Lecture Notes in Engineering

Edited by C. A. Brebbia and S. A. Orszag

40

R. Borghi, S. N. B. Murthy (Eds.)

Turbulent Reactive Flows

Springer-Verlag
New York Berlin Heidelberg
London Paris Tokyo

Series Editors
C. A. Brebbia · S. A. Orszag

Consulting Editors
J. Argyris · K.-J. Bathe · A. S. Cakmak · J. Connor · R. McCrory
C. S. Desai · K.-P. Holz · F. A. Leckie · G. Pinder · A. R. S. Pont
J. H. Seinfeld · P. Silvester · P. Spanos · W. Wunderlich · S. Yip

Editors
R. Borghi
Université de Rouen
Faculté des Sciences et Techniques
Laboratoire de Thermodynamique (U.A.C.N.R.S. No. 230)
BP 118
76134 Mont-Saint-Aignan
France

S. N. B. Murthy
Thermal Sciences & Propulsion Center
School of Mechanical Engineering
Purdue University
Chaffee Hall
West Lafayette, IN 47907
USA

"The volume is based on the proceedings of the U.S.A.-France Joint Workshop on Turbulent Reactive Flows, cosponsored by the C.N.R.S. of France and the N.S.F. of U.S.A., held in July, 1987."

ISBN 0-387-96887-3 Springer-Verlag New York Berlin Heidelberg
ISBN 3-540-96887-3 Springer-Verlag Berlin Heidelberg New York

Library of Congress Cataloging-in-Publication Data

Turbulent reactive flows / R. Borghi, S.N.B. Murthy, eds.
(Lecture notes in engineering ; 40)
"The volume is based on the proceedings of the U.S.A.-France Joint
Workshop on Turbulent Reactive Flows, cosponsored by the C.N.R.S. of
France and the N.S.F. of U.S.A., held in July, 1987."
ISBN 0-387-96887-3
1. Fluid dynamics. 2. Turbulence. I. Borghi, R. (Roland)
II. Murthy, S.N.B. III. U.S.A.-France Joint Workshop on Turbulent Reactive Flows (1987)
IV. Series.
TA357.T886 1989
620.1'064–dc19   88-38767

This work is subject to copyright. All rights are reserved, whether the whole or part of the material is concerned, specifically those of translation, reprinting, re-use of illustrations, broadcasting, reproduction by photocopying machine or similar means, and storage in data banks.
© Springer-Verlag New York Inc. 1989
Printed in Germany

The use of registered names, trademarks, etc. in this publication does not imply, even in the absence of a specific statement, that such names are exempt from the relevant protective laws and regulations and therefore free for general use.

Printing: Mercedes-Druck, Berlin
Binding: B. Helm, Berlin
2788-2161/3020-543210

PREFACE

Turbulent reactive flows are of common occurrance in combustion engineering, chemical reactor technology and various types of engines producing power and thrust utilizing chemical and nuclear fuels. Pollutant formation and dispersion in the atmospheric environment and in rivers, lakes and ocean also involve interactions between turbulence, chemical reactivity and heat and mass transfer processes. Considerable advances have occurred over the past twenty years in the understanding, analysis, measurement, prediction and control of turbulent reactive flows. Two main contributors to such advances are improvements in instrumentation and spectacular growth in computation: hardware, sciences and skills and data processing software, each leading to developments in others.

Turbulence presents several features that are situation-specific. Both for that reason and a number of others, it is yet difficult to visualize a so-called solution of the turbulence problem or even a generalized approach to the problem. It appears that recognition of patterns and structures in turbulent flow and their study based on considerations of stability, interactions, chaos and fractal character may be opening up an avenue of research that may be leading to a generalized approach to classification and analysis and, possibly, prediction of specific processes in the flowfield. Predictions for engineering use, on the other hand, can be foreseen for sometime to come to depend upon modeling of selected features of turbulence at various levels of sophistication dictated by perceived need and available capability.

Chemically reactive flows, involving in general mass and heat exchange, present additional complexities due to density change, compressibility effects and, most significantly, interactions between turbulence and chemistry. Such flows may be initially unmixed or mixed and are partially mixed in the course of development in all cases. The flows in reactors and combustors are generally multi-component and often multi-phase. The presence of discrete solid or liquid entities in the flow also gives rise to interactions with flow, chemical action and heat and mass transfer processes. The main governing parameters in chemically reactive flows in general are the local strain (in the structures and the interconnections between them) and the characteristic time scales for convection, transport and chemical reaction. Once again, the approach to the problem is becoming focussed at two levels: examining such processes as mixing and flame stabilization at the physical-structural level and developing prediction methods with information in state space and chosen modeling of turbulence processes.

One of the major problems in turbulent reactive flows, therefore, pertains to the manner and the extent to which advances in the determination and understanding of the structure of turbulence can be incorporated into the rationalization of predictive schemes.

The research communities in France and the United States, it was found through discussions, are variously engaged in the problem of developing structure-based predictions in turbulent reactive flows. It was also established that it may be fruitful and timely to have a joint workshop on the subject between the two countries. The Centre National de la Recherche Scientifique of France and the National Science Foundation of the United States agreed to cosponsor such a workshop (a) to provide an opportunity to bring together interested researchers in the two countries for an open discussion on current status of developments and (b) to generate ideas for future work, especially through collaborative efforts. The Workshop was held in July, 1987 with 66 participants who contributed 46 invited papers for discussion. The participants included specialist invitees from Germany, Spain and the United Kingdom in addition to those from France and the U.S.A. The current volume is an outcome of that Workshop and presents the contributions that were subsequently rewritten and reviewed for publication.

The Workshop was arranged with a series of sessions devoted to (a) analysis of structure, (b) experimental studies, (c) prediction methods and (d) four areas of application, namely engine combustion, supersonic combustion, liquid state reactors and multi-phase fluid systems. The contributions have been gathered into a single volume with two parts: one, devoted to structure: diagnostics and analysis, and the other, to structure and predictions. In addition to the main themes, Part I includes the papers on liquid flows, and Part II includes those on multi-phase flows as well as those on supersonic combustion.

The volume is not merely an introduction to the type of researches being conducted in the two countries in the subject of turbulent reactive flows. It should serve as a major review of developments in our understanding of the structure of such flows and of the predictive schemes as they are developing based upon our appreciation of the underlying physical features of the flows. The papers on diagnostics, quantification and interpretation of results in Volume I provide a discussion of the major tools of non-intrusive measurement and their use that points to several avenues for future research. The contributions in the four areas selected to illustrate the applications each present a comprehensive review of the status of the subjects.

The Workshop and, therefore, the volume have been possible through the generous assistance and valuable advice of Dr. J. C. Carpentier of the C.N.R.S. and Mrs. Christine Glenday, supported by Dr. R. Goulard, of the N.S.F. Several other organizations and our own institutions in France and the U.S.A. have also provided very valuable assistance in many ways.

It has been the greatest pleasure to organize the Workshop and to work with each of the authors. The authors have given generously their time and expertise and have been patient and forgiving in the inevitable needling for revisions and promptness. We would also like to take this opportunity to thank the various reviewers who read the contributions.

R. Borghi
S.N.B. Murthy

Part 1

Table of Contents

Part I
Structure: Diagnostics and Analysis

M.B. Long, B. Yip, M. Winger, and J.K. Lam
Measurement of the Topology of Large-Scale Structures in
Turbulent Reacting Flows.. 1

P. Magre and R. Dibble
Finite Chemical Kinetics Effects in a Subsonic Turbulent
Hydrogen Flame.. 10

P. Magre, P. Moreau, G. Collin, R. Borghi, M. Pealat,
and J.P. Taran
Cars Study of Premixed Turbulent Combustion in a High
Velocity Flow... 33

W.M. Roquemore, L-D. Chen, L.P. Goss, and W.F. Lynn
The Structure of Jet Diffusion Flames.................................... 49

D. Stepowski, K. Labbaci, and R. Borghi
Instantaneous Radial Profiles of OH Flourescence and Rayleigh
Scattering Through a Turbulent H_2 - Air Diffusion Flame............... 64

T.A. Baritaud
Flame Structure in Spark Ignited Engines, From Initiation to
Free Propagation.. 81

D. Durox, T. Baritaud, J.P. Dumont, and R. Prud'Homme
Comparison Between Two Highly Turbulent Flames Having Very
Different Laminar Burning Velocities...................................... 93

I. Gökalp, A. Boukhalfa, R.K. Cheng, and I.G. Shepherd
Structure of Turbulent Premixed Flames as Revealed by Spectral
Analysis... 112

P. Goix, P. Paranthoën, and M. Trinite
Turbulent Flow Field and Front Position Statistics in V-Shaped
Premixed Flame With and Without Confinement.............................. 131

P.O. Witze and F.E. Foster
Two-Component Velocity Probability Density Measurements
During Premixed Combustion in a Spark Ignition Engine.................... 158

M.J. Cottereau, J.J. Marie, and P. Desgroux
On The Accuracy of Laser Methods for Measuring Temperature and
Species Concentration in Reacting Flows........................... 169

F.A. Williams
Structure of Flamelets in Turbulent Reacting Flows and Influences
of Combustion on Turbulence Fields................................ 195

P. Calvin and G. Joulin
Flamelet Library for Turbulent Wrinkled Flames.................... 213

A. Linan
Diffusion Flame Attachment and Flame Spread Along Mixing Layers... 241

N. Peters
Length and Time Scales in Turbulent Combustion.................... 242

J.E. Broadwell
A Model for Reactions in Turbulent Jets: Effects of Reynolds,
Schmidt and Damköhler Numbers..................................... 257

F.C. Gouldin
A Fractal Description of Flamelets................................ 278

F. Gaillard and I. Gökalp
Some Results on the Structure of the Temperature Field in Low
Damköhler Number Reaction Zones................................... 307

D. Escudie
Interaction of a Flame Front with Vortices: An Experiment......... 323

A.R. Karagozian, Y. Suganuma, and B.D. Strom
Experimental Studies in Vortex Pair Motion Coincident with a
Liquid Reaction... 340

J.L. Lievre and J.N. Gence
On an Attempt to Measure the Decay of Concentration Fluctuations
in a Quasi-isotropic Grid by Use of the Flourescene of the
Solution.. 373

R. David
Turbulent Reactive Flows of Liquids in Isothermal Stirred Tanks... 381

MEASUREMENT OF THE TOPOLOGY OF LARGE-SCALE STRUCTURES IN TURBULENT REACTING FLOWS

Marshall B. Long, Brandon Yip, Michael Winter and Joseph K. Lam
Department of Mechanical Engineering and Center for Laser Diagnostics
Yale University, New Haven, Connecticut 06520

Turbulent combustion involves a complex interplay between fluid motion and chemistry. A widely held view is that large-scale structures play an important role in this interaction. Experimental work that concentrates on providing new information on the spatial characteristics of large-scale structures in flames has been done. Laser imaging techniques that can provide two-, and more recently, three-dimensional data on quantities such as species concentrations, temperatures, and densities in turbulent flames have been developed.

I. INTRODUCTION

In the past decade, the importance of large-scale structures in turbulent reacting and nonreacting flows has generally been accepted.[1,2] The need for more specific information on these structures has led to the development of multipoint measurement techniques that can provide direct quantitative information on the spatial characteristics of the structures in gas flows.

Many different interactions between laser light and the flow medium are possible, and it has been demonstrated that a number of these light scattering mechanisms can provide information on the state of the flow. Lorenz-Mie, Rayleigh, fluorescence, and Raman scattering have all been used in turbulent flows and flames, and multipoint measurements of species concentration,[3,4,5] temperature,[6,7] and density[8,9] have been demonstrated. Each scattering process has a different range of applicability, the specifics of which have been discussed elsewhere.[3-10]

For two-dimensional light scattering measurements, these different techniques share a common basic experimental configuration. A laser beam is formed into a thin sheet that intersects the flow in the region to be studied, and a two-dimensional detector is used to record the scattered light from the flow. The recorded intensity distribution is then used to determine the value of a scalar, such as gas concentration, temperature, or density, in the flow.

The two-dimensional measurement of a scalar quantity in a turbulent flow gives a much clearer picture of the nature and role of structures in the flow than does the corresponding single-point measurement. Because of the complexity of the flows, however, many questions remain unanswered. To completely specify a turbulent reacting flow would be extremely difficult, requiring the simultaneous measurement of the concentration of all species, temperature, pressure, and velocity at all points in three dimensions as a function of time. While this type of measurement is not currently possible, some progress has been made toward obtaining information that more completely characterizes the structures in turbulent reacting flows.

In the next sections, results are presented from recent light scattering experiments that seek to extend the basic two-dimensional measurement techniques to allow (1) simultaneous measurement of more than one scalar quantity, (2) measurement of some three-dimensional characteristics of the flow, or (3) measurement of the time development of structures. The techniques will be summarized and specific results that relate to the structure of turbulent flames will be presented.

II. MEASUREMENT OF MULTIPLE SCALARS

Figure 1 shows a schematic of the experimental configuration used to measure two scalar fields simultaneously. As usual in planar measurements, a thin sheet of laser illumination intersects the flow in the area to be measured. The scattered light of all wavelengths from the flow is collected by a lens and separated into two parts by a beam splitter. In each portion of the split beam, an optical filter is used to isolate the wavelength associated with a particular scattering mechanism and the image corresponding to that wavelength is focused onto a separate detector. The two detectors are operated synchronously and the data are stored in a single computer that controls the entire experiment. The particular scalars measured depend on the flow, the scattering mechanisms detected (i.e., the optical filters), and the laser wavelength. Thus far, results have been obtained in turbulent flames for several combinations including: (1) Rayleigh and

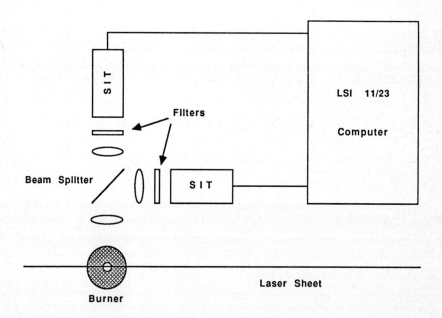

FIG. 1. Experimental configuration for the simultaneous measurement of two scalar fields in a flame or jet. A thin sheet of laser light illuminates a vertical plane in the flow field. Radiation (of all wavelengths) scattered by the molecules of the flow is collected by a forward lens and separated into two parts by a beam splitter. Secondary lenses image the light onto two SIT vidicon detectors. Each detector measures a different scalar field, since optical filters positioned before them ensure that they are sensitive to different spectral components of the scattered light. Imaging of fluorescence, Raman, or Rayleigh scattering provides a measurement of specific species concentrations or flame temperatures. When two closely spaced parallel laser sheets of different wavelength are used (instead of one), the detectors measure a scalar field in adjacent planes of the flow.

Raman scattering for temperature/fuel concentration measurements,[11] (2) Rayleigh scattering and fluorescence for temperature/radical concentrations,[12] and (3) Raman scattering and fluorescence for fuel gas concentration/radical concentration measurements.[13]

A pair of temperature/methane concentration distributions obtained in this manner is shown in Fig. 2. Rayleigh scattering was used to measure the temperature (right), and Raman scattering was used to measure the concentration of CH_4 (left) in a turbulent nonpremixed H_2/CH_4 flame.[11] The information provided by the two measurements is complementary. The CH_4 concentration distribution clearly shows the characteristics of the internal (mostly cold) fuel jet, while the temperature distribution better shows the length scales and associated characteristics of the flame front.

 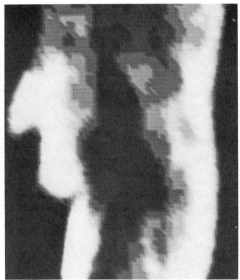

FIG. 2. A pair of simultaneous temperature and concentration distributions measured in a turbulent CH_4/H_2 diffusion flame. The left frame shows the CH_4 concentration as determined from Raman scattering from a single laser shot, with higher CH_4 concentration represented by lighter shading. The right frame depicts the corresponding temperature distribution as determined from Rayleigh scattering, ranging from 300 K (darkest shading) to 2200 K (lightest shading). The imaged region is 5 nozzle diameters in length (d = 1.9 mm) centered 7.5 diameters downstream of the burner, and digitized in a 70 x 80 pixel format for each frame. The spatial volume corresponding to a single pixel was 150 μm x 140 μm x 200 μm, where the last dimension corresponds to the thickness of the laser illumination sheet. The (upward) gas velocity at the nozzle opening was 69 m/sec.

Since both scalars are measured simultaneously, it is possible to obtain information on the spatial relationships of the two quantities. For example, the instantaneous temperature/concentration pair shows a close correspondence between the cool regions in the center of the flame and the regions of high fuel gas concentration. This close correspondence is not unexpected in a nonpremixed flame of this type. However, this correspondence does not always occur since a region can be seen in which the temperature is relatively low and yet the CH_4 concentration is also low. This indicates that there are regions in which unreacted cold fuel and air may be mixed. Such premixing can be expected to alter the burning characteristics of the flame considerably.

III. MEASUREMENTS IN THREE DIMENSIONS

A two-camera setup similar to that shown in Fig. 1 can also be used to obtain information on the three-dimensional characteristics of a flow by measuring a scalar and its full three-dimensional gradient. The main difference in the experimental configuration is that, instead of a single illumination sheet, two closely spaced parallel illumination sheets of different wavelength are used. The optical filters in front of each detector ensure that each camera will record the scattered light from only one of the sheets. Since the distribution of the scalar is recorded in two adjacent sheets, the gradient vector associated with the scalar can be determined at all points within a plane. From within a single illumination sheet, two gradient components can be calculated from the difference between adjacent pixels in two orthogonal directions. The third gradient component is obtained from the change in the scalar value from one sheet to the next.

To date, this measurement approach has been used with Rayleigh scattering in nonreacting jets of Freon mixing with air.[14] In this case, the Freon concentration and its gradient are measured. From this information, the mixture fraction and scalar dissipation rate can be determined. Figure 3 shows the mixture fraction (left) and associated scalar dissipation (right) from a Freon jet of Reynolds number 3850.

If a large number of instantaneous shots are recorded, the statistics of the scalar and its gradient can be determined. In particular, independent pdfs, as well as the joint pdfs of the scalar and its gradient, can be obtained. This information, even in a cold jet,[15] is of value for the modeling of turbulent combustion.[16,17] Work is currently underway to acquire reliable statistics over a range of flow conditions.

FIG. 3. Instantaneous mixture fraction and associated scalar dissipation field in a cold Freon jet of Reynolds number 3850 measured by imaging Rayleigh scattering from one Nd:YAG laser shot. The flow direction is from bottom to top, and the imaged region is between 14 and 16 nozzle diameters (d = 3.5 mm) downstream of the jet nozzle. The spatial resolution in the 65 x 78 pixel format is 90 μm x 90 μm (x 100 μm corresponding to the laser sheet thickness). On the left, higher values of the mixture fraction are represented by lighter shading, while on the right, lighter shading represents larger values of the scalar dissipation in the same plane.

The two-sheet measurements described above can provide large quantitative data sets from which pdfs can be determined. Also, a visualization of the interrelationship between the scalar and its full gradient is obtained as shown in Fig. 3. From these measurements, however, it is still difficult to visualize the topology of the structures in the flow. Further, while two sheets can provide a measurement of the gradient, a minimum of three parallel sheets would be required to determine the curvature of constant-property surfaces and many more would be needed to adequately describe the topology.

Recently, progress has been made toward obtaining instantaneous three-dimensional measurements of a scalar in turbulent reacting and nonreacting flows. One way to acquire such data is to make many independent two-dimensional measurements from closely-spaced parallel cross sections in a time frame that is short compared to the flow time scales.

An experimental configuration for performing instantaneous three-dimensional measurements is shown in Fig. 4. A two-dimensional laser sheet is swept through a volume of a turbulent flow by a rapidly

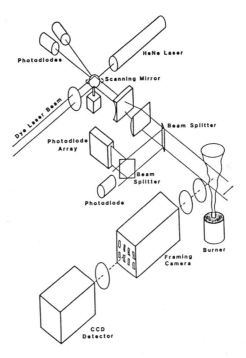

FIG. 4. Experimental configuration for measuring the three-dimensional gas concentration field in a jet or flame. A dye laser beam reflected from a resonant scanning mirror is formed into a thin illumination sheet which sweeps through the flow volume. Illuminated planes of the flow are imaged onto an electronic framing camera which records up to 18 parallel planes in the flow as the sheet traverses it. The output phosphor screen of the framing camera is demagnified onto a CCD array detector for data digitization. The speed and position of the dye laser sheet are monitored with a HeNe laser beam reflected from the rear of the scanning mirror and a pair of sensing photodiodes. Beam splitters allow the temporal and spatial energy profiles of the scanning laser sheet to be monitored by a fast photodiode and a linear diode array, so that nonuniformities in the illumination can be compensated for during data corrections. The experiment is controlled by a DEC MicroVAX II computer. The temporal resolution of the measurement is 0.6 μsec.

oscillating mirror. The scattered light corresponding to many different locations of the moving sheet is recorded by a high speed electronic framing camera, which is capable of recording up to 18 two-dimensional frames at rates up to 2×10^7 frames/sec. Images corresponding to consecutive frames (and thus to different flow planes) appear on different portions of a phosphor screen. The resulting sequence of frames can be recorded on film or directly digitized by a high resolution (600 x 400 element) CCD detector.

This technique has been applied to both nonreacting[18] and reacting[19] flows. In the most recent work, it has been possible to record the scattering from 12 different closely-spaced two-dimensional sheets in less than 1 μsec using both Rayleigh scattering and fluorescence.[19] Once the data corresponding to the different planar cross sections are recorded in the computer, the slices can be assembled so that the entire three-dimensional scalar field is known. At any point within the measured volume, the scalar, its gradient, and the curvature can all be determined.

There are many ways in which the data obtained can be visualized. One means of displaying the topology of the flow structures is to represent constant-property surfaces. Figure 5 shows a constant fuel gas concentration surface in a turbulent premixed flame. Computer graphics are used to represent the surface as a solid shell with hidden surface elimination, shading, and highlights added to convey as much three-dimensional information as possible. With this information, the topological characteristics of the flow structures can be investigated quantitatively. Properties currently being studied include surface areas, surface to volume ratios, and possible fractal dimensionalities.

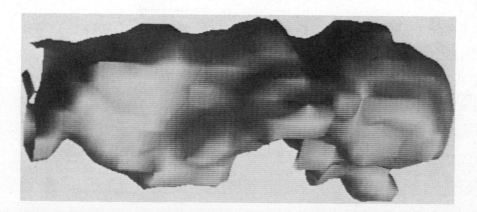

FIG. 5. Instantaneous constant fuel gas concentration surface in a turbulent premixed flame. The surface corresponds to a concentration of 64% of the maximum in this region, and has been generated from actual measured data by computer graphics software. The three-dimensional measurement was performed within the duration of a single dye laser pulse, giving a temporal resolution of 0.6 μsec. The flame was a stoichiometric CH_4-air mixture seeded with biacetyl vapor as a fluorescent marker, with a jet exit velocity of 19.5 m/sec. The imaged region is 2 x 1 x 0.5 nozzle diameters (d = 6 mm) in dimension, centered 5 diameters downstream of the burner. The data were digitized in a 128 x 64 x 12 pixel format with a resolution volume of 94 μm x 94 μm x 250 μm. The flow direction is from left to right.

Figure 6 shows a constant gas concentration contour from a nonreacting jet. In this figure, the surface has been shaded according to the magnitude of the concentration gradient vector at each point on the surface. Even though the data are from a cold flow, this representation should be well suited for providing insight into the interaction of turbulence and combustion in nonpremixed flames. In the flame sheet model of nonpremixed combustion, reactions are assumed to take place within a thin flame sheet located wherever the stoichiometric concentration of fuel and oxidizer exists.[20] In this model, the surface shown in Fig. 6 can be considered to be a hypothetical flame sheet for a chemical reaction with no heat release. The scalar dissipation rate for this cold flow can be obtained from the square of the mixture fraction gradient magnitude. Therefore, the value of the gradient (indicated by the shading level in Fig. 6) gives information on the reaction rate at each location on the flame sheet and, for high enough values, indicates regions where flame extinction would be likely.

FIG. 6. Instantaneous constant gas concentration surface in a turbulent Freon jet (Reynolds number 13,800) shaded according to the values of the magnitude of the concentration gradient vector on the surface. Lighter shading represents higher local gradients. The surface corresponds to a concentration of 73% of the maximum in this region, and has been generated from actual measured data by a computer graphics program. Rayleigh scattering from flow planes illuminated by a dye laser sheet sweeping through the gas was recorded to make the measurement. The imaged region is 3.4 x 1.7 x 0.7 nozzle diameters (d = 3.5 mm) in dimension, centered 6.9 diameters downstream of the burner. The data were digitized in a 128 x 64 x 10 pixel format with the resolution volume being 94 μm x 94 μm x 250 μm. The temporal resolution of the measurement was 0.5 μsec. The flow direction is from left to right.

IV. HIGH SPEED TWO-DIMENSIONAL MEASUREMENTS

The measurements described thus far can provide useful information on the spatial characteristics of turbulent reacting flow. Because of the limitations of the detectors and lasers used however, a continuous record of changes in the flow is not obtained. Sequential shots are essentially independent events in all of the techniques described above. From the standpoint of characterizing the flow statistically, this lack of time information is not critical, since many statistical quantities can be ensemble averaged rather than time averaged. To fully understand the flow, however, a measurement of the temporal evolution of structures is important.

For subsonic gas phase flows, the framing rate (i.e., the speed at which each image is recorded) must be in the range of 1 to 50 kHz to adequately resolve the movement of structures in the flow. Thus the large amount of information corresponding to each two-dimensional measurement has to be read from the detector and stored very rapidly. In addition, if a pulsed laser is used, the laser repetition rate must be synchronized with the detector's framing rate. These requirements represent severe constraints on the detector, laser, and computer system that can be used in the experiment. However, if each component is optimized for high speed operation, quantitative two-dimensional information can be obtained at rates sufficient to follow the temporal evolution of structures in some gas phase flows.[21]

Figure 7 shows several frames from a sequence of two-dimensional gas concentration measurements in a Freon jet mixing with air. The framing rate for the measurement was 1100 Hz and the evolution of structures from one frame to the next is clearly seen. In a qualitative sense, these data can provide insight into the formation and evolution of structures in the flow. Furthermore, the data can be used to calculate quantities such as convection velocities, shedding frequencies, and space-time correlations of the measured scalars.

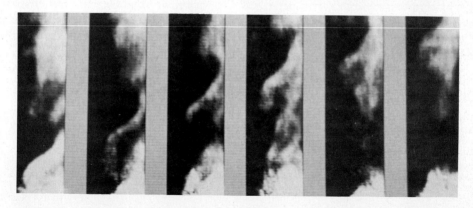

FIG. 7. Six gas concentration measurements (from a sequence of 320) made 0.8 msec apart in a cold Freon jet by imaging Rayleigh scattering from a flow plane illuminated by an argon-ion laser sheet. The flow direction is from bottom to top with time increasing from left to right. The Reynolds number of the flow was 4880. The elastically scattered light from a region between 6 and 9.5 nozzle diameters (d = 4 mm) downstream of the nozzle was imaged onto a 128 x 32 portion of an intensified photodiode array that was gated on for 50 μsec. The shading levels correspond to different nozzle gas concentration with lighter regions representing higher concentration.

V. CONCLUSION

Advances have been made in two-dimensional imaging techniques that allow a more complete view of the complex processes involved in turbulent reacting flows. Some advances, such as the measurement of a scalar and its full gradient, are directly applicable to the testing of existing theories of turbulent mixing and combustion. Other advances, such as the high speed two-dimensional and instantaneous three-dimensional measurements, provide new information on the large-scale structures that should be useful for developing and verifying new models that directly incorporate the characteristics of the structures.

ACKNOWLEDGMENTS

We gratefully acknowledge the partial support of this research by the U.S. Department of Energy (grant DE-FG02-85ER13427) and the National Science Foundation (grant MSM 83-51077).

REFERENCES

[1] A. R. Ganji, R. F. Sawyer: AIAA J. **18**, 817 (1980).
[2] G. L. Brown, A. Roshko: J. Fluid Mech. **64**, 775 (1974).
[3] M. B. Long, D. C. Fourguette, M. C. Escoda, C. B. Layne: Opt. Lett. **8**, 244 (1983).
[4] M. J. Dyer, D. R. Crosley: Opt. Lett. **7**, 382 (1982).
[5] G. Kychakoff, R. D. Howe, R. K. Hanson, J. C. McDaniel: Appl. Opt. **21**, 3225 (1982).
[6] D. C. Fourguette, R. M. Zurn, M. B. Long: Combust. Sci. Technol. **44**, 307 (1986).
[7] J. M. Seitzman, G. Kychakoff, R. K. Hanson: Opt. Lett. **10**, 439 (1985).
[8] A. J. R. Lysaght, R. W. Bilger, J. H. Kent: Combust. Flame **46**, 105 (1982).
[9] M. C. Escoda, M. B. Long: AIAA J. **21**, 81 (1983).
[10] R. K. Hanson: In *Proceedings of the 21st Symposium (International) on Combustion* (The Combustion Institute, Pittsburgh, 1986).
[11] M. B. Long, P. S. Levin, D. C. Fourguette: Opt. Lett. **10**, 267 (1985).
[12] R. W. Dibble, M. B. Long, A. Masri: In *Dynamics of Reactive Systems Part II: Modeling and Heterogeneous Combustion* (Vol. 105 of Progress in Astronautics and Aeronautics 1986) p. 99.
[13] M. Namazian, R. L. Schmitt, M. B. Long: submitted to Appl. Opt.
[14] B. Yip, M. B. Long: Opt. Lett. **11**, 64 (1986).
[15] K. N. C. Bray: In *Turbulent Reacting Flows*, ed. by P. A. Libby and F. A. Williams (Springer-Verlag, Berlin 1980).
[16] R. W. Bilger: In *Turbulent Reacting Flows*, ed. by P. A. Libby and F. A. Williams (Springer-Verlag, Berlin 1980).
[17] F. A. Williams: In *AGARD Conference Proceedings*, AGARD-CP-164 (1975).
[18] B. Yip, J. K. Lam, M. Winter, M. B. Long: Science **235**, 1209 (1987).
[19] B. Yip, R. L. Schmitt, M. B. Long: Opt. Lett. **13** (1988) to appear.
[20] N. Peters: In *Proceedings of the 21st Symposium (International) on Combustion*, (The Combustion Institute, Pittsburgh 1986).
[21] M. Winter, J. K. Lam, M. B. Long: Exp. Fluids **5**, 177 (1987).

FINITE CHEMICAL KINETICS EFFECTS IN A SUBSONIC TURBULENT HYDROGEN FLAME

P. Magre
ONERA, 92322 Chatillon, France

R. Dibble
Combustion Research Facility
Sandia National Laboratories, Livermore, CA, 94550

ABSTRACT

Departures from chemical equilibrium appear in nonpremixed turbulent flames at very high mixing rates, as shown by dimensional analysis based on Damköhler number (characteristic time of mixing over characteristic time of chemical reaction). This paper presents an experimental study that shows departures from chemical equilibrium in a hydrogen-air flame, which is often erroneously considered to have an infinitely fast chemical rate and therefore to be at chemical equilibrium.

These departures from chemical equilibrium are measured with nonintrusive laser diagnostics. Instantaneous and spatially resolved measurements of major combustion species (H_2, O_2, H_2O, and N_2), density, and temperature are performed by means of Raman and Rayleigh scattering in a turbulent jet flame with a fuel of 22 mole percent argon in hydrogen. From these measurements we infer the local fuel mixture fraction f. Departures from chemical equilibrium are manifested by the comparison between the measured temperature and the equilibrium temperature deduced from the value of f.

We vary the Damköhler number by adjusting either the aerodynamic conditions or the chemical rate. In the first case, a range of Reynolds numbers is explored: Re=8,500, Re=17,000, and Re=20,000, using the same fuel. The experimental results show a dramatic effect of the Reynolds number on the extent of departure from the limit of chemical equilibrium. Differences between the measured temperature and the inferred equilibrium temperature are as large as 450K as the flame approaches blow-off conditions. In the second case, we hold the aerodynamic conditions constant, and alter the chemical reaction rate by diluting the fuel with increasing amounts of nitrogen. These last experiments also show a difference between measured temperature and the inferred equilibrium temperature. Consequently, departures from the limit of chemical equilibrium are achieved through increasing the rate of mixing or by decreasing the rate of chemical reaction.

[*] Dr. P. Magre is a Combustion Research Facility Visiting Scientist sponsored by DRET, Direction des Recherches et Etudes Techniques, French Ministry of Defense.

[*] This research is supported by the United States Department of Energy, Office of Basic Energy Sciences, Division of Chemical Sciences.

1.0 INTRODUCTION

Much of our understanding of combustion in turbulent nonpremixed systems is a result of explorations with an idealized model that assumes that (A) the chemical kinetics are fast relative to the rate of mixing (large Damköhler number, equilibrium flow), (B) the mixing is dominated by turbulence and, hence, all species and enthalpy are transported at the same rate (unity Lewis number), and (C) the flow is adiabatic. With these assumptions all of the concentrations and temperatures can be related to a single conserved scalar [1-4].

Although some combustion systems closely approach this idealized model [5-7] most practical devices deviate from these assumptions. For example, the idealized model has no mechanism for predicting phenomena such as flame blowoff, and incorrectly predicts equilibrium values for exhaust pollutants such as oxides of nitrogen, carbon monoxide, and particulate carbon. Accordingly, devising modifications to the above model has become a major research topic in recent years. For example, nonadiabaticity has been explored [8], and implications of nonunity Lewis numbers [9,10] have been investigated.

Turbulent mixing also has a profound effect on flame chemistry [5,11-18]. Departures from chemical equilibrium occur when mixing rates are comparable to chemical kinetic rates. Insufficient time for complete combustion results in a depressed flame temperature, which, in turn, reduces the kinetic reaction rates; hence, products of incomplete combustion begin to appear in the exhaust. As the mixing rate is further increased, the temperature is further depressed until ultimately the kinetic rates become so slow (frozen) that the flame blows out. This evolution in mixing rate is characterized by the Damköhler number, which is the dimensionless ratio of the mixing time τ_{flow} divided by the chemical kinetic time τ_{chem} (Williams [19]; Strehlow [20]) :

$$Da = \tau_{flow}/\tau_{chem} \tag{1}$$

The relationship between the temperature and the Damköhler number is the well known 'S' shaped curve reproduced in Fig. 1. As the figure illustrates, the temperature approaches its equilibrium value as Damköhler number approaches infinity.

Departures from chemical equilibrium are easily shown in flame systems by comparing the actual temperature of the combustion products to the expected equilibrium temperature. Simultaneous measurements of temperature and concentrations of major species are necessary for this purpose. From the measurement of those species we can infer the local mixture fraction of fuel f, which is the point value of fuel mass divided by total mass; in the nozzle, f is unity and in the coflow air, f is zero. From f, we predict the equilibrium temperature that we compare to the measured temperature. The use of laser Raman scattering for thermometry and concentration measurements has been reviewed by several authors including articles by Eckbreth [21], and by Rahn, Mattern, and Farrow [22]. The spontaneous vibrational Raman technique is ideally suited for measurements in the turbulent nonpremixed flame of hydrogen in air. Unlike hydrocarbon flames, hydrogen flames do not produce species that fluoresce when exposed to visible laser light; in addition, hydrogen flames do not produce soot. Mie scattering and laser induced incandescence from the soot particles often are larger than the vibrational Raman scattering. Such positive features of hydrogen flames have allowed the accumulation of a wide variety of measurements in this flame by this technique alone or coupled with laser Doppler velocimetry (Dibble et al. [23,24]).

We explore two different ways to vary the Damköhler number. By increasing the velocity of the jet we are able to modify τ_{flow} while keeping τ_{chem} constant. Another way consists in changing τ_{chem} by using different kinds of fuels, *i.e.* by using hydrogen diluted with increasing amounts of nitrogen. For these different conditions we study the departures from the chemical equilibrium. These departures are quantified in terms of the Damköhler number representing the aerothermochemical conditions in the flame.

2.0 EXPERIMENTAL
2.1 Turbulent Nonpremixed Flame

All measurements were made in the Sandia Turbulent Combustion Tunnel Facility, which is depicted in Fig. 2. The Facility is a forced draft, vertical wind tunnel with an axisymmetric fuel jet located at the upstream end of a test section. A fully windowed test section has a 30-cm-square cross section and is 200 cm long. The test section empties into an exhaust hood which draws air from the room in addition to flow from the test section. The fuel nozzle consists of two concentric tubes with inside diameter D of 0.52 cm and an outside diameter of 0.95 cm; the tube walls are 0.07 cm thick. The annular void region has no gas flow. The fuel tube is straight for more than 500 diameters, and therefore a fully-developed velocity profile in the fuel tube is assumed. The coflow air has consistent temperature (20 ± 2 degrees Celsius) and humidity (31 ± 9 percent). Well upstream of the test section, the coflow air is passed through a bank of high efficiency particle arrest (HEPA) filters that remove particles larger than 0.3 μm.

2.2 Raman Scattering Techniques

The mixture fraction is determined from the Raman scattering from a single laser pulse from the Combustion Research Facility flashlamp-pumped dye laser (1 J/pulse, 3 pulses/s, 2 μs pulsewidth, λ=514.5nm, $\Delta\lambda$=0.3nm), which is located 15 m from the test facility in an adjacent room, is focused through a lens (focal length = 50cm). As Fig. 3 depicts, the Raman scattering from this focused laser beam is collected at right angles by a 6-element, 300-mm-focal length, f/2 collection lens and is relayed at X3 magnification to the entrance slit of a 3/4m grating spectrometer. The width of the entrance slit (3mm) determines the length of the Raman probe volume, while the height of the probe volume is determined by the laser beam diameter (< 1mm), since the slit height is 6mm. With this slit arrangement, the measured Raman signal is not sensitive to small movements of the laser beam, which can be caused by beam steering in the turbulent flame (Starner and Bilger [6]). At the exit plane of the spectrometer, five photomultiplier tubes are positioned to receive the Stokes-shifted vibrational Raman scattered light from N_2, O_2, H_2, H_2O, and the anti-Stokes vibrational Raman scattering from N_2. In addition, the elastically scattered (Rayleigh and Mie) laser light is also measured. The laser pulse energy is measured with a vacuum-photodiode that receives attenuated laser light before it passes through the test section. The electrical outputs of the phototubes are connected to a twelve-channel charge integrator (CAMAC), which is gated for the duration of the laser pulse. As an indication of the overall efficiency of the Raman collection system, 6000 photoelectrons per Joule of laser light are collected from vibrational Raman scattering of N_2 in room air.

The salient feature for the use of laser Raman scattering is that the integrated charge Q_i from a given photomultiplier tube is linearly related to laser energy Q_l and species concentration $[N_i]$ as follows:

$$Q_i = k_i Q_l [N_i] f_i(T) \qquad (2)$$

The proportionality constant k_i is dependent on the vibrational Raman cross section, the wavelength, geometry, and optical collection efficiency and is ultimately determined by calibration. The bandwidth factor $f_i(T)$ accounts for the temperature-dependent distribution of molecules in their allowed quantum states. Also convolved into the bandwidth factor $f_i(T)$ are the spectral location, shape, and bandwidth of the spectrometer and the bandwidth of the laser. The bandwidth factor can be calculated for N_2, O_2, and H_2; however, the calculation is less reliable for triatomics such as H_2O. In the work presented here, the bandwidth factors are experimentally determined in laminar, adiabatic, flat-flames of hydrogen premixed with air. The fuel flow rates determine the equivalence ratio from which equilibrium values of the species concentrations calculated. These concentrations are used for calibration of the Raman signals. Thus, after these calibrations, a measurement of a given species concentration from each laser pulse requires that laser pulse energy, Raman signal, and temperature be recorded.

2.3 Test Conditions

A wide range of Damköhler numbers are explored by systematically adjusting either the numerator or the denominator in Eq. (1). Five test conditions, Cases A-E, are enumerated in Table 1. For each of these cases the coflow air velocity remains constant at 9.2 m/s.

In the first three cases, the Damköhler number variation is achieved by changing the mixing time while holding the chemical time constant by using the same fuel. This fuel is a mixture of 22% argon-in-hydrogen. This choice of composition allows direct measurement of density by means of Rayleigh scattering (Dibble and Hollenbach [25]). For Cases A, B, and C, the respective values of the Reynolds numbers are Re=8,500, Re=17,000 and Re=20,000. At a Reynolds number of 21,500, the flame detaches from the nozzle and subsequently blows off, resulting in no flame.

The Case A flame, which is extensively documented by Dibble, et al. [23,24], is our reference case. The previously mentioned studies have shown that there was not any departure from equilibrium for those aerodynamical conditions.

In the last two cases in Table 1, the Damköhler number variation is achieved by changing the chemical time while holding the mixing time relatively constant. The chemical time is changed by diluting the hydrogen fuel with increasing amounts of nitrogen: 60 mole percent for Case D and 70 mole percent for Case E. At these flow rates, flame blow off occured for dilutions greater than 70 mole percent. As a consequence of this dilution, the adiabatic equilibrium flame temperatures, Table 1, are reduced by as much as 500 K relative to Cases A. This reduction in temperature will reduce the chemical reaction rate and thus achieve the expected variations of Damköhler number.

3.0 DATA REDUCTION
3.1 Measurements of Mixture Fraction

The Raman signals provide the concentrations of H_2, O_2, H_2O, and N_2. When the chemical system is assumed at equilibrium, the mixture fraction f can be determined from the unique relation between N_2 concentration and mixture fraction (Dibble et al. [5,24]).

When the chemical system is not at equilibrium, all of the above Raman signals are required for determination of the mixture fraction as follows:

$$f = \frac{2[H_2O] + 2[H_2] + 40[Ar]}{18[H_2O] + 2[H_2] + 32[O_2] + 28[N_2] + 40[Ar]} \quad (3)$$

For the argon-in-hydrogen flame (Cases A, B, and C) the concentration of argon $[Ar]$ is not directly measured. Instead, we assume all species have the same diffusivity and consequently the Lewis number for each species is the same. These assumptions are reasonable since the effect of unequal diffusivities scales as Re^{-1} (Bilger [9]) and Drake et al. [26] show the effect to be negligible, for $Re \geq 8000$. The equal diffusivity assumption allows the concentration of argon to be related to the concentration of hydrogen and water: $[Ar] = \frac{22}{78}([H_2O] + [H_2])$.

3.2 Measurements of Temperature

The Raman scattering apparatus, illustrated in Fig. 3, provides the intensities of the Stokes and Anti-Stokes lines of nitrogen. The ratio of these two signals are related to the temperature, but due to the weakness of the Anti-Stokes signal, the accuracy of this thermometry is low. As an alternative route, the temperature can be inferred from the total number density N_{total} and the ideal gas law:

$$T = T_o \frac{N_o}{N_{total}} = T_o \frac{N_o}{[H_2O] + [H_2] + [O_2] + [N_2] + [Ar]} \quad (4)$$

where N_o is the total number density of air at atmospheric pressure and temperature T_o (when $T_o = 273$ Kelvins, $N_o = 2.687 \times 10^{19}$ moles/cm^3). For hydrogen flames, it is fortunate that the total number density is adequately determined by the sum of concentrations of only four species; namely H_2, O_2, H_2O, and N_2; all of which are here measured. For hydrocarbon flames, the measurements of several additional species are required [27].

Temperature deduced from the various methods (ratio of anit-Stokes to Stokes signals, total number density, and equilibrium values from mixture fraction f) have been compared on an adiabatic flat-flame buner. The mean of these measurements agree within 100 Kelvins; they also agree with a radiation corrected thermal couple. The standard deviation of the total-number-density thermometry is 3 percent at 300 Kelvins and 7 percent at 2000 Kelvins.

3.3 Presentation of Data

For each location in the flame a series of 1000 laser shots provides an ensemble of 1000 individual measurements of the temperature T, of the mixture fraction f, and of the equilibrium temperature T_{eq} deduced from f. From this ensemble, the probability distribution P of any of these variables can be generated; Fig. 4, presents such a radial profile of $P(f)$ at x/D=50 for the Case A flame. Prediction of the evolution of probability distribution of mixture fraction $P(f)$, as shown by Fig. 4, is a challenging problem that is inherent in most models of turbulent combustion.

Rapid rates of mixture fraction dispersal may result in departures from chemical equilibrium. This departure is evident when the measured temperature T is less than the equilibrium temperature T_{eq}, which is determined from the measured mixture fraction f. Figure 5 is a plot of T versus f; each individual point is from the ensembles used to generate the radial profile in Fig. 4, around 8,000 points in total. In addition, Fig. 5 includes a curve that is the equilibrium relationship between T_{eq} and f. The significance of these results are discussed in the following paragraphs.

4.0 RESULTS AND DISCUSSION

4.1 Influence of Reynolds Number

The radial profile of the probability distribution $P(f)$ for Case A at x/D=50 (Fig. 4), shows features that are characteristic of these distributions. On the center line (y/D=0), a unimodal distribution is centered around an average value of $f_{avg} = 0.23$. With increasing radial position y/D this distribution shifts towards lower values of the mixture fraction due to mixing with air. The stoichiometric composition for the argon-in-hydrogen flame is f_{st}=0.166. The mean stoichiometric contour occurs at y/D=2.5. For y/D>3.5, zero values of f become more probable. The probability of this peak $P(0)$ increases and narrows as the region of pure coflow air is approached. The same series of data is now displayed on Fig.5 in terms of temperature T versus mixture fraction f. The symmetrical location of the set of points around the adiabatic equilibrium curve justifies the assumption of chemical equilibrium for the Case A flame; we define Case A at x/D=50 as our reference case.

The data for Case A at x/D=30 (Fig. 6), show a slight departure of the measured temperatures from the equilibrium curve. This phenomenon is enhanced for Case B at x/D=30 (Fig. 7), and even more so for case C at x/D=30 (Fig. 8). In this latter case, all of individual measurements are located below the curve of equilibrium temperature.

In order to quantify departures from chemical equilibrium, we imagine a smooth curve drawn through the data in Figs. 6-8; this imagined curve defines the average flame temperature T_{flame}; which is a function of mixture fraction f. Departures from chemical equilibrium are indicated by the difference between the equilibrium temperature and the measured temperature that is determined from the imagined curve. The difference is zero for Case A x/D=50, the reference Case, to as much as 450 Kelvins for the highest velocity case, Case C at x/D=30. For a given case, the maximum temperature difference occurs when the mixture fraction is at its stoichiometric value, $f = 0.166$; it is this temperature difference that will be scaled with the Damköhler number.

A relative estimation of the Damköhler number corresponding to the different cases presented by estimating τ_{flow} for each case while assuming a constant τ_{chem} since the fuel is the same for all cases A, B and C. Classical results on the development of a turbulent jet (Hinze [28], Everett and Robins [29]) indicate that the velocity on the axis of a high-momentum-turbulent jet decreases as 1/x (x distance from the nozzle):

$$U_j \propto \frac{d_0 \cdot U_0}{x} \qquad (5)$$

where d_0 and U_0 are characteristic length and velocity, respectively, and where $d_0 \cdot U_0/\nu$ is a Reynolds number.

From this relationship, and the definition of velocity as the time derivative of distance it follows:

$$dt = \frac{xdx}{d_0 \cdot U_0} . \qquad (6)$$

Integrating Eq.(6) from $t = 0$ to $t = \tau_{flow}$ and from $x = 0$ to $x = x$ we obtain:

$$\tau_{flow} \propto \frac{x^2}{2d_0 \cdot U_o} \propto \frac{x^2}{Re}. \qquad (7)$$

(Equation 7, which is based on macroscopic scales, can also be arrived at by consideration of the large-eddy-turnover time:

$$\tau_{large-eddy-turnover} \propto \delta/U_{cl} \propto \frac{x^2}{Re}$$

where δ and U_{cl} are the jet width and centerline velocity, respectively.) The ratio between the Damköhler number for the reference case, arbitrarily fixed at a value of $Da_{A50} = 100$, and the Damköhler number for the other cases can be calculated from Eq.(7). For instance:

$$\frac{Da_{B30}}{Da_{A50}} = \left(\frac{\tau_{chemA}}{\tau_{chemB}}\right)\left(\frac{U_{oA}}{U_{oB}}\right)\left(\frac{30}{50}\right)^2 = 0.18. \qquad (8)$$

As previously noted, we have assumed $\tau_{chemA} = \tau_{chemB}$; an assumption we will consider in the next paragraph. Table 2 summarizes the values of the Damköhler number and its effect on T_{flame}. Figure 9 presents the evolution of the maximum flame temperature versus this relative Damköhler number.

Any difference between τ_{chemA}, τ_{chemB}, and τ_{chemC} will be due to temperature differences, since the fuel composition for these Cases are identical. This temperature influence has not been taken into account. Consequently, the Da given here are slighly overestimated since we expect that τ_{chemC} will increase as T_{flame} is reduced.

4.2 Influence of Fuel Composition: Nitrogen Dilution.

In the previous section, 4.1, the Damköhler number was changed primarily by altering the fuel velocity while holding the fuel composition constant. In this section, the Damköhler number is changed by altering the fuel composition. The effect of this change is studied with cases D and E (see Table 1). For these cases, the results concerning x/D=30 are shown in Figs. 10 and 11. The features of these curves are similar to the ones obtained when only the aerodynamical conditions were changed. One difference appears in Case E (the highest dilution case), where sporadic breaks in the flame front are visually observed at the flame attachment point on the nozzle. As a consequence of this sporadic burning, fuel and air are, from time to time, mixing without burning; this explains the appearance in Fig. 11 of points with mixture fraction values greater than zero but with temperatures of unburned reactants at 300 Kelvins. As these room temperature points do not occur for mixture fractions near stoichiometric, where the temperature T_{flame} is determined, the appearance of these mixed but unburned points do not enter into the Damköhler scaling.

For comparison of Cases D and E to Cases A, B and C, an estimation of Da required. As Eq.(8) indicates, such an estimate demands an estimate of the ratio of $\frac{\tau_{chemA,B,orC}}{\tau_{chemDorE}}$. On the most elementary level, one would expect that the chemical time should increase simply because the concentration of the reactants are reduced by the dilution of the fuel with the inert gas. However, the largest effect on the chemical time results from reduction in the Arrhenius factor due to the reduction in maximum flame temperature that is also a consequence of addition of inert diluents to the fuel. Both effects of reduced reactant concentration and reduced Arrhenius factors are included into an estimation of τ_{chem} by assuming that the chemical time in these flames is related to the minimum residence time in a perfectly stirred reactor containing a stoichiometric mixture of the corresponding fuel and air. The minimum residence time τ_{min} for which mixtures A, D and E burn in a perfectly stirred reactor is readily calculated with the stirred reactor numerical code of Glarborg and Kee [30]. We expect τ_{min} to scale with τ_{chem}. The perfectly stirred reactor code produces the following: $\tau_{chemA} = 20\mu s$, $\tau_{chemD} = 58\mu s$, and $\tau_{chemE} = 112\mu s$. The corresponding Damköhler numbers deduced from inclusion of these chemical times into Eq.(8) are presented in Table 2. The two additional data points are plotted on the curve of Fig. 9; they are consistent with the other points of Cases A, B, and C. This qualitative agreement suggests that the molecular transport effects on τ_{chem} in these flames are nearly the same for Cases A-E since the molecular transport effects are not modelled by the stirred reactor.

4.3 Influence of Fuel Composition: Pure Hydrogen.

Drake [31] has measured mixture fraction of fuel and temperature at several axial locations in a turbulent nonpremixed flame of pure hydrogen. Since the geometries of the Drake[31] experiment and the present work are similar, the pure hydrogen data provides an additional data set with which to compare our Damköhler scaling, Eq.(8). Using the perfectly stirred reactor code, as done in Section 4.2, we estimate the ratio $\tau_{chemA}/\tau_{pureH_2}$ as $20\mu s / 18\mu s = 1.1$; the flow times τ_{flow} are implicitly handled by Eq.(8). With these assumptions, the Damköhler-temperature pairs are generated from the Drake[31] data and they are plotted with the present data in Fig. 9. The agreement is gratifying, especially in view of our uncertainty as to the effect of flame temperatrue on τ_{chem}.

4.4 Discussion

Historically, one of the main motivations for investigations into turbulent hydrogen combustion is that the fast chemistry nature of the flame allows one to concentrate on the turbulent aspects of the flow without the additional complication added by finite rate chemistry. The evolution from these large Damköhler number limits to near unity Damköhler numbers is logical from a fundamental point of view. However, with the renewed interest in hydrogen propulsion (e.g. Northam and Anderson [32]), the predictive capability that Fig. 9 and Eq.(8) provides, may guide development of combustion of hydrogen where the rates of mixing and of reaction are competitive.

Progress toward this predictive capability is an active research area; such progress requires that the time scale in Eq.(8) be generalized from jet mixing flows to a mixing time scale τ_{flow} that is independent of the system geometry and is based on length and velocity scale consistent with the microscopic dimensions of the reaction zone. Such a time scale, the Gibson scale, was recently proposed for premixed flames by Peters[33]; these new concepts have yet to be extended to nonpremixed flames. In addition to a generalized mixing time scale, an absolute chemical time scale is needed. Progress in this area is illustrated by the recent work on the combustion of hydrogen with fluorine combustion by Mungal and Frieler [34]. In spite of the existence of several elementary reaction steps, each of which has an associated time scale, a single time scale was developed from the rate of product appearance as determined from computer simulations of a premixed combustion bomb. Unfortunately, the direct extension of this determination of a chemical time scale to hydrogen (or hydrocarbon) in air combustion is questionable since the premixed hydrogen in air reactants have an induction time that would be measured in years; although the ignition rise time may still be a useful parameter. In order to overcome this ignition problem, it may be satisfactory to equate τ_{chem} to the computed transient response time of a perfectly stirred reactor that evolves from a near blowout residence time to arbitrarily large residence time such that the contents of the stirred reactor relax to chemical equilibrium.

5.0 CONCLUSION

The use of laser Raman scattering techniques in a turbulent hydrogen flame have provided instantaneous and local measurements of major combustion species (H_2, O_2, H_2O, and N_2) and temperature. These measurements have been performed in conditions where effects of finite chemical rates are observed. These conditions have been produced when the flame is burning with very high mixing rates (Reynolds numbers from 8,500 to 20,000). An analogous situation is generated when the fuel is diluted with nitrogen, depressing the chemical reaction rate from its large value for pure hydrogen. Those two types of conditions have been linked by means of the Damköhler number. Our data show that the reduction of the flame temperature can be plotted on a same curve, scaled by the Damköhler number whether the departures from chemical equilibrium are from increased mixing rates or reduced chemical kinetic rates.

These quantitative measurements will guide development of combustion of hydrogen where the rates of mixing and of reaction are competitive.

6.0 ACKNOWLEDGMENTS

This research is supported by the Department of Energy, Office of Basic Energy Sciences, Division of Chemical Sciences.

Dr. P. Magre is a Combustion Research Facility Visiting Scientist sponsored by DRET, Direction des Recherches et Etudes Techniques, French Ministry of Defense.

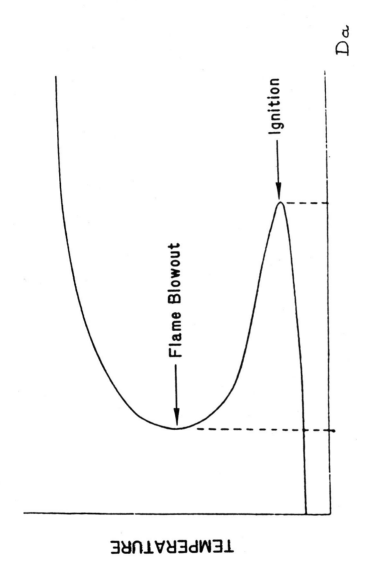

Fig.1 The flame temperature versus Damköhler number illustrates the 'S' shaped curve that is typical of most combustion systems.

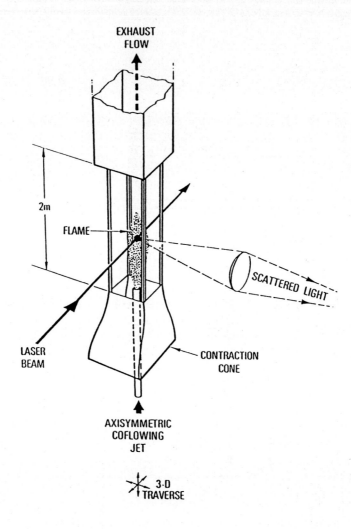

Fig.2 This illustration of the Sandia combustion tunnel shows that the tunnel moves relative to fixed optics.

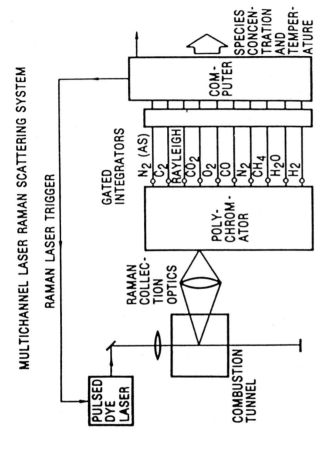

Fig.3 This figure illustrates the laser Raman scattering system.

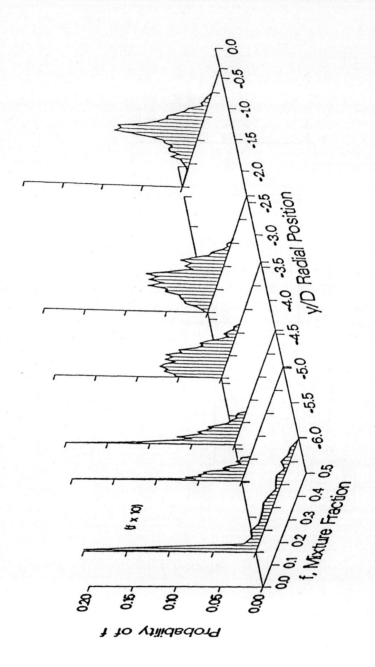

Fig.4 Radial profile of mixture fraction PDF $P(f)$, Case A50.

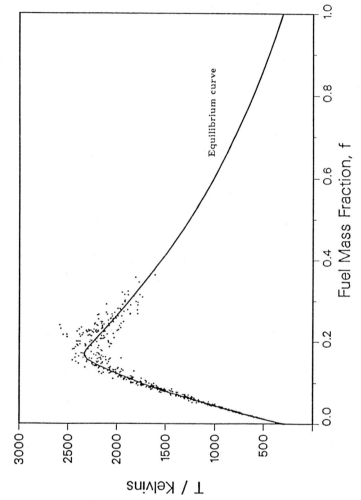

Fig.5 Ensemble of instantaneous temperature versus mixture fraction for Case A50.

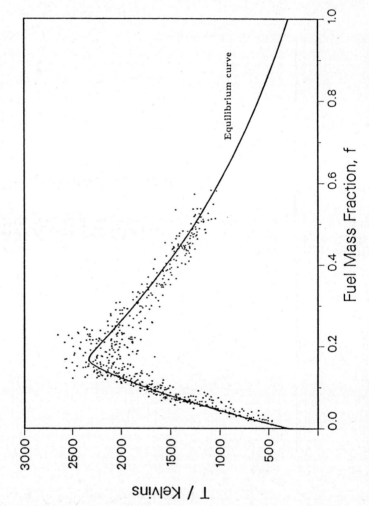

Fig.6 Ensemble of instantaneous temperature versus mixture fraction for Case A30.

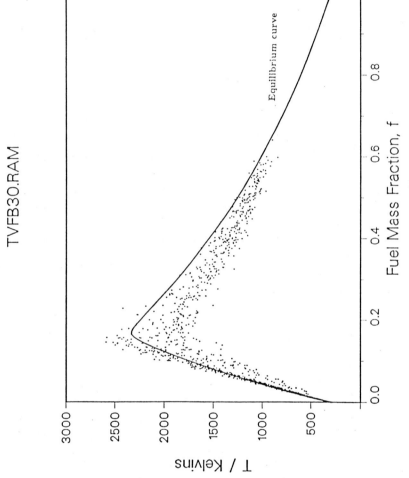

Fig.7 Ensemble of instantaneous temperature versus mixture fraction for Case B30.

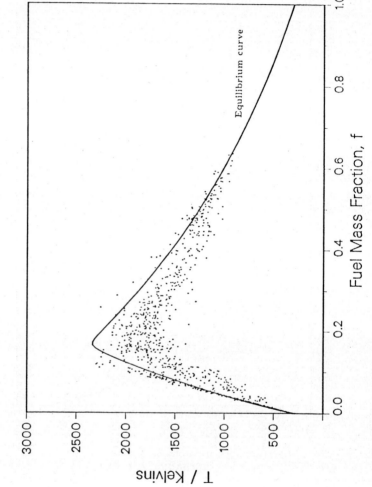

Fig.8 Ensemble of instantaneous temperature versus mixture fraction for Case C30.

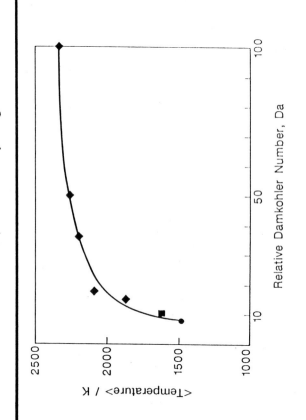

Fig.9 Measured temperature T_{flame} versus Damköhler number; this is the top branch of the general curve depicted in Fig. 1.

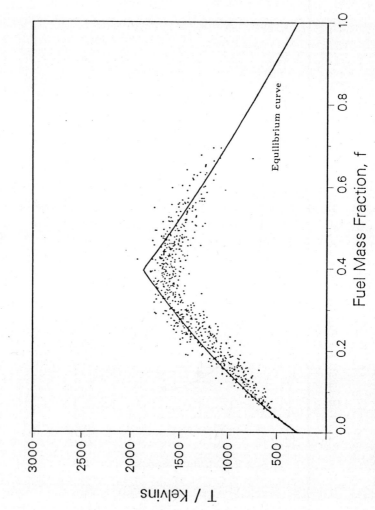

Fig.10 Ensemble of instantaneous temperature versus mixture fraction for nitrogen diluted fuel. Case D30.

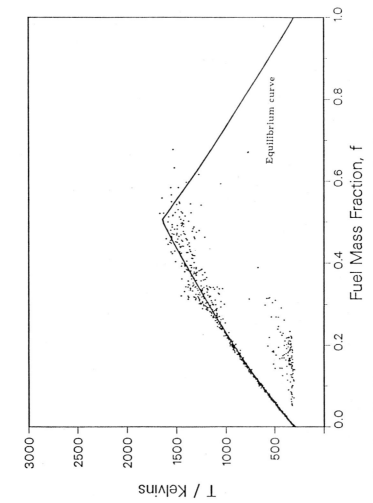

Fig.11 Ensemble of instantaneous temperature versus mixture fraction for nitrogen diluted fuel, Case E30.

7.0 REFERENCES

[1] Bilger, R. W., "Turbulent Jet Diffusion Flames", Prog. Energ. Combust. Sciences,1 87-109,1976

[2] Mitchell, R. E., "Chemical Element Diffusion Factors for Use in the Conserved Scalar Approach to Diffusion Flame Modeling", Western States Section/The Combustion Institute, Fall Meeting, 1980.

[3] Libby, P. A. and Williams, F. A., "Fundamentals Aspects", in *Turbulent Reacting Flows*, (P. A. Libby and F. A. Williams, Eds.), Springer-Verlag, New York,1980.

[4] Jones, W. P. and Whitelaw, J. H., "Modeling and Measurements in Turbulent Combustion", *Twentieth Symposium (International) on Combustion*, Pittsburg, The Combustion Institute, pp 233-249, 1984.

[5] Dibble, R. W., Kollman, W., and Schefer, R. W., "Conserved Scalar Fluxes Measured in a Turbulent Nonpremixed Flame by Combined Laser Doppler Velocimetry and Laser Raman Scattering", Comb. and Flame 55:3, pp 307-321, 1984.

[6] Starner, S. H. and Bilger, R. W. (1983). "Differential Diffusion Effects on Measurements in Turbulent Diffusion Flames by the Mie Scattering Technique", *Flames, Lasers, and Reactive Systems* (Bowen, J.R. et al Eds.), Prog. in Astronautics and Aeronautics, Vol. 88, American Institute of Aeronautics and Astronautics, New York, p. 81.

[7] Mungal, M. G., Dimotakis, P. E., and Broadwell, J. E., "Turbulent Mixing and Combustion in a Reacting Shear Layer", *AIAA J.* 22, pp 797-800, 1984.

[8] Wall, T. F., Duong, H.T., Stewart, I. M., and Truelove, J. S., "Radiative Heat Transfer in Furnaces: Flame and Surface Models of the IFRF M1- and M2-Trials", *Nineteenth Symposium (International) on Combustion*, The Combustion Institute, pp 537-547, 1982.

[9] Bilger, R. W., "Molecular Transport Effects in Turbulent Diffusion Flames at Moderate Reynolds Numbers", *AIAA J.* 20:962, 1982.

[10] Bilger, R. W., and Dibble, R. W., "Differential Molecular Diffusion Effects in Turbulent Mixing", Combustion Science Technology, Vol.28, pp 161-172, 1982.

[11] Liew, S. K.,Bray, K. N. C., and Moss, J. B., "A Stretched Laminar Flamelet Model of Nonpremixed Combustion", Comb. Flame 51, 199, 1984.

[12] Bilger, R. W., "Turbulent Flows with Nonpremixed Reactants," *Turbulent Reacting Flows*, (P.A.Libby and F.A.Williams,Eds.), Springer-Verlag, New York, pp. 65-113, 1980.

[13] Janicka, J. and Kollmann, W., "A Two-Variables Formalism for the Treatment of Chemical Reactions in Turbulent H_2-Air Diffusion Flames", *Seventeenth Symposium (International) on Combustion*, The Combustion Institute,pp. 421-430, 1979.

[14] Pope, S., "Monte Carlo Calculations of Premixed Turbulent Flames", *Eighteenth Symposium (International) on Combustion*, The Combustion Institute, pp 1001-1010, 1981.

[15] Eickhof, H., "Turbulent Hydrocarbon Jet Flames", Prog. Energ. Combust. Sciences, 8 159, 1982.

[16] Peters, N., and Donnerhack, S., "Structure and Similiraty of Nitric Oxide Production in Turbulent Diffusion Flames", *Eighteenth Symposium (International) on Combustion*, The Combustion Institute, pp 33-42, 1981.

[17] Janicka, J. and Peters, N., "Prediction of Turbulent Jet Diffusion Flame Lift-Off Using a PDF Transport Equation", *Nineteenth Symposium (International) on Combustion*, The Combustion Institute, pp 367-374, 1982.

[18] Peters, N., "Laminar Diffusion Flamelet Models in a Nonpremixed Combustion" Prog. Energy Combust. Sci., 10 p.319, 1984.

[19] Williams, F. A., "Combustion Theory", Second Ed., Benjamin/Cummings, Menlo Park, CA, 1985.

[20] Strehlow, R., "Combustion Fundamentals", McGraw-Hill, 1984.

[21] Eckbreth, A. C., "Recent Advances in Laser Diagnostics for Temperature and Species Concentration in Combustion", *Eighteenth Symposium (International) on Combustion*, The Combustion Institute, pp 1471-1488, 1981.

[22] Rahn, L. A., Mattern, P. L., and Farrow, R. L. (1981), "A Comparison of Coherent and Spontaneous Raman Combustion Diagnostics", *Eighteenth Symposium (International) on Combustion*, The Combustion Institute pp 1533-1542, 1981.

[23] Dibble, R. W., Schefer, R. W., Chen, J. Y., Hartmann, V., and Kollmann, W., "Velocity and Density Measurements in a Turbulent Nonpremixed Flame with Comparison to a Numerical Model" presented to the Spring Meeting of the Western States Section of the Combustion Institute, Banff, Cananda, 28-30 April 1986. to be submitted to Comb. and Flame.

[24] Dibble, R. W., Magre, P., Schefer, R. W., Chen, J. Y., Hartmann, V., and Kollmann, W., "Simultaneous Mixture Fraction and Velocity Measurements in a Turbulent Nonpremixed Flame", paper AIAA 86-1666, AIAA/SAE/ASME 22nd Joint Propulsion Conference, Huntsville, AL, 16-18 June 1986, to be submitted to Comb. Science and Tech.

[25] Dibble, R. W., and Hollenbach, R. E., "Laser Rayleigh Thermometry in Turbulent Flames", *Eighteenth Symposium (International) on Combustion*, The Combustion Institute, pp 1489-1499, 1981.

[26] Drake, M. C., Bilger, R. W. and Starner, S. H.,"Raman Measurements and Conserved Scalar Modeling in Turbulent Diffusion Flames", *Nineteenth Symposium (International) on Combustion*, The Combustion Institute, 1983, pp. 459-467.

[27] Masri, A., Dibble, R. W., and Bilger, R. W., "Simultaneous Concentration Measurements in Nonpremixed Flame of Methane", to be submitted to Combustion and Flame, 1987.

[28] Hinze, J. O., "Turbulence", 2nd Edition, McGraw-Hill, 1975.

[29] Everett, K. W., and Robins, A. G., "The Development and Structure of Turbulent Plane Jets", JFM 88, pp 563-583, 1978.

[30] Glarborg, P., and Kee, R. J., Sandia Report available from NTIS as SAND86-8209.

[31] Drake, M. C., *Twenty First Symposium (International) on Combustion*, 1986.

[32] Northam, G. B., and Anderson, G. Y., paper AIAA86-0159, AIAA 24th Aerospace Sciences Meeting, Reno, 6-9 Jan. 1986.

[33] Peters, N., *Twenty First Symposium (International) on Combustion*, 1986.

[34] Mungal, M. G., and Frieler, C. E., GALCIT Report FM85-01, 1985. Submitted to Combustion and Flame.

CARS STUDY OF PREMIXED TURBULENT COMBUSTION IN A HIGH VELOCITY FLOW

P. Magre, P. Moreau, G. Collin, R. Borghi, M. Péalat and J.P. Taran
Office National d'Etudes et de Recherches Aérospatiales,
BP 72, 92322 Châtillon Cedex, FRANCE

INTRODUCTION

Premixed turbulent flames offer interesting possibilities to study turbulent combustion and permit confrontation between experiments and calculations. In addition, there is direct application to the design of afterburners for turbojet engines, if the flow velocity is fast enough. For both reasons ONERA have undertaken continously over the last ten years an experimental and theoretical study of such a turbulent flame, stabilized either by a planar hot jet parallel to the main stream, or by a sudden enlargment. The studies began with shadowgraphic visualizations and wall pressure measurements, followed by gas sampling and analysis by chromatography in order to know the mean concentration profile inside the flame. The velocity profile was studied in detail later, with velocity and turbulence measurements by laser Doppler anemometry. An attempt to measure some aspects of the temperature profile was made also by means of a pneumatic probe and emission-absorption spectroscopy. Very recently, a programme of measurements by CARS (Coherent anti-Stokes Raman Spectroscopy) has led to measurements of the fluctuating temperature. These results are presented here.

1 - COMBUSTOR

The experimental device is a two dimensional duct, shown in fig. 1. Most of the tests have been performed with a flame stabilized by a 2D hot jet, parallel to the main stream, but we have also studied the case when the pilot jet is replaced by a simple downward facing step, playing the role of a flame holder. The fuel is methane, at different equivalence ratios. Different velocities of the jets have been used.

Actually, the main characteristics of the device is the high level of turbulence in the flow, even without combustion. Indeed the shear between the two jets can be very large when the velocities on both sides of the separating plate differ by 55 m/s; in addition, the turbulence in the incoming premixed stream is itself very large: 4 to 8 m/s of velocity fluctuation (R.M.S.) has been measured just upstream of the shear. By comparison with the laminar flame speed of methane and air, which is of the order of 0.5 m/s, we can see that the turbulence is very high.

The turbulent flame has been studied firstly by wall pressure measurements; the results are presented in [1]. The main result has been that a *half flame length*, defined from the pressure gradient, depends both on the velocity of the main flow and on its equivalence ratio. This flame length could be related to a *global turbulent flame speed*, which is shown here to be very large (5 m/s is an order of magnitude), and to depend on the equivalence ratio as well as the inlet velocity. Therefore, one can suspect that turbulent combustion is controlled by both turbulence and chemistry.

A detailed presentation of our current understanding of this flame is given in [2]. The structure of the flame is revealed by high speed shadowgraphy, appears to be a thick grainy zone with slow undulations. When the turbulence intensity becomes very high, small scale fluctuations begin to appear within the flame front itself. From the analysis of [3], this phenomenon could lead to the thickening of the turbulent flame by small scale turbulence while this flame remains wrinkled by the largest fluctuations. In effect, the Kolmogorov microscale η is found to be smaller than the laminar flame thickness e_L; the Kolmogorov time scale τ_K is shorter than the chemical time τ_c. The estimated figures are $0.07 \leq \eta \leq 0.14$ mm, $e_L \simeq 0.25$ mm, 10^{-5} s $\leq \tau_K \leq 1.4 \, 10^{-5}$ s, $\tau_c \simeq 10^{-3}$ s.

Using the estimates of [3] one finds a thickness between 1 and 2 cm. Measurements of temperature fluctuations, with a sufficient spatial resolution, would be of interest to verify this point. The CARS measurements will be discussed later with this in mind. The study of Ganji and Sawyer [4] seems to show a thinner flame; this can readily be explained, as shown in [3], because η is larger than in our experiment. The flame studied in detail by Hill, Talbot et al. [5,6] is also likely to be more wrinkled since η is here larger than e_L.

CARS system

The CARS spectrometer consists of two compact optical benches 1.5m x 0.5 m, weighing about 50 kg. These characteristics are particularly important when the CARS apparatus has to be installed to study combustion facilities of industrial character which are often built with no concern to optical access.

The main bench is the laser source assembly. It is virtually identical to that described in [7]. The core of the system is a stable frequency, single mode, passively Q-switched Nd-YAG laser, with two amplifiers. The energy of the frequency-doubled output is 170 mJ, delivered in 15 ns pulses at a rate of 10 Hz (ω_1 beam). However, the detector carry-over necessitates limiting the repetition rate at a maximum of 1 Hz to avoid overlap of consecutive spectra on the target. Part of the ω_1 beam also pumps a broadband oscillator-amplifier dye chain (ω_2 beam). To study N_2, the broadband spectrum is centered around 607 nm. A 40 μm air-spaced Fabry-Pérot etalon facilitates tuning of the dye laser and reduces its spectral width to 60 cm⁻¹ (FWHM). Several optics allow different beam arrangements: BOXCARS, parallel beams; crossed or parallel polarizations. At the output end of the laser source bench, the ω_1 and ω_2 beam energies are respectively 75 mJ and 3 mJ.

The other bench is the spectrometer bench. It supports several handling optics, dichroic filters, the spectrometer components and the detector. The spectrometer consists of a concave, 50 mm diameter, 2100 gr/mm holographic grating. Magnification optics (x2) image the spectrometer output on the target of the ISIT vidicon detector. The resolution and the dispersion of the spectrometer-detector system are respectively 1.2 cm⁻¹ (FWHM) and 0.235 cm⁻¹/ channel. The detector controller is connected to a PDP 11/23 minicomputer.

The computer is assigned two tasks. During the experiment, it controls the scanning procedure of the ISIT detector and fires the laser synchronously with the detector scanning. At the end of the experiment, it processes the spectra to obtain temperatures. This work takes 5 s per spectrum. The scanning procedure and the fitting program have been described in a previous paper [8].

CARS optical arrangement

Spatial resolution and measurement bias

It is evident that the instantaneous flame front does not present any particular symmetry even though the experimental burner has a 2D geometry. However the gradients are thought to be less stiff in a direction orthogonal to the symmetry plane (i.e. the plane of fig. 1), and, as a consequence, we have aligned the axis of the cylindrical probe volume along this direction.

Concerning the length of this probe volume, two reasons impose the arrangement giving the best spatial resolution: i.e. the BOXCARS arrangement [9]. We are interested in resolving the flame structure as finely as possible and must try to avoid simultaneous presence of hot and cold gases inside the probe volume. When such a situation occurs, the CARS spectrum created is the result of the complex blend of spectra of hot and cold gases. The problem is numerically simulated by adding the anti-Stokes fields created by a cold pocket of unburnt gases and a hot one of burnt gases, separated by a very thin flame front (typically 300 μm) in the measurement volume. The calculation has been made by mixing p molecules at T_{hot} = 2200 K and (1 - p) molecules at T_{cold} = 600 K, these values being close to the extrema expected to be found in the burner. For simplicity, we assume the focal volume to be cylindrical with total length L. We have noted that despite the non-canonical shape of these synthetic spectra, our data processing program is able to find a temperature T_{CARS} which gives the best fit between the simulated spectrum and the theoretical ones using our least squares criterion [8]. As expected, this temperature T_{CARS} is underestimated with respect to T_{th} the thermodynamic temperature of the mixture which is the temperature that the gases in the focal volume would have after being homogeneized, their enthalpy being conserved. The prevalence of the cold molecule contribution in the cold band of the Raman spectrum explains this result. Figure 6a presents T_{CARS} and T_{th} versus l_R, the ratio of the length of the measurement volume occupied by hot gases to its length L. The quantity l_R is derived from the parameter p using the simple relationships:

$$l_R = \frac{p}{(1 - p)\frac{600}{2200} + p}$$

The discrepancy shown in figure 2a makes the use of CARS in turbulent combustion questionable. We have to point out that the results depend strongly on the computer code used for the data processing. More precisely, since experimental and theoretical spectra are fitted in log scale [8], hot gases contribute more significantly than if the comparison were made in a linear scale.

However, the conditions of the calculation are pessimistic for two reasons. Firstly, the comparison between the Kolmogorov time scale and the chemical time suggests that the actual temperatures in the microstructures are distributed between the 600 and 2200 K extrema. If we reestimate the previous bias $T_{th} - T_{CARS}$ for a mixture of gases at T_{cold} = 1000 K and T_{hot} = 1800 K, we see in figure 2a that it is reduced significantly and remains lower than 100 K. Secondly, this potential problem is acute mainly in the flame front and poses itself whenever the flame front is located within the probe volume. We are not able to estimate the probability of this phenomenon but we tried to see how a temperature histogram formed with two Dirac peaks of equal probability at T_{cold} = 600 K and T_{hot} = 2200 K (the most severe conditions encountered in our burner) would be biased by spatial averaging. For this purpose we arbitrarily suppose that the measurement volume lies in cold gases (l_R = 0) 30% of the time, in hot gases (l_R = 1) 30% of the time, and that the rest of the time both hot and cold gases are in the probe volume. In that case, we assume the distribution function of l_R to be flat between 0 and 1. Figure 2b displays the distorsion of the original distribution. The distribution of the thermodynamic temperature is also shown for the same statistical distribution of l_R. It is important to point out that the two peaks remain located at the same place even if some intermediate values of T_{CARS} are now found, specially in the cold regions. Apparently, the hypothetical spatial averaging does not obscure the bimodal shape of the temperature distribution as can also be seen in the results presented in section 2.

For our experimental conditions (i.e. with 500 mm focal length focussing, beam diameter 6 mm and separation 8 mm) the total length of the probe volume is 12 mm. This value was measured experimentally. Increasing beam separation would improve it, but would also reduce signal strength below acceptable levels. Obviously, this length is not short enough to resolve either the Kolmogorov microscale or the laminar flame front.

Background suppression

For the present experiment, the non resonant background was not suppressed [10,12] to preserve the best signal to noise ratio. Fortunately, the non resonant contribution was shown to be constant whatever the advancement of the reactions. An estimate of the non resonant background can be made using the values of [13,14]. One calculates the same value for a fresh (air + methane) $\varphi = 1$ mixture and for a representative mixture of burnt gases [N_2, CO_2, H_2O]. This value for the non resonant susceptibility is ~ 1.15 times that of air. It is realistic to assume that the value is about the same in a combusting mixture. It would be strongly different if heavier hydrocarbons were used for the fuel [15]. Numerical simulation shows that a $\pm 20\%$ error on the non resonant susceptibility causes an error of ± 10 K at 1500 K; overestimating the background causes the temperature to be underestimated.

General layout

High noise levels and wide temperature changes in the combustion facility prevented installation of the CARS set up close to the burner. The two optical benches have been set up in the remote control room which implies a 15 m long optical path from the laser source bench to the burner and back to the spectrometer. The optical path is diagrammed in fig. 3.

A reference channel is mounted on a 1.5 m x 0.5 m cast aluminium table installed under the burner to support the optics of the signal channel. The need for a reference channel has been discussed in [8]. The signal and reference channels are arranged in series: the laser beams pass through the reference cell and then through the burner. The focusing and recollimating optics are 500 mm focal length, A.R. coated, air-spaced achromats.

The usual windowed reference cell is replaced by a small duct flushed with a stream of argon. The pressure at the nozzle exceeds the pressure in the test room by 1-3 mbar. This system, which is really convenient because it suppresses windows, is applicable only with the BOXCARS beam arrangement; in the collinear beam arrangement some resonant N_2 CARS signal is generated out of the argon stream, far from the focus of the achromat.

Optical access to the combustion chamber is done through two 1 x10 mm horizontal slits machined in the side walls of the burner. The use of windows is avoided. These would be damaged by the high power density ω_1 beams, the walls being only 5 cm away from the focus of the achromat. Ideally, two 1 mm diameter holes should suffice for passing the beams through the walls. However there is a specific problem with the combustion facility which makes it necessary to use the slits; this is because the combustion chamber is installed at the end of a pipe which expands when warm air is supplied, causing its horizontal motion. The slits then allow us to collect CARS data and perform checks on the alignment before and during the warm up periods.

ANALYSIS OF THE RESULTS

The purpose of the experiments performed with CARS in the conditions of fig. 1 was twofold:

i) *To gain more insight* into the structure of the developing flame, *qualitatively as well as quantitatively*. The shape of the spatially resolved temperature p.d.f. should help us to establish if the flame is purely wrinkled or also thickened by small scale turbulence. This structure being possibly not the same at each point in the flow, measurements at different locations are in order. In addition, transverse profiles of mean temperature (Favre averaged and conventionally averaged) as well as the variance of temperature fluctuations are useful quantities for comparisons with calculations.

ii) *To study the temperature field within the recirculation zone itself.* No measurements can be made within the recirculation zone by means of probes, and interesting questions remain unanswered:
- is the recirculation zone similar to a well-stirred reactor, with a constant mean temperature within the major part of it?
- what is the intensity of the fluctuations within it? Are two peaks clearly visible on the p.d.f., showing a large unmixedness?

The results will be discussed with respect to these points. Prior to this discussion, we recall that temperature accuracy deduced from measurements in a furnace [8] ranges from 20 K at 300 K to 70 K at 1600 K. In the present experiment, there are two locations in the burner where the temperature should be constant: at the exit of the pilot flame duct and in the fresh mixture. We found respectively 2000 K ± 170 K and 560 ± 35 K. The slight increase with respect to the furnace values is explained by the turbulence in the flows.

Figure 4 shows the p.d.f. obtained for the pure mixing of hot gases from the pilot with the main stream without methane. As the observation point is moved through a vertical section, the p.d.f. changes and shows one peak and progressively another one. In the middle, the p.d.f. displays a very wide shape, in which it is possible, sometimes, to distinguish two broad bumps. The p.d.f. can be compared with the p.d.f. measured by Brown and Roshko [16], in their well known study of an air-helium mixing zone. It is to be noticed that, when two bumps appear, the location of their maxima is not stable; the low temperature and high temperature bumps are shifted towards each other as they are interacting.

Figure 5 shows the p.d.f. found at the same positions, when methane is added to the main stream and the combustion occurs. The comparison with figures 4a and 4b is interesting.

In the first section, 4.2 cm after the beginning of the mixing, only a very small amount of fuel has been burnt: the p.d.f. at y = 2.5 cm and 1 cm look very similar; the one at 2 cm shows clearly a decrease of the peak of fresh gases; the same tendency, but less pronounced, is visible for the one at 1.5 cm. These reductions in peak intensity are due to combustion; consequently, an increase of probability density is noticed at a higher temperature, but this increase does not appear at the adiabatic temperature which is here 2350 K; for y = 1.5 cm, a peak appears at 1800 K, and for y = 2 cm, no well defined peak is seen and the probability density is increased over all the range between 1000 and 1800 K.

In the second section, the influence of the combustion is quite significant. The low temperature peaks are shifted towards high temperatures in a manner quite comparable with the non combusting case, but their areas are clearly smaller. High temperature peaks are appearing also, but they are again very wide and their location is not the adiabatic flame temperature; they are still centered at a lower temperature, yet higher than for the previous section.

Two explanations can be proposed:
1/ Heat losses
This can be invoked close to the walls, for instance for the y = 10 mm p.d.f. But this effect is not expected to be pronounced in the center of the channel at y = 20 mm and higher. Moreover radiation losses are not likely to be strong.
2/ Too small a reaction rate
In fact the hot peaks in the section x = 42 mm are found at about 1800 K, and at x = 122 mm they are at 2000 K. They would probably reach closer to 2350 K further downstream.

In order to quantify how close the flame is to a wrinkled turbulent flame, we have plotted: $\overline{T'^2}$ and $\left(T_{max} - T\right)\left(T - T_{min}\right)$ as a function of T; the ratio θ of these functions has to always be smaller than 1 [1,17]. Here, T' is the conventional temperature fluctuation. T_{min} is the minimum temperature of the fresh mixture; T_{max} must be taken as either the adiabatic temperature of the combustion with $\varphi = 0.8$, or the maximum temperature of the pilot flames; these two values are very close, and we will choose $T_{max} = 2250$ K. The curves are shown figure 6, for both sections, and for $\varphi = 0$ or $\varphi = 0.8$; it is seen that θ is about 0.5 at the first section, and closer to 1 in the second one. In both sections, the characteristic time of the chemistry is then not short enough to ensure a purely wrinkled flame; this tends to be less true in the second section, where the turbulence characteristic time is probably larger (the length scale being larger and the turbulence kinetic energy not larger).

The shapes of the p.d.f.'s in figure 5 suggest one more comment. We have noted that the peak corresponding to unburnt gases is not situated at the same temperature for the different positions and that it moves slightly toward the high temperatures, as the mixing and combustion proceed. Both phenomena complicate the choice of good "presumed" shapes for a turbulent combustion model. With this in mind, the shape obtained with the Lagrangian Eulerian model of [18] seems to be more realistic than that of [19].

CARS results can also be compared with mean and fluctuating temperatures measured previously in the same burner. The first measurements of mean temperatures have been done with a pneumatic probe [20]; later, the mean and the fluctuating temperatures have been obtained with an I.R. method working on both emission and absorption of CO_2. This method has a very good time resolution but is not able to give us a point measurement since the signal was integrated along the beam path through the flow. Then, we can expect that only the large, two dimensional fluctuations were properly detected. Figure 7 shows a comparison between probe, IR, and CARS results. The discrepancies are not negligible, specially with combustion: differences as large as 200 K between probe and CARS can be noticed: CARS and IR give closer measurements, except within the pilot jet where the IR seems to strongly underestimate the temperature. Compared to the two methods used previously, CARS appears to us as the most reliable technique: it is not intrusive, unlike probe technique, and affords a better spatial resolution than both classical techniques. The IR method also has the drawback to necessitate the injection of $\sim$ 5% CO_2 in the fresh mixture.

Figure 8 shows the p.d.f.'s obtained in the case of the downward facing step. Within the recirculation zone, a single peak p.d.f. is displayed without ambiguity. The mean temperature is sligthly lower than previously, with the pilot: this is probably caused both by heat losses through the wall and by incomplete combustion. Within the mixing layer, the pdf is bimodal, as expected.

The figures 9 and 10 show the profiles of mean temperature and variance; a very similar behavior is seen for the pilot and the downstream facing step. In particular, the advancement of the combustion in the first section is probably quite small in both cases.

The two questions of the beginning of Chapter 2 now can be answered without ambiguity:
i) The recirculation zone deviates clearly from a well stirred reactor:

the mean temperature is constant only in a third of its total volume and the shape of the p.d.f. is clearly varying from point to point.
ii) The fluctuations are clearly bimodal in the mixing layer between the recirculation zone and the external flow: this result looks like what is observed in the case of the pilot jet, but ten centimeters downstream. Within the core of the recirculation, the p.d.f. is singly peaked, although fairly thickened. We conclude, from this, that flamelets are not seen within the recirculation zone, with our spatial resolution.

CONCLUSIONS

The CARS measurements presented here, in the large experimental set up and under adverse conditions, have been made possible only thanks to a special organization and because great precautions were taken. The interest of the results is twofold:
i) they shed light on the structure of the turbulent premixed flame at high velocity, stabilized either by a pilot jet or by a flame holder. The analysis of instantaneous temperature p.d.f. has shown that the turbulent flame is wrinkled by the large scale turbulence but thickened by small scale turbulence. The characteristic chemical time is too long to make the flame purely wrinkled. When the flame is stabilized by a recirculation zone, we found similar features for the temperature p.d.f. (bimodal aspect) in the mixing layer. The recirculation zone does not exhibit a constant temperature profile and cannot be considered as a well-stirred reactor,
ii) they will allow a detailed comparison with numerical calculations, which include turbulent combustion models. Such a comparison requires extensive measurements at several points within the flame, and some features of the combustion shown in this study have already been used in order to improve the turbulent combustion models. For instance, an ignition temperature has been introduced in the model [21] to take into account the experimental fact of the non infinitely fast chemistry. Finite rate chemistry would allow a better prediction of the temperature fluctuations. This could be done via Eulerian Lagrangian models [18]. Such models would be validated by joint p.d.f.'s of scalar quantities such as temperature and concentration of one of the major species. We are planning further experiments in this area using CARS.

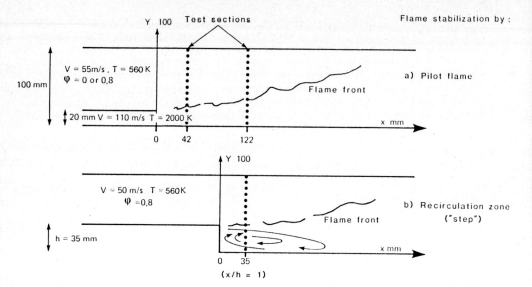

Figure 1 - The test set up

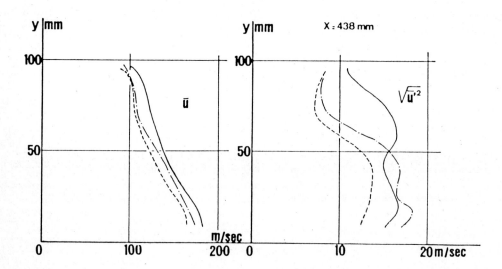

Figure 2 - a) Effect of spatial averaging on: T_{th} (----) and T_{CARS} (____) versus vol (% of measurement volume with hot gases for different couples of values T_{cold} and T_{hot}).
b) Distorsion of an original p.d.f. of T: p.d.f. of thermodynamic temperature (---), p.d.f. of temperature T_{CARS} deduced from the spectra (____).

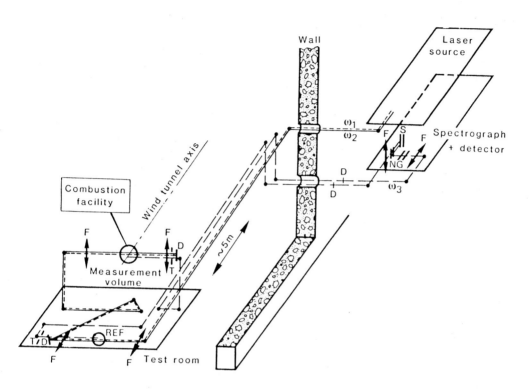

Figure 3 - Experimental arrangement. Code: •, mirror; D, dichroic; F, lens; T, trap for the collinear anti-Stokes beam; REF, reference probe volume; NG, neutral glass filter; S, slit of the spectrograph.

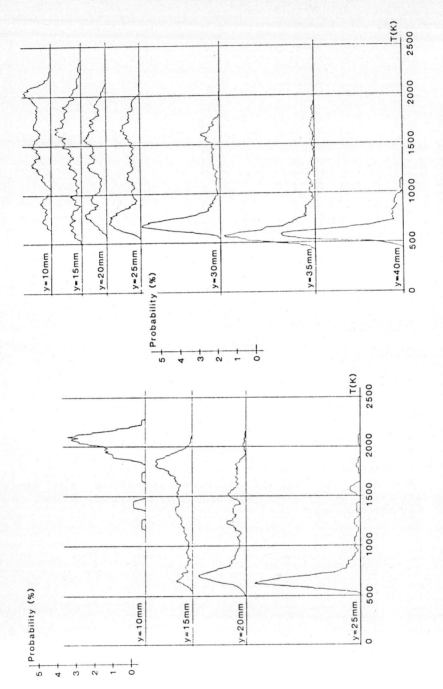

Figure 4 - Temperature p.d.f. obtained for the pure mixing of hot gases from the pilot with the main stream:
a) x = 42 mm;
b) x = 122 mm.

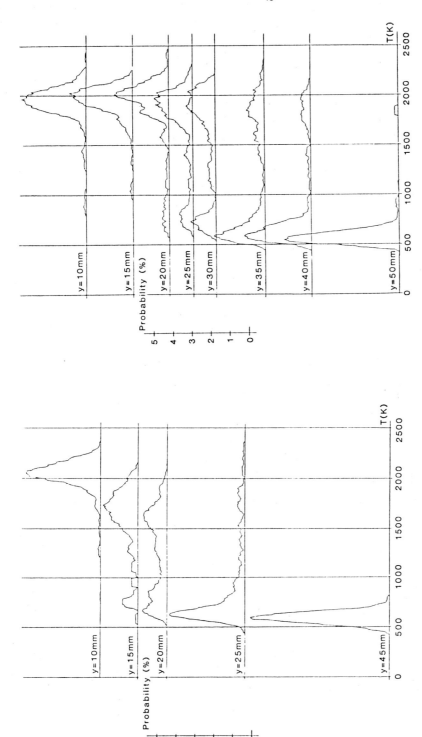

Figure 5 - Temperature p.d.f. obtained in the turbulent flame stabilized by the pilot flame. $\varphi = 0.8$ in the main stream a) x = 42 mm; b) x = 122 mm.

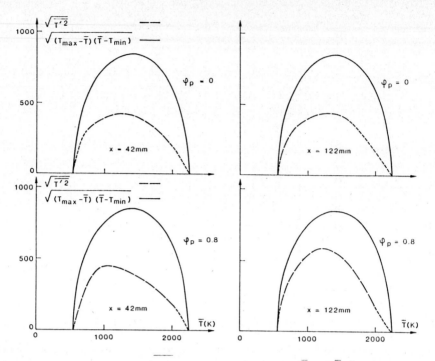

Figure 6 - Comparison of T'^2 (—) and $(T_{max} - \bar{T})(\bar{T} - T_{min})$ (---) for different conditions.

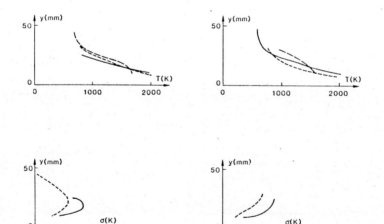

Figure 7 - Mean temperatures $\bar{T}$ and variance σ at x = 42 mm obtained by CARS (———), I.R. (__ __) and probe (----)
  a) for the pure mixing of the pilot flame with the main stream
  b) in the turbulent flame (φ = 0.8) stabilized by the pilot flame.

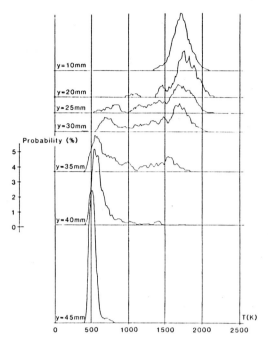

Figure 8 – Temperature p.d.f. obtained when the flame is stabilized by the downward facing step (x = 35 mm).

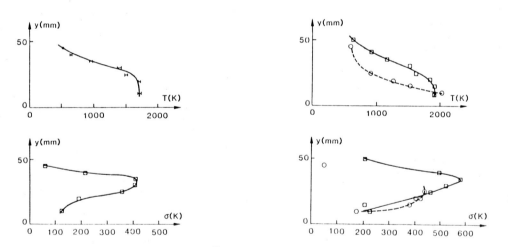

Figure 9 – Mean temperatures $\overline{T}$ and variance σ for the downward facing step experiment (x = 35 mm).

Figure 10 – Mean temperature $\overline{T}$ and variance σ for the turbulent flame (φ = 0.8) stabilized by the pilot flame: x = 42 mm (-o-); x = 122 mm (-□-).

REFERENCES

[1] - Singh, V.P., Borghi, R., Moreau P., *2ème Symp. Int. sur la Dynamique des Réactions chimiques : "flammes comme réactions dans les écoulements"*, Padoue, Déc. 1975.

[2] - Magre, P., Moreau, P., Collin, G., Borghi, R. and Péalat, M., Combust. Flame (to be published).

[3] - Borghi, R,. in *Recent Advances in the Aeronautical Sciences*, (C. Bruno, C. Casci Ed.), Plenum Press, 1984.

[4] - Ganji, A.R., Sawyer, R.F., AIAA Paper 79-0017, 17th Aerospace Science Meeting, New Orleans, 1979.

[5] - Bill, R.G., Namer, I., Talbot L., Cheng, R.K., Robben, F., Combust. Flame, 43:229 (1981).

[6] - Bill, R.G., Namer I., Talbot, L., Robben, F., Combust. Flame, 44:277 (1982).

[7] - Péalat, M., Taran, J.P., Moya, F., Opt. and Laser Tech. 12:21 (1980).

[8] - Péalat, M., Bouchardy, P., Lefèbvre, M., Taran, J.P., App. Optics., 24:1012 (1985).

[9] - Eckbreth, A.C., Appl. Phys. Letters., 32:421 (1978).

[10] - Song, J.J., Eesley, G.L., Levenson, M.D., App. Phys. Letters 29:567 (1967).

[11] - Bunkin, A.F., Ivanov, S.G., Koroteev, N.I., Sov. Tech. Phys 3:182 (1979).

[12] - Rahn, L.A., Zych, L.J., Mattern, P.L., Optics Commun. 30:249 (1979).

[13] - Lundeen, T., Hov, S.Y., Nibler, J.W., J. Chem. Phys. 79:6301, (1983).

[14] - Greenhalgh, D.A., Hall, R.J., Porter, F.M., England, W.A., J. Raman Spect., 15:71 (1984).

[15] - Hall, R.J., Boedeker, L.R., Appl. Optics 23:1340 (1984).

[16] - Brown, G.L., Roshko, A., J. Fluid Mech., 64:817 (1974).

[17] - Bray, K.N.C., Libby, P.A., Phys. Fluids 19:1681-1701 (1976).

[18] - Borghi, R., Pourbaix, E., Recherche Aérospatiale 4:29 (1983).

[19] - Borghi, R., Dutoya, D., 17th International Combustion Symposium, Leeds, August 20-25 1978, p. 235-244, the Combustion Institute, 1979.

[20] - Moreau, P., *15th AIAA Aerospace Science Meeting*, Los Angeles, 1977.

[21] - Dupoirieux, F., Workshop on the gas flame structure, Novosibirsk, July 27-31 1986.

The Structure of Jet Diffusion Flames

W. M. Roquemore
AFWAL/Aero Propulsion Laboratory
Wright Patterson AFB, OH 45433-6563

L-D. Chen
The University of Iowa
Iowa City, IA 52242

L. P. Goss and W. F. Lynn
Systems Research Laboratories, Inc.
Dayton, OH 45440-3696

Abstract

This paper presents the structural characteristics of free, round, jet diffusion flames as obtained using a new 2D laser sheet lighting visualization technique referred to as the RMS (Reactive Mie Scattering) method. The results of analyzing photographs and high speed movies of flames using the RMS method are discussed in terms of the visible flame structure. The fuel velocity is varied from 0.16 to 17 m/s. The presence of large toroidal vortices formed outside the visible flame zone have been known for many years but their importance in determining the dynamic structure of free jet diffusion flames has not been fully appreciated. The influence of the outer vortices on flame structure is prevalent for near laminar and transitional flames and diminishes for near turbulent flames. They are believed to result from a Kelvin-Helmholtz type instability formed by a buoyantly driven shear layer. They appear to be responsible for flame flicker defined by the separation of the flame tip or the oscillations of the flame surface and for determining the shape of the mean, rms, and pdf radial profiles of temperature. Vortex structures have also been observed inside the visible flame zone. In transitional flames established by a contoured nozzle, these structures are shown to be on the scale of the 10 mm diameter nozzle, toroidal, and coherent for a long distance downstream. However, they may have only a minimal impact on the mean temperature characteristics of transitional flames. Their impact on the visible flame structure of near turbulent flames is large. At high fuel velocities, coalescence of the large vortices appear to be correlated with the formation of small 3D vortices which are randomly distributed in size and space. Collisions of the small vortices with the visible flame front produce small localized flamelets which are responsible for the wrinkled appearance of the visible flame surface. The localized stretching of the flame surface is believed to invoke finite rate chemistry effects. Indeed, collisions are observed where the flame stretch is large enough to cause localized holes to form in the flame surface. This appears to occur when the radial velocities of the inner vortices are large. Holes formed near the lip of the jet are postulated to be one mechanism that induces flame lift-off.

Introduction

The Reactive Mie Scattering laser sheet-lighting technique has demonstrated the capability of providing information on bluff-body stabilized[1,2] and jet diffusion flames[3,4] in more detail than obtained by conventional Schlieren and shadowgraphic techniques. The technique involves adding titanium tetrachloride (TiCl$_4$) vapor to both the dry co-flowing air and the fuel. The TiCl$_4$ mixes with the water product of the flame and reacts spontaneously, near-instantaneously and near-isothermally to form 0.7 to 5 micron diameter titanium dioxide (TiO$_2$) particles. The particles are visualized by Mie scattering from a sheet of laser light. This provides a 2D view of the flow instead of a path integrated result.

Much of the detail provided by the technique results because the particles tend to highlight the regions where the water product mixes with the fuel and air on a molecular level. Since the scattering cross-section is significantly reduced at higher temperatures,[5] the cooler regions of the flame are marked. Also, the diffusion of the TiO$_2$ particles is considerably less than that for gas molecules. Thus, the particles tend to highlight the convection rather than molecular diffusion processes. The 2D flow visualization technique will be referred to as the RMS (Reactive Mie Scattering) method since the technique uses a reactive media (TiCl$_4$) to form the Mie scattering particles.

This paper discusses some of the insights into the visible structure of jet diffusion flames provided by examining flame photographs obtained using the RMS technique. Photographs were taken of near laminar, transitional and lifted flames. CARS temperature measurements were made in a transitional flame and the results are interpreted in terms of the observed flame structure.

Experimental Methods

The experimental set-up for visualizing the flame by the RMS method is shown in Fig. 1. TiCl$_4$ vapor was added to both the dried annulus air and the fuel. The co-flowing air stream had a velocity of 15 cm/s which was sufficient to reduce the room air disturbances while not causing a significant effect on the visible flame structure. The annulus jet had a diameter of 245 mm. An 11 mm diameter tube was used for the flame shown in Fig. 3a, all other photographs were made with a 10 mm diameter nozzle which was contoured to provide a flat velocity profile at the exit. The fuel type, either methane or propane, is indicated for each figure.

Mie scattering from the TiO$_2$ particles were observed at right angles to the laser sheet. A pulsed Nd:YAG laser was used as a light source and was electronically triggered when the camera shutter was opened. Both the orange colored blackbody radiation from the soot particles and the green Mie scattered laser light were recorded on the film. The flame luminosity was integrated over the 2 ms camera shutter speed, while the Mie scattered light was captured during the 15 ns duration of the laser pulse. Additional information on the experimental set-up is given in Ref. 3.

The luminous flame surface appears to be 2D in the photographs which can be misleading, since it is really a 3D surface superimposed on the 2D photograph. However, using the color discrimination between the luminous flame and the 2D Mie scattered light, one can normally obtain a clear interpretation of the flame photographs. Unfortunately, the reader will be hampered by the published black and white photographs. To partly negate this problem, the flame, Mie scattered light, and other interesting features have been marked on some of the photographs.

The Coherent Anti-Stokes Raman Spectroscopy (CARS) system used to make temperature measurements was developed by Goss, et al.,[6] and only the salient features will be mentioned here. The system used the frequency doubled output from a pulsed Nd: YAG laser to: pump a broadband dye laser and as the pump beam in the CARS process. A temperature measurement was made during the 15 ns firing of the Nd: YAG laser. Fifteen hundred measurements were made at each spatial location. This sample size was shown to be sufficient to obtain stable mean, rms and pdf temperature measurements. The precision of each single shot CARS measurement in a constant temperature environment is about ±75 K. The precision of the mean is estimated to be ±10 K.

A folded BOXCARS configuration was used to achieve a spatial resolution of less than 2 mm in the direction of laser beam propagation. There is undoubtedly some spatial integration occurring in the measurement volume. This could broaden the pdf's, reduce the rms values, reduce the probability of measuring near stoichiometric temperatures and reduce the sharpness of the mean temperature gradients. The uncertainties in the temperature measurements resulting from the size of the measurement volume will depend on the characteristics of the flame being studied and are unknown. The impact of the measurement volume size was somewhat reduced by traversing the burner in a direction perpendicular to the propagation direction of the laser beams when making the radial profile measurements.

Results and Discussion

General Flame Structure

Yule, et al.[7] made detailed studies of the structure of a transitional premixed (propane+air) jet flame having an equivalence ratio of 10.4. This flame was stabilized on a contoured nozzle with a diameter of 25.4 mm. The mass averaged fuel exit velocity was 640 cm/s and Reynolds number was 10^4 (based on air properties). A 100 cm/s sheath of air surrounded the flame. Their sketches of the flame structure, obtained from studies of color Schlieren photographs, are illustrated in Fig. 2. Also included in Fig. 2, is a photograph of a transitional propane jet diffusion flame (Re = 5318) obtained with the RMS technique. This flame is stabilized on a 10 mm diameter contoured nozzle.

A very good qualitative agreement between the Schlieren and RMS results are shown in Fig. 2. Yule, et al.[7] identified two regions containing organized structures. One region, located outside of the flame surface, consisted of large slow moving vortices that occurred at a frequency of about 30 Hz. They referred to this region as an outer preheat zone. Another region, referred to as the inner preheat zone, was located inside the flame at the potential core interface. These type of structures are clearly identified in the flame photograph in Fig. 2. The structures inside the visible flame surface represent the fuel/H_2O product interface and the large structures outside the flame surface mark the air/H_2O product interface. The frequency of the outer structures in our experiments was about 15 Hz. We observed that the frequency of the outer structures increased with an increase in annulus air velocity. This may account, in part, for the 30 Hz frequency observed by Yule, et al. since their annulus flow was considerably larger than ours. The characteristics of the outer and inner structures and the influence they have on the flame properties will be discussed in the next few paragraphs.

Outer Structures

The outer structures are believed to be established by a natural convection, buoyancy-driven shear layer. The flow field outside the stoichiometric flame surface consists of three components, two forced components and a natural convection component. The forced components are due to the momentum of the annular and fuel jets. From studies of high speed movies it appears, for the methane and propane flames under study, that most of the forced flow from the fuel jet was inside the visible (luminous) flame surface and did not contribute greatly to flow field outside the flame surface. The annulus flow velocity was significant in the lower regions of the flame but appeared to be overcome by the constant acceleration of the buoyancy driven flow at a height of about 70 mm from the nozzle exit. This height is noteworthy because it identifies the location where an instability in the outer flow had developed into a larger toroidal vortex. This location was nearly constant for all of our experiments which suggest that it was weakly dependent on fuel nozzle design, fuel type and fuel velocity. It was, however, strongly dependent on the annulus air velocity. Thus, we believe that the flow field outside the flame surface was dominated by natural convection induced by buoyancy.

The outer vortices result from an instability in the outer shear layer. Their impact on the flame structure at low and intermediate fuel velocities is significant if not dominating in the flames studied. The outer vortices are believed to be responsible for flame flicker as defined by the separation of the flame tip from the main body of the flame or the natural low frequency oscillation of the flame surface. This assessment is in agreement with the analysis reported by Buckmaster and Peters[8] who predicted that the flame flicker frequency results from an instability in the outer shear layer. The clipping of the flame tip is shown in Fig. 3a. The height of this flame was just above the height where the outer vortices are formed. This appeared to be a necessary condition for making the flame tip susceptible to the stretching and quenching actions of the outer vortices. Schlieren and flame photographs of the flame clipping process were taken by Kimura[9] and Toong et al.[10] but the photographs were not sufficiently revealing to allow a clear explanation of the phenomena.

Another form of flame flicker is the oscillation of the luminous flame surface. This occurs for longer flames (higher fuel flow rates) than that shown in Fig. 3a. Figure 3b illustrates the flame oscillation process. The flame surface is pulled radially outwards as the toroidal vortices rotate. This results in the formation of the bulge in the flame surface shown in Fig. 3b. To an observer in the laboratory, the upwards convective motion of the oscillating flame surface will have a flickering appearance. If velocities or temperatures are measured as a function of time at a point in the flow field, near sinusoidal oscillations will result as observed by Toong et al.[10] and Durao and Whitelaw.[11] The frequency of oscillation will correspond to the frequency of the outer vortices which was about 15 Hz in our experiment. This frequency was nearly independent of fuel type and fuel nozzle design (diameter, tube or contoured nozzle) for a wide range of fuel velocities. This is consistent with previously reported observations.[9,10,11] With an additional increase in fuel velocity, the flame has less time to be affected by the rotation of the outer vortices and the flame bulge is reduced in size, as illustrated by the methane jet flames shown in Figs. 3c and 3d. Thus, the flame flicker, resulting from large scale oscillations in the flame surface, appears to diminish as the relative velocity between the flame surface and the outer vortices increases. At the high fuel flow rates shown in Fig. 4, the outer structures are small and not easily recognizable. However, studies of high speed movies indicate that the outer structures still have some impact on the flame even at these relatively high velocities. They appear to be responsible, in part, for a long wavelength oscillation in the visible flame surface.

Entrainment of air is another important effect of the outer vortex structures. As the vortices rotate, they entrain air which is preheated by mixing with combustion products picked up from the flame surface. The preheated air and products are transported to the flame surface by rotation of the vortices where the oxygen reacts with the fuel which is supplied, in part, by the convective motion of vortex structures internal to the flame. At high fuel flow rates, the outer vortices are transformed to a layer of hot gases with very little rotational motion (See Fig. 4). However, air is still entrained, mixed and preheated in this hot layer and transported to the flame surface by diffusive and convective actions.

The transport of air and products from outside the high temperature flame surface into the interior of the flame could be a two step process. The first step is the entrainment of air by the outer vortices and the second is the transport of outside fluid into the interior of the flame. The rates of these two processes may also be different. The entrainment of air into the outer vortices is probably large judging by the volume they encompass; whereas, the transport across the high temperature surface of the flame will most likely be inhibited the by the high viscosity layer of the flame. The processes are complicated by a dependence on such parameters as air and fuel velocities and nozzle diameter. The entrainment processes are very important to our understanding of jet diffusion flames and are not clearly understood.

We have approached the discussions of the outer vortices from the view that they are causing the oscillation in the flame surface and not the motion of the flame surface inducing an instability in the outer shear layer. This view is suggested theoretically by the instability analysis performed by Buckmaster and Peters.[8] Experimental evidence also suggests that this view is correct. For example, the outer vortex structure is not changing greatly in Figs. 3b, 3c, and 3d (the photograph in Fig. 3b is not at the same spatial scale as those in Figs. 3c and 3d due to a difference in camera location); but, the outward extent of the flame bulge is reduced with an increase in fuel flow rate. This is difficult to understand if the flame was inducing the outer vortex motion. If the outer vortices are driving the flame oscillations, then eliminating the vortices should eliminate the oscillations. This requires replacing the natural buoyancy-induced convection by a dominating forced flow field. We performed such an experiment by adding a sufficient quantity of N_2 to a H_2 jet diffusion flame so that the stoichiometric surface lies well within the body of the momentum forced shear layer of the jet. When this was done the outer structures were not present and the bulges in the flame surface did not occur. It should also be possible to eliminate the outer vortices by forcing the annular air flow at a sufficiently high velocity to mask the natural convection. This experiment was not performed because our air supply was not large enough. Such an experiment was however performed by Kimura,[9] as well as an experiment where air was added to the primary jet (which would bring the stoichiometric flame surface into the body of the jet shear layer so the impact of the natural convection of the outer shear layer would be reduced just as for our H_2/N_2 flame). The result was that the flame flicker was not present. Kimura did not comment on whether the large vortex structures were absent in the above experiment since he did not attribute this effect to the elimination of the natural convection in the outer shear layer.

Inner Structures

Vortex structures located inside the flame surface also develop as the fuel velocity is increased from near laminar conditions to near turbulent conditions. The development of the inner vortex structures can couple to the outside structures under certain conditions, which are not clearly understood. This was not as evident in this study as it was in studies using larger fuel jets.[4,7]

In contrast with the outer vortex structures, the structures of the inner vortices are very dependent on the magnitude and radial profile of the fuel velocity at the nozzle exit. This dependence is shown in Figs. 3 and 4 for a contoured nozzle. At a low fuel flow rate such as that illustrated by the flame in Fig. 3b, an initial wave develops in the shear layer between the potential core and the flame. This instability develops into coherent vortices as the fuel velocity is increased (Figs. 3c). At the higher fuel flow rate shown in Fig. 3d, the inner vortices coalesce at some distance downstream. Shortly after the vortices coalesce, they lose their coherence and small-scale, unorganized vortex structures are observed. The height above the nozzle, where coalescence starts, decreases as the fuel velocity increases. Also, the flame spreading angle increases at the axial location where the vortices start to coalesce, much like a nonreacting shear layer grows when vortices coalesce.[12] These characteristics are shown in Figs. 3d, 4a and 4b.

A small scale wrinkling in the visible flame occurs at higher fuel flow rate which could also be interpreted as flame flicker.[13] However, the mechanism of flicker is very different from that discussed earlier. Color photographs and high speed movies of the flame show that small scale wrinkles or protrusions of the flame surface, which we will refer to as flamelets, are the result of localized collisions of the small-scale vortices with the flame surface. Small, localized flamelets are evident on close examination of Fig. 4b and the lifted flame in Fig. 4c. When a vortex (with a very large radial velocity component) collides with the flame surface, the flame can stretch to the extent that it is quenched, presumably in the fashion predicted by Peters and Williams,[14] and a localized hole appears in the flame surface. Normally, if a hole is formed high-up in the flame, the surrounding flame will reignite the unburned fuel in the hole. If the hole is formed near the jet exit where the flame is already highly stretched, the hole will open up and the flame will lift. The association of the abrupt lifting of the flame with the formation of a localized hole was first observed by Eickhoff, Lenze and Leucke[15] using Schlieren photography. This process was captured in Fig. 4d where the right side of the flame is attached and the left side is lifted to a height of about 3 nozzle diameters.

There are several interesting features associated with the lifting process captured in Fig. 4d. The inner vortices on the attached side maintain their coherence for a downstream distance of about 5 diameters; whereas, the inner structures on the lifted side are washed-out by the air entrained into the base of the flame. This flame has a Reynolds number of 22,600 based on nozzle diameter and cold flow inlet conditions and would normally be considered turbulent in a classical sense. However, the coherence of the inner vortices near the nozzle exit raises a question about whether it is indeed turbulent within the first five diameters of the nozzle exit. It is not known whether the coherent structures are the result of a change in fuel viscosity due to heating or whether they result from an inviscid instability due to the characteristics of the shear layer. This is an important point that needs additional study.

CARS Temperature Measurements

To ascertain the temperature field in a jet diffusion flame, a detailed series of CARS temperature profile measurements were made on a propane flame with an exit velocity of 1.8 m/s. A view of this flame, as obtained using the RMS technique, is shown in Fig. 5. Six axial locations were chosen for study: 10, 50, 100, 150, 200 and 250 mm above the nozzle exit. Radial temperature profiles were obtained at each of these axial locations at 1 to 0.2 mm increments, depending on the flame characteristics. Measurements inside the fuel jet were limited to those areas where N_2 from the surrounding air had penetrated.

To aid in interpreting the single point measurements in terms of the dynamic characteristics of the flame, radial profiles of mean (plots right of centerline) and rms (left of centerline) temperatures are plotted to the same spatial scale as the flame photograph in Fig. 5. Probability density functions (pdf's) can be interpreted as the fraction of time that the flame spends in a given temperature state. Values of pdf's at three axial locations are also shown in Fig. 5. At the 10 mm location the mean peak temperature occurs at a radial location of about 8 mm. The temperature pdf at this location is single mode and Gaussian-like. The low rms values indicate that the flame surface is relatively stable at this axial location. This is also supported by the result that the peak temperature occurs just outside of the visible flame surface. The rms temperature peaks well inside the flame surface very near the shear layer of the inner vortices.

At an axial location of 100 mm, the mean temperature peaks at a radial location of about 10 mm with an inflection point occurring at a location of 20 mm. The photograph of the flame shows that the visible flame surface expands outward at some distance downstream. This bulging effect results from the rotation of the outside vortices pulling the flame outward. It is a dynamic effect in that the bulge moves upward at the axial velocity of the flame. At a stationary measurement the movement of the flame surface would appear as a sinusoidal oscillation. A temperature measurement at the 20 mm position will depend on the location of the flame bulge at the time the measurement is made. If the flame surface is at the 20 mm position, a high temperature will be measured. If the flame surface is located radially at 10 mm, essentially room air with some combustion products is present at the 20 mm position and a low temperature will be measured. The bimodal shape of the pdf's at these radial locations support this description. The shape of the mean temperature profile is determined by the fraction of time that the flame surface spends at a given location. The inflection point, at the 20 mm location, probably represents the maximum spatial extent of the flame at the 100 mm axial location. The rms profile reflects the dynamic oscillation of the flame surface with one peak occurring where the hot and cold modes are equally probable and another peak occurring in the inner shear layer. It should be noted that if one were making spectral measurements of temperature or velocity that at least two discrete frequencies would result, one associated with the flame oscillation and the other associated with the inner vortex.

At the 200 mm axial station, the direction of rotation of the inner vortices changes. Observation of this effect was recently reported by Eickhoff and Winandy.[16] A vortex always rotates from the high to the low velocity side. The inner structures rotate outwards towards the flame surface until a height of about 175 mm at which point they rotate inward toward the centerline. This indicates that above 175 mm the flame surface has been accelerated by buoyancy to the point where it is moving faster than the inner, cooler fuel. This apparent change in direction of rotation does not have any obvious impact on the mean, rms or pdf temperature profiles at the 200 mm axial location. Indeed, the temperature profiles appear to be primarily determined by the time weighting of the flame oscillations induced by the outside vortices.

Summary and Conclusions

A systematic examination of jet diffusion flames, from very low velocities to lift-off conditions using the RMS technique, shows the presence of large toroidal vortices outside the visible flame surface. These vortices are believed to develop from a Kelvin-Helmholtz type instability driven by buoyancy. The vortices are responsible for flame flicker associated with the separation of the flame tip from the main body of the flame or oscillations of the flame surface. The flicker, resulting from the clipping of the flame tip, occurs when the height of the flame is a little larger than the height at which the outer vortices are formed. The oscillations of the flame surface occur in near laminar and transitional flames with heights that are appreciably larger than the formation height of the vortices. Flame flicker also results form the rotation of the outside vortices which creates a dynamic bulging motion in the flame surface. This oscillation of the flame surface is responsible for the shapes of the mean, rms and pdf temperature profiles. Thus, for transition flames, the temperature characteristics of the flame are determined to a large extent by buoyancy driven structures that are outside the flame surface.

The wrinkled, turbulent appearance of jet diffusion flames is primarily determined by small scale vortex structures formed inside the visible flame surface. Well organized vortex structures are formed for transitional flow conditions. Turbulent like characteristics, involving the formation of small, unorganized, 3D vortices in the flame, starts at the axial location just downstream of where the organized vortex structures start to coalesce. The coalescence starts high in the flame for low flow rates and moves toward the nozzle as the flow rate is increased. An increased spreading of the flame is observed at the axial location where coalescence occurs. Thus, the growth in flame width by coalescence is similar to the growth of a nonreacting shear. The appearance of the flamelets, normally associated with turbulence, is a result of collisions of small scale vortices with the flame surface. If the collision involves a vortex with a very high radial velocity component, the flame can stretch to the point that it is quenched locally around the vortex. Flame lift-off can occur when a localized hole, due to the collision process, is formed near the lip of the nozzle. This is most likely to occur at the axial location where the organized inner vortices start to coalesce.

Acknowledgements

The authors wish to express their appreciation to the Technical Photographic Division (ASD/RMVT) for their support in making and printing the still and high speed movies. We also wish to thank R. L. Britton, J. G. Lee and J. P. Seaba for their assistance in performing the experiments and to C. R. Martel, T. A. Jackson, C. A. Obringer, B. P. Botteri, Pei Lin and Shawn P. Heneghan for their helpful suggestions on the paper.

References

1. Roquemore, W. M., et al., Experimental Measurements and Techniques in Turbulent Reactive and Non-Reactive Flows, Eds. R. M. C. So, J. H. Whitelaw and M. Lapp, New York: The American Society of Mechanical Engineers, pp. 159-175,1984.

2. Roquemore, W. M., et al., "A Study of a Bluff-Body Combustor Using Laser Sheet Lighting," Experiments In Fluids, Vol. 4, pp. 205-213, 1986.

3. Chen, L.-D. and Roquemore, W. M., "Visualization of Jet Flames," Combustion and Flame, Vol. 66, pp. 81-86, 1986.

4. Chen, L.-D. and Roquemore, W. M. "Two-Dimensional Visualization and Single-Point Frequency Measurements of Low Reynolds Number Jet Flames," L.I.A. 58, ICALEO, pp.16-23, 1986.

5. Ebrahimi, I. and Kline, R., "The Nozzle Flow Concentration Fluctuation Field in Round Turbulent Free Jets and Jet Diffusion Flames", 16th Symposium (International) on Combustion, The Combustion Institute, pp. 1711-1723, 1977.

6. Goss, L. P., Trump, D. D., and Switzer, G. L., "Laser Optics/Combustion Diagnostics", Air Force Technical Report AFWAL-TR-86-2023, 1986.

7. Yule, A. J., Chigier, N. A., Ralph, S., Boulderstone, R., and Ventura, J., "Combustion-Transition Interaction in a Jet Flame", AIAA Journal, Vol. 19, pp. 752-760, 1981.

8. Buckmaster, J. D. and Peters, N. "Infinite Candle and its Stability- a Paradigm for Flickering Diffusion Flames," 21st Symposium (International) on Combustion, The Combustion Institute (in press).

9. Kimura, I., "Stability of Laminar-Jet Flames," 10th Symposium (International) on Combustion, The Combustion Institute, pp. 1295-1300, 1965.

10. Toong, T.-Y. Salant, R. F., Stopford, J. M., and Griffin Y. A., "Mechanisms of Combustion Instability", 10th Symposium (International) on Combustion, The Combustion Institute, pp. 1301-1313, 1965.

11. Durao, D. F. and Whitelaw, J. H, "Instantaneous Velocity and Temperature Measurements in Oscillating Diffusion Flames", Proc. R. Soc. Lond. A., 338, 479-501, 1975.

12. Wygnanski, I. and Fieldler, H., "Some Measurements in the Self-Preserving Jet," Journal of Fluid Mechanics, Vol. 38, pp. 577-612, 1969.

13. Scholefield, D. A. and Garside, J. E., "The Structure and Stability of Diffusion Flames," 3rd Symposium (International) on Combustion, The Combustion Institute, pp. 102-110, 1949.

14. Peters, N. and Williams, F. A., "Liftoff Characteristics of Turbulent Jet Diffusion Flames," AIAA Journal, 21, 3, 423-429, 1983.

15. Eickhoff, H., Lenze B., and Leuckel, W., "Experimental Investigation on the Stabilization Mechanism of Jet Diffusion Flames," 20th Symposium (International) on Combustion, The Combustion Institute, pp. 311-318, 1984.

16. Eickhoff H., and Winandy, A., "Visualization of Vortex Formation in Jet Diffusion Flames," Combustion and Flame, 60, 99-101, 1985.

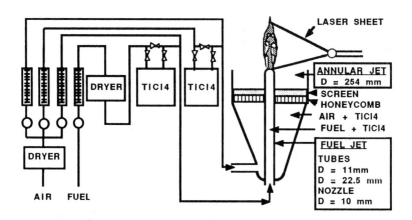

Figure 1. Jet diffusion flame experimental set-up.

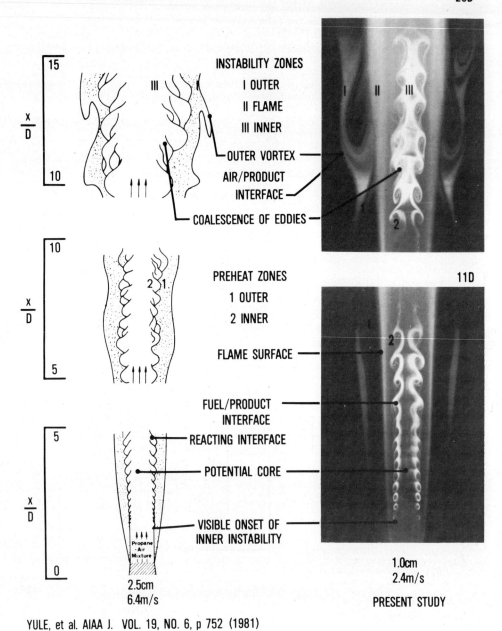

Figure 2. General structure of transitional propane jet flames.

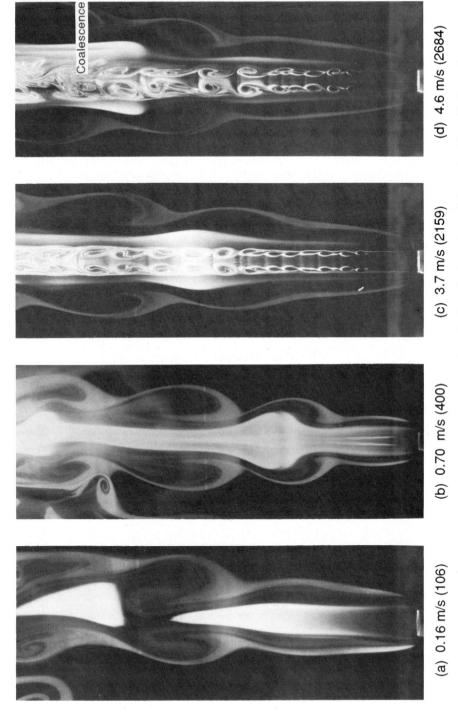

(a) 0.16 m/s (106) (b) 0.70 m/s (400) (c) 3.7 m/s (2159) (d) 4.6 m/s (2684)

Figure 3. Methane jet diffusion flames with an annulus air velocity of 15 cm/s and jet velocities (Re) as shown: Flame (a) is stabilized on an 11mm diameter tube. A 10mm diameter contoured nozzle is used for all other flames.

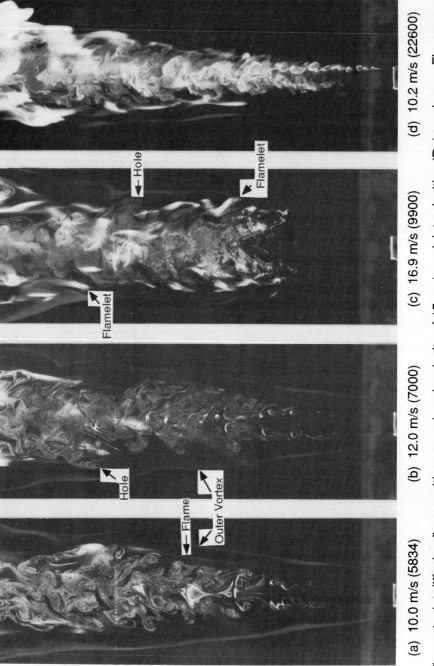

(a) 10.0 m/s (5834) (b) 12.0 m/s (7000) (c) 16.9 m/s (9900) (d) 10.2 m/s (22600)

Figure 4. Jet diffusion flames with an annulus air velocity of 15 cm/s and jet velocities (Re) as shown: Flames a-c are methane fueled and flame (d) uses propane. All flames are stabilized on a 10mm diameter contoured nozzle.

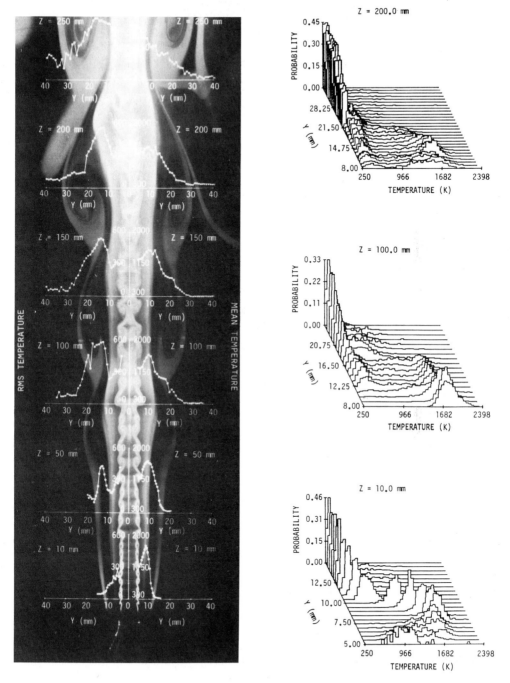

Figure 5. Mean, rms, pdf temperatures in a jet propane diffusion flame with an annulus air velocity of 0.15 m/s and a fuel velocity of 1.8 m/s (Re = 4000).

INSTANTANEOUS RADIAL PROFILES OF OH FLUORESCENCE AND RAYLEIGH SCATTERING THROUGH A TURBULENT H_2- AIR DIFFUSION FLAME

D. STEPOWSKI, K. LABBACI, R. BORGHI

U.A. CNRS 230 - Faculté des Sciences et Techniques
B.P. 118 - 76134 MONT SAINT AIGNAN Cedex (France)

I - INTRODUCTION

We have obtained instantaneous radial profiles of OH fluorescence and then Rayleigh scattering through a turbulent hydrogen-air diffusion flame by imaging a pulsed laser beam onto a gated linear diode array. Such data are helpful to determine what kind of flame front structure [1] can be expected in the flow : either a wrinkled flapping flame front if the turbulence scale is larger than the flame front thickness or a thickened flame front due to an internal micromixing by the small turbulence scales. Although turbulent flames must have one of the two structures above, the turbulence scale spectrum is spread over a considerable range and can be modified by combustion so that both structures may be alternatively found in a given flame.

We have investigated the lower zone (z/D < 20) of a hydrogen-air diffusion flame where the two flame fronts are sufficiently separated so that their position and thickness can be statistically studied. Similar diffusion flame have been studied by the teams of Bilger [2], Lapp [3], Dibble [4] and Takagi [5] ; most of these experiments were performed using point measurements and complementary to the few 2D imaging realizations [6, 7, 8] statistical multipoints data are still limited in such flows. In addition to the classical time average and fluctuation radial profiles our single shot profiles enable the structure of the instantaneous flame front to be investigated independently of flame motion. The two measurement sets -OH fluorescence and then Rayleigh scattering-provide informations on two different aspects of the flame front ; the radical and the thermal flame front. The behaviour of these two structures under the turbulent field will be compared.

II - APPARATUS (see Fig. 1)

We have built a vertical diffusion flame combustor with the classical configuration of a central jet of hydrogen diluted with nitrogen surrounded by a coflowing stream of air : dilution of the fuel makes it possible to choose the position of the reaction zone with respect to the shear zone[9]. The central pipe (diameter $\phi = 1$ cm) is fed with a stream of hydrogen and nitrogen (50 %-50 % volumetric) the velocity of which (u $\simeq$ 40 m/s) is controlled by sonic apertures. The coflowing air stream (u = 4 m/s) flows from a coaxial pipe ($\phi = 10$ cm) fed by a fan. The turbulent (Re $\simeq$ 15000) diffusion flame (1.2 m length) is stabilized at the central pipe tip.

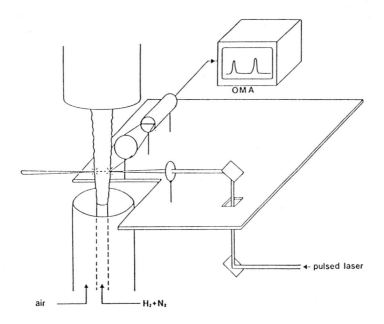

FIG. 1 : Apparatus

The pulsed ($\Delta t = 7$ ns) laser beam is horizontally focused through the pseudo-cylindrical flame along a diameter. The emitted light due to fluorescence or Rayleigh scattering is collected at 90 degrees and focused onto a gated intensified photodiode array (Δt gate = 20 ns) synchronized with the laser pulse at a repetition rate of 10

Hz. Radial exploration of the flame is made possible at different heights by setting all optics on a movable table. The 25µm pixels of the diode array are grouped by 8 or 16 (depending on the collected light intensity) so that the radial resolution of the flame is in the range .2 mm to .4 mm over a field of 20 mm to 40 mm depending on the magnification used.

III - OH FLUORESCENCE PROFILES

A) Procedure

OH radical has been measured due to its well known spectroscopic structure [10] and the important role it plays in combustion. Laser induced fluorescence is a very sensitive tool for radical detection in flames but accurate determination of absolute concentration is still very difficult to derive from fluorescence signals because the rate of fluorescence quenching due to collisional deexcitation is not sufficiently well known in flames [11]. First J.W. Daily [12] had suggested saturation of the transition to avoid the quenching dependence of the fluorescence. This method has been developed by Lucht [13] and the first statistical study of OH saturated LIF has been published [14] in 1984. In a complementary way, following Alden et al [15], we have chosen to use linear LIF to obtain instantaneous radial profiles in a similar flame.

Since the greatest absorption (and thus highest fluorescence signal) occurs in the $^2\Sigma^+(v'=0) \leftarrow {}^2\Pi(v''=0)$ transition, we have excited the $Q_1 6$ line (λ = 308.7 nm) of this band. The tuned laser pulse is derived from the second harmonic of a double stage dye laser pumped by a nitrogen laser. The laser spectral width ($\Delta\lambda \simeq .12$ A) is four times that of the absorption line. The OMA (PAR 1215) screen displays two peaks corresponding to the OH fluorescence emitted from the two intersections of the flame by the laser. From each instantaneous profile we register the position and the half-values width of the OH fluorescence peaks ; instantaneous OH concentration cannot be derived from fluorescence peak intensity because of the quenching dependence, nevertheless we have found a method to derive time average concentration :

When summing the instantaneous profiles to obtain time average fluorescence profile (see Fig. 2) it appears that right average peaks are always higher than left ones ; this asymmetry is only due to the noticeable absorption of the laser beam during its path through the first flame front.

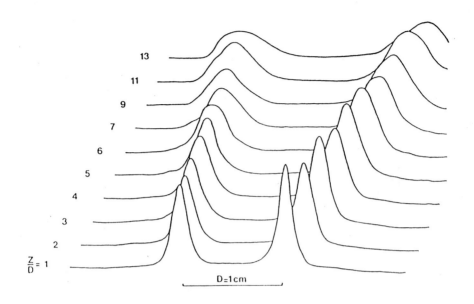

FIG. 2 : Set of mean radial OH fluorescence profiles (average over 100 laser shots) for increasing heights above the burner

Correspondingly the fluorescence trapping is no longer negligible but assuming our time average flame to be axisymmetric this trapping (likewise fluorescence quenching) does not modify the relative intensity difference between the two average fluorescence peaks. After its path through the first flame front the laser intensity is reduced (optical depth $K_{\nu 1} l_1$) so that the area under the first time average fluorescence peak

$$\overline{S_1} = C \int_{line} \overline{(1 - e^{-K_{\nu 1} l_1})} d\nu$$

is proportional to the energy lost by absorption. Likewise the area under the second fluorescence peak is

$$\overline{S_2} = C \int_{line} \overline{(1-e^{-(K_{\nu 1}l_1 + K_{\nu 2}l_2)})-(1-e^{-K_{\nu 1}l_1})} d\nu$$

Measuring the relative intensity difference between the two average fluorescence peaks we obtain:

$$\overline{\alpha} = \frac{\overline{S_1} - \overline{S_2}}{\overline{S_1}} = 2 - \frac{\int \overline{(1 - e^{(-K_{\nu 1}l_1 + K_{\nu 2}l_2)})} d\nu}{\int \overline{(1 - e^{-K_{\nu 1}l_1})} d\nu}$$

Assuming the two flame fronts not to be correlated (this will be corroborated), Reynolds decomposition of the $K_\nu l$ terms into their mean and fluctuating components, expansion of the exponential functions up to the third order and then average show that the mean relative attenuation is related to the mean optical depth over the flame front

$$\overline{\alpha} = 2 - \frac{\int (1-e^{-2\overline{K_\nu l}}) d\nu}{\int (1-e^{-\overline{K_\nu l}}) d\nu} \quad \text{if } \overline{\alpha} < .25$$

with an accuracy

$$\frac{\Delta\overline{\alpha}}{\overline{\alpha}} = \frac{\overline{K_0 l}}{2} \frac{\overline{K_0 l'^2}}{\overline{K_0 l}^2}$$

The curves $\overline{K_0 l} = g(\overline{\alpha})$ are plotted in Fig. 3 where K_0 is the maximum value of the absorption coefficient profile $K_\nu = K_0 f(\nu)$.

The time average population integrated over the flame front is given by the spectroscopic relation:

$$\overline{\int_{FF} |OH| dr} = \overline{K_0 l} \int_{line} f(\nu) d\nu \cdot \frac{4 \epsilon_0 \, mc \, Z_R Z_v}{e^2 \, S_{J'J''} \, F_{00}} \cdot e^{\frac{(E_{J''} + E_v)}{KT}}$$

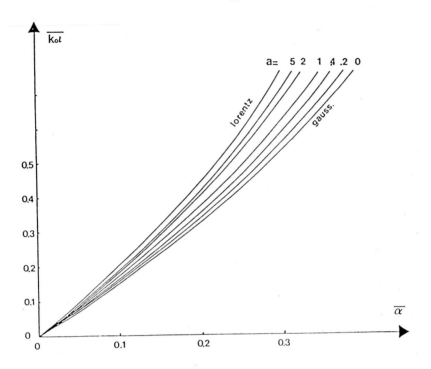

FIG. 3 : Optical depth of the medium versus relative attenuation of fluorescence peak for different values of the damping ratio a of the Voigt absorption line
(in the present flame a=.3)

involving band oscillator strength F_{oo}, Honl-London factor S and the partition functions of rotation Z_R and vibration Z_V. So we can realize a local absorption measurement over the time average flame front using the system -absorption line and subsequent linear fluorescence emission- as an in situ spectrometer with a resolving power of 10^5 [16]

B) Results and discussion

Measurement of α attenuation coefficient between the peaks of the time average fluorescence profiles provides a nearly constant integrated population $\overline{\int_{FF} [OH] \, dr} = .75 \; 10^{16} cm^{-2}$ independent of the height above the burner. Dividing by the width of the time average fluorescence peaks we find that peak average concentration, $\overline{([OH])}_{max}$ decreases from $3 \cdot 10^{16}$ cm^{-3} at the burner to 10^{16} cm^{-3} at 13 cm above. This result is in good agreement with the values previously measured in hydrogen-air turbulent diffusion flames [14,17].

Furthermore we are able to know whether the average broadening and decrease of the $\overline{[OH]}$ peak is due to a real broadening of instantaneous peaks or to an increase of peak position fluctuation or both. The position and the half-value width of the two instantaneous fluorescence peaks have been registered so that PDF and correlations can be calculated :

PDFs of Fig. 4 indicate that fluctuation of peak position increases with the level of probing at a same rate as the mean peak broadening. No correlation has been found between right and left peak position.

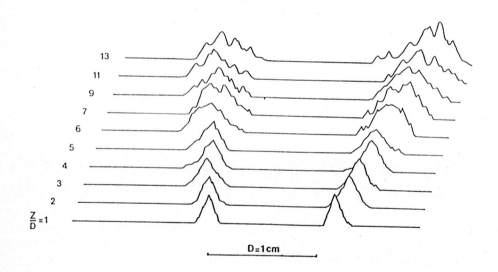

FIG. 4 : PDFs of OH fluorescence peak position for increasing heights above the burner

PDFs of half value width (see Fig. 5) show that the width fluctuation is very large (more than 60 %) but the average width ($\simeq 1.5$ mm) and its fluctuation are nearly independent of the height above the burner. Thus the apparent broadening of the average fluorescence peak is mostly due to the increase of position fluctuation in the lower flame zone investigated.

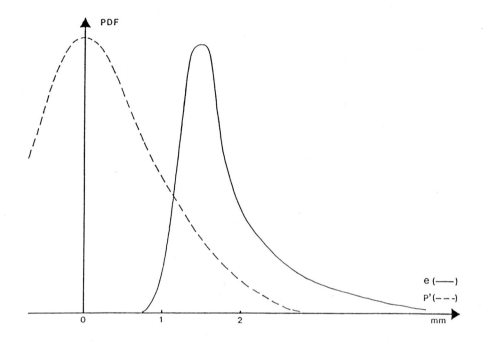

FIG. 5 : PDFs of OH fluorescence peak width e and relative position p' at 5cm above the burner

Dividing the mean integrated OH population by the actual mean value of instantaneous peak width we deduce that the mean maximum value of OH instantaneous peaks ($\overline{[OH]_{max}}$) is $5\ 10^{16} cm^{-3}$ whatever the height above the burner. This constant value (which holds about five times higher than equilibrium) is in agreement with the results of Drake et al [14] and with predictions of adiabatic partial equilibrium calculations [14,18] for hydrogen-air flames.

The range of fluorescence peak widths from 1.2mm to 4 mm is consistent with recent instantaneous OH flame front visualization by planar LIF [3]. The asymmetry of our PDF where a minimum width can be evaluated could be consistent with the intersection of the laser beam by a wrinkled laminar flame front structure ; expansion of the distribution towards high values being due to reaction sheets not perpendicular to the laser beam. We have investigated the correlation between position and width of fluorescence peaks. Under the above assumption of a wrinkled flamelet the probability of finding a thin fluorescence peak should be higher for peaks being far from the average peak position for which the intersection angle is statistically higher. The regression diagram plotted in Fig. 6 shows an inclination in this direction but does not allow a conclusive answer to be given about the radical flame front structure.

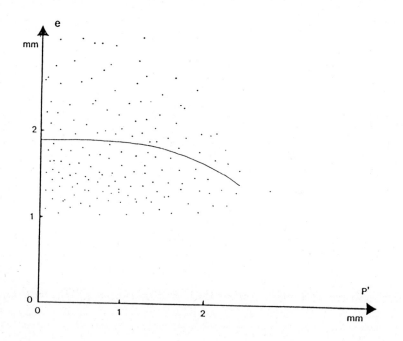

FIG. 6 : Regression diagram of OH fluorescence peak width e versus relative peak position p' at 5cm above the burner

IV - RAYLEIGH SCATTERING PROFILES

A) Procedure

For a single molecular species the scattered intensity in a

solid angle Ω due to Rayleigh interaction with a molecular concentration N over a path l is given by $W = W_0 \Omega l N \sigma$ where the Rayleigh scattering cross section σ depends on the transparent real refractive index μ through $(\mu-1)^2$ [19,20]. For a mixture of gases an effective Rayleigh scattering cross section can be used [20] $\bar{\sigma} = \Sigma\, x_i \sigma_i$ involving mole fractions x_i. In a medium with a constant effective Rayleigh cross section the scattered light intensity is proporitonal to the density and absolute temperature can be derived by inversion. Difficulties arise from the very low intensity of the Rayleigh scattered light (*) and from its dependence on a cross section which is not usually constant (**) :

(*) The excitation source is a pulsed nitrogen laser (λ=333 nm, W_{peak}=450KW, Δt = 9 ns). A filter has been interposed in the flow to avoid Mie scattering by dust. Ambiant light and flame emission are rejected by the gated intensification of the detector. Great care must also be taken to prevent the light due to parasitic reflection of the laser beam which cannot be rejected by time or wavelength filtering ; free optical access without window and long focal lenght optics with diaphragmes have been used. Nevertheless registration performed in Helium ($\sigma_{He} \approx 0.015\, \sigma_{N_2}$) shows that 2 % of the detected light is due to such parasite light. This background signal which induces noticeable errors particularly for high temperature must be substracted from each Rayleigh scattering profile.

(**) The best situation is to have the same cross section in the fuel and oxydant channels as realized by Dibble et al [21]. Under the situation that we had to maintain for comparison with prévious fluorescence measurements the Rayleigh cross section is 30 % lower in the fuel channel than outside. Our stoechiometric mixture fraction is 0.295, thus at the temperature peak where the stoechiometry is assumed the Rayleigh cross section should be only 10 % lower than outside. A first order correction for this variation is brought by dividing each instantaneous Rayleigh scattering profile by the average Rayleigh profile registered at the same level without combustion ; the growth rate of the cold jet is probably higher than that of the burning jet but the average mixture fraction profiles are not very different in the lower investigated zone of the flow. In addition this normalization procedure corrects for non uniform sensitivity of the photodiode array. The cross sections of reactants, intermediates and products of the combustion are slightly different but Namer and Scheffer [22] have shown that for our mixture the effective cross section variation is lower than 1.5 %.

Finally the accuracy of our temperature measurement is estimated to be about 5 %.

B) Results and discussion

As shown in Fig. 7 the flame expansion is the same as for $\overline{OH}$ but the maximum average temperature is nearly constant in the investigated zone and the half-value width is about twice that of $\overline{OH}$ maxima. The fluctuation of instantaneous peak position is 35 % smaller than for OH and increases proportionally to the height above the burner.

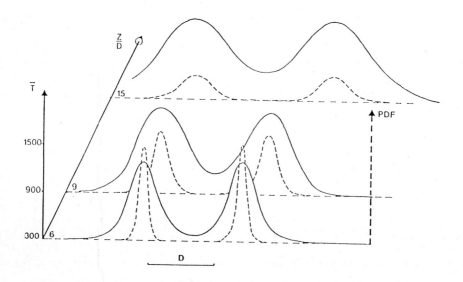

FIG. 7 : Set of mean temperature profiles (average over 256 single shot profiles)(———) and PDFs of peak temperature position (-----) for increasing heights above the burner

PDFs of instantaneous peak width (see Fig. 8) show that the average width is much larger than that of OH peaks (x 3 to x 8) and increases with the height above the burner ; similarly the width fluctuation are larger than the values found for OH. Notice that contrary to the case of OH these width distributions are nearly symmetrical. Finally the thermal structure behaves like an envelope of the radical structure as if its response time to the turbulence was longer.

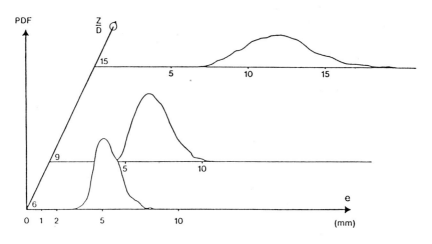

FIG. 8 : PDFs of temperature peak width e for increasing heights above the burner

PDFs of instantaneous peak temperature (see Fig. 9) for different distances above the burner are quite similar and are normal distributions spreading from 900 K to 2300 K (Tad ≃ 1950 K).

The surprising low temperature values cannot be explained by the poor spatial resolution ($\Delta x \simeq 0.4mm$, Δz and $\Delta y = 1mm$) provided by the nitrogen laser because the induced distortion would mostly alter the high temperature values rather than the lower ones. In addition, no correlation has been found between temperature peak value an width. According to Dixon-Lewis et al [23] 1300K should be the limit temperature of hydrogen-air steady diffusion flame under maximum strain. Rejecting the possibility of an error of 50 % in our temperature measurement we attribute our low temperature data to unsteady flame structures such as extinguishing flamelets ; although flame blow-off never occurs in the large range of flow rate allowed by our combustor some local and intermittent near extinctions due to turbulent stretching |24| could explain these low temperature data.

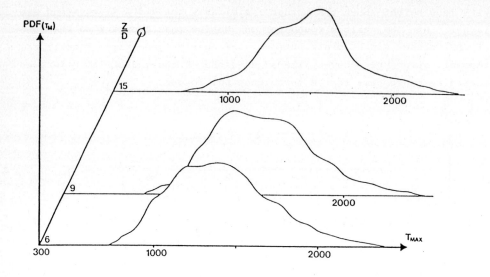

FIG. 9 : PDFs of temperature peak value T_{max} for increasing heights above the burner

V - INSTANTANEOUS RATE OF HEAT RELEASE ESTIMATE

The measurement of the radial second derivative of the temperature at the peak position where temperature gradient vanishes can provide information on the instantaneous reaction rate. We are aware that uncertainty in absolute temperature (and moreover in its second derivative) does not enable accurate results to be obtained but it should be interesting to know what can be derived from such procedure.

In unsteady form, assuming unity Lewis numbers, the Shvab-Zeldovich equation for enthalpy conservation may be written [25,26]

$$\rho \frac{\partial h}{\partial t} + \rho V \cdot \nabla h - \nabla (\rho D \nabla h) = \dot{\omega}_T$$

with

$$h = \int_{T_0}^{T} c_p \, dT \simeq c_p (T - T_0) \qquad D = D_0 (\frac{T}{T_0})^{7/4}, \qquad \rho = \rho_0 \frac{T_0}{T}$$

For a quasi-steady temperature structure in cylindrical coordinates neglecting θ and Z variations of temperature we obtain :

$$\frac{\rho_0 T_0 c_p}{T} v_r \frac{\partial T}{\partial r} - \frac{\rho_0 T_0^{-3/4} c_p D_0}{r} \left[\frac{3r}{4} T^{-1/4} (\frac{\partial T}{\partial r})^2 + T^{3/4} r \frac{\partial^2 T}{\partial r^2} + \frac{7}{4} T^{3/4} \frac{\partial T}{\partial r} \right] \simeq \dot{\omega}_T$$

so that at the instantaneous maximum where $\frac{\partial T}{\partial r} = 0$

$$\boxed{\dot{\omega}_T \simeq - \rho_0 D_0 c_p (\frac{T}{T_0})^{3/4} \frac{\partial^2 T}{\partial r^2}}$$

For each instantaneous radial temperature profile we have registered the temperature and its second derivative at the maximum so that the instantaneous heat release rate can be derived (see Fig. 10). Such diagram has also been set for higher levels above the burner showing similar average increase but with more and more scattering of the points as θ and Z temperature gradients may no longer be neglected downstream. $\dot{\omega}_T$ can be controlled either by kinetics or by turbulence depending on the Damköhler number ; we dont know in our case what is the controlling mechanism but we have compared our results with the kinetics of an equivalent single step reaction proposed for hydrogen-air diffusion flame 27

Semi-log plot of our value versus 1/T leads to a quasi-linear decay with a slope of 6600 K ± 800 K. Direct comparison with the above equivalent reaction rate (with 242 KJ per H_2O mole) provides a mean residual H_2 mole fraction of 1.5% at the temperature peak which is a reasonable order of magnitude.

A first glance result of Fig. 10 seems to confirm the surprising possibility of steady reacting structures at very low temperature (down to 800 K) ; but on the other hand if we attribute our low temperature data to unsteady extinguishing flamelets locally $\rho C_p \frac{\partial T}{\partial t}$ and terms involving θ and Z variation of temperature are no longer negligible and decrease the actual reaction rate possibly down to zero. Finally the uncertainty in our temperature measurements and the numerous assumptions and simplifications that we have made contrains us to circumspection and does not allow further analysis.

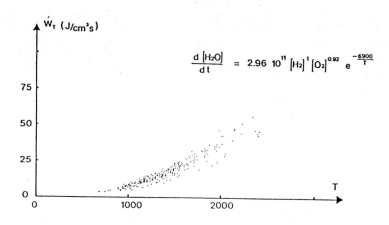

FIG. 10 : Heat release rate at the instantaneous temperature peak position versus peak temperature measured at 6cm above the burner

CONCLUSIONS

In the lower investigated zone ($Z/D < 15$) of the reacting flow the hydroxyl flame front is flapped by the turbulence field with amplitude increasing with height above the burner. Average thickness is constant despite large apparent fluctuations that cannot be only due to fluctuations of the angle between the laser beam and a wrinkled flame structure. A nearly constant mean value of OH peak concentration of $5 \cdot 10^{16}$ cm^{-3} has been found which is 3-4 times higher than equilibrium value : such a degree of super equilibrium is consistent with previous calculations and measurements in H_2-air flames.

The thermal structure is three times thicker than the OH flame front and is not so sharply perturbed by the turbulence. The surprising large distributions of peak temperature value and profile width suggest that turbulent stretching is mainly responsible for the fluctuations of this thermal structure possibly up tolocal and intermittent near extinction.

In addition a reasonable order of magnitude estimate for the instantaneous rate of heat release has been derived from the measurement of the second derivative of the temperature at the peak.

REFERENCES :

[1] BORGHI, R., Journal de Chimie Physique, 81, p. 361 (1984).

[2] KENT, J.H. and BILGER, R.W., Fourteenth Symposium (International) on Combustion, p. 615, The Combustion Institute (1973).

[3] DRAKE, M.C., PITZ, R.W. and LAPP, M., Laser Measurements on Non premixed Hydrogen-Air Flames for Assessment of Turbulent Combustion Models, 22nd Aerospace Sciences Meeting, Reno, Nev, January 1984, A.I.A.A. 84 (1984).

[4] DIBBLE, R.W. and HOLLENBACH, R.E., Eighteenth Symposium (International) on Combustion, p. 1489, The Combustion Institute (1981).

[5] TAKAGI, T., SHIN, H.D. and ISHIO, A., Combustion and Flame, 41, p. 26 (1981).

[6] CATTOLICA, R.J. and STEPHENSION, D.A., 9th ICOGERS, Poitiers, France, July 1983, A.I.A.A. Progress series, 95, p. 714 (1984).

[7] KYCHAKOFF, G., HOWE, R.D., HANSON, R.K., DRAKE, M.C., PITZ, R., LAPP, M. and PENNEY, C.M., Science, 224, p. 382 (1984).

[8] LONG, M.B., LEVIN, P.S. and FOURGUETTE, D.C., Optics Letters, 10, p. 267 (1985).

[9] MASRI, A.R., STARNER, S.M. and BILGER, R.W., Transition and Transports in the Initial Region of a Turbulent Diffusion Flame, 9th ICOGERS, Poitiers, France, July (1983), A.I.A.A. Progress Series, 95, p. 293 (1984).

[10] DIEKE, G.M. and CROSSWHITE, H.M., J.Q.S.R.T., 2, p. 92 (1962).

[11] FAIRCHILD, P.W., SMITH, G.P. and CROSLEY, D.R., J. Chem. Phys., 79, p. 1975 (1983).

[12] DAILY, J.W., Applied Optics, 16, p. 569 (1977).

[13] LUCHT, R.P., SWEENEY, D.W. and LAURENDEAU, N.M., Combustion and Flame, 50, p. 189 (1983).

[14] DRAKE, M.C., PITZ, R.W., LAPP, M., FENIMORE, C.P., LUCHT, R.P., SWEENEY, D.W. and LAURENDEAU, N.M., Measurement of Superequilibrium Hydroxyl Concentrations in Turbulent Non premixed Flame Using Saturated Fluorescence, General Electric Report N° 84 CRD 195, Schenectady, New-York 12345, and Twentieth Symposium International on Combustion, Ann Arbor, Michigan, The Combustion Institute (1984).

[15] ALDEN, M., EDNER, H., HOLMSTEDT, G., SWANBERG, S. and HOGBERG, T., Applied Optics, 21, p. 1236 (1982).

[16] STEPOWSKI, D., and GARO, A., Applied Optics, 24, p. 2478 (1985).

[17] JANICKA, J. and KOLLMANN, W., Seventeenth Symposium (International) on Combustion, p. 421, The Combustion Institute (1978).

[18] WARNATZ, J., Combustion Science and Technology, 26, p. 203 (1981).

[19] ROBBEN, F., in "Combustion Measurements in Jet Propulsion System", (R. Goulard, Ed.), Project Squid, ONR (1975).

[20] MULLER-DETHLEFS, K. and WEINBERG, F.J., Seventeenth Symposium (International) on Combustion, p. 985, The Combustion Institute (1978).

[21] DIBBLE, R.W., HOLLENBACH, R.E. and RAMBACH, G.D., Temperature Measurement in Turbulent Flame via Rayleigh Scattering, in "Laser Probe for Combustion Chemistry, A.C.S. Symposium Series 134 (D. Crosley, Ed.), p. 435, Washington, D.C. (1979).

[22] NAMER, I. and SCHEFER, R.W., Experiments in Fluids, 3, p. 1 (1985).

[23] DIXON-LEWIS, G., DAVID, T. and GASKELL, P.H., "Structure and Properties of methane-air and hydrogen-air counterflow diffusion flames", Archivum Combustionis in Press.

[24] PETERS, N., Combustion Science and Technology, Vol. 30, pp. 1-17 (1983).

[25] BILGER, R.W., "Turbulent Jet Diffusion Flames" in Energy and Combustion Science, Student Ed. 1 (N. Chigier, Ed.) Pergamon Press (1979).

[26] WILLIAMS, F.A., Combustion Theory, Comb. Science and Engineering Series, Benjamin/Cummings Pub. Comp. Inc. (1984).

[27] MARATHE, A.G., MUKUNDA, M.S. and JAIN, V.K., Combustion Science and Technology, 15, p. 49 (1977).

FLAME STRUCTURE IN SPARK IGNITED ENGINES, FROM INITIATION TO FREE PROPAGATION

T.A. BARITAUD
Institut français du Pétrole
BP 311, 92606 Rueil-Malmaison Cedex

ABSTRACT

Flame visualization techniques, Schlieren and tomography, are used in experimental engines running with lean propane- air mixtures. They reveal that engine flames are turbulent as soon as created, and that this turbulence favors initiation. At very low engine speed, the engine flames are of wrinkled type. When going to higher rpm, the turbulent overall flame thickness increases, being around 1 cm at 1500 rpm. The study of the inner structure of the turbulent flame brush shows that the combustion zone is made of a very corrugated flame fronts, with peninsula and pockets. The distribution of the wrinkle scales widens when increasing engine speed, ranging from a few millimeters to tenth of mm.

1 Introduction

The state of the art of multi-dimensional engine flow modeling has now reached an accuracy level allowing the insertion of a detailed combustion submodel. The choice of the latter requires a knowledge of engine flame structure and propagation mechanisms. Although spark- ignited engines have been studied for a long time, it is only since "transparent experimental engines" exist, along with the use of powerful new laser diagnostics, that one can actually have an idea of the relevant combustion regimes. A theoretical analysis of the turbulent engine flow properties, and of the laminar flame characteristics, led Abraham et al. (1) to locate the engine combustion regimes between reaction sheets for low engine speed, and something closer to distributed reactions for usual running conditions. The first regime can be viewed as a region where a thin corrugated flame front separates the fresh from the burnt gas. The second would be more like a region with a small scale mixing of reactants and products. The combustion zone structure between these two states is not well known, although certainly relevant for most of engine combustion. The questions are the followings: what are the turbulence scales in the engine ? How does the flame respond to these scales, modifying its overall shape but also its local internal structure ? Noticeable is the lack of turbulence scale measurements inside the engine to guide a theoritical work. One of the first attempts at making these measurements done by Fraser et al. (2) gave an integral length scale of about 2 mm for low engine speed. The flame front behavior and structure is still not well known, even if considereble insight has been gain recently, as will be shown in this paper.

Moreover, these considerations about flame propagation are only one part of the whole combustion problem in engine, because it deals with the fraction of the charge burning "freely", away from the spark plug, and far from the wall. A typical time evolution of the charge in a lean operating engine is shown on Figure 1, as given by a thermodynamic analysis of the pressure trace from an individual cycle. The ignition time is 330 crank angle degrees (cad), 30 degrees before top dead center (tdc). One can see that the combustion process lasts 65 degrees. During the first 20 degrees, almost nothing burns, although a flame is certainly propagating. Ninety per cent of the charge burns within the next 35 degrees, and the combustion ends smoothly during the last 10 cad. The flame contours from the single cycle previously analyzed can be seen on Figure 2, as seen through a transparent piston with a method described later. Only a 58 mm disk can be seen out of the 86 mm bore. The visualization field is filled up by the flame almost completely after the first 30 cad past ignition, when only a negligible part of the charge is burnt. The reason for this is the density difference between fresh and burnt gases. Assuming a turbulent reaction zone thickness of about 5 mm, one can obtain the bar chart of Figure 3, showing the fraction of the charge concerned by each of the three engine combustion phases: ignition, free propagation, and wall combustion. The ignition phase duration is taken as the time needed by the burnt kernel to reach a 1 cm diameter. It appears that the three combustion phases are of equal importance: ignition deals with less than 1/1000 th of the charge, but lasts about 25% of the total burn duration, free propagation must be able to propagate the flame across chamber and wall combustion concerns two third of the charge. A good combustion model, or models, should be able to correctly reproduce these three stages of different natures. A reasonable goal before choosing a relevant combustion model should be to identify the kind of flame encountered in engines, along with their controlling mechanisms. We tried in the present study to do that using laser visualisation techniques on engine flames for the two first combustion phases, ignition and free propagation.

2 Flame initiation

This study is limited to lean combustion, since it is the situation of interest for the engine designers. No attempts are made to understand the initial phenomena occuring during the beginning of the spark discharge. The work of Maly and Vogel (3) or Sher and Keck (4) bring a valuable contribution to the knowledge of this stage. A common idea is that the flame starts as a laminar kernel, with the laminar flame properties. The recent turbulence integral scales measurements we mentioned earlier seem to preclude the existence of this. Schlieren visualisation of the spark gap done by Smith (5) have already shown the turbulent nature of the early flame. Keck and co-authors (6) with a similar technique in a square engine found that the flame speed is linearly increasing when going away from the spark plug for burnt kernels of some millimeters to a few centimeters. The goal of the present work is to provide more information about the very early flame (between 0 and 4 millimeters radius). The role of turbulence and equivalent ratio is investigated for lean propane-air mixtures.

2.1 Experimental set-up

There are currently no techniques giving the three dimensional shape of a flame, especially if a high repetition rate of a few khz is required, as necessary to follow engine flames. Here, high speed Laser Schlieren cinematography is used. This technique integrates along a line of sight. Looking at early spark-ignited flames, this means that the flame expansion is vizualised along two directions. Because the flow field in our engine has no preferential directions, the properties of the flame kernel should be statistically isotropic. Hence, the analysis of our two dimensional integrated flame pictures gives a good idea of the overall flame growth process. A schematic of the optical apparatus is shown on Figure 4. The green ray of a five watt Argon ion Laser is expanded to form a 5 centimeter diameter beam. This beam crosses the engine combustion chamber through quartz windows inserted in a spacer. The beam is then focussed through a pinhole selecting the sensivity of the system. A lens recollimates the beam and image it on a ground glass plate. The scene is recorded with a high speed 16 mm camera running at 6000 frames per second. A Bragg cell is used to strobe the Laser beam at a given engine crank angle when tuning the optics. The engine has a disc chamber. The flat head is a production one with two valves. The compression ratio is 6.0 and the clearance height is 1.6 cm. The cylinder has an 86 mm bore. The spark plugs are extended to locate the spark at chamber mid-height and 2 cm from the axis. Some electronic devices allow the impression on the film of the ignition timing for each cycle, and another reference point for every crank revolution. This is done when the camera has sent a signal attesting a stabilized film speed.

After the film photographic processing, the frames are projected on a tablet digitizer. The Schlieren contours are manually traced and acquired by a microcomputer. From the surface enclosed by the contour, an equivalent burnt kernel radius is extracted. The evolution of this radius as a function of the time, R(t), when varying the engine parameter is the topic of this initiation study, along with the visual observations.

2.2 The early flame characteristics

The photographs sets of the early flame propagation from four individual cycles are shown on Figure 5. Three were taken for three engine speeds from 500 to 1500, and the fourth for a very lean combustion at 1040 rpm. The first clear feature is the turbulent nature of the early flame. Except for 500 revolutions per minute (rpm), the flame front is wrinkled at the first or second frame available. This is for burnt kernel diameter of 2 to 3 mm. After that, for normally operating engines, with higher rpm, one can conclude that the flame front is always turbulent. The kernel diameter of the region preheated by the spark is already of the order of 1 mm, since this is the electrode gap with the used spark-plug. This is not far from the presumed flow integral length scale. For this experiment, the spark energy has been measured to be 80 mJ. When going from low to higher rpm, one can see that the amount of wrinkling increases. Since Shlieren visualization integrates, this means that the minimum wrinkle size is decreasing. The second feature is the fact that the flame is faster from the beginning of its growth when increasing the engine speed, i.e. for diameters of only 2 to 3 mm. For leaner combustion, the fine structure of the flame is not obviously modified, but the overall flame kernel is more distorted by the flow field.

2.3 Evolution of the burnt kernel radius

For 1040 rpm, Figure 6 shows the evolution of the burnt kernel radius for individual cycles. The cyclic variations for lean mixture are large. Even if a part of this variability is due to measurement uncertainties, most of it is a real feature of engine lean combustion. Aerodynamic (turbulence and mean flow), mixture composition (local equivalence ratio and dilution with residual gases) and spark characteristic changes are the reason for these cycle to cycle variations. The mean evolution of the flame radius with equivalence ratio .85 is plotted on Figure 7. The camera framing speed of 6000 causes an incertainty on the real frame timing as compared to ignition time. It also prohibits a reliable study at engine speeds much higher than 1000 rpm. However, one can see that the flame reaches a diameter of 1.5 to 2 mm with about the same speed for the three rpms. The evolutions of the different cases vary only later, although there are clearly more turbulent structures on the pictures. At the very early stages, the kernel expansion is certainly driven by the spark energy release. This is confirmed by Figure 8. Looking at the same kind of curves, but for a constant engine speed, and varying the equivalence ratio, there is still this 1.5 to 2 mm to reach before having distinct evolutions. Looking at Figures 7 and 8, it is not obvious to see which, from the mixture strengh and the turbulence associated with engine speed, controls most flame velocity. These two parameters modify strongly the rate of expansion as soon as the burnt kernel radius reaches 3 mm. More data acquisition and processing would be necessary, especially having in mind the large cyclic variations. That would be very time consuming. Moreover, the Schlieren visualisation technique used to study the small scales involved has shown to have experimental limits for high engine speed and very lean mixture.

2.4 Expansion velocity behavior

In order to understand better the mechanisms of the early flame growth, the curves R(t) are derived with respect to the time using polynomial fits. This is done in detail at 500 rpm. For a set of cycles, the expansion velocity is shown on Figure 9. One can see that almost all these curves exhibit a minimum for a radius slightly larger than 2 mm, with a velocity of approximately 1.5 m/s. This seems to be supported by the work of Champion et al. (7), where both numeric calculations and asymptotic analysis predict the existence of a critical flame radius at which the expansion velocity should present a minimum value for heavy fuels. The observation of a single cycle evolution, thus avoiding the cyclic variation smoothing shows more clearly this trend (Figure 10). The velocity behavior on this curve is typical of very low turbulence level cycle at 500 rpm: the turbulence is not accelerating the flame right after the minimum, allowing a stabilized expansion for a while. Figure 11 shows the evolution of the expansion velocity with respect to the equivalence ratio. Clearly, more data should be required to give smooth and accurate results. However, it can be seen that the level of the minimim speed is an increasing function of the equivalence ratio and the radius at this minimum is a decreasing function of the equivalence ratio. The effect of the turbulence on early flame speed is to increase it, as shown on Figure 12. It is not clear if this is associated with a critical radius modification. Then, if the critical radius mechanism is the relevant one, the added effect of turbulence would be to increase the velocity levels.

3 Flame propagation

Two techniques were used to study this phase. Schlieren visualization gave the overall shape of the propagating flame with the same engine as for the ignition study, but benefiting from a transparent piston and from a mirrored head with the same geometry,. A two dimensional flame visualization method, tomography, was used to provide an insight in the inner structure of the flame front in the Sandia engine simulator. Although conducted for moderate rpm, not more than 1500, they are comparable to real engine running conditions, differing from them by a factor of 2 to 3 for the turbulence levels.

3.1 Global flame visualisation

The technique is the same as for the initiation study. However, since the enlarged collimated beam has to go through a mirror under the piston, to cross the test section (the combustion chamber), and to go back by the same path, the optical set-up turns out to be quite different, as seen on Figure 13. The field of view has a 58 mm diameter, out of a 86 mm bore. Results are presented for a volumetric efficiency of 0.5 and an equivalent ratio of 0.75.

The pictures of Figure 14 are flames taken at 500, 1040 and 1550 rpm. They were taken out of movies, but for the sake of brevity, we will discuss only the 'still' features of the photographs. The two circles are the contours of the valves, and the dark spot on the disk side is caused by the spark plug in the head. The first trend is that the combustion region thickness increase with rpm. In engines, the turbulence has been measured to vary roughly linearly with engine speed in various configurations, for instance by Witze et al. (8). In an engine with a very similar geometry, the turbulence was measured to be 1.7 m/s at 1000 rpm at tdc by Baritaud et al. (9). Since the Schlieren technique integrates along the optical path, the flame thickness is certainly less than it appears. One can suppose that the flame is curved towards the spark plug within the combustion chamber clearance height. If the combustion is slower near the walls, that would also increase the apparent thickness. Having all that in mind, one can estimate the combustion region zone to be around .5 cm at 500 rpm, and to reach 1 to 2 cm at 1500 rpm. Another feature of these flames is the fact that they are faster when increasing the thickness, i.e. rpm. This is not studied here, but requires a better understanding of the inner structure of the combustion region. The question is if the speed increase is due to a small scale mixing or if a flame sheet survives up to high turbulence regimes, although extremely corrugated. This is the reason for the study of the next section dealing with 2-D visualization.

3.2 Inner turbulent flame structure

The engine tomographic study was carried out in the Sandia engine simulator. We will recall briefly that this technique uses the Mie scattering of particles, well mixed with the fluid, within a Laser sheet of light sent through the combustion chamber and the flame. The density difference between fresh and burnt gas is used, along with a scattering efficiency reduction by combustion. An image of the 2-D section is captured with an electronic camera. A subsequent image processing gives the flame contours. A schematic of the experimental apparatus is shown on Figure 15. The engine side window being 9 mm wide, the images correspond to a 9x13 mm rectangle. A complete description of the apparatus is done by Baritaud and Green (10) with a detailed analysis. This experiment was extended by Zur Loye et al. (11) to give tomographies of the whole chamber.

In the Sandia engine, the turbulence is higher than in regular engines (8) since the intake is done with side located valves. For 1000 rpm, the velocity fluctuation level was found to be around 3.5 m/sec at tdc. This is a value about twice that measured in the engine studied in the previous sections. We report on Figure 16 three representatives shots. At 300 rpm, a smooth wrinkled flame front is observed, and corresponds quite well to the Schlieren structure seen at 500 rpm in the IFP engine. For 600 rpm, the flame front is a lot more corrugated, and can even be disconnected. The overall turbulent combustion thickness is already larger than the field of view of one centimeter for some of the tomographies. A further increase of the engine speed leads to large structures resulting in peninsulas or islands of unburned mixture in the burnt gas. Of interest is also the local radius of curvature of the front. It can be less than a fraction of a millimeter. This has to be compared to engine flame thicknesses of some tenths of millimeter measured by Smith (12) for methane-air mixture in engines at low rpm. This means that the local flame speed can differ strongly from the laminar burning velocity, as shown by Clavin and Joulin (13). At high engine speed, with the decreasing Kolmogorov length scale, the question can be raised of the importance and the nature of the interaction between small scale turbulence and inner 'flamelet' structure. This corresponds to the poorly known combustion regime between reaction sheets and distributed reaction. Obviously, more measurements are needed, capable of giving high resolution tomographies, but also to explore the inside of the flame front.

3.3 Conclusion

Flame visualisation techniques used in spark-ignited engines and applied to the study of initiation and free propagation for lean combustion have shown:

- the early flame is turbulent as soon as created

- the early flame speed is a strong function of the turbulence level (rpm) and of the equivalent ratio.

- for low engine speed, the burnt kernel growth velocity goes through a minimum for diameter of a few millimeters suggesting the existence of a critical radius with heavy hydrocarbons.

- the 'free' turbulent flame looks like a smoothly wrinkled surface at very low engine speed, and becomes a thick combustion region when increasing the rpm to usual engine values.

- the inner structure of the flame brush reveals that the combustion zone is made of an extremely corrugated flame front, certainly disrupted. When increasing the turbulence, both small and large scale wrinkles are created, with curvature radius ranging from tenth of mm to cm.

ACKNOWLEDGMENTS

These works were supported by Groupement Scientifique Moteur (GSM), and the US Department of Energy, and led at IFP (France) and Sandia Ntl. Labs (USA).

REFERENCES

1. J. Abraham, F.A. Williams, and F.V.Bracco, "A Discussion of Turbulent Flame Structure in Premixed Charge Engine," SAE Paper 850345, (1985)
2. Fraser, R.A., Felton, P.G., Bracco, F.V., and Santavicca, D.A., "Preliminary Length Scale Measurements in a Motored Engine," SAE Paper 860021, (1986)
3. Maly, R., and Vogel, M., "Initiation and Propagation of Flame Fronts in Lean CH4-air Mixtures by the Three Modes of the Ignition Spark," Seventeenth Symp. (Int.) on Combustion, The Combustion Institute, p. 821, (1978)
4. Sher, E., and Keck, J.C., "Spark Ignition of Combustible Gas Mixture," Combustion and Flame, Vol.66, pp. 17-26, (1986)
5. Keck,J.C., Heywood, J.B. and Noske, G., "Early Flame Development and Burning Rates Spark Ignition Engines and Their Cyclic Variability", SAE Paper 870164, (1987)
6. Smith, J.R., "The Influence of Turbulence on Flame Structure in an Engine", in Flows in Internal Combustion Engines, T. Uzkan Ed. (ASME, New-York), pp. 67-72, (1983)
7. Champion, M., Deshaies, B., Joulin, G., and Kinoshita, K., "Spherical Flame Initiation: Theory versus Experiments for Lean Propane-Air Mixtures," Combustion & Flame, Vol. pp. 319-338, (1986)
8. P.O. Witze, J.K. Martin and C.Borgnakke, "Measurements and Predictions of the Precombustion Flow Fluid Motionand Combustion Rate in a Spark IgnitionEngine, " SAE Trans., (92), pp.786-796, (1983)
9. Baritaud, T.A., Trapy, J., and Monreal J., "Effect of Swirl and Combustion on the Fluid Motion in a Spark ignited Engine," Third Int. Symp. on Appl. of Laser Doppler Anemometry, July 6th to 8th, 1986, Lisbon
10. Baritaud, T.A., and Green, R.M., "A 2-D Flame Visualisation Technique Applied to the Engine," SAE Paper 860025, (1986)
11. Zur Loye, A.O., and Bracco, F.V., "Two-Dimensional Visualization of Premixed-Charge Flame Structure in an IC Engine," SAE 870454, (1987)
12. J.R. Smith, "Turbulent Flame Structure in an Homogeneous-Charge Engine," SAE Trans., (91), pp 150-164, 1982
13. P. Clavin and G. Joulin, "Premixed Flames in Large Scale and High Intensity Turbulent Flow," Le Journal de Physique-Lettres, (44), pp. 1-12, 1983

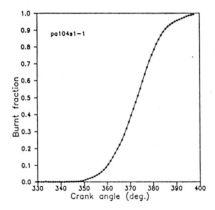

Figure 1 -   Evolution of the burnt gas fraction

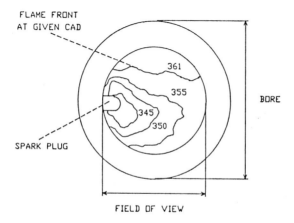

Figure 2 -   Evolution of the flame front position

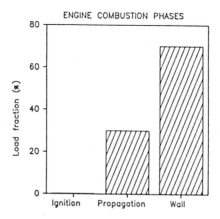

Figure 3 -   Fraction of the charge concerned by each of th three combustion phases

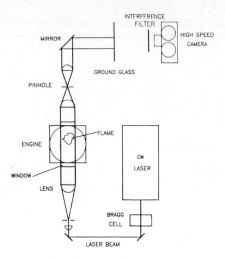

Figure 4 - Initiation study experimental apparatus

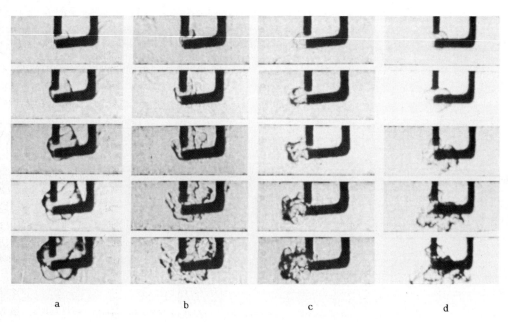

Figure 5 - Films of the ignition: a) 500 rpm, time between frame of 0.5 msec, b) 1040 rpm, time between frame of 0.33 msec, c)1550 rpm, time between frame of 0.17 msec, d) 1040 rpm, time between frame of 0.33 msec , equivalence ratio 0.75 for a) b) c and 0.58 for d)

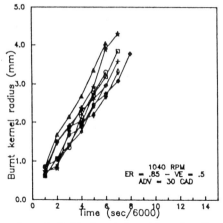

Figure 6 - Individual cycle burnt kernel radius evolution

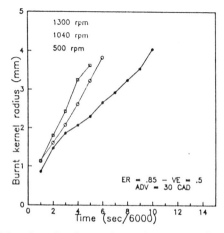

Figure 7 - Mean kernel radius evolution for an equivalence ratio of 0.85

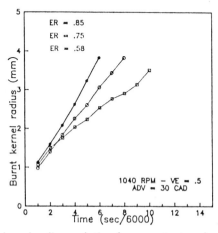

Figure 8 - Mean kernel radius evolution for a constant engine speed of 1040 rpm

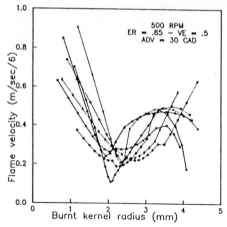

Figure 9 - Individual cycles for flame expansion speed

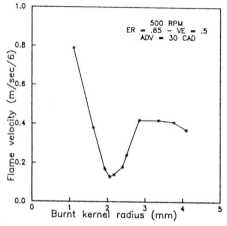

Figure 10 - Individual cycle for flame expansion speed

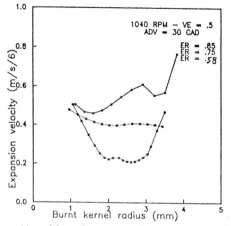

Figure 11 - Mean flame expansion speed at 1040 rpm

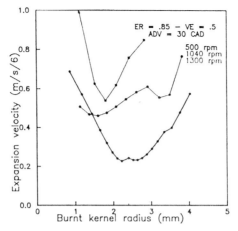

Figure 12 - Mean flame expansion speed at equivalence ratio 0.85

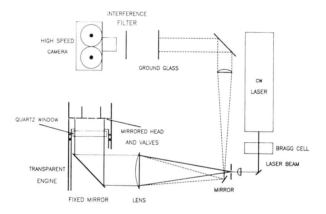

Figure 13 - Propagation study experimental apparatus

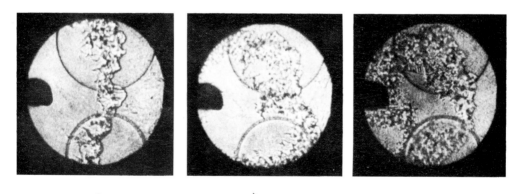

Figure 14 - Schlieren photograhs of the propagating flame:
a) 500 rpm, b) 1040 rpm, c) 1550 rpm

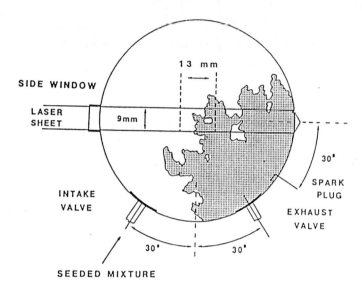

Figure 15 - Engine tomography set-up

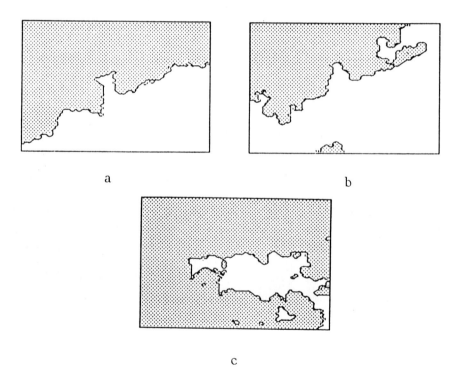

Figure 16 - Typical tomographies: a) 300 rpm, b) 600 rpm, c) 900 rpm

COMPARISON BETWEEN TWO HIGHLY TURBULENT FLAMES HAVING VERY DIFFERENT LAMINAR BURNING VELOCITIES

D. DUROX, T. BARITAUD*, J.P. DUMONT** and R. PRUD'HOMME

Laboratoire d'Aérothermique du C.N.R.S.
4 ter Route des Gardes, 92190 MEUDON, FRANCE

*Presently at IFP 92500 Rueil-Malmaison, France
**Presently at SEP 27200 Vernon, France

1. INTRODUCTION

Stationary jet burners, with Bunsen or V-shaped type flame are widely used for the study of premixed gas combustion. Since they are supposed to be simple configurations, they appear to the modeler scientist as a good test case to validate turbulent combustion equations closure assumptions. Modern diagnostic tools, such as laser Doppler anemometry [1] or fine wire thermocouple temperature measurements [2] have pointed out two essential features: a turbulence increase when crossing the combustion region and a counter-gradient diffusion. This helped to build the concept of "flamelet" combustion, where the combustion zone is seen as a region where a moving thin front, more or less corrugated, separates the fresh from burnt gas. To get a better knowledge of this flame front displacement, along with turbulence-combustion interaction, a statistical data processing involving the probability density function of the variables (PDF) was introduced /3,4,5,6/, and taken into account by the models, the most famous being the BLM model /7/.

Conditional sampling techniques brought new information about the respective contribution of the intermittency, between the fresh and burnt gas, and of the "real" turbulence /8,9,10/. Then, the models had to be adapted to take into account the last experimental results /11,12/.

However, it remains difficult to explain the large flame front movements shown by combustion measurements or visualizations /13,14,15,16/. Generally, the models try to express this as a simple function of the upstream turbulence. It may be that large scale structures have large effects on the overall flame behavior, even if experimental precautions have been taken, such a surrounding flow, or a flame location in the potential core. This importance of these large structure effects can be greatly changed by varying slightly the turbulence level or the laminar flame speed.

To show the influence of these effects two very close Bunsen flames have been studied. They have similar mean exit velocity and turbulence level ($\tau = (1/3\ (\overline{u'^2} + 2\overline{v'^2}))^{1/2} / \overline{U} \simeq 10\%$). They mainly differ by their laminar burning velocity.

2. EXPERIMENTAL SET-UP

Two flames are studied. The flame "A" is a flame of acetylene and air, and the flame "M" is a flame of methane and air; both are of Bunsen type. Flame "M" is set at atmospheric pressure and flame "A" is set at reduced pressure (0.25 atm.), which helps anchoring and allows high velocities with reduced flow rates. This thickens the flame front. A ring pilot flame holds "M" at its burner outlet.

The experimental conditions are:

	"A"	"M"
pressure	0.25 atm	1 atm
outlet diameter D	22 mm	11 mm
outlet velocity $\overline{U}$	6.7 m/s	7.2 m/s
turbulence rate $((1/3\,(\overline{u'^2}+ 2\overline{v'^2}))^{1/2} / \overline{U})$	9.3 %	9.7 %
$u'/\overline{U}$	4.2 %	9.3 %
equivalence ratio	1.42	1.2
laminar burning velocity S_l	1.15 m/s	0.39 m/s
laminar flame front thickness	1 mm	0.5 mm
$(1/3\,(\overline{u'^2} + 2\overline{v'^2}))^{1/2} / S_l$	0.54	1.72

Turbulence rates are very close although the axial fluctuation of "A" is lower. On the other hand, the chemistry clearly distinguishes "A" from "M", the laminar flame velocity ratio being of the order 3.

Some visualizations were made for both flames. Using tomography with a pulsed laser, Dumont and Borghi /17/ showed fresh gas zones in the flame "M". They used several kinds of seeding powders with various sublimation temperatures to visualize isotherms in the thermal diffusion zone of the flame. From their photographic results, it is possible to clearly show some caracteristic wrinkle sizes. The flame "A" is simply filmed with a high speed camera (1500 frames per second), the flame light being sufficient.

Velocity measurements are made with a laser anemometry. For flame "A", a one component LDA was used. Rotating the optics on its, axis it was possible to measure two velocity components (axial: $\overline{U}$ and u' and radial: $\overline{V}$ and v'), along with the correlation $\overline{u'v'}$. In the case of flame "M", dual component LDA was available, with dual Bragg cells. Signals coming from photomultipliers are processed by velocity counters. The velocity value and the RMS fluctuation are obtained from the analog signal of counters sent to integrator voltmeters /18/. A magnetic tape recorder permits to simultaneously store the velocity analog signals coming from the counters.

To draw the velocity histograms the analog signal is sent to a digital oscilloscope which can sample it at regular intervals. A micro computer sorts the data according to the velocity and gives histograms. The analog signal is also sent on a real time spectrum analyzer. A normalisation is made by the micro computer along with to take into account a specific bias due to high frequencies /19/. For flame "M", crossed terms $\overline{u'v'}$ are calculated with an oscilloscope card set inside the micro computer. This card simultaneously samples both channels and allows the correlation.

The seeding is made with zircone powder. A heavy seeding is achieved to get a high data rate much larger than the Shannon frequency. Particles are put in suspension into the air pipe. Gas flowrates are measured with massic flowmeters. The burners are set up on a motorized displacement.

3. VISUALIZATIONS

Time resolved photographies of "M" show a very different aspect of those of "A" obtained with film (fig. 1).

"A":-it does not present small scale wrinkles excepted at the top where tongues appear mainly oriented along the burner axis,
-some pockets can occur,
-a rough analysis of the cine-film recorded at a high speed shows a vertical pumping movement of this flame which can be related to the large turbulent jet structures. The pumping frequency is centered around 110 Hz and corresponds to a Strouhal number ($St = F * D / \overline{U}$) of 0.5.

"M":-higher than "A" due to a lower laminar flame speed,
-above one diameter, formation of small wrinkles with a distance between crests of the order of 1/2 diameter. This zone widens until a height varying between 2.5 times and 4.5 times the diameter,
-higher there is a zone with very corrugated wrinkles in all

directions. The characteristic size of the wrinkles is about diameter, as if the previous wrinkles have been paired,

-some formations of packets appears,

-it is difficult to see wrinkles with small curvature radii,

-the height reached by the fresh gases is fluctuating and may be related to outside jet type structures created by the shear zone between the jet and the ambiant air. If we measure on some photographies the wavelength of these swellings and if we relate it to the jet velocity, we find a frequency of 190 Hz corresponding to a Strouhal number of 0.29 (values found in the literature are about 0.3).

4. VELOCITY MEASUREMENTS

A velocities map is presented for "A" and for "M" on the figure 2. The deflection of velocity vectors is markedly greater for "A" than for "M" due to the large increase of temperature with acetylene. Flame "A" being shorter, velocity vectors are lined up again with the burner axis at the height $z = 5$ cm when $z = 7$ cm must be reached to observe the same for "M".

We present on figure 3 velocity measurements when crossing the flame front "A". By crossing radially or along verticals we can see an increase of the velocity $\overline{U}$ and of the velocity $\overline{V}$, excepted on the axis for the latter. Fluctuations u' and v' increase strongly, reaching turbulence rates of 20 % in comparison with $\overline{U}$ in the zone where the velocity gradient is maximum. Farther they decrease to reach a total turbulence rate similar to the one measured at the burner outlet.

Velocity results for flame "M" are plotted figure 4. There is almost no increase of the mean velocity $\overline{U}$, radially and axially. The velocity $\overline{V}$ grows at the crossing of the flame and shows the streamlines deflections. This can be explained by the mean flame front position, which is almost vertical. The density jump across the flame modifies only the radial velocity component, the burnt gas being pushed only towards the jet sides.

Going along parallels with the axis there is only a slow decrease of fluctuations u' and v'. Radially, we can distinguish a light raising of the fluctuations level by crossing the flame front but far less than for flame "A". The fluctuations caused by the flame front movements are small as compared to the upstream turbulence level. These results are comparable with other works /1, 2, 20/

The product $\overline{u'v'}$ is interesting to observe for flame "A" (fig.3). It takes high values in the flame center and involves positive correlations in this zone. At a distance r = 6 mm from the axis, by crossing the the flame, we can see at z = 25 mm that $\partial \overline{U} / \partial r > 0$, that $\partial \overline{V} / \partial z \gg 0$ and that the correlation is about 0.7 showing a countergradient diffusion. This can be explain by a phenomenon of intermittency. As a matter of fact "A" is mainly wrinkled type with flapping directions fairly oriented, and this can induced the creation of a pseudo-turbulence at the level of velocity measurements /4,9/.

For the methane flame the correlation is more difficult to analyze. On figure 4 three radial crossings are presented : z = 10, 30 et 50 mm. Results are always positive, even if between r = 0 and 4 mm the correlation is very weak. On the other hand a raising can be mentionned to the outside. At z = 10 mm,that is to say at about 1 diameter above the burner and at r = 5 mm, we have $\partial \overline{U} / \partial r \ll 0$,

$\partial \overline{V} / \partial z > 0$ and a positive correlation. In other words this result better corresponds to this one of a zone where the turbulent diffusion is of gradient type. At z = 30 mm two phenomena seem mixed : between r = 4 and 6 mm, $\partial \overline{U} / \partial r = 0$ and $\partial \overline{V} / \partial z > 0$ when the correlation is lightly positive and indicates a countergradient diffusion zone, farther to the outside we find again a diffusion zone of gradient type. At z = 50 mm the correlation remains almost zero until z = 5 mm while we observe deflections of streamlines almost since the flame axis. Farther the correlation reaches 0.5 in a diffusion zone of gradient type.

It seems that for the methane flame, the front flapping is too weak at one diameter from the outlet to induce an intermittency phenomenon on the velocity measurements. At 4 diameters, it seems that there is a small zone where this phenomenon is noticeable. This is confirmed by the front displacements observed on photographies. Above this height the front movements are too tormented and too irregular to let appear an intermittency with two states of velocity. The outside zone, where there is a diffusion phenomenon of gradient type apparently corresponds to the shear layer between the jet and the ambiant air.

5. PDF OF VELOCITY

PDF of velocity U for the flame "A" are presented on the figure 5. These histograms show a lightly bimodal shape at the crossing front and a very large spreading out of the velocity range. This well corresponds to what was visible on photographies, with important flame movements and directions of velocity vector varying from an instant to the other around the average direction.

For the flame "M" we present the results at a radial crossing at z = 30 mm (fig. 6) where there is an appearing of diffusion zone at countergradient. A lightly bimodal shape is visible about r = 3 or 4 mm for U as well as V. That confirms an intermittency phenomenon at two states of velocity in this area. Except for this zone the PDF are monomodal everywhere, without dissymetry and with a very few spreading of histograms. This last phenomenon, along with the continuous decrease of v' when going away from the axis is more difficult to explain. Some interactions with the pilot flame are suspected.

6. VELOCITY SPECTRA

Velocity spectra bring supplementary information on the movements for each flame. For the acetylene flame, spectra of U are presented on the figure 7. For axial crossings and radial crossings, there is a raising of spectra in the low frequency part, axially by going from z = 2 to z = 3 cm and radially by going from r = 0 to r = 12 mm. If we bring together this constatation with the below made at

§ 3, we can attribute this raising in the frequency range inferior to 200 Hz to the vertical pumping movement of the flame (the intermittency frequency of the front crossing is centred about 100 Hz). After the flame crossing the damping of spectra in high frequencies is caused by the effect of viscous dissipation on small eddies.

A behavior almost similar appears on spectra of flame "A" at $z = 30$ mm (fig. 8). About $r = 5$ mm, we can see a raising of the low frequency part. It is very difficult to observe modifications of slopes elsewhere excepted in burned gases where the viscosity strongly acts on high frequencies.

7. CONCLUSION

Flame visualization techniques and velocity measurements in two different Bunsen type flames with similar turbulence but different laminar flame speeds have demonstrated the influence of large scale structures on the flame behavior. These structures are related to the shear layer between the jet flow and the outer air. They induces large scale motions of the flame front at low frequencies centered around the jet typical Strouhal frequency. These pumping movements have a large influence on the velocity measurements and statistics. They induce a large part of the turbulence increase in the combustion region, modify strongly the velocity spectra and probability distributions shapes and may change the flux terms. The influence of these jet structures is also important on the flame front wrinkle sizes. Changing the chemistry time, i.e. the laminar flame speed and the density ratio has a strong influence on the nature of the jet structures and flame interactions by changing the location of the combustion region. Even for fast flames located in the jet potential core, the flame motion can be related to the jet structures. This study points out that results of works in jet flames aiming to determine the interaction between turbulence and combustion should be carrefully interpreted when compared with numerical model computations.

REFERENCES

/1/ Durst F. and Kleine R., "Velocity Measurements in Turbulent Premix Flames by Means of Laser Beams Anemometer", Gas Warme International, 1973, Bd22, Nâ12, p484.

/2/ Yoshida A. and Tsuji H., "Measurements of Fluctuating Temperature and Velocity in a Turbulent Premixed Flame", Seventeenth Symposium (International) on Combustion, The Combustion Institute, 1979, p945.

/3/ Moss J.B., "Simultaneous Measurements of Concentration and Velocity in an Open Premixed Turbulent Flame", Combustion Science and Technology, 1980, 22, p119.

/4/ Yoshida A. and Tsuji H., "Characteristic Scale of Wrinkles in Turbulent Premixed Flames", Nineteenth Symposium (International) on Combustion, The Combustion Institute, 1983, p403.

/5/ Cheng R.K. and Ng T.T., "Velocity Statistics in Premixed Turbulent Flames", Combustion and Flame, 1983, 52, p185.

/6/ Moreau P., "Experimental Determination of Probability Density Functions within a Turbulent High Velocity Premixed Flame", Eighteenth Symposium (International) on Combustion, The Combustion Institute, 1981.

/7/ Bray K.N.C., Libby P.A. and Moss J.B., "Flamelet Crossing Frequencies and Mean Reaction Rates in Premixed Turbulent Combustion", Combustion Science and Technology, 1984, 41, p143.

/8/ Cheng R.K., "Conditional Sampling of Turbulence Intensities and Reynolds Stress in Premixed Turbulent Flame, 1983, 52, p185.

/9/ Cheng R.K. and Shepherd I.G., "Interpretation of Conditional Statistics in a Open Oblique Premixed Turbulent Flames", Combustion Science and Technology, 1986, 49, p17.

/10/ Gulati A. and Driscoll J.F., "Velocity-Density Correlations and Favre Averages Measured in a Premixed Turbulent Flame", Combustion Science and Technolgy, 1986, 48, p285.

/11/ Bray K.N.C. and Libby P.A. "Passage times and Flamelet

Crossing Frequencies in Premixed Turbulent Combustion", Combustion Science and Technology, 1986, 47, p253.

/12/ Borghi R. and Pourbaix E., "Une modélisation eulérienne lagrangienne pour la combustion turbulente", La Recherche Aérospatiale, 1983, 4, p245.

/13/ Ganji A.R. and Sawyer R.F., "An Experimental Study of the Flow Field and Pollutant Formation in a Two Dimensional Premixed Turbulent Flame", AIAA paper 79-0017, 1979, 17th Aerospace Science Meeting.

/14/ Yule A.J., Chigier N.A., Ralph S., Boulderstone R. and Ventura J., "Combustion-Transition Interaction in a Jet flame", AIAA J., 1981, 19, p752.

/15/ Hertzberg J.R., Namazian M. and Talbot L., "A Laser Tomographic Study of a Laminar Flame in a Karman Vortex Street", Combustion Science and Technology, 1984, 38, p205.

/16/ Escudié D. and Charnay G., "Experimental Study of the Interaction Between a Premixed Confined Laminar Flame and Coherent Structures", Turbulent Shear Flow, 5, 1987.

/17/ Dumont J.P. and Borghi R., "A qualitative Study by Laser Tomography of the Structure of Turbulent Flames", Combustion Science and Technology, 1986, 48, p107.

/18/ Durox D. and Baritaud T., "Statistical Bias in LDA Measurement", Dantec Information, 1987, 4.

/19/ Boyer L. and Searby G., "Random Sampling : Distorsion and Reconstitution of Velocity Spectra from Fast Fourier-transform Analysis of the Analog Signal of a Laser Doppler Processor", J. Applied Physics, 1986, 60 (8), p2699.

/20/ Yanagi T. and Mimura Y., "Velocity-Temperature Correlation in Premixed Flame", Eighteenth Symposium (International) on Combustion, The Combustion Institute, 1981, p1031.

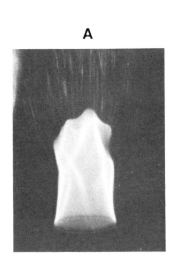

Fig. 1 - Instantaneous photographies

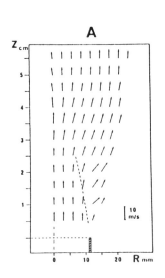
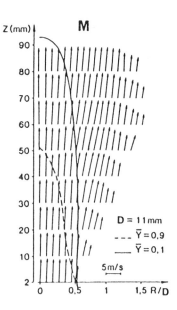

Fig. 2 - Velocities Map

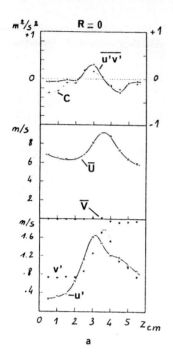
a

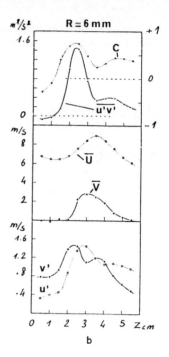
b

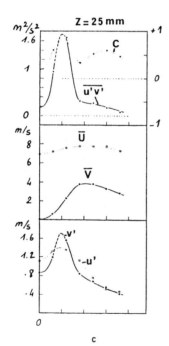
c

Fig. 3 - Velocities evolution. Flame "A"

    a. on the axis
    b. parallel with the axis at r=6mm
    c. radial crossing at z=25mm

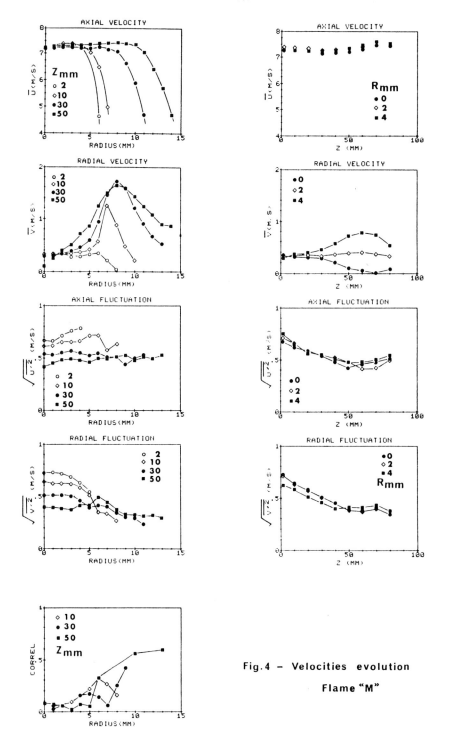

Fig.4 — Velocities evolution

Flame "M"

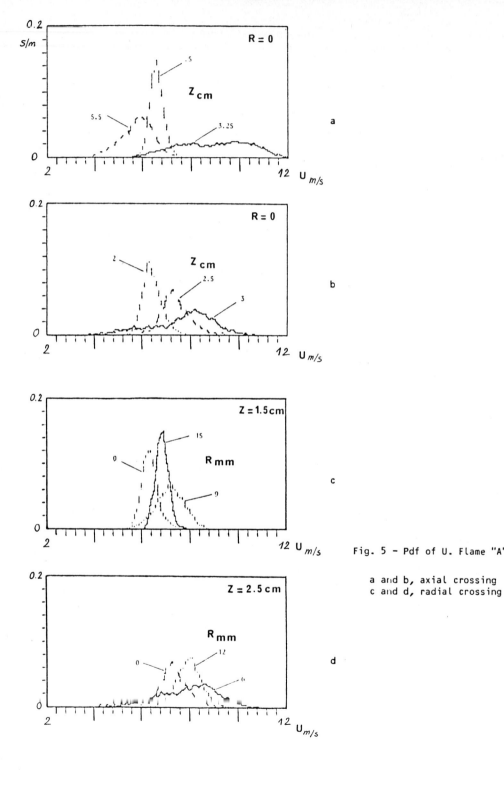

Fig. 5 - Pdf of U. Flame "A"

a and b, axial crossing
c and d, radial crossing

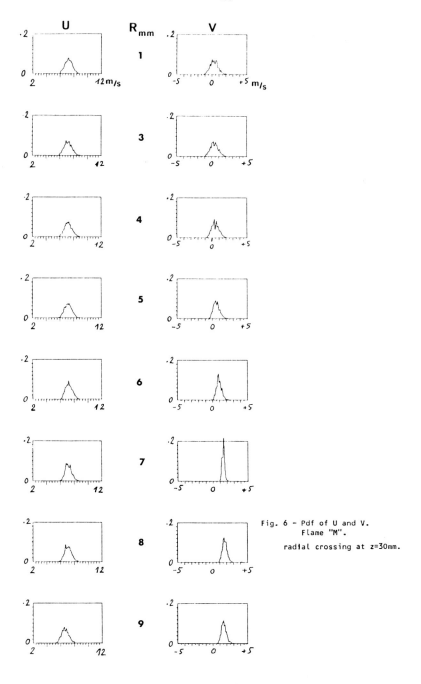

Fig. 6 - Pdf of U and V.
Flame "M".
radial crossing at z=30mm.

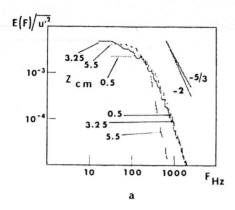

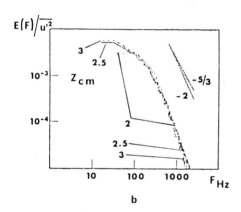

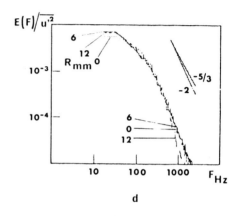

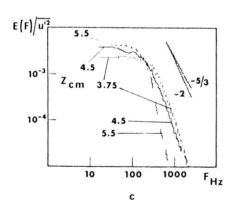

Fig.7 – Velocity spectra of U
Flame "A"
a,b,c axial crossing
d, radial crossing

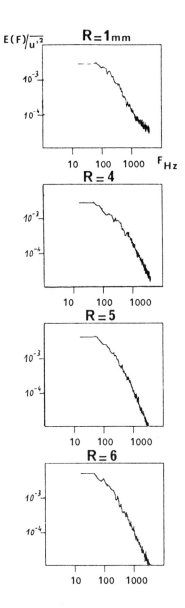
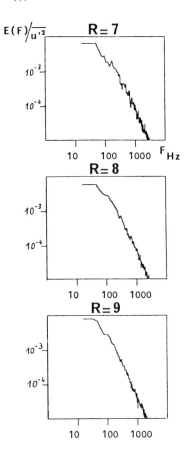

Fig. 8 - Velocity spectra of U. Flame "M" radial crossing at z=30 mm.

STRUCTURE OF TURBULENT PREMIXED FLAMES
AS REVEALED BY SPECTRAL ANALYSIS

I. Gökalp[1], A. Boukhalfa[1], R.K. Cheng[2] and I.G. Shepherd[2]

[1] Centre National de la Recherche Scientifique
Centre de Recherche sur la Chimie de la Combustion
et des Hautes Températures
45071 Orléans Cedex 2, France

[2] Lawrence Berkeley Laboratory
Applied Science Division
Berkeley CA 94720, USA

INTRODUCTION

Although spectral analysis has proved a powerful tool for studying non-reacting turbulent flows, its application in turbulent reacting flow investigations has been limited. With the development of recent theoretical models which place more emphasis on the temporal characteristics of the scalar and velocity fluctuations/1/, experimental investigation of their spectral behaviour would be useful to further the development of turbulent combustion models and to infer the physics of the turbulence-combustion interaction mechanisms which control the overall reaction rate.

In this paper, the spectra of dynamic and scalar fields in three different turbulent premixed flame configurations are explored. The three configurations are I) rod-stabilized oblique vee flames, II) turbulent flames stabilized in a stagnation flow and III) Bunsen type open conical flames. Configuration I and II are essentially used here to compare the spectra of dynamic and scalar fields in either frequency or vawe-number spaces. Configuration III is essentially used to perform a parametric study of the scalar field spectral behaviour.

The presentation of the results is organized in the following way. First, dynamic spectra in two vee flames having the same characteristic chemical time t_c and the same cold flow turbulence Reynolds number, but different u'/S_L ratios and different Damköhler numbers are compared. In a second step, the comparison of dynamic and scalar spectra in both vee and stagnation flow flames is presented. Finally, in a third step, the influence of the Damköhler number on the scalar spectra in Bunsen type flames is explored.

EXPERIMENTAL SET-UP, MEASUREMENT TECHNIQUES AND DATA ANALYSIS

The rod stabilized v-flames and the stagnation flow stabilized flames are investigated at Lawrence Berkeley Laboratory by using the same experimental facilities as in /2/ and /3/. The vertically placed burner produces an unconfined circular inner fuel/air jet of 50 mm diameter, surrounded by an annular coflowing air jet of 100 mm diameter. Incident turbulence is generated either by a square grid or a perforated plate placed 50 mm upstream of the nozzle exit. The stagnation plate is placed 75 mm downstream of the nozzle exit. To stabilize the vee flame, the stagnation plate is removed and a 1 mm diameter rod is placed at the nozzle exit.

The open conical turbulent premixed flames are investigated on the Bunsen type burner facility of CRCCHT at Orléans. They are stabilized by an annular pilot flame at the rim of a 16 mm inner diameter and 0.8 mm thick brass tube of 600 mm long. As no turbulence grids are used, the cold flow turbulence structure corresponds to that of a fully developed turbulent pipe flow, whose characteristics are varied by varying the exit Reynolds number of the mixture.

The velocity statistics and the spectra of flames from configurations I and II are measured by using a four beam two color laser Doppler anemometry system. All four beams are frequency shifted by Bragg cells. The velocity components measured are those parallel (U) and perpendicular (V) to the burner axis. The doppler bursts are analyzed by two frequency counters. The velocity spectra in the stagnation flow stabilized flames are obtained by fast Fourier transformation directly from the analog output digitized at 4 kHz. To obtain the spectra for the velocity components normal (U_n) and tangential (U_t) to the vee flames, the outputs of the two counters are digitized simultaneously at 4 kHz and stored on magnetic tape for subsequent coordinate transformation and spectral analysis. At each measurement point, 20480 pairs of instantaneous velocity vectors in two dimensions are recorded. The flame coordinate for transformation is deduced from the joint probability density function of the velocity components/4/.

For these two flame configurations, the scalar spectra are obtained from time resolved measurements of the flame intermittency with laser light Mie scattering intensity from silicon oil aerosol seeded in the flow. Since the oil droplets evaporate and disappear through the thin flame sheet, each passage of the flame is represented by a jump from burned to unburned conditions, or vice versa. The Mie scattering signal is digitized at 10 kHz. For both velocity and scalar spectra, 50000 samples are used to obtain 100 spectra for averaging.

In configuration III, the cold flow turbulence stucture is determined by laser Doppler anemometry. Laser induced Rayleigh scattering is used in the determination of the instantaneous density field. The Rayleigh signal is sampled, digitized and stored at a rate 10 kHz until 32768 points are accumulated for each measurement point. Signals due to background and scattered light and flame radiation are compensated for by measuring, averaging and removing them from the Rayleigh signal before the normalization by the cold flow value is carried out. The fluctuating density signal is also corrected for the shot noise. But the variation of the mean scattering cross section is not compensated for.

INFLUENCE OF THE DAMKOHLER NUMBER ON THE VELOCITY SPECTRA IN VEE FLAMES

Figure 1 shows a schematic of the burner and the coordinate system for the vee flame configuration. The statistics of three velocity components U, U_1 and U_2 shown on this figure are obtained from unshifted single component laser Doppler anemometry measurements/5/. For this particular study ethylene/air mixtures are used. Spectra for the velocity component U_2 in two flames are presented here. Cold flow conditions and flame parameters for the two flames are compared on Table I.

	u' cm/s	L cm	Re_L	τ_t ms	τ_c ms	Da	$K-W$
Flame V_1	30	0.57	110	19.0	0.18	106	0.14
Flame V_2	49	0.35	110	7.1	0.21	34	0.42

Table I

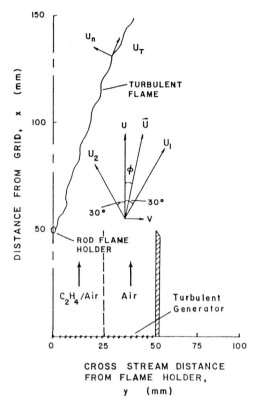

Figure 1. Schematic of the burner and the coordinate system for the vee flame configuration.

The derivation of these parameters and their relevance as turbulent flame citeria are fully discussed in /6/. Table I shows that flame V1 is much closer to a pure wrinkled flamelet regime than flame V2. The spectral data to be presented here have been selected to compare the influence of the large eddy Damköhler number Da on the spectral behaviour of the dynamic field. It may be noted that as the mean flame brush angle for flame V1 (32°) and flame V2 (30°) are very similar, the results from the two flames are directly comparable. These comparisons are presented in frequency and wave-number spaces on figures 2 and 3. Tables IIa and IIb present also the comparison of spectral slopes in different regions of these two flames.

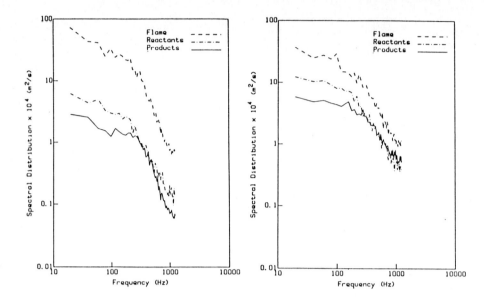

Figure 2. Comparison of the velocity spectra in three regions in the flame brush. (a) Flame V1; (b) Flame V2. For these spectra in the frequency space, the area under each curve is equal to the r.m.s. value of the corresponding velocity fluctuation.

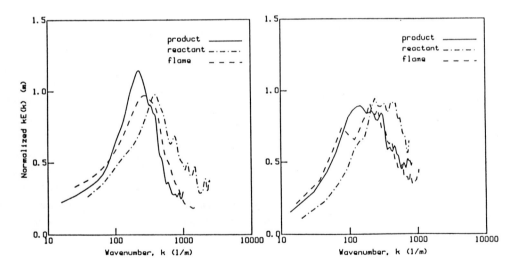

Figure 3. Comparison of the velocity spectra in three regions in the flame brush. (a) Flame V1; (b) Flame V2. For these spectra in the wave number space, the area under each curve is equal to unity.

Two classes of results arise from these comparisons /7/. The first class concerns the common features of both flames. The most important of these is the evolution of the relative energy content of the spectra when the measuring point moves from the reactants through the products. With respect to this evoution, both flames respond similarly, i.e. the energy content of the spectra is shiftedtowards larger structures as a coupled consequence of increased dilatation and viscous dissipation. In the center of the flame brush (called flame zone on the figures), increases in the spectral energy occur at all frequencies.

The second class of results concerns the differences in the flame responses. For example, the increase of the spectral energy (i.e. the rms value of the fluctuating velocity) within the flame zone is much larger for flame V1 which has the highest Damköhler number and the lowest u'/S_L ratio, so that the masking effect suggested by Klimov is smaller /8/. The spectral slopes are different for the two flames, although the corresponding cold flow turbulence Reynolds numbers are the same and their spectral dynamic consistent with the Kolmogorov slope.

These results indicate then that the classical spectral theory for non-reacting flows is not sufficient for the interpretation of many features of velocity spectra in premixed flames. This can in part be attributed to the fact that turbulence fluctuations within the flame consist of contributions from the incident turbulence, fluctuations in the products and the intermittent passage of the flamelet(s). Consequently, the determination of spectra of scalars such as temperature or density is necessary for the interpretation of the velocity spectra in premixed turbulent flames.

COMPARISON OF VELOCITY AND SCALAR SPECTRA IN VEE FLAMES AND STAGNATION FLOW STABILIZED FLAMES

For these comparisons, four sets of data have been obtained, two for each configuration and using two different turbulence generators. The mean velocity and the equivalence ratio of the methane/air mixtures used in all flames are the same and equal respectiveley to 5 m/s and 0.98. A schematics of the stagnation flow stabilized flame configuration and the data acquisition system is shown on figure 4. Table II gives the cold flow turbulence characteristics of each configuration.

flame	configuration	$\frac{u'}{S_L}$	turbulence	l	Re_l generator
V3	v	.80	plate	3.0	84
V4	v	.53	grid	2.0	32
S1	stagnation	.80	plate	3.0	84
S2	stagnation	.55	grid	2.0	32

Table II

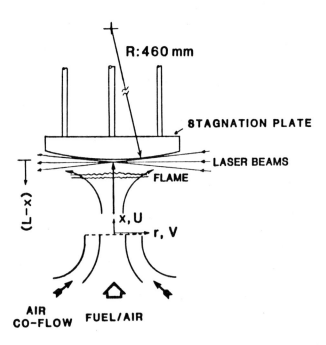

Figure 4. Schematics of the stagnation flow stabilized turbulent flame

One can easily calculate from Table II that the large eddy Damköhler number values of these four flames are very close to each other and only very slightly scattererd around 30. The spectral comparisons of this chapter emphazise then the influences of the flame configuration, the u'/S_L ratio and the turbulence Reynolds number. These comparisons are fully discussed in /9/; consequently only an overview is outlined here.

In order to present globally the flames investigated here, figures 5 and 6 show representative mean and rms profiles measured in flames V3 and S1.

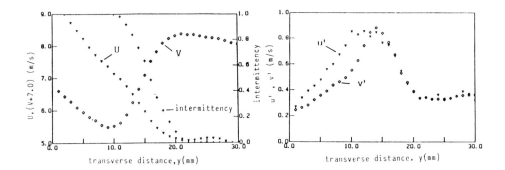

Figure 5. Mean and rms tranverse profiles of velocities and intermittency in Flame V3 at x=40 mm above the nozzle.

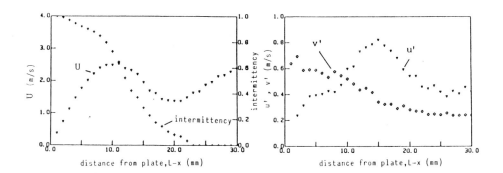

Figure 6. Mean and rms axial profiles of velocity and intermittency in Flame S1 along the centerline of the burner.

Comparison of the frequency weigthed spectra for Flame V4 at x=40 mm above the stabilizing rod is presented on figures 7 and 8. In figure 7, the velocity spectra are obtained with reference to laboratory coordinates and in figure 8 with respect to the flame coordinates. The corresponding spectral slopes for Flames V3 and V4 at two heights for each of them are presented on Table III.

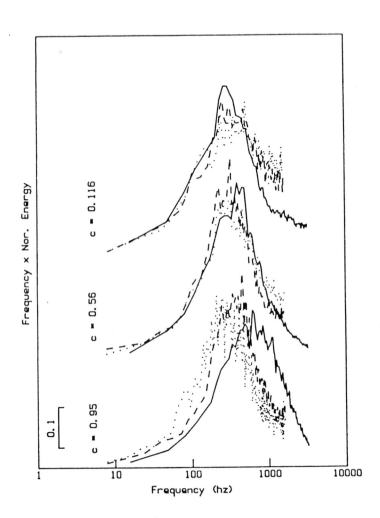

Figure 7. Comparison of the scalar (intermittency) spectra with the velocity spectra obtained with reference to the laboratory coordinates at three locations in Flame V4 at x = 40 mm. .... u' ; ------- v'; ─────── Intermittency.

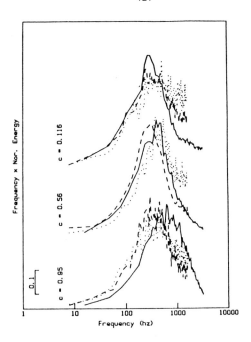

Figure 8. Comparison of the scalar(intermittency) spectra with the velocity spectra obtained with respect to the flame coordinate at the same positions as those of Fig. 7. .... u_t; ---- u'_n; ———— Intermittency.

Table II

	x (mm) (flame angle)	$\bar{c}$	Spectral Slopes				
			U	V	U_t	U_n	$\bar{c}$
V3	30.0 (35°)	0.145	-1.79	-2.11	-1.53	-2.01	-2.07
		0.48	-1.90	-2.12	-1.77	-2.14	-1.85
		0.86	-2.11	-2.20	-2.05	-2.24	-1.56
V3	40.0 (35°)	0.35	-2.09	-1.73	-1.73	-1.85	-2.02
		0.634	-1.99	-2.01	-1.55	-2.33	-1.50
		0.92	-2.22	-2.22	-1.89	-2.28	-2.17
V4	30.0 (18°)	0.13	-1.50	-1.74	-1.29	-1.75	-2.10
		0.31	-1.64	-2.02	-1.35	-2.01	-2.27
		0.8	-1.70	-2.20	-1.50	-2.20	-1.72
V4	40.0 (26°)	0.116	-1.48	-1.63	-1.24	-1.67	-2.14
		0.56	-1.86	-2.40	-1.87	-2.40	-2.07
		0.95	-2.07	-2.18	-1.57	-1.94	-1.31

Table III

The comparison of the most energetic frequencies of the scalar spectra and those of the velocity spectra with flame coordinates for Flames V3 and V4 is shown on figure 9.

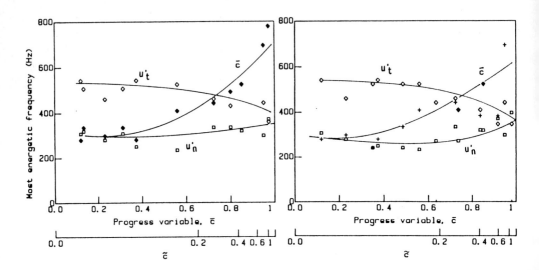

Figure 9. Comparison of the most energetic frequencies of the scalar and the velocity spectra respective to flame coordinates.
(a) Flame V3 ; (b) Flame V4.

Parallely, the spectral comparisons concerning the stagnation flow stabilized flame (Flame S1) are presented in the frequency weighted form on figure 10. The spectral slopes corresponding to both Flames S1 and S2 are displayed on Table IV.

		Spectral Slopes		
	$\bar{c}$	U_t	U_n	$\bar{c}$
flame S1	.148	-1.44	-2.01	-2.06
	.424	-1.96	-1.96	-2.09
	.727	-2.12	-1.94	-1.96
flame S2	.096	-1.52	-1.41	-2.27
	.35	-1.67	-1.83	-2.34
	.89	-2.11	-1.93	-2.04

Table IV

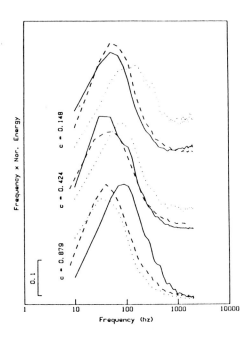

Figure 10. Comparison of the scalar (intermittency) spectra with the velocity spectra in three locations in Flame S1. ---- u'; ...... v'; ———— intermittency.

Finally, the comparison of the most energetic frequencies of the scalar and velocity spectra for Flames S1 and S2 is presented on figure 11.

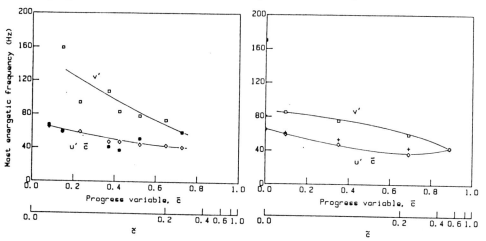

Figure 11. Comparison of the most energetic frequencies of the scalar and the velocity spectra. (a) Flame S1; (b) Flame S2.

These comparisons between the scalar and velocity spectra in two different turbulent premixed flame configurations indicate that the effect of the coordinate transformation on the velocity spectra is more apparent in the velocity component tangential to the mean flame front. These effects include changes in the spectral slope and shifting of the most energetic frequencies to higher levels. Near the middle of the flame brush, where c = 0.5, the scalar spectrum is closest to the normal velocity spectrum. The reason for this is that the combustion induced velocity jump which occurs after each flame passage is mostly in the normal direction. Consequently, the normal velocity spectrum also contains the flame passage (scalar/intermittency) spectrum.

This result is streghtened by the comparisons from the stagnation flow stabilized flame where, except in the hot boundary, the normal velocity and the scalar spectra behave similarly; whilst the tangential velocity spectrum is shifted towards higher frequencies. The parallel evolution of the most energetic frequencies of the scalar and the normal velocity spectra confirms the strong analogy between these two spectra and consequently between the time and length scales of the two fields.

INFLUENCE OF THE DAMKOHLER NUMBER ON THE SCALAR SPECTRA IN BUNSEN TYPE CONICAL TURBULENT PREMIXED FLAMES

The schematic of the experimental arrangement for the Bunsen flame facility is presented on figure 12.

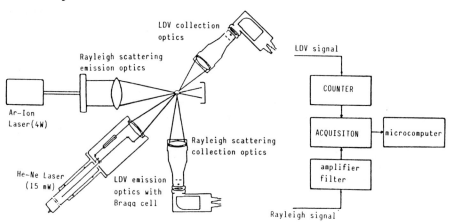

Figure 12. Schematics of the experimental arrangement for the Bunsen flame experiments.

Six methane/air flames, corresponding to combinations of three exit Reynolds numbers and four equivalence ratios have been investigated. The main parameters of these flames are summarized on Table V.

	u'/S_L	Re_ℓ	ℓ/δ_L	Da	t_c/t_η
C1	0.85	40	15	17	0.37
C2	1.54	110	21	14	0.76
C3	1.98	130	19	10	1.16
C4	2.01	110	17	9	1.24
C5	1.03	110	30	30	0.35
C6	1.25	110	26	21	0.51

Table V

For each flame, instantaneous density data from Rayleigh scattering are taken at one axial and four transversal traverses. The centerline evolutions of the mean and fluctuating density are used to determine the total and inner cone heights of the flames. The transversal mean density profiles are used to determine, by the maximum gradient method, the average turbulent flame brush thickness /10/.

Spectra of the density fluctuations for Flames C4 and C5 are presented below. Table V shows that these two flames correspond to extreme values of the Damköhler number range explored here; but they have the same cold flow turbulence Reynolds number. So, the variation of the Damköhler number is obtained by varying the equivalence ratio, which is equal to 0.7 for Flame C4 and to 1 for Flame C5.

Three spectra are presented in logarithmic coordinates for each flame on figures 13. They correspond to $z/D = 2$ above the burner exit and are representative of the mostly unburnt region, the mostly burnt region and the middle of the average flame brush where the density fluctuations peak. Both flames exhibit the same spectral trends. Density spectra corresponding to cold and hot boundaries of the flame have a weaker slope than that for the middle of the flame brush. These spectral slopes are summarized on Table VI. For both flames, the cold boundary spectral slope is very close to the -5/3 Kolmogorov slope, which indicates that in this part of the average flame brush the fluctuations of the instantaneous flame front are essentially triggered by the cold flow velocity fluctuations. The slopes increase slightly and approach -2 in the hot boundary of the flame front. For this evolution also, it may be assumed that the spectral dynamics of the velocity field dominates the temporal structure of the flame front movements. But, in this region the increased viscous dissipation gives rise to a steeper decay rate at high frequencies.

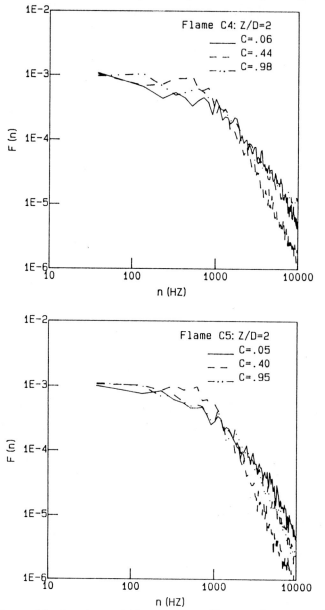

Figure 13. Spectra of the density fluctuations at z/D = 2.
(a) Flame C4; (b) Flame C5.

In the middle of the flame brush, the density spectra fall off with the highest slope, which is -2.5 for Flame C4 and -2.9 for Flame C5. These spectra are wholly representative of the dynamics of the instantaneous flamelet(s) passages. It is interesting to note that these slopes are very close to those obtained by Mie scattering in the middle of a vee flame brush/11/, but they do not present the low frequency hump obseved by Rayleigh scattering measurements in low speed vee flames/12/. One may note also that these slopes are higher than that predicted by the model based on time series representation of the flame passages and which gives a slope of -2. Another interesting feature of these spectra is that, for each value of the progress variable c, they tend to the same plateau value for low frequencies. This behaviour is indicative of the fact that the integral time scale associated with density fluctuations is constant through the flame brush. This may be considered as a confirmation of the same result established from the evaluation of the integral time scales from the density auto-correlation functions/11, 13/.

	Flame C4			Flame C5		
c	0.06	0.44	0.98	0.05	0.40	0.95
n_e	1300	720	1010	1200	710	1110
slope	-1.66	-2.46	-1.87	-1.57	-2.92	-2.04

TABLE VI

The same spectra weighted by the frequency are presented on figures 14 with semi-logarithmic coordinates. According to the normalization ,the area under each curve is equal to unity. They show clearly that the high frequency components are relatively less energetic for the spectra corresponding to the middle of the flame brush. As shown on Table VI, for both flames, the most energetic frequency range, n_e, in the middle of the flame brush is almost half of its values for the cold or hot flame boundaries. This again shows that in this region of the average flame brush, the spectral dynamics is dominated by the instantaneous fluctuations of the flamelet(s), and not by the intrinsic fluctuations of the dynamic turbulence field. Another important result which arises from this spectral comparison is that the large eddy Damköhler number does not appear as a major regulating parameter of the temporal characteristics of the scalar structure of the conical flames investigated in this study. As the Damköhler number is varied via the equivalence ratio for flames C4 and C5, this implies also that the temporal characteristics of the scalar field are insensitive to changes in the equivalence ratio.

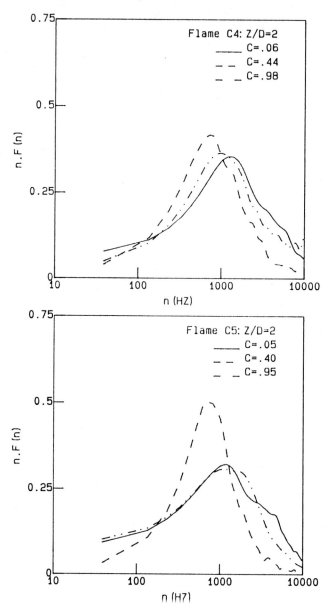

Figure 14. Frequency-weighted scalar spectra at z/d =2.
(a) Flame C4; (b) Flame C5.

CONCLUSION

Spectral information have been obtained for dynamic and scalar fields of premixed turbulent flames by investigating three different flame configurations: vee flames, stagnation flow stabilized flames and conical flames. For each configuration the influence of an important global parameter is explored. The most basic result which arises from this study is that the instantaneous flamelet(s) imposes their own spectral dynamics on to the flow field. This means that the spectral dynamics of the velocity field is strongly influenced by the passages of the flamelet(s). This is convincingly shown by the strong analogy between the spectra of the scalar field and the spectra of the velocity component normal to the mean flame brush. This analogy indicates that the time scales of the two fields are also strongly linked. On the other hand, the results obtained from the open conical flames show that the scalar field temporal characteristics are not strongly influenced by the Damköhler number when this parameter is varied via the equivalence ratio of the reactive mixture.

REFERENCES

/1/ Bray K.N.C. and Libby P.A., Combustion Science and Technology, 47:253, (1986)

/2/ Cheng R.K., Combustion Sciences and Technology, 41:109 (1984)

/3/ Cho P., Law C.K., Hertzberg J.H. and Cheng R.K., to appear in the 21st. Symposium (International) on Combustion, Munich, August 1986

/4/ Cheng R.K. and Shpeherd I.G., Combustion Science and Technology, 49:17 (1986)

/5/ Cheng R.K. and Ng T.T., Combustion and Flame, 57:156 (1984)

/6/ Gökalp I., Combustion and Flame, 67:111 (1987)

/7/ Gökalp I., Shepherd I.G. and Cheng R.K., to appear in Combustion and Flame (1987)

/8/ Klimov A.M., Progress in Astronautics and Aeronautics, 88:146 (1983)

/9/ Cheng R.K., Shepherd I.G. and Gökalp I., paper presented at the Sixth Symposium on Turbulent Shear Flows, Toulouse, September 1987.

/10/ Boukhalfa A. and Gökalp, I., to appear in Combustion and Flame, 1988.

/11/ Shepherd I.G. and Cheng R.K., to appear in **Combustion Science Technology**, (1988)

/12/ Gouldin F.C. and Halthore R.N., **Experiments in Fluids**, 4:269 (1986)

/13/ Boukhalfa A., Sarh B., Debbich M. and Gökalp I., paper presented at the **Eleventh ICDERS**, Warsaw, Poland, August 1987.

TURBULENT FLOW FIELD AND FRONT POSITION STATISTICS IN V-SHAPED PREMIXED FLAME WITH AND WITHOUT CONFINEMENT

Ph. Goix, P. Paranthoën, M. Trinité

Université de Rouen - Faculté des Sciences et des Techniques
Laboratoire de Thermodynamique (U.A. C.N.R.S. N° 230)
B.P. 118 - 76134 Mont-Saint-Aignan Cedex (FRANCE)

I - INTRODUCTION

Premixed turbulent flames have received considerable attention during the past few years both from an experimental and theoretical point of view. However our understanding of the mutual interaction of combustion and turbulence is still incomplete due to the large number of mechanisms involved in such a problem.

With the new development of laser based diagnostics (LDV, Rayleigh, tomography...), some results concerning the flame structure and the combustion mechanisms are now available in the litterature. Recent measurements of velocity, temperature, density performed in a V shaped turbulent flame concern several studies [1,2,...8]. Although these studies have been carried out in an approach turbulence generated by a grid, comparison of experimental results is often difficult due to differences linked to experimental combustor configurations (confinement, flame-holder diameter,...) and gas mixture compositions (fuel nature, equivalent ratio). In most of these studies the flame thickness is (generally) lower than turbulent scales of the approach turbulence and the structure of the instantaneous flame-front is wrinkled in agreement with the classification proposed by BORGHI [9].

For this kind of flame it is then important to study the wrinkling of the flame which is known from the work of DAMKOHLER [10] to be the main reason of the increase in the burning rate. From simple geometrical considerations the average length of the wrinkles can be related to the value of the root-mean-square displacement of a flame element from the mean flame position $\left(\overline{Y'^2}\right)^{1/2}$. Understanding of effects involved in the increase (or the decrease) of $\left(\overline{Y'^2}\right)^{1/2}$ can then lead to a better knowledge of the turbulent flame propagation mechanism. As analysed by KARLOVITZ et al. [11], SCURLOCK and GROVER [12], $\left(\overline{y'^2}\right)^{1/2}$ is increased by eddy diffusion associated with turbulence of the approach flow while propagation of the flame into the unburned gases tends to reduce $\left(\overline{y'^2}\right)^{1/2}$. Furthermore, amplification or stabilization of disturbances in the flame-front can be associated with diffusion-heat instability effects or hydrodynamic instability effects (WILLIAMS [13], CLAVIN [14]).

The purpose of this paper is to present and discuss some experimental results concerning both velocity fluctuations and instantaneous flame-front positions measurements in the case of a lean V shaped turbulent H_2-Air flame in presence of grid turbulence flow-field. First, complete results concerning longitudinal and transversal mean and fluctuating velocity are presented for three cases of pressure gradient (0 ; - 40 Pa/m ; - 72 Pa/m).

The second part of these results consists in quantitative visualizations of instantaneous flame-fronts of the unconfined V flame by means of a tomography laser technique.

Turbulence characteristics of the approach flow have been extensively studied for two grids (M = 5 mm ; 10 mm) in cold conditions.

Modification of turbulent kinetic energy is found to be mainly influenced by the longitudinal mean pressure gradient.

From visualizations, the turbulent flame brush characterized by $\left(\overline{y'^2}\right)^{1/2}$ and the correlation coefficient between the positions of the two flame-fronts have been determined. Their evolution versus flame-holder downstream distance is particularly studied. Lengthening rate of the flame is found to be proportionnal to the burning rate. These results have been interpreted in a satisfactory way using a rescaling scheme based on the Lagrangian approach early proposed by KARLOVITZ et al. [11], SCURLOCK et GROVER [12].

II - EXPERIMENTAL ARRANGEMENT

Experimental apparatus

A schematic diagram of the experimental apparatus is shown in figure 1. The V flame is stabilized, without recirculation effect, on a 0.4 mm-diam catalytic wire. The wire-Reynolds number is always lower than the critical Reynolds number ($Re_d \simeq 15$). Turbulence of the approach premixed Air-Hydrogen flow is generated by a grid located at the output of the settling chamber. The wire is successively located at different distances from the grid (12 M_G or 24 M_G where M_G is the mesh size) in order to modify turbulence parameters at ignition.

Experiments are carried out successively in an unconfined free flow or in a confined flow. The confinement is obtained by means of a combustion chamber of 8x8 cm^2 in cross section and 25 cm in length. Two rectangular quartz windows are installed for optical measurements. The longitudinal mean pressure gradient is controlled by adjusting the walls of the combustion chamber.

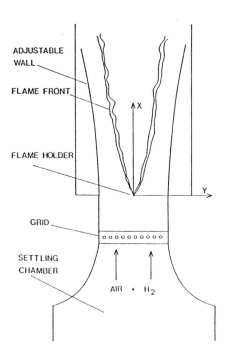

Figure 1 - Diagram of experimental apparatus

Laser Doppler Anemometry

Two classical LDV systems have been successively used : a one-channel real-fringe forward scatter system and a two-channel real-fringe back scatter system (with a 3.75 beam expander in order to increase signal/noise ratio). The probe volume is positionned in the three directions through a traverse-periscopic system driven by stepping motors connected to the computer so that automatic scanning of the flame is performed.

A connection between the counters and a PDP 11/34 is used for data acquisition and reduction. In order to detect and to reduce bias [15] statistics are obtained by two averaging procedures : classical ensemble averaging and time averaging.

The flame is seeded at the entrance of the settling chamber with ZrO_2 particles dispersed by a rotary brush system. In addition, for conditional sampling, the flow can also be seeded with an oil spray which is used for tomography.

Tomography

The well known tomography method initiated by Boyer [16] and used in premixed combustion by Dumont [17] and Escudié [18], is chosen for flame visualization.

An instantaneous thin laser sheet is obtained with a pulsed ruby laser and a lens system as shown in Figure 2. The upstream flow is seeded with silicon oil droplets with a diameter less than 1 μm. These droplets disappear within the flame front imaging the sheet of light with a CCD camera. This camera is connected to a videotape recording system and to an image processing device. A 75mm height turbulent flame is visualized. The dark zone represents burned gas and the bright zone represents a fresh mixture. The flame front position is located at the boundary between the two zones.

For each laser shot, the video signal is stored and digitized in a memory matrix of 512 x 512 pixels and on 256 grey levels. This image memory is connected to a microcomputer (figure 2). Digital analysis is then performed line by line in three steps : filtering, digitizing by thresholding, flame front position detection and storage. Then different statistics can be obtained such as 1st and 2nd moment of the front position, correlation between the left and the right front, burned zone width, rate of lenghtening of the front and curvature. These statistics are performed on 240 flame samples. Accuracy of the method and stability of the flame is systematically tested in the laminar case (the same flow without grid).

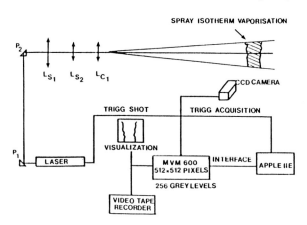

Figure 2 - Experimental device

Flow parameters

Experimental conditions, without combustion, are summarized in table I. In the coordinate system used throughout, X coordinate is taken to be parallel to the flow and Y coordinate is normal to the flow and the flame holder. The origin is at the catalytic wire location. G_1 and G_2 correspond respectively to the two values of the mesh size : 5 mm and 10 mm. C_0 indicates the case without confinement, C_1 and C_2 are respectively the cases with $\frac{dp}{dx} = -40$ pa/m and $\frac{dp}{dx} = 72$ Pa/m. Subscript 0 corresponds to the parameters values measured at the wire location (A : $\frac{X_0}{M_G} = 12$, B : $\frac{X_0}{M_G} = 24$, C : $\frac{X_0}{M_G} = 22$).

Decay of turbulence, behind the grid in cold conditions, is studied in details with the two colors LDV. The law $\frac{\overline{U^2}}{u'^2} = C\left[\frac{X}{M_G} - \frac{X_0}{M_G}\right]^n$ is verified with n=1 (fig. 3 and 4). Comparison of the RMS velocities downstream the grid G_1 with and without confinement is shown in fig.5. It is worth to note the influence of the stagnant surrounding on the longitudinal velocity fluctuation for X > 60 mm. However this effect is negligible on the lateral velocity fluctuation.

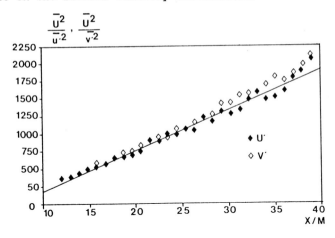

Figure 3-Decay of fluctuations u' and v' in cold conditions (grid n°1)

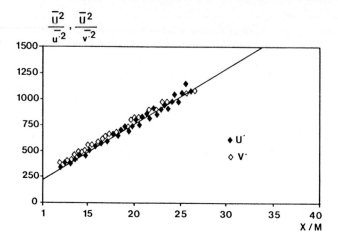

Figure 4-Decay of fluctuations u' and v' in cold conditions (grid n°2)

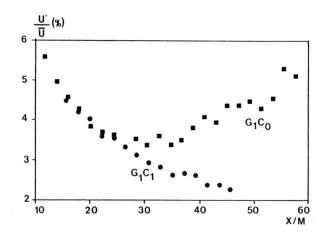

Figure 5 - Influence of the stagnant surrounding on the turbulence decay

Equivalent ratio, φ^*, of hydrogen in the mixture is 0.4 and the maximum temperature of the burned gases does not exceed 1200 K. In order to test the sensitivity of the flame speed to the equivalent ratio, the value $\varphi = 0.43$ is also considered. Under these conditions, flame angle is about 4° in the confined case and smaller always smaller thanin the unconfined case.

reference	M_G (mm)	$\frac{dP}{dX}$ (pa/m)	$\frac{X_0}{M_G}$	$\overline{U_0}$ (m/s)	I_0 (%)	φ
A $G_1 C_1$	5	-40	12	4.84	5.2	0.4
A $G_1 C_2$	5	-72	12	4.84	5.25	0.4
A $G_1 C_0$	5	0	12	4.9	5.3	0.4 - 0.43
B $G_1 C_0$	5	0	24	4.9	3.3	0.4 - 0.43
A $G_2 C_0$	10	0	12	5.1	4.9	0.4
C $G_2 C_0$	10	0	22	5.	3.3	0.4

Table I

$$* \; \varphi = \frac{N_{H_2}}{N_{H_2} + N_{O_2}} \bigg/ \left(\frac{N_{H_2}}{N_{H_2} + N_{O_2}} \right)_{stochio}$$

III - EXPERIMENTAL RESULTS

Velocity measurements within turbulent flame

. **Mean velocities** : Evolution of longitudinal and transversal velocity, $\overline{U}$ and $\overline{V}$ is presented in figure 6 a, b, c for the three cases of longitudinal mean pressure gradient (0, - 40 pa/m ou - 72 pa/m). The mean flame position defined at the location of $\left(\frac{\partial \overline{V}}{\partial y}\right)_{max}$ as suggested by Cheng [21] is indicated on the velocity profiles by an arrow.

In the first studied section (X = 4.5 cm) the mean velocity profiles present a decay in the central zone (burned gases) resulting from the wake of the flame holder. It is worth to note that the wake is less important than in similar experiments owing to the use of a catalytic wire.

Far downstream, an important increase of $\overline{U}$, located in the central zone, is observed in the <u>confined cases</u> resulting from two different phenomena :

- the mean longitudinal pressure gradient which accelerates preferentially the burned gases ;

- the flow convergence of stream lines in the central zone and their divergence in the external zone of the flame where the flow is deflected outside.

In the unconfined case only this last effect is present. The divergence of the external flow is maximum and a small increase of velocity is observed in the reaction zone.

. **RMS velocities** : Evolution of longitudinal and transversal RMS velocity, u' and v' is presented in Figure 7 a,b,c,. Close to the flame holder, a strong reduction of the velocity fluctuations is observed within the flame, resulting from increase of dissipative terms with temperature increase. This decrease is more accentuated for v' due to the strong divergence of the flow in this zone ($\partial \bar{V}/\partial y < 0$).

Far downstream a strong increase of u' is observed. This effect is more accentuated for the higher mean pressure gradient case and inexistant for the unconfined case. In the confined case, maxima are centered on the maximum values of $|\partial \bar{U}/\partial y|$. The v' profiles are more complex due to competition between the decrease in the reaction zone and the increase in the all-burned gases. In all cases, with zero pressure gradient, increase of turbulent kinetic energy is not observed.

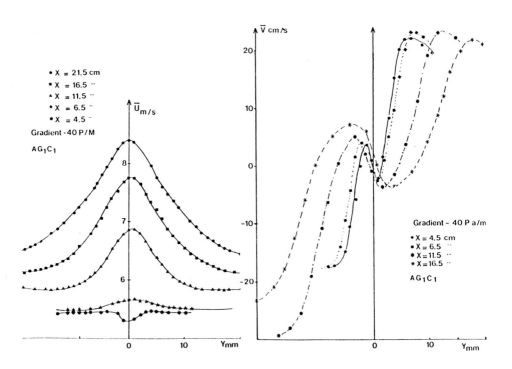

Figure 6-a - Evolution of mean velocity profiles (gradient -40pa/M)

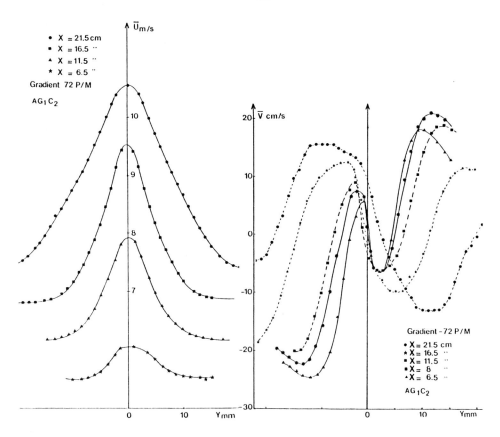

Figure 6-b - Evolution of mean velocity profiles (gradient -72pa/M)

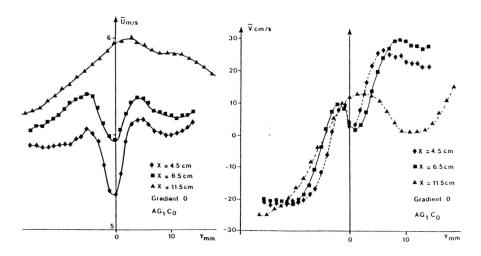

Figure 6-c - Evolution of mean velocity profiles (gradient 0pa/M)

. **Reynolds Stresses** : Reynolds stresses profiles are plotted in figure 8 a,b,. An important increase of $\overline{u'v'}$ is observed, corresponding to the maxima of the mean velocity gradient (when it is present). A counter gradient region is not evidently detected as in the case studied by Cheng [21].

Visualization and results on digitized images

Visualizations and statistical computations have been performed on different turbulent flames summarized in the Table I. Visualizations of laminar and two turbulent flames of equivalent ratio 0.4 are shown in Figure 9. The turbulent intensities at the ignition are 5.3 % and 3.3 % respectively. For this case the turbulence is generated by the same grid and the turbulent Reynolds number is quite the same in the two configurations and about 30. However the development of these two flames is rather different. The referential used for statistical analysis is shown in figure 10. All statistics are performed on 240 samples.

Statistical computations

Y_1, Y_2 : mean front position ; $\left(\overline{Y_1'^2}\right)^{1/2}, \left(\overline{Y_2'^2}\right)^{1/2}$ R.M.S. front position

$$R_{Y_1 Y_2} = \frac{\overline{Y_1' Y_2'}}{\left(\overline{Y_1'^2}\right)^{1/2} \left(\overline{Y_2'^2}\right)^{1/2}} \; ; \; Y_1 - Y_2 \; : \text{represents the burned zone width}$$

$$A_T = \int_{\text{front}} ds \; : \text{represents the front length} ;$$

$$R_c = \frac{\overline{Y_1'(x) Y_1'(x+dx)}}{\left(\overline{Y_1'^2(x)}\right)^{1/2} \left(\overline{Y_1'^2(x+dx)}\right)^{1/2}} \; : \text{for one front}$$

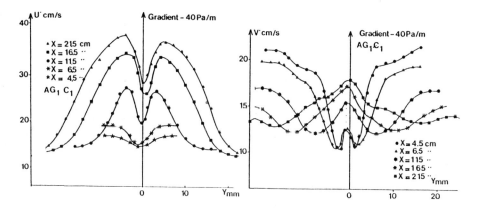

Figure 7-a - Evolution of RMS velocity profiles (gradient -40pa/M)

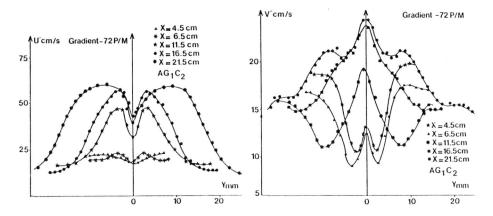

Figure 7-b - Evolution of RMS velocity profiles (gradient -72pa/M)

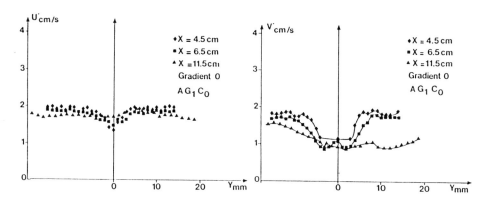

Figure 7-c - Evolution of RMS velocity profiles (gradient 0pa/M)

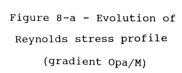

Figure 8-a - Evolution of Reynolds stress profile (gradient 0pa/M)

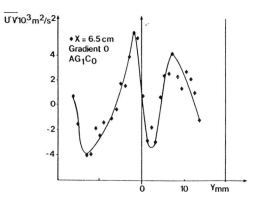

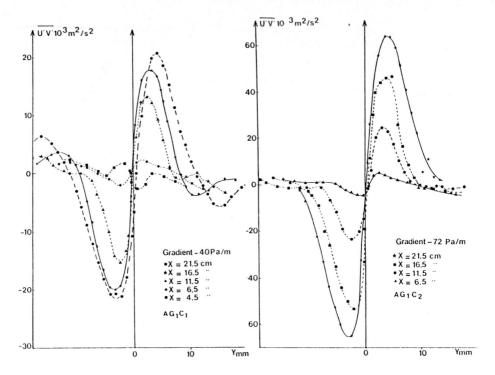

Figure 8-b - Evolution of Reynolds stresses profiles
(Gradient -40pa/M and -72pa/M)

Longitudinal evolution of the root-mean square of displacement of flame element $\left(\overline{Y'^2}\right)^{1/2}$ is presented in Figure 11 for the 5 mm mesh grid and in Figure 12 for the 10 mm mesh grid.

For each case results are shown in the unconfined case when the flame holder is located at 12 M and 24 M from the grid. Comparison with the laminar case shows that turbulence strongly increases the value of $\left(\overline{Y'^2}\right)^{1/2}$ and its evolution is approximately proportionnal to the turbulence intensity at the flame holder. This influence of the approach turbulence level clearly appears in Figure 13 where $\left(\overline{Y'^2}\right)^{1/2}$ is plotted versus $\frac{X}{M_G}$ for the two situations corresponding to the 5 mm mesh grid. Over the studied range a linear spreading region is observed as in similar experiments [2], [4], [5].

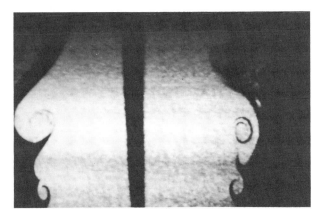

Laminar Flame

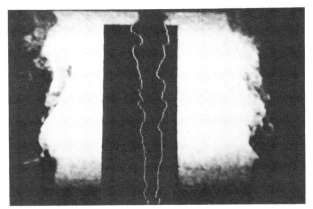

Turbulent Flame A $G_1 C_0$

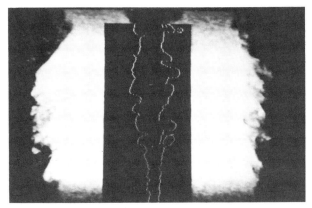

Turbulent Flame B $G_1 C_0$

Figure 9

Examples of visualizations

Figure 10 - Referential

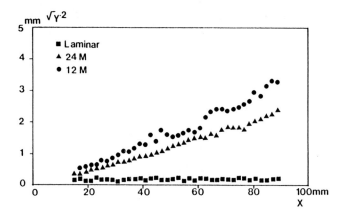

Figure 11 - Evolution of $(\overline{Y'^2})^{1/2}$ (G_1)

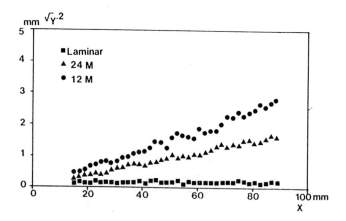

Figure 12 - Evolution of $(\overline{Y'^2})^{1/2}$ (G_2)

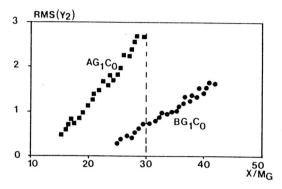

Figure 13 - Influence of turbulent level at ignition on $(\overline{Y^2})^{1/2}$

Evolution of the burned zone width ΔY is presented in Figures 14 and 15. For each case in the initial stage, the burned zone width of the turbulent flame is found equal to that of the laminar case. For each grid when the flame is generated at 12 M_G, this first stage is observed over a smaller distance. It is followed by a second stage where a strong increase of $\Delta \overline{Y}$ appears.

Information about the correlation of positions of the two flame-fronts of the V flame is shown in Figure 16 and 17. For each grid the two flame-fronts are found correlated over a larger distance when the flame-holder is located at 24 M_G. The initial value of the correlation coefficient $R_{Y_1 Y_2}$ seems lower than 1 in relation with the width of the ignition zone around the catalytic wire. It is worth to note that over the region where $R_{Y_1 Y_2}$ is maximum the burned zone width is the same as in the laminar flame without modification.

Mean values of the flame-front element length :

$l(\Delta x) = A_T (\Delta x + \Delta x_0) - A_T (\Delta X)$

measured between two points distant of $\Delta x_0 = 3.5$ mm are presented in Figure 18 for the 5 mm mesh grid case. The slope of the curve, $\overline{l}$, versus Δx represents the elongation of the flame-front. In agreement with the burned zone width results the maximum values of $\overline{l}$ are found when the flame-holder is located at 12 M. It is worth to note that the rate of lengthening of the flame front is proportional to the burned zone width evolution for the two cases of turbulence (12 M_G and 24 M_G). It indicates a local velocity propagation equal to the laminar one.

An example of the correlation of the flame-front positions in the longitudinal direction is shown in Figure 19. This coefficient is characteristic of the wavy form of the instantaneous flame-front. A smaller length scale is found when the flame holder is located at 12 M_G

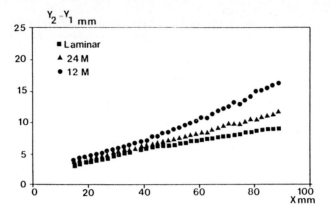

Figure 14 - Burned zone width evolution (G_1)

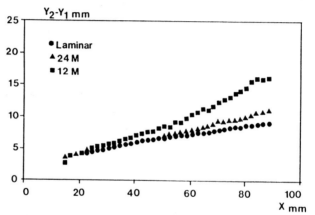

Figure 15 - Burned zone width evolution (G_2)

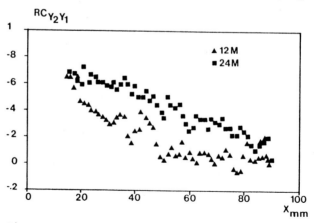

Figure 16 - Two fronts correlation evolution (G_1)

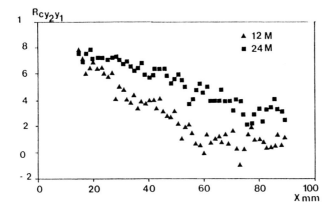

Figure 17 - Two fronts correlation evolution (G_2)

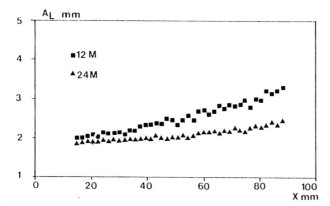

Figure 18 - Mean value evolution of the flame front element length (G_1)

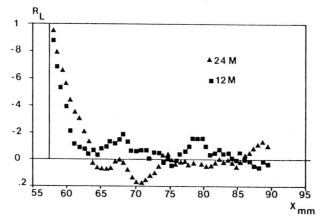

Figure 19 - Evolution of the flame front positions correlation in the longitudinal direction (G_1)

IV - ANALYSIS

Turbulent diffusion of a flame-element

The mechanism of flame propagation in a turbulent flow with an integral scale which is very large by comparison with the thickness of the laminar flame is considered in relation with the Lagrangian approach developed by KARLOVITZ et al. [11]. In this approach small positions of the flame-front are transported at random by the turbulent flow field when flame elements diffuse away from the flame holder position. Simultaneously each flame element moves at its normal burning velocity S_L from one small gas volume to the next. The turbulent flow field at the flame holder can be characterized by :

- its intensity $\left(\overline{v'^2}\right)^{1/2}$ where v' is the instantaneous value of the fluctuating velocity component perpendicular to the average turbulent flame ;

- the correlation coefficient $R(\tau) = \dfrac{\overline{v'(t_0) \, v'(t_0 + \tau)}}{\overline{v'^2}}$ where $v(t_0)$ is the fluctuating velocity component at the time $t_0 + \tau$;

- the scale of turbulence $l_1 = \left(\overline{v'^2}\right)^{1/2} \int_0^\infty R(\tau) d\tau$;

- the characteristic time $T_L = \int_0^\infty R(\tau) d\tau$

When the flame propagates at a very low velocity $\left(\dfrac{S_L}{\left(\overline{v'^2}\right)^{1/2}} \ll 1\right)$ a diffusing flame element is bound during the diffusion process to the same fluid particle and the displacement of a flame element coinciding with a fluid particle at zero time (at the flame holder location) is deduced from the displacement Y of the fluid particle. The root mean square value of the turbulent displacement $\left(\overline{Y'^2}\right)^{1/2}$ can be calculated according to the theory of G.I. Taylor given by the following equation :

$$\frac{d\overline{Y'^2}}{dt} = 2 \, \overline{v'^2} \int_0^t R(\tau) d\tau$$

When the relation $\dfrac{S_L}{\left(\overline{v'^2}\right)^{1/2}} \ll 1$ is not verified the flame-front during the diffusion process is not bound to the same marked fluid particle. In this situation SCURLOCK and GROVER [16] defined a mixed correlation coefficient $R(\tau,y)$ which characterizes the correlation between the fluctuating velocity of a flame element at zero time and that of the same flame element at a later time after the flame has propagated into the unburned mixture a distance $y = S_L t$.

The equation becomes :

$$\frac{d\overline{y'^2}}{dt} = 2 \overline{v'^2} \int_0^t R(\tau,y)d\tau \quad (1)$$

Assuming $R(\tau,y) = R(\tau) R(y)$ and a correlation function $R(y) = \exp - (y/l_2)$ where l_2 is an eulerian length scale of v'.

A mixed scale l_i and a mixed characteristic time T'_l may now be defined as follows :

$$l_i = \frac{l_1}{1 + \frac{S_L}{2\left(\overline{v'^2}\right)^{1/2}}}$$

$$T'_l = \frac{1}{1/T_L + \frac{S_L}{2 \ l_2}}$$

For time intervals t which are small by comparison with the characteristic time T'_l, $R(\tau,y) \sim 1$ and $\left(\overline{y'^2}\right)^{1/2} = \left(\overline{v'^2}\right)^{1/2} t$.

For time intervals t which are very large with respect to T_L,

$$\left(\overline{y'^2}\right)^{1/2} = \left\{2 \ l_i \left(\overline{v'^2}\right)^{1/2} t\right\}^{1/2}.$$

For intermediate time $\left(\overline{y'^2}\right)^{1/2}$ depends on the shape of the correlation curve $R(\tau,y)$. Assuming an exponential form :

$$\left(\overline{y'^2}\right)^{1/2} = \left\{2 \ l_i \left(\overline{v'^2}\right)^{1/2} t\right\}^{1/2} \left\{1 - \frac{T'_l}{t} \left(1 - \exp - \left(\frac{t}{T'_l}\right)\right)\right\} \quad (2)$$

Using the Taylor hypothesis and relation (1), $\left(\overline{y'^2}\right)^{1/2}$ can be expressed as a function of downstream distance :

$$\left(\overline{y'^2}\right)^{1/2} = \frac{\left(\overline{v'^2}\right)^{1/2}}{U} \Delta x$$

By defining $L_l' = \bar{U} T_l'$ relation (1), (2) can be written in a dimensionless form :

$$\frac{\left(\overline{Y'^2}\right)^{1/2}}{l_i} = \frac{\Delta X}{L_l'} \qquad \text{when } \frac{\Delta X}{L_l'} \ll 1$$

and

$$\frac{\left(\overline{Y'^2}\right)^{1/2}}{l_i} = \left(\frac{\Delta X}{L_l'}\right)^{1/2} \qquad \text{when } \frac{\Delta X}{L_l'} \gg 1$$

Application to our results

At this point accurate measurements of the Lagrangian time-scale T_L are not available and we choose to calculate T_L from the relation proposed by CSANADY [24] :

$$T_L(x) = 1.1 \frac{\overline{v'^2}(x)}{\varepsilon(x)}$$

where $\varepsilon(x)$ is the turbulent dissipation rate at the distance x from the grid.

In our experiment the turbulent dissipation rate $\varepsilon(x)$ can be calculated as :

$$\varepsilon = -\frac{3}{2} \bar{U} \frac{d\overline{u'^2}}{dx}$$

Examples of values of l_i and L_l' integrating the influence of flame propagation are presented in Figures 20 and 21 and Table II. As shown in these figures l_i and L_l' are found to be dependent on the location of the flame holder downstream the grid. In order to take into account this evolution results concerning $\left(\overline{Y'^2}\right)^{1/2}$ are presented by plotting $\left(\overline{Y'^2}\right)^{1/2} /l_L(\Delta X)$ versus $\int_0^{\Delta X} \frac{d(\Delta X)}{L_L(\Delta X)}$.

By using this rescaling scheme results concerning the root-mean square of flame displacement, $\left(\overline{Y'^2}\right)^{1/2}$, are shown in figures 22 and 23. It is worth to note that for the four studied cases results are gathered in a satisfactory way. In our experiment, the studied range is confined to small and intermediate values of $\overline{\Delta X}/L_L$ (0.5 < $\overline{\Delta X}/L_L$ < 1.5). Over this range an approximative linear spread is observed. Comparison with the theoretical curve deduced from (2) indicates that $\left(\overline{Y'^2}\right)^{1/2}_{meas.}$ is always lower than $\left(\overline{Y'^2}\right)^{1/2}_{theor.}$ in the first part of the flame development. This could be reasonably explained by the fact that propagation speed of angular points exceeds the normal laminar speed of the flame owing to the interaction of two flame fronts [11]. This influence appears on photographs presented in figure 9 where sharp edges are preferentially found on the burned gas side. This is also confirmed by lower values of $\left(\overline{Y'^2}\right)^{1/2}/l_i$ found when increasing the equivalent ratio as shown in figures 24 and 25. However, in a second stage the measured $\left(\overline{Y'^2}\right)^{1/2}$ values tend to exceed the theoretical value only based on eddy diffusion. This behavior is accentuated in figures 24 and 25 where the equivalent ratio increases. This could be related to instability effects but a better understanding of this aspect will need more systematic study.

Rescaling of the correlation of displacement of adjacent flame fronts can be achieved using a similar scheme based on mixed scales l_1'' and T_L'' defined as follows :

$$l_1'' = \frac{l_1}{1 + \dfrac{S_L}{\left(\overline{v'^2}\right)^{1/2}}} \quad \text{and} \quad T_L'' = \frac{1}{\dfrac{1}{T_L} + \dfrac{S_L}{\left(\overline{v'^2}\right)^{"2}}}$$

As shown in Figures 26-29, a satisfactory agreement is observed for each grid for the two locations of the flame holder and the two values of the equivalent ratio.

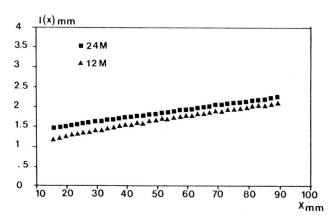

Figure 20 - Evolution of l'_1 (G_1)

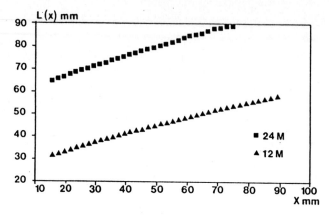

Figure 21 - Evolution of L'_1 (G_1)

	U m/s	I %	$(v')^{2\,1/2}$ m/s	C	ξ^2_2 m²/s²	T_L MS	L_L MM	$L_{1.0}$ MM	$L'_{1.0}$ MM	$L'_{1.0}$ MM	$L''_{1.0}$ MM	$L''_{1.0}$ MM	$\eta=(\nu^3/\varepsilon)^{1/4}$ MM
$AG_1\ C_0$ $S_L=.16$M/s	4.90	5.3	0.26	55.50	15.30	4.40	21	1.15	16	0.90	13	0.7	0.13
$BG_1\ C_0$ $S_L=.16$M/s	4.90	3.3	0.16	55.50	2.20	11.60	57	1.90	38	1.25	28	0.95	0.20
$AG_2\ C_0$ $S_L=.16$M/s	5.10	4.9	0.25	51.75	6	10.40	53	2.60	40	1.95	32	1.60	0.16
$CG_2\ C_0$ $S_L=.23$M/s	5.10	3.3	0.17	51.75	1.30	23	118	3.90	81	2.65	61	2	0.25
$AG_1\ C_0$ $S_L=.23$M/s	4.90	5.3	0.26	55.50	15.30	4.40	21	1.15	15	0.80	11.5	0.60	0.13
$BG_1\ C_0$ $S_L=.23$M/s	4.90	3.3	0.16	55.50	2.20	11.60	57	1.90	32	1	23	0.75	0.20

TABLE II

Turbulent Lagrangian Scales Modified by S_L at the Flame Holder Location

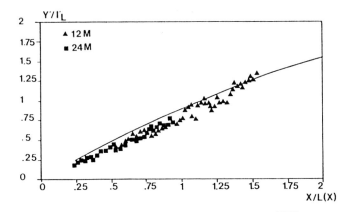

Figure 22 - Dimensionless evolution of $(Y'^2)^{1/2}/'l_1$
(G_1, solid line : theoretical curve, $\Phi = 0.4$)

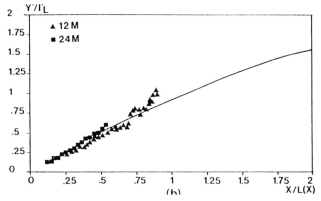

Figure 23 - Dimensionless evolution of $(Y'^2)^{1/2}/l'_1$
(G_2, solid line : theoretical curve, $\Phi = 0.4$)

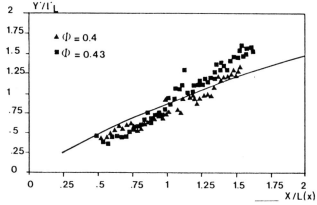

Figure 24 - dimensionless evolution of $(Y'^2)^{1/2}/l'_1$
(G_1, solid line : theoretical curve, $\Phi = 0.43$)

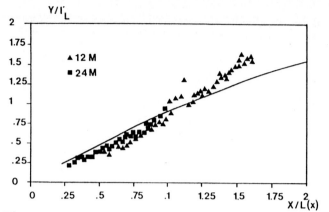

Figure 25 - Influence of equivalent ratio (G_1, 12M)

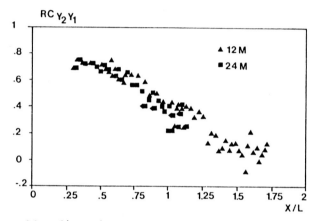

Figure 26 - dimensionless evolution of $R_{Y_2 Y_1}$ (G_1, $\Phi = 0.4$)

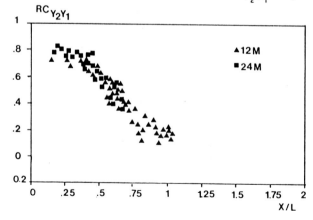

Figure 27 - Rescaling evolution of $R_{Y_2 Y_1}$ (G_2, $\Phi = 0.43$)

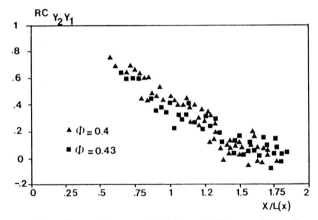

Figure 28 - Rescaling evolution of $R_{Y_2Y_1}$ (G_1, $\Phi = 0.43$)

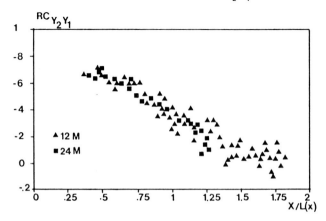

Figure 29 - Incidence of equivalent ratio on $R_{Y_2Y_1}$ (G_1, 12M)

V - CONCLUSION

In this paper the two aspects of the mutual interaction between combustion and turbulence have been investigated in the case of a wrinkled V shaped flame.

From LDV measurements, modification of the turbulence of the approach flow is found to be strongly dependent on the mean stream lines field which is controlled by the mean pressure gradient. In absence of confinement, no significative increase or decrease of turbulent kinetic energy is observed. When increase exists it seems more related to the difference between the velocity of fresh and burned gases.

The influence of turbulence on flame propagation can be thoroughly studied by coupling the tomography technique and digital image processing. By this method, a direct observation of the flame allows us to qualitatively define the combustion regime. Analysis of digital images leads to statistics of the flame geometry. Use of this new technique has allowed us to study the limit of the eddy diffusion model proposed by Karlovitz et al. [11] and Scurlock and Grover [12] in the Fifties for the initial development of the flame.

The results presented show the importance to rescale measurements obtained in a premixed flame by means of flame modified Lagrangian scales. The interest of this approach is to try to separate the respective parts of laminar propagation, approach turbulence and instability effects on the flame.

Further development concerning flame front lenghtening and the longitudinal correlations of the flame front position are in progress and could explain the second stage of the flame evolution (presence of pockets).

REFERENCES

[1] R.K CHENG, I.G. SHEPHERD, Interpretation of Conditionnal Statistics in open oblique Premixed flames - CST 1986, vol. 49.

[2] J.R. SMITH and F.C. GOULDIN, Turbulence effects on Flame Speed and Flame structure - 17th Aerospace Sciences Meeting, 1979.

[3] F.C. GOULDIN, DANDEKAR, Time Resolved Density Measurement in Premixed Turbulent Flame - AIAA Journal, 1984.

[4] J.R. SMITH, G.D. RUMBACH, S. RAJAN, Instantaneous Rayleigh Measurements in Turbulent Flame - Combustion and Flamme, 1984.

[5] N. NAMAZIAN, I.E. SHEPHERD and L. TALBOT, Characterization of the Density Fluctuations in Turbulent V shaped premixed Flame.

[6] S. RAJAN, Effect of equivalence Ratio and Approach Flow Turbulence proprieties on Internal Structure of Premixed Rod Stabilized Flame - Combustion and Flame, 1987.

[7] D. ESCUDIE, M. TRINITE, P. PARANTHOEN, "Modification of turbulent flow field by an oblique premixed Hydrogen-Air Flame" - Progress in Astronautics and Aeronautics, vol. 88, 1983.

[8] A. RACHID, P. GOIX, P. PARANTHOEN and M. TRINITE, Modification of a Turbulent Flow Field by an Oblique Confined Flame in the case of a small Angle - Turbulent Shear Flow V - Cornell University - ITHACA, 1985.

[9] R. BORGHI, Mise au point sur la structure des flammes turbulentes - Journal Chimie Physique, 1984.

[10] G.Z. DAMKHOLER - Elecktrochem 46, 601, 1940.

[11] BELA KARLOVITZ, D.W. DENNISTON, F.E. WELLS - The Journal of Chemical Physics, May 1951.

[12] SCURLOCK and GROVER, Propagation of turbulent Flames - 4th Symposium, 1953.

[13] F.A. WILLIAMS - Combustion Theory, 2^d Edition.

[14] P. CLAVIN, Combustion des gaz prémélangés - Structure et Dynamique des fronts de flamme, 7ème Congrès de Mécanique, 1955.

[15] K. LABBACI, M. TRINITE, D. STEPOWSKI, Comparison of Velocity Statistics as Obtained by Different Averaging Procedures in a Turbulent Diffusion-Flame - Saint-Louis, Mai 1987.

[16] L. BOYER, Laser Tomographic Method for Flame-Front mouvement studies - Combustion and Flame, 1980.

[17] J.P. DUMONT, R. BORGHI, A qualitative study by laser tomography of the structure of turbulent flames - Combustion Science and Technology, 1986.

[18] D. ESCUDIE, An experimental study of the effect of Vortex street on the stability of a laminar flame-front - Joint Meeting of the french and italian sections of the Combustion Institute, 1987.

[19] BOUKHALFA, B. SARH, DEBBICH and I. GOKALP, Spatial and temporal characteristics of the density field in premixed turbulent conical methane-air flame - Joint Meeting of the french and italian sections of the Combustion Institute, 1987.

[20] HINZE, Turbulence, 2^d edition Mc Graw Hill Series in Mechanical Engineering.

[21] R.K. CHENG and T.T.NG, Velocity statistics in premixed turbulent flame - Combustion and Flame, 1983.

[22] P. PARANTHOEN, A. FOUARI, A. DUPONT, J.C. LECORDIER, Dispersion measurements in turbulent flows - Internal Journal Heat Mass Transfer, to be published, 1987.

[23] CSANADY, Turbulent diffusion in the environment, D. Redel, Publishing Company, 1973.

[24] K.O. SMITH, F.C. GOULDIN, Experimental investigation of flow turbulence effects on premixed methane-air flame - AIAA 15th Aerospace Science Meeting, 1977.

TWO-COMPONENT VELOCITY PROBABILITY DENSITY MEASUREMENTS DURING PREMIXED COMBUSTION IN A SPARK IGNITION ENGINE*

Peter O. Witze
Combustion Research Facility
Sandia National Laboratories, Livermore CA 94550

David E. Foster
Department of Mechanical Engineering
University of Wisconsin, Madison WI 53706

ABSTRACT

A two-component laser Doppler velocimeter is used to investigate turbulence during confined premixed combustion in an internal combustion engine. Of particular interest is the effect of unidirectional compression on the turbulence structure of the preflame gas, and the effect of the density jump across the flame on the turbulence in the burned gas. Measurements are presented for two components of turbulence intensity and their probability density functions, the correlation coefficient, and the joint probability density function. It is concluded that for low levels of preflame turbulence, both compression in front of the flame and expansion through it cause an increase in the turbulence intensity. For high initial levels of turbulence, however, compression has a negligible effect on the preflame turbulence, and expansion across the flame actually results in reduced turbulence levels.

INTRODUCTION

The combustion of premixed reactants in a confined volume is heavily influenced by combustion-induced fluid motion. Illustrated in Fig. 1a is the very simple case of a plane laminar flame propagating down a closed channel; it is assumed that the burning velocity is constant, the pressure is uniform, and the heat transfer is negligible. In the laboratory reference frame the flame velocity U_f is the sum of the laminar burning velocity S_L and the unburned gas velocity at the flame surface, U_{u_f}. The gas velocity immediately behind the flame is U_{b_f}.

Figure 1b shows these two gas velocities at the flame surface as a function of the flame position in the channel. At the time of ignition at the left end, the expansion velocity induced on the unburned gas is equal to the density ratio ρ_u/ρ_b across the flame surface, arbitrarily assumed here to be 5. As combustion progresses the unburned gas velocity decreases until it stagnates when the flame reaches the end of the channel. On the burned gas side of the flame the flow direction is back toward the ignition plane; when the flame reaches the end of the channel the velocity ratio is equal to the current density ratio, which is less than the initial value because of the pressure increase.

* This paper was first published in the Proceedings of the Conference on *Combustion in Engines - Technology and Applications*, 1988. It is reproduced by permission of the Council of the Institution of Mechanical Engineers.

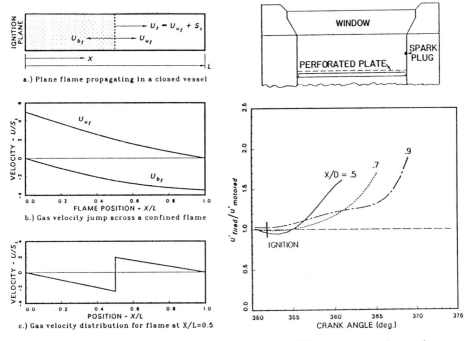

Fig. 1 - Characterization of the gas velocities induced by the propagation of a plane laminar flame from the end wall of a closed vessel.

Fig. 2 - Effect of compression on the preflame turbulence intensity measured above a stationary perforated plate in an engine operated at 300 rpm. From Martin et al. [1].

Finally, shown in Fig. 1c is the gas velocity distribution in the channel for the time when the flame is located at $X/L = 0.5$. The velocity field is linearly distributed between the flame and the end walls. Martin et al. [1] have shown that a similar velocity distribution can be created for turbulent conditions in an idealized internal combustion engine. Of particular interest is the effect the normal strain fields shown in Fig. 1c have on the turbulence intensities in front of and behind the flame. Because the turbulent flame itself is highly wrinkled and possibly multiply connected, it is also unclear what changes turbulence may undergo as the gas passes through the turbulent flame brush and experiences rapid expansion.

The study by Martin et al. used a fixed perforated plate installed parallel to the piston surface at a location 1 mm from the piston at top-dead-center. The purpose of the plate was to remove large bulk fluid motions from the combustion volume. This greatly simplified the bulk flow, but at the same time removed a lot of energy from the turbulence field, such that the turbulent burning velocity was too low to be characteristic of practical engines. The very controlled flow field that was established revealed some interesting features, however, as shown in Fig. 2. Ignition was at the side-wall of a disc-shaped combustion chamber of diameter D. Laser Doppler velocimeter (LDV) measurements were made of the velocity component normal to the average flame surface at three locations along the cylinder diameter passing through the spark. The results presented in the figure clearly show that the normal strain field imposed on the unburned gases resulted in a significant increase in turbulence intensity, on the order of fifty percent. Additional measurements made parallel to the average flame surface showed no similar increase, indicating an anisotropic response to the normal strain field.

An increase in turbulence intensity of this amount is obviously important to our understanding and subsequent modeling of confined turbulent combustion. It is not apparent, however, that these results can be extrapolated to more realistic engine conditions with higher turbulence levels. Hall and Bracco [2] have recently observed, at most, only a small increase in turbulence ahead of the flame in a ported engine, although they did note a measurable increase in the turbulence behind the flame. Their result does not agree with the data in Fig. 2, but is consistent with the $k-\epsilon$ turbulence model results of Morel and Mansour [3]. They performed a sensitivity study to show that the production of turbulence by the unidirectional compression of a moving piston is only noticeable under conditions of low initial turbulence levels.

To gain a qualitative understanding of the predictions of Morel and Mansour, we exercised the engine simulation model of Borgnakke et al. [4]. The model is based on a two zone (burned and unburned gases) thermodynamic analysis of the combustion process that includes an averaged $k-\epsilon$ turbulence model that gives the bulk values of the turbulent kinetic energy k and the dissipation rate ϵ in both the unburned and burned zones. Because the model is zero-dimensional the turbulence is assumed to be isotropic, and the only production terms P in the simulation are those due to the dilatation effect, which are formulated according to Reynolds [5] to be

$$P_k = \frac{2}{3} k \frac{d\rho}{dt} \quad \text{and} \quad P_\epsilon = \frac{4}{3} \epsilon \frac{d\rho}{dt}$$

To demonstrate the trends predicted by the model, we performed a parametric study of premixed combustion in a closed vessel. The initial turbulence intensity was varied, heat transfer was neglected and, based on previous model correlations with turbulence decay rates [6], a constant integral length scale of 1.5 mm was assumed. The results for three initial levels of turbulence intensity u'_o are presented in Fig. 3. For each condition a constant turbulent burning velocity S_T was assumed that had the form

$$S_T = S_L + u'_o$$

with the laminar burning velocity S_L prescribed at 1.0 m/s. The heavy solid line in each figure represents the natural decay of turbulence without combustion during the time interval T corresponding to the combustion duration; the dashed line shows the turbulence intensity of the reactants, and the dotted line is for the products. These results agree with the observation of Morel and Mansour that the production of turbulence ahead of the flame becomes less significant with increasing initial turbulence. The model also predicts a large decrease in turbulence intensity across the flame surface; recall, however, that the model does not include a mechanism for production by the flame itself.

The model results presented in Fig. 3 are only intended to provide a qualitative introduction into some of the trends that may occur during confined turbulent combustion. The LDV measurements to be presented in the present paper will look further into the issues of dilatation effects and flame-induced turbulence. Because we have been able to resolve some particle seeding problems that greatly restricted our data rates in the past [4], we are now able to make simultaneous two-component velocity measurements in both the unburned and burned gas regions. Beam steering problems within the flame zone itself still preclude our making reliable measurements in this region, however.

The same research engine used by Martin et al. was employed, except that much higher turbulence levels were achieved by not using the perforated plate. At the time of ignition, the earlier study measured the turbulence intensity to be 0.25 m/s for an engine speed of 300 rpm; here, the initial turbulence levels are on the order of 1.0 and 8.0 m/s for corresponding engine speeds of 300 and 1200 rpm. A stoichiometric propane-air mixture was used for both studies.

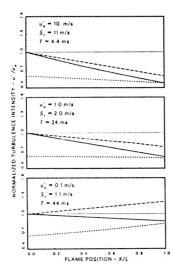
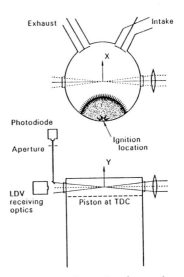

Fig. 3 - Behavior of confined turbulence as a function of initial turbulence intensity levels u'_o. The dashed line is for the reactants, the dotted line is for the products, and the solid line is without combustion. T is the burn duration for $S_T/S_L = 1 + u'_o/S_L$.

Fig. 4 - Schematic of experimental setup, showing the forward-scatter LDV optics and the laser-beam-refraction flame arrival detector.

THE EXPERIMENT

The forward-scatter LDV measurements reported here were made in the combustion chamber illustrated in Fig. 4. The side-wall location of the unshrouded intake valve produces a tumbling motion during the induction stroke that stores kinetic energy that is recoverable as turbulence during compression. Because of the directed nature of the intake port, however, there is also a small amount of residual swirl that increases with engine speed. The interaction of these two large-scale fluid motions leads to a bulk flow that becomes increasingly complex with increasing engine speed.

The LDV measurements were made at the center of the clearance volume having a 76.2 mm diameter and 17.5 mm vertical height. The spark plug was located 90-degrees from the optical axis of the LDV system. Dual Bragg cells were used for both velocity components, set for frequency shifts of 5.0 MHz. The zirconium fluoride (ZrF_4) used as the seeding material was introduced into the intake flow by atomizing a water dispersion. In the engine, a low temperature reaction during compression converts the zirconium fluoride to zirconium oxide (ZrO_2), which is a high temperature refractory material that survives combustion with no measurable loss in scattering efficiency. Details of the experimental setup [7] and the seeding system [4] have been reported previously.

Conventional ensemble-averaging procedures were used to analyze the data. To reduce bias errors caused by cyclic variations in the combustion rate, we conditionally sampled the engine cycles included in the average by employing a flame-arrival-time criterion. Figure 4 illustrates an adaptation of the laser beam refraction technique developed by Dyer [8] in a combustion bomb and by Swords et al. [9] in an engine. One of the spent LDV beams for the velocity component normal to the piston surface is directed onto a photodiode. When the flame crosses the path of this beam, gradients in the index of refraction cause the beam to deflect off the photodiode,

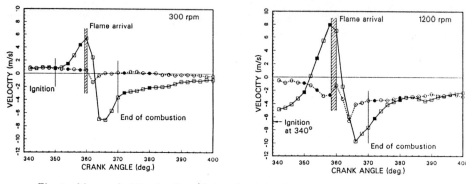

Fig. 5 - Mean velocities for the x (□) and y (○) velocity components measured at $X/D = 0.5$. The closed symbols indicate crank-angle positions for which pdf data will be given.

producing a distinct signal. The technique is, of course, insensitive to where along the beam path the flame is first encountered, such that some error can be introduced by asymmetric flames. However, a detailed comparison between direct flame arrival measurements by ionization probe and this technique showed good agreement.

Based on the ensemble statistics of the flame arrival time at the LDV probe volume, we selected conditional criteria of 359.4-360.2° for use at 300 rpm, and 358.4-360.0° for 1200 rpm. This conditioning was done on-line, such that only the LDV data from engine cycles that satisfied the criteria were recorded.

RESULTS

Figure 5 summarizes the mean velocity measurements that were made. The barred regions indicate the band of the flame-arrival conditions. Ignition timing was selected to place the average end of combustion, defined as the point of maximum pressure, at 370°. The solid symbols indicate those crank angles for which probability density distributions (pdf's) and joint-pdf contour plots are given in a later figure.

The x-component velocity histories clearly show the effect of the expansion velocity on the unburned gas, and the velocity jump across the turbulent flame brush. The large negative velocity at the time of ignition for 1200 rpm is indicative of the complicated bulk flow that exists. The 300 rpm results appear less complex initially, but the large negative velocity that exists at the completion of combustion at 370° suggests that this is not the case. Simple one-dimensional analysis dictates that the velocity should return to zero at this point. Instead, there is a strong flow directed toward the spark plug that decays slowly with time, suggesting a well-ordered secondary flow set up by the geometry of the unburned gas volume, such as the "tulip" flame phenomena of Dunn-Rankin et al. [10].

The y-component mean velocity results shown in Fig. 5 are not particularly informative. At times we have speculated that this measurement could give an indication of a preferred "tilt" orientation of the flame surface, but it would seem to be risky to put too much faith in a single point measurement.

The ensemble-averaged turbulence intensity measurements are given in Fig. 6. The thin lines without symbols are curve fits to LDV measurements made without combustion. The data are not shown for the period between flame arrival and the end of combustion because of uncertainties in their accuracy. This is mainly a result of increased statistical error due to fewer data, but also

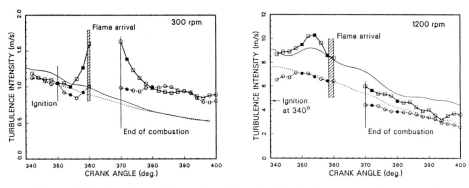

Fig. 6 - Turbulence intensities for the x (□) and y (○) velocity components measured at $X/D = 0.5$. Lines without symbols represent motored results. The data of questionable accuracy from the combustion period are not shown.

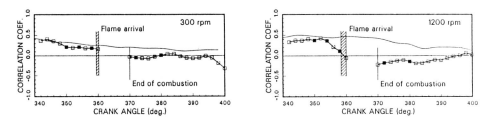

Fig. 7 - Correlation coefficients measured at $X/D = 0.5$. Lines without symbols represent motored results.

because of bias errors caused by index-of-refraction-gradient beam steering [11]. The data shown within the flame-arrival window are also uncertain because of bias. More than 90 percent of the measurements contributing to these averaged results were obtained from the unburned gases; however, because some measurements were obtained from within and behind the flame, these data represent a mixture of unburned and burned gas velocities. We chose to report these results because they consistently reflect the trends seen in the data as the flame approaches the LDV probe volume.

There are several important features to Fig. 6: 1) At the time of ignition, the turbulence is isotropic at 300 rpm and anisotropic at 1200 rpm. 2) At 300 rpm the preflame x-component of turbulence is enhanced by approximately 50 percent, whereas there is no similar effect at 1200 rpm (although the maximum in the x-component of turbulence at 352-354° may be a flame-induced effect). 3) After the completion of combustion the turbulence intensities at 1200 rpm are slightly lower than the corresponding motored values, whereas at 300 rpm the turbulence quickly reaches an isotropic state that is about 60 percent higher than the motored measurements.

The simultaneous measurement of two components of turbulence permits the calculation of the correlation coefficient C_{xy}, defined as

$$C_{xy} = \frac{\overline{uv}}{u'v'}$$

such that $-1 \leq C_{xy} \leq 1$. Here u and v are the instantaneous velocity fluctuations in the x and y directions, respectively, and u' and v' are the corresponding turbulence intensities. The correlation

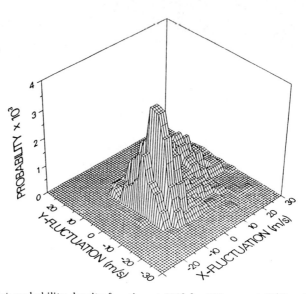

Fig. 8 - Joint probability density function at 350° for 300 rpm at $X/D = 0.5$.

coefficient is a measure of the asymmetry of the joint pdf of the orthogonal velocity fluctuations. The results presented in Fig. 7 suggest that at the lower engine speed the approaching flame does not influence the correlation coefficient, whereas at 1200 rpm there is a linear decay toward zero, followed by a sign reversal after the completion of combustion. Because the correlation coefficient is dependent on the orientation of the coordinate system chosen, it may be that the effect seen at 1200 rpm is a result of tilting of the flame surface. Our coordinate system was selected assuming the x- and y-directions would be aligned normal and parallel to the average flame surface. The y-component of mean velocity given in Fig. 5 does suggest that the flame is tilted, with a portion of the expansion velocity directed down toward the piston. A flame of this shape may also be expected because of the earlier ignition timing for the 1200 rpm case, which in effect places the ignition site above the midplane of the combustion volume.

Presented in Fig. 8 is a typical joint pdf, which gives the probability of simultaneously observing any pair of u and v values. By taking five equally spaced slices through this joint pdf, parallel to the uv plane, contour plots of the type presented in Figs. 9 and 10 were generated. Also shown are the individual pdf's for the two turbulence components measured, compared with a Gaussian distribution represented by the solid curve in each figure. (Note that different scales are used for the two engine speeds.)

Many of the features of the turbulence structure noted earlier in Figs. 6 and 7 can be seen in Fig. 9. Isotropy of the turbulence at the time of ignition (350°) is evident, as is the Gaussian nature of the pdf. The approaching flame increases the width of the x-component pdf at 358°, but the shape does remain Gaussian. Behind the flame, this component appears to become skewed to the high velocity side. The corresponding contour plots are not particularly interesting, with perhaps the exception of an relatively orderly pattern at 358°, just in front of the flame.

The 1200 rpm results given in Fig. 10 show some very different features. The anisotropy before the flame is characterized by a highly skewed x-component that appears to become more nearly Gaussian with the approach of the flame. The corresponding contour plots clearly show the initially large positive correlation coefficient, and its evolution toward symmetry with the approach of the flame. Behind the flame the narrow pdf's indicative of lower turbulence intensities are evident, and perhaps there is renewed skewness in the x-component at 372°.

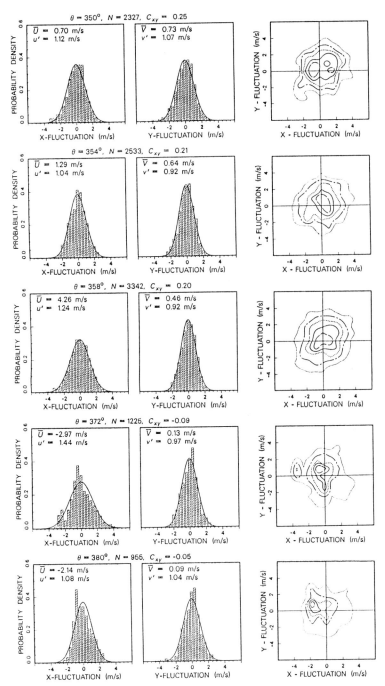

Fig. 9 - Probability density functions and contour plots of the joint pdf's for $X/D = 0.5$ and 300 rpm. N is the number of velocity measurements used in the pdf's.

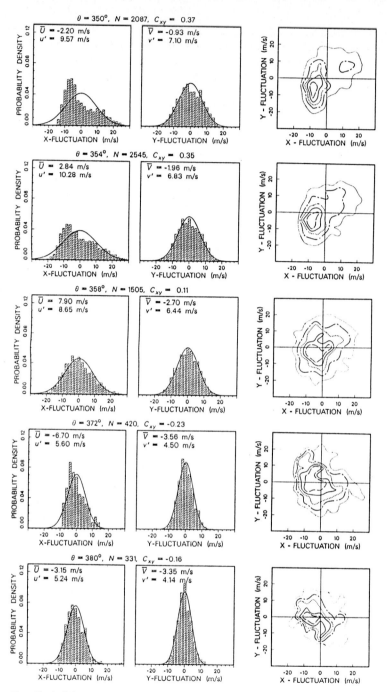

Fig. 16 - Probability density functions and contour plots of the joint pdf's for $X/D = 0.5$ and 1200 rpm.

CONCLUSIONS

The preflame turbulence measurements presented here, together with the previous results of Martin et al. [1], support the conclusions of the Morel and Mansour [3] study, which were that unidirectional compression will amplify turbulence when the initial intensity is low, but that high levels of preflame turbulence will exhibit a continual decay during compression. This conclusion is also consistent with the turbulence measurements of Hall and Bracco [2]. Because practical internal combustion engines operate in a high turbulence regime, it would seem that detailed consideration of this amplification mechanism is of marginal priority to the immediate needs of predictive analytic models.

The flame has a similar effect on the gases that pass through it; for low unburned gas turbulence levels the turbulence in the burned gases is amplified, while high unburned gas turbulence levels are actually reduced when crossing the flame. Because of heat transfer implications, a better understanding of the mechanisms governing the postflame turbulence could be important to improved engine design.

It is not clear at this time whether the measurements of the correlation coefficient are of particular value. The trends that they show are somewhat surprising in that compression by the approaching flame does not alter the coefficient for a low initial turbulence level, while for higher turbulence the coefficient linearly decays to zero at the flame surface. This result appears to be inconsistent with the turbulence intensity measurements. Since it may be that this behavior is due to tilting of the average flame surface relative to the reference coordinate system, however, further investigation should be pursued.

The joint pdf contour plots reflect the trends seen in the correlation coefficient for the high turbulence case, and the accompanying individual pdf's identify skewed deviations from Gaussian distributions. However, because only the joint pdf's and the correlation coefficient measurements require the more difficult (and expensive) simultaneous measurement of two velocity components, it is not evident to us at this time that enough can be learned from the joint measurements to justify the added complexity.

ACKNOWLEDGEMENT

This work was performed at the Combustion Research Facility, Sandia National Laboratories, and was funded by the U. S. Department of Energy, Energy Conversion and Utilization Technologies Program, and the University of Wisconsin-Madison Graduate School.

REFERENCES

1. Martin, J. K., Witze, P. O., and Borgnakke, C., "Combustion Effects on the Preflame Flow Field in a Research Engine," Trans. SAE 94, 722, 1985.
2. Hall, M. J., and Bracco, F. V., "A Study of Velocities and Turbulence Intensities Measured in Firing and Motored Engines," SAE Paper No. 870453, 1987.
3. Morel, T., and Mansour, N. N., "Modeling of Turbulence in Internal Combustion Engines," SAE Paper No. 820040, 1982.
4. Witze, P. O., and Baritaud, T. A., "Particle Seeding for Mie Scattering Measurements in Combusting Flows," Proceedings of the Third International Symposium on Applications of Laser Anemometry to Fluid Mechanics, Lisbon, 5.1, 1986.
5. Reynolds, W. C., "Modeling of Fluid Motions in Engines: An Introductory Overview," in Combustion Modeling in Reciprocating Engines, J. N. Mattavi and C. A. Amann (Eds.), Plenum Press, New York, 41, 1980.

6. Witze, P. O., Martin, J. K., and Borgnakke, C., "Measurements and Predictions of the Precombustion Fluid Motion and Combustion Rates in a Spark Ignition Engine," *Trans. SAE* **92**, 786, 1983.

7. Witze, P. O., "An Engine Simulator for Laser Diagnostic Studies of One-dimensional Turbulent Flame Propagation," in *International Symposium on Diagnostics and Modeling of Combustion in Reciprocating Engines*, JSME, Tokyo, 305, 1985.

8. Dyer, T. M., "Characterization of One- and Two-dimensional Homogeneous Combustion Phenomena in a Constant Volume Bomb," *Trans. SAE* **88**, 1196, 1979.

9. Swords, M. D., Kalghatgi, G. T., and Watts, A. J., "An Experimental Study of Ignition and Flame Development in a Spark Ignition Engine," *Trans. SAE* **91**, 3826, 1982.

10. Dunn-Rankin, D., Barr, P. K., and Sawyer, R. F., "Numerical and Experimental Study of "Tulip" Flame Formation in a Closed Vessel," *Twenty-First Symposium (International) on Combustion*, The Combustion Institute, Pittsburgh, 1987.

11. Witze, P. O., and Baritaud, T. A., "Influence of Combustion on Laser Doppler Velocimeter Signal Quality in a Spark Ignition Engine," in *International Symposium on Laser Anemometry - FED Vol. 33*, A. Dybbs and P. A. Pfund (Eds.), ASME, New York, 3, 1985.

ON THE ACCURACY OF LASER METHODS FOR MEASURING TEMPERATURE
AND SPECIES CONCENTRATION IN REACTING FLOWS

M.J. COTTEREAU - J.J. MARIE - P. DESGROUX

U.A. CNRS 230 - Faculté des Sciences et Techniques de Rouen
B.P. 118 - 76134 - Mont-Saint-Aignan Cedex (France)

INTRODUCTION

During the past decade, following the improvements of both the lasers and the detectors and associated processors, a number of studies have been devoted to temperature and species concentration measurements by laser-based methods. In their prospective paper (1) in the 1975 Project Squid Workshop on Combustion Measurements in Jet Propulsion Systems, R. GOULARD., A.M. MELLOR and R.W. BILGER presented the type of measurements needed in combustion studies as following :

-A) *Discrete (but simultaneous) Measurements of Temperature, Velocity and Concentration*

Spatial resolution to 0.1mm, temporal resolution to $10 \mu s$ (100KHz), and accuracy to 5 percent or better are desirable. Data accumulation at each point sufficient so that means, variances, convariances, p.d.f.s and joint p.d.f.s can be determined to better than 5 percent. Sufficient simultaneous information must be obtained so that instantaneous density can be computed and temperature or other corrections applied. Good coverage (mapping) of the combustion field is required.

-B) *Time and Space Correlated Data*

Instantaneous temperature, velocity and concentration data separated by variable time and spatial amounts is desirable to determine fine scale flame structure turbulence length scales, local reaction rates etc. Resolution and accuracy as in A above but only limited coverage of the combustion field is required.

-C) *Spray Characteristics*

Droplet sizes, velocities and frequencies at one point and Lagrangian measurements of droplet velocity and size as a function of time are needed. Instantaneous spatial distribution of droplets (holograms?) are necessary.

-D) *Imaging*

Schlieren, shadowgraph and other integrated path imaging, as well as Ramonograph or other single scattering plane imaging are useful. Instantaneous and sequential (movie) images are required. Spatial resolution to 0.1mm and temporal resolution to 10µs are needed but information does not have to be quantitative as it is required only as a diagnostic for conditional sampling work on coherent structures and recirculation zone.

-E) *Coherent Structure Investigation*

If data under A and B are obtained in a continuous analog fashion or the equivalent by rapidly pulsing it is possible to do conditional sampling on the data at a later time ; simultaneous imaging with accurate cross-referencing is highly desirable. If data under A and B are obtained in a more or less random discrete manner, then simultaneous imaging and or CW signal (Raman N_2?) is essential if quantitative investigation (conditional sampling etc.) of coherent structures is to be carried out : the data processing problem is then enormous but the measurements are not dependent on the experimenter. Alternatively the obtaining of discrete date (A & B) can be triggered by some continuous monitoring device (e.g. CW Raman N_2) with appropriate time delays, the selection of the trigger criterion making the measurements experimenter subjective.

After ten years of active research how deeply are these requirements fullfilled ? This paper is an attempt to answer this question, at least partially.

Unquestionably the imaging is the domain in which the most dramatic improvements have been achieved. Laser tomography techniques associated with Spontaneous Raman Scattering, Rayleigh Scattering or Mie Scattering, or with a laser induced fluorescence have been successfully used to look at instantaneous turbulent flames structures. The state of art of these techniques is presented elsewhere in this meeting and will not be discussed here. Therefore this paper is mainly concerned with point-measurements. The Coherent Anti-Stokes Raman Spectroscopy (CARS) or the Laser Induced Fluorescence Spectroscopy (LIFS) which were still in their infancy as diagnostic tools in 1975, have been widely studied in the meanwhile and will be mainly presented here.

II - LASER INDUCED FLUORESCENCE SPECTROSCOPY

II-A - Generalities

As soon as LIFS was proposed as a method for measuring concentration in reactive media the difficulties due to the collisional relaxation have been clearly pointed out. Two main paths of research have been followed to overcome these problems. The saturated LIF method (which, in principle, need not the knowledge of the collisional relaxation rates) appeared early (2,3) as the most promising. Unfortunately a number of experimental problems actually limit the method, as explained later.

Accurate measurements allowing the modelisation of the various collisional processes appeared also as the means to perform accurate measurements by LIFS. Low-pressure flames measurements proposed at the end of seventies (4,5,6) have been used in a number of laboratories to obtain data on these collisional processes.

II-B - Multishot measurements

1) Saturated Fluorescence

The rate equation which is generally valid for the usual experimental conditions easily describes the principle of the method. It is, for a two level model :

$$\frac{dN_2}{dt} = N_2 B_{12} u_\nu - N_2 B_{21} u_\nu - N_2 (A_{21} + Q_{21}) \qquad (1)$$

For a constant spectral energy u_ν of the exciting laser pulse, the population N_2 of the upper level increases during the excitation according to the relation :

$$N_2(t) = \frac{N_1^0 B_{12} u_\nu}{(B_{12}+B_{21})u_\nu + A_{21} + Q_{21}} (1-\exp-[(B_{12}+B_{21})u_\nu + A_{21}+Q_{21}]t) \quad (2)$$

where N_1 and N_2 are the population of the two level $N_1 + N_2 = N_1^0$ is the initial population of the lower state, the upper level being nearly empty at flame temperatures B_{12}, B_{21} and A_{21} are the Einstein coefficients, respectively, for absorption, stimulated and spontaneous emission and Q_{21} is the quenching rate. After a sufficient time the steady state can be reached and :

$$N_2 = \frac{N_1^0 B_{12} u_\nu}{(B_{12}+B_{21}) u_\nu + A_{21} + Q_{21}} \quad (3) \text{ or :}$$

$$N_2 = N_1^0 \frac{B_{12}}{B_{12}+B_{21}} \frac{1}{1 + \frac{A_{21}+Q_{21}}{B_{12}+B_{21}} \frac{1}{u_\nu}} \quad (4)$$

The saturation energy u_ν^s being defined as :

$$u_\nu^s = \frac{Q_{21}+A_{21}}{B_{12}+B_{21}} \qquad \text{one obtains}$$

$$N_2 = N_1^0 \frac{B_{12}}{B_{12}+B_{21}} \frac{1}{1 + \frac{u_\nu^s}{u_\nu}} \quad (5)$$

For $u_\nu \gg u_\nu^s$ the steady state population of the upper level is therefore independant of both the quenching rate and the laser energy, like the fluorescence signal :

$$S_F = \frac{h\nu A_{21}}{4\pi} \Omega V N_2 \quad (6)$$

which is collected with a solid angle Ω from a volume V defined by the detector optics.

As far as the laser energy u_ν is always, within the shot-to-shot fluctuations, much greater than the saturation energy, accurate shot-to-shot measurements can therefore be performed. In relative values the accuracy is limited only by the detector noise. In absolute value the accuracy of the calibration of the optical detection system is preponderant.

The electrical signal out of the detector can be written :

$$V_F = C \cdot \frac{h\nu \, A_{21}}{4\pi} \, \Omega \, V \, N_2 \qquad (7)$$

where C is the efficiency of the detection system. Usually the overall factor $C \frac{\Omega}{4\pi} V$ can be accurately determined with Rayleigh scattering measurements.

When the energy of the laser is not large enough to fully saturate the transition it can be shown from relation (5) that reciprocal of the fluorescence signal is linear vs reciprocal of laser energy.

As proposed by Baronavski and Mc Donald (7,8) in stationary media, it is possible to draw this straight line by progressively attenuating the energy of the laser. Extrapolation to $\frac{1}{u_\nu} = 0$ gives the value of the fully saturated signal.

Unfortunately with the actual experimental conditions the measurement is not so straightforward. First $u_\nu(t)$ has not a rectangular shape and the measurements have to be made with a sampling system having a 1.ns or less window and looking at the maximum of the laser pulse. Therefore the number of collected photons is smaller than that of the whole fluorescence pulse and the signal-to-noise ratio is worse. Secondly, and that is much more difficult to take into account, the spatial intensity distribution accross the laser beam is not a perfect top hat profile. Perfect truncation of pulsed, multimode lasers commonly used is very difficult to achieve and wing effects have to be considered as first pointed out by Rodrigo and Measures.

Actually u_ν in relations 1 to 4, then $N_2(t)$ have to be considered as local values within the laser beam. The fluorescence signal collected become :

$$S_F = \int \int \int_\nu \frac{h\nu}{4\pi} A_{21} \, N_2 \, (x,y,z) \, dv \qquad (8)$$

Different degrees of saturation appear in the collection

volume. The variation of S_F vs the laser intensity depends on the laser beam profile and on the geometrical collection arrangement used.

Recently Kohse-Hoïnghaus et al (10) and Salmon and Laurendeau (11 a,b) proposed methods for calibrating LSF measurements taking wing effects into account. These two methods, while formally different, have the same basis i.e define an effective fluorescence collection volume to be used in relation (6). References (10) and (11) present the two different procedures used to experimentally determine this effective volume. The measurements have been performed in low-pressure flame. These calibrations need a number of laser shots in a constant medium. As the effective volume depends on both the laser intensity and the quenching rate, single-shot measurements are not possible. Application to atmospheric pressure is difficult.

Either of these two methods appear to give the most accurate absolute measurements. Following the authors the accuracy does not exceed 15% in the best case for OH with a number density of $10^{15}/cm^3$ and for values averaged on a number of laser shots.

Two main causes of further inaccuracy occur when the measurements are performed, as usual, on molecular species (C_2, CH, CN, NH, NO, OH ...). First the two-level system description is no longer valid and interpretation of the fluorescence signal is not straightforward unless the pseudo two level model, the so-called cross-rate model (12) is valid. It is assumed in this model that the sum $N_1 + N_2$ of the two laser coupled vibrational-rotational (V-R) level is constant. This is questionable (13). Secondly the measurements give the population of a given (V-R) level and the total population is calculated using the Boltzmann factor. The temperature has to be known. However the measurements can usually be performed using energy levels for which the temperature dependance is weak (as example the Boltzmann factor of the F_1 (8) level of OH varies by only 3% over a temperature range of 1400-2000K).

2) Linear Fluorescence

As pointed out by Stepowski and Cottereau (5) relation 2 becomes :

$$N_2(t) = N_1^0 \, B_{12} \, u_\nu \, t \qquad (9)$$

for $[(B_{12}+B_{21})\,u_\nu + A_{21} + Q_{21}]\,t \ll 1 \qquad (10)$

Then the fluorescence signal, under this weak and short excitation, is at a given time t :

$$S_F(t) = \frac{h\nu}{4\pi} A_{21} \, \Omega \, V \, N_2(t) \qquad (11)$$

It does not depend on the quenching rate and is a direct measure of N_1^0.

As $N_2(t) \ll N_1$ the condition $N_1 + N_2 = N_1^0$ is still valid for a multilevel system. In that case Q_{21} in relation 10 represents the total collisional deexcitation rate of the V.R. excited level (rotational, vibrational and electronic energy transfers).

The accuracy of the measurements is directly dependent on the determination of the volume probed V (this can be done with Rayleigh scattering measurements) and, especially, on the determination of the laser spectral energy u_ν. Due to the presence of close adjacent lines it is difficult to have a laser beam spectral width much larger than the width of the absorption line chosen. The convolution between the two spectral profiles has to be taken into account to accurately define or measure u_ν. Therefore the method is not very accurate for absolute number density measurements. The fluorescence signal which, moreover, has to be sampled at a given time, is weak and the S/R ratio on the detector is not good enough for single-shot measurements.

For relative and time-averaged measurements the accuracy of the method is comparable to that of the saturation method and can probably lowered to a few percent.

On a practical point of view the available pulsed lasers and detection systems limit the minimum value of t in relation 10 to be t $\sim 10^{-9}$ s. Then in order to fulfil relation 10 the method is only applicable for $Q_{21} < 10^{-8}$ s, i.e in low pressure flames (p < 0,1atm) for the usual species.

The feasability of measurements in atmospheric pressure flames has been demonstrated (14), using a subnanosecond laser pulse and a streak camera. It still appears difficult to actually apply the method at this pressure.

II-C Single-shot measurements :

As previously mentioned accurate single-shot measurements are not achievable with the methods described in the above paragraphs.

In our laboratory we are strongly involved in researches aimed at obtaining methods for such single-shot measurements at atmospheric or higher pressure.

Both linear and saturated fluorescence methods have been studied.

Linear fluorescence :
In a premixed low pressure hydrocarbon flame the variation of the quenching rate for OH through the flame front was found to be less than 25 percent over a range of temperature from 1000 to 2200 K (6) ; this is probably true in atmospheric pressure flames. Within this uncertainty it is interesting to perform measurements independant of the laser intensity fluctuations.

At atmospheric pressure the fluorescence decay is faster than the laser pulse decay. The fluorescence signal collected during the total pulse duration can be written for linear excitation

$$S_F = \eta \frac{A_{21} B_{12}}{A_{21} + Q_{21}} N_1^0 \int_{pulse} u_\nu(t) \, dt$$

where η is the light collection efficiency of the optical detection system.

To perform single-shot measurements independant of the laser intensity the fluorescence signal has to be normalized by the laser intensity $\int u_\nu(t) \, dt$.

At first sight, the simplest is to simultaneous measure the

fluorescence signal and the laser intensity. Such measurements have been performed in our laboratory (15) with the experimental set up shown on Fig. 1.

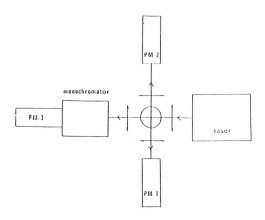

Figure 1 - Experimental Arrangment

The u_ν laser light for the fluorescence excitation of the OH molecule is provided by a nitrogen pumped dye laser system with subsequent frequency doubling. The porous disk burner is supplied with carefully regulated flows of methan and air ; measurements made with LDV have shown that the velocity is constant within 1%. It is therefore assumed that the flat flame properties are constant. The $Q_1(7)$ OH line is excited and the fluorescence is detected through broadand filters or by two symetrical optical paths with two photomultipliers. The laser pulse energy is measured with a third photomultiplier through a monochromator to eliminate remaining few amount of undoubled red light of the dye laser beam.

The r.m.s. of the laser signal is 10% either after absorption by the flame or without absorption. The r.m.s. of the fluorescence signals on path 1 or 2 is higher (20 to 30%). The first line of Table 1 gives the correlation

| Laser-Fluo | 0,56 | 0,32 | 0,35 | 0,35, | 0,18 | 0,40 | 0,95 | 0,54 | 0,71 | 0,60 |
| Fluo 1-Fluo 2 | 0,86 | 0,63 | 0,68 | 0,64 | | 0,72 | 0,62 | 0,82 | 0,80 | 0,94 | 0,82 |

Table 1 - Correlation coefficients

coefficient between the laser signal and one of the fluorescence. The second line gives the correlation coefficient between the two fluorescence signals. 500 events have been analysed in each case. The figures in the first line are always much smaller than those of the second line. The maximum values, respectively 0,94 and 0,71 corresponds to the hightest signal values for which the detectors noise contribution to the fluctuations are the smallest. The explanation is the following : as already noted the spectral width of the laser beam cannot be much larger than the width of the absorption line. Therefore small shot-to-shot variations in the laser wavelength can induce large fluctuations in the absorption efficiency at constant laser energy. It is concluded that normalization by the laser pulse energy can be not sufficient and that it is better to normalize with a fluorescence signal produced in a stable reference source.

Saturated fluorescence

We have proposed (16) an original method to perform single-shot concentration measurements.

The bases of the method are very simple.

Relation (8) :

$$S_F = \int \int \int_v \frac{vh}{4\pi} A_{21} \, N_2(x,y,z) \, dv$$

when : $N_2(x,y,z)$ is given by relation (5)

$$N_2 = N_1^0 \frac{B_{12}}{B_{12} + B_{21}} \cdot \frac{1}{1 + \frac{u_\nu^s}{u_\nu}}$$

shows that the fluorescence signal for a non-uniform laser illumination depends on the geometry of the probe volume V. There are two typical focal arrangements in common use. As illustrated in Fig. 2, they correspond to (a) a slit cutting out a cylindrical shapped portion of the beam and (b) a slit cutting out a slab along the beam.

Writting the local spectral energy of the laser beam as $u_\nu(x,y,z) = u_\nu^0 f(x,y,z)$ where u_ν^0 is the spectral energy at a given point in the beam (on the laser axis for example), $f(x,y,z)$ is a geometrical function representing the spatial distribution of the laser irradiance. This function is assumed not to vary shot by shot.

Then the fluorescence signals are :

for case a :

$$S_{F(a)} = \frac{h\nu \, A_{21} \, \Omega}{4\pi} N_1^0 \frac{B_{12}}{B_{12} + B_{12}} \int\int\int_{V_a} \frac{dv}{1 + \dfrac{u_\nu^s}{u_\nu^0 \, f(x,y,z)}}$$

for case b :

$$S_{F(b)} = \frac{h\nu \, A_{21}}{4\pi} N_1^0 \frac{B_{12}}{B_{12} + B_{21}} \int\int\int_{V_b} \frac{1}{1 + \dfrac{u_\nu^s}{u_\nu^0 f(x,y,z)}}$$

These signals are distinct functions of the laser power and the collisional transfer rates only through the ratio $\dfrac{u_\nu^0}{u_\nu^s}$. The ratio of the two signals is also a function of $\dfrac{u_\nu^0}{u_\nu^s}$ the variation of which represents the different contributions in the two geometries of the wings of the laser beam distribution, where the signals are not satured.

This function can easily be determined experimentally in simultaneously recording the signals of the two arrangements using two spectrometers facing on opposite sides of the laser beam and having their slits parallel and perpendicular to the laser beam axis, respectively, as shown of the schematic diagram of the experimental device we use (Fig. 3). The laser is a Quantel "Data Chrom" YAG pumped dye laser modified to obtain a convenient spatial distribution not too far from a gaussian shape. The spectrometers are two Jobin Yvon HRS 2 model. Two sampling leads (Tektronix S 5) with a 1ns aperture, or the Lecroy Camac pulse integrator have been used.

The principle of the method has been tested using radical OH as the studied species. The experiments have been performed on stoechiometric propane-oxygen low-pressure flames burning at 50 and 100 mbar. By doubling the dye laser frequency the Q_1 (6) line of the $2\Sigma+$ (v'=0) $\leftarrow$ 2 (v"=0) OH band line is excited. To avoid interference due to scattering of the laser beam, the fluorescence is collected on the P_1(7) line issued from the same excited level.

With that process, the two level system description given above if strictly speaking not applicable can be used as far as the balanced cross-rate model (5) is valid. N_1 and N_2 are in that case the populations of the directly coupled rotational levels, the quenching rate Q_{12} has to be replaced by an effective rate coefficient for collisional transfert out of level 2, taking into account both the electronic quenching and the rotational transfers

Fig. 4 deals with the results obtained with the sampling heads.

It shows the variation in arbitrary units of averaged values (100 single shots) of signal $S_{F(a)}$ and $S_{F(b)}$ as functions of the energy of the exciting beam (arbitrary units) obtained in the burnt gas zone of a flame burning at 50 mBar. To vary the energy of the laser without changing the spatial distribution function, the laser beam has been successively attenuated by an increasing number of partially absorbing glass plates. As the flame conditions remains constant u_ν^s is constant and the abscissa is also proportional to $\frac{u_\nu^0}{u_\nu^s}$.

Fig. 5 shows same curves for the flame burning at 100 mBar.

Taking into account that the collisional transfer rates must be proportional to the pressure, the saturation energy at 50 mBar is twice the one at 100m Bar. Therefore the curves of Figures 4 and 5 can be related to a same arbitrary abscissa. The abscissa of data at 100m Bar is half the abscissa of data at 50 mBar. Figure 6 which shows that the experimental values of ration $\frac{S_{F(a)}}{S_{F(b)}}$ are as expected gathered along a single curve, is a good indication of the validity of the assumption made.

Probably due to a failure in one of the sampling leads single-shot data were of poor quality.

Fig. 7 deals with results obtained with the Lecroy Camac pulse integrator. The measurements have been performed in an atmospheric pressure flames. The result demonstrate the ability of the method to work at this pressure.

We are now pursing the work to test the accuracy of single-shot measurements.

III - ACCURACY OF COHERENT ANTISTOKES RAMAN SPECTROSCOPY TEMPERATURE MEASUREMNTS

III-A : Literature data

If coherent Anti-Stokes Raman Spectroscopy has also the capablity for performing major species concentration measurements with good spatial and temporal resolution, most of the published works deal with temperature measurements regarding time resolved (e.g single shot) determination.

A number of recent papers (17-22) have been devoted to the precision of multiplex CARS temperatures. Our purpose here is only to resume the main conclusions of these papers and to compare with the experimental results we have obtained in measuring the temperature by CARS in the combustion chamber of an IC engine.

The theory of CARS is described in recent review articles (25-26). Briefly CARS is a nonlinear optical phenomenon involving three wave mixing. Two lasers beam of frequencies, respectively w_p for the so-called pump beam and w_s for the so-called stokes beam, interact and a coherent beam, the anti-stokes beam with the frequency $w_{as} = 2w_p - w_s$ is generated.

The intensity of the CARS signal is, for monochromatic wave :

$$I(w_{as}) \sim I_p^2 \, I_s \, |x^{(3)}|^2$$

where I_p, I_s and $I(w_{as})$ are the pump, Stokes and CARS intensities, respectively $x^{(3)}$ is the third-order non linear electric susceptibility ; it has a non-resonant x_{NR} (real) and a resonant x_R (complex) part

$$x^{(3)} = x_{NR} + x_R$$

the resonant contribution x_R occurs when $w_p - w_s$ approaches a Raman frequency of the sample medium and gives a contribution to the CARS signal much higher than the non-resonant contribution.

Roughly speaking x_{NR} is proportional to the total number density and x_R is proportional to the number density of molecules on the particular level of the Raman line used.

Therefore the measurement of the CARS intensity is a measurement of the particular level population, provided that I_p, I_s and the relation between x_R and the number density are known. If the Stokes-beam is moved to a number of Raman lines the populations of the various energy levels are measuredand the temperature can be determined, as long as a Boltzmann equilibrium occurs. This describes the scanned CARS method applicable for stationary medium. Clearly the accuracy of such measurements depends on the accuracy of the models relating x_R and x_{NR} to the population. Improvements of the models have been made in recent years, taking into account, in particular, the collisional narrowing occuring when interferences occurs between adjacent transitions (see for example 27,28,29) for important diatonic species like N_2, O_2, CO, H_2 it can be throught that the models are now almost perfect for monochromatic laser beams.

Actually the pump beam is either monochromatic (single-mode

pump laser) or not (multi-mode pump laser)

Improvements have also been made in the theory for taking into account the effect of laser bandwidths. It has been shown (30) that cross-coherence effects have to be taken into account when the pump bandwith is comparable to the Raman linewidth, which is the case with multimode pump laser.

For performing single-shot measurements, one excites simultaneously a number of Raman lines with a broadband dye laser. The generated anti-stokes beam is dispersed by a spectrograph and detected with an optical multichannel analyser. The temperature is determined by fitting the experimental spectrum with theorical ones (which take into account the instrumental slit function). Only the shapes of the spectra, not their absolute intensities are needed.

The main causes of inaccuracy are :

- shot-to-shot variation in the spectral energy profile of the broadband dye dye laser

- shot-to-shot variation in the spectral energy of the multimode pump laser

- noise of the detector

The first cause of inaccuracy can be reduced by shot-to-shot ratioing the experimental CARS spectra with a nonresonant reference spectrum.

At first sight it seems that single-mode pump laser would give better signal-to-noise ratio than a multimode pump laser. This is yet a subject of controversy, both on theoretical and experimental points of view.

Obviously the detector noise is inavoidable, the greater is the number of photons collected by each channel, the better is the S/N ratio.

A pointed out by Pealat et al. (17) a simple least-squares

routine using a linear law should be less sensitive to the noise but gives too much weight to portions of spectra which change little with the temperature. After their analysis it is better to minimize the quantity $\sum_i (\text{Log } T_i - \text{Log } E_i) T_i$ where T_i is the theoretical spectral density in channel i and E_i the corresponding experimental density. Snelling et al (19) follow a similar approach and minimize the quantity $\sum (T_i - S_i)^2 \times \sigma_i^2$ where σ_i^2 is the experimentally determined variance of the signal of channel . They show that the standard deviation of repeated measurements slightly decreases when σ_i takes into account a shot-to-shot CARS noise and not only the detector noise.

Snelling et al's results confirm the experimental results of Pealat et al and does not significantly modify the figures given in previous papers.

To summarize, CARS mean temperature measurements performed at atmospheric pressure (either in flames, reacting flows or heated cells) are within less than 50K those obtained by other techniques over the range from 300 to 1500K (the actual figure is likely to be 20-30K). Snelling et al found that the standard deviation of repeated single pulse measurements can be lowered with the best processing to 59 ± 6K with a single-mode pump laser to 42 ± 5K with a multi-mode pump laser (the measurements being performed in a flame a ~ 1600K). Unlike Pealat et al who report analog figures they do not use shot-by-shot referencing. According to Pealat et al the effect of shot-by-shot referencing appeared to be marginal at the best in their measurements. The refinements of Snelling et al's processing deserve perhaps such referencing

III-B - Measurements in an SI engine

As a practical example of what actually can be the accuracy of CARS temperature measurements in severe conditions we will present in this paragraphg typical results we obtained in the combustion chamber of a S.I. engine.

Not yet discussed in the previous paragraph the spatial

resolution, i.e., the size of the volume probed is, of course, an important parameter.

As the intensities of the laser beams have to be limited to avoid breakdown of the medium, the number of emitting molecules, i.e. the volume probed, must be large enough to give not too bad S/N ratio. A compromise as to be found between the accuracy of single-shot measurements and the spatial resolution. In the usual 2D BOXCARS configuration the length of the probed volume cannot probably be reduced below 1-2 millimeters. The focal length of the focusing lens to be used is therefore 50-100cm. In our engine to avoid damages on the windows it was impossible to use so large a focal length, therefore the focal volume was to weak to allow single-shot measurements with the BOXCARS configuration. Otherwise it was impossible with our laser to use the 3D phase matching technique where the Stokes beam passes coaxially through the annular pump beam which has been described (31) as well suited for measurements in engine. Therefore we used the colinear beams configuration which gives a very poor resolution : with the 20cm focal length lens the probe volume is cylindrical, approximately 1cm long and less than 100μm in diameter. (The bias due to a poor resolution, compared to the flame thickness, is discussed by Taran in this workshop (32).

The CARS assembly is the commercial instrument described in (23) and available from SOPRA. The shot-by-shot referencing is used. The data reduction is that of Pealat et al. improved to take into account the collisional narrowing.

Nitrogen CARS spectra are recorded at different crank angle the engine operating at a constant speed of 1,600 RPM.

Accuracy of temperature determinations in absolute values can be affected by wrong estimations of non resonant background, especially before combustion where the nonresonant susceptibility of hydrocarbons are not well-known. We estimate that this systematic error does not exced 50K. Available techniques for elimination of nonresonant background were not used as they reduce the number of collected photons. We report here some significant results already published in ref. (33).

Fig. (8) shows the temperature evolution versus crank angle. Each symbol is the average temperature and each error bar is the calculated standard deviation. Therefore error bars reflect both the measurement uncertainty and the cycle-to-cycle variation which is large in the engine used.

The pressure maximum reaches in the cycle can be taken as an indicator of the cycle-to-cycle variation.

The individual measurements used on Fig. (8) are reported Fig. (9), for two crank angles as functions of the peak pressure.

At 180 c.a.d., before combustion, the cycle-to-cycle variation d. is not important and the temperature is independant of the peak pressure. The standard deviation is 25K, comparable to that obtained in less severe experimental conditions. At 660K the temperature decreases slightly when the peak pressure increases. Here the remaining scatter of the values is larger than the scatter which would be due to the measurement itself. The peak pressure cannot reflects alone all the causes of cycle-to-cycle fluctuations.

We have also performed measurements with the engine knocking. The knock intensity (defined as the sum of successive extrema in the characteristic knock oscillations of the pressure history) is then the best indicator of the cycle-to-cycle variations. Among the results we obtained, the following appear to be very representative of the accuracy of the measurements. Fig. (10) shows that the mean temperature (described by the regression straight line drawn) decreases by about 6.5% on the knock intensity range explored. This is mainly due to a better thermal transfer to the walls. Fig. (11) shows that the mean pressure (computed from individual measurements performed at the same time that the temperature measurements) decreases also as it could be expected from the laws of thermodynamics. However this decrease is only 4%.

The difference in the two figures could eventually be explained by larger losses of load by the rings at heavy knock. Anyhow, these results shows that in mean values, a variation of 100K can be measured by CARS in very severe experimental conditions.

IV - CONCLUSION

Due to continual efforts, unquestionable improvements have been made in the knowledge of the fundamentals on which diagnostic techniques are based, in particular LIFS and CARS methods. However a number of problems already raised ten years ago and some newly appeared problems have not yet been satisfactorily solved. The largest improvements have been made in the imaging techniques, following the improvements of lasers and, especially, the development of laboratory computers.

Requirement A reported in the introduction i-e need for discrete (but simultaneous) measurements of temperature, velocity and concentration can be, formally fullfilled. Hovewer the increasing complexity of the techniques and the cost have still limited such works. This is certainly the field open for the next years.

REFERENCES :

(1) R. Goulard, A.M. Mellor and R.W. Bilger "Review and Suggested Experiments" in "Combustion Measurements" R. Goulard Editor, Academic Press 1976.

(2) E.H. Piepmeir,"Theory of Laser Saturated Atomic Resonance Fluorescence", Spectrochim. Acta Part B 27, 431 (1972).

(3) J.W. Daily, "Saturation Effects in Laser Induced Fluorescence Spectroscopy", Appl. Opt. 16, 568 (1977).

(4) M. Mailander, "Determination of Absolute Transition probabilities and Particle Densities by Saturated Fluorescence Excitation,J. Appl. Phys. 49, 1256 (1978).

5) D. Stepowski and M.J. Cottereau, "Direct Measurement of OH local Concentration in a Flame from the Fluorescence Induced by a single Laser Pulse", Appl. Opt. 18, 354 (1979).

(6) D. Stepowski and M.J. Cottereau, "Time Resolved Study of rotational "Energy Transfer in the $A^2\Sigma^+$ (v=0) State of OH in a flame by Laser Induced Fluorescence" J. Chem. Phys. 74,6674 (1981).

(7) A.P. Baronavski and J.R. Mc Donald, "Measurement of C_2 Concentration in an Oxygen-Acetylene Flame : An Application of Saturation Spectroscopy", J. Chem. Phys. 66, 3300 (1977).

(8) A.P. Baronavski and J.R. Mc Donald, "Application of Saturation Spectroscopy to the Measurement of C_2 $^3\Pi_u$ Concentration in Oxy-Acetylene Flames", Appl. Opt. 16, 1897 (1977).

(9) A.B. Rodrigo and R.M. Measures, "An Experimental Study of the Diagnostic Potential of Laser Selective Excitation Spectroscopy for a Potassium Plasma", IEEE J. Quantum Electron.9, 972 (1973)

(10) K. Kohse-Hoïnghaus, W. Perc and T. Just, "Laser-Induced Saturated Fluorescence as a Method for Determination of Radical Concentration in Flames", Ber. Bunsenges Phys. Chem. 87, 1052 (1983).

(11 a) J.T. Salmon and N.M. Laurendeau, "Calibration of Laser Saturated Fluorescence Measurements Using Rayleigh Scattering", Appl. Opt. 24, 65 (1985).

(11 b) J.T. Salmon and N.M. Laurendeau, "Analysis of Probe Volume Effects Associated with Laser-Saturated Fluorescence Measurement". Appl. Opt. 24, 1313 (1985).

(12) R.P. Lucht, D.W. Sweeney and N.M. Laurendeau", Balanced Cross-rate Model for Saturated Molecular Fluorescence in Flames using a Nanosecond Pulse Length Laser", Appl. Opt. 19, 3295 (1980).

(13) D.H. Campbell, "Collisional Effects on Laser-Induced Fluorescence Measurements of Hydroxide Concentrations in a Combustion Environment.

1 : Effects for $v'=0$ Excitation", Appl. Opt. 23, 689 (1984).

(14) N.S. Bergano, P.A. Jaanimagi, M.M. Salour and J.H. Bechtel, "Picosecond Laser-Spectroscopy Measurement of Hydroxyl Fluorescence Lifetimes in Flames", Opt. Lett.8, 443 (1983).

(15) J.F. Dionet, Rapport de D.E.A. Université de Rouen, 1984.

(16) M.J. Cottereau, "Single-shot Laser-Saturated Fluorescence Measurements : a new method", Appl. Opt. 25, 744 (1986).

(17) M. Pealat, P. Bouchardy, M. Lefebvre and J.P. Taran, "Precision of Multiplex CARS Temperature Measurements", Appl. Opt. 24, 1012 (1985).

(18) D.R.Snelling, R.A. Sawchuk and R.E. Mueller, "Single-Pulse CARS Noise : A comparison between Single-Mode and Multi-mode Pump Lasers", Appl. Opt. 24, 2771 (1985).

(19) D.R. Snelling, G.J. Smallwood, R.A. Sawchuk and T. Parameswaran,"Precision of Multiplex CARS Temperatures Using Both Single-Mode and Multimode Pump Laser", Appl. Opt. 26, 99 (1987).

(20) S. Kröll, M. Alden, T. Berglind and R.J. Hall, "Noise Characteristics of Single Shot Broadband Raman-Resonant CARS with Single and Multimode lasers", Appl. Opt. 26, 1068 (1984).

(21) D.A. Greenhalgh and S.T. Whittley, "Mode Noise in Broadband CARS Spectroscopy", Appl. Opt. 24, 907 (1985).

(22) R.J. Hall and D.A. Greenhalgh, "Noise Properties of Single-Pulse Coherent Anti-Stokes Raman Spectroscopy with multimode Pump Sources", J. Opt. Soc. Am. B 3, 1637 (1986).

(23) S.A.J. Druet and J.P.E. Taran, "CARS Spectroscopy", Prog. Quantum Electron. 7, 1 (1981).

(24) R.J. Hall and A.C. Eckbreth, 'Coherent Anti-Stokes Raman Spectroscopy (CARS) : Application to Combustion Diagnostics" in Laser Applications, vol. 5, J.F. Ready and R.K. Erf, Eds. (Academic, New-York, 1984), pp. 213-309.

(25) A.B. Harvey, Ed., Chemical Applications of Nonlinear Raman Spectroscopy (Academic, New-York, 1981).

(26) J.W. Nibler and G.V. Knighten, "Coherent Anti-Stokes Raman Spectroscopy", in Raman Spectroscopy of Gases and Liquids, A.Weber, Ed. (Springer, Berlin, 1979), pp. 253-299.

(27) M.L. Kozykowski, R.L. Farrow and R.E. Palmer, "Calculation of Collisionnally Narrowed Coherent Anti-Stokes Raman Spectroscopy Spectra", Opt. Lett. 10, 478 (1985).

(28) A.D. May, J.C. Stryland and G. Varghese, "Collisionnal Narrowing of the Vibrational Raman Band of Nitrogen and Carbon Monoxide", Can. J. Phys. 48, 2331 (1970).

(29) J. Bonamy, L. Bonamy and D. Robert, "Overlapping Effects and Motional Narrowing in >Molecular Band Shapes : Application to the Q-Branch of HD", J. Chem. Phys 67, 4441 (1977).

(30) R.E. Teets, "Accurate Convolutions of Coherent Anti-Stokes Raman Spectra", Opt. Lett. 9, 226 (1984).

(31) D. Huck, K.A. Marko and L. Rima, "Broadband Single-Pulse CARS Spectra in a Fired Internal Combustion Engine", Appl. Opt. 20, 1178 (1981).

(32) J.P. Taran, "CARS in premixed Turbulent Combustion Problems, Biases and Results. This Workshop.

(33) J.J. Marie and M.J. Cottereau, "Single-Shot Temperature Measurements by CARS in an I.C. Engine for Normal and Knocking Conditions". SAE Technical paper n°870458 (1987).

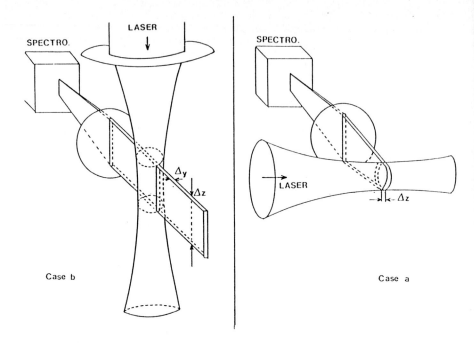

Figure 2 - Geometrical Arrangements for LIF Measurements

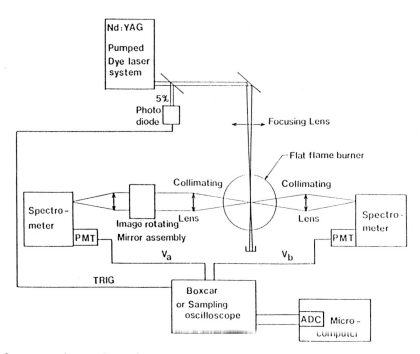

Figure 3 - Experimental Device Used for Saturated LIF Measurements

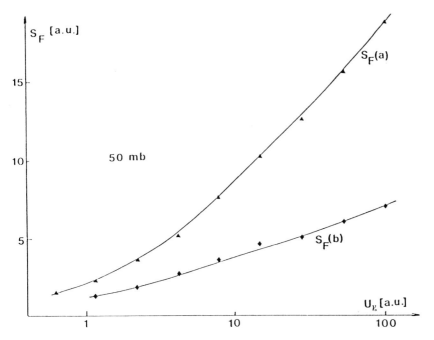

Figure 4-Experimental Results Values at the Fluorescence peak

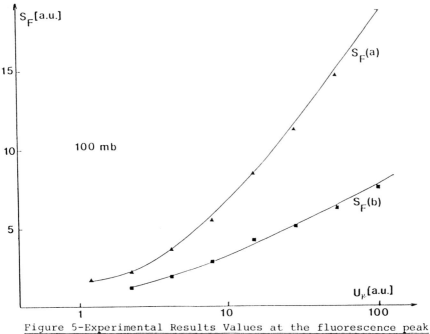

Figure 5-Experimental Results Values at the fluorescence peak

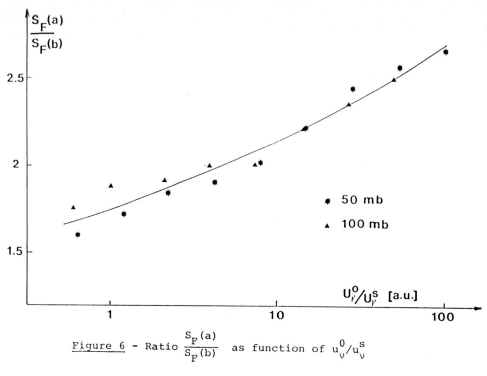

Figure 6 - Ratio $\frac{S_F(a)}{S_F(b)}$ as function of u_ν^0/u_ν^s

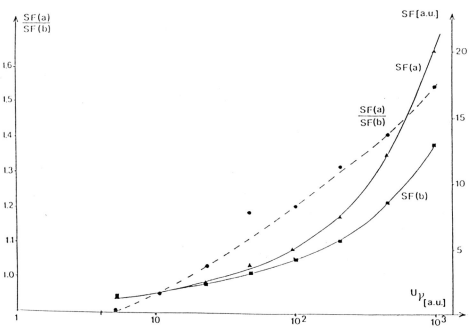

Figure 7 - Experimental results integrated fluorescence signal at 1atm

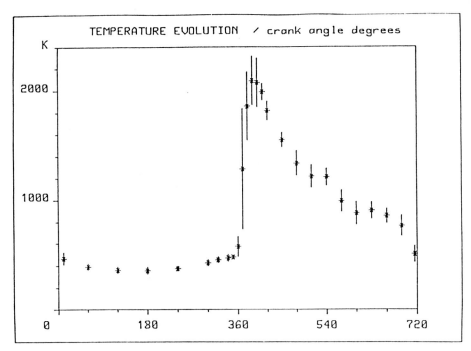

Figure 8 - Temperature measured in an SI engine by CARS (Mean values and RMS)

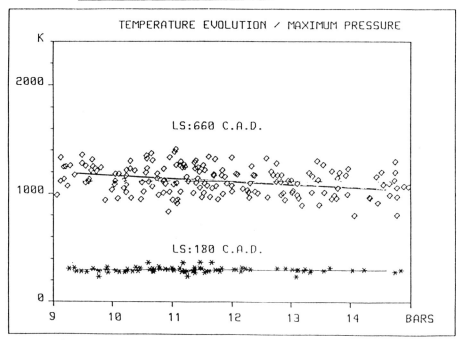

Figure 9 - Single shot temperatures as function of the maximum of the pressure in the cycle

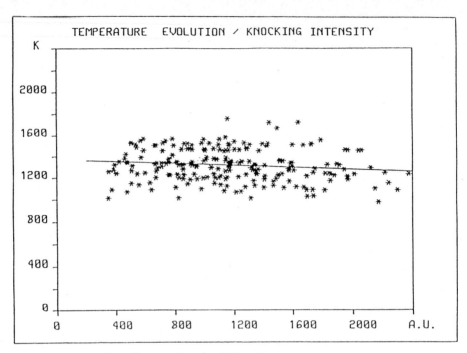

Figure 10 - Temperature of the burnt gases as function of the knock intensity

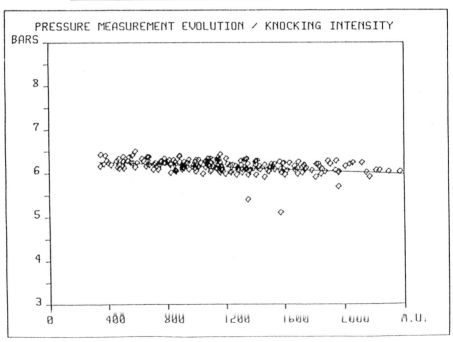

Figure 11 - Pressure of the burnt gases as function of the knock intensity

STRUCTURE OF FLAMELETS IN TURBULENT REACTING FLOWS AND
INFLUENCES OF COMBUSTION ON TURBULENCE FIELDS

F.A. Williams
Princeton University
Princeton, NJ 08544

ABSTRACT

Attention is focused on the reaction-sheet regime of turbulent reacting flows. In this regime the chemistry occurs in thin sheets convected and distorted by turbulent motions. Consideration is given to structures of reaction sheets, to their stability, and to influences of these sheets on the turbulence in turbulent combustion. These considerations are pursued separately for premixed and nonpremixed combustion.

In the past, analyses of reaction-sheet structures have been based largely on activation-energy asymptotics. These analyses have addressed influences of strain and curvature on sheet structures and extinction. More recently there have been reaction-sheet analyses that account for more detailed chemistry, including hydrocarbon-air chemistry. These analyses are reviewed and evaluated with respect to their utility in turbulent-flame calculations.

To address influences of reaction-sheet combustion on turbulence fields, attention is restricted mainly to large-scale, low-intensity turbulence and to premixed systems. It is then possible to calculate adjustments to the turbulence that occur in the preheat zone and in larger hydrodynamic zones of essentially constant-density fluid dynamics outside the flame. The adjustments in the hydrodynamic zones are significant and affect measured turbulence characteristics appreciably. The nature of the modifications introduced by the hydrodynamic zones is discussed.

1. INTRODUCTION

It is well known that turbulent combustion can occur in two principal regimes, the reaction-sheet regime and the distributed-reaction regime. These regimes may be indicated [1] in a plane of Damkohler number Da and Reynolds number Re. In terms of the root-mean-square velocity fluctuation u' and the integral length scale ℓ, the turbulence Reynolds number is Re_ℓ and $u'\ell/\nu$, where ν is a kinematic viscosity. A corresponding Damkohler number is $Da_\ell = \ell/(u'\tau_L)$, where τ_L is a representative laminar chemical time. With these definitions, the plane can apply equally for premixed and diffusion flames. Reaction-sheet combustion occurs in the upper part of the plane and distributed-reaction combustion in the lower part. Precise boundaries of the regimes are unclear, and it is possible that there are intermediate, possibly even overlapping, regimes in the plane.

Careful studies are needed to identify just where any flow of interest falls in the plane. These studies must be pursued separately for premixed and diffusion flames. For premixed flames the parameters in the plane can be related to the laminar burning velocity S_L and the laminar flame thickness δ_L by using the rough formulas $\nu = \delta_L S_L$, $\tau_L = \delta_L/S_L$. Thus, lines of constant values of ℓ/δ_L and of u'/S_L can be identified and drawn. The Kolmogorov scale η is related approximately to ℓ by $\eta = \ell/Re^{3/4}$, and lines of constant values of η/δ_L also may be shown. Physical interpretations for premixed flames then are clear: u'/S_L is the nondimensional turbulence intensity, measured in units of the laminar burning velocity; ℓ/δ_L is the largest turbulence scale, measured in units of the laminar flame thickness; and η/δ_L is the smallest turbulence scale, measured in these same units.

Spark-ignition engines for automotive applications have been investigated [2] to identify their combustion regimes. The results indicate that most of these engines have combustion in the reaction-sheet regime. Through continuing experimental studies there has recently been growing acceptance of the idea that combustion in spark-ignition engines belongs in the reaction-sheet regime [3]. Thus, it seems clear that there is at least some practical interest in reaction-sheet turbulent combustion.

More studies are needed for identifying the turbulent combustion regimes in practical applications. Large furnaces, fires and marine diesels probably involve diffusion flames in the reaction-sheet regime. Turbojet afterburners and ramjet engines perhaps involve premixed reaction-sheet combustion, depending on their size. Turbojet and rocket engines especially need further detailed study to identify their regimes well; some claim turbojets to involve distributed-reaction combustion and other claim them to involve reaction-sheet combustion.

2. PREMIXED FLAMELETS

The basic structures of premixed flamelets are derived from analyses of steady, planar, adiabatic flames. Numerical integrations and analytical approximations are used to describe these structures. Irrespective of the approach, achievement of human understanding of the structure requires the introduction of simplifications. Asymptotic methods are analytical approximations in which simplifications are derived in a systematic manner. Significant improvements in understanding of basic flamelet structures have been obtained recently, largely through asymptotic methods.

The physical ideas underlying the most well-known asymptotic analysis for premixed flames involve an asymptotic expansion for large values of the Zel'dovich number $\beta_1 = E_1(T_\infty - T_0)/R^0 T_\infty^2$, where T_0 is the initial temperature, T_∞ the equilibrium temperature of the burnt gas, and E_1 the overall activation energy for the reaction, which is treated in a one step Arrhenius approximation. This "activation-energy asymptotics" leads to the existence of a convective-diffusion or "preheat" zone followed by a thin reaction zone. A representative specific heat at constant pressure c_p may be introduced and a nondimensional distance variable defined in terms of the mass flow rate m and thermal conductivity λ as $\xi = \int^x (mc_p/\lambda)dx$, along with a nondimensional temperature $\tau = (T-T_0)/(T_\infty-T_0)$. Solutions for τ and the mole fractions X_i tend to be exponential in ξ in the preheat zone. From the translational invariance the reaction zone may be placed at $\xi = 0$ in the outer variable, and then appropriate inner-variable scalings identify $\beta_1\xi$ and $\beta_1(1-\tau)$ as variables of order unity in the reaction zone, whereas ξ and τ are of order unity in the preheat zone. Differential equations in the inner variables are derived and solved subject to matching conditions to determine profiles in the reaction zone as well as the burning velocity S_L, which is an eigenvalue of these equations. In this way, relations for flamelet structure and burning velocity are obtained that can be used as an element in a premixed-flamelet library for turbulent-flame computations. General flamelet solutions of this type for two-reactant flames with Stefan-Maxwell transport have been published recently [4].

Results for flamelet structures are becoming available for real flames with full chemistry. The simplest real flame is the ozone decomposition flame, for which the full structure has been identified [5-8]. This flame exhibits two regimes of propagation, a steady-state regime at small initial ozone mole fractions X_{3O} and a merged regime at large X_{3O}, in which the step $O + O_3 \to 2O_2$, which follows $O_3 + M \to O_2 + O + M$, is not fast enough to maintain a steady state for the O atom in the thin reactive-diffusive zone. In both regimes, the burning-velocity predictions of the asymptotic analysis agree well with those of numerical integrations [8]. A distinguishing attribute of this flame is the existence of a large recombination zone downstream in which $2O + M \to O_2 + M$ and $O + O_2 + M \to O_3 + M$, $O_3 + O \to 2O_2$ lead to exothermicity without any influence on the burning velocity. Under most practical conditions the recombination is found to be convective-reactive, so that in turbulent-flame computations, this chemistry should be included in the field equations downstream from the flame.

One way that nonuniformities in flows may affect laminar burning velocity is by introducing variations of the temperature T_0 of the fresh mixture, with corresponding variations of the adiabatic flame temperature T_∞. In the simplest description, the Arrhenius factor gives the resulting change in the laminar burning velocity. For the ozone decomposition flame, the situation remains similar to this. The temperature of the reaction zone is $T_f = T_\infty - (1+\mu)(T_\infty-T_0)X_{1f}/X_{3O}$, where μ is the ratio of the heat of dissociation of O_3 to its heat of combustion, and X_{1f} is the maximum O mole fraction, which in the steady-state regime is simply k_{1f}/k_{2f}, the ratio of the specific reaction-rate constant, k_1, for O_3 dissociation to that, k_2, for $O + O_3 \to 2O_2$, evaluated at T_f. The burning-velocity formula is like that for a one-step process, with the kinetic parameters of O_3 dissociation, but with T_∞ replaced by the lower temperature T_f. Since the formulas show that the influences of temperature variations on T_f are similar to those on T_∞, there is not an appreciable difference between an ozone-flame response and that of a flame described by a one-step reaction. The effective activation energy involves not only the E_1 for the dissociation but also the rate parameters in k_1 and k_2, and in the merged regime its calculation necessitates integration for the reaction-zone structure with two independent rate equations. The resulting effective specific reaction-rate constant varies from k_1 in the steady-state limit of the merged regime to k_1^3/k_2^2 in the opposite limit of the merged regime, where $T_\infty - T_f$ is $[\mu(1+\mu)/(2+\mu)](k_{1f}/k_{2f})(T_\infty - T_0)/X_{3O}$. This opposite limit approaches a new regime (never applicable to the ozone flame) in which flow nonuniformities may exert more profound influences on laminar-flame structure.

There is a theoretical possibility of a two-zone regime [7] in which in the reaction zone ξ becomes of order unity, so that convection becomes important at leading order, and a convective-reactive-diffusive balance is maintained. A steady-state approximation becomes applicable in the differential equation for thermal enthalpy in this zone. The O concentration in this reaction zone is now so large that the intermediate diffuses into the preheat zone, where it liberates heat by the second reaction. The preheat zone therefore becomes an exothermic zone, and what was formerly the only reaction zone may now be called a generation zone, the zone in which the active intermediate is generated. The label "two-zone" was introduced because of the two (convective-reactive-diffusive) reaction zones. In this regime there is a mechanism whereby diffusion of an active intermediate can play an important role in the flame structure and dynamics. Diffusion coefficients for intermediates affect burning velocities already in merged regimes by influencing radical distributions within the thin reaction zone, but direct influences of moderate strain or curvature on structure through modification of intermediate diffusion are not encountered until the two-zone regime. If the activation energy of the second step is too high, then the preheat zone must remain inert, but if it is small enough then the two-zone structure exhibits exothermicity all the way to the cold boundary (without compromising activation-energy asymptotics). Parameter estimates suggest that two-zone structures like this might occur for H_2-Cl_2 or H_2-F_2 flames.

Simplified asymptotic descriptions are becoming available for hydrocarbon-air flames with full chemistry [9-11]. Thus far the principal results pertain to methane-air flames [10], but the principles seem readily extendable to other hydrocarbons and alcohols [11] and to oxygen-enriched or near-limit mixtures. Steady-state and partial-equilibrium approximations are introduced systematically to reduce the full reaction mechanism to a small number of overall steps. The overall fuel-consumption step is found to be $CH_4 + 2H + H_2O \rightarrow CO + 4H_2$ and to proceed at the rate of step 3 of Table I. The water-gas shift, $CO + H_2O \rightleftarrows CO_2 + H_2$, which is reversible, with a steady state for OH has the rate of step 4 for its forward rate; the qualitative essence of the structure can be obtained by assuming water-gas equilibrium in the reaction zones. The overall oxygen-consumption, radical-production process is found through steady states for OH and O to be $O_2 + 3H_2 \rightleftarrows 2H_2O + 2H$, reversible but with the forward rate of step 5. Finally, if a steady state is assumed for HO_2, the overall recombination process is found to be $2H + M \rightarrow H_2 + M$, at the rate of step 6. A test mechanism with four overall steps thus is derived. If a steady state is assumed for H, then a three-step mechanism is obtained, the fuel-consumption step indicated above being replaced by $CH_4 + O_2 \rightarrow CO + H_2 + H_2O$, and the last two steps combining into the oxidation step $O_2 + 2H_2 \rightarrow 2H_2O$, both with readily derivable rates dependent, for example, on concentrations of intermediates. Introduction of water-gas equilibrium gives a two-step mechanism, $(1+r)CH_4 + (1+r)O_2 \rightarrow 2(H_2+rCO) + 2rH_2O + (1-r)CO_2$, $2(H_2+rCO) + (1+r)O_2 \rightarrow 2H_2O + 2rCO_2$, where r is the ratio of CO to H_2 concentration at water-gas equilibrium.

Table I
Some Elementary Reaction Steps

1. $O_3 + M \rightarrow O_2 + O + M$
2. $O_3 + O \rightarrow 2O_2$
3. $CH_4 + H \rightarrow CH_3 + H_2$
4. $CO + OH \rightarrow CO_2 + H$
5. $H + O_2 \rightarrow OH + O$
6. $H + O_2 + M \rightarrow HO_2 + M$
7. $CH_4 + OH \rightarrow CH_3 + H_2O$

The asymptotic analyses of methane flames relate to the four-step, three-step or two-step mechanism. For the premixed flame, in the structure according to the three-step mechanism, following the preheat zone there is a thin layer in which the fuel-consumption step occurs, and this is followed by a broader (but still reactive-diffusive) layer of oxidation of intermediates, in the upstream part of which the water-gas reaction departs from equilibrium. The ratios of the thicknesses of these three reaction layers to that of the preheat zone will be denoted below by the small parameters δ, ϵ and ω, respectively in discussing structures of the diffusion flame. The quantitative agreement between predicted and measured burning velocities for stoichiometric methane flames, obtained with this three-step mechanism, is not very good. The difficulty lies in the specified chemical mechanism for the fuel-consumption zone. For example, reaction 7 of Table I has not been included in the kinetic scheme. By beginning with the mechanism defined, it is straightforward to systematically test for the importance of new steps in the hydrocarbon chemistry [10] when their rates are given. As a result, it is found that additional steps are likely to be important. Since the new chemistry is confined to the fuel-consumption layer, it becomes straightforward to revise the burning-velocity computations [10]. Resulting revised burning velocities [10] agree well with results of full numerical integrations employing the same rate coefficients and also with experiment. For methane flames, the structure defined here is better than that of one-step activation-energy asymptotics as an element in a premixed-flamelet library.

3. DIFFUSION FLAMELETS

Asymptotic analyses have also contributed to our understanding of structures or diffusion flamelets. It is convenient to transform to the mixture fraction Z as the independent variable in describing diffusion flames, where Z = 0 in the oxidizer (O) stream and Z = 1 in the fuel (F) stream. The resulting well-known structure for one-step, irreversible reactions in the limit of infinite Damkohler number has an infinitesimal reaction sheet at
$Z = Z_s$, the stoichiometric mixture ratio. Activation-energy asymptotics reproduce this structure and the frozen structure in limiting cases [12-14]. Unlike unstrained premixed flames, diffusion flames possess a one-parameter family of structures. The parameter can be taken to be a Damkohler ratio of a diffusion time to a chemical time. A diffusion time is $t = (D_{12}|\nabla Z|^2)^{-1}$ where D_{12} is a representative diffusion coefficient. The strain rate in a counterflow and the rate of scalar dissipation in a turbulent flow are inversely proportional to t. Extinction occurs if t becomes small enough to decrease the Damkohler number below its critical value. Activation-energy asymptotics provide the critical Damkohler number for extinction [12]. The predicted structures near extinction are close to the equilibrium profiles. Thus relatively few Z-profiles are allowed by activation-energy asymptotics; elements of the flamelet library could be the equilibrium and frozen solutions, the solution just before extinction, and perhaps one between that and equilibrium. The region between near-extinction and frozen conditions is excluded entirely.

For laminar diffusion flames there is ample experimental evidence for structures qualitatively resembling those just described [15]. Some corresponding evidence has been obtained recently for turbulent hydrogen diffusion flames [16,17]. Representative results [17] of simultaneous, pointwise measurements of T and Z cluster around the equilibrium curve at low strain rates and fall into two distinct bands, one corresponding to near equilibrium and the other to flame-extinguished mixing, at high strain rates. At present there is controversy concerning the extent to which results of this kind can be obtained for hydrocarbon-air flames. Current understanding of hydrocarbon-air diffusion flamelets can be addressed by considering the methane-air flame.

Measurements [18-23] and numerical integrations of the conservation equations [22-26] for the structures of methane diffusion flames with detailed chemistry show that increasing strain produces increasing leakage of O_2 through the reaction zone but no CH_4 leakage. This behavior is contrary to predictions of activation-energy asymptotics with one-step chemistry. However, the two-step mechanism of the premixed flame qualitatively yields the correct behavior [27]. In the structure with the two-step mechanism there is a thin fuel consumption zone, with thickness in Z of order δ, on the fuel-rich side of a thicker (but still thin) zone of H_2 and CO oxidation, with thickness of order ϵ. The intermediates H_2 and CO are produced along with H_2O and CO_2 in the fuel-consumption zone, whose structure is essentially identical to that for the premixed flame, so that δ is given by the same formula. For the oxidation zone, ϵ is like that of the premixed flame and is proportional to $t^{-1/4}$; the structure of this zone differs from that of the premixed flame only in that now the temperature decreases and the O_2 concentration increases with increasing distance from the fuel-consumption zone. An increase in ϵ (a decrease in t) increases the O_2 leakage, according to the asymptotic analysis. The predictions agree qualitatively with reality.

Quantitative predictions of structure and extinction from the two-step mechanism are inaccurate, but substantial improvements are achieved by introducing corrections for water-gas nonequilibrium. This gives the three-step mechanism and results in a thin water-gas nonequilibrium layer of thickness ω on the rich side of the oxidation zone, adjacent to the fuel-consumption zone. The zone arrangement then is entirely analogous to that of the premixed flame; the kinetic schemes are the same, and activation energies play correspondingly negligible roles.

An alternative viewpoint currently being explored suggests that there may be significant differences between premixed-flame and diffusion-flame kinetics [28]. Estimates indicate that the full steady-state approximations for H and OH may not be very accurate and that partial equilibrium for step 5 of Table I in fact may be poor over key ranges of Z. It therefore may be necessary to revert to a four-step mechanism to obtain an improved approximation to the flame structure. Simplifications motivated by asymptotic analysis may be introduced into the four-step kinetics to obtain a readily manageable formulation from which extinction conditions can be computed. If, for simplicity, water-gas equilibrium is introduced, then a reduced four-step mechanism is obtained, which can be written as

$$CH_4 + (\frac{2}{1+s})OH + (\frac{2s}{1+s})H \rightarrow \frac{r(3+5s)}{(1+r)(1+s)} CO$$
$$+ \frac{(3+5s)}{(1+r)(1+s)} H_2 + \frac{3r(1+s)-2s}{(1+r)(1+s)} H_2O + \frac{(1+s)-2r(1+2s)}{(1+r)(1+s)} CO_2 , \quad (I)$$

$$O_2 + \frac{(1+3s)}{(1+r)(1+s)} H_2 + \frac{r(1+3s)}{(1+r)(1+s)} CO \rightleftarrows (\frac{2}{1+s}) OH + \frac{2s}{1+s} H$$
$$+ \frac{2s-r(1+s)}{(1+r)(1+s)} H_2O + \frac{r(1+3s)}{(1+r)(1+s)} CO_2 \quad (II)$$

$$(\frac{2}{1+s}) OH + (\frac{2s}{1+s}) H + \frac{(1-s)}{(1+r)(1+s)} H_2 + \frac{r(1-s)}{(1+r)(1+s)} CO \rightarrow$$
$$\frac{2+r(1+s)}{(1+r)(1+s)} H_2O + \frac{r(1-s)}{(1+r)(1+s)} CO_2 , \quad (III)$$

where s is the ratio of H to OH concentrations at equilibrium of $H_2 + OH \rightleftarrows H_2O + H$. Step (I) represents fuel consumption by radicals, step (II) oxidation of H_2 and CO to consume oxygen and produce radicals, and step (III) radical consumption by recombination. All steps are bimolecular except for (III), which is termolecular. The rate of (I) involves k_3 and k_7, that of (II) k_5, and that of (III) k_6.

Step (I) occurs in a narrow fuel-consumption zone at $Z = Z_f$ and involves a diffusion flame between CH_4 and radicals. Step (III) occurs in a narrow recombination zone at a temperature T_c, defined by $k_5 = k_6\ (p/R^0T)$, at a mixture fraction $Z = Z_c < Z_s < Z_f$; for $p \leq 1$ atm, it is found that $T_c < 1000$ K, and so three-body processes are negligible throughout the main high-temperature reaction zones, although they still influence flame structure through the boundary conditions that they impose at $Z = Z_c$. Between $Z = Z_c$ and $Z = Z_f$ is the radical-production, oxygen-consumption zone in which step (II) occurs; this is the only reversible step of the three, and it is found to maintain partial equilibrium at low T while having a negligible reverse rate at high T. To approximate this behavior, a point of sudden freezing for (II) is introduced, $Z = Z_i$, such that there is equilibrium for $Z_c < Z < Z_i$, while the reverse reaction is negligible for $Z_i < Z < Z_f$. The relatively high activation energy of step 5 is an aspect of the motivation for the sudden-freezing approximation. Extinction occurs either through nonexistence of a solution for the structure of the oxygen-consumption zone or through activation-energy asymptotics applied to the finite rates in the fuel-consumption zone. The roles of the radicals thus differ appreciably from their roles in the premixed flame with this model.

The substantial differences between the two models just described illustrate the large uncertainties that remain concerning the structures and the extent to which methods of activation-energy asymptotics may play a role. Further research is needed to better identify accuracies and ranges of validities of the models. Extensions to calculations of soot production at sufficiently large values of t by perturbation methods in the region $Z > Z_f$ also are needed. Much remains to be done in asymptotic analyses of diffusion-flame structures to provide improved elements in flamelet libraries for nonpremixed turbulent combustion.

4. PREMIXED-FLAME INSTABILITIES AND TURBULENT EFFECTS

The preceding discussions have focused on the simplest flamelet configurations to expose most easily the considerations involved in describing flamelet structures. In turbulent flows planar flamelets are subjected to strain and curvature, and they may experience heat losses and instabilities, irrespective of whether the flow is turbulent. The instabilities of premixed flames (excluding system-dependent phenomena, such as acoustic instabilities) may be classified as body-force instabilities, hydrodynamic instabilities, and diffusive-thermal instabilities [12]. The order listed here typically goes from the largest to the smallest scales. Acceleration or buoyancy produce body-force instabilities that exist, for example, for upward-propagating flames. There are many flames for which diffusive-thermal effects are known to be stabilizing, and in favorable configurations body-force effects are stabilizing; hydrodynamic effects always are destabilizing, and stability of the planar premixed flame is achievable only with the help of diffusive-thermal effects at small scales plus body-force effects (or boundary influences) at large scales [29]. It is of interest to investigate how the new developments in understanding of laminar-flame structure, reviewed above, affect these conclusions.

Ideas about premixed-flame instabilities have been based on activation-energy asymptotics. For the two-zone structure, or for the methane flame structure, these ideas require revision. From the scales involved, it seems clear that body-force and hydrodynamic effects remain the same; diffusive-thermal phenomena, however, may be influenced by modifications in diffusion of reaction intermediaries. Good understanding of aspects of these changes associated with radical diffusion has been obtained in recent model studies of Linan and others. In need of further investigation are stability for the two-step model of methane flames and influences of somewhat extended downstream zones of finite-rate heat release on diffusive-thermal stability. It might be conjectured that these additional phenomena would tend to enhance stability, so that the current ideas will not be in need of substantial revision.

The strain in turbulence fields modifies the structures of planar premixed laminar flamelets. If a is the negative of the element of the rate-of-strain tensor normal to the flamelet, then a nondimensional flame stretch may be defined as $\kappa = a\nu_0/S_L^2$, which is the inverse of a Damkohler number. This κ is small for weak stretch and of order unity for moderate stretch. In any given configuration, there thus exists a one-parameter family of structures, dependent on κ. Many of these families of structures have been calculated for one-step chemistry with activation-energy asymptotics [30-35], most recently for general strain fields [35]. Numerical integrations with more complete chemistry also have been made, as summarized by Rogg [36]. If the turbulence is described in terms of its strain-rate distribution, then the results of these analyses can contribute to the flamelet library needed in turbulent combustion calculations. The previously described asymptotic descriptions of real laminar flames have not yet been extended to address influences of stretch, although some ideas are available from model studies (e.g., by Seshadri and Peters).

A significant aspect of strained-flame analyses concerns the prediction of abrupt extinction. As κ increases, the structure may change abruptly to one in which there is little heat release. Sufficient cooling of the product stream leads to the existence of abrupt extinction, in comparison with the gradual extinction found in adiabatic and superadiabatic cases at strong stretch. Systems with gradual extinction might be expected to exhibit continuous variations in turbulent combustion regimes as the parameters are varied. Systems with abrupt extinction, on the other hand, may exhibit abrupt transitions in regimes and might support more than one regime for the same parameter values.

Experimentally, flamelet extinctions seem to be abrupt, but results of activation-energy asymptotics suggest that gradual extinctions should not be uncommon. The discrepancy may be associated with radiant heat losses from the hot combustion products or with specific chemical-kinetic mechanism effects not described by activation-energy asymptotics. Clarifications of the degree of prevalence of the latter could be obtained from asymptotic analyses of stretch influences on the real-flame kinetic models previously described.

Preceding candidates for premixed-flamelet libraries are planar flames. The best selection depends on the strain-rate distribution of the turbulence, concerning which there is conflicting evidence at present [37,38]. If the turbulence is like spaghetti, i.e., stretched vortex tubes, then it seems likely that the principal flames will be cylindrical and wrapped around the vortex lines. Numerical [39] and asymptotic [40,41] analyses of premixed cylindrical flames of this kind for one-step, irreversible, Arrhenius chemistry have been developed recently. At sufficiently high turbulence Reynolds numbers these flamelets (which do exhibit abrupt extinctions) may be more prevalent than planar flamelets. Additional ideas include flame propagation along the nonuniform cores of vortices [42], balances between flame propagation and strain [43], and eddy sizes to quench reaction zones [44]. More study of turbulence structure and its influences on flamelet configurations certainly is needed.

5. DIFFUSION-FLAME INSTABILITIES AND TURBULENCE EFFECTS

Laminar diffusion flamelets exhibit fewer instabilities than premixed flames. Body-force instabilities of course may occur and may be involved in transition to turbulence in large fires with buoyant plumes; since the low-density region is in the interior of the diffusion flame, in contrast to adiabatic premixed flames there is no entirely buoyantly stable configuration. Most significant is that, since diffusion flames are nonpropagating, hydrodynamic instability is absent for them; there is no need to identify stabilizing effects to counterbalance hydrodynamic instability. Finally, diffusion flames appear to admit a smaller variety of diffusive-thermal instabilities; banded near-limit hydrogen flames, supported by preferential diffusion of fuel, can be identified under suitable conditions [12], but this sort of thing is rare, and no pulsating diffusive-thermal instabilities have been discovered.

Although analysis of influences of radiant heat losses from diffusion flames by asymptotic methods is possible [45], these effects seem negligible compared with those of strain. Unlike analyses of premixed flames, analyses of all types of common diffusion flamelets predict abrupt extinction. The reason is that both reactants are appreciably colder than the adiabatic flame temperature. Conductive heat loss from the reaction region is always significant, and increasing strain increases this loss rate, eventually giving abrupt extinction. While abruptness thus is characteristic of reaction-sheet regimes, there is less of a tendency for it to occur (and extinctions may be gradual) in distributed-reaction regimes for nonpremixed combustion; this is in need of further investigation.

Descriptions of turbulent diffusion flames for reaction-sheet regimes can be based on a distribution of strain rates and employ a flamelet library parameterized by strain [46-51]. So long as all strain rates are too small for extinction, this procedure is relatively straightforward; the principal difficulties lies in the modeling of the strain-rate distribution. There certainly are experimental conditions under which flamelet extinctions are not approached [52,53]. However, there are also important conditions, e.g. for lifted flames, in which extinctions occur. Simulations have shown [54] that extinctions tend to occur in the braids between vortices in mixing layers, where the strain rates are highest. Recent procedures for including effects of flamelet extinctions involve the introduction of partially premixed diffusion flamelet structures into the libraries [49-51], under the idea that premixing and reignition will occur after extinction. Although this certainly has some appeal, there clearly are difficulties in estimating extents of premixing.

A significant fundamental question involves the dynamics of the edge of a hole in a diffusion-flamelet sheet. This question has received recent attention [55,56], and it now appears that the edge can propagate so as to close the hole. The propagation velocity, however, is not large; with activation-energy asymptotics, it seems to be on the order of β_1^{-1} times the laminar burning velocity of the stochiometric adiabatic premixed laminar flame [56]. Further development of these analyses could provide improvements over percolation-theory concepts introduced in the past [48].

6. VORTICITY GENERATION IN DIFFUSION FLAMES

We turn now from questions of influences of turbulence on flames to address some aspects of influences of flames on turbulence. Improved understanding of influences of diffusion flames on turbulence has been obtained from experimental [53] and computational [57] investigations of diffusion flames in two-dimensional mixing layers at large Damkohler numbers. Beyond the obvious effect of increased rates of dissipation through the increase of viscosity with increasing temperature, how does the heat release in thin, nonextinguishing flamelets influence the flow? This question can be considered on the basis of the conservation equation for vorticity ω, in two-dimensional flow with dissipation neglected, $\partial \omega / \partial t + \nabla \cdot (\underline{v}\omega) = (\nabla p) \times \nabla(1/\rho)$, the right-hand side of which is the baroclinic torque, which is zero without heat release.

Vorticity contour plots have been obtained from simulations [57] of two-dimensional time-dependent mixing layers by pseudospectral methods. Corresponding contours of constant values of the baroclinic torque can be understood by realizing that ∇p is roughly radially outward from the vortex center while $\nabla \rho$ is outward from the stochiometric contour of the flame. The baroclinic torque therefore vanishes where the flame and vortex are concentric but has a nonzero contribution in the wings, positive outside the stoichiometric contour and negative (of nearly the same magnitude) inside. Thus the flame generates additional vorticity in the inner part of the vortex wings and generates some vorticity of the opposite sign in the outer wings. In the simulation this gives small overshoots of mean velocity at the outer edges of the mixing layer [57], which have not been seen experimentally [53]; it also steepens the maximum gradient of mean velocity, in agreement with experiment.

This baroclinic effect, along with the thermal expansion ($-\omega\nabla\cdot\underline{v}$), spreads the vortex, decreasing its vorticity in the center, and thereby decreases the rate of entrainment into the mixing layer, decreasing the overall rate of product formation. Simulations and experiments are in qualitative agreement in these respects. The effects are observable but not dramatically large [53]; if the tendencies towards cancellations of effects are not considered, then it can appear that increasing the heat release experimentally has surprisingly little influence on the mixing layer. The reduced entrainment could contribute to observed flame lengthening in fire whirls.

While there is some understanding of the two-dimensional mixing layer, three-dimensional effects need further exploration. Simulation [58] has suggested that in three-dimensional turbulence, vorticity tends to align roughly with the smaller of two extensive components of the rate-of-strain tensor, while the gradient of a conserved scalar tends to align with the compressive component of the rate-of-strain tensor. If this is true, then the three-dimensional situation locally may not be too different from the two-dimensional situation described above.

7. INFLUENCES OF PREMIXED FLAMES ON TURBULENCE

The baroclinic effects described above can also be applied to premixed flames wrapping around vortices. Again the larger effects are in the tails, generating vorticity of opposite sign, but now the regions of generation of like-sign vorticity are absent because the burnt-gas density is practically constant; thus a greater net rate of change of vorticity is expected in the premixed example.

Premixed turbulent flames in high-intensity, large-scale turbulence pose challenging problems that are difficult to address. There is an interesting new approach to the calculation of turbulent burning velocities in this regime [59,60]. Fractal concepts can play a role in describing flames in this regime [61], but their role seems fundamentally descriptive and not prone to elucidation of dynamics. Because of the difficulties in describing large-scale, high-intensity situations, the discussion here will be restricted to large-scale, low-intensity turbulence.

Many experimental studies have now been completed on rod-stabilized vee-shaped premixed flames in turbulent flows of large scale and relatively low intensity [62-69]. Reasonable correlations between measurements and predictions of the Bray-Libby-Moss theory have been obtained; (successes achieved by going to second order in this theory have been remarkable in overcoming difficulties presented by countergradient diffusion). A difficulty in these experiments is their evolutionary character; scales continue to increase with increasing distance from the rod for reasons that have not been fully explained. Thus there is only partial success, and more research is needed. Experiments with flames normal to the approach flow can be designed to maintain homogeneity in transverse directions [70], thereby removing some of the complexities of the vee-flames. Comparisons between theory and experiments of this last type can provide the most stringent tests for predictions of premixed turbulent flame theories.

Influences of premixed flames on large-scale, low-intensity turbulence can be calculated by methods that are straightforward in principle if there is transverse homogeneity and stationarity, and if the planar laminar flame is stable. It is essential to observe that in these configurations there are zones of hydrodynamic adjustment, having thicknesses comparable with the integral scale, located immediately upstream and downstream from the wrinkled laminar flame. The dynamics of the wrinkled flame itself have been analyzed [71-73], and changes in turbulence quantities across the wrinkled flame have been calculated [71]. Changes across the zones of hydrodynamic adjustment have been addressed only more recently [74,75]. The changes encountered can be understood on physical grounds.

Quantities of interest are the turbulent kinetic energy per unit mass, q, and the turbulent kinetic energy per unit volume, Q. The wrinkled flame modifies these quantities mainly through continuity of locally transverse velocity components and a fixed increase in the locally normal velocity component by the density ratio $R > 1$. For weak wrinkling it is then found that in the first approximation the flame does not influence the component of
the fluctuating velocity normal to the turbulent flame but through tilt affects components transverse to the turbulent flame. The isotropy of the turbulence therefore is modified by the wrinkled flame.

If u_- and v_- are the fluctuating longitudinal and transverse velocities just upstream from the wrinkled flame, then the upstream kinetic energies are $q_- = (\overline{u_-^2} + \overline{|v_-|^2})/2$ per unit mass and $Q_- = \rho_-(\overline{u_-^2} + \overline{|v_-^2|})/2$ per unit volume, while those just downstream are found to be $q_+ = \overline{[u_-^2 + |v_- - \nabla f\, S_L(R-1)|^2]}/2$ and $Q_+ = \rho_+ q_+$, where the instantaneous wrinkled flame shape is described by $x = f(y,z,t)$, x being the coordinate normal to the turbulent flame. When R is large it can dominate the downstream kinetic energy, giving $q_+ \approx R^2 S_L^2 \overline{|\nabla f|^2}/2$ so that the downstream fluctuations are mainly transverse, and the flame increases the turbulent kinetic energy appreciably. Continuity and transverse homogeneity give $\overline{v_-\cdot\nabla f} = \overline{f\partial u_-/\partial x}$, from which stationarity and a Taylor hypothesis give $\overline{v_-\cdot\nabla f} = \overline{u_-^2}/S_L$, an approximation which results in $q_+-q_- = \overline{|\nabla f|^2}\, S_L^2(R-1)^2/2 - \overline{u_-^2}(R-1)$ and in $Q_+-Q_- = \rho_+[\overline{|\nabla f|^2} S_L^2(R-1)^2/2 - (\overline{|v_-|^2} + 3\overline{u_-^2})(R-1)/2]$. Under these conditions, the turbulent burning velocity is approximately $S_T = S_L(1+\overline{|\nabla f|^2}/2) = S_L A$ (where A is the ratio of the wrinkled flame area to the cross-sectional area); the first of these equalities enables $\overline{|\nabla f|^2}$ to be eliminated from the q and Q expressions in terms of S_T/S_L. If the turbulence just upstream from the wrinkled flame were isotropic, then the Taylor hypothesis would give $\overline{|\nabla f|^2}/2 = \overline{u_-^2}/S_L^2$, so that $q_+-q_- = \overline{u_-^2}(R-1)(R-2)$ and $Q_+-Q_- = \rho_+\overline{u_-^2}(R-1)(R-7/2)$, in which it may be recalled that $\overline{u_-^2} = \overline{u_+^2}$; the ratio of a transverse to longitudinal component of downstream turbulent kinetic energy would then be R^2-3R+3, which is unity at $R=1$, achieves a minimum value of 3/4 at $R=3/2$, and exceeds unity for $R>2$.

These changes across the wrinkled flame are further modified in the upstream and downstream zones of hydrodynamic adjustment [74]. So long as the decay of turbulent kinetic energy is negligible over the distance of an integral scale, the changes across the hydrodynamic zones mainly involved isotropy adjustments. Thus the anisotropy introduced by the wrinkled flame need not be seen outside the zones of hydrodynamic adjustment, but the changes in kinetic energy that it produces tend to persist.

The analysis of the zones of hydrodynamic adjustment involves linearization in the ratio of the velocity fluctuations to S_L [74]. Nonlinear hydrodynamic effects can be studied by proceeding to second order in an expansion treating this ratio as a small parameter; the necessary analysis can be performed and has been initiated. Completion of the analysis could provide improved descriptions of influences of premixed flames on turbulence and better expressions for turbulent burning velocities at small and moderate turbulence intensities. The linear analysis [74] already had revealed mechanisms whereby the flame response can amplify or diminish turbulent fluctuations through diffusive-thermal and buoyant effects.

Conditions under which planar flames are not stable need more attention. In real combustors buoyancy seldom is an effective stabilizing agent. There are hydrodynamic instabilities at the larger scales whose influences therefore need to be considered. The turbulence itself involves hydrodynamic adjustments and therefore interacts with these flame instabilities. The turbulence thus may control the instabilities, engendering statistically stationary finite-amplitude structures. Investigations of nonlinear hydrodynamic effects are required for understanding of turbulent flame propagation under such conditions.

8. CONCLUDING COMMENTS

From these discussions it may be seen that significant progress has been made recently in describing flamelet structures of real flames for both premixed flames and diffusion flames. Research in this area continues to be very active, and there is much more to be done, not only with respect to the basic structures, but also with respect to instabilities and influences of strain and curvature.

In recent years some useful new ideas on interactions between reaction sheets and turbulence have been developed. There have been some important new theoretical and experimental results on turbulent burning velocities and on influences of heat release on turbulence. More study is needed on the difficult problem of calculating effects of combustion on turbulent fields.

REFERENCES

1. F.A. Williams, "Turbulent Combustion," Chapter III of The Mathematics of Combustion, (J.D. Buckmaster, editor), Society for Industrial and Applied Mathematics, Philadelphia, PA, 1985, 97-131.

2. J. Abraham, F.A. Williams and F.V. Bracco, "A Discussion of Turbulent Flame Structure in Premixed Charges," Engine Combustion Analysis: New Approaches (S.M. Shahed, editor), Society of Automotive Engineers, Warrendale, PA, 1986, 27-42; SAE Transactions 94, 128-143 (1985).

3. A.O. zur Loye and F.V. Bracco, "Two-Dimensional Visualization of Premixed-Charge Flame Structure in an IC Engine," Society for Automotive Engineers, Paper #870454, SAE Congress, Feb., 1987.

4. H.K. Chelliah and F.A. Williams, Combustion Science and Technology 51, 129-144 (1987).

5. B. Rogg and I.S. Wichman, Combustion and Flame 62, 271-293 (1985).

6. A. Linan and M. Rodriquez, "Combustion and Non Linear Phenomena," (P. Clavin, B. Larrouturou, and P. Pelce, editors), Les Editions de Physique, Les Ulis, 1986, pp. 51-69.

7. B. Rogg, A. Linan and F.A. Williams, Combustion and Flame 65, 79-101 (1986).

8. B. Rogg, Combustion and Flame 65, 113-116, (1986).

9. N. Peters, "Numerical Simulation of Combustion Phenomena," R. Glowinski, B. Larrouturou, and R. Teman, editors, Springer-Verlag, New York, 1986, pp. 90-108.

10. N. Peters and F.A. Williams, Combustion and Flame 68, 185-207 (1987).

11. G. Paczko, P.M. Lefdal, and N. Peters, Twenty-First Symposium (International) on Combustion, The Combustion Institute, Pittsburgh, 1987, to appear.

12. F.A. Williams, "Combustion Theory," Second Edition, Benjamin/Cummings Publishing Co., Menlo Park, CA, 1985.

13. N. Peters and F.A. Williams, "Dynamics of Flames and Reactive Systems," J.R. Bowen, N. Manson, A.K. Oppenheim and R.I. Soloukhin, editors, Vol. 95, Progress in Astronautics and Aeronautics, American Institute of Aeronautics and Astronautics, New York, 1984, pp. 37-60.

14. F.A. Williams and N. Peters, "Dynamics of Reactive Systems, Part I: Flame and Configurations," J.R. Bowen, J.-C. Leyer and R.I. Soloukhin, editors, Vol. 105, Progress in Astronautics and Aeronautics, American Institute of Aeronautics and Astronautics, New York, 1986, pp. 152-166.

15. F.A. Williams, "A Review of Flame Extinction," Fire Safety Journal 3, 163-175 (1981).

16. M.C. Drake, "Stretched Laminar Flame Analysis of Turbulent H_2 and $CO/H_2/N_2$ Diffusion Flames," Twenty-First Symposium (International) on Combustion, The Combustion Institute, Pittsburgh, 1987, to appear.

17. R. Dibble and P. Magre, "Finite Chemical Kinetic Effects in a Subsonic Turbulent Hydrogen Flame," AIAA Aerospace Sciences Meeting, Reno, Nevada, January 1987.

18. H. Tsuji and I. Yamaoka, "The Structure of Counterflow Diffusion Flames in the Forward Stagnation Region of a Porous Cylinder," Twelfth Symposium (International) on Combustion, The Combustion Institute, Pittsburgh, 1969, pp. 997-1005.

19. H. Tsuji and I. Yamaoka, "Structure Analysis of Counterflow Diffusion Flames in the Forward Stagnation Region of a Porous Cylinder," Thirteenth Symposium (International) on Combustion, The Combustion Institute, Pittsburgh, 1971, pp. 723-731.

20. S. Ishizuka and H. Tsuji, "An Experimental Study of Effect of Inert Gases on Extinction of Laminar Diffusion Flames," Eighteenth Symposium (International) on Combustion, The Combustion Institute, Pittsburgh, 1981, pp. 695-703.

21. I.K. Puri and K. Seshadri, "Extinction of Diffusion Flames Burning Diluted Methane and Diluted Propane in Diluted Air," Combustion and Flame 65, 137-150 (1986).

22. M.D. Smooke, I.K. Puri and K. Seshadri, "A Comparison Between Numerical Calculations and Experimental Measurements of the Structure of a Counterflow Diffusion Flame Burning Diluted Methane in Diluted Air," Twenty-First Symposium (International) on Combustion, The Combustion Institute, Pittsburgh, 1987, to appear.

23. I.K. Puri, K. Seshadri, M.D. Smooke and D.E. Keyes, "A Comparison Between Numerical Calculations and Experimental Measurements of the Structure of a Counterflow Methane-Air Diffusion Flame," Combustion and Flame, submitted, 1987.

24. G. Dixon-Lewis, T. David, P.H. Gaskell, S. Fukutani, H. Jinno, J.A. Miller, R.T. Kee, M.D. Smooke, N. Peters, E. Effelsberg, J. Warnatz, and F. Behrendt, "Calculation of the Structure and Extinction Limit of a Methane-Air Counterflow Diffusion Flame in the Forward Stagnation Region of a Porous Cylinder," Twentieth Symposium (International) on Combustion, The Combustion Institute, Pittsburgh, 1984, pp. 1893-1904.

25. J.A. Miller, R.J. Kee, M.D. Smooke and J.F. Grcar, "The Computation of the Structure and Extinction Limits of a Methane-Air Stagnation Point Diffusion Flame," Western States Section, The Combustion Institute, Paper No. WSS-CI84-20, April 1984.

26. N. Peters and R.J. Kee, "The Computation of Stretched Laminar Methane-Air Diffusion Flames Using a Reduced Four-Step Mechanism," Combustion and Flame, to appear, 1987.

27. K. Seshdri and N. Peters, "Asymptotic Structure and Extinction of Methane-Air Diffusion Flames," submitted to Combustion and Flame, 1987.

28. C. Trevino and F.A. Williams, "An Asymptotic Analysis of the Structure and Extinction of Methane-Air Diffusion Flames," to be presented at the Eleventh International Colloquium on Dynamics of Explosions and Reactive Systems, Warsaw, Poland, August 1987.

29. P. Clavin, "Dynamical Behavior of Premixed Flame Fronts in Laminar and Turbulent Flows," Progress in Energy and Combustion Science 11, 1-59 (1985).

30. P.A. Libby and F.A. Williams, "Structure of Laminar Flamelets in Premixed Turbulent Flames," Combustion and Flame 44, 287-303 (1982).

31. P.A. Libby and F.A. Williams, "Strained Premixed Laminar Flames Under Nonadiabatic Conditions," Combustion Science and Technology 31, 1-42 (1983).

32. P.A. Libby, A. Linan and F.A. Williams, "Strained Premixed Laminar Flames with Nonunity Lewis Numbers," Combustion Science and Technology 34, 257-293 (1983).

33. P.A. Libby and F.A. Williams, "Strained Premixed Laminar Flames with Two Reaction Zones," Combustion Science and Technology 37, 221-252 (1984).

34. V. Giovangigli and S. Candel, "Extinction Limits of Premixed Catalyzed Flames in Stagnation Point Flows," Combustion Science and Technology 48, 1-30 (1986).

35. P.A. Libby and F.A. Williams, "Premixed Laminar Flames with General Rates of Strain," Combustion Science and Technology, to appear, 1987.

36. B. Rogg, "Response and Flamelet Structure of Stretched Premixed Methane-Air Flames," Combustion and Flame, submitted, 1987.

37. C.H. Gibson, "Fossil Turbulence in the Denmark Strait," J. Geophys. Res. 87, 8039-8096 (1982).

38. W.T. Ashurst, A.R. Kerstein, R.M. Kerr and C.H. Gibson, "Examination of a Three Dimensional Turbulent Flow Field Obtained by Direct Navier-Stokes Simulation," submitted to J. Fluid Mech., 1986.

39. A.M. Klimov and V.N. Lebedev, "Limiting Phenomena in Turbulent Combustion," Fiz. Gor. Vzr. 19, 7-9 (1983).

40. A. Linan, "Structure and Extinction of Cylindrical Flames," unpublished.

41. P.A. Libby, N. Peters and F.A. Williams, "Cylindrical Premixed Laminar Flames," to be submitted, 1987.

42. J. Chomiak, "The Uniform Distortion of a Turbulent Flame," Combustion and Flame 22, 99-104 (1974).

43. N. Peters, "Laminar Flamelet Concepts in Turbulent Combustion," Twenty-First Symposium (International) on Combustion, The Combustion Institute, Pittsburgh, 1987, to appear.

44. N. Peters, personal communication, 1987.

45. S.H. Sohrab, A. Linan and F.A. Williams, "Asymptotic Theory of Diffusion-Flame Extinction with Radiant Loss from the Flame Zone," Combustion Science and Technology 27, 143-154 (1982).

46. S.K. Liew, K.N.C. Bray and J.B. Moss, "A Flamelet Model of Turbulent Nonpremixed Combustion," Combustion Science and Technology 27, 69-73 (1981).

47. N. Peters and F.A. Williams, "Liftoff Characteristics of Turbulent Jet Diffusion Flames," AIAA Journal 21, 423-429 (1983).

48. N. Peters, "Laminar Diffusion Flamelet Models in Nonpremixed Turbulent Combustion," Progress in Energy and Combustion Science 10, 319-339 (1984).

49. S.K. Liew, K.N.C. Bray and J.B. Moss, "A Stretched Laminar Flamelet Model of Turbulent Nonpremixed Combustion," Combustion and Flame 56, 199-213 (1984).

50. B. Rogg, F. Behrendt and J. Warnatz, "Modeling Turbulent Methane-Air Diffusion Flames: The Laminar-Flamelet Model," Ber. Bunsenges. Phys. Chem. 90, 1005-1010 (1986).

51. B. Rogg, F. Behrendt and J. Warnatz, "Turbulent Nonpremixed Combustion in Partially Premixed Diffusion Flamelets with Detailed Chemistry," Twenty-First Symposium (International) on Combustion, The Combustion Institute, Pittsburgh, to appear, 1987.

52. L.D. Chen and W.M. Roquemore, "Two-Dimensional Visualization and Single-Point Frequency Measurement of Low Reynolds Number Jet Flames," Laser Institute of America ICALEO 58, 16-23 (1986).

53. J.C. Hermanson, M.G. Mungal and P.E. Demotakis, "Heat Release Effects on Shear-Layer Growth and Entrainment," AIAA Journal 25, 578-583 (1987).

54. P. Givi, W.H. Jou and W.R. Metcalfe, "Flame Extinction in Temporally Developing Mixing Layers," Twenty-First Symposium (International) on Combustion, The Combustion Institute, Pittsburgh, to appear, 1987.

55. W.J. Dold, unpublished work (1987).

56. A. Linan, unpublished work (1987).

57. P.A. McMurty, W.H. Jou, J.J. Riley and R.W. Metcalfe, "Direct Numerical Simulation of Reacting Mixing Layer with Chemical Heat Release," AIAA Journal 24, 962-970 (1986).

58. W.T. Ashurst, A.R. Kerstein, R.M. Kerr and C.H. Gibson, "Alignment of Vorticity and Scalar Gradient with Strain Rate in Simulation of Navier-Stokes Turbulence," Physics of Fluids, to appear, 1987.

59. A.R. Kerstein, "Computational Study of Propagating Fronts in a Lattice-Gas Model," Journal of Statistical Physics 45, 921-931 (1986).

60. A.R. Kerstein, "Pair-Exchange Model of Premixed Turbulent Flame Propagation," Twenty-First Symposium (International) on Combustion, The Combustion Institute, Pittsburgh, to appear, 1987.

61. F.C. Gouldin, Combustion and Flame 68, 249-266 (1987).

62. K.O. Smith and F.C. Gouldin, AIAA J. 17, 1243-1250 (1975).

63. K.V. Dandekar and F.C. Gouldin, AIAA J. 20, 652-659 (1981).

64. R.G. Bill, Jr., I. Namer, L. Talbot, R.K. Cheng and F. Robben, Combustion and Flame 43, 229-242 (1981).

65. R.G. Bill, Jr., I. Namer, L. Talbot and F. Robben, Combustion and Flame 44, 277-285 (1982).

66. R.K. Cheng and T.T. Ng, Combustion and Flame 52, 85-202 (1983).

67. F.C. Gouldin and K.V. Dandekar, AIAA J. 22, 655-663 (1984).

68. S. Rajan, J.R. Smith and G.D. Rambach, Combustion and Flame 57, 95-107 (1984).

69. S. Rajan, Combustion and Flame 68, 221-229 (1987).

70. G. Searby, F. Sabathier, J. Monreal, P. Clavin and L. Boyer, "The Feedback of a Flame Front on Turbulent Flows," Dynamics of Flames and Reactive Systems, J.R. Bowen, N. Manson, A.K. Oppenheim and R.I. Soloukhin, editors, Vol. 95, Progress in Astronautics and Aeronautics, American Institute of Aeronautics and Astronautics, New York, 1984, pp. 103-114.

71. P. Clavin and F.A. Williams, J. Fluid Mech. 116, 251-282 (1982).

72. P. Clavin and G. Joulin, J. Physique-Lettres 44, L-1 - L-12 (1983).

73. P. Clavin and P. Garcia-Ybarra, J. Mecanique Teorique et Appliquee 2, 245-264 (1983).

74. G. Searby and P. Clavin, Combustion Science and Technology 46, 167-193 (1986).

75. P. Clavin and G. Joulin, "Theoretical Approaches for the Flamelet Library," France-USA Workshop on Turbulent Reactive Flows, July 1987.

FLAMELET LIBRARY FOR TURBULENT WRINKLED FLAMES

Paul Clavin
Laboratoire de Recherche en Combustion,
CNRS / Université de Provence, S.252 Centre St Jérôme, 13397 Marseille Cedex 13

and

Guy Joulin
Laboratoire d' Energétique et de Détonique,
CNRS / ENSMA, Rue Guillaume VII, 86034 Poitiers Cedex

ABSTRACT

The characteristics of a turbulent combustion in premixed gases are depending on the physico chemical properties of the reactive mixture and on the scales of the approaching turbulent flow compared to the thickness d and the speed u_L of the laminar flame. All the properties of the reactive mixture that are necessary to characterize the turbulent wrinkled flame regime are included in few parameters such as the gas expansion parameter γ, the Markstein number Ma (describing stretch and curvature effects upon the local combustion rate) and the Froude number (when gravity effects are important). The theory of this regime is reviewed with a special emphasis on derivations of local laws and expressions of the Markstein number. Both stable and unstable flame fronts are considered in the wrinkled flame regime. New experimental data concerning the Markstein number are reported.

I- INTRODUCTION

Premixed flames may be considered as nonlinear reaction diffusion waves propagating into a reactive gaseous mixture which is frozen at a composition far from the chemical equilibrium. The final state of the burned gases is in a stable equilibrium state. Temperature and composition varie across the flame thickness d. In planar geometry, the laminar combustion rate is characterized by the flame speed u_L defined as the front velocity of a flame propagating in a quiescent reactive mixture. Typical orders of magnitude are: $d \approx (D_{th} \tau_t)^{1/2} \approx 0.3$ mm, $u_L \approx (D_{th}/\tau_t)^{1/2} \approx 0.5$ m.s^{-1} with a diffusivity $D_{th} \approx$ 1cm^2.s^{-1} (at the flame temperature $T_b \approx 2000$ K) and a transit time $\tau_t \equiv (d/u_L) \approx 6\ 10^{-4}$ s which corresponds to a characteristic time of reaction at T_b of the order of magnitude of 10^{-6}s (see Eq. I.13 below).

Because of the gas expansion across the flame, the density of burned gases ρ_b is smaller than the density of the fresh mixture ρ_u ($\rho_b < \rho_u$) and the streamlines crossing a tilted front are deflected. As a consequence, the dynamics of wrinkled flames involves hydrodynamical phenomena which develop on characteristic lengthscales of the same order of magnitude as the wavelength Λ of the wrinkles of the front. In the case of a flame thickness d sufficiently small compared to the size Λ of the wrinkles ($\varepsilon \equiv d/\Lambda \ll 1$), the variations of density, temperature and concentrations stay confined inside the flame thickness and the flame can be considered as a surface of discontinuity. The combustion problem is thus reduced to an hydrodynamical free boundary problem for incompressible flows with a different density in the upstream and downstream sides of the flame.

1. Local laws

The boundary conditions which are satisfied by the flow fields at the flame front, may be obtained as a result of an analysis of the local structure of the wrinkled flame. As an example, the normal burning velocity u_n which may be properly defined in the limit $\varepsilon \to 0$ as the normal component of the relative velocity of the unburned gases crossing the flame front, may be computed by a perturbation analysis in the framework of simplified flame models. Let $x=\alpha(y,t)$ denote the equation of the flame front (written in the two dimensional case for simplicity), kinematic considerations lead by definition to:

$$u_n = \{ u - \partial\alpha/\partial t - v\partial\alpha/\partial y \}/\{1 + (\partial\alpha/\partial y)^2\}^{1/2} \qquad (I.1)$$

where u and v are the values at the front of the x and y components of the velocity of the upstream flow of fresh mixture. Let Σ be the area of the flame surface, the combustion rate is

$$\int_\Sigma \rho_u u_n d\sigma \qquad (I.2)$$

where $d\sigma$ is the differential elementary area of the flame front. For $\varepsilon = 0$ one has $u_n = u_L$, but as anticipated by Markstein /1/ and Karlowitz /2/, the flame speed is modified ($u_n \neq u_L$) by front curvature and flow inhomogeneities and these corrections may be obtained at the following orders of a perturbation analysis of the wrinkled flame structure in power of ε. Such studies have been carried out recently in the framework of an asymptotic expansion of the solution of simplified models /3-6/. The other jump conditions concerning the flow fields (pressure and velocity) are also obtained as results of the analysis /12-13/. As usual, when a multiscale analysis of an internal structure is used to describe the effects of non zero thickness of an interface upon local relations, the form obtained for such laws depends on the arbitrary choice of the origin inside the thin layer of the interface. Obviously, the final solution for the front dynamics does not depend on such an arbitrary choice. For example, the choice of the origin may be imposed by a prescribed form of one of the local law which is kept valid at all order in the perturbation expansion in powers of ε. For simple flame models with a large reduced activation energy represented by a Zeldovich number large compare to unity, $\beta \gg 1$, a natural choice is to take the reactive sheet as origin /6/. At the first non trivial order, the local analysis of the flame stucture yields the modification of the local burning rate in the form:

$$(u_n/u_L - 1) = \text{Ma} \{ 2d/R + \vec{t}_t \vec{n} . \nabla \vec{u} . \vec{n} \} + O(\varepsilon^2) \qquad (I.3)$$

where R is the mean curvature of the reaction sheet defined to be positive when the front is concave towards the fresh mixture and **n** is the unit vector normal to the reaction sheet oriented towards the fresh gases. $\nabla \mathbf{u}$ is the extrapolation to the reaction sheet of the rate of strain tensor of the upstream flow outside the total flame thikness. The first term in the r.h.s. of Eq. (I.3), describes the curvarure effect. The second is the stretch effect. The dimensionless coefficient Ma, called the Marsktein number (L = d.Ma is the Markstein length), caracterizes the intensity of stretching and curvature effects. Its value is found as a result of the analysis and depends on the diffusive and reactive properties of the mixture. It is predicted to be positive in ordinary cases. Eq. (I.3) can also be written as:

$$(u_n/u_L - 1) = Ma \{\sigma^{-1}\tau_t (d\sigma/dt)\} + O(\epsilon^2) \tag{I.4}$$

where σ is the elementary area of the flame surface. Such results have also be extended to complex chemistry /6-7/.

2. Hydrodynamical effects

Eqs. (I.1) and (I.3) put together yield an equation for the motion of the flame front. But because of the hydrodynamical effects which are produced by the gas expansion, this equation has not a closed form. More precisely, if u_e and v_e are the components of the flow velocity in the absence of the flame, one has :

$$u = u_e + u_i, \qquad v = v_e + v_i, \tag{I.5}$$

where u_i and v_i are the components of the flow velocity which are induced upstream by the gas expansion of the burnt gases. By solving the corresponding hydrodynamical problem in which the flame is considered as a surface of discontinuity located at the reaction sheet, the induced flow u_i and v_i is computed for small ϵ as a functional of the dynamical characteristics of the flame front $x = \alpha(y,t)$. Such a calculation was first carried out indepently by Darrieus (1938) and Landau (1944) in the linear approximation of a front weakly wrinkled around a planar shape and in the limiting case $\epsilon = 0$ where the effects associated with the flame thickness are completly neglected, $u_n = u_L$. As a result, a planar flame front standing perpendicular to a uniform flow ($u_e = u_L$, $v_e = 0$) is found to be unstable for all wavelengths Λ with a linear growth σ of the form:

$$\sigma = a_1(\gamma) u_L K, \tag{I.6}$$

where K is the modulus of the wave vector, $K \equiv 2\pi/\Lambda$. The nondimensional coefficient $a_1(\gamma)$ has the same sign as the gas expansion coefficient $\gamma \equiv (\rho_u - \rho_b)/\rho_u \approx (T_b - T_u)/T_b$. Working in perturbation of ϵ, relations such as Eq. (I.3) must be used, and the first correction in non zero thickness of the wrinkled flame front ($u_n \neq u_L$, ...) introduces a K^2 term in the r.h.s. of Eq. (I.6) with a coefficient depending on the reactive and diffusive properties of the reactive mixture. When this coefficient is positive, the corresponding effects stabilize the short wavelengths and the so called Michelson-Sivashinsky (M.S.) model equation for the front dynamic has been derived in the limit of a weak gas expansion ($\gamma \ll 1$) /8/. At the leading order in a γ expansion this equation takes the following form :

$$\partial\alpha/\partial t = \mathbb{L}(\alpha) - |\nabla\alpha|^2/2 \qquad (I.7)$$

where when conveniently nondimensionalized, the linear differential operator $\mathbb{L}$ can be written in the Fourier representation ($\alpha \rightarrow \int \alpha(Y) \exp(+iKY) dY$) as:

$$\mathbb{L}_k = k - k^2, \qquad (I.8)$$

where k ($\approx$ Kd) is the modulus of the dimensionless wavevector scaled in such a way that k=1 is the marginal case and k=1/2 corresponds to the most unstable wavevector. The nonlinear term in Eq. (I.7) is of a geometrical nature: it comes from a expansion in powers of the gradients of the geometrical term $\{1+(\partial\alpha/\partial y)^2\}^{1/2}$ appearing in the definition of the normal flame speed given in Eq. (I.2). Numerical and analytical studies of this equation have been carried out /8-10/. Cells much larger than the marginal wavelength are observed with sharp cusps. For periodic boundary conditions, a single cell is observed at long times. A defect of the M.S. equation is to not include nonlinearities coming from the advection of the front by the induced flow which are of same order of magnitude as $|\nabla\alpha|^2$ for ordinary values of γ. Nevertheless no new terms appear in (I.7) at the following order in the γ expansion /11/ and one must go further in this expansion. This tedious calculation has not been yet carried out.

When the flame propagates downwards, the gravity acceleration g is a stabilizing mechanism /1/. A detailed analysis of this case has been carried out /12/ and a model equation similar to (1.7) but with a modified linear operator can be obtained for conditions close to the instability threshold:

$$\mathbb{L}_k = -r + k - k^2 \qquad (I.9)$$

where the dimensionless number r is proportional to the inverse of the Froude number $Fr^{-1} = (gd/u_L^2)$. For sufficiently low flame speed ($r > 1/4 \Leftrightarrow U_L < 10$ cm/s), a planar flame front is unconditionally stable. A steady and regular pattern of cells with a size corresponding to the marginal wave number, k =1/2, appear in the unstable domain close to the stability limits. In the stable domain, the linear response of the flame front to an external exitation (u_e, v_e) as a weak incoming turbulence, has been computed /14/.

3. The different combustion regimes of turbulent premixed gases.

In turbulent flows there are different combustion regimes and flame fronts are meaningful only for particular conditions. Several authors have supplied phase-diagrams to illustrate the different combustion regimes of turbulent premixed gases. We will present here the Barrère-Borghi diagram /18-19/ which is based on the Kolmogorov dimensional analysis of the fully developed turbulence. According to this analysis, the following relations hold between the integral and the Kolmogorov scales at which dissipation takes place

$$l_{Ko}/l_\Lambda = Re^{-3/4}, \quad u'_{Ko}/u'_\Lambda = Re^{-1/4}, \quad \tau_{Ko}/\tau_\Lambda = Re^{-1/2}, \tag{I.10}$$

l, u' and $\tau \equiv l/u'$ are the characteristic length, velocity fluctuation and time. Subscrits Ko and Λ refer to Kolmogorov and integral scales respectively. The Reynolds number based on the integral scales is

$$Re = u'_\Lambda l_\Lambda / \upsilon. \tag{I.11}$$

Different cases may be considered depending on the magnitude of the scales of the turbulence compared to those of the laminar flame structure d, u_L and τ_t. Two limiting cases are particularly instructive:

-a) When the longest characteristic timescale in the spectrum of the turbulent flow (i.e. the integral time τ_Λ) is shorter than the reaction time τ_r at the combustion temperature,

$$\tau_\Lambda \ll \tau_r, \tag{I.12}$$

the reaction zone is homogeneously distributed all over the volume of the combustion chamber as in well stirred reactors. Here the integral time τ_Λ plays the role of the resident time. In the framework of simplified models the transit time τ_t and the reaction time τ_r are related together through

$$\tau_r \approx \beta^{-2} \tau_t, \tag{I.13}$$

where β is the Zeldovich number ($\beta \approx 10$). According to the small values of the reaction time ($\tau_r \approx 10^{-6}$s.) at the temperature of combustion T_b such an extreme case is not easily produced in ordinary combustion chambers.

-b) When the shortest timescale in the turbulent spectrum i.e. the Kolmogorov time, is larger than the laminar transit time ($\tau_{Ko} > \tau_t$), the smallest structures of the turbulence are larger than the flame thickness d,

$$\tau_{Ko} > \tau_t \Rightarrow l_{Ko} > d \quad \text{and} \quad u_L > u'_{Ko}. \tag{I.14}$$

As the Prandtl number is of order unity in gases, $Pr = D_T/\upsilon \approx 1$, (I.14) is a direct consequence of the scaling of the flame structure, $u_L d / D_T \approx 1$, and of the definition of Kolmogorov scales:

$$l_{Ko} u'_{Ko}/\upsilon = 1. \tag{I.15}$$

When (I.14) is verified, laminar flamelets must exist. These flamelets are flapping back and forth under the action of the fluctuations of the velocity field. This regime is called the flamelet regime.

The different regimes of turbulent combustion may be presented in the Barrère-Borghi diagram

(see Fig. 1) in which the reduced integral scales of the turbulence are plotted on the coordinate axes, $X = (l_A / d)$ and $Y = (u'_A / u_L)$. According to (I.10) and (I.11) and to the scaling of the flame structure, one has

$$(l_{Ko}/d) = (X/Y^3)^{1/4} \quad \text{and} \quad (\tau_A/\tau_t) = (X/Y). \tag{I.16}$$

-i) The domain of the flamelet regime defined by (I.14) corresponds to

$$Y < X^{1/3} \quad \text{and} \quad X > 1. \tag{I.17}$$

The corresponding curves are plotted in figure 1. The flamelet regime is subdivised into two parts
   -i-a) For sufficiently weak turbulence $u'_A < u_L$ (Y<1), the interaction time l_A/u_L between a coherent structure and the flame front is smaller than the turnover time l_A/u'_A and one is in the wrinkled flame regime characterized by a single wrinkled flame sheet without pockets and corrugations which can be represent by an equation $x = \alpha(y,t)$.
   -i-b) For a sufficiently high turbulence intensity, $u'_A > u_L$ ($X^{1/3}>Y>1$), corrugations and pocklets appear. This is the corrugated flame regime.

   -ii) According to (I.13), the well stirred reactor regime with a distributed reaction zone which is defined by (I.12) corresponds in figure 1 to

$$Y > \beta^2 X. \tag{I.18}$$

   -iii) In between the flamelet and the well stirred reactor regimes (see Fig. 1), one may represent the turbulent combustion as thick fronts whose internal structure depends on the turbulent mixing associated with vortices of the turbulent flow smaller than the flame thickness. This regime is called the thick turbulent flame regime.

   Typical conditions of industrial combustion chambers are $u'_A \approx 10$m/s, $l_A \approx 0.1$m (Re $\approx 10^4$ for $\upsilon \approx 1$ cm^2/s at the combustion temperature). This correspond to the boundary between the flamelet and the turbulent thick flame regimes:

$$l_{Ko} \approx 10^{-2} \text{cm} \approx d, \quad \text{and} \quad \tau_{Ko} \approx 10^{-4}\text{s} \approx \tau_t. \tag{I.19}$$

4. Turbulent wrinkled flames
   In the flamelet regime (l >> d) the flame may be considered as surface of discontinuity. In order to predict the overall turbulent combustion rate, one must be able to relate the statistical properties of the flame front to those of the incoming turbulent flow (u_e, v_e). One has to solve a free boundary problem of an hydrodynamical nature. The problem is simpler in the wrinkled flame regime ($u'<u_L$). As explained before, one of the main difficulties is due to the gas expansion: the wrinkled flame front

produces an induced flow (u_i, v_i) which participates to the motion of the front described by Eqs. (I.1) and (I.3).

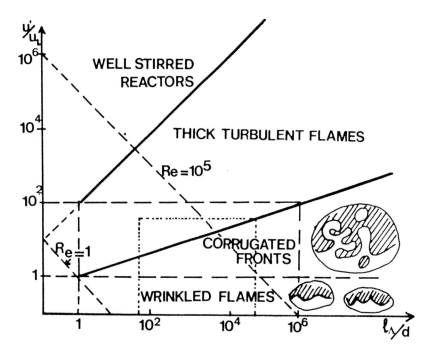

Figure 1. Barrère-Borghi diagram in coordinates Log (u'/u_L) versus Log (l_Λ/d). The accessible domain in ordinary conditions of industrial combustion chamber is delimited by a dotted line

In the case of a turbulent flame front with a planar shape on an average and standing perpendicular to the mean flow, it may be shown /3-4/ that, for homogeneous incomming turbulence, the turbulent flame speed U_{tur} is given by the product of the laminar flame speed with the mean relative increase of the flame surface as anticipated by Damköhler in 1940,

$$U_{tur} = u_L \langle \Sigma \rangle. \tag{I.20}$$

where Σ is the area of the reaction sheet by unity of cross-sectional area and where the brackets $\langle \rangle$ denote the time average. In the wrinkled flame regime, Eqs (I.1-4) yield

$$U_{tur} = u_L \langle (1 + |\nabla \alpha|^2)^{1/2} \rangle. \tag{I.21}$$

On an average, the effects of front curvature and flame stretch described by Eq. (I.3-4) are, most of the time, negligible compared to the surface increase in the expression of the turbulent flame speed. Nevertheless even weak, these effects control the dynamics and the shape of the flame and are important to determine the mean flame surface and thus the turbulent flame speed. This is particularly clear for the model equations (I.7-9). Thus, in the wrinkled flame regime one of he main problem is to relate the averaged area of the flame front to the characteristics of the incoming turbulence.

The problem has been completly solved in the linear approximation for stable fronts /14/ as those described by Eq.(I.9) for which gravity is a stabilizing mechanism. Experiments have been devoted to this case /15/ as well as to the determination of the stability limits /16-17/. The relative magnitude of the characteristic lenght scale of the incoming turbulent flow compare to the Markstein length is essential for the turbulent combustion rate. For sufficiently large turbulent length scales, the turbulent intensity is damped out when approaching the flame front, and the combustion rate is the same as for the planar laminar flame. At the opposite, the turbulent intensity can be amplified when the characteristic turbulent length scale becomes of the same order of magnitude as the Markstein length. When the Froude number approaches its critical value, a resonance appears for a length scale close to the wavelength corresponding to the maximum of the r.h.s. of Eq.(15). Such experiments are useful to obtain experimental data concerning the Markstein number Ma (see § III).

It is one of the purpose of the present paper to review recent experimental works concerning measurements of the Markstein number carried out at Poitiers /20/ and at Marseilles /21/. Preliminary results are presented in § III. Before, a general derivation of the local laws free from detailed chemical kinetic models is presented in § II.

When an unstable premixed flame propagates into a weakly turbulent flow, irregular and chaotic cellular patterns appear on the flame front as a result of an interaction between the Darrieus Landau instability described by Eqs.(I.7-8) and the external noise constituted by the fluctuations in the velocity of the approaching turbulent flow. This regime which we will called the "cusped flame regime" is indeed the generic case of the turbulent wrinkled flame regime. It was studied experimentally for sherical flames expanding in turbulent flows /22/ and in a V-shape turbulent flame stabilized by a flame-holding device /23/. A rough analysis of the cusped flame regime which is based on the ideas developed by Zeldovich et al./24/ for the anomalous stability of curved fronts in channel, is presented in § IV.

II - LOCAL EFFECTS

1. The mechanisms involved

In the wrinkled flame regime, a flame front propagating in a turbulent flow is curved and encounters a non-uniform flow. Furthermore, as happens in internal combustion engines, the thermodynamical properties of the fresh mixture - albeit uniform - may change with time.

To begin with, we consider the effects of curvature, by assuming that the umpstream flow is uniform and (provisionnally) that the streamlines are not deflected when crossing the flame (Fig.2).

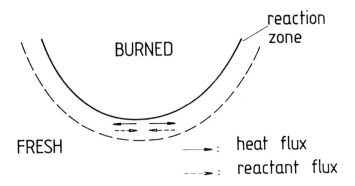

Figure 2. Sketch of a curved flame front.

As a consequence of the transverse heat and reactant fluxes generated by the corresponding gradients (via the curvature), the mixture is locally cooled (or heated) or enriched (or depleted) depending on the sign of the curvature and on the departure of the Lewis numbers (Le_i) from unity. The reaction temperature, hence the chemical rates and u_n, are modified accordingly. Even when $Le_i=1$, the heat flux at the entrance of the reaction zone is no longer balanced by normal convection, since geometry is also involved; when geometry changes, so does u_n.

Similar effects appear in flat flames stabilized in stagnation flows, even if the combustion is assumed not to affect the fluid mechanics. Owing to the streamline divergence, the velocity component normal to the flame depends on the normal distance; heat and species are then bound to behave in different ways, if $Le \neq 1$, with the same consequence as above, i.e. $u_n \neq u_L$. Even if $Le \neq 1$, u_n has to differ from u_L, because the heat flux issuing from the reaction zone is a constant and the streamline divergence modifies the convection/conduction balanced.

These influences of the shape of the front and of the structure of the upstream flowfield on u_n would exist if there were no feed-back of combustion on fluid mechanics. In actual flames, which are nearly-isobaric, the changes in mixture density due to the temperature rise induce streamline deflections. A deflection which would be constant along a locally flat flame could be absorbed locally by an adequate change in frame of reference. Only tangential variation in flow deflection can affect the flame structure, and this is possible only if the flame is curved and/or the upstream flowfield is non-uniform (Figs.3a and 3b). Then the non-uniform hydrodynamical effects taking place inside the flame lead to further inbalances between convection, conduction and the diffusive processes, there by modifying the response of u_n to curvature and to upstream rates of strain.

Finally unsteadiness can also contribute to changes in u_n, in three ways. The first, obvious, consequence of temporal changes in the upstream pressure and density (as occurs in internal combustion engines) is a modification of u_L itself. Furthermore, unsteadiness implies transient storages of mass, heat and reactant inside the flame thickness. Finally, a change in the ambiant pressure leads to a work inside the flame (the dP/dt term remains in the energy equation even for low - Mach number flames), thereby modifying the energy balances, the temperature, the chemical rates... and u_n.

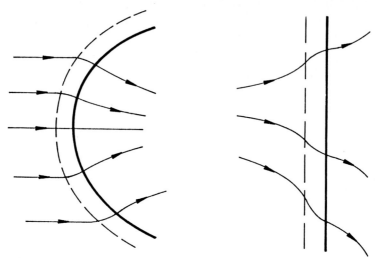

Figures 3. a) Deflection of the streamlines for a curved front in a uniform flow of fresh mixture
b) Deflection of the streamlines for a planar front in a sdtagnation point flow.

2. Phenomenological laws

As explained in the preceding paragraph, a "local law" for the combustion rate is expected to hold along the flame with the following typical form:

$$u_n = u_L \ F\{ \text{ flame shape, flowfield, time-dependence, mixture composition } \}. \qquad (II.1)$$

The functional $F\{\ \}$ is not exactly known in the general case...and is bound to remain so for a while. We present here a phenomenological way of getting an approximation to it, upon use of the following assumptions :
- i) all the lengths involved in the flame sheet geometry and in the outer flow structure are $O(d/\varepsilon)$
- ii) the corresponding time-scales are $O(d/\varepsilon u_L)$
- iii) the physico-chemical properties of the upstream mixture are uniform and isotropic
- iv) the result cannot depend on any particular frame of reference.

Here, $\varepsilon = d/\Lambda$ is a small, non-dimensional, paranemeter, and $d = D_{th}/u_L$ is the conduction / convection length based upon u_L and the heat diffusivity D_{th}. We shall restrict ourselves to finding $u_n - u_L$ within $O(\varepsilon^2 u_L)$ errors. Being a scalar, u_n can be written as :

$$u_n = u_L + (S_1 + S_2 + ...) + O(\varepsilon^2 u_L) \qquad (II.2)$$

where the S_i's are $O(\varepsilon u_L)$ scalars ; computing u_n within the desired accuracy is equivalent to finding them from the "entities" bracketed in (II.1).

Let us begin with the flame-sheet geometry. It is locally characterized by the rank-2, curvature tensor C and by the higher order tensors deduced from it by covariant differenciations. Because according to assumption -i), the latter have $o(\varepsilon/d)$ elements, only C can contribute. However it cannot appear as a tensor. From C it is nevertheless possible to extract two independent scalars, viz : $\det(C) = O(\varepsilon^2/d^2)$ and $\operatorname{trace}(C) = O(\varepsilon/d)$; hence the first scalar :

$$S_1 \propto \operatorname{trace}(C) = (1/R_1) + (1/R_2)$$

where the R_i's are the flame-sheet principal radii of curvature. This gives the "Markstein" contribution to u_n.

We next consider the upstream flow structure, which is characterized by the velocity vector field $\mathbf{u}$, the rate-of-strain tensor $\nabla \mathbf{u}$ and higher order tensors obtained from it by differentiation. Because of assumption ii), only $\nabla \mathbf{u}$ can contribute , but not as it is. From the rank-3 tensor $\nabla \mathbf{u}$ one can extract three independent scalars, upon noticing that if fulfills its own characteristic equation (the so-called Cayley-Hamilton theorem). Among them, only $\operatorname{trace}(\nabla \mathbf{u}) = O(\varepsilon u_L/d)$ has the required magnitude, hence :

$$S_2 \propto \operatorname{trace}(\nabla \mathbf{u}) = \operatorname{div} \mathbf{u}$$

We further note that scalars can be built from the time variations in the density of the upstream frozen medium, viz $d\operatorname{Log}\rho/dt = - \operatorname{div} \mathbf{u}$, and from the change in ambiant pressure $P(t)$, hence :

$$S_3 \propto d\operatorname{Log}P/dt$$

We also note that the scalars obtained upon time - or space - differentiations of $u_n - u_L$ are too small (assumption ii) ; those coming from du_L/dt can be expressed in term of S_2 and S_3. We finally have to consider the scalars which one can build from $\nabla \mathbf{u}$ and vectors, by contraction. $\mathbf{u}$ itself can not be used, since otherwise the result would not be Galilean-invariant, nor the vectors obtained from $\mathbf{u}$ by differentiations (e.s. $\partial \mathbf{u}/\partial t$, **curl u**), since they are too small (assumptions (i)(ii)). The only vector one is left with is the unit vector n normal to the flame sheet ; hence :

$$S_4 \propto \mathbf{n}.\nabla \mathbf{u}.\mathbf{n}$$

which gives the "Karlowitz" contribution. Putting all the pieces together in (II.2), one obtains

$$(u_n - u_L) = L_1 u_L (R_1^{-1} + R_2^{-1}) + L_2 \mathbf{n}.\nabla \mathbf{u}.\mathbf{n} + L_3 \operatorname{div} \mathbf{u} + L_4 d\operatorname{Log}P/dt \qquad (II.3)$$

as an approximation to (II.1).The phenomenological coefficients $L_1...L_4$ have the dimension of a length, and we call them the Markstein lengths. One introduces the Markstein numbers $Ma_1....Ma_4$

defined as

$$Ma_i = L_i / d \quad (II.4)$$

Along with u_L, they cannot be computed phenomenologically, and have to he regarded as physico-chemical properties of the mixture. The L_i's and Ma_i's are configuration-independent, they may however depend on time, through the variations in pressure and in u_L. For flames propagation in a constant-ρ and constant-P fresch medium, the last two contributions to u_n-u_L disappear. As for confined flames, such as in internal combustion engines, they may however be worth taking into account.

3. Results from models

To the best of our knowledge, the problem of giving theoretical estimates of Ma_3 and Ma_4 has so far not been addressed. We thereafter restrict our attention to flames propagating in constant-density, constant-pressure mixtures. Finding Ma_1 and Ma_2 from realistic, detailed flame models has been done only on particular cases /7/, mainly because the realistic chemical kinetic schemes are too indricate. Estimates of the Markstein numbers Ma_i have theoretically been provided by considering simple, analytically tractable models. Even though the results have been generalized to account for two-step or few-step chemical schemes /6/, most of them have been derived by assuming that the whole combustion process can be modelled by a one-step overall reaction of the Arrhenius type, and by assuming that the activation energy is large compared to the thermal one. The results obtained in this way have the form given in Eq. (I.3) /1/ i.e. the same structure as (II.3), but with Ma_1 = Ma_2 (= Ma). Furthermore, such theories provide with theoretical estimates of the Markstein number Ma. First obtained in the context of one reactant model whith assigned constant diffusive properties of the mixture /3//4/, the estimates of Ma furnished by the asymptotic theories have later been improved to account for : the depletion of the abundant primary reactant, such as O_2, the temperature - and comparition-dependence of the transport coefficients, the dilution effects, the thermal diffusion of the reactants /5/ and, recently, non chain branching reactions /7/. As far as adiabatic flames were concerned, the form (I.3) was always obtained. Non adiabatic case has also been considered /30/.

We present here a way of obtaining estimates of the Markstein numbers Ma_i, which includes the results of the corresponding asymptotic theory but is a bit more "transparent" and at least as accurate. Constitutive laws such as (II.3) hold for moving flames, but it is convenient to look at steady configurations to comute Ma_1 and Ma_2. We thus have to study :

$$\nabla.(\rho \mathbf{u}) = 0 \quad (II.5)$$
$$\rho C_p \mathbf{u}.\nabla T = \nabla.(\lambda \nabla T) + Qw \quad (II.6)$$
$$\rho \mathbf{u}.\nabla y = \nabla.(\rho D \nabla y + \nabla.(\rho Dsy \nabla(\log T))- w \quad (II.7)$$

Besides the usual symbols ρ, T, C_p, λ we use **u** to denote the current velocity vector and w for the reaction term. y is the reactant mass fraction, D is its coefficient of binary diffusion into the diluent, and the Soret coefficient s accounts for its thermal diffusion into the diluent. As suggested by asymptotic theories /5/, the composition-induced changes in molecular weight, transport coefficients and heat capacity can be numerically neglected in calculation of Ma, when the reactive mixture is higly diluted in nitrogen. We therefore only retain the variations with temperature, and assume that the Lewis number and the Soret coefficient are constant. We further assume that all the chemical activity is concentrated along a reaction-sheet, which we use to define the flame front location. Specifically, we replace w by a suitable surface δ function, such that the normal heat flux undergoes the jump :

$$\left[\lambda \frac{\partial T}{\partial n}\right] = -\left(2 Le.B.C_p \lambda \rho_b^2 \sqrt{T_b} \frac{R^2 T_b^4}{E^2} e^{-\frac{E}{RT_b}}\right)^{1/2} \quad \text{(II.8)}$$

accross the reaction-sheet, along which y vanishes. Here B is a temperature - independent pre-exponential factor, Le = $\lambda/\rho DC_p$ is the Lewis number of the reactant - diluent pair and the subscript "b" refers to quantities evaluated as the local reaction temperature T_b. The above expression has been selected to yield the same value of u_L as that given by the leading order result of an asymptotic method. We note that the $T^{1/2}$ dependence of the collision frequency is accounted for and that a quadratic dependence on ρ has also been retained to yield a value of u_L which is pressure-independent (as is nearly the case in experiments on lean propane-air flame /3/. Although (II.8) leads to the same u_L as leading order analysis which postulates $E/RT_b \gg 1$, and the form of (II.8) is suggested by such analyses and rely on them, we do not need here to make particular assumptions on the values of E/RT_b, hence on Le-1 and s. If needed (II.8) could he improved by including the contribution of the next order result of an asymptotic analysis of the reaction zone. Usint (II.8) can therefore give information on the accuracy of asymptotic analyses when neither RT_b/E nor Le-1 nor s are very small quantities and thus could be useful for mixtures involving heavy fuels and/or hydrogen. To compute the two Markstein number M_1 and M_2, it is enough to study two simple, independent steady configurations.

We first examine an adiabatic spherical steady flame stabilized around a point-source of fresh mixture. Then the radical velocity component u is given by $\rho u = m/r^2$, where r is the current radius and m is the mass flow rate of injected mixture. Upon integration of (II.6, II.7), with w=0, over the entire upstream zone, the following results are found :

$$\lambda \frac{\partial T}{\partial r}\bigg|_{r_F^-} = \frac{m\, C_p(T_b - T_u)}{r_F^2}$$

$$\rho D \frac{\partial y}{\partial r}\bigg|_{r_F^-} = -\frac{m\, y_u}{r_F^2}$$

where y=0 has been used at the reaction-sheet location $r=r_F$. From the above fluxes, it is readily shown that T_b is the adiabatic combustion temperature $T_{ad} = T_u + Qy_u/C_p$, and that the burning velocity $u_n = m/r_F^2 \rho_u$ equals u_L for any value of the flame radius r_F. For this to be compatible with (II.3) for $r_F \gg D/u_L$, one should have :

$$u_L = u_L + \frac{2u_L}{r_F}(L_1 - L_2) + o\left(\frac{u_L d}{r_F}\right)$$

because $(1/R_1 + 1/R_2) = -2/r_F$ and $\mathbf{n}.\nabla U.\mathbf{n} = -2u_L/r_F$ for the present spherical flame. Therefore :

$$L_1 = L_2 \quad (= L) \tag{II.9}$$

and the form (I.3) is recovered, provided the flame front is identified with the reaction-sheet.

To compute L, we next study an adiabatic flat flame stabilized far enough upstream of a stagnation point. By using the momentum equations, we may then write :

$$\frac{d(\rho u)}{dn} = \rho \left.\frac{du}{dn}\right|_{-\infty} + \ldots\ldots \tag{II.10}$$

where n is the a streamwise coordinate normal to the flame and counted from the reactio-sheet location, and $du/dn|_{-\infty} = o(u_L/d)$ is the far-upstream) normal rate of strain. The equations to be solved are now :

$$\rho u C_p \frac{dT}{dn} = \frac{d}{dn}(\lambda \frac{dT}{dn})$$

$$\rho u \frac{dy}{dn} = \frac{d}{dn}\left(\rho D \frac{dy}{dn} + \rho D s \frac{dLogT}{dn}\right) \tag{II.11}$$

in the upstream zone we can integrate (II.11) and use (II.10) to give :

$$\rho_u u_n = (\rho u)_{n=0} - \left.\frac{du}{dn}\right|_{-\infty} \int_{-\infty}^{0} (\rho - \rho_u) dn \ldots\ldots \tag{II.12}$$

$$(\rho u)_{n=0} C_p(T_b - T_{ad}) = \left.\frac{du}{dn}\right|_{-\infty} \int_{-\infty}^{0} \rho(C_p(T-T_u) + (y-y_u)Q) dn + \ldots$$

$$\left.\lambda \frac{dT}{dn}\right|_{n=0_-} = (\rho u)_{n=0} C_p(T_b - T_u) - \left.\frac{du}{dn}\right|_{-\infty} \int_{-\infty}^{0} \rho C_p(T-T_u) dn + \ldots$$

We further realize that all the above integrals can be evaluated at the leading order in the limit $(du/dn)_{-\infty}/(d/u_L) \to 0$, ($\varepsilon \to 0$) by using the structure of the planar, stretch-free flame. Indeed, in the latter case, one may write $dn = \lambda dT / \{\rho_u u_L C_p(T-T_u)\}$ and $y = y_a \psi(\theta)$, where $\theta = (T-T_u)/(T_{ad}-T_u)$ and ψ is given by :

$$\frac{d\psi}{d\theta} = Le\frac{\psi-1}{\theta} - \frac{s\psi\gamma}{1+\gamma(1-\theta)} \qquad (II.13)$$

$$\psi(1) = 0$$

Here, and in all the subsequent formulas, $\gamma=(T_{ad}-T_u)/T_{ad}$. Using this in (II.12) then in (II.8), finally yields an expression for u_n-u_L as a function of $du/dn|_{-\infty}$. Upon identification with (I.3) one deduces the Markstein number, owing to (II.9)

$$Ma = J(\gamma) + \frac{\beta_{eff}}{2}\kappa(Le, s, \gamma) \qquad (II.14)$$

where:

$$J(\gamma) = \int_0^1 \frac{\Lambda(\theta,\gamma)}{1+\gamma(1-\theta)} d\theta$$

$$\kappa(Le,s,\gamma) = \int_0^1 \frac{\Lambda(\theta,\gamma)(1-\gamma)}{1+\gamma(1-\theta)}\left(1 + \frac{\psi-1}{\theta}\right)d\theta$$

$\psi(\theta)$ being given by (II.13) and Λ standing for λ/λ_u when it is expressed in terms of θ and γ. The "effective" Zeldovich number measures the fractionnal changes in $[\lambda\partial T/\partial n]$ when T_b varies about T_{ad}:

$$\beta_{eff} = \beta - 2 + \frac{T_{af}T_a}{T_{ad}}\left(\frac{\partial Log\lambda/\lambda_u}{\partial Log T_{ad}}\right) + \frac{3}{2}$$

where $\beta = (T_{ad}-T_u)E/RT^2_{ad}$ is the "usual" Zeldovich number.

The above expression of Ma can be compared to the of the corresponding asymptotic analysis performed in the distinguished limit $\beta\to\infty$, $\beta(Le-1) = O(1)$, $\beta s = O(1)$, namely /5/:

$$Ma|_{asymp.} = J(\gamma) + \frac{\beta(Le-1)}{2}\frac{\partial\kappa}{\partial Le}(1,0,\gamma) + \beta s\frac{\partial\kappa}{\partial s}(1,0,\gamma) \qquad (II.15)$$

which is merely an expansion of (II.14) in the above mentionned distinguished limit. Thus it may be expected that (II.14) could be more accurate than (II.15) when not-so-small values of Le-1 or/and s are involved (e.g. lean C_3H_8/air or H_2/air flames) and when β is not that large. Aboud the Zeldovich number it is worth realizing that $\beta_{eff}-\beta$ is often comparable in magnitude to the uncertainty to which β_{eff} is known from the values of E available in the litterature. Improving (II.8) by the inclusion of $O(1/\beta)$ corrections may thus be unwarranted; another consequence of the uncertainty on β is that asymptotic theories - which neglect the difference between β and β_{eff} - are not, much in error about this particular point. One may also note that any variation of λ/λ_u or of ρ/ρ_{ad} ($\equiv (1+\gamma(1-\theta))^{-1}$ here) with composition could easily be implemented in (II.14) by introducing extra equations analogous to

(II.13) for the abundent reactant and the main products. Similarly, (II.14) can be generalized so as to include the stoichiometry effects on the reaction "intensity" ; the analysis is available, but the results are too lengthy to be reproduced here and will be published elsewhere when very accurate experimental tests will be available. Here, we only quote the corresponding value of Ma as is given by asymptotic methods /5/ :

$$Ma = J(\gamma) + \beta(Le_g - 1)\frac{\partial \kappa}{\partial Le}(1,0,\gamma) + \beta s_g \frac{\partial \kappa}{\partial s}(1,0,\gamma)$$

in which any "global" quantity q_g is defined by :

$$q_g = q_A + (q_B - q_A)(2 + \beta(\phi - 1))$$

ϕ being the equivalence ratio and the subscript "A" (resp."B") referring to the deficient (resp. abundant) reactant.

In any cases, both Eqs. (II.14) and (II.15) indicate that Ma is positive and larger than unity for most hydrocarbon/air mixtures, by contrast to what follows from thermal-diffusional ($\gamma \to 0$) models of adiabatic flames, as a result of the hydrodynamical effects induced by the density changes within the flame front. This means that as anticipated in Eq. (I.8), the second order effects $O(\varepsilon^2)$ are usually stabilizing at short wavelength. Apart from some "exotic" mixtures specially prepared to test the functional form of (II.14), the only "natural" exception are lean H_2/air mixtures.

4. Stretch-dependent u_n, or not ?

We consider Eq.(I.3) once again :

$$u_n = u_L - u_L L \left(\frac{1}{R_1} + \frac{1}{R_2}\right) + L\,\mathbf{n}.\nabla\mathbf{u}.\mathbf{n} + \ldots\ldots$$

where $u_n = \mathbf{n}.\mathbf{u} - D$, D being the normal flame speed in the laboratory frame. Recall that $\nabla\mathbf{u}$ and $\mathbf{u}$ are the upstream rate-of-s/rain tensor and the upstream flow velocity evaluated at the flame front. It has also to be recalled that (I.3) holds when the flame-front is identified with the reaction-sheet ; we thereafter refer to this as "definition #1".

We now consider a "definition #2" of the front location : The "new" flame front is defined by locally shifting the reaction sheet by an amount L, counter streamwise and along the normal $\mathbf{n}$. In other words, to a point labelled by the vector $\mathbf{F}$ according to definition # 1, there corresponds a point labelled by $\mathbf{F}-L\mathbf{n}$ in the second case (Fig.4). Provided $\nabla\mathbf{u} = O(\varepsilon u_L/d)$, with $\varepsilon \ll 1$, and the higher order tensors are smaller, one may Taylor-expand $\mathbf{u}$ and $\nabla\mathbf{u}$ to show that :

$$\mathbf{u}|_F = \mathbf{u}|_{F-L\mathbf{n}} + L\,\mathbf{n}.\nabla\mathbf{u}|_{F-L\mathbf{n}} + \ldots$$

$$\nabla u|_F = \nabla u|_{F-L\,n} + \ldots$$

One can further show that **n**, D and $(1/R_1 + 1/R_2)$ are unaffected by the change in definition, at the relevent leading order(s). Therefore :

$$(u.n - D)|_{F-nL} = u_L - u_L\left(\frac{1}{R_1} + \frac{1}{R_2}\right)L + \ldots \qquad (II.16)$$

and the new u_n turns out to be stretch-independent !

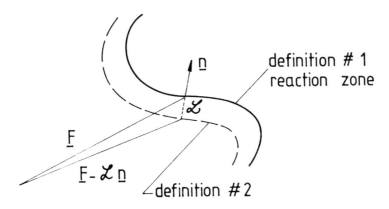

Fig. 4 Definitions #1 and #2 of the front location

Any other definition (1000-K isotherm...) could be used as well, but would lead a different result from (I.3) and (II.16). Using the locus of minima of **u.n** (when any) would even alter the structure of (II.3) by bringing in "stretch-dependent" Ma. It has to be stressed that the above, strange property holds in the limit where ∇u is weak. It cannot be ignored in any theoretical, numerical and experimental accurate determination of $u_n - u_L$, and of the Markstein length(s) by direct measurements. Increasing the value of $n.\nabla u.n$ would not make life easier. The next question has of course to be answered : at which speed does the flame move in the laboratory of reference, given that u_n is definition-dependent ? It must be realized that definitions #1 and #2 refer to different discontinuous idealizations of the same flame : they have different local u_n's but, at each instant, they encounter the upstream flow at different locations (Fig.5), so that the value of D is unchanged. A last remark is in order. Knowing u_n is not enough to compute D and the flame motion at the desired accuracy. One also has to find the upstream flowfield to the same accuracy. To this end, one needs jump relations across the flame front ; in the case of definition #1, they have been evaluated in references /13/ end /14/, and they differ from the usual Rankine-Hugoniot relationship. For the sake of argument, we consider the jump in normal mean flow. Let :

$$J = [\rho(u.n-D)]_-^+$$

denotes its value corresponding to definition #1. When definition #2 is used instead, one obtains :

$$[\rho(u_n-D)]_-^+ = J - L n.[\rho\nabla u]_-^+.n + ...$$

The interpretatin of this result is obvious on figure (5) is considered. It is true that one may choose u_n as stretch-dependent (e.g. in /32/) or stretch-independent (e.g. in /33/), according to one's own convenience, but once the choice is made, the other jump relations are fully determined. This should be kept in mind when using (I.3) or (II.16) in the numerical simulations of discontinuous flame propagating in turbulent media.

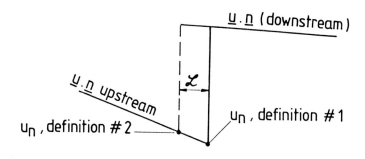

Figure 5. Skech of the outer flow structure along the normal to a stretched flame

III. EXPERIMENTAL TESTS

Two classes of tests of (I.3) or (II.3) can be distinguished, namely :

- <u>indirect tests</u> : they consist of determining the Markstein number by a comparison between a Ma-dependent overall property of the flame, and the same Ma-dependent property as is given by an extra theory which uses (I.3) and the corresponding jump relations /12//14/.

- <u>direct tests</u> : they consists of measuring the three quantities u_n, $(1/R_1 + 1/R_2)$, $n.\nabla u.n$ in different configurations, then trying to correlate them to test (I.3) or (II.3).

1. Indirect test

As explained in details in §I, the local laws giving u_n are important ingredients of the flame dynamics. As a consequence, the Markstein length(s) play a role in the flame stability or response to the incoming flow. At the threshold corresponding to the appearance of cells on a flat flame propagating downward, both the flame velocity u_L and the cell wavelength depend on the Markstein length /12/. Measuring the former can give estimates of the latter. Quinard /16/ performed such measurements of cell wavelengths by assuming that the one step model and the stability theory are accurated enough. For C_3H_8-N_2-O_2 mixtures, he obtained the results presented in fig. 6. It must be recalled that the corresponding procedure is applicable only at the threshold for instability. Another kind of indirect measurement has very recently been developed /21/. It consists of studying the shape of a steady, stable, 2-dimensional flame propagating downward, when a steady parallel, spatially-periodic shear flow is superimposed to the uniform incoming flow (see pictures in reference /14/). Specifically, the Ma-dependent transfer function, as is analytically computed /14/, is used to estimate the Markstein number by a comparison with the experimental data. The preliminary results are not yet completely satisfactory and new experiments are in process.

2. Direct tests

No additional theory seems to be needed here : measuring $(1/R_1 + 1/R_2)$, $\mathbf{n}.\nabla\mathbf{u}.\mathbf{n}$ and u_n in different configurations is apparently enough to determine whether (I.3) or (II.3) applies, and to determine the unknown parameters L_1, L_2, u_L. The following remarks must be put forward, however : L and Ma are definition-dependent, thus, the same flame front definition must be used in the theories and the experiments. In particular, if the flame front is defined as the reaction zone of the simplified one step model in the asymptotic limit $\beta \to \infty$ (definition #1), the experimentalists have to use the experimental data and the model to reconstruct the equivalent chemistry - free outer profiles of temperature and velocity. It is a kind of "circular" procedure since a part of the model used to show that $L_1 = L_2$ and to obtain a estimate of their commun value, is also used to locate the flame in the experiments ; one has so far been unable to avoid that.

Preliminary results of a direct test have recently been obtained in Marseille /21/. Using laser diagnostics (LDA, tomography) the needed quantities at the crests and the troughs of the near sinusoidal, two-dimensional flame already described previously, are measured. Because of the small number of configurations considered for a given mixture (typically 2), the accuracy on the determination of Ma is yet rather poor, typically 40% . Another direct test of (I.3) and (II.3) has recently been performed in Poitiers by Cambray, Deshaies and one of as /20//34/, at the tips of a larger series of axially-symmetric steady flames (Fig.7).

By also using LDA and Laser Tomography, the meridian shape (hence $1/R_1+1/R_2 = 2/R$) and the axial velocity profile along the symmetry axis were measured. Then by using the chemistry-free form of (II.5-7) along the axis, the measured value of R and boundary values given by measurements, the reaction zone is located via numerical integration and u_n is determined for each configuration. By changing the inlet conditions and/or the burner obstacle distance, several (typically 10) triple measures $(u_n, 2/R, \mathbf{n}.\nabla\mathbf{u}.\mathbf{n})$ were obtained for a given mixture. For each mixture (the jth, say), the results were found to be compatible with a linear law such as (I.3) and with $Ma_1=Ma_2$. They also give the values $Ma^{(j)}$ of the Markstein number and $u_L^{(j)}$ of the reference burning velocity, upon least-square fitting of

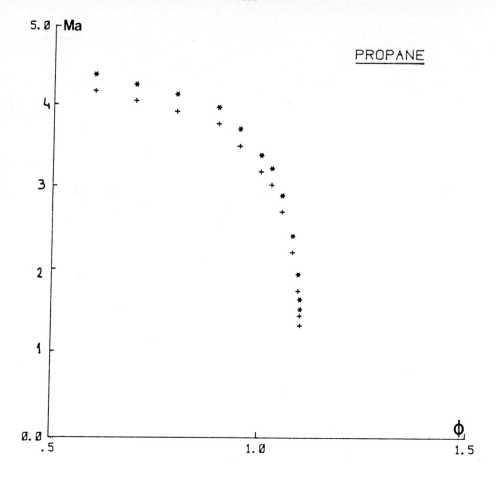

Fig. 6. Indirect measurements by J. Quinard of Markstein numbers for propane-oxygen-nitrogen flame propagating downward at the threshold of stability. ϕ is the equivalent ratio. Crosses are data obtained with the simplified model used in reference /12/. Stars correspond to the more detailed model of /5/.

the data and an iteration on $u_L^{(j)}$ which is involved in the defintion of the "total" rate of strain $\mathbf{n}.\nabla\mathbf{u}.\mathbf{n} - 2u_L/R$ (see Eq. I.4). Once repeated for different mixtures this finally yields the results presented in Fig.(8), in which all the measurements corresponding to three similar mixtures of C_3H_8-N_2-O_2 ($\phi=0.8\pm0.01$ dilutions about 0.160) are gathered. Within the experimental uncertainties and given that all the mixtures have approximately the same γ, Le, s, all the measurements are compatible with the same Markstein number Ma (but not the same Markstein lengths because the $u_L^{(j)}$'s differ from each other) viz

$$\langle Ma \rangle = 8.0 \pm 1.5,$$

since different combinations of geometrical and hydrodynamical stretches were found to give the same value of u_n/u_L when the total adimensinal stretch is kept constant.

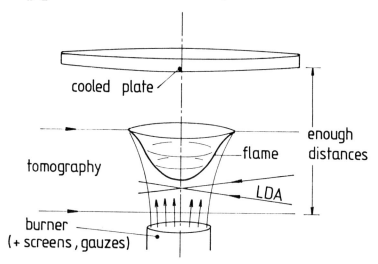

Fig. 7 Sketch of the configuration used in references /20/ and /34/

Here $\langle Ma \rangle$ is given by a least-square fit of all the measurements $u_n^{(j)}/u_L^{(j)}$, the quoted uncertainty being the RMS variations of the various $Ma^{(j)}$. Given the uncertainties on each measurement of u_n, of the total stretch and of u_L, the quoted uncertainty on $\langle Ma \rangle$ is fully compatible with that of the slope of a least-square fit drawn through 10 points (as was typically used for each mixture) which have been randomly displaced (by the experimental uncertainties) about an exact straight line of slope $\langle Ma \rangle = 8$. The above value of Ma can be compared with that given by the theory and one-reactant model. With Le=1.64, s=0.48 (recommended in /5/ for the C_3-H_8 - air), E=37.7 kcal/mole ±5 kcal/mole (from /31/ and the litterature), and the value of T_{ad} given by a thermodynamical code, eq.(II.14) gives /20/ :

$$Ma = 8.0 \pm 10\%$$

provided the data of /35/ are used to evaluate the variations in thermal conductivity and the specific heat is taken as constant. Accounting for the specific heat change with temperature lowers Ma by about 10%. Finally, accounting for the stiochiometry effects, one obtains a value of Ma which is lowered further, by about -0.5 for $\phi=0.8$ if the Lewis number and the Soret coefficient of oxygen are taken from /5/. Given the concertainty on both the experimental and the theoretical estimates of Ma, theory and experiment are compatible.

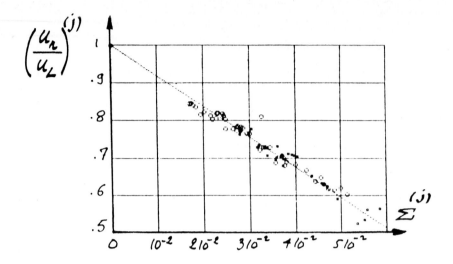

Fig. 8 Experimental values of u_n/u_L plotted versus the total reduced stretch (see Eq. I.4) obtained by direct measurements /20/ /34/ in propane flames.

It is yet unexplained why the indirect method of Quinard /34/ gives substantially lower values than in /20//34/. May be the Markstein number is very sensitive to the slightest inaccuracy of the theories which relate it to the observed cell wavelength or critical velocity. Despite this discrepancy between different kinds of preliminary experimental tests, it is fairly resonnable to expect that the local effects in a weakly wrinkled and/or stretched premixed flame, will be fully understood in the very near future. One has finally to remark that the linear law (I.3) has originally been derived in cases where $u_n \approx u_L$, however it is apparently valid down to $u_n \approx u_L/2$ (Fig.8), and could be used to simulate a much larger than expected number of wrinkled and turbulent flames by introducing a cutt off to describe the flame quenching mechanism by stretch.

IV. THE TURBULENT COMBUSTION REGIME OF CUSPED FLAMES

When an unstable flame front propagates into a weakly turbulent flow ($u'_e < u_L$ where u'_e is the velocity fluctuation in the approaching flow), the Landau-Darrieus instability of the front is nonlinearly coupled with the turbulent motion of the approaching flow to produce irregular cells with a chaotic motion. A detailed analysis of such patterns is presently out of the question. Studies of the M.S. equation (I.7-8) excited with an external noise u'_e could be of interest. Such numerical studies were recently developed /10/. The attention is focused here on global properties such as the average size of these cells which determines the turbulent flame speed. A rough analysis is provided for the cell size selection by a weak turbulence of the approaching flow. The Markstein number Ma is found to play an important role. The analysis is based on the stability analysis of curved fronts /24/ which is briefly recalled in the following section.

1. Curved fronts

Flames in channel exhibit a stationary curved front with a size R much larger than the most unstable wavelength of the planar front (k = 1/2 in the notation of Eq.I.8). An approximate steady state solution of the corresponding curved front with a simple explanation of its anomalous stability was presented by Zel'dovich *et al.* /24/. The corresponding analysis has been extended recently to other types of fronts /25-26/. A detailed presentation of these questions can be found in a recent book by P. Pelcé /27/. As sketched in fig.9 , the analysis consists in evaluating the growth rate of an initially localized perturbation initiated close to the tip, stretched and carried along the front by the tangential component of the flow. During its travel, such a perturbation is submitted locally to a linear growth rate similar to the one in the planar case (I.7-8) but with a varying wavelength which is increasing under the flow stretch. The final amplitude is the result of a time integration in the Lagrangian frame of the perturbation. The most amplified perturbations are initiated near the tip with an initial size of the same order of magnitude as the marginal wavelength Λ_m of the dispersion relation (k $_m$=1 in the notation of Eq. I.8). Λ_m is proportional to the flame thickness d and its magnitude depends on the Markstein number. In the absence of gravity the simplest model yields /12/ /28/:

$$\Lambda_m = 2\pi d \{ 1 + (2+\gamma)Ma/\gamma - (2/\gamma)\ln(1/1-\gamma) \} \tag{IV.1}$$

The following developments are limited to the cases for which, as in Eq. (I.8), the second order term is stabilizing. In the framework of Eq (IV.1), this is the case for Ma > 1 for which $\Lambda_m > 0$ as soon as $\gamma >$ 0.8. The total growth rate Γ of a perturbation can be computed in its Lagrangian frame by a W.K.B. method for stationary fronts of curvature radius R much larger than Λ_m as those observed in experiments /24/. For curved flames, the leading order of the WKB expansion of Γ is found to be determined by the vicinity of the tip whose the configuration is similar to a stagnation point flow and where the tangential velocity is small and the transit time is long enough to let the instability grow. The overall growth rate Γ turns out to be of the order of R/Λ_m which corresponds to a Reynolds number Re based on the size of the channel R

$$\Gamma \approx \text{Re} \quad \text{with} \quad \text{Re} \approx R/\Lambda_m, \tag{IV.2}$$

and where by definition

$$A_f/A_i \equiv \exp\Gamma, \tag{IV.3}$$

where A_f is the final amplitude in the wing of a perturbation initiated close to the tip with an initial amplitude A_i. A stability criterion is roughly obtained by limiting the admissible amplitude A_f to the radius R* of the tube:

$$(R^*/\Lambda_m) - \text{Log}(R^*/\Lambda_m) = \text{Log}(\Lambda_m/A_i). \tag{IV.4}$$

For a given amplitude A_i of the perturbation of the front at the tip produced by the external noise, this

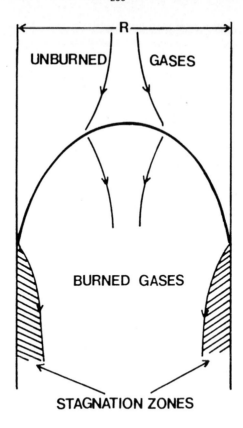

Figure 9. Sketch of the streamlines associated with a curved flame front propagating in a tube. The gas flow in the fresh mixture presents the form of a local stagnation point flow at the tip. Local perturbations which are initiated near the tip are swept toward the wings and stretched by the induced flow. Stagnation zones appear in the burned gases close to the walls.

gives an estimate of the maximum size of the tube R^* in which a single cell propagates. When $R > R^*$ the flame front breaks up in at least two cells. The present theory is limited to $A_i \ll \Lambda_m$ in such a way that R^* can be much larger than Λ_m. A typical order of magnitude is $\Lambda_m \approx 0.5$ cm.

As an instructive example, an exact calculation of Γ, can be carried out in the framework of a simple model where the effect of a stagnation point flow is added to the operator $\mathbb{L}$ in (I.7)

$$\partial\alpha/\partial t = \mathbb{L}(\alpha) - (2\pi Re)^{-1} y \partial\alpha/\partial y, \qquad (IV.5)$$

where $\mathbb{L}$ is defined in the Fourier representation by Eq. (I.8). By introducing the following change of variables $\tau = t$, $Y = y\, e^{-t/(2\pi Re)}$, Eq. (IV.5) can be written in the Fourier representation ($a_K = \int \alpha(Y) \exp(-ikY)\, dk$) as:

$$da_K/d\tau = \{|k|e^{-\tau/(2\pi Re)} - |k|^2 e^{-2\tau/(2\pi Re)}\} a_K. \tag{IV.6}$$

The long time asymptotic behavior of the solution is

$$\tau \gg 2\pi Re \Rightarrow a_K(\tau)/a_K(0) \approx \exp\{2\pi Re(|k| - |k|^2/2)\} \tag{IV.7}$$

from which it appears that the most amplified wavelength is the one corresponding at t=0 to the marginal condition (k=1 i.e. $\Lambda = \Lambda_m$) with a final amplification similar to (IV.2).

2. The cusped flame regime

Consider for simplicity an unstable free flame front propagating in the direction normal to its mean position which is assumed to be planar. Consider also for simplicity a two dimensional geometry. On such a front, each cell can be considered as a curved front whose stability analysis is outlined in the preceding section (see fig. 10.). The analogy is clearly exhibited when one mentally substitutes the symmetry axis of the cusps for the walls of the channel /29/. To each of these cells is associated a self consistent induced flow u_i the amplitude of which may be of the order of magnitude of u_L. At the tip of the cells, this flow has a stagnation point. When curvature effects are neglected at the cusps, stagnation zones must be introduced in the burnt gases.

The question we want to address now is the selection of the size of these cells in a weakly turbulent flow. According to Eq. (IV.4) in the absence of external noise ($A_i=0$), large cells much larger than Λ_m are stable whatever be their size. In this case, the size of the cells is determined by the dimension R of the vessel. As suggested by the numerical calculation of the model (I.7) equation /8-10/, the basin of attraction of the largest possible cell is expected to be very large and the corresponding solution is selected in a long time limit. When the flame propagates in a weakly turbulent flow characterized by $u'_e < u_L$, the mean size of the cells $\langle \Lambda \rangle$ is expected to be controled by this external noise according to an equation obtained in a straightforward manner from Eq. (IV.4) :

$$(\langle\Lambda\rangle/\Lambda_m) - \text{Log}(\langle\Lambda\rangle/\Lambda_m) = \text{Log}(\Lambda_m/\alpha_e') \tag{IV.8}$$

where $A_i = \alpha_e'$ is the amplitude of the fluctuation of the flame position at the tip produced by u'_e with a wave length equal to Λ_m. Notice that $\alpha_e' = \Lambda_m$ is a critical value for the solution $\langle\Lambda\rangle$ of Eq. (IV.8) corresponding to $\langle\Lambda\rangle = \Lambda_m$: there is no more solution for for $\alpha'_e > \Lambda_m$. In fact Eq.(IV.8) is well justified only for $\alpha'_e \ll \Lambda_m$ and for initial perturbations of the front with a wavelength equal to Λ_m. When the external noise is produced by a turbulent flow, one must, according to the results presented above, focus the attention upon those vortices with a size Λ_m. As $u_L > u'_e$, the front perturbation due to these vortices is determined by the time of interaction Λ_m / u_L between the vortex and the front in

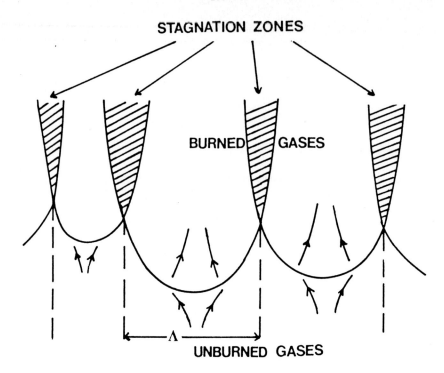

Figure 10. A cusped flame front is constituted of cells separated by sharp edges. Each cell can be considered as the pattern represented in Fig.9 for the propagation in a channel.

such a way that

$$\alpha'_e = \Lambda_m (u'_e/u_L), \tag{IV.9}$$

so that

$$(\langle\Lambda\rangle/\Lambda_m) - \text{Log}(\langle\Lambda\rangle/\Lambda_m) = \text{Log}(u_L/u'_e) \tag{IV.10}$$

which determines the average size $\langle\Lambda\rangle$ of the cells in terms of the amplitude u'_e ($< u_L$) of the velocity fluctuations associated with the length scale $l = \Lambda_m$ in the approaching turbulent flow. Notice that the solution depends upon the Markstein number Ma through Λ_m (see Eq. IV.1). The situation described by Eq. (IV.10) is rather unusual in the sense that small scale structures of size Λ_m in the approaching turbulent are found to control the average size of the cells on the flame front which in turn control large scale structures of size $\langle\Lambda\rangle$ in the burnt gas flow. Using the result obtained for the curved fronts propagating in channel /24/, Eq. (IV.10) may be used to determine the turbulent flame speed. Notice also that Eq. (IV.10) delimits unambiguously the cusped flame regime by the relation $u'_e = u_L$. Numerical studies /10/ of the model equation (I.7 0) triggered by an external periodic force have roughly confirmed the above predictions. But in addition, they have pointed out a mechanism of tip splitting.

V. CONCLUDING REMARKS

The second order effects such as those associated with flame stretch and front curvature are essential for the determination of the combustion rate in the wrinkled flame regime. Whether or not the situation is somehow similar in the corrugated flamelet regime which has been considered recently in reference /36/, is an open question. Detailed theoretical analysis are now available for wrinkled flames both in stable and unstable cases. The corresponding experiments as those carried out for measuring Markstein numbers, have provide us already with interesting data which have to be confirmed by further investigations. In most of the ordinary reactive mixtures far from the limits of flamability, the detailed chemical kinetics and/or transfer processes do not seem to have a drastic effect upon the Markstein number at least for its order of magnitude, and excepted for rich hydrogen oxygen mixtures, no new dynamical behaviours of wrinkled flame fronts are introduced.

Acknowledgements

We wish to acknowledge partial financial support of this work by the D.R.E.T. under contract No 85/130 and C.E.C. under contracts No. EN3E-0084-F. P. Cambray , B. Deshaies from Poitiers and L; Boyer, J. Quinard and G. Searby from Marseilles are greatly acknowledged for providing experimental data and for many fruitfull discussions.

References

1. G.H. Markstein "*Nonsteady Flame Propagation*" Pergamon Press (1964)
2. B. Karlovitz et al. ,IV Combustion Symposium, 613-620. The Williams & Wilkins Company (1953)
3. P. Clavin, F.A. Williams. J. Fluid Mech. **116**, 251-282 (1982)
4. P. Clavin, G. Joulin, J. Phys.Lettre, **44**, L 1-12 (1983)
5. P. Garcia-Ybarra, C.Nicoli, P. Clavin. Combust. Sci. Technol.**42**, 87-109 (1984)
   P. Clavin , P. Garcia-Ybarra, J.Mécanique Theorique et Appliquée. **2**, 245-263 (1983)
6. P. Clavin. Prog. Energy Combust. Sci. **11**, 1-59 (1985)
7. A. Linan, P. Clavin, Combust. Flame **70**, 137-160 (1987)
8. G.I. Sivashinsky, Acta Astronautica, **4**, 1177-1206 (1977)
   D.M. Michelson, G.I. Sivashinsky, Acta Astronautica **4**, 1207-1221 (1977)
9. O. Thual, U. Frisch and M. Henon, J. Physique. **46**, 1485-1494 (1985)
10. B. Denet. Thèse Université de Provence, Aix-Marseille I (1988)
11. G. Sivashinsky, P.Clavin, J. Physique, **48**, 193-198 (1987)
12. P. Pelcé, P. Clavin, J. Fluid Mech. **124**, 219-237 (1982)
13. M. Matalon, B.J. Matkowsky, J.Fluid. Mech. **124**, 239-259 (1982)
14. G. Searby, P. Clavin, Combust. Sci. Tech.**46**, 167-193 (1986)
15. G. Searby, F. Sabathier, P. Clavin, L. Boyer, Phys. Lett. **51**, 1450 (1983)
16. J. Quinard, Thèse d'Etat, Université de Provence, Aix-Marseille I (1984)
17. J. Quinard, G. Searby and L. Boyer - Lecture Note in Physics **210**, 331 (1984)
    - Progress. Astro. Aero. **95**, 129-141 (1984)

18. M. Barrère, J. Chimie Physique, **81** (9) 519-531 (1984)
19. R. Borghi, in *"Recent advances in the Aerospace Sciences"* Ed. Corrado Casci, 117-138, Plenum Press (1985)
20. P. Cambray, B. Deshaies, G. Joulin, Proceedings of the AGARD P.E.P.70th Symposium. Preprint No 422, pp.36.1-36.5 (1987)
21. J. Quinard, G. Searby, in preparation (1988)
    R. Hadef, Thèse de l'Université de Provence, Aix-Marseille I (1988)
22. A. Palm-Leis, R. Strehlow, Combust. Flame **13** (1) 111-129 (1969)
23. D.R. Ballal, A.H. Lefebvre, Proc. Roy. Soc. London, **A344**,217-234 (1975)
24. Y.B. Zeldovich, A.G. Istratov, N.I.Librovich, Combust. Sci.Technol. **24**, 1 (1980)
25. P. Pelcé, Thèse de Doctorat d'Etat, Université Aix-Marseille I (1986)
26. P. Pelcé, P. Clavin, Europhysics Letters, **3** (8) 907-913 (1987)
27. P. Pelcé, *"Dynamics of curved fronts"*, Academic Press, à paraître (1988)
28. M.L. Frankel and G.I. Sivashinsky. Combust. Sci. Technol. **29**, 207 (1982)
29. Y.B.Zeldovich, Combust. Flame, **40**, 225-234 (1981)
30. C. Nicoli, P. Clavin, Combust. Flame, **68**, 69-71 (1987)
31. J.C. Keck, C.R. Fergusson, Combust. Flame, **28**, 197-205 (1977)
32. A. Giovannini, These Proceedings, paper No 11, Vol II.
33. A.F. Ghoniem, O.M. Knio, **21th** Symposium (Int.) on Combustion, Munich (1987)
34. P. Cambray, B. Dehaies, Proceedings of the Joint Meeting of French and Italian section of the Combustion Institute, Amalfi p 6.19 (1987)
35. A.J. Reynolds, *" Thermofluid Dynamics"* Wiley Interscience (1971)
36. P. Clavin, Invited lecture at the NATO Conference *"Disorder and Mixing"* Cargèse August 1987, NATO Asi Series, M. Nijhoff to appear (1988)

DIFFUSION FLAME ATTACHMENT AND FLAME SPREAD ALONG MIXING LAYERS

Amable Liñán
E. T. S. de Ingenieros Aeronáuticos, Univ. P. de Madrid
Pza. Cardenal Cisneros 3, 28040 Madrid

U.S.-France Workshop on Turbulent Reacting Flows
Rouen, France, July 7-10, 1987.

ABSTRACT

An analysis is presented for the description of the diffusion flame attachment region to the splitter plate separating the fuel and air, which should provide the flame lift-off speed of the fuel jet in laminar and turbulent diffusion flames.

The region of flame attachment and the flow velocity there are small enough to allow for a balance of convection and upstream diffusion, so that the complete quasi-steady Navier-Stokes equations must be used to describe the flow in this region. The problem that must be solved is posed and it is conjectured that for the typical values of the overall activation energy there is a diffusion controlled solution with a flame attached close to the splitter plate, only for values of the jet flow velocity below a critical value; otherwise the flame is lifted-off.

For those small values of the jet velocity there is a second unstable solution, which may also be attached to the wake region of the plate, and a third nearly frozen solution. The transition from the nearly frozen solution to the stable attached flame solution is only possible after flame ignition by a hot source in the mixing layer and subsequent flame propagation along the local laminar mixing layers or vortex cores. The problem of flame propagation along the local laminar mixing layers is also posed.

LENGTH AND TIME SCALES IN TURBULENT COMBUSTION

N. Peters

Institut für Allgemeine Mechanik
RWTH Aachen
Templergraben 64, 5100 Aachen
West-Germany

Abstract

The different regimes of premixed turbulent combustion may well be illustrated in a diagram, initially proposed by R. Borghi, where the ratio of the turbulence intensity to the laminar flame speed is plotted over the ratio of the turbulence integral length scale to the flame thickness. In this diagram four different regimes of turbulent combustion are specified:

1. the regime of wrinkled flamelets,
2. the regime of corrugated flamelets,
3. the regime of distributed reaction zones and
4. the regime of the well-stirred reactor.

While the first and fourth regime, representing limiting conditions, have been analysed in the past rather successfully, the second and third regime describe a more intense interaction between turbulence and combustion. Using arguments based on Kolmogorov's energy cascade, a new length scale is identified in each of these two intermediate regimes.

In the regime of corrugated flamelets, the Gibson scale $L_G = v_F^3/\varepsilon$ is derived. Here v_F is the flame velocity which by definition is equal to the characteristic turn-over velocity of the eddy of size L_G. Eddies much larger than L_G which have a larger turn-over velocity than v_F will convect the flame front within the flow field as if it was a passive surface. Eddies much smaller than L_G, having a smaller turn-over velocity than v_F, are consumed by the flame front very rapidly and therefore cannot corrugate the front. The Gibson scale therefore is the lower cut-off of all scales that appear in the corrugated flame surface.

In the regime of distributed reaction zones the mean turbulent flame thickness is of the order of the quench scale δ_q, which is predicted to be proportional to $(\varepsilon\, t_q^3)^{1/2}$, where $t_q = 1/a_q$ is the quench time, a_q is the stretch rate at quenching of a premixed flame and ε the dissipation of turbulence. The quench scale presents the largest eddy within the inertial range, which is still able to quench the thin inner reaction zone of a premixed flame. Smaller eddies, inducing a larger stretch, will quench this thin layer more readily and therefore will try to homogenize the scalar field locally. Therefore a thickened flame front will appear. Larger eddies, inducing a weaker stretch that will not be able to quench the inner reaction zones, will only wrinkle this thickened flame front.

Introduction and outline

A fully developed turbulent flow can be characterized by a continuous range of length and time scales. This range, which is called the inertial range, extends from the large energy containing eddies characterized by the length scale ℓ_t at which the turbulent kinetic energy is fed into the system, down to the smallest eddies of size ℓ_K, the Kolmogorov scale, where the energy is dissipated. Associated to the length scales are particular time scales, a turn-over time of the eddies, which shall be denoted by t_t for the large eddies and t_K for the Kolmogorov eddies; and a particular turn-over velocity, which is v' (the turbulence intensity) for the large eddies and v_K for the Kolmogorov eddies.

Combustion reactions taking place within a turbulent flow field introduce their own time scales, the chemical times, and in the case of a propagating flame their own length scale and velocity, the flame thickness and flame velocity. These scales may interact with the scales of the inertial range in a very specific way, leading to different regimes in premixed turbulent combustion. These regimes often are displayed in phase-diagrams. We will discuss these regimes in terms of Borghi's phase-diagram, where the turbulence intensity v', normalized by the flame velocity, is plotted over the integral length scale ℓ_t, normalized by the flame thickness.

Based on the classical theory of Kolmogorov two characteristic length scales will be derived for two central regimes in the phase-diagram, the regime of corrugated laminar flamelets and the regime of distributed reaction zones. These two regimes are the most important in many practical applications, particularly in engine combustion. In these two regimes there is a close interaction between turbulence and combustion, because neither turbulence nor combustion is completely dominating.

Length and time scales in premixed combustion

The most important property of a premixed flame is its ability to propagate normal to itself. This leads to a characteristic velocity scale - the flame speed. Flame propagation is due to a diffusive-reactive mechanism, which may be illustrated by considering the structure of a plane steady flame in fig. 1. It consists of a chemically inert upstream preheat zone, a thin reaction zone and a downstream burn-out zone. The details of the structure depend on the kinetic model that is used, but from a recent asymptotic analysis of methane-flames /1/ it seems clear now that there always exists a thin reaction zone, at least in hydrocarbon flames, in a similar way as in one-step model flames with a large activation energy (cf. /2/, /3/).

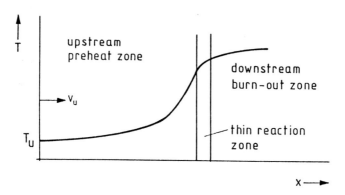

Fig. 1: Structure of a plane steady premixed flame.

In order to have a steady configuration, the flame velocity v_F of the flame illustrated in fig. 1 must be equal to the velocity v_u of the oncoming flow. The heat generated in the thin reaction zone diffuses against the flow into the preheat zone. There is a reactive-diffusive balance in the reaction zone and a convective-diffusive balance in the preheat zone. Reaction provides only a time scale, the chemical time, which may be considered as a characteristic measure of the time required for all the various elementary reactions to transform the reactants entering into the reaction zone into products. The velocity scale and the chemical time scale are linked to each other by the diffusion process. Therefore, from dimensional analysis one may write

$$v_F = \sqrt{a/t_c} \; , \qquad (1)$$

where $a = \lambda/(\rho c_p)$ is the thermal diffusivity, ρ is the density, λ the thermal conductivity and c_p the heat capacity at constant pressure. Equation (1) defines the chemical time t_c, if the conditions, at which the thermodynamic properties are to be evaluated, are defined. Since the temperature varies considerably within the flame structure, it is convenient to evaluate a at a well defined reference temperature T_{ref}. The reference temperature should be that of the inner reaction layer, which occurs typically between 1600 K and 2000 K.

Associated with the flame velocity and the diffusive nature of the propagation process, a flame thickness can be defined

$$\ell_F = a_{ref} / v_F \quad . \tag{2}$$

With the simplifying assumption of a Prandtl number $Pr = \nu/a$ of unity, where ν is the kinematic viscosity, the flame thickness may be defined

$$\ell_F = \nu_{ref} / v_F \quad . \tag{3}$$

Then the time required for the flame to traverse its own thickness, the flame time, is equal to

$$t_F = \ell_F / v_F \quad . \tag{4}$$

The flame time has the same physical significance as the chemical time, and from the definition, in view of eqs. (1)-(4), with $\nu_{ref} = a_{ref}$ it is equal to

$$t_F = t_c \quad . \tag{5}$$

In addition, premixed flames may be subject to extinction due to flame stretch. Flame stretch, which was introduced by Karlovitz et al. /4/, is the local fractional increase of flame surface area. In a steady flow field positive stretch is an addition of the effect of straining by flow divergence and of flame curvature (cf. /3/). For a one-step reaction an essential additional factor is differential diffusion of heat and the deficient reactant characterized by its Lewis number

$$Le = \lambda / (\rho c_p D) \quad . \tag{6}$$

Here D is the diffusion coefficient of the deficient reactant. Differential diffusion is enhanced by flame stretch and induces a temperature change at the thin reaction layer. Due to the large activation energy of a single global reaction, that was assumed in all the theoretical work based on activation energy asymptotics, the flame speed is very sensitive to temperature changes. Positive stretch increases the enthalpy and thereby the temperature in the thin reaction layer if $Le < 1$ and decreases it if $Le > 1$. The flame responds therefore by an acceleration if flame stretch is positive and $Le < 1$ or negative flame stretch and $Le > 1$, and by a deceleration if flame stretch is negative and $Le < 1$ or positive flame stretch and $Le > 1$. It follows that cellular patterns will form for $Le < 1$, because an initial perturbation of the flame front is enhanced, while the $Le > 1$ case is stabilizing.

As far as extinction of a steady flame is concerned, different answers are obtained depending on the value of the Lewis number or, more precisely, its deviation from unity. If the Lewis number is of an order $O(1)$ amount larger than one, Sivashinsky /5/ obtains for a plane flame in a stagnation point flow a maximum value of the stretch rate k, at which the flame extinguishes

$$k \frac{Le-1}{Le} Ze = e^{-1} \quad . \tag{7}$$

In this case k is the non-dimensional velocity gradient across the flame $k = (du/dx)/(v_F/\ell_F)$ and Ze is the Zeldovich number

$$Ze = \frac{E}{R} \frac{T_b - T_u}{T_b^2} \qquad (8)$$

assumed to be large (typically of order ten), where E is the activation energy, R the universal gas constant and T_b und T_u are the temperatures in the burnt and unburnt gas, respectively.

Equation (7) shows that the stretch rate k is small of order Ze^{-1} if Le-1 is of order O(1). This case is called weak stretch. Another possibility, called strong stretch, was analysed by Buckmaster /6/ for the case of a plane constant-density stagnation point flame. Here k is of O(1) but

$$\lambda = \frac{Le - 1}{Le} Ze \qquad (9)$$

is also of order O(1), requiring that the difference of the Lewis number from one is small. In this case the flame speed is reduced to zero with increasing stretch, but there is not necessarily a maximum value of k as for weak stretch which would be typical for abrupt extinction. The values $v_F = 0$ correspond to the stretch rates

$$k = \frac{\pi}{2} \exp\left(-\frac{\lambda}{2}\right) . \qquad (10)$$

If we identify the inverse of the stretch rate in physical units with a time scale t_q, which is typical for flame quenching, we can write

$$t_q = k^{-1} t_F \qquad (11)$$

since the stretch rate was non-dimensionalized with the flame time. For the two cases of weak or strong stretch the quench time is either an order of magnitude larger or of the same order of magnitude as the flame time. Therefore, in addition to the chemical time, which is equal to the flame time, we obtain an additional time scale t_q which, in principle, differs from the chemical time.

Regimes of turbulent combustion

Several authors (Bray /7/, Williams /2/, Borghi /8/) have supplied phase-diagrams to illustrate different regimes in premixed turbulent combustion as a function of dimensionless quantities. Those are the turbulent Reynolds number

$$\text{Re} = \frac{v' \ell_t}{\nu}, \qquad v' = \sqrt{\overline{v'^2_K}/3}, \tag{12}$$

the turbulent Damköhler number

$$\text{Da} = \frac{t_t}{t_F} = \frac{v_F \ell_t}{v' \ell_F} \tag{13}$$

and the turbulent Karlovitz number

$$\text{Ka} = \frac{\gamma \ell_F}{v_F} = \frac{t_F}{t_K}, \qquad \gamma = \frac{1}{t_K} = \sqrt{\frac{\varepsilon}{\nu}}. \tag{14}$$

Here γ is the inverse of the Kolmogorov time $t_K = \ell_K/v_K$ and describes the straining by the smallest eddies of the Kolmogorov size $\ell_K = (\nu^3/\varepsilon)^{1/4}$ which have a turn-over velocity $v_K = (\nu\varepsilon)^{1/4}$. Also, ε is the dissipation of the turbulent kinetic energy in the unburnt gas. We have defined the Reynolds number and the Kolmogorov scale with $\nu = \nu_{ref}$. With v' and ε prescribed, the integral length scale of the energy containing large eddies may be defined by

$$\ell_t = v'^3/\varepsilon. \tag{15}$$

These definitions can be used to derive the following relations between the ratios v'/v_F and ℓ_t/ℓ_F in terms of the three non-dimensional numbers Re, Da and Ka as

$$\frac{v'}{v_F} = \text{Re}\left(\frac{\ell_t}{\ell_F}\right)^{-1}, \qquad \frac{v'}{v_F} = \text{Da}^{-1}\frac{\ell_t}{\ell_F}, \qquad \frac{v'}{v_F} = \text{Ka}^{2/3}\left(\frac{\ell_t}{\ell_F}\right)^{1/3} \tag{16}$$

as well as the relation

$$\text{Re} = \text{Da}^2 \text{Ka}^2. \tag{17}$$

In the following we will adopt a modified version of Borghi's /8/ phase-diagram and plot the logarithm of v'/v_F over the logarithm of ℓ_t/ℓ_F in fig. 2. In this diagram the lines Re = 1, Da = 1 and Ka = 1 represent boundaries between the different regimes of premixed turbulent combustion. Another boundary of interest is the line $v'/v_F = 1$ which separates wrinkled from corrugated flamelets.

The regime of laminar flames (Re < 1) in the lower left corner of the diagram is not of interest in the present context. Among the remaining four regimes the wrinkled flamelets and the corrugated flamelets belong to the flamelet regime which is characterized by the inequalities Re > 1 (turbulence), Da > 1 (fast chemistry) and Ka < 1 (sufficiently weak flame stretch). The boundary to the distributed reaction regime given by Ka = 1 may be expressed in view of

$$\text{Ka} = \frac{t_F}{t_K} = \frac{\ell_F^2}{\ell_K^2} = \frac{v_K^2}{v_K^2} \tag{18}$$

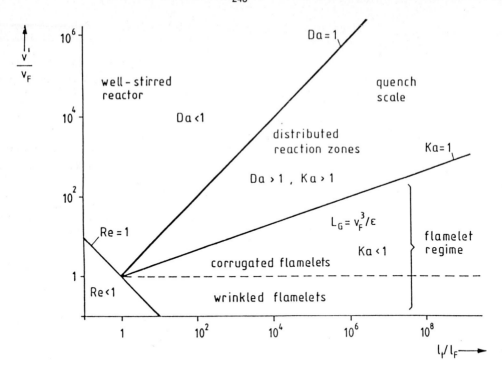

Fig. 2 : Phase-diagram showing different regimes in premixed turbulent combustion.

as the condition where the flame thickness is equal to the Kolmogorov scale (this is the Klimov-Williams criterion). But also, since viscosity as a molecular transport process relates velocity, length and time scales to each other at the Kolmogorov scale in the same way as the velocity, length and time scales for flame propagation, the flame time is equal to the Kolmogorov time and the flame velocity equal to the Kolmogorov velocity. This will be important in the discussion below.

The distributed reaction zones regime is characterized by $Re > 1$, $Da > 1$ and $Ka > 1$, the last inequality indicating that flame stretch is strong and that the smallest eddies can enter into the flame structure since $\ell_K < \ell_F$, thereby broading the flame structure. The smaller eddies produce the largest strain rates and may lead to local extinction of the inner reaction layer. Finally, the well-stirred reactor regime on the upper left of the diagram is characterized by $Re > 1$, $Ka > 1$ but $Da < 1$, indicating that the chemistry is slow compared to turbulence.

We will now enter into a more detailled discussion of the various regimes. The flamelet regime is subdivided into the regimes of wrinkled flamelets and corrugated flamelets. This boundary is viewed by Williams /2/ as the one between single and multiple flame sheets. Clearly, if $v' < v_F$ and v' is interpreted as the turn-over velocity of the large eddies, even those eddies cannot convolute the flame front enough to form multiply connected reaction sheets. Flame propagation is dominating and there is no strong interaction between turbulence and combustion in this regime. In the regime of wrinkled flamelets, asymptotic methods using large activation energy have been a very powerful tool to describe the interaction between weak turbulence and the flame front. An excellent review on theoretical as well as on experimental results was given by Clavin /9/.

The regime of corrugated flamelets is much more difficult to analyse analytically or numerically. In view of eq. (18) we have with Ka < 1 within this regime

$$v' \geq v_F \geq v_K . \qquad (19)$$

Since the velocity of the large eddies is larger than the flame speed, these eddies will push the flame front around, causing a substantial convolution. On the other hand the smallest eddies, having a turn-over velocity less than the flame speed, will not wrinkle the flame front. We may construct a discrete sequence of eddies within the inertial range by defining

$$\ell_n = \frac{\ell_t}{2^n}, \quad \ell_n \geq \ell_K \qquad (n = 0, 1, 2, ...) . \qquad (20)$$

Then, energy cascade arguments require that ε is independent of n and dimensional scaling laws lead to a turn-over velocity v_n of the eddy of size ℓ_n as

$$v_n^3 = \varepsilon \ell_n \qquad (21)$$

indicating that the velocity decreases as the size of the eddy decreases.

Now, as illustrated in fig. 3, we want to determine the size of the eddy which interacts locally with the flame front by setting the turn-over velocity v_n equal to the flame speed v_F. This determines the Gibson scale

$$L_G = v_F^3 / \varepsilon . \qquad (22)$$

It is the size of the burnt pockets that move into the unburnt mixture, try to grow there due to the advancement of the flame front normal to itself, but are reduced in size again by newly arriving eddies of size L_G. Therefore, there is an equilibrium mechanism for the formation of burnt pockets, while unburnt pockets that penetrate into the burnt gas will be consumed by the flame advancement. It is worth noting that L_G increases with v_F if the turbulence properties are kept constant. At sufficiently low turbulence levels, the mean thickness of a turbulent flame should be influenced by this mechanism and therefore also increase with v_F. This is observed in the V-shaped flame by Namazian et al. /10/, where the mean flame thickness increases by a factor between 2 and 3, as the equivalence ratio is changed from $\phi = 0.6$ to $\phi = 0.8$, thereby increasing v_F. On the contrary, the size of cellular wrinkles due to the instability mechanism

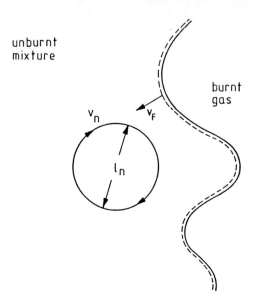

Fig. 3: Interaction between an eddy of size ℓ_n and turn-over velocity v_n with a flame front.

described above decreases with v_F, since it is proportional to ℓ_F which decreases with v_F according to eq. (2). Using eq. (15), one may also write eq. (22) in the form

$$L_G / \ell_t = (v_F / v')^3 \quad . \tag{23}$$

An illustration of the kinematics of the interaction between a premixed flame and a turbulent flow field may be found in fig. 9 of the paper by Ashurst and Barr /11/. In this numerical study the characteristic integral length scale ℓ_t was kept constant, while the turbulence intensity was increased, showing corrugations of smaller and smaller size. A similar effect is observed in the 2-D visualisations of the flame front in I.C. engines by Baritaud and Green /12/ and zur Loye and Bracco /13/ with increasing engine speed. Here it may be argued that the integral length scale is determined by the geometrical dimensions of the combustion chamber and the turbulence intensity increases linearly with engine speed. Again a stronger corrugation with finer scales is observed at increasing engine speed in agreement with eq. (23).

A graphical determination of the Gibson scale L_G is shown in fig. 4. Here the logarithm of the velocity v_n is plotted over the logarithm of the length scale ℓ_n according to eq. (21). If one enters on the ordinate with the flame velocity v_F equal to v_n into the diagram, one obtains L_G as the corresponding length scale on the abscissa. This diagram also illustrates the limiting values of L_G: If the flame velocity is equal to v', L_G is equal to the integral length scale ℓ_t. This case

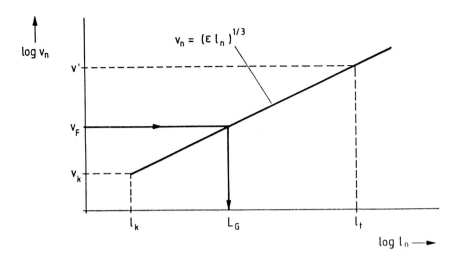

Fig. 4: A graphical determination of the Gibson scale.

corresponds to the border line between corrugated and wrinkled flamelets in fig. 2. On the other hand, if v_F is equal to the Kolmogorov velocity v_K, L_G is equal to ℓ_K. This corresponds to the line Ka = 1 in fig. 2. Therefore L_G may vary between ℓ_K and ℓ_t in the corrugated flamelet regime.

The next regime of interest in fig. 2 is the regime of distributed reaction zones. As noted before, the small eddies can enter into the flame structure and even destroy it, since $\ell_K < \ell_F$ in this regime. Therefore, the notion of a well defined flame structure and therefore of a flame velocity as a relevant velocity scale has no meaning in this regime. Another scale, however, that of the chemical time scale t_c, remains still meaningful since reactions may occur independently of flame propagation. Also, since there is an inner reaction zone smaller than the flame thickness, the quenching of this inner zone by flame stretch is the important physical process. Therefore, the quench time t_q is the more relevant time scale. In analogy to eq. (22) we may relate the turn-over time t_n to the size ℓ_n of an eddy with $t_n = \ell_n/v_n$ as

$$t_n^3 = \ell_n^2 / \varepsilon \quad . \tag{24}$$

Then, by setting $t_q = t_n$, one obtains the quench scale

$$\delta_q = (\varepsilon t_n^3)^{1/2} \quad . \tag{25}$$

This scale may be interpreted as the size of the largest eddy within the inertial range that is still able to quench a thin reaction zone. Smaller eddies up to δ_q, inducing a larger stretch, will quench the thin reaction layers within the flow field more readily and thereby try to homogenize the scalar field locally over a distance up to δ_q. Therefore, δ_q may be interpreted as the thickness of a turbulent flame brush or the overall thickness of a turbulent flame. This is illustrated in fig. 5. Larger eddies than δ_q will only wrinkle this thickened turbulent flame front.

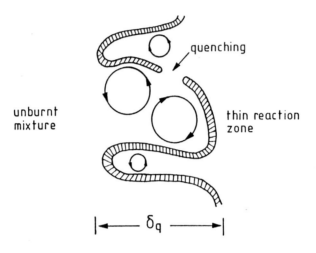

Fig. 5: Illustration of the quenching of the thin reaction zone by eddies smaller than δ_q.

It should be noted that in the distributed reaction zones regime the Damköhler number, which is based on the integral time scale t_t and the chemical time scale $t_c = t_F$ is large. Therefore, at the order of these scales, chemistry is fast and therefore thin reaction zones may be generated locally. These zones are quenched by flame stretch such that an equilibrium mechanism exists between the generation and the destruction of thin reaction zones over a region of thickness δ_q.

On the other hand the Damköhler number based on the turbulent time $(\delta_q^2/\epsilon)^{1/3}$ and the quench time t_q is unity by definition. Over any smaller length scale than δ_q the scalar field appears like a localized well-stirred reactor. This justifies the characterisation of this regime as distributed reaction zones regime.

Again, the derivation of δ_q may be illustrated in a diagram in fig. 6, showing eq. (24) in a log-log-plot of t_n over ℓ_n. If one enters the time axis at $t_q = t_n$, one obtains the quench scale δ_q on the length scale axis. It should be noted that all eddies of size ℓ_n, where $\ell_K < \ell_n < \delta_q$, have a larger stretch than δ_q and therefore are able to quench thin reaction zones locally.

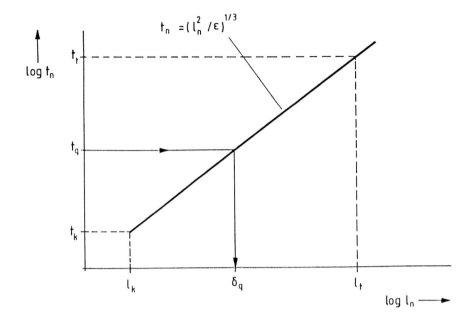

Fig. 6 : A graphical determination of the quench scale δ_q.

If t_q is equal to the Kolmogorov time t_K, fig. 6 shows that δ_q is equal to the Kolmogorov scale ℓ_K. In this case, corresponding to

$$Ka = k \qquad (26)$$

according to eqs. (11) and (14), the border line between the distributed reaction zones regime and the corrugated flamelet regime at $Ka = k$ is not equal to but parallel to the line $Ka = 1$. This is shown as a dotted line in fig. 7 for the case $k < 1$. In this case the distributed reaction zones regime covers part of the corrugated flamelet regime, a part where now quenching is possible. If this occurs flamelets and the Gibson scale are not well defined anymore. Since the characterization of premixed flames in diagrams like fig. 2 or fig. 7 is based on order-of-magnitude estimates, the case $k > 1$ is not meaningful because k is always of order $O(1)$ or smaller according to eqs. (10) or (7).

Similarly, one sees from fig. 6, that if the quench time t_q is equal to the integral time t_t, the quench scale is equal to the integral length scale. This would correspond to $Da = 1$ in fig. 2 only if t_q was equal to t_F. According to eq. (11), the case $t_q = t_t$ is equal to

$$Da = k^{-1}, \qquad (27)$$

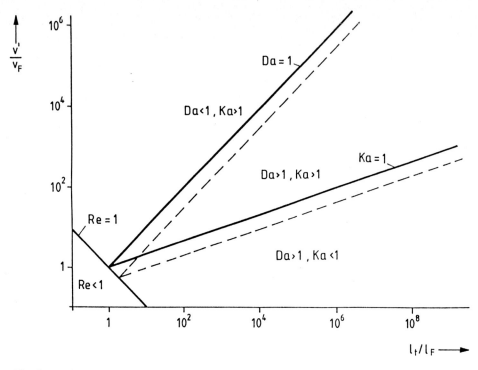

Fig. 7 : Modification of the boundaries of the distributed reaction zones regime for $t_q > t_F$.

which corresponds to a line parallel to $Da = 1$ in fig. 2. This border line between the distributed reaction zones regime and the well-stirred reactor regime is also shown in fig. 7 for the case $k < 1$ as a dotted line.

As a final remark related to the distributed reaction zones regime it may be noted that turbulence in real systems is not homogeneous and ε is not a local constant but has a distribution. This refinement of Kolmogorov's theory has led to the notion of intermittency, or "spottiness" of the activity of turbulence in a flow field /14/. This may have important consequences on the physical appearence of turbulent flames at sufficiently large Reynolds numbers. One may expect that the flame front shows manifestations of distributed reaction zones with local quenching events as well as of regions, where corrugated flamelets appear. Therefore, the regimes discussed above may well overlap each other in an experimentally observed turbulent flame.

In the well stirred reactor regime chemistry is slow as noted before. Now all eddies up to ℓ_t can quench the inner reaction zones, since $Da \leq k^{-1}$, where the order of magnitude of k is $O(1)$ or smaller. Therefore, turbulence homogenizes the scalar field by rapid mixing leaving the slow chemistry to be the rate determining process. Therefore, no specific interaction between turbulence and combustion can occur in this regime and no specific interaction scale is to be defined.

Concluding remarks

In talking about turbulent combustion, it is important to specify the length and time scales of turbulence as well as of combustion, and thereby the regime of interaction. In the four regimes discussed above, the corrugated flamelet regime and the distributed reaction zones regime are the most interesting, not only from the fundamental point of view. It is in these regimes that most practical combustion devices operate. Since strong turbulence is difficult and costly to generate in practical combustion devices and since it dissipates very rapidly, one only generates as much turbulence as necessary to promote combustion and to maintain it. If quenching events appear too frequently, the continuation of the combustion process is threatened and it becomes uncertain. Therefore it is not surprising that most turbulent flames under engine conditions fall into the corrugated flamelet regime or the distributed reaction zones regime /15/.

It should be emphasized, however, that the two interaction scales derived in these two regimes, while illustrating the important physical processes quite nicely, should not be taken as readily measurable quantities. They have been derived essentially by dimensional analysis and various specific influences like gas expansion, detailed chemistry and local variations of the flame velocity due to flame stretch, have not been taken into account. At this stage they only represent a concept of turbulent combustion. Their existence illustrates, in particular, that turbulent combustion cannot be described by a single length scale and a single time scale, but that the entire range of scales enters into the process.

References

/1/ Peters, N., Williams, F.A.: Combust. Flame $\underline{68}$, 185 (1987).

/2/ Williams, F.A.: Combustion Theory, 2nd Edition, The Benjamin/Cummings Publishing Company, Menlo Park, 1985.

/3/ Buckmaster, J.D., Ludford, G.S.S.: Theory of laminar flames, Cambridge University Press, 1982.

/4/ Karlovitz, B., Denniston, D.K., Knappschafer, D.H., Wells, F.E.: 4th Symposium (International) on Combustion, p. 163, The Combustion Institute, 1953.

/5/ Sivashinsky, G.I.: Acta Astronautica $\underline{3}$, 889 (1976).

/6/ Buckmaster, J.: 17th Symposium (International) on Combustion, p. 835, The Combustion Institute, 1979.

/7/ Bray, K.N.C.: Turbulent Flows with Premixed Reactants, Turbulent Reacting Flows, P.A. Libby, F.A. Williams (Eds.), p. 115, Springer, Berlin, 1980.

/8/ Borghi, R.: On the structure of turbulent premixed flames, Recent Advances in Aeronautical Science, C. Bruno, C. Casci (Eds.), Pergamon, 1984.

/9/ Clavin, P.: Progr. Energy Combust. Sci. 11, 1 (1985).

/10/ Namazian, M., Shepherd, I.G., Talbot, L.: Combust. Flame 64, 299 (1986).

/11/ Ashurst, W.T., Barr, P.K.: Combust. Sci. Technol. 34, 227 (1983).

/12/ Baritaud, T.A., Green, R.M.: SAE-paper 860025 (1986).

/13/ zur Loye, A.O., Bracco, F.V.: International Symposium on Diagnostics and Modeling of Combustion in Reciprocating Engines, Tokyo, p. 249, 1985.

/14/ Kolmogorov, A.N.: J. Fluid Mech. 13, 82 (1962).

/15/ Abraham, J., Williams, F.A., Bracco, F.V.: SAE-paper 850345 (1985).

A MODEL FOR REACTIONS IN TURBULENT JETS:
EFFECTS OF REYNOLDS, SCHMIDT, AND DAMKÖHLER NUMBERS

James E. Broadwell

Graduate Aeronautical Laboratories
California Institute of Technology
301-46, Pasadena, California 91125

ABSTRACT

Data from several recent experiments on mixing and chemical reactions in turbulent shear layers and jets is discussed in detail and used to formulate a picture of the path from the freestream to the molecularly mixed state. A model is proposed which incorporates the essential steps in this path and which appears to provide a framework for understanding the major effects of Reynolds, Schmidt, and Damköhler numbers on the chemical reaction. A simplified version of the model reproduces the observed Reynolds number dependence of nitric oxide production in turbulent fuel jets.

INTRODUCTION

It is a tenet of classical turbulence theory that in the limit of Reynolds number approaching infinity, dissipation is independent of Reynolds number and takes place at the Kolmogorov scale. The same assumptions concerning scalar mixing are the basis for a model of molecular mixing in shear layers and jets that has been under development for some time, Broadwell and Breidenthal (1982), Broadwell and Mungal (1986,1987). The approach was suggested by the experiments that revealed large scale organized motions in shear layers and jets, primarily those of Brown and Roshko (1974), Konrad (1976), and Breidenthal (1982) in shear layers and of Dimotakis, Papantoniou and Miake-Lye (1983), in jets. In the shear layer, these experiments and the computations of Corcos and Sherman (1976) showed that instabilities lead to concentrated regions of vorticity and that free stream fluids enter the layer in streams with dimensions of order of the layer thickness, δ. It is a basic postulate in the model that in the limit $Re \rightarrow \infty$, no mixing takes place until the scale of these entering streams is reduced, by inviscid motions, to the Kolmogorov scale, $\lambda_K \sim \delta/Re^{3/4}$, where Re is the Reynolds number. When the scale of the concentration fluctuations in the streams reaches λ_K, it is shown by arguments given later that the time required for diffusion to "homogenize" the mixture is negligible compared to the time to reach λ_K from δ, i.e. compared to δ/U, where U is a characteristic large scale velocity. The quantity, or volume fraction, of molecularly mixed fluid so formed is, therefore, independent of both the Reynolds number, and the Schmidt number, $Sc = \nu/D$, where ν and D are the viscosity and diffusion coefficients. In this limit then, the rate of molecular mixing is a constant as is the rate of entrainment into the layer.

When the Reynolds number is finite, but such that $(Re\,Sc)^{1/2} \gg 1$, diffusion layers form at the boundaries of the entering streams and, because their thickness, λ_τ, scales with the large scale variables, U and δ it is given by

$$\lambda_\tau \sim \delta/(Sc\,Re)^{1/2} \sim \delta/Pe^{1/2}$$

where Pe is a Peclet number, equal to $U\delta/D$. Since the surface area per unit volume, S, of these diffusion sheets scales only with δ^{-1}, their fractional volume ($S \cdot \lambda_\tau$) is proportional to $Pe^{-1/2}$.

In this case, also, when all scales in the entering stream are reduced to the Kolmogorov scale the mixture is homogenized as before. In this Lagrangian description of the path from the free stream to the molecularly mixed state, the same quantity of fluid is involved whatever the Reynolds number. Therefore, the fractional volume of mixed fluid produced by diffusion at the Kolmogorov scale is independent of the Reynolds number. The diffusion at the scale λ_τ causes mixing that is "early" or upstream, in the Eulerian viewpoint.

The above described picture of shear layer mixing is, by the same arguments, applicable to jets, in which case the entering stream of reservoir fluid mixes at a constant rate with jet fluid. Differences between the two flows emerge only when chemical reactions are considered.

At any axial station in either shear layers or jets, then, the molecularly mixed volume fraction, V_m, can be written,

$$V_m = A + B/(Sc\,Re)^{1/2} \qquad \begin{array}{c} Re \gg 1 \\ (Sc\,Re)^{1/2} \gg 1 \\ Re \gg (\ln Sc)^2 \end{array} \qquad (1)$$

in which A is the average volume fraction generated when the entrained fluid reaches the scale λ_κ and B is the average non-dimensional surface area per unit volume, i.e., S measured in terms of δ. The notation Sc Re is used instead of Pe because of the need to state the conditions of applicability of the model. The origin of the restriction, $Re \gg (\ln Sc)^2$ is discussed later.

In the above referenced discussions of the model, the diffusion layers were called strained laminar flames. Connection is made with the preceding discussion by noting that if these layers are in equilibrium with the local large scale strain $\varepsilon \sim U/\delta$, their thickness is proportional to $(D/\varepsilon)^{1/2}$ and $\lambda_\tau \sim \delta/Pe^{1/2}$ as above. See Carrier, Fendell, and Marble (1975) for a discussion of these flames. When Sc = 1, λ_τ is the Taylor scale, a fact brought to our attention by H. W. Liepmann who points out that they are, in the mathematical sense, internal boundary layers.

The discussion so far has been concerned only with molecular mixing with no restrictions on the nature of the chemical reactions; in particular, reaction rates of any magnitude can be considered. This flexibility, also a characteristic of the coherent flame model, Marble and Broadwell (1977), may be useful in interpreting the recently measured finite kinetic effects in hydrogen flames, Dibble and Magre (1987), and in studying lift-off and blow-out of fuel jets, Broadwell, Dahm and Mungal (1984).

With regard to the Taylor diffusion layers, notice that in the jet they form between newly entrained reservoir fluid and jet fluid, i.e. fluid that is, or was, "turbulent". For them to persist it appears to be necessary for the small scale turbulent motions in the jet fluid to have been destroyed, by momentum diffusion, when the concentration variations were destroyed. In the shear layer, this picture is consistent with the observations of Coles (1983) and Hussain (1983), that turbulence is generated in the braids and dissipated in the cores of the vortices. Furthermore, if the small scale motion is dissipated, the vortices would be left in solid-body-like rotation, evidence for which is reported by Wygnanski and Fiedler (1979). In both shear layers and jets, the motions are unsteady on the largest spatial and temporal scales in the flow. These unsteady motions, as illustrated by the flame length fluctuations observed by Dahm and Dimotakis (1985), for instance, make possible, if not plausible, the idea that the small scale motions and concentration fluctuations are successively generated and destroyed in the jet fluid as it moves along the axis.

While, as just stated, large scale unsteadiness is an inherent feature of the flows being considered, the model attempts to deal only with their averaged consequences. In addition, in most of the discussion that follows, the density will be taken to be constant. The modifications that are needed when the jet and reservoir fluid have different densities, or when heat release is considered, are discussed after the basic ideas are presented.

THE HIGH REYNOLDS NUMBER LIMIT

In the above cited references to earlier work on the model, an upper limit for the time required for diffusion to homogenize the fluids at the Kolmogorov scale was estimated from the time to diffuse across that scale, i.e., from $\lambda^2_K/D \sim (d/u)\, Sc/Re^{1/2}$ where d is the local jet radius and u the axial velocity. Batchelor (1959) shows that local straining reduces λ_K to $d/(Re^{3/4}Sc^{1/2})$ in a time $(d/u)(1/Re^{1/2}) \ln Sc$, a result brought to the author's attention by P. E. Dimotakis. Since the time for diffusion across the smaller scale is only $(d/u)/Re^{1/2}$, the larger time d/u $(1/Re^{1/2}) \ln Sc$ is controlling and is the restriction given relative to Equation (1). For ordinary substances, the condition $Re \gg 1$ is sufficient to ensure that $Re \gg (\ln Sc)^2$.

When the conditions, $Re \gg 1$, $(Sc\, Re)^{1/2} \gg 1$, and $Re \gg (\ln Sc)^2$ are satisfied, it is a simplification to consider the scales to be divisible into the three separate categories:

d, $d/Re^{1/2}$ and $d/Re^{3/4}$,

and to assume instantaneous mixing at the scale λ_K.

Consider first the limiting case $Re \to \infty$ for any fixed Schmidt number, in which case mixing takes place only at λ_K. (Note that the requirement $Re > (\ln Sc)^2$ precludes the case $Sc \to \infty$ for fixed Reynolds number. It seems clear, intuitively, that the mixed volume fraction goes to zero in this limit.)

A critical check on the model in the high Reynolds number limit is whether or not the mixing rate is independent of Re and Sc. Very little evidence is available concerning this point, but the experiments of Breidenthal (1981) and Koochesfahani and Dimotakis (1986) (in the same facility and over a limited Reynolds number range, $10^4 < Re < 10^5$) show that for $Sc \sim 600$, there is no dependence on Re. These results suggest that $Sc\, Re > 10^7$ is sufficient for the limiting condition to apply. In gases, $Sc \sim 1$, at $Re \sim 10^5$, however, Mungal, Hermanson, and Dimotakis (1985) find Reynolds number effects; 10^5 is apparently not high enough.

Indirect support for the proposed mixing mechanisms in the high Reynolds number limit is in the composition and spatial distribution of the molecularly mixed fluid. If there is a delay, after entrainment, during which the fluids are distributed by large scale motion throughout the shear layer or the jet, the mixed fluid composition should be independent of the lateral coordinate in shear layers and of the radius in axi-symmetric jets.

Konrad (1976) noted the approximate flatness, in his gas shear layer, of the mixed fluid composition and, in fact, this observation was one of the motivations of the present model. The most remarkable results, however, are those obtained in shear layers in water ($Sc \sim 600$) and $Re \sim 2 \cdot 10^4$ by Koochesfahani and Dimotakis (1986).
Figure 1 shows their mean mixed fluid composition deduced from chemical reaction experiments that eliminated probe resolution errors. (They show, incidentally, the rigorous requirements for accurate measurements of this quantity by conventional instrumentation). We see that there is no indication whatever of a lateral gradient. It is a convenient and useful additional approximation for modeling, used the following, to assume that the mixed composition has a single value, but what is to be noted here is lateral distribution. It is difficult to escape the conclusion that the fluids are distributed throughout the layer before they mix.

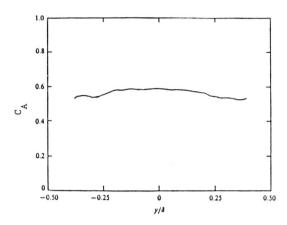

Figure 1. Dependence on lateral coordinate, y/δ, of average concentration of mixed fluid. C_A denotes high speed fluid. Koochesfahani and Dimotakis (1986).

The picture is not as simple for jets. Dahm and Dimotakis (1985), again in experiments in water in the Re range $10^3 - 10^4$, conclude from laser sheet photographs and composition profiles that there are large region of nearly uniform composition but that they are so arranged that radial scans sometimes intersected more than one. Figure 2 contains samples of this measurement. More data is needed but the few results are in accord with the notion that large scale motions distribute reservoir fluid throughout the jet before it mixes molecularly. Such motions are dramatically illustrated in high speed laser sheet images of Long (1987). Likewise, temperature-time traces in a plane jet, Antonia et al (1986), have been shown by them to be evidence of large scale motions.

JET MODEL OF THE LIMIT Sc Re $\to \infty$, Re $\gg 1$

To put the above description in approximate quantitative form, we consider only the far field where the jet spreads linearly and similarity in the velocity distribution, $u(x/d_o, \eta)$ has been attained. Here, x is the axial distance, d_o, the jet nozzle diameter, d, the local jet radius, and $\eta = r/d$. Likewise the molecularly mixed fraction, α, is taken to depend only on η. Recall that the justification for the independence of α from x is that the entrained fluid reaches the Kolmogorov scale and mixes after it moves downstream a distance that scales only with the radius at which it was entrained.

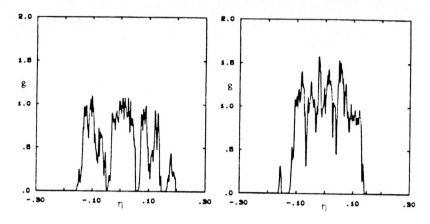
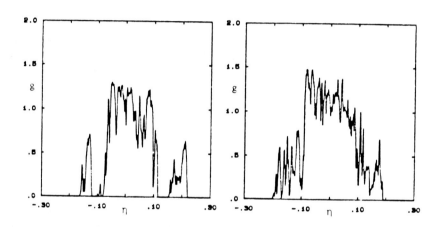

Figure 2. Representative instantaneous radial profiles of concentration in the jet, $x/d_o = 300$, $Re = 5000$. Dahm and Dimotakis (1985).

The connection between α and the rate, v_h, at which reservoir fluid "enters" the molecularly mixed flux is

$$v_h = 2\pi \frac{d}{dx} d^2 \int \alpha(\eta) u(x/d_o, \eta) \eta \, d\eta \tag{2}$$

Let

$$d = C_1^{1/2} x \text{ and } u = (u_o d_o/x) g(\eta). \tag{3}$$

Then,

$$v_h = 2\pi C_1 u_o d_o E \tag{4}$$

where

$$E = \int \alpha(\eta) g(\eta) \eta \, d\eta. \tag{5}$$

The independence of α from x implies, therefore, that the molecularly mixed fluid flux rises linearly with x as does the total jet flux. As is well known, the latter result follows simply from the conservation equations; the numerical rate has been measured by Ricou and Spalding (1961).

In the shear layer, as was noted above, the mean concentration in this limit is independent of the lateral coordinate. If, in addition, the concentration is taken to have a single value, then the integral $\int \alpha(\eta) \, d\eta$, the average mixed volume fraction, can be determined from chemical reaction experiments to be 0.28. (See Broadwell and Mungal, 1986.) Dahm (1985) computed the mean mixed concentration from profiles like those in Figure 1, finding, of course, a radial variation in this quantity, Figure 3. Such variation, as well as probe resolution questions, makes the accuracy of the determination of the corresponding value for the jet questionable. It is instructive nevertheless, to use the approximate expressions derived from the data of Dahm for the mean mixed jet fluid concentration, $\overline{C}_m$, and the mean concentration, $\overline{C}$, approximated by the equations

$$\overline{C}_m = 3.2(d_o/x) \tag{6}$$

$$\overline{C} = 1.7(d_o/x)$$

to find

$$A = \int \alpha(\eta) \eta \, d\eta \approx 0.5.$$

This value, twice that in the shear layer, is in approximate accord with that found by Waddell in his use of acid-base reactions in water to study mixing in jets, Hottel (1953). (This early work provided the starting point for the liquid shear layer and jet

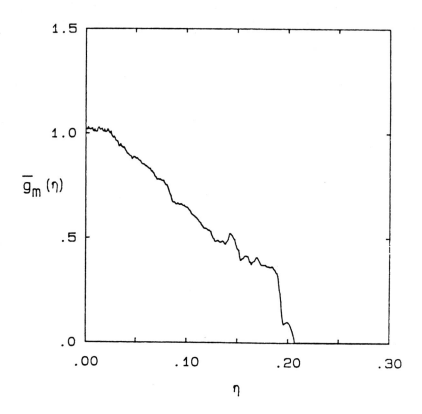

Figure 3. Mean radial profile of mixed fluid concentration $x/d_o = 300$, Re = 5000. Dahm (1985).

experiments discussed above.)

With the assumption that the mixed fluid composition is independent of η and has a single value, the fuel conservation equation becomes

$$2\pi \frac{d}{dx} (C_F)_h d^2 \int \alpha(\eta) u(x,\eta) \eta d\eta = -2\pi k (C_F)_h (C_O)_h d^2 \int \alpha(\eta) \eta d\eta \qquad (7)$$

The oxidizer, carried in the reservoir fluid, is governed by

$$2\pi \frac{d}{dx} (C_O)_h d^2 \int \alpha(\eta) u(x,\eta) \eta d\eta = -2\pi k (C_F)_h (C_O)_h d^2 \int \alpha(\eta) \eta d\eta + v_h (C_O)_\infty \qquad (8)$$

where $(C_F)_h$ and $(C_O)_h$ denote the fuel and oxidizer concentrations <u>in the mixed fluid</u>, $(C_O)_\infty$ is the reservoir oxidizer concentration, and k the reaction rate coefficient for the reaction:

$$O + F \xrightarrow{k} P$$

in which the stoichiometric ratio is taken to be unity. For definiteness, the reaction has been taken to be a single step forward binary reaction with a constant rate coefficient. As will be seen, this simpification can be dropped when numerical computations and complex reactions are considered.

Introducing the variables $C^*_O = (C_O)_h/(C_O)_\infty$, $C^*_F = (C_F)_h/(C_F)_i$, and $z = x/d_0$, where $(C_F)_i$ in the initial fuel concentration, and using Equations (2)-(5), we can write Equations (7) and (8) in the form

$$\frac{d}{dz} (z C^*_F) = -Da\, z^2 C^*_F C^*_O \qquad (9)$$

$$\frac{d}{dz} (z C^*_O) = -Da\, \phi\, z^2 C^*_F C^*_O + 1 \qquad (10)$$

in which ϕ is the equivalence ratio, $(C_F)_i/(C_O)_\infty$, and Da the Damköhler number defined by

$$Da = \frac{A(C_O)_\infty k\, d_0}{E\, u_0}$$

and $A = \int \alpha(\eta)\, \eta\, d\eta$.

When there is no reaction, $Da = 0$, Equation (9) yields

$$C^*_F = c/z = 3.2/z \qquad (11)$$

where the constant is taken from Equation (6).

Since $(C^*_F + C^*_H) = 1$ for $Da = 0$,

$$C^*_F = 1 - 3.2/z$$

Multiplication of Equation (9) by ϕ and subtraction from Equation (10) yields

$$\frac{d}{dz}[z(C^*_O - \phi C^*_F)] = 1,$$

from which we find,

$$(C^*_O - \phi C^*_F) = \frac{z - 3.2(\phi+1)}{z}$$

in which the constant comes from Equation (11). For $Da \to \infty$, $C^*_O \to 0$ in the flame; therefore

$$C_F^* = \frac{3.2(\phi+1) - z}{\phi z}$$

and the flame ends, $C^*_F = 0$, at

$$z_e = 3.2(\phi+1) \qquad (12)$$

Since Equation (12) is valid for any form of the reaction term, as long as $Da \to \infty$, it may be compared with the "flame length" measurements of Waddell and of Dahm et al (1984) for reactions in water where $R_e \sim 10^3 - 10^4$ and $Sc\,Re \sim 10^6 - 10^7$.

The experimental results are shown in Figures 4 and 5, the first establishing the independence of z_e from Reynolds number for $Re > 3000$, and the second the linear dependence of z_e on ϕ. Since in the experiments z_e was taken to be the axial location of the last indication of <u>any</u> unreacted jet fluid and since the flame length fluctuations are large, it is understandable that the measured lengths are much larger than that given by Equation (12). The important point, however, is that the linear dependence of z_e on ϕ in the experiment is consistent with the Equation (12), and hence consistent with the assumption that α is independent of axial distance.

FINITE REYNOLD NUMBER JET

At Reynolds numbers for which the jet is turbulent but low enough for the Taylor layers or strained flames to make significant contributions to the molecular mixing, the procedure is as follows. To simplify the notation, the Schmidt number will be taken to be unity in this section.

Recognizing that the strained flames form between the newly entrained reservoir fluid and the continuously reacted jet fluid, we may sketch the process as shown in Figure 6. The equality of

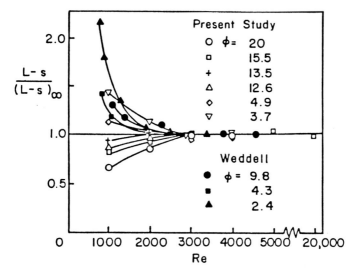

Figure 4. Mean turbulent flame length normalized by its asymptotic value versus Reynolds number and equivalence ratio. Dahm et al. (1984).

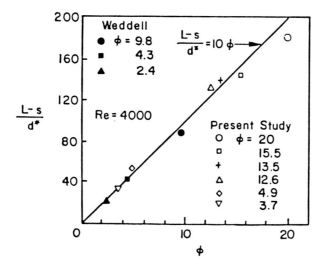

Figure 5. Mean turbulent flame length at high Reynolds number versus equivalence ratio. Dahm et al. (1984).

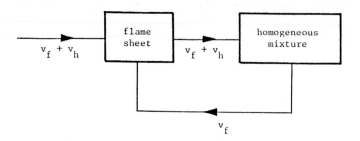

Figure 6. Schematic diagram for reactions in the jet model.

Figure 6

the net volume flux to the flame sheets, v_f, from the homogeneous mixture and the free stream comes from the assumption of constant molecular transport properties, an assumption leading to sheets constituted of equal amounts of fluid from the two streams. Notice that the net flux to the homogeneous mixture, v_h, is unchanged; the presence of the flame sheets only changes the constituents in the entering stream. The fluid in the flame sheets at any axial station becomes part of the homogeneous fluid further downstream.

The net volume addition to the flame sheets, v_f, can be written

$$v_f = 2\pi \frac{d}{dx} d^2 \int \frac{\beta(\eta)}{R} \cdot \frac{bR}{Re^{1/2}} u \eta \, d\eta \tag{13}$$

in which $\beta(\eta)/R$ is the flame sheet area per unit volume and the sheet thickness is taken to be that of a strained flame in equilibrium with the large scale strain u/d, i.e.

$$\lambda_\tau = b \, (D/\varepsilon)^{1/2} = b \, (Dd/u)^{1/2} = b \, d/(Re)^{1/2}$$

where b is a constant of order one (See Carrier, Fendell, and Marble, 1975.) With Equations (3), Equation (13) becomes

$$v_f = 2\pi \, C_1 \, u_0 \, d_0 \, F/Re^{1/2} \tag{14}$$

where

$$F = b \int \beta(\eta) \, g(\eta) \, \eta \, d\eta \, .$$

As before

$$v_h = 2\pi \, C_1 \, u_0 \, d_0 \, E \tag{3}$$

and

$$E = \int \alpha(\eta) g(\eta) \eta \, d\eta$$

With the help of Equations (3) and (14), fuel and oxidizer conservation equations similar to Equations (9) and (10) can be written for both the flame sheets and the homogeneous mixture. Now, however, the flame sheet chemical reactions must be treated numerically even for the simple reaction considered in the preceding section. The proposed scheme for solution, currently being carried out for a simple reaction, is to consider at each integration step, the flame sheet properties to be those of a strained flame in equilibrium with the local strain, u/d, and to use either numerical solutions, such as those of Dixon-Lewis et al (1984), or approximations to them. One side of the flame is pure reservoir fluid and the other the fluid of changing composition from the homogeneous mixture.

In the absence of such solutions, however, several useful conclusions can be drawn from the nature of the model. First, since the flame ends when all the fuel is consumed in both regions, and since the addition rate to the homogeneous mixture is unaffected by the flame sheets, the flame length is independent of Reynolds number in this case also. The same argument shows the independence from Schmidt number. Experimental information on this point is discussed later.

Julian Tishkoff called this paradox to our attention, i.e., the apparent contradiction between the influence of Reynolds and Schmidt numbers on molecular mixing and the independence of flame length from these parameters. The arguments just given appear to resolve the paradox: the upstream molecular mixing in the strained flames influences the mixing rate but does not reduce the distance required for _every_ element of jet fluid to be mixed with ϕ parts of reservoir fluid.

Next, the model provides a basis for understanding the dependence of the nitric oxide production rate on Reynolds number. The experiments of Bilger and Beck (1974) on hydrogen-air flames and of Peters and Donnerhack (1981) on methane-air showed that when the effect of residence time was removed, there remained an effect of Reynolds number; the amount produced appeared to decline approximately with $Re^{1/2}$. Since NO is produced by a slow chemical reaction when the temperature is high, in the model it appears in the flame sheets only at small x/d_o, where, in the absence of radiation, the adiabatic flame temperature is reached. The forward NO reactions stop and NO may be destroyed when the flame sheet material returns to the homogeneous mixture. Near the end of the flame NO is produced in both regions. As Equation (14) illustrates explicitly, the flame sheet NO production also depends on $1/Re^{1/2}$ and the model, therefore, has a mechanism for explaining the experimental results.

To examine these ideas in more detail, an approximate calculation was carried out for both fuels with a full set of kinetic equations. A representative list of the reactions and species can be found in Heap, et al (1976), and the actual updated set used in the calculations in Bowman and Corley (in press).

For the model computation, Tyson, Kau and Broadwell (1982), the strained flame "reactor" was considered to contain a homogeneous mixture like that generated at the Kolmogorov scale, but v_f from the homogeneous mixture was varied to maintain a stoichiometric mixture in the flames and thus to yield the adiabatic flame temperature reached in strained flames. The constant fixing the flow rate, v_h, was inferred, as described above, from the flame length data of Waddell (see Hottel, 1953) and the remaining constant chosen to match the model to the data at one point. Approximate account was taken of radiation in both flames.

The model-experiment comparison for the hydrogen flame is shown in Figure 7. Only the NO emission index was measured for the methane flame and while the agreement, with the same constants, in that case is equally good, the Reynolds number range for the high Froude number runs was limited. In fact, the Reynolds numbers of both experiments are so low that the good quantitative agreement between the model predictions and the experiments is likely only fortuitous. The qualitative trend with $1/Re^{1/2}$, however, is clear in the model and, having been observed in different experiments in two gases, seems believable. There are plans to repeat the experiment and to make a more complete model calculation.

A more straight-forward check of the ability of the model to deal with molecular effects is a comparison of model predictions with experiment of the differences between gas and water reactions, Mungal and Dimotakis (1984). The satisfactory agreement in that case as well not only gives more confidence in the model but lends credence to the idea that nitric oxide production in flames is, indeed, Reynolds number dependent.

EFFECTS OF DENSITY CHANGE

In the discussion so far the density has been taken to be constant. However, Ricou and Spalding (1961) show that when the jet density, ρ_j, differs from that in the reservoir, ρ_∞, the ratio of the total jet flux to the initial value is

$$m/m_0 = c(x/d_0)(\rho_\infty/\rho_j)^{1/2} \tag{15}$$

Since this relation is consistent with the idea that the entrainment is controlled by the jet momentum, it may be expected to hold when the jet is reactive. Avery and Faeth (1974) develop a theory for such jets and with it are able to correlate the flame lengths of a remarkable variety of reactants by a relationship of the form of Equation (15) in which d_0 is replaced by $d^* = d_0(\rho_\infty/\rho_j)^{1/2}$. Dahm & Dimotakis (1985) present their data together with that collected by Avery and Faeth in this form as shown in Figure 8.

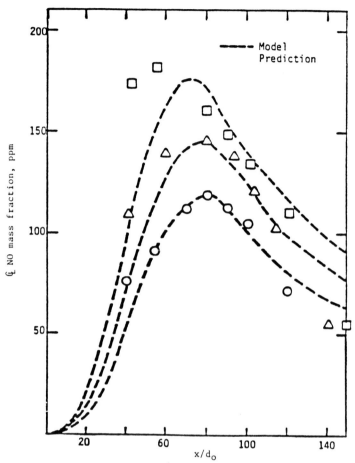

Figure 7. Comparison of calculated nitric oxide concentration with measurements for hydrogen-air flame. □, Re = 1,540; △, Re = 4,350; ○, Re = 12,300. (Froude number ~$6 \cdot 10^5$. Data from Bilger and Beck (1974). From Tyson, et al. (1982).

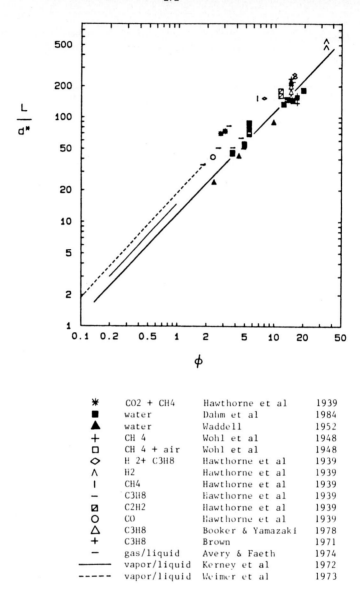

Figure 8. Mean flame length versus equivalence ratio for various reacting jets. Dahm (1985).

The results in this figure are explained by the present model if the distance after entrainment for the jet and reservoir fluid to reach the Kolmogorov scale continues to depend, in this case also, only on the local jet radius. The arguments leading to Equation (11) appear to be valid in this more general situation.

DAMKÖHLER NUMBER EFFECTS

In the shear layer, it is at least a reasonable approximation to consider the flame sheets to be formed between the fixed composition free stream fluids. In this case, approximate analytical treatment of simple chemical reactions with arbitrary reaction rates is possible. See Broadwell and Mungal (1986) for a comparison of such analytical results with the observations of Mungal and Frieler (1987) for the hydrogen-fluorine reactions.

In the jet, the varying jet fluid composition appears to preclude a similar treatment of the flame sheets. Hence, only the high Reynolds number situation, described by Equations (9) and (10), has been considered. C. F. Frieler (private communication) notes that an analytical solution of these equations is possible but that it is of such complexity that numerical results, which he has also kindly provided, are more useful. These are shown in Figure 9, in which we see that the limiting concentration distribution is attained for $Da > 1$. The flame length for $Da \to \infty$, i.e. > 1, is 32 in agreement with Equation (11).

CONCLUDING REMARKS

As has been pointed out in the foregoing, an implication of this analysis is that extremely high values of the parameter Sc Re are required to eliminate explicit molecular effects on molecular mixing. For a Schmidt number of unity, it appears that Reynolds numbers of the order of $10^6 - 10^7$ are necessary. This would mean that most laboratory combustion experiments are significantly influenced by "low Reynolds number" effects and that analyses neglecting them are likely to be inadequate.

With regard to flame sheets, the present analysis emphasizes the notion that the fuel side consists of fuel mixed with a linearly rising amount of products, from $x/d_o \sim 10$, and that the flame sheet concept is appropriate for any Damköhler number provided the Peclet number is sufficiently high. The proposed model is simple enough that output from fairly elaborate strained flame calculations should be able to be incorporated fairly easily. Since there is no limitation on the Damköhler number, the same sort of calculations may help clarify the mechanisms controlling lift-off and blow-off.

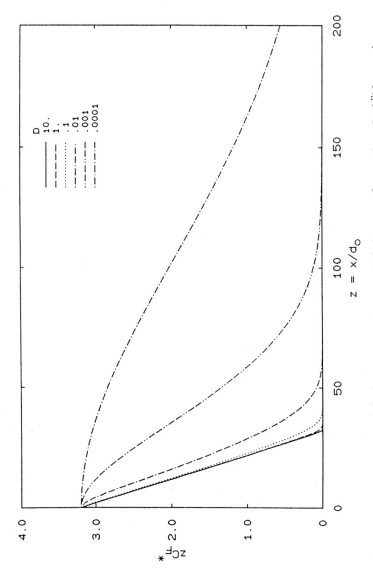

Figure 9. Dependence of fuel concentration on axial distance for various Damköhler numbers. Equivalence Ratio, $\phi_r = 10$.

ACKNOWLEDGEMENTS

This paper is extracted from a summary report on mixing and reactions in shear layers and jets that is in preparation in collaboration with Godfrey M. Mungal. His assistance as well as that of the author's GALCIT colleagues is gratefully acknowledged. The ideas of Roddam Narasimha and Anatol Roshko concerning the volume fraction in which dissipation occurs helped greatly to clarify the relationship between dissipation and scalar mixing. The author also thanks Werner Dahm and Manooch Koochesfahani for permission to use figures from their theses and for many instructive discussions of their work.

We wish to acknowledge support for this work by the Air Force Office of Scientific Research under Contract No. F49602-79-C-0159 and Grant No. AFOSR-83-0213, the Office of Naval Research under Contract No. 00014-85-k-0646, and the Gas Research Institute under Grant No. 5083-260-0878.

REFERENCES

ANTONIA, R. A., CHAMBERS, A. J., BRITZ, D. and BROWNE, L. W. B. [1986] "Organized structures in a turbulent plane jet: topology and contribution to momentum and heat transport", J. Fluid Mech. 172, 211-229.

AVERY, J. F. and FAETH, G. M. [1974] "Combustion of a Submerged Gaseous Oxidizer Jet in a Liquid Metal", 15th Annual Combustion Institute, 501-512.

BATCHELOR, G. K. [1959] "Small-scale variation of convected quantities like temperature in turbulent fluid. Part 1. General discussion and the case of small conductivity", J. Fluid Mech. 5, 113-133

BILGER, R. W. and BECK, R. E. [1974] "Further experiments on turbulent jet diffusion flames", Fifteenth Symposium (International) on Combustion (The Combustion Institute), 541.

BOWMAN, C. T. and CORLEY, T. C. [in press] "Chapter II", EPA Project Decade Monograph, Ed. A. F. Sarofim, T. Lester, G. B. Martin, W. S. Lanier.

BREIDENTHAL, R. E. [1981] "Structure in turbulent mixing layers and wakes using a chemical reaction", J. Fluid Mech. 109, 1.

BROADWELL, J. E. [1982] "A Model of Turbulent Diffusion Flames and Nitric Oxide Generation. Part I", TRW Document No. 38515-6001-UT-00, EERC Final Report, PO No. 18889.

BROADWELL, J. E. and BREIDENTHAL, R. E. [1982] "A Simple Model of Mixing and Chemical Reaction in a Turbulent Shear Layer", J. Fluid Mech. 125, 397-410.

BROADWELL, J. E., DAHM, W. J. A., and MUNGAL, G. [1984] "Blowout of Turbulent Diffusion Flames", Twentieth Symposium (International) on Combustion/The Combustion Institute, 303-310.

BROADWELL, J. E. and MUNGAL, M. G. [1986] "The Effects of Damköhler number in a turbulent shear layer", GALCIT Report FM86-01.

BROADWELL, J. E. AND MUNGAL, G. M. [1987] "Molecular Mixing and Chemical Reactions in Turbulent Shear Layers and Jets", GALCIT Report in preparation.

BROWN, G. L. and ROSHKO, A. [1974] "On Density Effects and Large Structure in Turbulent Mixing Layers", J. Fluid Mech. 64(4), 775-816.

CARRIER, G. F., FENDELL, F. E. and MARBLE, F. E. [1975] "The Effect of Strain on Diffusion Flames", SIAM J. Appl. Math. 28(2), 463-500.

COLES, D. [1983] "On one mechanism of turbulence production in coherent structures", Turbulence and Chaotic Phenomenon in Fluids. Ed. T. Tatsumi, IUTAM Elsevier Science Pub. B. V., 397-402.

CORCOS, G. M. and SHERMAN, F. S. [1976] "Vorticity Concentration and the Dynamics of Unstable Free Shear Layers", J. Fluid Mech. 73(2), 241-264.

DAHM, W. J. A. [1985] Experiments on Entrainment, Mixing and Chemical Reactions in Turbulent Jets at Large Schmidt Numbers, California Institute of Technology, Ph. D. thesis.

DAHM, W. J. A. and DIMOTAKIS, P. E. [1985] "Measurements of Entrainment and Mixing in Turbulent Jets", AIAA Paper No. 85-0056.

DAHM, W. J. A., DIMOTAKIS, P. E. and BROADWELL, J. E. [1984] "Non-premixed turbulent jet flames", AIAA 22nd Aerospace Sciences Meeting (Reno, Nevada), AIAA Paper No. 84-0369.

DIBBLE, R. and MAGRE, P [1987] "Finite chemical Kinetics Effects in a Subsonic Turbulent Hydrogen Flame", Presented at AIAA 25th Aerospace Sciences Meeting (Reno, Nevada) 12-15 January 1987, SAND87-8601.

DIXON-LEWIS, G., DAVID, T., GASKELL, P. H., FUKUTANI, F., JINNO, H., MILLER, J. A., KEE, B. J., SMOOKE, M. D., PETERS, N., EFFELSBERG, E., WARNATZ, J. AND BEHRENDT, F. [1984] "Calculation of the structure and extinction limit of a mathane-air counterflow diffusion flame in the forward stagnation region of a porous cylinder", from: Twentieth Symposium (International) on Combustion/The Combustion Institute, 1893-1904.

HEAP, M. P., TYSON, T. J., CICHANOWICZ, J. E., GERSHMAN, R., and KAU, C. J. [1976] "Environmental Aspects of Low Btu Gas Combustion", 16th Symposium (International) on Combustion, 535-545.

HOTTEL, H. C. [1953] "Burning in Laminar and Turbulent Fuel Jets", 4th (International) Symposium on Combustion (The Williams and Wilkins Co.), 97-113.

HUSSAIN, A. K. M. F. [1983] "Coherent structures and incoherent turbulence", Turbulence and Chaotic Phenomenon in Fluids, Ed. T. Tatsumi, IUTAM, Elsevier Science Pub. B. V., 453-460.

KONRAD, J. H. [1976] An Experimental Investigation of Mixing in Two-Dimensional Turbulent Shear Flows with Applications to Diffusion-Limited Chemical Reactions, Ph.D. Thesis, California Institute of Technology, and Project SQUID Technical Report CIT-8-PU (December 1976).

KOOCHESFAHANI, M. M. and DIMOTAKIS, P. E. [1986] "Mixing and chemical reactions in a turbulent liquid mixing layer", J. Fluid Mech. 170, 83-112.

LONG, M. B. [1987] "Measurement of the topology of large scale structure in reacting flow", These proceedings.

MARBLE, F. E. and BROADWELL, J. E. [1977] "The Coherent Flame Model for Turbulent Chemical Reactions", Project SQUID Technical Report TRW-9-PU.

MUNGAL, M. G. and DIMOTAKIS, P. E. [1984] "Mixing and combustion with low heat release in a turbulent mixing layer", J. Fluid Mech. 148, 349-382.

MUNGAL, M. G, HERMANSON, J. C. and DIMOTAKIS, P. E. [1985] "Reynolds Number Effects on Mixing and Combustion in a Reacting Shear Layer", AIAA J. 23(9), 1418-1423.

MUNGAL, M. G. and FRIELER, C. E. [1985] "Chemical reactions in a turbulent mixing layer: The effects of the reaction rate coefficient - Part I", GALCIT report FM85-01 (31-Dec-85).

MUNGAL, M. G. and FRIELER, C. F. [1987] "The Effects of Damköler Number in a Turbulent Shear Layer", Combustion & Flame, (in press).

PETERS, N. and DONNERHACK, S. [1981] "Structure and Similarity of Nitric Oxide Production in Turbulent Diffusion Flames", Eighteenth Symposium (International) on Combustion (The Combustion Institute), 33-42.

RICOU, F. P. and SPALDING, D. B. [1961] "Measurements of Entrainment by Axisymmetrical Turbulent Jets", J. Fluid Mech. 11, 21-32.

TYSON, T. J., KAU, C. J., and BROADWELL, J. E. [1982] "A Model of Turbulent Diffusion Flames and Nitric Oxide Generation. Part II" Energy & Environmental Res. Corp. (June 1982) Report.

WYGNANSKI, I. and FIEDLER, H. E. [1970] "The Two Dimensional Mixing Region", J. Fluid Mech. 41(2), 327-361.

A FRACTAL DESCRIPTION OF FLAMELETS

F C Gouldin

Sibley School of Mechanical and Aerospace Engineering
Cornell University
Ithaca, New York
USA

Abstract

Under appropriate conditions, reaction in premixed turbulent flames is known to occur in thin sheets called flamelets. For weak turbulence the local structure of these flamelets approaches that of unperturbed steady laminar flames. It has been hypothesized that the geometry of flamelets may be described by the techniques of fractals, and combustion models based on this hypothesis have been proposed.

In this paper fractal based combustion models for burning velocity and chemical closure are reviewed. Evidence for the fractal behavior of surfaces in turbulent flow and of flamelets is reviewed, and new experimental data on the fractal character of premixed flamelets obtained by laser tomography and related techniques are presented and discussed.

Several conclusions regarding the fractal description of flamelets are reached. These include: 1) If valid a fractal description of flamelets can be extremely useful in modeling turbulent combustion rates. 2) While the evidence is not conclusive it appears that flamelet surfaces can be represented by fractal surfaces and that this fractal wrinkling is related to turbulent velocity fluctuations in the inertial subrange. 3) Additional experiments are need to confirm the fractal behavior of flamelets, to determine the fractal dimension and its dependency on conditions and to study the cutoffs to fractal behavior.

I. INTRODUCTION

Mandelbrot [1] has coined the term fractal to denote a field of mathematics [2] which has developed in the last hundred years and which covers a wide range of subjects including the characterization of surfaces and curves with multiple scales of wrinkling and of the distribution of discrete points or mass in space. For many years these topics were of interest only to mathematicians. However Mandelbrot has pointed out that many physical processes can be characterized at least in part by fractal mathematics. In recent years fractals have been applied in the study of many different physical systems, and their study and application to physical problems are currently the subject of intense investigation [3].

One way to define fractal behavior for a wrinkled surface is to consider the variation of measured surface area with varying measurement scale [4, 5] Consider an isotropic, homogeneous fractal surface filling a volume L^3 in space. If the area of this surface is measured with a measurement scale ε which is commensurate with the scales of surface wrinkling it is found that

$$A/L^3 \sim |\varepsilon/L|^{-D} \varepsilon^2/L^3. \tag{1}$$

See also Fig. 1 and [1, 4, 5]. D is the fractal dimension and is defined by (1), and fractal character is reflected in the power law dependency of A versus ε. In many cases the scales of wrinkling of a fractal surface are self-similar, and in this case the power law behavior of (1) is attributed to this self-similarity. For a curve a similar method for determining fractal behavior can be employed except that A is replaced by measured curve length, and the exponent of ε is 1 - D instead of 2 - D. For a surface D falls between 2 and 3, while for a curve D is between 1 and 2. In either case, the larger D is the more "space filling" the curve or surfaces becomes.

Fractals provide new mathematical tools for the description of highly wrinkled surfaces and curves which, as noted by Mandelbrot [1], are complementary to other descriptions such as those obtained through harmonic analysis. There has been some criticism of fractals and their current popularity [6] because the application of fractals has provided little new physical knowledge and insight. This criticism is certainly premature in view of the novelty of the application of fractals to physical problems. In addition there are arguments over the exactness of a fractal description of a physical quantity or object. Frequently such arguments can lose sight of the important question. Namely, do fractals provide a **useful** description or characterization whether or not the description is **exact** ?

Under appropriate conditions, combustion reactions in turbulent flows occur in thin, wrinkled sheets called flamelets. If the geometry of these sheets can be represented by fractal sheets, expressions for mean flamelet surface area can be developed and models for turbulent burning velocity and mean reaction reaction rate can be proposed [5, 7]. These models give relatively simple expressions for quantities of interest in modeling turbulent combustion processes and, therefore, appear to be very attractive for application to a wide range of turbulent combustion problems. In this paper, these fractal-based, combustion models and a related model for jet mixing are summarized and discussed. Then evidence bearing on the validity of the assumption of fractal surfaces in turbulent flow is presented and discussed. It is concluded: 1) that the fractal assumption leads to attractive models for turbulent combustion; 2) that available evidence obtained from analysis and experiment indicate that surfaces in turbulent flow, including flamelets, do exhibit fractal character; but 3) that the evidence is not conclusive especially in regard to the exactness of the fractal representation of these surfaces.

II. MODEL DEVELOPMENT

Models for predicting the turbulent burning velocity [5] and the mean fuel consumption rate per unit volume of premixed, turbulent flames [7] have been proposed. In addition fractal ideas have been applied to the modeling of jet mixing [8]. This work is summarized and discussed in this section.

II.1 Model for Turbulent Burning Velocity
As noted above for a range of turbulence conditions, including conditions and

mixture compositions of practical interest, reaction in premixed turbulent flames occurs in thin reaction sheets referred to as flamelets. For modest levels of turbulence the structure of these sheets approaches that of steady laminar flames. For this condition Damkohler [9] has suggested that

$$u_t = u_0 \frac{A_T}{A_L} \qquad (1)$$

where u_t and u_0 are the turbulent and laminar burning velocities, A_L is the flamelet surface area for laminar flow, and A_T is the ensemble average flamelet area in turbulent flow.

Damkohler's model can be extended to higher levels of turbulence by allowing for perturbations of the flamelet structure by the turbulence, i.e., allowing for flamelet stretch.

$$u_t = <u_L> \frac{A_T}{A_L}, \qquad (2)$$

where $<u_L>$ is the mean, stretched-flamelet burning velocity.

Early attempts to use Damkohler's model suggestion were not successful. A major reason for this lack of success appears to be the failure of area ratio estimates to account for the multiple length scale wrinkling of flamelet surfaces by turbulence [5]. Rough surfaces with multiple, self-similar scales of wrinkling can be characterized with the mathematics of fractals [1], and an expression for the area ratio in (1) and (2) can be obtained.

Mandelbrot [4] has suggested that constant property surfaces in isotropic, homogeneous turbulence are fractal. This hypothesis is supported by experiment [10,11] and by analysis of cloud dispersion [12, 13]. Since there usually is a limit to the maximum and minimum length scales present in turbulent flow, these surfaces are expected to exhibit fractal character for a limited range of measurement scales corresponding to a limited range of scales of surface wrinkling. The experimental results of [11] indicate that the limits to fractal behavior are associated with an inner cutoff (ε_i), which, in turn, is related to the

Kolmogorov (η) scale of turbulence, and an outer cutoff (ε_o), which is related to the integral (l) scale. These findings suggest that the inertial subrange turbulence eddies are the agents of self-similar fractal wrinkling. A simple model of how an eddy could wrinkle a surface is also suggestive of this conclusion. Assume an eddy of length scale l_n and velocity scale u_n causes a wrinkle of wavelength l_n and amplitude $u_n \cdot \Delta t$, where Δt is the eddy life time. Typical scaling [14] gives $\Delta t \sim l_n / u_n$ and the aspect ratio of the wrinkles is seen to be one, independent of the wrinkle amplitude and wavelength, ie., the wrinkling is similar at all scales.

In [5] it is assumed that flamelet surfaces can be represented as fractal surfaces and the assumption is used to estimate the area ratio in (2). Also it is assumed that there are inner (ε_i) and outer (ε_o) cutoffs to the fractal behavior for flamelets and that the ensemble average of measured flamelet area will vary with measurement scale as shown in Fig. 1b. Further it is argued that A_i, the area at the inner cutoff, is the ensemble average flamelet area in L^3, while A_o, the area at the outer cutoff, would be the flamelet area in the absence of turbulence. Thus

$$u_t = <u_L> \frac{A_i}{A_o} = <u_L> \left(\frac{\varepsilon_o}{\varepsilon_i}\right)^D. \tag{4}$$

One sees that in this model the area ratio is given in terms of three parameters - two length scales and the fractal dimension. The assertion that $A_o = A_L$ warrants some discussion. It implies that when viewed at length scales larger than the outer cutoff the surfaces looks smooth and not space filling. Constant property surfaces in homogeneous, isotropic turbulence do not behave in this manner as they appear to be space filling at all measurement scales. On the other hand flamelet visualization data available (for example see below) indicate that flamelets do appear reasonably smooth for large ε at least for laboratory premixed turbulent flames. Since flamelets in this case are not homogeous, isotropic surfaces one might question the direct application, as in [5], of Mandelbrot's ideas which were developed for homogeneous, isotropic surfaces. Various aspects of this question are discussed in [5 and 7].

To complete the burning velocity model expressions for the cutoff length scales, a value for D and an expression for $<u_L>$ are required. In [5] it is argued that

$$\varepsilon_o = l , \text{ and} \tag{5}$$

$$\varepsilon_i = \eta \ (1-(1-A_t^{-1/4}R_l^{-3/4}) \exp(-A_t^{1/4}R_l^{-1/4} u' / <u_L>)). \tag{6}$$

A_t is an empirical constant obtained from data on noncombusting turbulent flow. u' is the root mean square of velocity fluctuations, the absolute turbulence intensity; and R_l is the turbulence Reynolds number ($R_l = u'l / \nu$). Also presented in [5] is an expression for $<u_L>$ based on an analytical expression for flamelet stretch effects developed by Law [5].

$$u_t = <u_L> \{[1-(1-A_t^{-1/4}R_l^{-3/4}) \exp(-(A_t/R_l)^{1/4} u' / <u_L>)]A_t^{1/4}R_l^{3/4}\}^{D-2}. \tag{7}$$

In [5] model predictions for burning velocity are compared to data obtained at Leeds University [15, 16]. A value for D of 2.37 is used based on the results of [10-12]. The comparison shows good agreement for a wide range of R_l (1,000 - 40,000) and of u'/u_0 (u'/$u_0 \le 40$) for various fuels; for an example see Fig. 2.

II.2. Chemical Closure Model

The arguments of the previous subsection can be extended to obtain a chemical closure model for premixed turbulent flames. The proposed closure model expression for fuel formation can be written as

$$<\omega_f> = C_R\rho_o<\Delta Y_f u_L>_f (l \ f/\eta)^{D-2} l \ F^{-1} <c> (1-<c>), \tag{8}$$

with f given by

$$f = [1-(1-A_t^{-1/4}R_l^{-3/4})\exp(-A_l^{1/4}R_l^{-1/4}u'/\langle u_L\rangle)].$$

C_R is a model constant; ρ_o is the reactant gas density; ΔY_f is the change in fuel mass fraction across the flamelet; l_F is the local flame brush thickness; and c is the reaction progress variable which is defined as the ratio of the fuel mass fraction less the initial fuel mass fraction and to ΔY_f. The f subscript on the angle bracket denotes a conditional average with conditioning on the flamelet.

Details of the model development and the analyses are presented in [7]. Here a brief description of the model and a summary of the findings are presented. The model expression for $\langle \omega_f \rangle$ is essentially the product of three terms: 1) the mean fuel consumption rate per unit flamelet area $\{\rho_o\langle \Delta Y_f u_L\rangle_f\}$, 2) the conditional average flamelet surface area per unit volume, conditional on the flamelet being in the volume $\{(l\,f/\eta)^{D-2}\}$, and 3) the probability of the flamelet being in the volume of interest $\{C_R\, l_F^{-1}\langle c\rangle(1-\langle c\rangle)\}$. The flame brush thickness appears in the model to appropriately scale the probability of finding the flamelet in a unit volume. This probability is given by $l_F^{-1}\langle c\rangle(1-\langle c\rangle)$ as discussed in [7] and [17].

The closure model can be used in simplified analyses of normal and oblique flames [7]. For the normal flame analysis, high Reynolds number, one-dimensional flow in the mean, and constant ΔY_f, $\langle u_L\rangle$, η, f and l are assumed. Integration of the equation for $\langle c\rangle$ across the flame brush gives

$$u_t/\langle u_L\rangle = (l\,f/\eta)^{D-2}$$

(which is identical to the burning velocity model prediction) provided one assumes that

$$C_R = l_F / \int_{-\infty}^{\infty} \langle c\rangle(1-\langle c\rangle)\, dx, \qquad (9)$$

which is reasonable. Thus the closure model for normal flames is consistent with the

The analysis can be extended to the oblique flame case provided several additional assumptions are made. The assumptions to be made include constant density, parabolic flow (the flame brush lies almost parallel to the flow direction), and flow similarity. A major point of interest in this analysis is to see if a turbulent burning velocity can be defined for the oblique flame case, and if so, how does it relate to u_t for a normal flame. A definable burning velocity is obtained with $u_t / <u_L> = (l\,f/\eta)^{D-2}$ provided (9) is valid and that either the flames brush thickness is constant or grows linearly with distance in the flow direction [7].

Numerical calculations using the transport closure of [18] and the new chemical closure model are performed for oblique flames with constant density in decaying grid turbulence. Flame results were obtained for a range of conditions with a single value for C_R; $C_R = 4.0$. Calculated burning velocities compare favorably with values obtained from the burning velocity model as would be expected from the results of the normal and oblique flames analyses discussed above. However flame brush thickness did not grow linearly and there were slight but noticeable variances between the burning velocity calculated and that obtained from the u_t model. These differences are attributed to a departure from similarity in the flows obtained by numerical calculations. In turn the departure from similarity is attributed to turbulence decay.

Overall the performance of the burning velocity model in predicting the Leeds data and the consistency of the closure model with the burning velocity model are extremely encouraging. While further development and testing are required, the representation of flamelet surface geometry by fractals appears to be a very promising and useful approach to modeling turbulent premixed combustion. If a flamelet surface can be represented by a fractal surface, an easily applied expression for mean flamelet area can be developed and used for modeling. The chemical closure model can be generalized to provide a rate expression for the species α by replacing ΔY_f with ΔY_α where ΔY_α is the change in α mass fraction across a flamelet. This step assumes that ΔY_α as well as $<\rho \Delta Y_\alpha u_L>_f$ can be defined

II.3 Modeling of Jets

The flamelet concept may also be applied to nonpremixed flames [19]. In this case reaction still is assumed to occur in thin sheet-like regions which are centered on the stoichiometric mixture fraction surface, $Z = Z_s$. (Z is the mass weighted mixture fraction.) Modeling ideas which are applicable to jet mixing and reacting jets can be developed based on the assumption that constant mixture fraction surfaces can be represented as fractal surfaces. In [8] such ideas are applied to jet mixing, and an expression for the fractal dimension is obtained.

An axisymmetric jet is modeled, and the jet structure is viewed as a set of constant concentration surfaces. Stationary turbulence and uniform density are assumed, and a balance equation for the ensemble mean flux of jet fluid across a Z constant surface is developed which states that this flux must equal the mean flow supplying the jet. Terms in

Consider a slice of the jet defined by x to x + dx, where x is axial distance. The

$$dA_Z = 2\pi r_{<Z>}(A_t^{1/4}R_f^{3/4})^{D-2} \, dx. \tag{10}$$

$r_{<Z>}$ is the local radius of the $<Z>$ contour with $<Z>$ = a particular Z value. The area of the $<Z>$ constant surface is identified as A_o and dA_Z is this area multilped by the A_i/A_o ratio which is given by $(A_t^{1/4}R_f^{3/4})^{D-2}$. The average flux of jet fluid per unit area of a Z constant surface is modeled as $C_d\rho Zv$ with v being the Kolmogorov velocity, ρ the density, and C_d a model constant. The average flux of jet fluid across a Z surface between x and x + dx is obtained by multiplying (10) by $C_d\rho Zv$. The resulting expression is integrated over x using a standard similarity form for $r_{<Z>}$ to obtain

$$L/d = C\, C_d\, R_l^{[3/(4-\mu)][D-2] + [1/2 - 3/(4-\mu)]}\, Z_0. \tag{11}$$

L is the axial distance at which $\langle Z \rangle$ on the jet axis falls to Z_0, and d is the initial jet diameter. C is an empirical constant [8], and μ is the intermittency exponent -- $0.25 \leq \mu \leq 0.5$. The form of (11) is that expected from similarity and experiment provided there is no Reynolds number dependency which in turn requires

$$D = 2 + \frac{(\mu + 2)}{6}. \tag{12}$$

Thus the jet mixing analysis gives an expression for the fractal dimension. The expression is identical to the one obtained by others [12, 13] from an analysis of cloud dispersion in homogeneous, isotropic turbulence. For the expected range of μ values, (12) gives $2.33 \leq D \leq 2.42$ which is consistent with experiment [10, 11] and the value used in the burning velocity model. All of these results support the idea of a universal expression for D valid for high Reynolds number.

As in the case of premixed flames the initial application of fractal concepts to modeling of jet mixing has proved useful and encouraging. Further work along these lines is to be encouraged. Also it is essential that the fractal character of flamelet surfaces and other surfaces in turbulent flow be evaluated if possible.

EVIDENCE FOR FRACTAL BEHAVIOR

The primary evidence for fractal behavior of surfaces in turbulent flow must come from experiment. However some support for fractal behavior can be obtained from modeling efforts. The jet mixing analysis of the previous section is an example of such a modeling study. In this section experimental evidence for fractal behavior will be discussed. In addition two models for cloud dispersion will be discussed as they shed some light on the question of fractal behavior.

Models for cloud dispersion

Hentschel & Procaccia [12] and Kingdon & Ball [13] both consider the dispersion of a cloud, e.g., a pollutant cloud, in homogeneous isotropic turbulence. Molecular diffusion is ignored. An expression for the structure function $S(l, t)$ is obtained from a model for $Q(l, t | l', 0)$ which is the probability that a pair of diffusing particles of fluid separated by l' at $t = 0$ will be a vector distance l apart at time t. The structure function, $S(l, t)$, of a scalar, θ, is defined as follows:

$$s(l, r, t) = <|\theta(r+l,t)-\theta(r,t)|^2>,$$

and

$$S(l,t) = \frac{1}{v}\int dr\; s(l,r,t),$$

where integration is over the volume v. Finally, the behavior of S as $|l| \to 0$ is related to D.

While the two works are similar in many ways, they differ in their model for Q and in their relationship between S and D. However both obtain the same expression for D

$$D = 2 + \frac{(\mu + 2)}{6}. \tag{12}$$

As before μ is the intermittency exponent.

Mandelbrot [4] made two suggestions for the fractal dimension of surfaces in isotropic, homogeous turbulence - 2 2/3 for Gauss-Kolmogorov turbulence and 2 1/3 for Gauss-Bergers turbulence. Both Mandelbrot and Hentschel & Procaccia assume Gaussian statistics, while Kingdon & Ball do not. In this regard it is interesting to note that if a Gaussian field is assumed, the Kingdon & Ball approach gives $D = 2\ 2/3$ in agreement with Mandelbrot for Gauss-Kolmogorov turbulence. At this time the Kingdon & Ball approach and result appear to be the most appropriate for the fractal dimension. Their result is identical to that obtained in [8] and presented in the previous section, namely (12).

Experimental evaluation of the fractal character of flamelets and surfaces

While it is not obvious how to assess surface fractal character directly, inferences can be draw from the application of laser tomography and point, scalar measurements; see for example [11]. Preliminary experiments on v-flames have been performed using tomography and point scattering measurements [20, 21], and these experiments are now described.

Measurements were performed on low Reynolds number v-flames. A schematic of the burner used is shown in Fig.3. An unconfined v-flame is stabilized in grid turbulence on a rod (1.5mm dia.) mounted across the exit of a cylindrical burner (50mm dia.). Turbulence is generated by woven-wire screens mounted 30 mm below the rod. Fuel (commercial grade methane) and air are mixed in a plenum chamber and flow into the burner. For details see [17, 22, 23].

The intersection of a flamelet surface with a plane can be visualized with laser tomography [24]. For these experiments the reactant flow is seeded heavily with an oil mist in the same manner as for velocity measurements [17]. The output of a high energy, pulsed Nd:Yag laser is frequency doubled into the green (300 mj per pulse in the green), and lenses used to form the laser beam into a sheet of light approximately 0.5mm in thickness. The flame brush is illuminated by the laser, and the reactant gas in the plane of illumination is made visible by scattering of light from the oil mist. Photographs of the flame brush are taken with a 35mm camera synchronized with the laser pulse. Because the oil evaporates in the flamelet (see [17] for a discussion of this.) there is no scattering from the products, and consequently the curve defined by the boundary between regions of scattering and of no scattering observed in the photographs --called tomograms -- is the intersection of the plane defined by the laser and the flamelet surface. A sample tomogram is presented in Fig. 4.

Curves formed by the intersection of a plane with a fractal surface are fractal, and for isotropic fractal surfaces the fractal dimension of the curve is one less than that of the surface from which it is formed [1, 25]. It is assumed here that the curve fractal dimensions which are measured by this technique are also one less than that of the fractal flamelet even though flamelet surfaces in these experiments are not isotropic. For curves, fractal dimension can be defined in terms of measured length, L, in a manner analogous to that for surfaces. Thus

$$L/L^2 \sim |\varepsilon/L|^{1-D} \varepsilon/L^2. \tag{16}$$

Here D is the fractal dimension of the curve; L is the scale of the surface area over which L is measured; and ε is the measurement scale.

In these studies tomograms for several different flame conditions -- see Table 1 -- are digitized and their fractal dimension determined by application of (16). Plots of L versus ε were made and slopes taken. To obtain L for a given ε the curved is fitted by a piecewise continuous curve composed of straight line segments each segment being ε in length. Then L is the the number of segments multiplied by ε.

To analyze the data a digitizing tablet is used to transfer flamelet curves from photographs to a computer. The digitized data are analyzed for L in the manner just described. There are several potential problems with this approach. The use of a digitizing tablet requires operator judgement, and bias may enter if the curves to be analyzed are too complex or not distinct enough. Digitization in the present study is not a problem. The algorithm for finding L suffers from round-off error as ε approaches L. To correct for this error a remainder term is added to L. This correction has been evaluated and found to give good results although not all round-off error is removed [20]. It is highly desirable to obtain large data records with good spatial resolution so that ε can be varied over several orders of magnitude. For the present measurements the resolution is approximately 0.5 mm (the thickness of the laser sheet), and the maximum picture dimension is approximately 10 cm.

A sample plot of L versus ε is given in Fig. 5. Note that length is scaled by the integral scale in the cold flow turbulence 7 cm above the turbulence generating grid; $l = 2.7$ mm. Since there are limits to the length scales of wrinkling a curve is expected which is similar in shape to that in Fig. 1b, two horizontal lines connected by a straight line with slope 1 - D. Furthermore for a surface fractal dimension of 2.37 ones expects a curve dimension of 1.37. In Fig. 5 one sees that the shape is in general what is expected. However the maximum ε is not large enough to unambiguously show behavior above the outer cutoff. Also the fractal dimension taken from the curve is well below the expected value. Low fractal dimensions are observed for all the flame conditions studied, Table 1. From the plots of logN versus logε, inner and outer cutoffs can be estimated. On physical grounds the sharp cutoff portrayed in Fig. 1 are not expected. In addition data for ε large compared to the outer cutoff are not available. Hence the determination of cutoffs is quite difficult. For the flame conditions studied it is concluded that: $\varepsilon_i/l \approx 0.25$ and $\varepsilon_o/l \approx 5$.

The low fractal dimension obtained in these measurements is attributed to two factors, low turbulence Reynolds number and low u'/u_0. Fractal wrinkling is most likely caused by the inertial subrange turbulent eddies. For low Reynolds number turbulence these eddies are not fully developed which may influence the character of surface wrinkling. In section **II.1** a simple model for surface wrinkling by turbulent eddies was presented. For this model, the amplitude of wrinkling depends on a time scale, Δt, which is the time a surface is exposed to an eddy and was taken, in Section II, as the eddy life time. Since a flamelet propagates relative to the reactant mixture, the exposure time may be less than an eddy life time. An alternate exposure time is given by l_n / u_0. For $u_0 > u_n$ this latter time is, most likely, the exposure time, and the amplitude of corrugation by this crude model is $l_n(u_n/u_0)$. Thus the condition $u_n = u_0$ marks a transition in behavior. The l_n for $u_n = u_0$ is defined as the Gibson scale [26] -- l_G -- and by standard scaling arguments $l_G = l\,(u_0/u')^3$. For the flame conditions studied $l_G > l$, i. e., the Gibson scale is undefined.. By our simple, wrinkling model all wrinkle amplitudes scale as $l_n\,(u_n/u_0)$ rather than l_n and it is reasonable to expect a variation in fractal behavior. By the above argument one expects D would approach 1.37 for large R_l and large u'/u_0.

The set of points formed by the intersection of a line with a fractal surface also exhibits fractal behavior, and for an isotropic fractal surface the fractal dimension of the set of points is two less than the fractal dimension of the surface. A fractal set of points is known as a fractal dust [1]. For points along a line the fractal dimension is determined by the variation in the measure of the set with measurement scale. Divide the line into segments of length ε. The number of segments in which there is one or more members of the set is N. Then for a fractal set

$$N \sim \varepsilon^{-D}. \tag{17}$$

D is again the fractal dimension, and its value for a dust falls between 0 and 1.

Time-series, point-scattering measurements were performed to study the fractal character of the distribution of flamelet crossing events in time. By Taylor's hypothesis this experiment is equivalent to intersecting a surface with a line in th flow direction. As before the reactants are seeded with an oil mist but scattering is excited by a He-Ne laser. Scattered photons from a small volume are detected and used to drive a logic circuit which generates a short duration, logic pulse at each flamelet crossing event. These pulses control a computer clock which counts the time between crossing events, and these passage times are stored in computer. The spatial resolution of these measurements is a few tenths of a millimeter and the temporal resolution is approximately 20 μsec.

From recorded data the measure of a set of crossing events along the time axis can be determined as a function of ε. Plots of N versus ε are then generated; see Fig.6. To generate such curves data sets of approximately 8000 entries are obtained and for analysis are divided into segments of approximatelty 100 ms in length. Each segment is evaluated to find $N(\varepsilon)$, and the $N(\varepsilon)$ for the several segments are averaged to obtain $<N(\varepsilon)>$ from which the fractal dimension is obtained. If data segments much longer than 100 ms are used in the analysis, no regions of fractal behavior are observed. A similar lack of fractal behavior for long time records is observed by Sreenivasan and Meneveau [11] who attribute this lack to a breakdown of Taylor's hypothesis. As in the analysis of curves, a round-off error is encountered in finding N when ε approaches the data segment length. As a result, for large ε, $N(\varepsilon)$ behaves like a step function. In this case D is found by considering only the left most point of log<N> versus logε at each step level, see Fig.6.

Fractal behavior is not obvious in the data shown in Fig. 6. However, these data may be represented by several straight line segments (see Fig. 6) and a physical interpretation can be given to each of the segments. With reference to Fig. 6, the horizontal line to the left is for the case where ε is below the inner cutoff and all members of the set are resolved. The right most line has a slope of approximately -1 which indicates these ε values are above an outer cutoff since a slope of -1 is expected for random events without fractal behavior. Fractal behavior in data of this type is associated with events being grouped into bursts of events. A slope of -1 for a range of ε indicates that there are set members in all time intervls into which the time segment being analyzed is divided, and therefore, N equals the number of intervals in the segment which in turn is proportional to $1/\varepsilon$. In between these two regions of limiting behavior are two regions of fractal behavior; one having a fractal dimension of approximately 0.1 and the other a dimension of approximately 0.5.

The tomographic results imply a fractal dimension for the dust of approximately 0.1, and therefore we associate the $D \approx 0.1$ segment of the curve with the fractal wrinkling of the flamelet surface. This interpretation is further corroborated by applying Taylor's hypothesis and comparing the regions of fractal behavior in the time and space measurements. From Fig. 6 with a flow velocity of 3.8 m/s one finds the limits of the region in which $D \approx 0.1$ correspond to the center of the region of fractal behavior observed for flamelet curve; note vertical lines in Fig. 5. More precisely, the cutoffs, both inner and outer, for the region $D \approx 0.1$ may be defined by the straight line segments representing the regions where $D \approx 0.1$, and 0.5. The results of this exercise are given in Table 2 where it is seem that the $\varepsilon_i \approx 0.3$ ms and $\varepsilon_o \approx 1$ ms.

To understand the portion of the curve where $D \approx 0.5$ consider a coin toss game in which the winner of each toss receives the coin. The coins are identical and fair. A plot of cumulative winnings for one of the players versus trial number is a random walk, while the distribution of zero winnings versus trial number is a fractal set with dimension 1/2 [1].

This observation suggests that the $D \approx 0.5$ segment may be the result of large scale flapping of the flamelet. In other words, a smooth, flapping flamelet would give fractal behavior with $D \approx 0.5$, while the fractal wrinkling of the flamelet contributes a region in the <N> versus ε plot with $D \approx 0.1$. With regard to the proposed burning velocity and chemical closure models the fractal character and fractal dimension of the surface wrinkling are the critical features to be tested by the experiments.

While estimates are possible, neither the tomographic data nor the time series data are adequate to establish clear cutoffs to fractal behavior. The tomographic data do not cover a large enough region of space to determine outer cutoffs. For both types of data the regions of observed fractal behavior are narrow which hinders a clear determination of cutoffs as well as determination of the fractal dimension. Broader regions of fractal behavior should be present for flames at higher turbulence Reynolds number and higher u'/u_0. Experiments at such conditions are sorely needed.

This section is concluded with a brief summary of additional experimental results not obtained at Cornell. Lovejoy [10] has reported a fractal dimension of about 2.34 for clouds inferred by examining the fractal character of cloud projections on the earth's surface as obtained from satellite pictures and from rain fall patterns observed by radar. Sreenivasan and Meneveau [11] have made measurements in nonreacting shear layers using tomography and hot-wire anemometry. The boundary between turbulent and nonturbulent fluid is visualized with tomography, while velocity measurements are used to generate an intermittency function which is analyzed as a fractal dust. For a boundary layer and axisymmetric jet they find a D between 2.32 and 2.4, while $\varepsilon_i \approx \eta$ and $\varepsilon_o \approx 3l$. Only intermittency measurement are made in a plane wake and a mixing layer where D is found to vary with location in the flow; 2.38 to 2.4 are typical values. The dimension of isovelocity surfaces is inferred from point velocity measurements and found to vary from about 2.2 to 2.8 depending on location in the flow and on the velocity surface observed. These large variations in D are not fully explained as yet. In general the results both of Lovejoy and of Sreenivasan and Meneveau indicate fractal behavior for surfaces in turbulent flows.

D measurements using tomography in premixed flames have been reported by Peters [26] and by North and Santavicca [27]. Peters has studied a v-flame similar to the one used at Cornell; only one flame condition is studied and $D \approx 2.1$ is inferred. North and Santavicca present results for unanchored, propagating flames. D values between 2.1 and 2.2 are observed. Also their data show that D increase as Reynolds number increases. This latter finding is extremely significant, and measurements are needed to confirm the observation for other flame configurations and to see if D approaches a limit as Reynolds number increases.

SUMMARY AND CONCLUDING COMMENTS

A fractal representation of surfaces in turbulent flows is found to be a very useful tool for modeling purposes. The fractal modeling of turbulent combustion introduces three new parameters, the fractal dimension and two length scales. Experiments support the hypothesis that flamelet surfaces and constant property surfaces can be represented by fractal surfaces, but more experiments are need before the evidence can be considered conclusive. Experimental evidence on the cutoffs to fractal behavior is not as strong. The data of Sreenivasan and Meneveau are most adequate in this regard. More measurements at high trubulence Reynolds number with good resolution and large samples are needed to adequately address the question of cutoffs.

Acknowledgement

This work was supported primarily by the Army Research Office (contract numbers DAAG29-82-K 0187 and DAAL03-87-K-0039).

Table 1. Tomographic Fractal Analysis Results

CH$_4$ - Air flames.
8 mesh screen turbulence generator

ϕ	mean axial velocity (m/s)	R_l	number of curves analyzed	mean fractal dimension	standard deviation
0.8	4.1	25	18	1.09	0.03
1.0	4.2	25	16	1.11	0.03
0.8	3.8	23.5	16	1.11	0.04

Table 2. Results from Point Scattering Measurements

CH4 - Air flames
8 mesh screen turbulence generator
All data taken 5.5 cm from turbulence generator;
$<U> = 3.8$ m/s; $l = 2.48$ mm; $R_l = 24.4$.

ϕ	$<c>$	$<ptime>$	flamebrush thickness, y	log $\varepsilon_i/<ptime>$	log $\varepsilon_o/<ptime>$	D for regions: 1	2	3
1.0	0.6	1.70 ms	6.6 mm	-0.81	-0.20	0.96	0.50	0.13
1.0	0.4	2.69	6.6	-0.89	-0.33	1.02	0.51	0.11
0.8	0.6	1.58	4.0	-0.78	-0.25	0.99	0.50	0.11
0.8	0.4	2.33	4.0	-0.93	-0.33	0.94	0.50	0.11

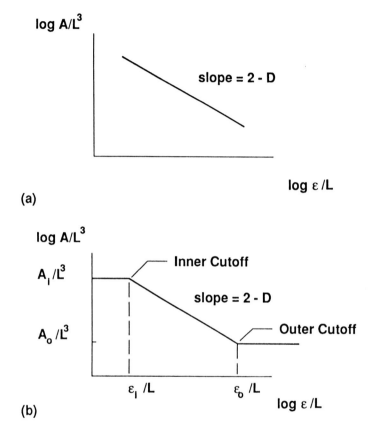

1. Variation of measured area with measurement scale for a fractal surface in a cubic volume of side L. a) No inner or outer cutoffs to fractal behavior. b) Cutoffs to fractal behavior.

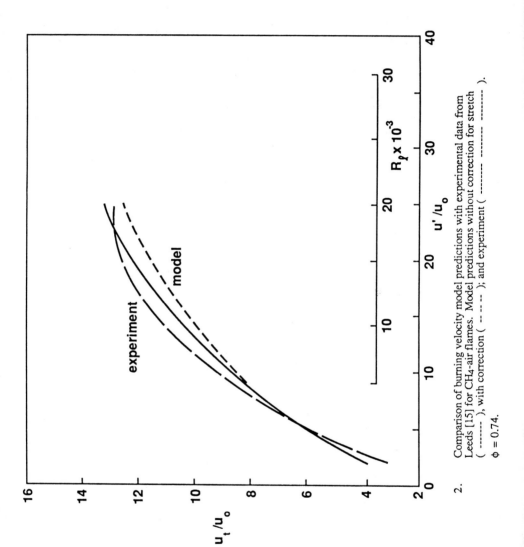

2. Comparison of burning velocity model predictions with experimental data from Leeds [15] for CH_4-air flames. Model predictions without correction for stretch (— — —), with correction (– – – –); and experiment (— — —). $\phi = 0.74$.

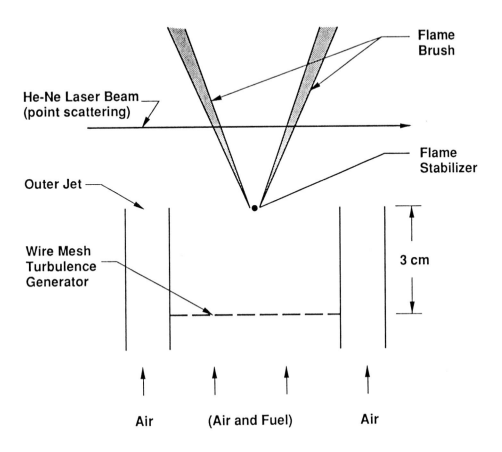

3. Schematic of burner showing location of wire mesh turbulence generator, stabilizer rod, and laser beam for flamelet passage measurements. The inner jet diameter is 5 cm, and the outer, annular jet diameter is 7.6 cm. The fuel is methane.

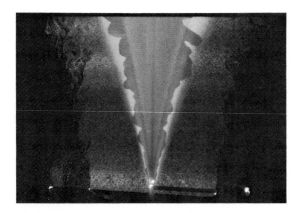

4. Tomogram of methane-air v-flame for conditions similar to those of Fig. 5.

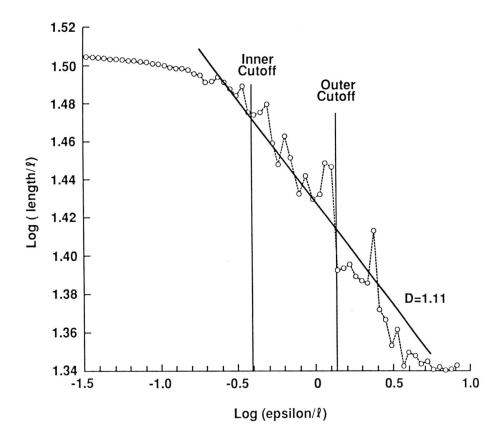

5. Typical length plot for fractal analysis of tomograms for methane-air v-flames [20]. $\phi = 0.8$; $<U> = 3.8$ m/s; $R_l = 23.5$; and $l = 2.7$ mm.

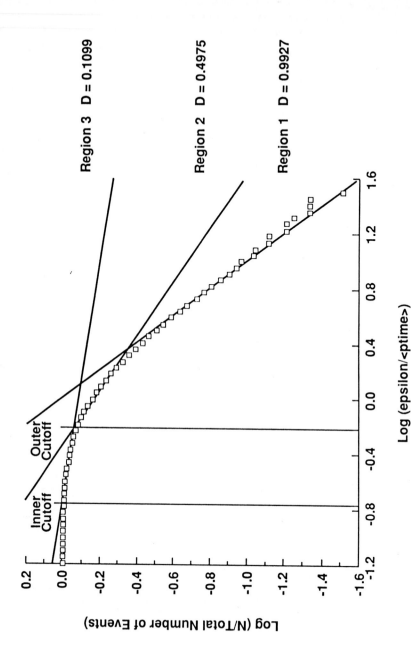

6. Typical plot for fractal dust analysis of the time series measurements of flamelet crossing events in methane-air flame [21]. Data taken 5.5 cm above turbulence generating grid. $\phi = 0.8$; $<c> = 0.6$; $<U> = 3.8$ m/s; $l = 2.48$ mm; and $R_l = 24.4$. $<ptime>$, the mean time between passage events, is 1.58 ms.

REFERENCES

1. Mandelbrot, B B: **The Fractal Geometry of Nature**, Freeman, New York, 1983.

2. Falconer, K J: **The Geometry of Fractal Sets**, Cambridge, Cambridge, 1985.

3. La Brecque, M: **Mosaic 14(4)**, 22-41, 1987.

4. Mandelbrot, B B: **J. Fluid Mech. 72**, 401-416, 1975.

5. Gouldin, F C: **Comb. Flame 68**, 249-266, 1987.

6. Kadanoff, L.: **Physics Today**, Feb. 1986.

7. Gouldin, F C, Bray, K N C, and Chen, J-Y: Chemical Closure Model for Fractal Flamelets, to be submitted to **Comb Flame**.

8. Gouldin, F C: An interpretation of jet mixing using fractals, submitted to **AIAA J**.

9. Williams, F A: **Combustion Theory**, 2nd ed., Benjamin/Cummings, Menlo Park, 1985, p.429.

10. Lovejoy, S: **Science 216**, 185, 1982.

11. Sreenivasan, K R and Meneveau, C: **J Fluid Mech. 173**, 356-386, 1986.

12. Hentschel, H G E and Procaccia, I: **Physical Review A - General Physics 29**, 1461-1470, 1984.

13. Kingdon, R D & Ball, R C: The fractal dimension of clouds, submitted to **J. Phys. A**, 1987.

14. Frisch, U, Sulem, P-L, and Nelkin, M: **J. Fluid Mech. 87**, 719-736, 1978.

15. Abdel-Gayed, R G, Al-Khishali, K J, & Bradley, D: **Proc. Roy. Soc. Lond. A391**, 393-414, 1984.

16. Abdel-Gayed, R G, Bradley, D, Hamid, M N, & Lawes, M: **20th Symposium (International) on Combustion**, The Combustion Inst., 505-512, 1984.

17. Miles, P & Gouldin, F C: College of Engineering, Energy Report E-86-03, Cornell University, Ithaca, NY, 1986. Presented at the 2nd ASME-JSME Thermal Engineering Conference, Honolulu, Hawaii, March 1987.

18. Chen, J-Y, Lumley, J L, & Gouldin, F C: to appear in the **21st Symposium (International) on Combustion**, The Combustion Inst., 1987.

19. Williams, F A: pp. 189-208, in **Turbulent Mixing in Non-reactive and Reactive Flows**, Plenum Press, 1975.

20. Hilton, S M: Measurements to Determine the Fractal Character of Premixed Turbulent Flames, Master of Engineering (Engineering Physics) Report, Cornell University, Ithaca, NY, May 1987.

21. Lamb, T: Point In Time Measurements of the Fractal Dimension of Premixed Turbulent Flames, Master of Engineering (Engineering Physics) Report, Cornell University, Ithaca, NY, June 1987.

22. Gouldin, F C & Dandekar, K V: **AIAA J. 22**, 655-663, 1984.

23. Gouldin, F C & Halthore, R: **Exp. in Fluids 4**, 269-278, 1986.

24. Boyer, L: **Comb. Flame 39**, 321-323, 1980.

25. Procaccia, I: **J. Stat. Phys. 36**, 649-663, 1984.

26. Peters, N: Laminar Flamelet Concepts in Turbulent Combustion, invited paper presented at the **21st International Symposium on Combustion**, Munich, August 3-8, 1986.

27. North, G L, and Santavicca, D A: Fractal Analysis of Premixed Turbulent Flame Structure, paper presented at the 19th Fall Technical Meeting of the Eastern Section: The Combustion Institute, December, 1986, San Juan, Puerto Rico.

SOME RESULTS ON THE STRUCTURE OF THE TEMPERTATURE FIELD IN LOW DAMKOHLER NUMBER REACTION ZONES

F. Gaillard and I. Gökalp

Centre National de la Recherche Scientifique
Centre de Recherche sur la Chimie de la Combustion
et des Hautes Températures
45071 Orléans Cedex 2, France

INTRODUCTION

Low Damköhler number premixed gaseous reaction zones, either thickened flame fronts or distributed reactions regimes, are much less investigated than high Damköhler number flames. Nevertheless, there are several reasons, ranging from the most fundamental ones to those of applied nature, which should motivate their study. Concerning the formers, low Damköhler number reaction zones constitute turbulent reacting flow situations where the chemistry, mixing and turbulence are the most profoundly intermingled due to the comparability of their respective time scales. They constitute then a challenging configuration to look for the interactions between chemical reactions and turbulence, as the combustion reactions are distributed throughout the volume occupied by the turbulent reaction zone /1/. Furthermore, due to their low exothermicity, the low Damköhler number reaction zones are most suitable from the perspective of a systematic comparison between the turbulence structure in reacting and non-reacting but strongly variable density flows.

The reasons of applied nature which may motivate the investigation of low Damköhler number reaction zones are, among others, the ignition and extinction periods in industrial combustion chambers, the pollutant production mechanisms, the knock phenomenons ,the design of industrial continuous stirred tank reactors, etc. Indeed, it is today generally accepted that in order to predict the performances of turbine engines or internal combustion engines, computation codes including fundamental aspects of different combustion regimes are needed.

For all these reasons, we have initiated an experimental work on the interactions between turbulence and chemistry during low oxidation reactions in a jet stirred reactor. Parallel to this work, another low Damköhler regime, constituted by a turbulent cool flame, is also investigated /2,3/. The study of low Damköhler regimes in well stirred reactors may be justified from two different but strongly linked perspectives. The first one is related to chemical kinetic studies for which this kind of technique is quite suitable. Indeed, owing to the capability of ensuring a good macromixing level in this kind of reactor, detailed experimental data have been collected in CRCCHT on hydrocarbon oxidation (C_1 to C_4) up to high conversion rates of the initial fuel and kinetic modelling has been developed/4-7/. On the other hand, the knowledge of the micromixing level (or the segregation index) is essential either to ensure a perfect mixing level or to evaluate the influence of the imperfect mixing on chemical conversion rates/8,9/. This knowledge may only be obtained from the experimental determination of the small scale scalar structure in the reactor.

The second perspective which motivates this study is related to the understanding of the interactions between turbulence and chemistry. Indeed, this turbulent reacting flow situation is most suitable for the investigation of this interaction for the following reasons:

* the mean thermodynamic parameters are homogeneous (perfect macromixing), eliminating then the turbulence generation processes due to gradients of mean quantities;

* the non-reacting turbulence structure is very close to the ideal homogeneous isotropic representation;

* for oxidation studies with very diluted mixtures, the heat release parameter is very low, which eliminates the strong variable density effects.

In order to address simultaneously these two perspectives, an experimental study of the instantaneous temperature field in a jet stirred reactor has been initiated. The temperature field is determined by fine wire thermometry, with and without chemical reactions and under several combinations of the major experimental parameters, such as the average residence time, the equivalence ratio or the intial reactant concentration. Ethylene has been choosen as a typical fuel for this study. Indeed, it correctly represents the light fragments arising from the degradation of practical fuels. Also, previous studies performed at CRCCHT have allowed to establish a detailed kinetic model for C2H4 oxidation, validated over a wide rage of operating conditions/5/.

EXPERIMENTAL SET-UP AND MEASUREMENT TECHNIQUES

The jet stirred reactor technique was first conceived by Matras and Villermaux to investigate hydrocarbon pyrolysis /10/ and low temperature hydrocarbon oxidation /11/, for average residence times around 1 second. Following the construction rules defined by these authors to ensure a good macromixing, two jet stirred reactors, specially designed to investigate the hydrocarbon oxidation in the intermediate temperature range, were built at CRCCHT /12/. One of them is located in a steel vessel which can be pressurized up to 10 bars. The second can be operated at atmospheric and subatmospheric pressures, with the main purpose of detecting molecular as well as radical species. In the investigation reported here, the second jet stirred reactor is used; consequently the further description will only concern this reactor.

The spherical reactor of 40 mm diameter is made of fused silica to prevent wall catalytic reactions. It comprises 4 nozzles of 0.5 mm diameter and located on the equatorial plane of the sphere. The four jets from the nozzles achieve turbulent mixing and recirculation of the gases within the reactor (figure 1).

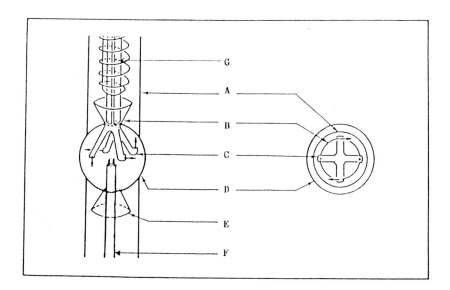

Figure 1. Detail of the reactor. A, external tube; B, convergent cone; C, injectors; D, spherical quartz reactor; E, divergent cone; F, fine wire thermometer; G, capillary surrounded by the preheating resistor.

The reactor is also equipped with four quartz windows to allow optical access for laser based measurements. The flow is preheated by means of electrical resistors before its admission into the reactor. The fuel, diluted in nitrogen, is introduced separately through a capillary and mixed with the principal flow composed of oxygen and nitrogen at the entrance of the nozzles. The reactor is surrounded by an electrical heater whose adequate adjustment allows to reduce the temperature gradient along the vertical diameter at a negligible level, provided the reactants is highly diluted in nitrogen. The residence time of the mixture in the nozzles are extremely lower (more than 300 times) than the average residence time in the reactor, thus preventing any pyrolysis or oxidation of the reactants inside the injectors.

The mean temperature in the reactor is measured by a chromel-alumel thermocouple and the molecular species are sampled through a quartz sonic probe and analyzed by gas-phase chromatography. The position of the injectors allows to investigate easily along the vertical diameter and to verify the macromixing of the system. In this way, it can be shown that the reactor is well macromixed for average residence times ranging between 0.02 s. and few seconds.

The fluctuating temperature field is determined by the fine-wire thermometry (or the so-called cold wire thermoanemometry). The probe supporting the 0.01 mm diameter platinum/rhodium fine wire is introduced from the bottom of the reactor, by using the probe position adjuster used for the sampling probe and the thermocouple. The fine wire is heated by a constant current of 1.5 mA. This low heating current minimizes any contamination of the temperature signal by the velocity fluctuations. At this stage of the investigation, the thermal inertia of the wire is not compensated for. However, as the mean thermal and dynamic conditions are very similar between all the reacting and non-reacting cases investigated, it may easily be assumed that the error introduced by the time constant of the wire is the same for each condition, so that the measured temperature fluctuations are comparable with each other. The cold wire is previously calibrated in a static oven against a thermocouple. The instantaneous temperature signal, delivered by the DISA thermoanemometric unit and the signal conditioner, is tape recorded, digitized at 2000 Hz. and stored on a micro-computer. The various moments of the instantaneous temperature field, its probability density function and power spectral density are then calculated numerically.

EXPERIMENTAL STRATEGY AND EXPERIMENTAL CONDITIONS

The fluctuating temperature field in the jet stirred reactor is explored for several conditions in order to compare the reacting and non-reacting scalar field structures and to investigate the influence of the following parameters: the position z along the vertical diameter of the reactor, the average residence time t_R, the equivalence ratio PHI and the initial concentration of the hydrocarbon $(C2H4)_o$. The non-reacting case is investigated with the reactor traversed with a nitrogen-oxygen mixture flow heated up to the reaction temperature which is kept equal to 1033 K in each reacting run. For both reacting and non-reacting cases, average residence times t_R=60, 100 and 200 ms are investigated. The influence of the position in the reactor (the spatial homogeneity of the micromixing) is investigated for two reacting cases. Three values of the equivalence ratio (PHI= 0.4, 1 and 1.5) and three values of the initial C2H4 volume concentration (0.15%, 0.3% and 0.4%) are explored at atmospheric pressure.

Each experimental run includes two parts. In a first part, gas flows are adjusted so as to produce a kinetically well known reaction for the given experimental conditions and the temperature measurements are performed. After the completion of the measurements the flow of ethylene is suppressed, while the flow rates of nitrogen and oxygen are kept unchanged and the temperature measurements are repeated for this non-reacting case. As the fuel concentration is very low, the average residence times are considered as identical in both cases. Experimental conditions under consideration correspond to fuel conversion stages exhibited on figure 2 . Concentration profiles of major species (C2H4, CO, CO2) are deduced from a simulation work taking into account the detailed kinetic model recently established for C2H4 oxidation/5/.

TIME SCALES IN THE JET STIRRED REACTOR

The determination of the different relevant time scales in the jet stirred reactor is a necessary step in the evaluation of the experimental results. The case for t_R=60 ms corresponds to sonic conditions and the compressibility effects may be neglected for t_R=100 ms and 200 ms. The mean flow velocities calculated with these approximations are used in the determination of the energy input per unit mass in the reactor, which is also put equal to the

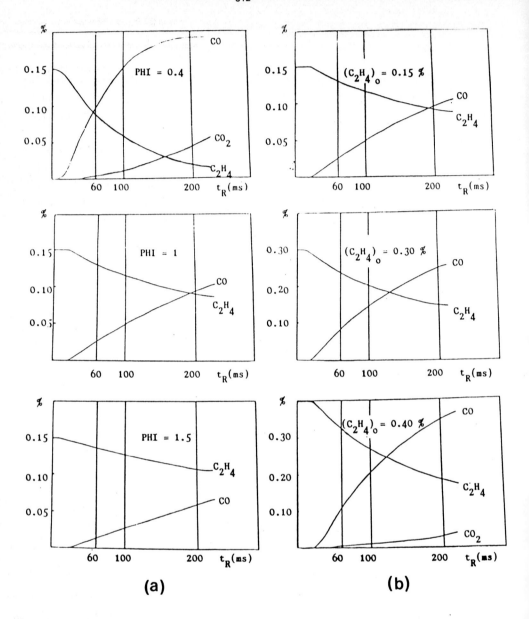

Figure 2. Simulated conversion rates for the oxidation of ethylene at atmospheric pressure and T = 1033 K. (a) influence of the equivalence ratio PHI for $(C2H4)_o$ = 0.15%; (b) influence of the initial ethylene concentration $(C2H4)_o$ for PHI = 1.

turbulence dissipation rate, ϵ. The average residence time, t_R, is calculated as the reciprocal of the flow rate per unit volume of the reactor. The micromixing or the diffusion time t_D, which is the inverse of the mixing intensity β, is estimated by using the Corrsin's relation /8/ $\beta = 0.5 \, (\epsilon/D^2)^{1/3}$, where D is the reactor diameter. The Kolmogorov length and time scales may also be estimated from the usual relations as $\eta = (\nu^3/\epsilon)^{1/4}$ and $t_\eta = (\nu/\epsilon)^{1/2}$, where is the kinematic viscosity of air evaluated at 1033 K. These different time scales are listed on Table I.

t_R	U	ϵ	t_D	t_η	η	t_D/t_R
ms	m/s	cm^2/s^3	ms	ms	mm	
60	655	3.7x10^{10}	1.5	0.006	0.029	0.026
100	376	7.1x10^9	2.6	0.013	0.043	0.026
200	188	8.8x10^8	5.2	0.038	0.072	0.026

Table I

The ratio t_D/t_R is the usual definition of the Damköhler number in well stirred reactors. However, it is worth to recall that in this reactor configuration, the average residence time does not represent the actual characteristic time of the reacting system, since it is related to an arbitrary conversion rate, which, for given values of the initial parameters, may be any value of this rate between very low and very high ones. This is why this definition of the Damköhler number depends only on the reactor design parameters. Therefore, it would be desirable to use a more appropriate definition of the characteristic chemical time in the estimation of the t_D/t_c ratio in order to ensure a better comparison of the micromixing and chemical time scales. In this study, we evaluated t_c, from the simulated conversion rate profiles, as the necessary time delay to observe 50% of fuel conversion. The values of the Damköhler number corresponding to this definition of the characteristic chemical time are listed in Table II. In general, these values are close to 0.026 which is the value of the Damköhler number obtained by taking the average residence time as the characteristic chemical time.

COMPARISON OF THE TEMPERATURE FIELD STRUCTURE BETWEEN THE NON-REACTING AND REACTING FLOWS.

The typical variation of the r.m.s. values of the temperature fluctuations T' corresponding to various experimental conditions are listed in Table II.

PHI	$(C_2H_4)_o$	t_R	t_C	t_D	z	t_D/t_C	T'
	%	ms	ms	ms	cm		K
0	0	60		1.5	0		0.98
0	0	100		2.6	0		0.86
0	0	200		5.2	0		1.32
0.4	0.15	60	76	1.5	0	0.020	0.96
0.4	0.15	100	76	2.6	0	0.034	1.32
0.4	0.15	200	76	5.2	0	0.068	1.88
0.4	0.15	200	76	5.2	-1	0.068	2.18
0.4	0.15	200	76	5.2	-1.5	0.068	1.96
1	0.15	200	285	5.2	0	0.018	1.24
1	0.30	200	205	5.2	0	0.025	0.81
1	0.40	200	180	5.2	0	0.029	0.68
1.5	0.15	200	340	5.2	0	0.015	2.23
1.5	0.15	200	340	5.2	-1	0.015	1.66
1.5	0.15	200	340	5.2	-1.5	0.015	1.75

Table II

The levels of the temperature fluctuations displayed on this table are only slightly scattered around 1 K. This low level of the temperature fluctuations is compatible either with the small spatial mean temperature gradient (5 to 10 K) and with the small temperature increase due to reactions (less than 20 K) Furthermore, in the cases without reaction, at the center of the reactor, T' appears as independent of the average residence time t_R. However, with reaction, T' increases with t_R. As shown on figure 2, when t_R increases, the fuel conversion rate is higher and the most exo-energetic stages of the reactions, due to carbon monoxide consumption, are reached. This increased exothermicity may then explain the increase of T' with t_R. On the other hand, as shown on Table II, the ratio t_D/t_C increases with t_R, for a given equivalence ratio and initial C_2H_4 concentration. This increase of the Damköhler number, due to the increase of the turbulent micromixing time t_D, may also be correlated with the increase of T'. Finally, Table II shows that T' is unchanged between the non-reacting and reacting cases for t_R=60 ms. For the two highest average residence times, T' with reaction is approximately 50% higher than the non-reacting values.

The probability density distributions of the temperature fluctuations are quasi-gaussian in the non-reacting cases, for the three average residence times investigated. Indeed, as shown on Table III for the center of the reactor, the skewness and flatness values of these distributions are very close to their gaussian levels.

t_R ms	S	F
60	0.048	3.68
100	0.077	3.70
200	0.025	3.34

Table III

An example of the spectral distribution of the temperature fluctuations, $E_T(n)$, is presented on figure 3 for the non-reacting case with t_R=100 ms and at the centre of the reactor. The flat portion of the spectrum extends up to approximately 100 Hz. For higher frequencies, the spectrum rolls off for almost three decades with a slope around -3.5. It is worth to recall here that as the thermal inertia of the fine wire is not corrected, the actual slope would be weaker. On the other hand, the spikes on this spectrum are the harmonics of 50 Hz and are due to the contamination of the temperature signal with the periodic frequency of the electrical heating of the reactor. However, this contamination influences only slightly the overall energy level of the spectrum up to 500 Hz, where the noisy portion of the spectrum is entered. The comparison of the reacting and non reacting cases shows that the temperature spectra are very similar for each case and that they are not strongly influenced by the variations of the parameters investigated here.

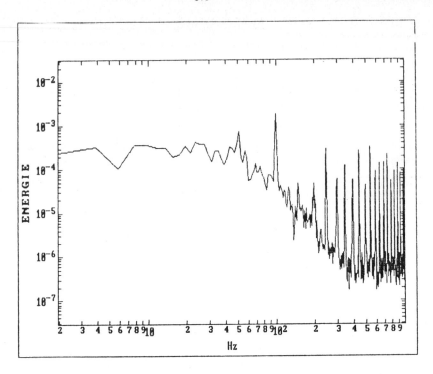

Figure 3. Spectral distribution of the temperature fluctuations for the non-reacting case with $t_R=100$ ms and at $z=0$.

When weighted with the frequency and plotted on a semi-logarithmic scale as $nE_T(n)$, the same spectrum may be used to estimate the time scale of the most energetic scalar eddies. For all the cases investigated here, this time scale ranges between 10 ms and 25 ms. It is then lower than the average residence time and the characteristic chemical time t_c. This means that if this turbulence time scale is used in the evaluation of the large eddy Damköhler number, the distributed reaction zone nature of the jet stirred reactor is again confirmed.

Spectral information about the scalar structure can also be used to estimate the time scale of the most dissipative eddies by plotting $n^2E_T(n)$; this function may be considered as representative of the scalar dissipation spectrum. For all of our reacting and non-reacting conditions, this time scale is found fluctuating around 10 ms, then lower than the micromixing time t_D calculated above. However, if the non-compensation of the thermal inertia of the fine wire is taken into account, it may be argued that the dissipation time scale should be smaller and then closer to t_D, which is considered here as a good estimation of the time scale of the turbulent inhomogeneities.

PRELIMINARY RESULTS ON THE INFLUENCE OF VARIOUS PARAMETERS ON THE STRUCTURE OF THE INSTANTANEOUS TEMPERATURE FIELD.

The above comments on the temporal characteristics of the temperature field have shown that the spectral distribution and the time scales of the temperature field were found only sligthly sensitive to variations of the operating conditions of the jet stirred reactor. The insufficient temporal resolution of the fine wire technique used here is probably partly responsible of this result. Therefore, the conclusive results about this part of the work should wait the application of time resolved laser based diagnostic techniques. However, the probability distribution functions of the temperature fluctuations are much less sensitive to the insufficiency of the temporal resolution of the diagnostic technique. The modification of the temperature pdfs with the variation of the operating parameters is then addressed below.

Influence of the average residence time

Probability density functions of the instantaneous temperature are presented on figures 4 for $(C2H4)_o$ = 0.15% and PHI = 0.4 and three average residence times; they correspond to the center of the reactor. For the lowest two residence times, the pdfs are very close to a gaussian distribution as were also their non-reacting correspondants. However, for the highest residence time, a low probability long tail appears on the positive side; it extends up to 4 r.m.s. value of the temperature signal, i.e., according to Table II, almost 8 K. This positively skewed distribution indicates the occurrence of some reactions which are more strongly exothermal than the others and which is again compatible with the observation of the last stages of C2H4 oxidation (CO comsumption).

Influence of the initial ethylene concentration

The pdfs of the temperature for three values of $(C2H4)_o$ at the center of the reactor and for PHI= 1 and t_R= 200 ms. are shown on figures 5. As listed on Table II, the r.m.s. temperature fluctuations decrease with increasing $(C2H4)_o$. For the highest $(C2H4)_o$ value, equal to 0.4 %, this decrease is accompanied by a shift of the corresponding pdf from the mono-modal, quasi-gaussian shape to a clearly bi-modal shape, very close to that observed in turbulent premixed flames with rapid chemistry or high Damköhler number. This bi-modal probability distribution would seem to indicate that at high initial fuel concentrations, a flamelet type combustion may locally occur, even when the global Damköhler number is very low.

Influence of the distance from the center of the reactor

Temperature pdfs for two distances from the reactor center ($z/D=-0.25$ and -0.38, where z is the azimuthal distance from the center taken positive upwards), for $t_R=200$ ms, PHI=0.4 and $(C2H4)_o=0.15\%$, are shown on figures 6. The correspondant pdf on the center is given on figure 4c. With increasing distance from the reactor center, the temperature pdfs are skewed towards lowest temperatures. For $z/D=-0.38$, a high probability low temperature peak appears. For the reacting case with PHI= 1.5 (not shown here), it is observed that the low temperature peak appears for smaller distances from the center.

CONCLUSION

The preliminary investigation of the interactions between ethylene oxidation and the instantaneous temperature field in a jet stirred reactor in the intermediate temperature range, has shown the feasibility of this kind of experimental work, even if the non-intrusive optical diagnostic techniques with high temporal resolution should increase the accuracy of the measurements. The r.m.s. values of the temperature fluctuations do not exceed 2 K in all the reacting cases; they are higher than the non-reacting case values only for the longest residence times, over 100 ms. These observations confirm then that the jet stirred reactor configuration we are using is most suitable for chemical kinetic studies. However, under some operating conditions, instantaneous temperature fluctuations as high as four times the corresponding r.m.s. value, thus approaching the overall temperature increase due to chemical reactions, are found. This type of observation is very useful to assess the limits of the operating conditions of the jet stirred reactor.

Concerning the structure of the instantaneous temperature field, this preliminary parametric investigation shows that the temporal characteristics of the scalar field are not strongly influenced by the variations of the parameters. However, the results strongly indicate that the instantaneous temperature field is more sensitive to the presence of chemical reactions for high conversion rates (high residence times) and for high initial fuel concentrations.

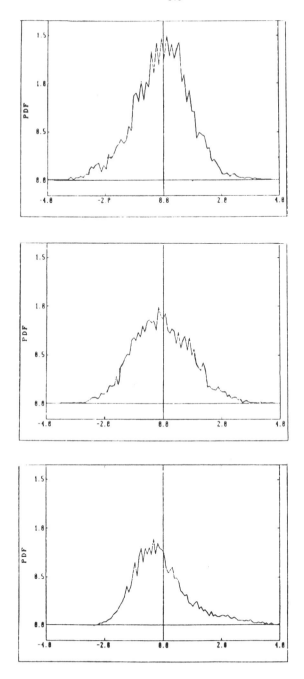

Figure 4. Temperature probability density functions at z= 0 for $(C2H4)_o$=0.15% and PHI=0.4. (a) t_R=60 ms; (b) t_R=100 ms; (c) t_R=200 ms.

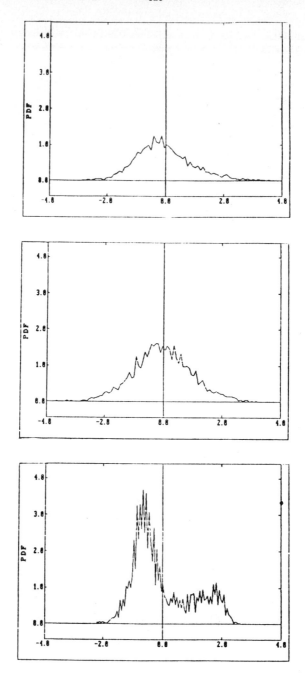

Figure 5. Temperature probabilty density functions at z=0, PHI=1 and t_R=200 ms. (a) $(C2H4)_o$=0.15%; (b) $(C2H4)_o$=0.3%; (c) $(C2H4)_o$=0.4%.

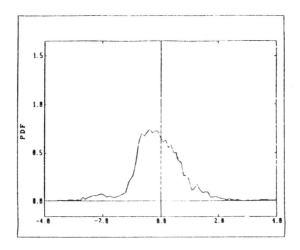

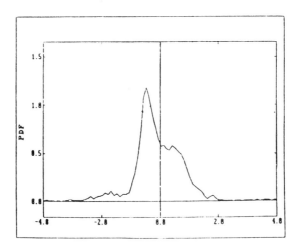

Figure 6. Temperature probability density functions for PHI = 0.4 and t_R = 200 ms. (a) z/D = -0.25; (b) z/D = -0.38.

REFERENCES

/1/ Bray K.N.C., 'Methods of including realistic chemical reaction mechanisms in turbulent combustion models', paper presented at the **Second Workshop on Modelling of Chemical Reaction Systems**, Heidelberg, August 1986.

/2/ Gökalp I., Dumas G.M.L. and Ben-Aïm R.I., **Eighteenth Symposium (International) on Combustion**, The Combustion Institute, p. 969, 1981.

/3/ Debbich M., **Etude expérimentale des champs dynamique et thermique dans les flammes froides laminaires et turbulentes de buthane-air**. Ph.D. Thesis in preparation, Université Pierre et Marie Curie, Paris, 1988.

/4/ Dagaut P., **Etude expérimentale et modélisation de l'oxydation de l'éthylène, du propane et du propène en réacteur auto-agité par jets gazeux**. Ph.D. Thesis, Université Pierre et Marie Curie, 1986.

/5/ Dagaut P., Cathonnet M., Boettner J.C. and Gaillard F., **Combustion Science and Technology**, 56:23, 1987.

/6/ Dagaut P., Cathonnet M., Gaillard F., Boettner J.C., Rouan J.P. and James H., **Progress in Astronautics and Aeronautics**, 105:377, 1986.

/7/ Dagaut P., Cathonnet M., Boettner J.C. and Gaillard F., to appear in **Combustion and Flame**, 1988.

/8/ Evangelista J.J., Shinnar R. and Katz S., **Twelfth Symposium (International) on Combustion**, The Combustion Institute, p. 901, 1969.

/9/ Nenninger J.E., Kridiotis A., Chomiak J., Longwell J.P. and Sarofim A.F., **Twentieth Symposium (International) on Combustion**, The Combustion Institute, p. 473, 1984.

/10/ Matras D. and Villermaux J., **Cemical Engineering Science**, 28:129, 1973.

/11/ Ferrer M., David R. and Villermaux J., **Oxidation Communications**, 4:353, 1983.

/12/ Dagaut P., Cathonnet M., Rouan J.P., Foulatier R, Quilgars A., Boettner J.C., Gaillard F. and James H., **J. Phys. E : Sci. Instrum.** 19:207, 1986.

INTERACTION OF A FLAME FRONT WITH VORTICES : AN EXPERIMENT

D. Escudié

Ecole Centrale de Lyon, Laboratoire de Mécanique des Fluides,
36, Avenue Guy de Collongue, BP. 163 _ 69131 Ecully Cedex (France)

INTRODUCTION

When chemical reactions occur in turbulent flows turbulence plays an important role along with thermodynamics and chemistry. This results in complex problems involving the interaction of turbulent motions with combustion. Many analytical and numerical analyses have been carried out to gain an understanding of the turbulent combustion and to predict the structure of turbulent reacting flows. However, progress has been impeded by the lack of experimental results due in a large part to the difficulties to provide accurate time-resolved velocity, density, temperature and concentration fields in combustion environment. But new developments in optical diagnostic techniques, and particularly the visualization techniques, now allowed us to examine the interaction of turbulence and combustion at a fundamental level.

This work focuses on the interaction between a premixed laminar flame with vortices of a Karman vortex street, an idealization of turbulent flow. Each vortex is considered as a "two-dimensional" organized parcel of turbulent fluid with its own dynamic behaviour that is a coherent structure. Such structures, arising also in boundary layers, jets or mixing layers, have received considerable attention from both the turbulence and the combustion communities. Turbulent fields can be decomposed into a spectrum of eddies with a distribution of length scales, thus the interaction between turbulence and combustion can be analyzed in terms of relative scale of an eddy and the thickness of a laminar flame. Damköhler /1/ was the first, in 1940, to put forward the idea that, in comparison with the thickness of the laminar flame, large scale turbulence wrinkles a premixed laminar flame without modifying its internal structure, while small scale turbulence primarly affects the transport processes inside the flame.

The first experimental study devoted to a detailed description of a non-confined premixed laminar flame (C2H4) in a Karman vortex street was that of I. Namer /2/. By means of non intrusive diagnostic techniques such as Rayleigh scattering and L.D.A, he showed that the vortex street does not persist beyond the flame, and that the eddies in the vortex street are "consumed" in the flame by combination of dilatation and increased rate of dissipation. Few years after Hertzberg et al /3/ using a visualization method originated by Boyer /4/, "Laser Tomography", found that the inboard vortex (closest to the flame holder) creates a perturbation which is damped due to the stabilizing effects of the flame holder and to the constant downward bulge caused by the velocity defect in the center of the vortex street. Only the outboard vortex creates a smaller incursion which develops into cups convected downstream at the flow velocity.

In 1986 the investigations by D. Escudié et al /5/ revealed the possible effect of the inboard vortices upon the other branch of a V-shaped flame and consequently the necessity to take into account stability problems besides the turbulent-combustion problem /6/.

The purpose of the present study is to use non intrusive techniques like tomography and Laser Doppler Anemometry (L.D.A.) to study the structure of a laminar flame front which interacts with vortices generated by a rod set upstream in the flow (some hot wire probe measurements have also been realized without combustion). Depending on the diameter of the rod, the characteristic scales of the flow could be greater than or of the same order of magnitude as the flame front thickness. The Damköhler's limiting regime, where the turbulence scale is smaller than the laminar flame thickness is also investigated.

EXPERIMENTS AND TECHNIQUES

In Figure 1, a schematic of the existing experimental set up is shown. The flame holder is a 0.4mm diameter catalytic wire situated in the lower part of a vertical quartz cylinder. The lean air-hydrogen premixed flame (6.5% H2 volumic concentration) is V expanding in the confined chamber and disturbed by vortices generated by a cylinder set 1cm upstream in the flow. The apparatus is such that the eccentricity E(Y/D) can vary from 0mm (right upstream of the platinum wire) to 12mm. The average upstream velocity is 5m/s and the residual turbulence about 0.4%.

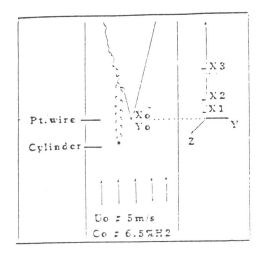

Figure 1 : Experimental set up

Two principal cases have been investigated. The first concerning *the laminar Karman vortex street*, and the second dealing with turbulent velocity fluctuations accompanying the periodic formation of vortices, *the turbulent Karman vortex street* /7/8/. This variation depends on the cylinder diameter but also on the fixing method. As a matter of fact Berger /9/ first observed that the transverse vibrations of the rod at and near the natural vortex shedding frequency extend the stable vortex shedding range from 40 < Re < 150 to Reynolds numbers about 300-400.

_ In our experiment *the laminar Karman vortex street* is obtained for three cylinder diameters : the fixed rods D0=0.2mm and D1=0.4mm, for which vortex street patterns are stable and well defined for long distances downstream, and the vibrating rod D2=1mm. As pointed out by Griffin /10/, in this last configuration, there is an increase in the vortex strength compared to the non oscillating case. The formation length could increase or decrease depending on the ratio : cylinder vibration frequency (F) / Strouhal frequency (Fs) ; when F/Fs < 1 the distance to formation is extended and when F/Fs > 1, there is an decrease of the formation length. For the D0, D1 and D2 cylinder diameters, the Reynolds numbers are Re=67, Re=133 and Re=330 respectively and the Strouhal frequencies Fs=3750Hz, Fs=2250Hz and Fs=1000Hz.

_ *The turbulent Karman vortex street* is generated by the fixed D2=1mm cylinder. Periodic vortex shedding also occurs for this Reynolds number (Re=330), but downstream, the discrete energy at the shedding frequency, is quickly dissiped or transferred to other frequencies. So that by 50 diameters the wake is completely turbulent and the energy spectrum of the velocity fluctuations approachs that of the isotropic turbulence.

The experimental set up and the techniques have been detailed in a previous paper /5/ : the mean axial velocity field is measured by hot wire probes (5μm) and to L.D.A technique (forward scatter method with a Bragg cell). There the flow has to be seeded by non-burning particles (Al_2O_3-SiO_2). Mean temperature measurements (coated chromel-alumel thermocouples) completes this quantitative part of the description. The tomographic method mentioned above is also used to observe the effect of the various vortex scales on the flame front. The main flow is then seeded by incense and the flame front appears at the interface between the bright and dark areas.

RESULTS AND DISCUSSIONS

The shapes of the flame differ owing to the different scales generated by the rods, and whether the eddies are with or without turbulent velocity fluctuations. But before developing this events it is fruitful to know what happens without combustion in the wake of the cylinders. Hot wire measurements have been realized for two cylinder diameters : D1=0.4mm and D2=1mm. Results concerning the mean velocity U and the fluctuations I%= $\sqrt{u'^2}/U$ versus Y position are plotted in Figure 2 and Figure 3, for the D2 and D1 rod respectively and at three stations downstream in the flow : X1=1cm, X2=4cm, X3=8cm.

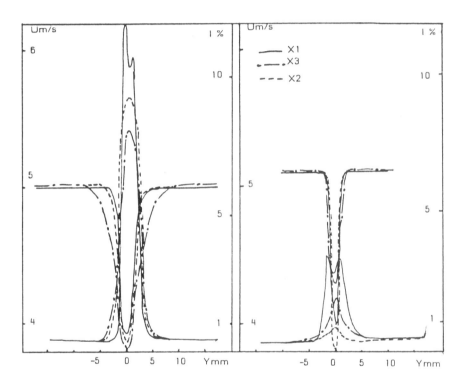

Mean axial velocity and turbulence intensity profiles in the wake of the cylinders, without combustion :

Figure 2: cylinder D2=1mm   *Figure 3* : cylinder D1=0.4mm

The defect in the mean velocity associated with the drag of the cylinder is considerable. The maximum intensity of velocity fluctuations was found to coincide with the end of the formation region /11/, we can also observe the swift decrease of this maximum as a function of the downstream distance. Nevertheless this values remains important compared to the residual turbulence of the flow and particularly in the vicinity of the laminar flame. As a matter of fact in a region close to the stabilizing wire, a relaminarization phenomenum occurs /12/ due to an increase in viscosity by a factor 8. Figure 4, obtained by means of the A.D.L. technique, presents for the three stations, the mean axial velocity profiles during combustion and the turbulence intensity profile.

Concerning the mean longitudinal velocity profiles, a global acceleration of the flow due to confinement and associated with the important decrease in density is noted.

With respect to turbulent fluctuations, in the general case of a premixed flame expanding in a grid induced turbulence (in a confined chamber), it was observed far downstream an important increase in turbulence intensity /12/13/. This can be explained by both the velocity gradient and the pressure gradient terms in the turbulent kinetic equation. There is only a slight increase in axial velocity fluctuations in this study, but in any case these results are not sufficient to understand how structure of the flame front is modified by the vortices. Tangential measurements are needed to improve on our understanding.

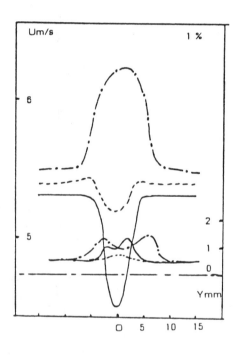

Figure 4: Mean axial velocity and turbulence intensity profiles across the laminar premixed flame without any disturbances

Hence L.D.A. transverse velocity measurements were also carried out, enabling us to determine the probability density functions. These results are presented in /6/. They reveal the possibility of deriving /14/ the tangential velocities values and the vortex characteristic scale, both by experiments and calculations. For the small cylinder (D1), the tangential velocity VT=0.8m/s and the viscous vortex radius is r*=0.425mm, while for the D2 cylinder VT=1,5m/s and r*=1.2mm. Moreover the laminar flame is defined by a laminar burning velocity SL=0.18cm/s and a thickness d=0.7mm, obtained due to the mean temperature profiles.

Till such investigations (transverse velocities) have been carried only in non combustion cases and the future ones should concentrate on combustion cases to complete visualization results.

For the sake of simplified interpretation, the velocity fluctuations (that is tangential velocity of the vortices) have to be compared to the laminar flame propagation and the characteristic scales of vortices to the thickness of the flame front. In such a context it could be possible to use the Borghi's diagram /15/ to intend to classify the various flame front obtained.

a. The laminar Karman vortex street

In this case there is only one particular frequency Fs on the power spectrum. Velocity fluctuations in the main flow are generated by the vortices and approximated by their tangential velocity.

_ 1. _ Δ --> d

When the characteristic scales of the flow (D) are of the same order of magnitude as the flame front thickness (d), and when their tangential velocity VT=0.8m/s is four times greater than the laminar burning velocity, the visualizations show a regular wrinkle of the flame surface. The wavelength (2.5mm) can be determined by the tomographic method and the same value is obtained by simple calculations from the shedding frequency. In the same way the evolution of the amplitude of the corrugation is a function of the stage of interaction between the vortex street and the flame front. There is an increase from the vortex scale D to the width value of the street and then a gradual diminution to reach again the viscous vortex diameter value.

When the cylinder is drawn towards the platinum wire, the length of the interaction (that is the wrinkled part of the flame) decreases and the cups becomes thicker (cf. Photo N°1 and Photo N°2). This seems to prove that the characteristic parameters of the vortex street are not perceptibly affected by combustion. On the contrary small perturbations would not only increase transfers, they also disturb the flame front by wrinkling it all along the interaction zone. But there is no an amplification of these disturbances, and the end of the interaction region is defined by the eddies crossing the flame front and being "consumed" as suggested by Namer.

To give an accurate description of what is happening in this configuration we have to remark that some packs of burned gases are flowing away at the interface with the mean flow velocity. Unfortunately there is no trivial explanation of this event. It can be probably due to a stability problem, either of the tilted flame or of the Karman vortex street, and various experiments have to be done to confirm or disprove these hypotheses.

_ 2. _ Δ < d

The study of the vortices, whose characteristic scale is smaller than the flame front thickness, required a specific experimental set up, not to have the cylinder 1cm upstream in the flow, but just close to the flame.

As in the last case the flame is wrinkled with a well defined wavelength (1.2mm) and all the typical parameters of this interaction are almost the same that in the previous experiment but divided by a factor two. So even if scales are smaller and the tangential velocity reduced to 2SL, the response of the flame to a laminar Karman vortex street still remains a succession of cups along the whole interaction length.

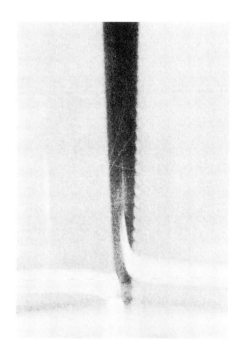 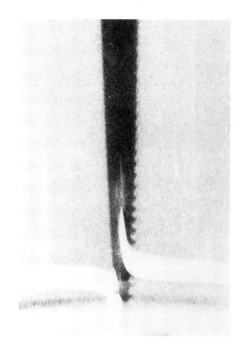

Tomographic view of the interaction between the laminar flame and the laminar Karman vortex street (D=0.4mm)

Photo N°1 : For a large eccentricity
E(Y/D)-->6

Photo N°2 : For a small eccentricity E(Y/D)-->3

_ 3. _ Δ >> d

The case of vibrating rod is a very interesting one because even though the amplitude and the tangential velocity are increased, compared to the fixed one, vortices are shed in a laminar Karman vortex street with only one particular frequency. However an indented structure due to the existence of large velocity gradients is noted. The vortex street sweeps the flame front down with it. In the same time, Photo N°3 points out the competing effects of the two processus : a well known deflection of the mean streamlines (that is the vortex street) at the flame front and in turn a singular deviation of the flame, shifted by the vortex street.

The waves of the flame front generated by this oscillating cylinder are created by alternating vortices of the vortex street and consequently the flame front is submitted to high shearing stresses (Photo N°4). As one could expect, these stretches can possibly lead to local extinctions of the flame front and the structure of this special turbulent flame could be composed of segments.

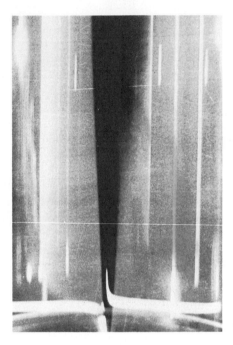 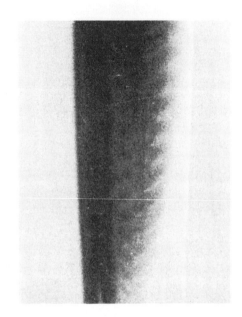

Tomographic visualization of the interaction between the laminar flame and the laminar Karman vortex street generated by the vibrating rod.

Photo N°3: A general view ; the flame is shifted by the vortex street

Photo N°4: A detail of this interaction ; the flame front is carried away by the vortex street

Compared to the fixed cylinder D1=0.4mm, the evolution of disturbance seems identical, but the strength and the scale of the vortices are greater and their effects are perceptible for a larger interaction range.

b. Turbulent Karman vortex street

The fixed diameter rod is used to generate the turbulent Karman vortex street. Depending on the eccentricity E(Y/D), the behaviour of the flame can be different and some special conditions reveal a stability problem, besides the classical turbulent-combustion analysis.

 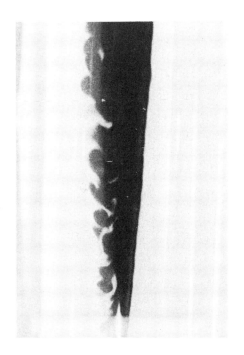

Tomographic view of the laminar flame in the turbulent Karman vortex street
Photo N°5 : Large eccentricity   Photo N°6 : Eccentricity E(Y/D)-->4
the flame is developping as in a   effect of the vortices on the opposed
turbulent flowfield   V_branch of the flame

Even for large eccentricity, the effect on the flame front is very sensitive and combustion reacts to the perturbation wrinkling as if it were a turbulent flow (cf. Photo N°5). When E(Y/D) is large, the interaction between the Karman vortex street and the flame front occurs far downstream in the flow, where the wake (as mentioned before) is fully turbulent.

As the cylinder is approached close to the stabilizing wire (Photo N°6) the cylinder influence is obvious on the other side of the V_flame ; this corresponds to a local disturbance created by one vortex in the lower part of the flame and propagating downstream with the tangential velocity of the flame front (--> the mean axial velocity of the flow). So the combustion front loses its linear aspect by waving and only one frequency (Fs) of the power spectrum is effective. This regular shape is opposed to the V_branch, just above the cylinder, teared by turbulent vortices. Some packs of burned gases are carried away by vortices and at the same time packs of unburned gases are trapped by running loops of burned gases.

If the cylinder is immediatly upstream of the platinum wire, the flame presents a regular front curvature (cf. Photo N°7) due to a local perturbation by one vortex of the street. Each branch of the V_flame is then disturbed only by vortices of the same side of the vortex street, always rotating in the same direction :

the vortices rotating clockwise interact with the left branch, and the ones rotating counter- clockwise with the right V_branch of the flame. Because of this particular phenomenum, the phase lag between two alternative vortices of the street was rechecked by examining the eddies at the flame front. Moreover the wavelength of the oscillations is very stable and eddies propagate downstream without evolution in scale exept a slight deformation. This is due to the various mean velocity gradient in which they travel. In the wake of the stabilizing wire, the slow down inside the flame gives rise to eddies orientated towards the top of the combustion chamber, while further downstream the acceleration of the burned gases involves a reversed orientation.

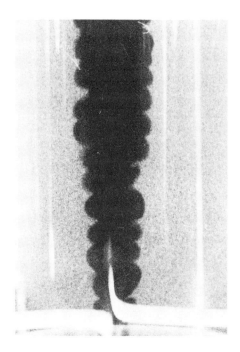

Photo N°7: *Visualization of the laminar flame in the turbulent vortex street E(Y/D)=0*

It appears interesting to summarize now these results by classifying them in the synthetic Borghi's representation of the morphology of turbulent premixed flames. The different ratios to be considered are the square root of the turbulent kinetic energy to the laminar flame speed $k^{1/2}/S_L$ and the integral length scale of turbulence to the laminar flame thickness l_t/d.

In this particular turbulent field, the velocity fluctuation is linked to the tangential velocity of the vortex assumed two-dimensional ; $k^{1/2}$ can be approximated by VT and lt by the characteristic scale of the vortices Δ. There lies an ambiguity to keep in mind for a reasonable comparison. Nevertheless the four case presented before could be reported in the diagram.

1 _ lt --> d and $k^{1/2}/S_L$ --> 4
2 _ lt < d and $k^{1/2}/S_L$ --> 2   laminar Karman vortex street
3 _ lt >> d and $k^{1/2}$ >> S_L
4 _ lt > d and $k^{1/2}/S_L$ --> 8   turbulent Karman vortex street

Figure 5 presents the various classifications obtained.

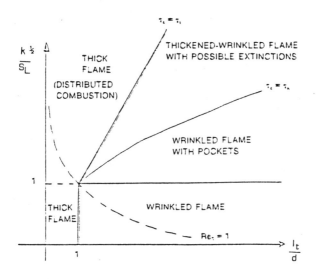

Figure 5: The different regimes of premixed turbulent combustion

There is some dicrepancy in this rough comparison between the structures predicted and the tomographic results. In the two first events (1,2) the flame is experimentally found to be wrinkled by the vortices whereas the diagram indicate a distributed combustion. This disparity could be due to a wrong evaluation of $k^{1/2}$, which is intented to be the turbulence kinetic energy including the influence of combustion on turbulence and because of the peculiar power spectrum obtained in a laminar Karman vortex street. Moreover it may due to a higher value of the flame front thickness derived from the mean temperature profile.

The agreement seems better when considering the two other cases (3,4). The vibrating cylinder gives rise to a thickened wrinkled flame with possible extinctions and the fixed one to a wrinkled flame with pockets which could explain the particular structures visualized.

But in these whole investigations, special care must be taken to compare them directly to the various type of turbulent combustion predicted.

CONCLUDING REMARKS

The laminar premixed flame, disturbed by the coherent structures of a Karman vortex street is an interesting problem in the field of turbulent combustion. Therefore understanding could be improved by tangential velocity measurements in the combustion zone which would lead to a better comprehension of the generation of peculiar waves.

For the various cases investigated, an attempt to compare the observed special patterns at the flame front with the Borghi's synthesis (on the structure of turbulent premixed flames) is certainly worthwhile but has to be done with caution.

To improve our understanding of turbulent combustion especially the stability associated with the interaction between turbulence and combustion, a more basic experiment in which only one vortex is created could be considered. This will provide a more fundamental understanding and experimental results needed by current researchers.

REFERENCES

/1/ G. Damköhler : Z Elektrochem, 46, p 601_626 (1940)

/2/ I. Namer : An experimental investigation of the interaction between a Karman vortex street and a premixed laminar flame ; Ph D, Thesis Berkeley CA (1980)

/3/ J.R. Hertzberg et al : A laser tomographic study of a laminar flame in a Karman vortex street ; Comb. Scien. and Techn. , vol 38, p 205_216 (1984)

/4/ L. Boyer : Laser tomographic method for flame front movements studies ; Comb. and Flame, 39, p 321 (1980)

/5/ D. Escudié and G. Charnay : Experimental study of the interaction between a premixed flame and coherent structures ; 5th Turbulent Shear Flows, p 347_310 (1986)

/6/ D. Escudié : Stability of a premixed laminar V_shaped flame ; accepted for publication in 11th ICDERS, Varsovie, Pologne (1987)

/7/ L.S.G. Kovaznay ; Hot wire investigation of the wake behind cylinders at low Reynolds numbers ; Proc. Roy. Soc. A, vol 198, 1053, p 174_190 (1949)

/8/ A. Roshko : On the development of turbulent wakes from vortex streets ; NACA, Tech. Note 2913 (1953)

/9/ E.W. Berger ; Phys. Fluids, 10, S191, S193 (1967)

/10/ O.M. Griffin : Effects of a sychronized cylinder vibration on vortex formation and mean flow ; Symp. on Flow Induced Structural Vibration, Karlsruhe, E. Naud. ed. Springer Verlag, p 455_470 (1972)

/11/ M.S. Bloor and J.H. Gerrard : Measurements on turbulent vortices in a cylinder wake ; Proc. Roy. Soc., A294, (1966)

/12/ D. Escudié, M. Trinité, P. Paranthoën : Modification of turbulent flowfield by an oblique premixed hydrogen air flame ; Prog. Astro. and Aero., 88, p 47 (1983)

/13/ R. Borghi and D. Escudié : Assesment of theoretical model of turbulent combustion by comparison with a simple experiment ; Comb. and Flame 56, p 149_164 (1984)

/14/ H. Lamb : Hydrodynamics, 6th Ed. Cambridge, University Press (1932)

/15/ R. Borghi : On the structure and morphology of turbulent premixed flame ; Recent advances in aerospace sciences, Pergamon Press (1985)

EXPERIMENTAL STUDIES IN VORTEX PAIR MOTION
COINCIDENT WITH A LIQUID REACTION

A. R. Karagozian, Y. Suganuma, and B. D. Strom
Mechanical, Aerospace, and Nuclear Engineering Department
UCLA
Los Angeles, California 90024-1597

Abstract

An experimental examination of the coincidence of a liquid reaction (acid-base) with the formation of a vortex pair structure is described, in which emphasis is placed on the evolution of the strained diffusion layer and reacted core structures. Flow visualization of the reaction process is achieved via the technique of chemically sensitive laser-induced fluorescence. The observed growth of reacted core structures, associated with each vortex, is compared with theoretically predicted behavior published recently (Marble, Adv. in Aero. Sci., 395 (1985) and Karagozian and Manda, Comb. Sci. and Tech., 49, 185 (1986)). Vortex pair separation is also compared with theoretical correlations, and the relevance of the analogy between a fast liquid reaction (Sc $\gg$ 1) and a gaseous reaction (Sc $\sim$ 1) is discussed.

Introduction

The formation and interaction of vortical flow structures is significant to the understanding and analysis of a variety of reacting and non-reacting flowfields. When a jet is injected normally into a crossflow, for example, a vortex pair is observed to dominate the cross-section of the jet, particularly in the jet's farfield[1]; in the case where the jet fluid (fuel) reacts with species in the crossflow (oxidizer), a bifurcated flame structure dominated by the vortices is observed[2]. These vortical structures are seen to have a significant influence on the trajectory of the nonreacting transverse jet[3,4] as well as the effective arclength of the deflected diffusion flame[5]. Vortical structures shed from burner lips or flameholders[6] in a combustion chamber have a particularly strong effect on the turbulent combustion process, occasionally resulting in very severe acoustical instabilities resulting from localized heat pulses[7,8].

In an effort to more fully understand the relationship between the fluid mechanics and combustion processes involved in these types of complicated flowfields, a number of analytical[9-12], numerical[13], and experimental studies[2,14-17] involving flame deformation through vortex interactions have been undertaken. The analytical/numerical studies have involved two-dimensional examinations of the distortions of diffusion flames by single vortex structures[9-11,13] and by vortex pair structures[12], while the experimental studies[14-16] have examined flames interacting with toroidal (axisymmetric) vortices which are shed from a burner by acoustical excitation. Comparison of these 2D analytical and approximately axisymmetric experimental studies is possible in terms of the qualitative nature of flame distortion and effects of heat release, but a quantitative comparison of reacting flow structures is less direct.

It is the purpose of the present study to experimentally examine the behavior of a two-dimensional vortex pair structure in a reacting flowfield, largely to serve as a basis for comparison with the 2D theoretical predictions of reaction structure, but in addition, to serve as a 2D analog to the toroidal vortex/flame structures observed in experiments to be shed from cylindrical burners. The present experiments are carried out in a liquid vortex pair generator facility, with a liquid reaction (acid-base) used to simulate a combustion process with fast chemistry. These types of reacting flowfields in the absence of heat release have been shown to adequately represent certain features of turbulent combustion processes (e.g., the dependence of effective reacted "flame" length on stoichiometry and Reynolds number) for a range of Schmidt numbers applicable to liquid as well as gaseous reactions[18-20]. Quantitative and qualitative observations in the present experiment are compared with specific analytical results for the deformation of diffusion

flames bounding a semi-infinite fuel strip by a vortex pair[12], a study which
isolates the effects of flow and diffusion processes and which neglects the
release of heat due to the combustion process. These studies indicate that, due
to the regions of high straining in the vortex pair flowfield, reacted "cores" of
combustion products are formed at the vortex centers, surrounded by winding
flame arms which continue to consume reactants. Later analytical studies[21]
have shown that, when heat release due to the chemical reaction is
represented in the vortex pair-diffusion flame interaction problem, the effect
is to shift the winding flame arms radially outward, away from the vortex
centers. While the diffusion flames are strained to a lesser degree than if
heat release were neglected, the same fundamental flame behavior in terms of
deformation about the vortices and formation of reacted cores is observed.

Description of the Experimental Facility and Procedure

A schematic diagram of the vortex pair generator facility used in these
experiments is shown in Figure 1. The tank, constructed of aluminum with a
5 cm thick plexiglass front plate, is 2.44 m high, 1.22 m long, and 15.2 cm
wide; it was originally constructed to simulate vortex pair motion in a
stratified environment[22,23]. The vortex generating apparatus is situated in the
lower portion of the tank, consisting of two aluminum side plates separated by
a distance of 12.7 cm, and narrowed at the top to 8.255 cm. The presence of
two horizontal plexiglass sheets allows separation of fluid in the lower
portion of the tank from that in the upper portion, and thus accomodates a two-
fluid (liquid) system. A counter-rotating vortex pair structure can be formed
in either of two ways: 1) by a rapid introduction of liquid pumped from
external tanks through the lower fill valves, or 2) by placement of a piston
spanning the aluminum side plates, with the piston accelerated then stopped
through the action of an external motor and clutch. Method 1) has been found
to be most effective in impulsively generating a relatively long-lived vortex
pair, and is used in the experiments described here.

In the liquid reaction experiments, a dilute base solution initially fills the lower portion of the tank to the top of the aluminum side plates, as indicated in Figure 1. A dilute acid solution, which has a slightly lower density than that of the basic solution, slowly fills the upper portion of the tank, above the horizontal plexiglass side plates. Due to the stable stratification, a uniform liquid interface between the two fluids is maintained at the top of the aluminum side plates. After the tank is filled, base is pumped into the tank through the lower fill valves, thus forming a vortex pair initially arising from the impulse of the two-dimensional flow of base through the side plates. At later times the vortex pair is influenced by the continual shedding of vorticity in the fluid passing between the side plates.

As the vortex pair is formed and propagates upward in the tank, the interface between the acid and base is distorted to wind about each of the vortices. This stretched interface region consists largely of the neutralized products of the liquid reaction, which are analogous to the combustion products formed at the diffusion flame "interface" in a fuel-oxidizer reaction process. The stoichiometry of the liquid reaction can be varied by changing the relative concentrations of the acid and base solutions in water. Thus, ϕ, the volume of acidic solution per volume of basic solution required for the reaction to go to completion, is analogous to $1/f^*$, the inverse of the stoichiometric fuel-oxidizer ratio in a combustion reaction, where the base is analogous to fuel and the acid is analogous to oxidizer in the present experiment.

Flow visualization is achieved in this experiment through the phenomenon of chemically sensitive laser-induced fluorescence. A 5W Argon Ion Laser (Coherent Radiation Innova 90-5) is used as a light source, situated at the top of the vortex generator tank. When the laser beam impinges on an optical scanner (General Scanning Model G115), the oscillating motion of the scanner mirror forms an effective laser "sheet", of reasonably uniform intensity, illuminating a two-dimensional slice of the tank and flowfield, as shown in Figure 1. In the present situation, fluorescein dye is mixed with the acid solution, while the basic solution contains no dye. This arrangement takes advantage of the fact that the fluorescence intensity of the dye in the acid is relatively low, since the higher concentration of H^+ ions in the acid interrupts the resonant absorption of photons by the fluorescein[20]. When the acid reacts with the base, forming neutralized products, a sharp increase in the pH of the solution as well as in the intensity of emitted light is observed; for a local pH above 7.0-7.5, the intensity does not change appreciably[19,20]. Figure 2, for example, indicates how the relative fluorescence intensity of fluorescein dye in solution measured in the present case varies with the local volumetric mixture ratio of acid in solution to base in solution, μ. When the mixture ratio $\mu = \phi$, the stoichiometric value, a sharp drop in intensity is observed, coincident with a drop in the pH of the solution (see Figure 3). As previously mentioned, the mixture ratio ϕ at which the sharp transition occurs is determined by the concentrations of acid and base in aqueous solution. The fluorescence intensity in a given experiment is thus a measure of the local pH as well as the local mixture ratio of the solution. The regions of the present reacting flowfield which are strongly illuminated consist mostly, due to the diffusion process at the acidbase interface, of neutralized products in an effective "combustion" process occurring at a diffusion flame. As the vortex pair is formed and the reaction zone winds about the vortex structures, the thickness of the effective diffusion layer depends both on the stoichiometric mixture ratio ϕ for the reaction and on the local straining of the reaction zone[9,12].

Photographic recording of the propagation of the reacting vortex pair is accomplished using a video camera and cassette recorder as well as a 35 mm SLR camera and motor drive. Video images are digitized and analyzed using a micro computer (IBM PC/AT) outfitted with a DT-IRIS image processing board (DT-2851) and arithmetic processor board (DT-2858). The local fluorescence intensity in the flowfield is thus determined with a high degree of accuracy (512x512 pixels, with approximately 5-10 pixels across the reaction front), and the local pH and mixture ratio of the solution may be determined.

Results and Comparison with Theory

$$\Gamma_{impulse} = \frac{P}{2\rho h_0} \tag{1a}$$

where P is the impulse per unit depth of the fluid introduced into the tank, passing through the side plates, and h_0 is the half-separation of the vortices at their initiation (i.e., $2h_0$ = 8.255 cm). If the rate of change of the impulse is equated to the force required to generate fluid motion from rest, which is equal to the net change in momentum of the fluid passing between the side plates, then the circulation $\Gamma_{impulse}$ is related to the volume flux of fluid introduced into the tank. Based on an impulse per unit depth of the tank P ≃ 1209 dyne-sec/cm, then, the circulation due to impulse, $\Gamma_{impulse}$, is approximated as 146.5 cm^2/sec, so that in this liquid system an effective Reynolds number $\Gamma_{impulse}/2\pi\nu$ is on the order of 2000.

Evaluation of the circulation $\Gamma_{b.l.}(t)$ arising from the shedding of vorticity from the boundary layer assumes that the local vorticity vector in each of the two boundary layers is $\underline{\zeta} \simeq -(\partial u/\partial y)\hat{k}$, where $\hat{k}$ is a unit vector lying perpendicular to the 2D flowfield illuminated. Hence the rate of change of the circulation associated with the layer takes the form

$$\frac{d\Gamma}{dt} = \frac{-U^2}{2},$$

where U is the velocity outside of each boundary layer, between the side plates. This yields

$$\Gamma_{b.l.}(t) \simeq \frac{\pm U^2 t}{2} \tag{1b}$$

for the circulation associated with vortices shed from each side plate, with the positive sign corresponding to the left side plate, and the negative sign corresponding to the right side plate. Finally, an effective Schmidt number (Sc ≡ ν/D) is determined by approximating the binary diffusivities of acid into products and base into products as identical, so that the Schmidt number here is on the order of 600.

A simple qualitative comparison of the reacting flow structures in these experiments can first be made with theoretical predictions of an equivalent diffusion flame in the absence of heat release by using the model described in Reference 12. In this model, the superposition of an inviscid vortex pair of constant strength and separation with a "semi-infinite" fuel strip is considered (a situation which is somewhat representative of the coincidence of the liquid reaction with the vortex pair here), shown in Figure 7. The fluid mechanical straining of the flame front causes an augmentation in the volumetric consumption rate of reactants and the formation of combustion products, especially in the high-strain region near the vortex centers. A crossflow of magnitude $U_\infty = \Gamma/4\pi h$ is superposed in order to consider the vortices to be stationary. According to inviscid theory, a vortex pair of constant circulation in crossflow will remain parallel[21], so that the half-spacing $h = h_0$ here. Fast chemistry is assumed in this reaction analysis, so that products are considered to be formed instantly; in addition, the heat release accompanying the formation of products is neglected so that the problem effectively represents the liquid reaction situation. In making the comparison between the theoretical situation and the present experiment, a constant circulation $\Gamma \simeq \Gamma_{impulse}$ is assumed. Typical results for the evolution of the flame structure and regions of reactants and products are shown in Figures 8a-8b for the stoichiometric mixture ratio $\phi = 10$, or, equivalently, the stoichiometric fuel-oxidizer ratio $f^* = 0.10$. The effective ratio of vortex strength to diffusivity is very high here, on the order of $\Gamma_{impulse}/D \simeq 10^5$, representing a typical liquid (rather than a gaseous) reaction. A structure of the reacted interface which is similar to experimental observations is observed (c.f. Figures 4 and 5), as is the formation of reacted cores of combustion products, in which where the flame has been extinguished, coinciding with the vortex centers and growing in size in time. This observation is described in detail below.

As mentioned, these reacted cores are comprised mostly of the products of the fast reaction, and as such are strongly illuminated in the experimental vortex pair flowfield. It should be noted, however, that the cores must also contain one of the reactants (either base or acid) depending on the

Reaction Zone Structure

Typical photographs of the vortex pair structure coincident with the liquid reaction for the stoichiometric ratio $\phi = 2$ ($f^* = 0.5$) are shown in Figure 4. The evolution in time of the vortices for the stoichiometric ratio $\phi = 10$ is shown in Figure 5. As is predicted by analytical modeling[12], a circular, reacted "core", comprised mostly of neutralized products, is observed to form quickly at each vortex center due to the excessive fluid mechanical straining of the acid/base interface at the vortices. A thin illuminated region surrounding the vortex pair structure represents the product diffusion layer which is formed at the interface. The actual acid/base interface or "flame" here is indistinguishable from the surrounding products in these experiments, due to the fluorescence technique used.

It is observed that the thickness of the winding diffusion layer actually decreases in time, particularly in the region close to the forward stagnation point of the vortex pair structure. This observation is consistent with the theory of the strained diffusion flame, in which diffusion thickness is determined to be proportional to $(D/\epsilon)^{\frac{1}{2}}$, where ϵ is the local strain rate[9,10,24]. As vorticity continues to be shed in time from the solid surfaces in this experiment, the effective circulation associated with each vortex is increased, and thus the strain rate ϵ associated with the "flame front" preceding the vortices increases, reducing the diffusion layer thickness.

It is also observed that the thickness of the illuminated part of the diffusion layer increases with the stoichiometric ratio ϕ; this phenomenon may be explained in terms of the local concentrations of reactants and products at the interface or flame. The analytically derived concentration profiles of acid, base, and products in the vicinity of the reaction zone, shown in Figure 6a for the case $\phi > 1$ (or $f^* < 1$), indicate that a greater portion of the product layer lies in a basic environment, situated above the x-axis or interface position. The $\phi > 1$ configuration thus tends to cause most of the product layer to become slightly basic in its composition, increasing the intensity of light emitted from the layer. The portion of the product layer which lies above the interface increases in thickness as ϕ increases above unity, thus explaining the present observations. When the stoichiometric ratio is less than unity, with concentration profiles shown in Figure 6b, the majority of the product layer lies in an acidic environment, tending to reduce the average pH of the products below the x-axis in the layer and reducing the intensity of light emitted from a large portion of the layer. For unity stoichiometry, exactly half of the product layer lies in a basic environment and half in an acidic environment (Figure 6c); the portion of the layer which is most strongly illuminated, though, above the flame, is thinner than in the $\phi > 1$ situation.

A Kelvin-Helmholtz instability is also observed to occur in these experiments, with successive vortices shed from the side plates, causing the formation of additional smaller reacted cores. These small reacted cores are observed to wind about the large reacted core structures, but are stretched and distorted in the successive spirals such that they do not appear to strongly affect the fundamental behavior of the large reacted cores for most of the vortex pair's trajectory, except in the overall addition of vorticity to the flowfield.

Qualitative and quantitative analyses of the behavior of this reacting flow structure is achieved by determining the effective strength of each vortical structure. Vorticity generation in this experiment is a process which continues as fluid passes between the vertical side plates, so that the effective vortex strength Γ of each vortical structure is actually a function of time. At time t = 0 (measured from the initiation of the vortex pair), the vortex strength is related to the impulse of the fluid passing through the vertical side plates, but as time elapses, Γ continues to increase due to the continual shedding of the boundary layer from each side plate. Hence, the effective vortex strength here is approximated by $\Gamma(t) = \Gamma_{impulse} + \Gamma_{b.l.}(t)$. As described in Lamb[25], for example, the circulation arising from the impulsive motion of the liquid is

stoichiometric mixture ratio at which the experiment is run. This can be explained in terms of the "initial" conditions prevalent in the tank, shown approximately in the schematic diagram of Figure 7 (after the analysis described in Reference 12). As the vortex pair coincides with the "corners" of the semi-infinite fuel strip, it is clear that there is a smaller volume of fuel (or base solution) contained within a given pair of closed streamlines than there is oxidizer (or acid solution). Thus, if equal parts of fuel and oxidizer are required for the fast reaction to go to completion (as in the case $\phi = 1$), then fuel will be completely consumed by adjacent flame elements much sooner than the oxidizer is consumed. According to the fast chemistry of the problem, the diffusion flame or reaction interface is locally extinguished when one of the reactants is consumed; this leaves a reacted "core" region where products and excess oxidizer wind about the vortices. In the case where the stoichiometric ratio is greater than 3, however, corresponding to $f^* = 0.33$ or lower, the geometry of Figure 7 indicates that since more oxidizer is required than fuel in the reaction, oxidizer is depleted first and the reacted core is comprised of products and excess fuel. Hence, by analogy, the fluorescence intensity in the core region by the fluorescein dye in the present experiment should increase as ϕ increases beyond 3, since the average pH in the core is higher than when ϕ is less than 3. These predictions are also borne out in a comparison of illuminated core structures for $\phi = 2$ and 10 in Figures 4 and 5.

Reacted Core Growth

The theoretical interaction of a vortex pair structure with a semi-infinite fuel strip also allows for a prediction of the time dependent growth of the reacted core. The radius of the reacted core, r^*, is found theoretically to depend on flow variables (the vortex strength Γ and the binary diffusivity D of reactants into products) as well as stoichiometry. This correlation takes the dimensional form (c.f. References 9, 10):

$$r^*(t) = C(f^*) \, \Gamma^{1/3} D^{1/6} t^{1/2} \qquad (2)$$

where the dependence of the function C on the stoichiometric fuel-oxidizer ratio f^* (equivalent to $1/\phi$) is shown in Figure 9, valid for the problem of the semi-infinite fuel strip-vortex pair interaction without heat release. This analysis assumes a constant circulation Γ, although the time-dependent circulation here will not alter the size of the recirculation cell of the vortex pair, nor will it significantly affect the shapes of the streamlines. Hence it is reasonable to postulate that relation (2) may be compared with results from the present experiment, with $\Gamma = \Gamma(t)$ as previously described.

An estimate of the reacted core radius in the present experiment is possible through determination of the regions of uniform, high intensity light emitted in the flowfield using the DT-IRIS digitizing boards and software. It is assumed that outside of these uniform circular regions, where spirals of high intensity light are observed, the interface or "flame" arms of the reaction process are continuing to consume reactants, as in the theoretical interaction. It should be noted that in these experiments, the diffusion layer separating the acid and base solutions has had time to grow in thickness before the vortex pair is generated, while the theoretical prediction assumes that the vortices are initiated at the same time as the reaction is started. Thus, at a given time measured from the initiation of the vortices (experimentally), the experimental core radius r^* should be slightly larger than the theoretically predicted core radius. Despite these slight differences between the experiments and the theoretical assumptions, however, the correspondence for reacted core radius is quite good, as shown in Figure 10. Again, a variable circulation $\Gamma(t) = \Gamma_{impulse} + \Gamma_{b.l.}(t)$ is incorporated in relation (2) to achieve this comparison.

Vortex Pair Separation

The separation of the reacting vortex structures with time in this experiment arises due to several complicated fluid mechanical processes. Vortex pair separation is influenced by the fact that the circulation associated with each vortex is largely due to an impulsive motion in the tank, but separation is also influenced by viscous effects. Because vorticity generation associated with the larger flow structures appears to be more strongly influenced initially by the impulse of the initially 2D flow rather than the shedding of the boundary layer, the assumption will be made in the separation analysis that $\Gamma \simeq \Gamma_{impulse}$.

While the vortex pair in the present experiment propagates upward in the tank due to the mutual induction by the vortices as well as the 2D jet-like passage of fluid between the side plates, the propagation distance z due to induction *only* varies with vortex strength and spacing according to the relation

$$\frac{dz}{dt} = \frac{\Gamma_{impulse}}{4\pi h} \qquad (3)$$

where h, the vortex pair half-spacing, is a function of time. According to the theory described in Reference 4 for the separation of two viscous vortices in the initial stages of incompressible viscous vortex pair development, the separation distance 2h is proportional to the square root of time, and hence is proportional to the square root of propagation distance z in the present situation. If we thus assume that $z = C h^2$ in (3), where C is an undetermined constant, the solution for h(t) from the governing equation (3) takes the form

$$h(t) = \left[h^3(t_0) + \frac{3}{8\pi C} \Gamma_{impulse}(t - t_0) \right]^{\frac{1}{3}} \qquad (4)$$

Here t_0 corresponds to the effective "flow time" at which the vortex pair and the viscous cores are **observed** to be shed from the side plates in the present experiment. This initial time does not, however, correspond to the time $t = 0$ at which the motion in the tank is actually initiated and at which the viscous cores begin to grow due to the growth and shedding of the boundary layers at the side plates.

In order to approximate this non-zero time t_0 at which motion is observed to be initiated, the asymptotic behavior of the separating viscous vortices[4] is incorporated. This analysis assumes that the two viscous cores, coincident with each vortex structure, have radii which grow in proportion to $(\nu t)^{\frac{1}{2}}$, in accordance with the classical theory of the viscous vortex[26]. As mentioned above, this produces the initial behavior of vortex pair separation near $t = t_0$, described by

$$h \simeq b \left[\Gamma t \right]^{1/2} \quad \text{where} \quad b \equiv \left\{ \frac{1}{2\pi^2 (k+2)} \right\}^{1/4} \quad (5)$$

The constant k above represents the virtual mass coefficient associated with motion of the viscous cores in a vortex pair flowfield, determined in Reference 4 to be of the order $k = 4$. Given that the initial separation of the vortices in the experiment is $2h_0 = 2h(t_0) = 8.255$ cm, the "initial" time t_0 corresponding to observed vortex formation (and the non-zero viscous core radius) is approximately 1.266 sec. Hence, in comparing experimental results with correlation (4), the factor 1.266 sec must be added to the time measured from the observed initiation of the vortex pair.

Comparison of the approximate theoretical correlation in (4) with experimental results for vortex separation is shown in Figure 11, where the constant C in (4) is determined to be equal to 0.159, according to a fit of the curve. As is postulated above, correlation (4), describing vortex separation due to the impulse of the jet, is more appropriate in the initial stages of vortex pair propagation. At later times ($t \gtrsim 5$ sec) the vortex pair appears to be more strongly affected by the dissipation in the secondary vortices, as well as possible three-dimensional effects in the flowfield. Thus, correspondence with theory at these later times is not as strong.

Summary and Conclusions

An experimental study of the coincidence of vortex pair generation with a liquid reaction allows qualitative and quantitative comparison of reacting flow structures with two-dimensional theory that was previously not possible. In addition, fundamental features of the liquid reaction process (at high Schmidt numbers) may be determined, and the basis for which comparison with gaseous reactions (at low Schmidt numbers) is valid may be made, as described below.

Our experiments reveal the formation of a thin reaction zone surrounding the generated vortex pair structures, with illumination of a diffusion layer achieved via laser-induced fluorescence, in addition to the formation of reacted cores comprised largely of neutralized products, coincident with the vortex centers. These reacting flow structures are qualitatively similar to the toroidal eddy structures shed from circular burners[14-16], coincident with flame structures. The reaction layer in the present experiment is most strongly illuminated for acid/base stoichiometric ratios which are fairly large ($\phi > 1$); in accordance with the combustion analogy, this corresponds to fuel/oxidizer stoichiometric ratios much less than unity. This observation is consistent with the theory of the strained diffusion flame in the absence of heat release, which predicts that the diffusion layer is thicker on the fuel (base) side for $\phi > 1$ or $f^* < 1$. The actual *variation* in the diffusion layer thickness predicted for a gaseous reaction (see Reference 12), resulting from variation in the local strain rate of the diffusion flame, is not observed in these liquid experiments, however, due to the lower effective diffusivities of reactants arising in a liquid. This variation in diffusion thickness is also not observed in theoretical predictions of the liquid reaction layer evolution, shown in Figure 8, for example. This indicates that some of the fundamental features of the diffusion process in gases (Sc ~ 1) cannot be simulated by a liquid reaction (Sc > 100). Specifically, reaction characteristics cannot be determined which are related to the effect of the diffusion layer on the surrounding flowfield, especially when the heat release due to the combustion process in the diffusion layer is significant and affects local flame extinction. Nevertheless, simulation of the combustion process using a liquid reaction is a useful means of isolating the effects of flame strain on the augmentation of the fast reaction process.

Quantitative comparison of the experimentally observed growth of the reacted cores with theory is quite good, and the dependence of reacted core size on stoichiometry is reaffirmed as well. In predicting vortex pair separation in the current experiment, a "nearfield" approximate analysis is fairly successful, but at later times in vortex pair evolution, the effects of secondary vortices and the development of a three-dimensional flowfield cause deviation of experimental results from the theory.

Acknowledgements

This work has been supported by the National Science Foundation under Research Initiation Grant MEA 83-05960 and by NASA Lewis Research Center under grant NAG 3-543. The authors wish to express appreciation to Mr. Bill Floud for his technical assistance and to acknowledge helpful discussions with Professors W. J. A. Dahm of the University of Michigan, P. A. Monkewitz of UCLA, and F. E. Marble of Caltech.

References

1. Kamotani, Y. and Greber, I., AIAA Journal, 10, 1425 (1972).

2. Brzustowski, T. A., Progress in Energy and Combustion Science, 2, 127 (1976).

3. Coelho, S. L. V. and Hunt, J. R., "On the initial deformation of strong jets in cross-flows", preliminary draft, 1987.

4. Karagozian, A. R., "AIAA Journal, 24, 1502 (1986).

5. Karagozian, A. R. and Nguyen, T. T., "Effects of buoyancy and flame distortion in the transverse fuel jet", to appear in the 21st Symposium (Intl.) on Combustion, 1987.

6. Oppenheim, A. K. and Ghoniem, A. F., "Aerodynamic features of turbulent flames", AIAA Paper No. 83-0470 (1983).

7. Zukoski, E. E. and Marble, F. E., "Experiments concerning the mechanism of flame blowoff from bluff bodies", Proc. of the Northwestern Gasdynamics

Symposium, 205 (1955)

8. Rogers, D. and Marble, F. E., Jet Propulsion, 26, 156 (1956).

9. Marble, F. E., "Growth of a Diffusion Flame in the Field of a Vortex", Advances in Aerospace Science, 395 (1985).

10. Karagozian, A. R. and Marble, F. E., Combustion Science and Technology, 45, 65 (1986).

11. Norton, O. P., "The effects of a vortex field on flames with finite reaction rates", Ph.D. Thesis, California Institute of Technology, 1983.

12. Karagozian, A. R. and Manda, B. V. S., Combustion Science and Technology, 49, 185 (1986).

13. Laverdant, A. M. and Candel, S. M., "A Numerical Analysis of a Diffusion Flame-Vortex Interaction", ONERA TP no. 1987-26, 1987.

14. Vandsburger, H., Lewis, G., Seitzman, J. M., Allen, M. B., Bowman, C. T., and Hanson, R. K., "Flame-flow structure in an acoustically driven jet flame", Paper 86-19, Proc. of the Western States Section/ The Combustion Institute, October, 1986.

15. Chen, L-D and Roquemore, W. M., Combustion and Flame, 66, 81 (1986).

16. Linevsky, M. J., Fristrom, R. M., and Smith, J. R., "Single eddy combustion: a new approach to turbulent flames", presented at the 20th Symposium (Intl.) on Combustion, Ann Arbor, MI, 1984.

17. Cattolica, R. and Vosen, S., Combustion and Flame, 68, 267 (1987).

18. Hottel, V. O. and Luce, R. G., Fourth Symposium (Intl.) on Combustion, Williams and Wilkins Co., 97 (1953).

19. Dimotakis, P. E., Miake-Lye, R. C., and Papantoniou, D. A., The Physics of Fluids, 26, 3185 (1983).

20. Dahm, W. J. A., "Experiments on entrainment, mixing, and chemical reactions in turbulent jets at large Schmidt number", Ph.D. Thesis, California Institute of Technology, 1985.

21. Manda, B.V.S., "A Study of the Interaction of Diffusion Flames with a Vortex Pair Flowfield", Ph.D. thesis, University of California, Los Angeles, 1987.

22. Barker, S. J. and Crow, S. C., Journal of Fluid Mechanics, 82, 659 (1978).

23. Tomassian, J. D., "The motion of a vortex pair in a stratified medium", Ph.D. thesis, University of California, Los Angeles, 1979.

24. Carrier, G. F., Fendell, F. E., and Marble, F. E., SIAM Journal of Applied Mathematics, 28, 463 (1975).

25. Lamb, H., Hydrodynamics, Sixth Edition, Dover, 1932.

26. Robertson, J. M., Hydrodynamics in Theory and Application, Prentice-Hall, 1965.

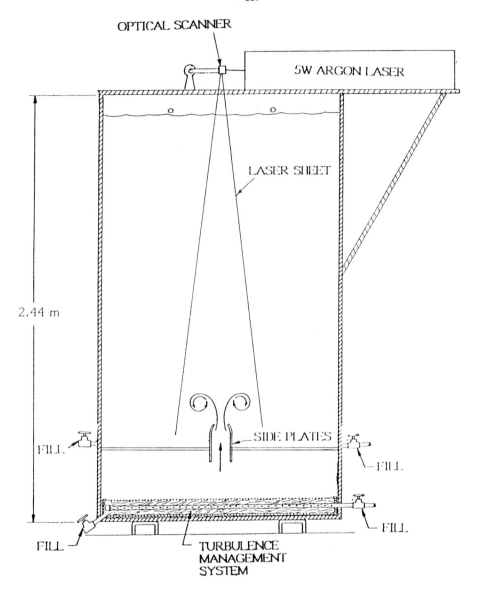

Figure 1. Schematic diagram of vortex pair generator tank.

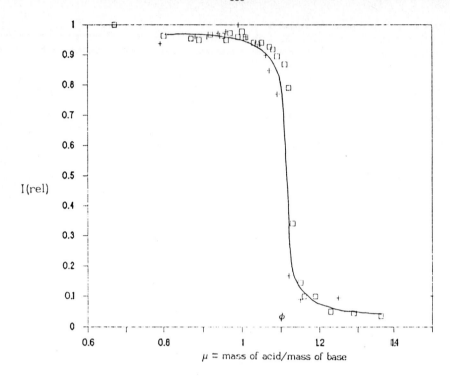

Figure 2. Relative intensity of light emitted from a solution of acid and fluorescein titrated with base (+) and from a solution of base and fluorescein titrated with acid (□) as a function of solution mixture ratio μ, for reactant concentrations producing a stoichiometric acid/base ratio $\phi \simeq 1$. The concentration of fluorescein dye in solution is 2.5×10^{-6} M.

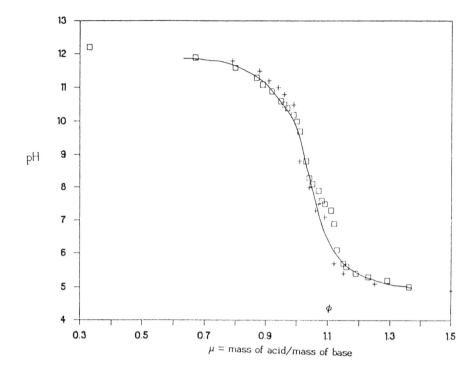

Figure 3. Variation in pH for a solution of acid titrated with base (+) and a solution of base titrated with acid (□) as a function of solution mixture ratio μ, for stoichiometric acid/base ratio $\phi \simeq 1$.

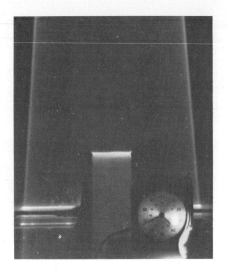 

4a.) 4b.)

4c.)

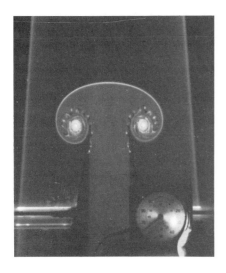

4d.)

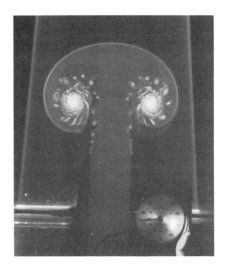

4e.)

Figure 4. Evolution of vortex pair structure coincident with liquid reaction for stoichiometric acid/base ratio $\phi = 2$, at times: a.) t = 0 sec, b.) t ≃ 0.5 sec, c.) t ≃ 1.5 sec, d.) t ≃ 3 sec, and e.) t ≃ 5.5 sec.

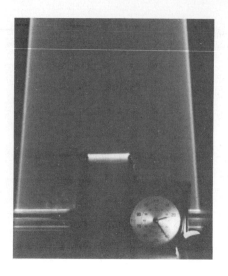

5a.)

5b.)

5c.)

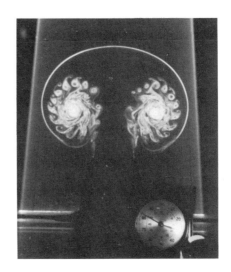

5d.) 5e.)

Figure 5. Evolution of vortex pair structure coincident with liquid stoichiometric acid/base ratio $\phi = 10$, at times: a.) t = 0 sec, b.) sec, c.) t ≃ 1.5 sec, d.) t ≃ 3 sec, and e.) t ≃ 5.5 sec.

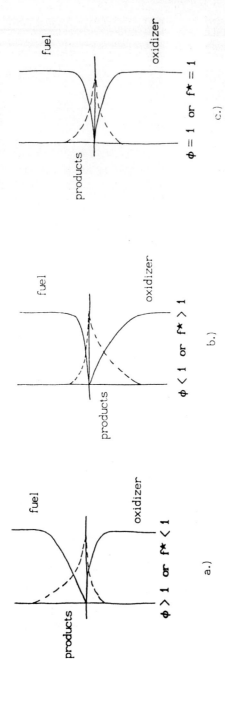

Figure 6. Theoretically calculated concentration profiles for diffusion flame with fast chemistry (at time t > 0). Profiles shown are for concentrations of fuel, oxidizer, and combustion products at fuel/oxidizer stoichiometric ratios a.) $f^* < 1$ (or $\phi > 1$), b.) $f^* = 1$ (or $\phi = 1$), and c.) $f^* > 1$ (or $\phi < 1$).

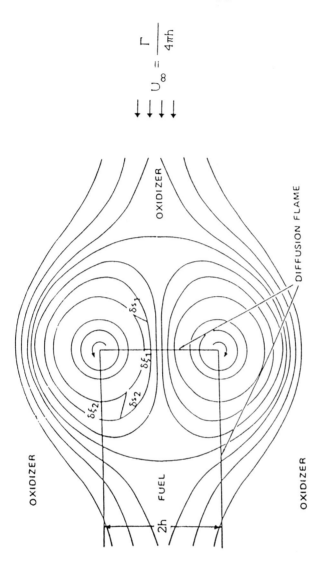

Figure 7. Schematic description of analytical framework for problem of semi-infinite fuel strip interacting with a vortex pair of strength $\pm\Gamma$, solved in detail in Reference 12.

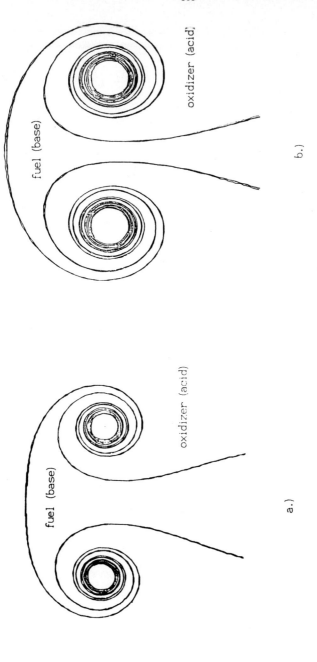

Figure 8. Theoretical prediction of deformation of diffusion flames bounding a semi-infinite fuel strip by the vortex pair flowfield, where $\Gamma/D \simeq 10^5$, $f^* = 0.10$ (or $\phi = 10$) at approximate times a.) $t \simeq 2$ sec, and b.) $t \simeq 4$ sec.

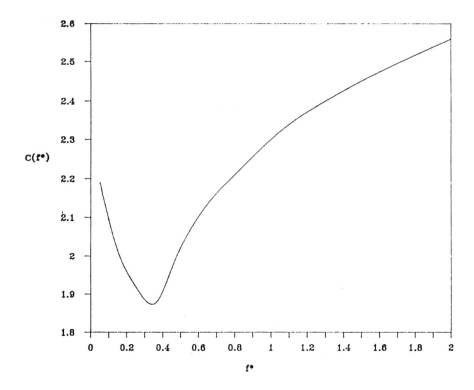

Figure 9. Theoretically computed dependence of function C (defined in equation (2) on stoichiometric fuel/oxidizer ratio $f^\star$.

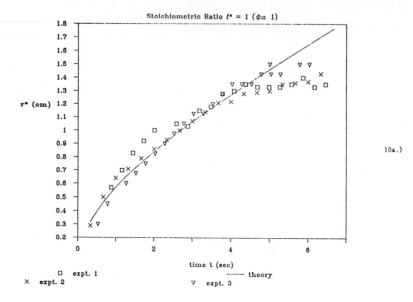

10a.)

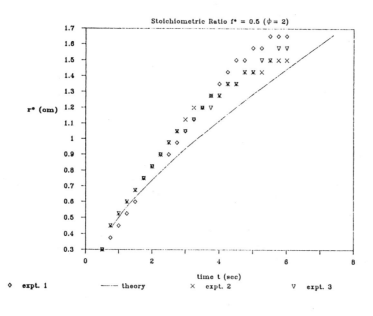

10b.)

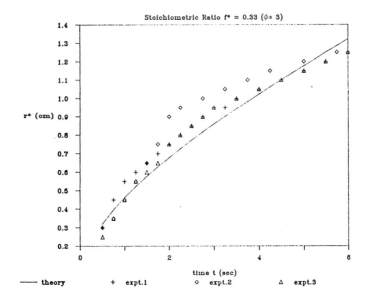

10c.)

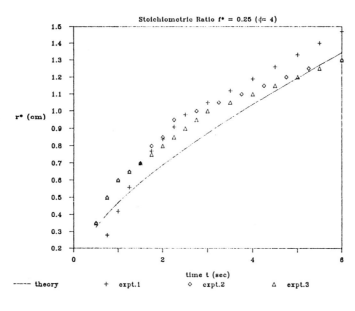

10d.)

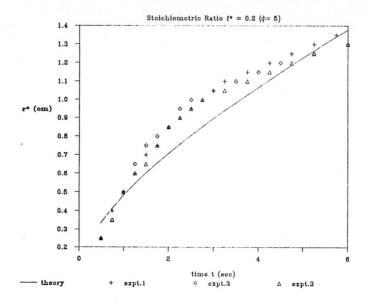

10e.)

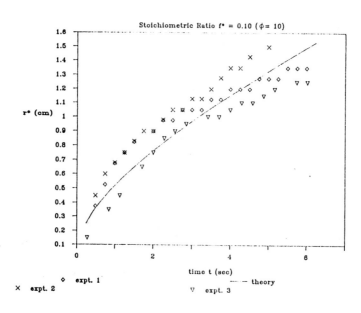

10 f.)

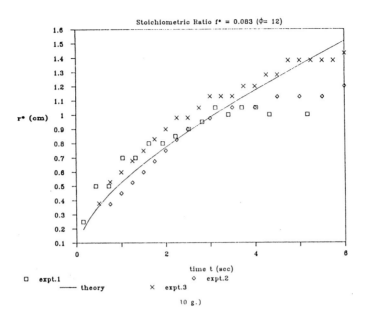

10 g.)

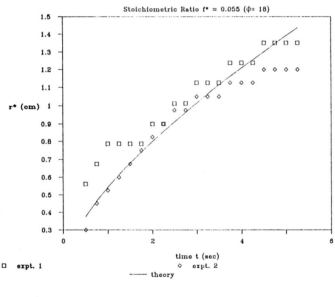

10 h.)

Figure 10. Comparison of measured reacted core radius r^* (in time) with theoretical prediction represented by equation (2). Comparisons shown are for fuel/oxidizer stoichiometric ratios a.) $f^* = 1$ (or $\phi = 1$), b.) $f^* = 0.5$ (or $\phi = 2$), c.) $f^* = 0.33$ (or $\phi = 3$), d.) $f^* = 0.25$ (or $\phi = 4$), e.) $f^* = 0.20$ (or $\phi = 5$), f.) $f^* = 0.10$ (or $\phi = 10$), g.) $f^* = 0.0833$ (or $\phi = 12$), h.) $f^* = 0.055$ (or $\phi = 18$).

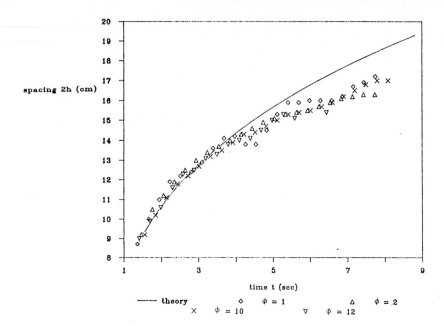

Figure 11. Comparison of measured vortex separation 2h with theoretical prediction according to equation (4) (——), as a function of time.

"This paper has been published in <u>The Physics of Fluids</u>, Vol. 31, No. 7, July, 1988, pp. 1862-1871. Printed with the permission of the American Institute of Physics."

Please let me know if there are any problems or questions.

ON AN ATTEMPT TO MEASURE THE DECAY OF CONCENTRATION FLUCTUATIONS IN A QUASI-ISOTROPIC GRID BY USE OF THE FLUORESCENCE OF THE SOLUTION.

J.L. LIEVRE, J.N. GENCE

Laboratoire de Mécanique des Fluides, Ecole Centrale de Lyon
36, avenue Guy de Collongue - 69130 - ECULLY, France

and I.S.I.D.T., Université de Lyon I
43, Boulevard du 11 Novembre 1918 - 69622 - VILLEURBANNE CEDEX, France

In collaboration with the L.S.G.C. and the G.R.A.P.P., E.N.S.I.C., NANCY, France

A main characteristic of turbulence is its ability to mix transported scalar fields such as temperature or concentration. In chemical engineering the different species having to be mixed are often diluted in liquids, and the Schmitt number

$$Sc = \nu/D$$

where ν is the kinematic viscosity and D the molecular diffusivity of one species, is much greater than in a mixture of gases. A typical value of this number for diluted species in water is about 10^3. As the molecular diffusivity is much smaller than ν, the turbulent concentration field exhibits much smaller "dissipative" scales than the velocity field. The classical measurement method of concentration fluctuations of ionic species in water uses a conductimetric probe whose measurement volume is too large to give satisfactory information about the smallest scales of the concentration field. Such a method was used by Gibson (1962) to study the evolution of a concentration field of Nacl in a quasi-isotropic grid turbulence. BENNANI, GENCE and MATHIEU (1985) used the conductimetric method to study the evolution of a slow chemical reaction in the same conditions.

This communication presents another measurement technics based on the fluorescence of the solution and which permits to reduce the measurement volume. It is applied to study the decay of concentration fluctuations of rhodamine B diluted in water in a quasi-isotropic grid-turbulence.

I - The measurement method :

The principles of the measurement method may be summed up as follows :
- a laser beam is focussed at one point P (or a small region) of a liquid containing one species which is fluorescent.
- a photomultiplier (P.M.) looking at this point P receives an intensity of fluorescence depending on the fluorescent species concentration at this point. If the concentration field is a random turbulent field $C(M,t)$, then the P.M. observes a random fluorescence intensity $I(P,t)$ directly connected to $C(P,t)$. More precisely, let us considere a cylindrical incident laser beam penetrating inside a tank with square section containing a concentration field $C(M,t)$ of

a fluorescent species (Figure 1).

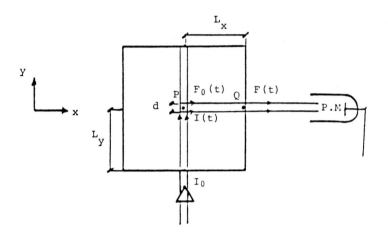

Figure 1

Let us considere also a P.M. looking at a small volume of characteristic length d located at the center P of the section. It receives a fluorescence intensity F(t) linked to the intensity $F_0(P,t)$ emitted from the measurement volume by the absorption law of Beer-Lambert, i;e; :

$$F(t) = F_0(P,t) \exp\left(-\varepsilon_f \int_0^{L_x} C(x,t).dx\right)$$

where ε_f is the absorption coefficient of the emitted fluorescent light and $C(x,t)$ the concentration field along the path PQ.

If I(t) is the incident intensity of the laser-beam penetrating the measurement volume, and I_0 its initial intensity, we may also write

$$I(t) = I_0 \exp\left(-\varepsilon_i \int_0^{L_x} C(y,t).dy\right)$$

where ε_i is the absorption coefficient of the laser intensity. A part δI of this intensity is absorbed in the measurement volume along length d, i.e.

$$|\delta I| = I(t).\varepsilon_i.C(P,t).d$$

where C(P,t) is the averadge concentration in the measurement volume. A fraction of this absorbed intensity is emitted by the fluorescence process, so that

$$F_0 = K.|\delta I|,$$

where K is a constant of the process.

Then, we may write that the intensity F(t) received by the P.M. is given by a functional of the concentration field

$$F(t) = K.d.\epsilon_i.I_0.C(P,t).\exp(-\epsilon_i\int_0^{L_y} C(y,t) - \epsilon_f\int_0^{L_x} C(x,t)dx)$$

As usual in turbulence description, we split the concentration field into its mean value $\bar{C}$ and a fluctuation c(M,t) so that in a statistically stationnary field :

$$C(M,t) = \bar{C}(M) + c(M,t)$$

Then, it may be written :

$$F(t) = K.d.\epsilon_j.I_0.(\bar{C}(P) + c(P,t))$$

$$\exp(-\epsilon_i\int_0^{L_y} (\bar{C}(y)+c(y,t)).dy - \epsilon_f\int_0^{L_x} (\bar{C}(x)+c(x,t)).dx)$$

Moreover, if the values of the concentration are small enough, the exponential term may be approximated by 1 so that F(t) becomes proportionnal to the concentration in the measurement volume, so that

$$F(t) = F(P,t) = K.d.\epsilon_i.I_0.(\bar{C}(P) + c(P,t)) = A.(\bar{C}(P) + c(P,t))$$

Putting

$$F(P,t) = \bar{F}(P) + f(P,t),$$

We have

$$\bar{F}(P) = A\,\bar{C}(P) \quad ; \quad f(P,t) = Ac(P,t),$$

which implies

$$f/\bar{F} = c/\bar{C}.$$

Such a measurement method have two main advantages :
- the concentration field is not influenced by the measure
- the measurement volume, that is to say the volume observed by the P.M. can be theorically reduced as we want, the limit being the minimum intensity detected by the P.M.

II - Application to the study of the decay of concentration fluctuations in a quasi-isotropic grid turbulence

The principle of the experiment is analogous to that of Gibson (1962) or Bennani,

Gence, Mathieu (1985) : a classical quasi-isotropic grid-turbulence is created in a "pure" water flow and a passive concentration of a fluorescent species (Rhodamine B) is injected through the grid nods so that a random and statistically isotropic and stationnary concentration field developped downstream the grid.

a) - Experimental set-up

The whole facility is composed of two circuits (Fig. 2) i) A main circuit corresponding to the "pure" water flowing through the grid. It is directly connected to a water duct feeding the laboratory so that no pump is used in order to avoid vibrations. Before entering the test section, the water flows through a diverging converging channel containing small grids and whose contraction ratio is 9. The grid generating the turbulence is located at the convergent outlet and is followed by the test-section with glass wall-sides. This test-tunel is 1m long and has a square 7cm side cross-section. The uniform mean velocity is equal to 0.5 m/s.

ii) A secondary circuit containing a concentrated solution of rhodamine which is injected through the grid nods. This solution is flowing by gravity effect from a tank located 2m over the grid.

The grid is biplane with round rods of 2mm and square mesh whose side are 6mm long. Its solidity is equal to 0,44. Each horizontal rods possesses 8 injectors of 6mm long. The flow rate through each rod is regulated by taps in order to obtain a good homogeneity.

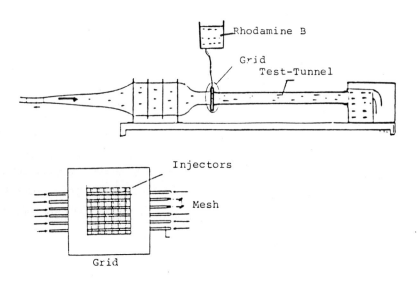

Figure 2

b) Measurement set-up :

The optical set-up is schemed on Figure 3. An enlarged laser beam of wave length 488 nm is concentrated at the center of the measurement cross section of the test tunnel and the P.M. observed this point through a 100 μm diameter pin hole. The source is an Argon laser Spectra physics whose power is 25 mw.

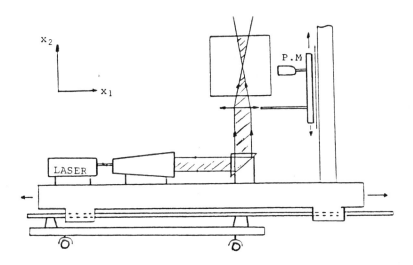

Figure 3

The signal of the P.M. is amplified by means of DISA units (which are commonly used in laser anenometry but have been modified in order to give information about low frequency fluctuations up to 15 Hz) and filtered by means of DISA units from 1 hz to 10 Khz. The high frequency cut-off is choosen as the double of the highest observable time frequency of the concentration fluctuations.

To measure the noise delivered by the P.M. in the absence of concentration fluctuations, the laser beam is focussed at the center of a cross-section of a reference tank having the same glass side-walls as the test tunnel and containing an homogeneous solution of rhodamine, whose concentration is equal to the mean concentration $\bar{C}$ in the test tunnel which, of course, is constant far enough downstream the grid (Figure 4).

It is assumed that this noise is not correlated with the concentration fluctuations in the test tunnel, so that the mean square noise fluctuations is substracted to the mean square of the concentration fluctuations. In the experiment, the ratio of these mean square values is equal to 100.

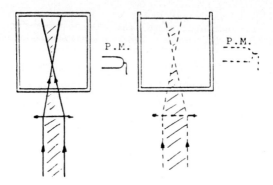

Figure 4

c) Results on the decay of concentration fluctuations

All the measurement were made on the center line of the test tunnel. The evolution of $\overline{c^2} / \overline{c_r^2}$ as function of x_1/M (where M is the mesh size) are represented in Figure 5 on a Log-Log plot. The mean square $\overline{c_r^2}$ is taken at x_1/M equal to 20. As usual in such physical situation, a power law of decay of the form

$$\overline{c^2} / \overline{c_r^2} = A.(x_1 / M)^{-n}$$

is observed with a value of n about 1.

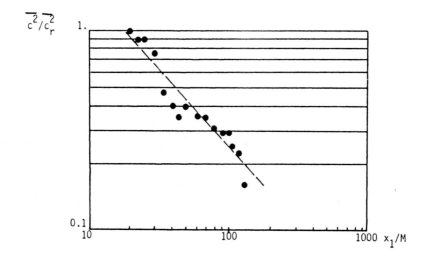

Figure 5

It is easy to deduce an order of magnitude of the Taylor's scale λ_c of the concentration fluctuation field, which is decaying according to the equation valid in quasi isotropic turbulence

$$\overline{U}.d\overline{c^2}/dx_1 = -12.D.\overline{c^2}/\lambda_c^2$$

where D is the molecular diffusivity of the rhodamine in water, which is equal to $10^{-9} m^2/s$ Using the power law of decay, we obtain

$$\lambda_c = (12.D.x_1 / n.\overline{U})^{-1/2}$$

Then, at a downstream distance from the grid x_1/M equal to 100 ($x_1 \simeq 60 cm$) we obtain λ_c equal to $120 \mu m$. It corresponds to the minimum spatial frequency which is observed by the P.M. in that experiment.

A typical power spectrum of $\overline{c^2}$ measured at $x_1/M = 130$ is given on Figure 6. It exhibits a kind of $-5/3$ power law which is often observed in passive scalar spectra but is not justified here by inertial effects because the mesh Reynolds number $Re=\overline{U}M/\nu$ is about 3000. (See for example Warhaft and Lumley (1978)). The Figure 6 clearly indicates the influence of the P.M. noise at high frequencies and of the measurement volume.

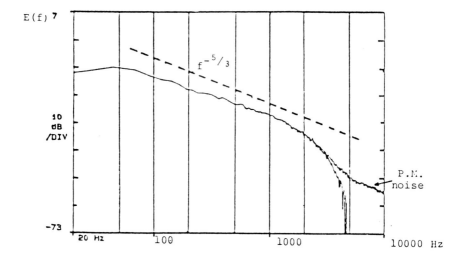

Figure 6

III - First conclusion

The measurement Technics based on the fluorescence of the solution gives a power-law of decay for concentration fluctuation downstream a grid, in agreement with the results obtained in analogous experiments carried out by Gibson (1962) or Bennani, Gence, Mathieu (1985) by a conductimetric method.

In this experiment, the measurement volume has a characteristic length scale of 100 µm. It may be reduced by diminishing the pin hole diameter. This volume is smaller than that of a conductimetric probe and this measurement technics does not perturb the flow.

Another advantage of such method is that it allows us to measure simultaneously and at the same point the fluctuations of concentration of two different fluorescent species whose wave-length of fluorescence are different, by means of optical filtering and of two P.M.

LITERATURE REFERENCES

- GIBSON Ch. : "Scalar mixing in Turbulent Flows" Ph. D. thesis, Stanford University (1962)
- BENNANI A., GENCE J.N., MATHIEU J. : "The influence of a Grid-Generated Turbulence on the development of Chemical Reactions" AIChE. Vol.31, No.7 (1985)
- WARHAFT Z., LUMLEY J.L., : "An experimental study of the decay of temperature fluctuations in grid-generated turbulence" J. Fluid Mech. Vol. 88 part. 4 (1978)

TURBULENT REACTIVE FLOWS OF LIQUIDS IN ISOTHERMAL STIRRED TANKS

by

René DAVID

Laboratoire des Sciences du Génie Chimique - CNRS-ENSIC-INPL
1, rue Grandville 54042 NANCY-CEDEX (France)

Reactive flows in the liquid phase are of major interest for the chemical engineers, but also for the scientist from a theoretical point of view.

Compared to the gas phase, the liquid phase shows three main simple characteristics :
* incompressibility
* dilute reactants in a dense solvent, leading to the appreciable reaction rates with minor thermal effects
* discoupling between the velocity and concentration microgradients, due to high values of Schmidt number

Reactions in the liquid phase are often done in stirred tank reactors which are designed for two different goals :
* reduce the average concentration (or temperature) gradients by macromixing by the velocity field.
* reduce the concentration (or temperature) fluctuations over all the thank by micromixing by the turbulent velocity field followed by diffusion.

So, stirred tanks allow mostly a chemical reaction to be achieved with uniform values of concentration and temperature. Consequently, they are widely used in industry for their simplicity of modeling and control. But, the reduction of the fluctuations occurs at a finite rate, so that a rapid reaction step can contribute to regenerate the microgradients of concentration and temperature. In turn, those microgradients change the yield or selectivity of the reactions.

It is clear that the most sensitive cases to such effects, are those where feeding maintains the macroscopic gradients of reactants in association with a multiple reaction system involving at minimum one rapid step whose extent influences the selectivity.

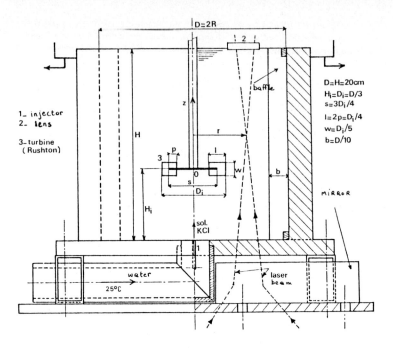

Figure 1. Stirred tank after Holland and Chapman [1]. Work of Mahouast et al. [5]

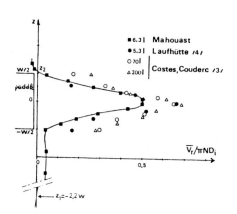

Figure 2. Average radial velocities [5] $[r-(D_j/2) = 1mm]$

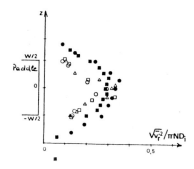

Figure 3. Radial velocity fluctuations [5].

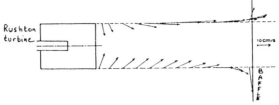

Figure 4. Average velocities in the discharge flow [5].

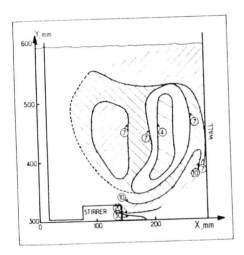

Figure 5. Map of turbulence intensity $10^{+2}u'/\pi ND$ in an upper quarter ($\beta=30°$) of the tank [2].

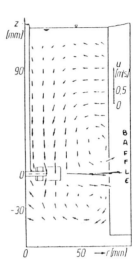

Figure 6. Radial velocity fluctuations [5]

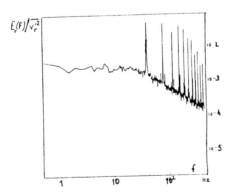

Figure 7. Spectrum of velocity fluctuations
($z=0$; $[r-(D_i/2)=1$ mm] ; $\beta=45°$).

The following paper is restricted to concentration fluctuations, which are the most frequent problem in the liquid phase, because the large values of the Lewis Number (>> 1) indicate that temperature fluctuations (if generated by the feed or reaction) are smoothed out more rapidly than concentration fluctuations.

1. MIXING AND REACTION IN STIRRED TANKS
1.1. - Hydrodynamics of stirred tanks

The characteristics of flow within stirred tanks have been studied by several authors (Costes and Couderc [3], Barthole et al. [2], Laufhütte and Mersmann [4], Mahouast et al. [5] have established the most recent results) in aqueous solutions.

The most investigated stirred tank was that after the standard of Holland and Chapman [1] (fig. 1) equiped by a six flat-blade Rushton turbine [2-6] . The velocities were measured by hot-film anemometry [[2-3] or LDA [4-6].

Figures 2-6 give some examples of results. The flow was found 3-dimensional, highly turbulent provided that the stirrer Reynolds number $N^2 D/\upsilon$ was higher than 10,000. Figure 6 shows clearly an internal "organisation" including recycle flows which are non-symmetric in the upper and lower parts of the tank. Besides, the influence of the angle on the profiles was smaller [5] . Turbulence intensities were found between 5 and 50 %.

The turbulence is anisotropic near the stirrer, but returned to isotropy in the rest of the flow. The spectral analysis of velocity signals (fig. 7) showed a periodic component in the Fourier spectra of the discharge flow of the Rushton turbine which disappeared stepwise away from the stirrer.

These conclusions hold also for other types of stirrers [6].

Macro-and microscales of velocity (fig. 8) were also determined in the whole tank [2-5]. Power dissipation maps have been drawn by several authors [2,4]. Placek et al. [7] have clearly shown the mechanism of power dissipation (fig. 10).

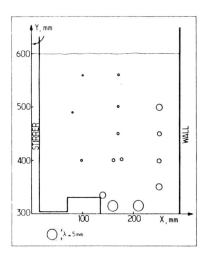

Figure 8. Map of Taylor microscale λ (mm) in the upper quater of the tank. [2]

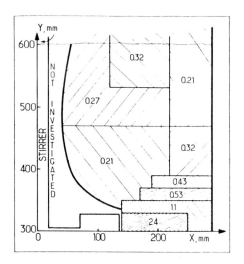

Figure 9. Map of $\varepsilon^* = \rho V \varepsilon / P$ values in the upper quarter of the tank. [2]

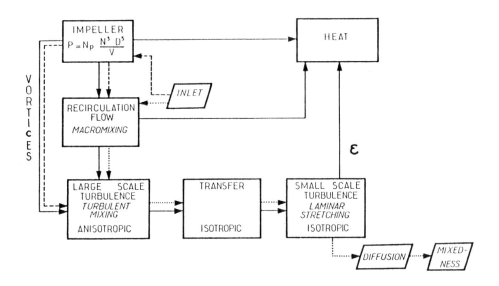

Figure 10. Model of dissipation of mechanical energy and segregation in an agitated vessel.

The circulation of flow within the tank is characterized by a circulation time t_c which has been mostly correlated by Bryant and coworkers [8]. This time represents the average duration of a recycle loop. The velocity field and the intensity of turbulence are such that the entering flow can reach after 3 times t_c every point of the tank. As t_c is in the order of magnitude 0.1-2.0 s, the tank can be considered as a perfect macromixer with uniform average concentration after a few seconds.

1.2. Mechanism of mixing

Within the tank, the mixing of inactive or reactive species proceeds via four consecutive (with respect to the age of the fluid) steps [9] :

* macroscopic distribution of the inlet flow, in the tank by the average velocity field,
* shearing and cutting by turbulence (so-called turbulent diffusion),
* laminar streching of smaller fluid eddies and formation of vortices,
* molecular diffusion when the eddies are small enough.

These steps are linked to the dissipation of mechanical energy in fig. 10.

We shall also consider the so-called concepts of macro-and micromixing.

These are often misunderstood. The distinction between these processes is sometimes ascribed to the scale of unmixed regions, or related to the cause of mixing : average velocity field for macromixing, and turbulence for micromixing.

Actually, macromixing in continuous or semi-batch stirred reactors should be defined as the process leading to equal values of the average concentration in space, whereas the term micromixing should be reserved to processes governing the decay of physical segregation, characterized by local concentration fluctuations. These processes have been discussed in several review papers on micromixing [10, 11, 12].

What are the interactions between chemical reaction, turbulence and average residence time in the tank ?

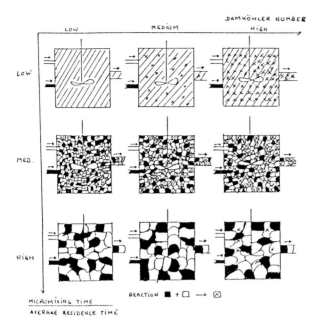

Figure 11. Interaction between micromixing, flow and reaction in a stirred tank.

The figure 11 gives a qualitative explanation of them for a simple reaction A (white) + B(black) → C (crosses) occuring at different 1st Damköler number (ratio of average residence time to reaction time) and 2nd Damköhler number (ratio of micromixing time to average residence time values). A good sensitivity to micromixing effects is achieved through the yield of reaction when the 1st Damköler number has an intermediate value.

The case without chemical reaction is the column on the left.

2. PHYSICAL METHODS

By mixing two solutions containing different amounts of an inert tracer, microscopic concentration gradients arise. By means of a local sensitive probe, the tracer concentration $C = \overline{C} + c(t)$ is recorded and treated as a probability distribution function, or as an autocorrelation function, or by Fourier transform (spectral analysis). It is also possible to deduce the standard deviation of fluctuations $\sqrt{\overline{c^2}}/\overline{C}$ and the different length scales (macroscale, dissipation scale) of the concentration field.

Until now, two different physical methods have been used for recording microfluctuations of concentration in liquids.

2.1. Microconductometry

Microconductometry relies on a pin electrode (approx. 100 μm^2 cross section) coupled with a second classical electrode. The surface of the active cross section is coated with platinum by electrolysis [5,13,14]. The response of this probe is limited only by the reaction rate of electrons with the Pt-coating (time-constant $\tau_1 \approx 5 \times 10^{-3}$ s). So it is possible to dectect fluid aggregates with size :

$$l = U\tau_1 \qquad (U = \text{average velocity})$$

or frequencies up to 2×10^2 Hz.

The method is thus limited and it is impossible to reach sizes of Kolmogorov (10-100 μm) or Batchelor (2-10 μm) microscales encountered in stirred tanks (see spectrum fig. 14).

An example of results obtained in the stirred tank is given on figures 12, 13 from Mahouast et al. [5] in the 6.25 dm^3 standard stirred tank described in § 11.

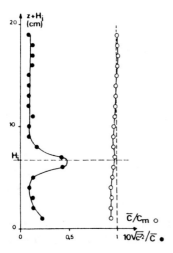

Fig. 12 : Average concentration and concentration fluctuation intensity = (r=0.5R, β=45°)

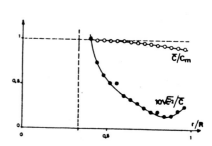

Fig. 13
Average concentration and concentration fluctuation intensity (z=0, β=45°)

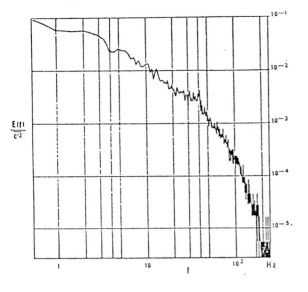

Fig. 14 : Concentration spectrum ($z = 0$; $[r-(D_i/2)] = 13$ mm ; $\beta=45°$) by microconductometry

However, these results and others [14] demonstrate that the average concentration is quite uniform over the tank and that the standard deviation of fluctuations $\sqrt{c^2}/C$ remains small. Concentration macroscales were observed in the range of 70-120 mm in the tank.

2.2. Fluorescence

The method consists, by focusing a laser beam in the flow, in the reemission of light by a fluorescent dye in a aqueous solution with different concentrations of the dye in the two feedstreams of the tank. The light reemitted is collected by a photomultiplier followed by an amplifier and by data acquisition and treatment by a "compatible PC" computer.

The method has been first tested by Patterson [15] without laser light source.

Mixing of both solutions (here of equal flowrate) is done in a stirred tank of 60 mm diameter provided with a Rushton turbine and 4 baffles. Other dimensions refer to the standard of Holland and Chapman [1].

The signal reemitted by the focus point of the incident laser beam ($\lambda = 488$ nm) has a wavelength of 515 nm with fluoresceine as dye. These signals should be proportional to the local concentration of the solute.

The sampling of the signal on the computer was donne up to 25 KHz frequencies. The microcomputer calculates the Fourier transform over 100 blocks of 2048 points each recorded by the computer. The spectrum obtained is the one of concentration fluctuations, but one has to take care of three main problems :

* due to absorption (Beer-Lambert's law) by the solute the light intensity decreases along the pathway of incident and reemitted beams, but this fact can be neglected because of the homogeneity of the average concentration and the small standard deviation $\sqrt{c^2}/C$ in the tank (the pathways are about 3 times the concentration macroscale).

* the presence of non-fluorescent solid impurities which give rise to an additional signal derived from the velocity spectrum, showing especially spectra including the peaks generated by stirring (frequency 6N and harmonics). In order to reduce these sources of error, the solutions were microfiltered.

* the reemission of photons by the solute excited by the laser beam is made through a Poisson process ; when the flux of photons is insufficient, the current recorded by the photomultiplier + amplifier contains additional fluctuations which modify in turn the Fourier spectrum of the signal. The flux of photons has to be reasonably increased by an higher power of the laser beam, or an extension of the volume observed, or using higher concentrations of solute. Actually, the frequency has to be cut by a low-pass filter (here 20 KHz) which leads to maximum frequency of 8 KHz approximately on the spectra.

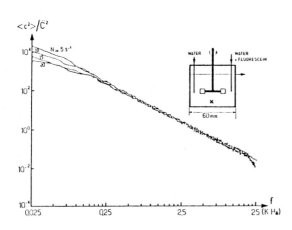

Figure 15. Fluorescence concentration spectrum in a stirred tank. (x = point of measurement)

The figure 15 shows how the spectrum looks like at a point located 10 mm over the bottom of the tank on the revolution symmetry axis. 99% of the standard deviation $\sqrt{c^2}/\bar{C}$ is contained in eddies whose size corresponds to frequencies less than 250 Hz. The influence of stirrer speed N is important in this region of the spectrum. Between 250 Hz and 5 KHz, all the spectra are the same which seems to indicate that the fluctuations of smaller eddies are independent of the behaviour of larger eddies. The slope is about -1.9 differing slightly from the theoretical value of -1.67 for isotropic homogeneous turbulence.

When plotting log $\sqrt{c^2}/\bar{C}$ vs log N, one observes a decreasing straight line :

$$\sqrt{c^2/\bar{C}} = N^{-0.75}$$

Finally, one may calculate that, when $N = 10 \text{ s}^{-1}$, the Batchelor and Kolmogoroff scales correspond to frequencies of 100 and 5 KHz respectively. So the method has to be developed towards higher frequencies if we want to reach the diffusive range predicted by the theory of turbulence equilibrium.

However, the method gives already interesting results in a channel flow, as reported by Gence and coworkers in this meeting [16].

3. CHEMICAL METHODS FOR EVALUATION OF THE MICROMIXING STATE

Let us examine the chemical reaction systems which are the most sensitive to micromixing effects. First, it is obvious that monomolecular reactions - generally 1st order with respect to the reactant - are not dependent on their chemical environment and, thus, not sensitive to micromixing. For testing such effects, we define the two extreme states of mixing in a perfectly macromixed tank.

3.1 Extreme states of micromixing in macromixed tanks

Microfluid is the physical state where the reactants are always mixed up to the molecular level immediately after entering the tank. The mass steady state balance is the written with one single concentration per species j :

$$QC_{jE} - rV = QC_{jS}$$

where $r(C_{1S}...C_{jS})$ is a function of concentrations.

Complete segregation, also called macrofluid, is the state where the reactants initially set into an eddy cannot escape of it. The reaction takes places within "batch" eddies (macrofluid state) :

$$\frac{dC_j}{d\alpha} = rV \text{ with } C_j(0) = C_{j,E}$$

The outlet concentration is averaged through the residence time distribution :

$$C_{jS} = \int_0^\infty C_j(\alpha) \frac{e^{-\alpha/\tau}}{\tau} d\alpha$$

If the reactants are not premixed, such a model leads to zero conversion of reactants.

3.2. Single step reactions

The simplest reactions which are sensitive to micromixing are the second order reactions between two reactants and sometimes the zero order reactions.

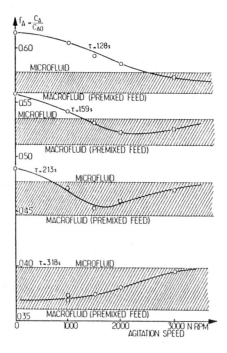

Figure 16. Experimental remaining fraction of nitromethane vs. stirring speed at different space times. Experimental conditions : alkaline hydrolysis of nitromethane; 5% polyethylene glycol, C_{AO} = 33.1 mol/m^3 C_{BO} = 49,7 mol/m^3 ; k_1 = 0.016 m^3 mol^{-1} s-1 ; ν = 2,32 x 10^{-6} m^{*2} s^{-1} T = 192°C. Microfluid means conversion obtained when the reactants are allowed to react in state of microfluid (whatever the kind of feed, unmixed or premixed) ; macrofluid means conversion obtained when then reactants are initially premixed and the allowed to react in a state of total segregation. The shaded area represents the conversion obtainable with premixed reactants and a variable state of segregation (premixed region). The region between y_A = 1 and the microfluid limit represents the conversions obtainable with unmixed feed and a variable state of segregation (unmixed region) [21].

Such reactions have been used e.g. by Paul and Treybal [17], Aubry and Villermaux [18], Villermaux and Zoulalian [19], Plasari et al... [20], Klein et al. [21]. Pohorecki and Baldyga have tested micromixing effects with and instantaneous acid-base reaction [22].

The figure 16 shows some results of Klein et al [21] mixing sodium hydroxide (A) and nitromethane (B) in a stirred tank at different average residence times τ and changing the stirring speed. Clearly, two effects of the stirring process are put here into evidence. At low values of τ, the mixing is essentially devoted to reduce the gradients of reactants entering the tank. At higher values of τ, the mixing changes the physical structure of the fluide and destroys segregation.

However, the study of mixing in stirred tanks by reactions which do not go to completion has three main disadvantages :
* the instantaneous conversion has to be deduced from the concentration of reactants and products and this requires specific sensors with short response times,
* segregation effects on conversion are often small and the kinetics have to be known with accuracy,
* the maximum sensitivity is achieved when the reaction time is in the same order of magnitude as the average residence time or space time and/or the micromixing time (fig. 11); consequently, the number of available reactions is limited and this may lead to unusual short space times in the liquid phase.

Thus, the following conditions should be fullfilled by a chemical system sensitive to micromixing :
* two or more reactants are involved so that contacting between these is controlled by micromixing ;
* the reactants are fed separately into the tank in order to initially create large concentration gradients ;
* at least one of the reactions must be much faster than the mixing process itself. This practically implies that reaction times for the fastest step are less than O.1 s in liquids ;
* a trace of mixing history must be kept in the system in the form of one or several stable products.

The last criterion eliminates single reactions which have to be stopped at the reactor outlet and can only be used with given space times. Conversely, fast multistep reactions leading to a distribution of products frozen by the consumption of one of the major reactants are well suited for micromixing studies. The space time is then generally much longer than the shortest reaction time and has no direct influence on the selectivity so that the method may be used as well in continuous as in semi-batch reactors.

3.3. Multistep reactions

Two examples will be treated here : consecutive-competing reactions and precipitation. A third classical example are polymerization and polycondensation reactions, which are not treated here. The reader will find details about micromixing and polymerization elsewhere [11,23-24].

331. Consecutive-competing reactions

They have been extensively used by several authors [19-25-26]. The general scheme is :

$$A + B \xrightarrow{t_{R1}} R$$

$$R + nB \xrightarrow{t_{R2}} S$$

The most interesting case is $t_{R2} \gg t_{R1}$ when a limited amount of B, in stoichiometric defect, is added into A, so that B is totally consumed at the end of both reactions.

If the mixing of B occurs instantaneously, no secondary product S is formed. If the fluid is partly segregated, R stays in contact with B and is reconverted to S. Both reactions are then stopped by total consumption of B. Therefore, the amount of S formed is some kind of segregation index. Bourne [25] and coworkers have defined this index as :

$$X_S = \frac{(n+1)C_S}{C_{Bo}}$$

Villermaux and David [27] have chosen :

$$\beta = \frac{X_S - X_{SM}}{1 - X_{SM}}$$

where X_{SM} is the yield one would observe in a well micromixed reactor (microfluid).

Two reactions have been especially used : azo-coupling of 1-naphtol (A) with diazotised sulphanilic acid (B) [25] and precipitation of barium sulphate complexed by EDTA in an alkaline medium (reactant A) under the influence of an acid (reactant B) [26]. In the second case, X_S writes :

$$X_S = \frac{(2n+1)C_S}{nC_{Bo}}$$

332. Precipitation reactions

A precipitation process can be represented by the following reaction scheme :

$$A + B \underset{}{\overset{K_e}{\rightleftarrows}} R \xrightarrow{\text{Crystallization}} C \downarrow \text{ solid crystal}$$

The solubility of the intermediate R, less soluble than A and B, is C_R^*. The concentrations of A, B and R are always linked by an equilibrium relationship :

$$K_e C_A C_B = C_R$$

In some cases, especially for inorganic salts, R doesn't exist at all. Then one writes the equilibrium in terms of the solubility product P_S :

$$P_S = C_A^* C_B^* = C_R^*/K_e$$

The driving force of crystallization is supersaturation :

$$S = \frac{C_R}{C_R^*} - 1 = \frac{C_A C_B}{P_S} - 1$$

The crystallization proceeds mainly via two processes :

* nucleation : creation of small crystals called nuclei

$$r_N = k_N S^i \quad + \quad k_N' S^{i_1} C_c^k \quad = k_N S^i (1 + K_N' S^{i_1-i} \frac{C_c^k}{P_S^{k/2}})$$

        primary   secondary
        nucleation   nucleation

with $1 \leq k \leq 2$.

The primary nucleation dominates when starting from a solution without crystals. In suspensions, however, secondary nucleation is preponderant.

* growth : we must distinguish between "chemical" and "diffusional" growth. The growth rate G is defined from the surface supersaturation S_S :

$$G = k_c S_S^j P_S^j$$

Generally, G is independent on the crystal size L.

But, to reach the surface, the reacting species (R or A+B) have to diffuse through the external layer surrounding the crystal. The balance between the chemical growth and the mass transfert fluxes leads to the equation :

$$h S_S^j + S_S - S = 0$$

where $h = \dfrac{k_c}{k_{DR} K_{ep} P_S}$ when R diffuses

and $h = \dfrac{k_c P_S^j}{k_{DAB} P_S^{1/2}}$ when both A and B diffuse

From the nucleation and the growth, it is possible to define time and length scales :

$$t^* = (k_c^3 k_N)^{-1/4} \quad ; \quad L^* = (k_c/k_N)^{1/4}$$

The chemical yield is here replaced by the crystal size distribution -CSD- ψ (L,t) which is the solution of the crystal population balance over a volume V for crystal size L :

$$\dfrac{\partial(\psi V)}{\partial t} + F_{OUT} + \dfrac{\partial(VG\psi)}{\partial L} = F_{IN} + V r_N \delta(L-L_N)$$

where F_{IN} and F_{OUT} are the inlet and outlet fluxes of crystals of size L, respectively.

The moments of the CSD are often used :

$\mu_{0,\psi}$ = number of particles per unit volume or mass of suspension

$L = \mu_{1,\psi} / \mu_{0,\psi}$ = average size (in number) of crystal

$$VAR = \sqrt{\frac{\mu_{2,\psi}\mu_{0,\psi}}{\mu_{1,\psi}^2}} - 1 = \text{standard deviation of } \psi$$

The concentration of C can be deduced from $\mu_{3,\psi}$:

$C = b\mu_{3,\psi}$ with $b = \phi_V \rho_C/M_C$

Garside and Tavare [28] have demonstrated that the crystal size distribution (fig. 17) is very sensitive to the micromixing state.

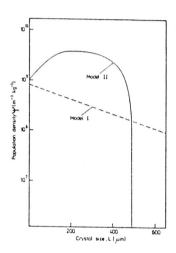

Figure 17. Comparison of crystal size distributions for models I (microfluid) and II (macrofluid premixed) [28]

4. MICROMIXING MODELS
4.1. Single parameter models

These models rely on following assumptions :

* perfect macromixing,
* uniform concentration fluctuations throughout the tank,
* the fluid is divided into small fluid particles or aggregates with uniform concentration inside of them. These particles undergo mass transfer between them,
* these aggregates are identified by their age (Lagrangian point of view).

The single parameter models assumes one micromixing process starting from two opposite sets of initial conditions :

* one premixed feed of reactants
* two or more unmixed feeds of reactants

4.1.1. Coalescence - Dispersion model

Proposed initially by Curl [29], the model was adapted by Spielman and Levenspiel [30]. It assumes aggregates of equal size, coalescing with neighbouring aggregates and producing by redispersion two aggregates of same size as the former ones with averaged concentration. The frequency of the coalescence-redispersion process ω is the parameter of the model. Chemical reaction occurs batchwise in the aggregates between two successive coalescences-redispersions. Obviously, $\omega = 0$ corresponds to complete segregation and $\omega \to \infty$ to the microfluid.

The distribution of concentrations obtained in the tank can be discrete or continuous (Curl). In open reactors, inlet and outlet feed streams are simulated by adding and removing aggregates from the coalescing population.

These models are often by the Monte-Carlo method, which consists of picking up aggregates at random in a large number of aggregates and making them coalesce and redisperse, the process being repeated ω times per simulated time unit.

4.1.2. IEM (Interaction by Exchange with the Mean) Model

As the rigorous treatment of coalescence-dispersion models leads to complicated equations, other authors [31-32] starting from the idea of Harada [33] introduced an exchange of the aggregates with a fictitious average concentration in the tank, which accounts for multiple contacting with other aggregates.

Villermaux and Devillon [31] and Costa and Trevissoi [32] proposed simultaneously the basic developments of this model. The mass exchange between aggregates is characterized by a time constant t_m generally taken as uniform throughout the tank ; $t_m = 0$ is the case of the microfluid, whereas $t_m \to \infty$ is the case of the macrofluid (complete segregation).

The basic equation for the variation of the concentration C_j of species j in an aggregate of age α is written :

$$\frac{dC_j}{d\alpha} = \frac{\bar{C}_j - C_j}{t_m} + R_j$$

R_j is the rate of production of j by reactions ; $\bar{C}_j$ is defined by the condition that the net sum of all exchange fluxes over the tank is zero.

For instance, in a CSTR, as the probability of encountering aggregates of age α is $e^{-\alpha/\tau}/\tau$, $\bar{C}_j$ writes :

$$\bar{C}_j = \frac{1}{\tau} \int_0^\infty C_j e^{-\alpha/\tau} d\alpha$$

An approximate equivalence could be found [11] between coalescence-dispersion and IEM-model when taking :

$\omega = 4/t_m$

Results of calculations by the IEM-model have been drawn in figures 18 and 19 in a CSTR [11] for second order and consecutive-competiting reactions, respectively.

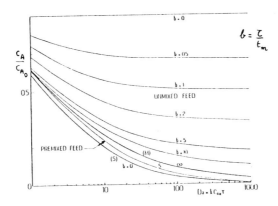

Figure 18.
Second order reaction A+B→products. Influence of micro-mixing on conversion. IEM-model.

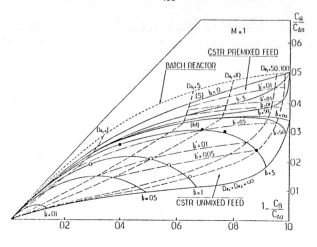

Figure 19.

Consecutive-competing reactions : A+B = R, R+B = S. Influence of micromixing on the yield of R. $K = k_2/k_1 = 0.5, M = C_{Bo}/C_{Ao} = 1$; Open and black circles are calculations by the C-D model showing excellent agreement with the IEM-model [11].

4.1.3. The concept of "mixing earliness" in continuous stirred tanks

Let us define the behaviour of fluid aggregates in a stirred tank by an internal age α (= time elapsed since entering the tank) and a life expectancy or residual lifetime λ (= time remaining until leaving the tank). The sum $\alpha + \lambda = t_s$, the residence time within the tank.

Obviously, all entering aggregates have the same α close to zero and all leaving aggregates have the same λ close to zero too. As the aggregates have different residence time t_s, they do not spend all their life in the tank having simultaneously the same α and the same λ. They start in an entering environment (E.E.) which is a state of minimum mixedness and switch to an leaving environment (L.E.) which is a state of maximum mixedness (all the aggregates have the same probability to leave and thus same λ). E.E. and L.E. should not be confused with macro- and microfluid which denote a physical structure.

On the figure 20, these phenomena are represented in form of a "bundle of parallel tubes". The reactor volume is reorganized in a bundle of small tubes of increasing length equal to t_s. The fluids flows with a flowrate

$$dQ = QE(t_s)dt_s = \frac{Q}{\tau} e^{-t_s/\tau} dt_s$$

through each tube. In the E.E. on the left, the tubes are piled in such a way that aggregates of the same α are on the same vertical line. On the right (L.E.), aggregates with same λ are on the same vertical.

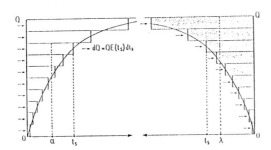

Figure 20. BPT (bundle of parallel tubes) model. Left : minimum mixedness state. Right : maximum mixedness state, mixing is achieved on a vertical line [11].

All the models differ by the way of transferring the aggregates from the E.E. to the L.E. Spencer and al. [34] have defined a general "segregation function" $s(t_s, \lambda)$ or $s(\infty, \lambda)$ which is in fact the fraction of fluid of a given λ and t_s (or ∞) which is in the E.E. ; (1-s) is in the L.E. Derived models have been proposed which are summarized in [10] and [11].

In all these models, the E.E. is assumed to be a microfluid and the L.E. a macrofluid, but one could imagine different combinations. Clearly, the microfluid can be obtained in eliminating the E.E. and taking microfluid structure (mixing up to molecular level) for the L.E. Conversely, the macrofluid can be obtained in suppressing the L.E. and taking complete segregation in the E.E.

All these phenomenological models (coalescence-dispersion, IEM, mixing earlyness) are approximately equivalent to predict conversions or yields with in the macro- and microfluid limits. But the major criticism which may be addressed to this kind of models is their lack of physical meaning in comparison with the structure of flow. In addition, the single parameter leads to ignore the spatial distribution of the micromixing intensity in the tank. Finally, they are unsuitable for scale-up of reactors, owing to the empirical character of involved parameters.

4.2. Models with segregation in physical space

We have distinguished above (§ 12) four stages in the physical mixing process. Starting from these stages, a comprehensive description should provide a quantitative model of the partial segregation and a physical interpretation of the parameters of the model.

Such models take into account one or more stages according to the importance of a given mixing stage for the conversion or yield of chemical species involved in the reactions.

4.2.1. single stage physical models

The micromixing process is then characterized by one time-constant depending on the mechanism involved. As the chemical process is also characterized by a reaction time (e.g. $t_R = 1/kC_0^{n-1}$ for n-th order reaction), the parameter controlling the process is the ratio t_R/t_m which allows to describe all the intermediate states between microfluid ($t_R/t_m \to \infty$) and macrofluid ($t_R/t_m = 0$).

4.2.1.1. Erosion model (2nd stage)

In these models the size l of aggregates generated by the inlet flow is supposed to be reduced with the age of the aggregate linearly [20] or exponentially [22].

$$l = l_0(1 - \alpha/t_e) \text{ or } l = l_0 e^{-\alpha/t_e}$$

t_e was correlated with the kinetic or mechanical energy dissipated in the tank [20, 38].

4.2.1.2. Stretching as major process (3rd stage)

In the theory of Ottino and coworkers [37-38], the striation thickness of δ of lamellae of fluid is determined by stretching process. δ is a function of a stretching time defined as :

$$t_\delta \equiv - \delta \left(\frac{d\delta}{dt}\right)^{-1}$$

These authors have completed their description in including reduction of the concentration gradients of the striation by diffusion when δ is small enough (4th stage).

4.2.1.3. Molecular diffusion (4th stage)

Here the main process is a diffusion process which can be modelized by a reaction-diffusion model in aggregates of size I as done by Bourne and coworkers in their first papers [25] starting from the classical mass balance with partial derivative equations.

The IEM model (§ 412) can be also used to this purpose. It has been shown to be almost equivalent to the reaction-diffusion model by Villermaux et al. [11,27]. The equivalence between time constants of both models is $t_D = \mu l^2/D$ (D = diffusivity, μ = shape factor).

An interesting property is revealed by the simulation reported on figure 21. When one assumes that a real tank with perfect macromixing can be represented by a macrofluid (volume fraction β) and a microfluid (volume fraction $(1-\beta)$ of the reactor), it comes out that the ratio $(1 - \beta)/\beta$ is close to t_R/t_D for several different reacting conditions (fig. 21). On the average :

$$\frac{1-\beta}{\beta} \cong 2 \left(\frac{t_R}{t_D}\right)^{0.8}$$

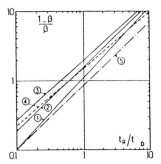

Figure 21. Micromixedness ratio against the ratio of the reaction time to the micromixing time. Curves (1) to (4) : simulations with the IEM model. Second order reaction A+B → products (premixed feed), $C_{Ao} = C_{Bo}$ (I) ; $kC_{Ao}\tau = 2$; (2) $kC_{Ao}\tau = 5$; (3) $kC_{Ao}\tau = 10$. Consecutive competing reactions :

$$A + B \xrightarrow{k_1} R, R + B \xrightarrow{k_2} S \ (C_{Ao} = C_{Bo}) \ (4) \ k_2/k_1 = 0.5, \ t_R = 1/(k_1 C_{Ao})$$

(5) reaction and diffusion in a slab [27].

4.2.2. Multistage physical models

Such models have been developed by Klein et al. [21], Pohorecki and Baldyga [22,38], Bourne and coworkers [25,39], and finally David and Villermaux [35,36]. We shall deal here with only the last one, which involves modelling of the four stages of the mixing process. The process of circulation and mixing of the model of David and Villermaux in a batch or semi-batch tank is explained on figure 22.

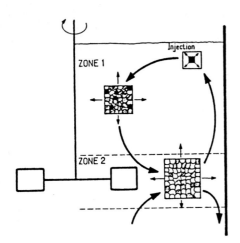
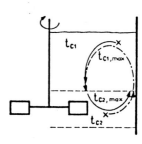

Figure 22. Four stage mixing model [35,36].

○ A Eddies ● ⬣ B Eddies (more or less concentrated)
___ Addition in zone 1. ---- Addition in zone 2

Mixing stage 1

An injected volume V_{Bo} is added in m fractions of volume $V_1 = V_{Bo}/m$, m being equal to 1 in the case of a pulse injection. Each fraction is initially associated to a volume V_2 of surrounding fluid to form the initial reacting cloud. The cloud is convected along the recirculation streams according to the trajectories determined by average velocities. It passes accross two zones, one far from the stirrer (zone 1, low mixing intensity), the other close to it (zone 2, high mixing intensity). The time for closing the recirculation loop is t_c. An injection point is thus defined by time required by the cloud to reach the next zone. In more complicated models, the tank is divided in a network of cells [42,43], each cell having different mixing parameters and exchange flows with neighbouring cells.

Mixing stages 2 and 3

The reaction cloud grows by two mechanisms. The velocity fluctuations cause the invasion of the whole tank by the cloud. The parameter is the turbulent diffusion D_T :

$$D_T \sim u'D$$

A second growth mechanism is that suggested by Badlyga and Bourne [39]. Due to vorticity, small eddies of fluid form striated structures which incorporate fresh fluid of equal volume at regular time intervals t_{inc} :

$$t_{inc} \sim \left(\frac{\nu}{\varepsilon}\right)^{1/2}$$

D_T and t_{inc} change according to the zone where the cloud passes through [35,36].

Actually, both mechanisms compete and the reacting cloud volume is determined by that which - at a given time - would lead to the smallest growth of the cloud volume.

Mixing stage 4

The reacting cloud consists of a collection of added volumes V_i and consequently has a microstructure. The microscopic mass exchange is represented by an IEM model. For species j, in the volume V_i, the mass balance writes :

$$\frac{dC_{ji}}{dt_i} = \frac{C_j - C_{ji}}{t_{mk}} + \sum_l \nu_{lj} r_{li}$$

t_i is the age of fluid element i, l is the reaction index, t_{mk} is the micromixing time in zone k = 1, 2.

Initial conditions are given by the composition of the volume V_1 and the concentration of species initially present in the tank.

The average concentration in the cloud is calculated as:

$$C_j = \frac{\sum_i C_{ji} V_i}{\sum_i V_i}$$

Overall concentrations in the tank are calculated from C_j and initial concentrations in the rest of the tank. The calculations can be stopped when the key reactant is totally consumed.

4.2.2.1. Results for consecutive-competing reactions

The test-reaction developed by Barthole et al. [26] is based on the precipitation of $BaSO_4$ (S) from a basic EDTA-barium complex [$(BaY^{2-})_n$, OH^- = A] by addition of H_3O^+ ions (B) in the presence of sulfate ions (U):

$$A + B \xrightarrow{\infty} R + 2H_2O$$

$$nU + R + 2nB \rightarrow nS + nYH_2^{2-} + 2H_2O$$

$$2A + YH_2^{2-} \xrightarrow{\infty} 2R + Y^{4-} + 2H_2O$$

where $R = (BaY^{2-})_n$

The segregation index X_S is defined as indicated in § 331 ; n is the initial ratio $(Ba^{2+})/(OH^-)$.

The reactor where the experiments were done was of 140 dm^3 volume stirred by a Rushton turbine with tank diameter/turbine diameter ratio equal to 2. It was also possible to insert a Mixel TT axial stirrer (diameter 0.36 m) instead of the Rushton turbine ; 0.1 dm^3 of 1.2 N hydrochloric acid was added either dropwise or as a pulse injection at five different places (fig. 23) in the tank. After completion of all three reactions which needs between 0.2 t_c and 0.5 t_c, the concentration of S, i.e. X_S, could be measured by taking a sample analyzed by spectrophotometry [26].

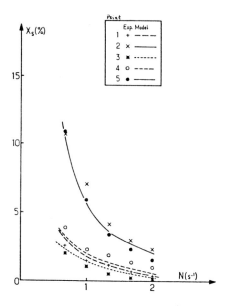
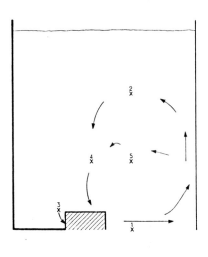

Figure 23. Segregation index X_S vs. rotation speed N in a stirred tank [40] for 5 different points of addition of reactant HCl.

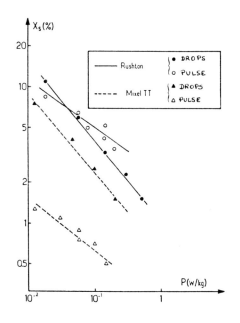

Figure 24. X_S vs. power dissipated per unit mass P for addition at point 5 (● ▲ dropwise addition [40]; ○ △ pulse addition [35]). Comparison of the performance of 2 stirrers.

In figure 23, X_S is plotted vs N for dropwise addition. The best mixing is achieved when injecting near the stirrer (point 3) at a given stirring speed. In figure 24, X_S is drawn vs the power delivered by the two stirrers to the liquid for both modes of addition of HCl.

Dropwise addition leads to

$$X_S \sim P^{-1/2} \sim N^{-1.5} \sim \text{(dissipated energy)}^{-1/2}$$

Pulse addition to

$$X_S \sim P^{-1/3} \sim N^{-1} \sim \text{(velocities in the tank)}^{-1}$$

The four free parameters (D_t, t_m, t_{inc}, V_2/V_1) of the model were fitted on the experiments. Simulations have shown that when using dropwise addition, the size of the reaction cloud was determined by vorticy (stage 3), whereas when using pulse addition, this size depended on turbulent diffusion (stage 2). This could explain the dependencies on P reported above [40].

The model explains well also all the X_S vs N experiments of figure 23. Further experiments were made in 0.1 dm^3 tank [40] and extrapolation rules could be set [36].

4.2.2.2. Simulations with precipitation of sparingly soluble salts

The same mixing model was applied to a second type of reaction, the precipitation of a sparingly soluble salt (see section 332) by adding the ion B dropwise in a solution of A making a so-called "single-jet precipitation" (fig. 25c) [41].

We took reasonable values of the reaction parameters and of the mixing parameters $\theta_{inc} = t_{inc}/t^*$, $\theta_m = t_m/t^*$; V_2/V_1 was set to its optimal value of 10 determined with consecutive-competing reactions. For the sake of simplicity, no distinction was made between two zones here.

Typical results could be found on figures 25a and 25b. Micromixing seems to be very important (fig. 25b). Efficient mixing leads to more crystals of smaller size and to distribution of crystal sizes which are more spread out. External transfer limitation (h increasing, see § 332) results in smaller crystals. Secondary nucleation (K'_N high) has the same effect.

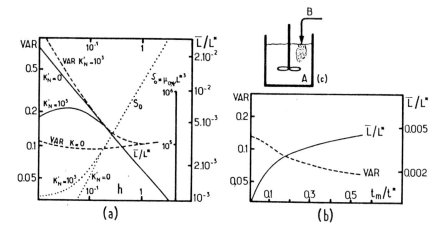

Figures 25a and 25 b. Mixing and precipitation A+B → C (solid). Data of the simulation [38] : j=2 ; i=6 ; k=2 ; i'=2 ; b= 10^4 ; $C_{Ao} = C_{Bo} = 100$ $P_S^{1/2}$; 1 $dm^3 = V_{Ao} = V_{Bo}$; $t_{inc}/t^* = 0.1$; $t_m/t^* = 0.1$ (fig. 25a) ; h = 1 (fig. 25b).

CONCLUDING REMARKS

The major concepts explaining micromixing processes are now well identified in stirred reactors and progress has been made towards an unified theory combining knowledge from the fluid mechanics and the reactions and allowing rapid predictions by means of phenomenological models.

Modes describing mixing in continuous reactors are superabundant, but those relying on a more physical description should be preferred. Macro- and micromixing should not be considered separately but in the frame of models integrating both phenomena.

In practice, the question for the chemist is often to determine the best operating conditions of mixing in order to optimize the selectivity of a wanted product or the distribution of products (e.g. polymerization or precipitation) and the role of the chemical engineer is to provide simple rules and models allowing this.

There are still important problems to be solved, concerning for instance micromixing in non-Newtonian fluids, multiphase reactors or scaling-up and down.

REFERENCES

1. HOLLAND F.A., CHAPMAN F.S., Liquid mixing and processing in stirred tanks, Rheinhold Publ. Co., New-York (1966)

2. BARTHOLE J.P. et al., Chem. Engng. Fundam., $\underline{1}$, 17 (1982)

3. COSTES J., COUDERC J.C., Proc. 4th European Conference on Mixing, Cranfield (U.K.), BHRA Fluid Eng., 14 (1982)

4. LAUFHUTTE H.S., MERSMANN A., Chemie Ingenieur Technik, $\underline{56}$(11), 862 (1984)

5. MAHOUAST M., DAVID R., COGNET G., Entropie, $\underline{133}$, 7 (1987)

6. MAGNI F., COSTES J., COUDERC J.P., Colloque sur l'agitation mécanique, Toulouse (France), 1-17 (June 1986)

7. PLACEK J., TAVLARIDES L.L., SMITH G.W., AIChE J., $\underline{32}$(11), 1771 (1986)

8. BRYANT J., SADEGHZADEH S., Proc. 4th European Conference on Mixing, Cranfield (U.K.) BHRA Fluid Eng., 325 (1982)

9. BEEK J., Jr. and MILLER R.S., Chem. Eng. Prog. Symp. Series, $\underline{55}$, 23 (1959)

10. VILLERMAUX J., Mixing in Chemical Reactors, ACS Symp. Series, $\underline{226}$, 135 (1983)

11. VILLERMAUX J., Micromixing phenomena in stirred reactors, Encyclopedia of Fluid Mechanics, Gulf Publishing Co., West Orange NJ, 707 (1986)

12. BOURNE J.R., Micromixing revisited, Proc. ISCRE 8, Edimburgh, (1984)

13. GIBSON C.H., SCHWARTZ W.H., J. Fluid Mechanics, $\underline{16}$, 357 (1963)

14. BRODBERGER J.F., VALENTIN G., STORCK A., Entropie, $\underline{112}$, 20 (1984)

15. PATTERSON G.K., BOCKELMAN W., QUIGLEY J., Proc. 4th European Conference on Mixing, BHRA Fluid Eng., 303 (1982)

16. GENCE J.N., LIEVRE J., Measurements of chemical reactions in turbulent liquid flows, (present workshop) (1987)

17. PAUL E.L., TREYBAL R.E., AIChE J., 17, 718 (1971)

18. AUBRY C., VILLERMAUX J., Chem. Eng. Sci., 30, 457 (1975)

19. ZOULALIAN A., VILLERMAUX J., Adv. Chem. Series, Chem. React. Engng., Evanston, 133, 348 (1974)

20. PLASARI E., DAVID R., VILLERMAUX J., ACS Symp. Series, 65, 125 (1978)

21. KLEIN J.P., DAVID R., VILLERMAUX J., Ind. Eng. Chem. Fundam., 19, 373 (1980)

22. POHORECKI R., BALDYGA J., Ind. Eng. Chem. Fundam., 22(4), 398 (1983)

23. VILLERMAUX J., BLAVIER L., Chem. Engng. Sci., 39(1), 87 (1984)

24. NAUMAN E.B., J. Macromol. Sci. Revs. Macromol. Chem., C10, 75 (1974)

25. BOURNE J.R., KOZICKI F., RYS P., Chem. Engng. Sci., 36, 1643 (1981)

26. BARTHOLE J.P., DAVID R., VILLERMAUX J., ACS Symp. Series, 196, 545 (1982)

27. VILLERMAUX J., DAVID R., Chem. Eng. Commun., 21, 105 (1983)

28. GARSIDE J., TAVARE N.S., Chem. Engng. Sci., 40, 1485 (1985)

29. CURL R.L., AIChE J., 9, 175 (1963)

30. SPIELMAN L.A., LEVENSPIEL O., Chem. Engng. Sci., 20, 247 (1965)

31. VILLERMAUX J., DEVILLON J.C., Chem. React. Engng., Amsterdam, Elsevier Publ. Co., B1-13 (1972)

32. COSTA P., TREVISSOI C., Chem. Engng. Sci., 27, 2041 (1972)

33. HARADA M. et al., Memoirs of the Faculty of Engineering, Kyoto (Japan), 24, 431 (1962)

34. SPENCER J.L., LUNT R., LESHAW S.A., Ind. Eng. Chem. Fundam., 19, 135 (1980)

35. DAVID R., BARTHOLE J.P., VILLERMAUX J., 5th European Conference on Mixing, Würzburg (FRG), BHRA Publ., 433 (1985)

36. DAVID R., VILLERMAUX J., Chem. Eng. Commun. (in press)

37. OTTINO J.M., AIChE J., 27, 184 (1981)

38. POHORECKI R., BALDYGA J., 5th European Conference on Mixing, Würzburg (FRG), BHRA Publ., 89 (1985)

39. BALDYGA J., BOURNE J.R., Chem. Eng. Commun., 28, 231 (1984)

40. DAVID R., GOUYE D., VILLERMAUX J., Chemie Ingenieur Technik, 59, 254 (1987)

41. VILLERMAUX J., DAVID R., Colloque sur la cristallisation industrielle, VII-2, Toulouse (France), (June 1987)

42. MIDDLETON J.C., PIERCE F., LYNCH P.M., Chem. Eng. Res. Des., 64, 18 (1986)

43. KNYSH P., MANN R., Symp. Series Inst. of Chem. Engineers, Fluid Mixing II, 89, 127 (1984)

Part 2

Table of Contents

Part II
Structure and Predictive Schemes

P. E. Dimotakis
Turbulent Shear Layer Mixing With Fast Chemical Reactions......... 417

J.J. Riley and P.A. McMurtry
The Use of Direct Numerical Simulation in the Study of Turbulent, Chemically-Reacting Flows.. 486

A.D. Leonard and J.C. Hill
Direct Numerical Simulation and Simple Closure Theory for a Chemical Reaction in Homogeneous Turbulence...................... 515

K.N.C. Bray, M. Champion, and Paul A. Libby
The Interaction Between Turbulence and Chemistry in Premixed Turbulent Flames... 541

R. Borghi, A. Picart, and M. Gonzalez
On the Problem of Modelling Time or Length Scales in Turbulent Combustion... 564

S.B. Pope
Statistical Modelling of Turbulent Reactive Flows................. 589

N. Darabiha, V. Giovangigli, A. Trouve, S.M. Candel, and E. Esposito
Coherent Flame Description of Turbulent Premixed Ducted Flames.... 591

A.F. Ghoniem, G. Heidarinejad, and A. Krishnan
Turbulence-Combustion Interactions in a Reacting Shear Layer...... 638

M.S. Anand, S.B. Pope, and H.C. Mongia
A PDF Method for Turbulent Recirculating Flows.................... 672

A. Giovannini
Dynamics of Cold and Reacting Flows on Backward Facing Step Geometry... 694

W. Kollmann
PDF - Transport Equations for Chemically Reacting Flows........... 715

M. Champion
Modelling the Effects of Combustion on a Premixed Turbulent Flow:
A Review.. 731

J. Boris, E. Oran, and K. Kailasanath
The Numerical Simulation of Compressible and Reactive Turbulent
Structures... 754

G.M. Faeth
Turbulent Multiphase Flows................................... 784

A. Berlemont, G. Gouesbet, and P. Desjonqueres
Lagrangian Simulation of Particle Dispersion................. 815

O. Simonin and P.L. Viollet
Numerical Modelling of Devolatilization in Pulverised Coal
Injection Inside a Hot Coflowing Air Flow.................... 824

M. Barrere
Flame Stabilization in a Supersonic Combustor................ 847

J. Swithenbank, F. Boysan, B.C.R. Ewan, L. Shao, and Z.Y. Yang
Mixing Problems in Supersonic Combustion..................... 863

T. Poinsot, A. Trouve, D. Veynante, S.M. Candel, and E. Esposito
Morphology of Flames Submitted to Pressure Waves............. 890

K.C. Schadow, E. Gutmark, T.P. Parr, D.M. Parr, and K.J. Wilson
Control of Turbulence in Combustion.......................... 912

F.E. Marble, G.J. Hendricks, and E.E. Zukoski
Progress Toward Shock Enhancement of Supersonic Combustion
Processes.. 932

TURBULENT SHEAR LAYER MIXING WITH FAST CHEMICAL REACTIONS

Paul E. DIMOTAKIS

Graduate Aeronautical Laboratories
California Institute of Technology
Pasadena, California 91125

ABSTRACT

A model is proposed for calculating molecular mixing and chemical reactions in fully developed turbulent shear layers, in the limit of infinitely fast chemical kinetics and negligible heat release. The model is based on the assumption that the topology of the interface between the two entrained reactants in the layer, as well as the strain field associated with it, can be described by the similarity laws of the Kolmogorov cascade. The calculation estimates the integrated volume fraction across the layer occupied by the chemical product, as a function of the stoichiometric mixture ratio of the reactants carried by the free streams, the velocity ratio of the shear layer, the local Reynolds number, and the Schmidt number of the flow. The results are in good agreement with measurements of the volume fraction occupied by the molecularly mixed fluid in a turbulent shear layer and the amount of chemical product, in both gas phase and liquid phase chemically reacting shear layers.

1.0 INTRODUCTION

Understanding chemically reacting, turbulent free shear flows is important not only for the obvious technical reasons associated with the engineering of a variety of reacting and combusting devices but also for reasons of fundamental importance to fluid mechanics and our perception of turbulence.

From a theoretical point of view, chemically reacting flows provide important tests of turbulence theories by adding to the dimensionality of the questions that can be asked of turbulence models. To compute chemical reactions in turbulent flow, the physics of reactant species turbulent transport and mixing need to be described correctly down to the diffusion scale level. This is a much more stringent specification than needs be imposed on momentum transport turbulence models.

From an experimental point of view, a fast chemical reaction provides a probe with an effective spatial and temporal resolution and sensitivity that is usually unattainable by conventional direct flow field measurement techniques in high Reynolds number turbulent flows. Chemically reacting turbulent flow experiments are therefore to be regarded as a complementary means of interrogation; a valuable adjunct to the more conventional probing of the behavior of turbulent flow.

A broad class of current efforts to understand chemically reacting turbulent flows is based on classical turbulence formulations founded on the Reynolds-averaged Navier-Stokes equations. In such formulations, species transport is conventionally modeled as proportional to the gradient of the corresponding mean species concentration, with an effective diffusivity that is prescribed to be some function of the flow. See Tennekes & Lumley (1972) for an introduction. Estimates of mixing at the molecular scale must be modeled separately, in these formulations, in a manner that unfortunately cannot be addressed without additional assumptions, that are essentially ad hoc. See Sreenivasan, Tavoularis & Corrsin (1981), the introduction in Broadwell & Breidenthal (1982) and the discussion in Broadwell & Dimotakis (1986) for a discussion of these issues.

A different approach is taken by modeling efforts based on attempts to write transport equations for the probability density functions (PDF) of the conserved scalars, or joint PDFs for scalars, and/or the (vector) velocity field and pressure. See Pope (1985) and related work by Kollmann & Janicka (1982) and Kollmann (1984), for example. These efforts, which are in principle capable of addressing the issues of transport and mixing in a unified manner, must nevertheless resort to essentially equally ad hoc assumptions to close the problem. In other words, while having the correct fluctuation statistics through the relevant PDFs, and conditional statistics through one-time joint PDFs, would undoubtedly permit the molecular mixing and resulting chemical product formation to be computed correctly, it would appear that those PDFs are no easier to obtain than the ab initio solution of the original problem.

Finally, a model was recently proposed by Broadwell & Breidenthal (1982) which is not based on gradient transport concepts. This model will be discussed below in the context of recent data on chemically reacting shear layers in both gas phase and liquid phase shear layers.

1.1 Recent experimental results

The aspirating probe (Brown & Rebollo 1972) measurements of Brown & Roshko (1974), and the measurements of Konrad (1976) of the probability density function of the high speed fluid fraction in a non-reacting, gas phase shear layer suggested that the mixed fluid composition does not vary appreciably across the width of the layer, even as the mean high speed fluid fraction varies smoothly from unity on the high speed side, to zero on the low speed side. Additionally, as Konrad recognized, the most likely values of the mixed fluid high speed fluid fraction seem to be clustered around a value dictated by the shear layer entrainment ratio. In the light of these results, the smooth variation of the mean is then to be understood as the variation of the local probability of finding:

a. pure high speed fluid,

b. mixed fluid,

and,

c. pure low speed fluid,

as we traverse the width of the layer.

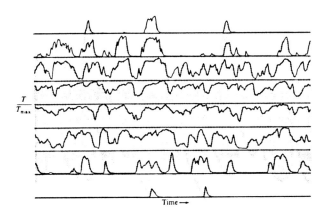

FIGURE 1. Temperature vs. time time traces for $\phi = 1$ ($\Delta T_{flm} = 93$ K). High speed ($U_1 = 22$ m/s) fluid (1% F_2 + 99% N_2) on top trace. Low speed ($U_2 = 8.8$ m/s) fluid (1% H_2 + 99% N_2) on bottom. Probe positions at y/x = 0.076, 0.057, 0.036, 0.015, -0.008, -0.028, -0.049, -0.070. Partial record of 51.2 ms time span ($\Delta T_{max} = 81$ K). From Mungal & Dimotakis (1984, figure 4b).

The near uniformity in the mixed fluid composition, apparent in Konrad's passive scalar non-reacting shear layer experiments, can be seen to have an important counterpart in the gas-phase, chemically reacting shear layer experiments (e.g. Mungal & Dimotakis 1984). Measuring the temperature field in the reaction zone of a mixing layer bringing together H_2 and F_2 reactants carried in a N_2 diluent, it is found that within the discernible regions that can be associated with the interior of the large scale structures the temperature was nearly uniform. See figure 1. The resulting mean temperature (chemical product) profile that peaks in the interior of the reaction zone is more a consequence of the variation of the fraction of the time a given fixed point is visited by the hot large scale cores (duty cycle), rather than the variation of the temperature field within a core. See figure 2.

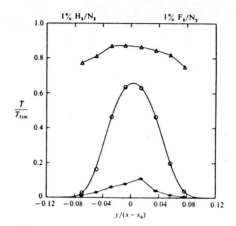

FIGURE 2. Peak, mean and minimum temperature rise observed for total data record at each station. Experimental parameters as in figure 1. Smooth curve least squares fitted through mean data points. From Mungal & Dimotakis (1984, figure 4c).

These results and conclusions are in good agreement with the results of Fiedler (1975), who measured the temperature with a fair temporal/spatial resolution at several points across a shear layer, one free stream of which was marked by a small temperature difference serving as a label for the passive conserved scalar. Measurements in both reacting and non-reacting liquid phase shear layers of the PDF of the high speed fluid fraction also corroborate these findings. See figure 3.

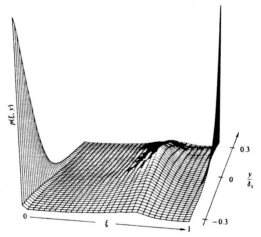

FIGURE 3. Probability density function of the high speed fluid mixture fraction in a liquid layer ($U_2/U_1 = 0.38$, Re $= 2.3 \times 10^4$). $\xi = 0$ corresponds to low speed fluid, $\xi = 1$ corresponds to high speed fluid. From Koochesfahani & Dimotakis (1986, figure 10).

An important conclusion can be drawn from these data, which is also consistent with the results of the flow visualization studies and the earlier pilot, liquid phase, chemically reacting experiments in Dimotakis & Brown (1976), as well as the study of liquid phase reacting layers by Breidenthal (1981), namely that the large scale motion within the cores of the shear layer vortical structures is capable of transporting a small fluid element from one edge of the layer to the other, before any significant change in its internal composition can occur. During this transport phase, initially unmixed fluid within the fluid element will mix to contribute to the amount of molecularly mixed fluid, but will do so to produce a range of compositions clustered around the value corresponding to the relative amounts of unmixed fluid originally within the small fluid element. This is the reason why the mixed fluid composition cannot exhibit a substantial systematic variation across the layer and, in particular, need not be centered about the value of the local mean. This observation represents an important simplification to the problem, as it suggests that it may be justified to treat the composition field in a uniform manner across the shear layer width.

In the gas phase, hydrogen-fluorine experiments of Mungal & Dimotakis (1984), the stoichiometric mixture ratio ϕ, defined by

$$\phi = \frac{c_{02}/c_{01}}{(c_{02}/c_{01})_s} \; , \tag{1.1}$$

was varied, where c_{02} and c_{01} are the low and high speed free stream reactant concentrations respectively, and the subscript "s" in the denominator denotes the corresponding chemical reaction stoichiometric ratio (unity for the $H_2 + F_2$ reaction). The quantity ϕ can be viewed as representing the mass of high speed fluid required (to be mixed and react) to exactly consume a unit mass of low speed fluid. For uniform density, chemically reacting shear layers (low heat release), ϕ can also be interpreted in terms of the requisite volumes of the free stream fluids for complete reaction.

For a given value of ϕ, the total amount of chemical product in the mixing layer can be expressed in terms of the integral product thickness

$$\delta_{P1} = \frac{1}{c_{01}} \int_{-\infty}^{\infty} c_P(y,\phi) \, dy \; , \tag{1.2}$$

where the subscript 1 in δ_{P1} denotes that c_{01}, the high speed stream reactant concentration, was used to normalize the mean chemical product concentration profile $c_P(y,\phi)$. Using the mean temperature rise $\Delta T(y,\phi)$ as the measure of product concentration, and normalizing the transverse coordinate y by the total width of the layer δ, we can also write

$$\frac{\delta_{P1}}{\delta} = \frac{1}{\delta} \int_{-\infty}^{\infty} \frac{\Delta T(y,\phi)}{\Delta T_{flm}(\infty)} \, dy \; . \tag{1.3}$$

If we keep c_{01} fixed and vary ϕ by, say, increasing c_{02}, also keeping the heat capacities for the free stream fluids matched, we find that the dependence of the adiabatic flame temperature rise on ϕ is given by

$$\Delta T_{flm}(\phi) = \frac{2\phi}{\phi + 1} \Delta T_{flm}(1) , \qquad (1.4)$$

where $\Delta T_{flm}(1)$ is the adiabatic flame temperature rise corresponding to a stoichiometric reactant concentration ratio. Note that, for a fixed high speed stream reactant concentration c_{01}, the normalizing temperature in equation 1.3 is given by $\Delta T_{flm}(\infty) = 2 \Delta T_{flm}(1)$. The experimental values for the product thickness δ_{P1}/δ, in such an experiment, are plotted in figure 4.

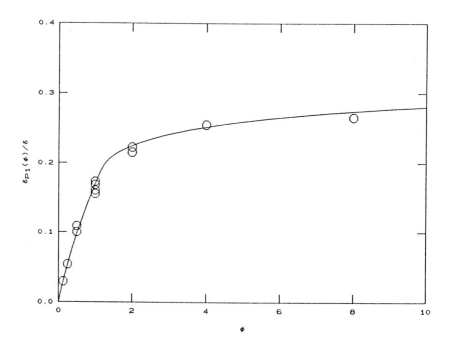

FIGURE 4. Normalized δ_{P1}/δ HF gas phase chemical product thickness data vs. stoichiometric mixture ratio ϕ (Mungal & Dimotakis 1984). $U_2/U_1 = 0.38$, Re $= 6.6 \times 10^4$. Smooth curve drawn to aid the eye.

In the data in figure 4, the width of the layer δ was estimated by δ_1, which is defined as the extent across the layer where the product concentration (mean temperature rise) has fallen to 1% of its peak mean value. To remove the small differences in the values of δ_1 computed from the temperature profiles measured for each ϕ (see Mungal & Dimotakis 1984, Table I), a fixed (average) value for δ/x ($= 0.165$) was used in normalizing the data in figure 4. We note that the 1% width δ_1, in both the gas phase reacting layer data and the liquid phase measurements of Koochesfahani & Dimotakis (1986), was found to be very close to the visual shear layer width δ_{vis} of Brown & Roshko (1974). As can be seen in the data in figure 4, as ϕ is increased from small values, the amount of chemical product at first increases rapidly. Beyond a certain value, however, a further increase in ϕ (increase of the low speed stream reactant concentration) does not result in a commensurate increase in the total chemical product, as the fluid in the shear layer is low speed reactant rich and much of the entrained high speed stream reactant has already been consumed. The smooth curve in figure 4 was drawn to aid the eye.

A slightly different definition of product thickness, which avoids the asymmetric choice of using one stream or the other as a reference, is to use the adiabatic flame temperature $\Delta T_{flm}(\phi)$ to normalize the temperature profile, corresponding to each value of ϕ. This yields a new normalized product thickness δ_P/δ, given by

$$\frac{\delta_P}{\delta} = \frac{1}{\delta} \int_{-\infty}^{\infty} \frac{\Delta T(y,\phi)}{\Delta T_{flm}(\phi)} \, dy \ , \qquad (1.5)$$

which represents the volume fraction occupied by chemical product. Note that the integrand is in the units of the normalized mean temperature rise profile, as plotted in figure 2, and that $\delta_P/\delta = (\delta_{P1}/\delta)/\xi_\phi$, where

$$\xi_\phi = \frac{\phi}{\phi + 1} \ . \qquad (1.6)$$

For equal density free streams, negligible heat release, and a given free stream reactant stoichiometric mixture ratio ϕ, the quantity ξ_ϕ represents the high speed fluid volume fraction, in the mixed fluid, required for complete consumption of both reactants. A volume fraction $\xi > \xi_\phi$ in the molecularly mixed fluid, for example, corresponds to an excess of high speed fluid, relative to that required by the stoichiometry of the reaction, and would result in complete consumption of the low speed reactant in the mixture, and a remainder of unreacted high speed fluid. A plot of the experimental values of δ_P/δ for the hydrogen-fluorine gas phase data, versus ξ_ϕ, appears in figure 5. The smooth curve through the gas phase data of Mungal & Dimotakis (1984) denoted by circles is the same curve that appears in figure 4, transformed to the coordinates of figure 5. The data point denoted by the triangle corresponds to the similarly defined chemical product volume fraction in a liquid phase two dimensional shear layer, as measured by Koochesfahani & Dimotakis (1986) at the same free stream speed ratio and comparable Reynolds number.

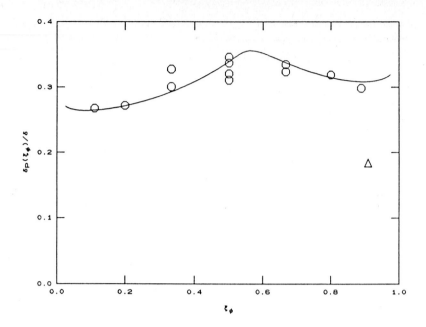

FIGURE 5. Chemical product δ_P/δ volume fraction data vs. stoichiometric mixture fraction ξ_ϕ. Circles from gas phase Mungal & Dimotakis (1984) data (see figure 4). Triangle from liquid phase Koochesfahani & Dimotakis (1986) data ($U_2/U_1 \approx 0.4$, $Re \approx 7.8 \times 10^4$). Smooth curve transformed from that of figure 4.

Since (for equal species and heat diffusivities) $\Delta T_{flm}(\phi)$ is the highest temperature that can be achieved in the reaction zone, the ratio δ_P/δ represents the volume fraction occupied by the chemical product within the mixing zone and is a measure of the shear layer turbulent mixing and chemical reactor "efficiency". If the two reactants were entrained from the two free streams in such a way as to produce molecularly mixed fluid everywhere within the layer at a single-valued composition corresponding to a mixture fraction ξ_ϕ, then the resulting temperature profile would be a top-hat of height $\Delta T_{flm}(\phi)$ and width δ, resulting in a value of δ_P/δ of unity. This clearly represents the highest possible total chemical product that can be formed within the confines of the shear layer turbulent region. If, on the other hand, the mean temperature rise profile was a triangle whose base was equal to δ and which reached $\Delta T_{flm}(\phi)$ at the apex somewhere within the layer, then δ_P/δ would be equal to $1/2$. It is interesting that, in these units, the gas phase data (circles) in figure 5, for all the values of the stoichiometric mixture ratio investigated, are in the relatively narrow range of $\delta_P/\delta \approx 0.31 \pm 0.03$.

Comparison of the total amount of chemical product measured in gas phase reacting layers (Mungal & Dimotakis 1984), and liquid phase reacting layers (Breidenthal 1981,

Koochesfahani & Dimotakis 1986), points out another important feature of these data; at comparable flow conditions, the amount of chemical product formed at high Reynolds numbers is a function of the (molecular) Schmidt number $Sc = \nu/D$ of the fluid, where ν is the kinematic viscosity and D is the relevant species diffusivity. In particular, roughly twice as much product is formed in a gas phase chemically reacting shear layer ($Sc \approx 0.8$) as in a liquid phase layer ($Sc \approx 600$).

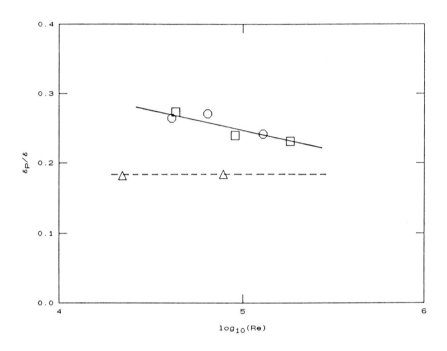

FIGURE 6. Chemical product δ_P/δ volume fraction versus Reynolds number. Circles and squares are for gas phase data (Mungal et al 1985) at $\phi = 1/8$. Circles are for initially laminar splitter plate boundary layers, squares for turbulent boundary layers. Triangles for liquid phase data from Koochesfahani & Dimotakis (1986), at $\phi = 10$.

Finally, in a further investigation in gas phase reacting shear layers, the Reynolds number was varied over a range of almost an order of magnitude, keeping all other conditions as constant as was feasible (Mungal et al 1985). The resulting data for δ_P/δ, for a fixed stoichiometric mixture ratio of $\phi = 1/8$, are plotted in figure 6. It can be seen that there is a modest but unmistakeable decrease in the total amount of product in the layer as the Reynolds number is increased. The authors estimate that, at the operating conditions for those experiments, a factor of 2 increase in the Reynolds number results in approximately a 6% reduction in δ_P/δ, the chemical product volume fraction. Also included in the same plot, for comparison purposes, are the reacting liquid layer data of Koochesfahani & Dimotakis (1986) at a stoichiometric mixture ratio of $\phi = 10$. As can be seen, the data indicate a much weaker Reynolds number dependence of the liquid phase product volume fraction δ_P/δ. We note, however, that the lower Reynolds number liquid data point may be at a value of the Reynolds number that is too close to the shear layer mixing transition (Konrad 1976, Bernal et al 1979, Breidenthal 1981) and the flow may not have attained fully turbulent behavior.

1.2 Entrainment ratio for a spatially growing shear layer

An important conclusion drawn by Konrad (1976) was that a spatially growing shear layer entrains fluid from each of the two free streams in an asymmetric way, even for equal free stream densities. In particular, for equal free stream densities ($\rho_2/\rho_1 = 1$) and a free stream speed ratio of $U_2/U_1 = 0.38$, Konrad estimated the volume flux entrainment ratio E to be 1.3. For a free stream density ratio of $\rho_2/\rho_1 = 7$ (helium high speed fluid and nitrogen low speed fluid), and the same velocity ratio, he estimated an entrainment ratio of $E \approx 3.4$.

This behavior can be understood in terms of the upstream/downstream asymmetry that a given large scale vortical structure sees in a spatially growing shear layer. Simple arguments suggest that the volume flux entrainment ratio can be estimated and is given by

$$E = \left(\frac{\rho_2}{\rho_1}\right)^{1/2} (1 + \ell/x) , \qquad (1.7a)$$

where ℓ/x is the large structure spacing to position ratio. See Dimotakis (1986) for the arguments leading to this result.

Konrad's data support the hypothesis that $\langle \ell/x \rangle$, the ensemble averaged value of ℓ/x, is independent of the free stream density ratio ρ_2/ρ_1. Fitting available data for ℓ/x, one finds that the relation

$$\langle \ell/x \rangle = 0.68 \frac{1-r}{1+r} , \qquad (1.7b)$$

where $r = U_2/U_1$ is the free stream speed ratio, is a good representation for this quantity. It can be verified that equations 1.7 produce estimates for E that are in good agreement with Konrad's measurements. Finally, we note that to the extent that ℓ/x is a fluctuating quantity, we would expect, on the basis of equation 1.7a, that the entrainment ratio E should exhibit corresponding fluctuations. We will develop this idea in the discussions to follow and incorporate its consequences in the proposed model calculations.

In the context of chemically reacting flows, it is important to recognize that fluid homogenized at the entrainment ratio E produces a (high speed fluid) mixture fraction ξ_E given by,

$$\xi_E = \frac{E}{E+1} . \qquad (1.8)$$

For $E > 1$, as is always the case for matched density free streams, this corresponds to a value for ξ_E that is greater than $1/2$. The resulting mixture fraction ξ_E has a special significance in the shear layer, as Konrad recognized, and helps explain the large differences in the composition fluctuations between his equal free stream density data and his helium/nitrogen free stream data. See sketch and discussion on page 27 in Konrad (1976).

This picture suggests a zeroth order model for mixing in a two-dimensional shear layer in which the reactants are entrained at the ratio E, as dictated by the large scale dynamics, and eventually mixed to a (nearly) homogeneous composition in which the distribution of values ξ of the resulting mixed fluid mixture fraction is clustered around ξ_E by the efficient action of the turbulence. A useful cartoon is that of a bucket filled by two faucets with unequal flow rates, as a laboratory stirring device mixes the effluents. For all the complexity of the ensuing turbulent motion, we would expect to find a distribution of mixed fluid compositions in the bucket clustered around the value of the mixture fraction given by equation 1.8, where E, in our cartoon, would correspond to the ratio of the flux from each of the two faucets. In fact, as the the faucet flow rate is decreased relative to the mixing rate, the mixed fluid composition probability density function is tightened around the value ξ_E, with $p(\xi) d\xi \to \delta(\xi - \xi_E) d\xi$ in the limit.

The asymmetric entrainment ratio also helps explain the outcome of the chemically reacting "flip" experiments, as they have been coined. In particular, it is known that if the concentration of the reactants carried by the two free streams corresponds to a stoichiometric mixture ratio $\phi \neq 1$, then one obtains more or less total chemical product, depending on whether or not the lean reactant is carried by the free stream fluid that is preferentially entrained. This can be seen in the gas phase reacting shear layer data pairs for $\phi = (1/4, 4)$ and $\phi = (1/8, 8)$, which correspond to "flipping" the side on which the lean reactant is carried. Compare the coresponding pairs of values for δ_P/δ in the data in figure 5. See figures 9 and 17, and related discussions in Mungal & Dimotakis (1984), and also the liquid phase "flip" experiments documented in Koochesfahani et al (1983), and in Koochesfahani & Dimotakis (1986) for additional information and discussions.

1.3 The Broadwell-Breidenthal model

In the Broadwell-Breidenthal (1982) mixing model for the two-dimensional shear layer, the entrained fluid is described as existing in one of three states:

1. recently entrained, as yet unmixed fluid from each of the two free streams,

2. homogeneously mixed fluid at a composition ξ_E corresponding to the entrainment ratio E (equation 1.8),

and,

3. fluid mixed at strained laminar interfaces (flame sheets).

In this picture, the total chemical product is computed as the sum of the contributions corresponding to the homogeneously mixed fluid, and the contribution from the flame sheets.

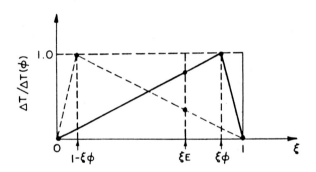

FIGURE 7. Normalized temperature rise for free stream fluids at a stoichiometric mixture fraction $\xi_\phi = \phi/(\phi+1)$, as a function of the high speed mixture fraction ξ. Dashed triangle indicates correspondding function for a "flip" experiment and the resulting temperature rise for a mixture at the entrainment mixture fraction $\xi_E = E/(E+1)$.

The volume fraction in the reaction zone, corresponding to the homogeneously mixed fluid at $\xi = \xi_E$, has experienced a temperature rise (product concentration) $\Delta T_H(\xi_E, \xi_\phi)$,

$$\Delta T_H(\xi_E, \xi_\phi) = \Theta_H(\xi_E, \xi_\phi) \times \Delta T_{flm}(\phi) , \qquad (1.9a)$$

where, for a fixed low speed stream reactant concentration, $\Delta T_{flm}(\phi)$ is given by equation

1.4. $\Theta_H(\xi_E,\xi_\phi)$ is the dimensionless temperature rise, normalized by the adiabatic flame temperature rise, that results when the two fluid elements at a stoichiometric mixture ratio ϕ are homogenized to form a mixture fraction equal to the entrainment mixture fraction ξ_E. This is given by

$$\Theta_H(\xi_E,\xi_\phi) = \begin{cases} \dfrac{\xi_E}{\xi_\phi}, & \text{for } \xi_E \leq \xi_\phi \\ \\ \dfrac{1-\xi_E}{1-\xi_\phi}, & \text{for } \xi_E > \xi_\phi, \end{cases} \quad (1.9b)$$

corresponding to the complete consumption of the lean reactant as a function of the resulting composition ξ_E. See figure 7.

The heat released (amount of product) in the strained laminar interfaces (flame sheets), for equal species and heat diffusivities, is found proportional to

$$(Sc \cdot Re)^{-1/2} F(\xi_\phi) \Delta T_{flm}(\phi), \quad (1.10)$$

where $F(\xi_\phi)$ is the Marble flame sheet function (Marble & Broadwell 1977), and given by

$$F(\xi_\phi) = \frac{e^{-z_\phi^2}}{\sqrt{\pi}\,\xi_\phi(1-\xi_\phi)}. \quad (1.11a)$$

In this expression, z_ϕ is implicitly defined by the relation

$$\text{erf}(z_\phi) = \frac{2}{\sqrt{\pi}} \int_0^{z_\phi} e^{-\zeta^2} d\zeta = \frac{\phi-1}{\phi+1}, \quad (1.11b)$$

where $\text{erf}(z)$ is the error function. The flame sheet function $F(\xi_\phi)$ is plotted in figure 8. We note here that in the original discussion (Broadwell & Breidenthal 1982), the exponent for the Reynolds number dependence could be taken as $-1/2$ or $-3/4$, depending on whether the appropriate flame sheet strain rate was estimated from the large scales of the flow or the small (Kolmogorov) scales, respectively. The Reynolds number exponent is taken here (equation 1.10) as $-1/2$, corresponding to the large scale strain rate, following the recommendation in the revised discussion of this model in Broadwell & Mungal (1986).

The contributions from the homogeneously mixed fluid and the mixed fluid on the flame sheets should be added. Normalizing the total amount of product with $\Delta T_{flm}(\phi)$, as in equation 1.5, we obtain the Broadwell-Breidenthal expression for the product volume fraction, i.e.

$$\frac{\delta_P}{\delta} = c_H \Theta_H(\xi_E, \xi_\phi) + c_F (Sc \cdot Re)^{-1/2} F(\xi_\phi) , \tag{1.12}$$

where c_H and c_F are dimensionless constants to be determined by fitting the data.

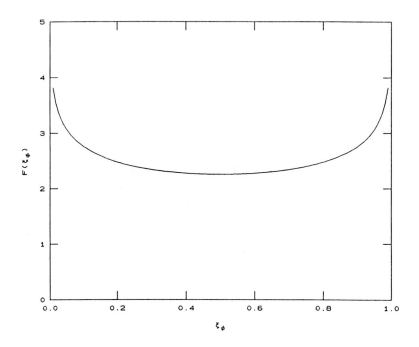

FIGURE 8. Marble (Marble & Broadwell 1977) flame sheet function $F(\xi_\phi)$.

In the more recent discussion of this model, Broadwell & Mungal (1986) recommend that the coefficients c_H and c_F in equation 1.12 should be determined by fitting the experimental value for δ_P/δ at $\phi = 1/8$, $Sc \approx 0.8$ and $Re \approx 6.6 \times 10^4$, derived from the gas phase data of Mungal & Dimotakis (1984), and the experimental value for δ_P/δ at $\phi = 1/10$, $Sc \approx 600$ and $Re \approx 2.2 \times 10^4$ derived from the liquid phase data of Koochesfahani & Dimotakis (1986). It should be mentioned, however, that in the latter discussion (which also models finite chemical kinetic rate effects in two-dimensional shear layers using the Broadwell-Breidenthal model) Broadwell & Mungal concluded, on the basis of their model calculations, that the gas phase data of Mungal & Dimotakis (1984) and Mungal et al (1985) were not quite in the fast chemistry limit. In fitting the two coefficients to the data, however, we will ignore such effects for the purposes of the discussion, noting that the differences in the resulting estimates for the model coefficients are not large. In the notation of equation 1.12, we then obtain

$$c_H = 0.27, \quad c_F = 11.5, \tag{1.13}$$

for the Broadwell-Breidenthal model constants.

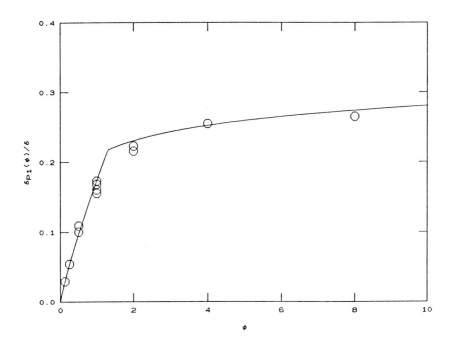

FIGURE 9. Broadwell-Breidenthal model predictions for the δ_{P_1}/δ gas phase product thickness data of Mungal & Dimotakis (1984) at $Sc \approx 0.8$ and $Re \approx 6.6 \times 10^4$.

The curve in figure 9 represents the resulting model predictions for the gas phase product thickness $\delta_{P_1}(\phi)/\delta$ data that were plotted in figure 4. The δ_P/δ product volume fraction data and the corresponding Broadwell-Breidenthal model curves are plotted in figure 10 versus ξ_ϕ. The top solid curve in figure 10 is computed for the gas phase data (circles; $Sc \approx 0.8$, $Re \approx 6.6 \times 10^4$). The dashed curve is computed for the lower Reynolds number $\phi = 1/10$ and $\phi = 10$ (inverted triangles; $Re \approx 2.3 \times 10^4$), while the dot-dashed curve is computed for the higher Reynolds number experimental value at $\phi = 10$ (upright triangle; $Re \approx 7.8 \times 10^4$) of the liquid phase data ($Sc \approx 600$) of Koochesfahani & Dimotakis (1986).

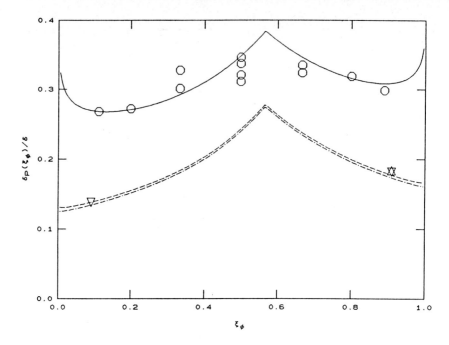

FIGURE 10. Broadwell-Breidenthal model predictions for $\delta p/\delta$ vs. ξ_ϕ data. Solid line for gas phase data (circles; $Sc \approx 0.8$, $Re \approx 6.6 \times 10^4$, Mungal & Dimotakis 1984). Dashed line for liquid phase ($Sc \approx 600$, Koochesfahani & Dimotakis 1986) data (inverted triangles; $Re \approx 2.3 \times 10^4$). Dot-dashed line for higher Reynolds number point (upright triangle; $Re \approx 7.8 \times 10^4$).

It can be seen that several features of the reacting shear layer data can be accounted for by this model. For a given Reynolds number, the $1/\sqrt{Sc}$ Schmidt number dependence of the flame sheet part renders its contribution in a liquid ($Sc \sim 600$) negligible (~ 25 times smaller) as compared to that in a gas ($Sc \sim 1$). Secondly, we can see that even though the flame sheet contribution is symmetric with respect to a change from ϕ to $1/\phi$, i.e. $F(\xi_\phi) = F(1-\xi_\phi)$, the homogeneous mixture contribution is not, since $\Theta_H(\xi_E, 1-\xi_\phi) = \Theta_H(\xi_E, \xi_\phi) / E$ (compare the solid triangle function with the dashed triangle function in figure 7). This allows the outcome of the "flip" experiments to be accommodated. We note here that, for values of the stoichiometric mixture ratio ϕ close to the entrainment ratio E, the model predicts a relatively smaller difference for the product volume fraction between gases and liquids, than for small (or high) values of ϕ. Unfortunately, no relevant chemically reacting liquid phase data are available at present to provide a direct assessment of Schmidt number effects in this regime.

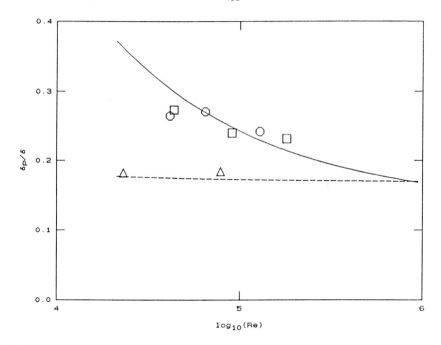

FIGURE 11. Broadwell-Breidenthal model predictions for δ_p/δ dependence on Reynolds number. Solid curve for gas phase data (Mungal et al 1985) at $\phi = 1/8$. Dashed curve for liquid phase data (Koochesfahani & Dimotakis 1986) at $\phi = 10$. Note that curves cross as a result of the larger homogeneously mixed fluid contribution for the liquid phase data at large ϕ.

Plots of the Broadwell-Breidenthal model predictions for the product volume fraction versus Reynolds number are depicted in figure 11 along with the corresponding gas phase data at $\phi = 1/8$ of Mungal et al (1985) and the liquid phase data at $\phi = 10$ of Koochesfahani & Dimotakis (1986). The predicted curves start at a Reynolds number of 2×10^4, based on the velocity difference and the local visual width of the layer, estimated to be the minimum Reynolds number for the quasi-asymptotic behavior to have been attained, following the shear layer mixing transition (Konrad 1976, Bernal et al 1979, Breidenthal 1981). The solid line is the model prediction for the gas phase data. The dashed line corresponds to the model prediction for the liquid phase data. Note that the predicted curves for the gas and the liquid phase product thickness curves are computed for the values of the stoichiometric mixture ratio corresponding to the one used in the experiments ($\phi = 1/8$ and $\phi = 10$ respectively) and will cross at some Reynolds number as a consequence of the larger homogeneous fluid contribution for the ($\phi = 10$) liquid data. There would, of course, be no crossing of the model predictions at the same ϕ, as the gas phase product volume fraction would always be larger than the corresponding liquid phase estimate for each value of the Reynolds number.

It can be seen that an additional important feature of the data is well represented by the model. Namely, the Reynolds number dependence of the product thickness for the gas phase data is predicted to be stronger than that for the liquid data. In fact, the model prediction is that at a Schmidt number of 600 the liquid product thickness will be almost independent of the Reynolds number. On the other hand, it would appear that the Broadwell-Breidenthal model predicts a Reynolds number dependence for the gas phase product thickness that may be too strong (algebraic), when compared to the dependence of the experimental data for the product thickness versus Reynolds number of Mungal et al (1985), which suggest a dependence on Reynolds number that may be closer to logarithmic (recall that those authors suggest a 6% drop in δ_P/δ, per factor of two in Reynolds number, for the range of Reynolds numbers investigated). It may be interesting to note, as was pointed out by these authors, that the model dependence on Schmidt number and Reynolds number is through the product Sc×Re (Peclet number) considered as a single variable. Lastly, in the limit of infinite Reynolds number, the model prediction is that gas phase shear layers should behave like liquids, with an asymptotic value of δ_P/δ, the chemical product volume fraction, given by $c_H \theta_H(\xi_E, \xi_\phi)$.

From a theoretical vantage point, the Broadwell/Breidenthal model considers the mixed fluid as residing in strained flame sheets, as would be appropriate for interfaces separated from each other by distances large enough such that the composition ξ (mixture fraction) swings from 0 to 1 across them, and as homogeneously mixed fluid, as would perhaps be appropriate at scales of the order of the (scalar) diffusion scale λ_D, after the diffusion process has homogenized adjacent layers of the entrained fluids. This partition of the mixed fluid states is an idealization, as the actual dynamics of this process would be expected to result in a smooth transition from one regime to the other. The authors argue that the Lagrangian time associated with that transition is short and, therefore, intermediate states can be neglected. It can also be argued, however, that the volume fraction associated with the molecularly mixed fluid in this intermediate state is not small, increasing rapidly as the diffusion scales are approached by the force of the same arguments, and is consequently not necessarily negligible.

Another related difficulty of the Broadwell/Breidenthal model, in my opinion, is the assignment of the volume fraction given to the homogeneously mixed fluid at $\xi = \xi_E$, i.e. the value of the coefficient c_H in equation 1.12. According to the model, c_H is a constant that, in particular, is independent of both Schmidt number and Reynolds number. It is reasonable to expect, however, that the fraction of the mixed fluid generated at the scalar diffusion scales of the flow will be a function of the ratio λ_D/δ, i.e. of the scalar diffusion scale λ_D to the overall transverse extent of the flow δ.

We shall return to these issues in the discussion of the model proposed in this paper and the comparison of its predictions with those of the Broadwell-Breidenthal model.

2.0 THE PROPOSED MODEL

The approach that is adopted in the model proposed here is that of viewing an Eulerian slice of the spatially growing shear layer, at a downstream station in the neighborhood of x, and imagining the instantaneous interface between the two interdiffusing and chemically reacting fluids as well as the associated strain field imposed on that interface. It is recognized that both the Eulerian state and the local behavior of that interface are the consequence of the Lagrangian shear layer dynamics from all relevant points upstream of the station of interest at x. It is assumed, however, that this upstream history acts in such a manner as to produce a self-similar state at x, whose statistics can be described in terms of the local parameters of the flow. In particular, it is assumed that a Kolmogorov cascade process has been the appropriately adequate description of the upstream dynamics, leading to the local Eulerian spectrum of scales and associated strain rate field at x.

The justification for this approach is that while the large scale dynamics are all important in determining such things as the growth rate and entrainment ratio into the spatially growing shear layer, the predominant fraction of the interfacial area is associated with the smallest spatial scales of the flow, which can perhaps be adequately dealt with in terms of universal similarity laws. The large scales, therefore, are to be viewed as the faucets in our cartoon, feeding the reactants that are entrained at some upstream station into the smaller scale turbulence at the appropriate rate. These reactants subsequently get processed by the evolution of the cascade processes upstream to produce the local spectrum of scales at x (see discussion in Broadwell & Dimotakis 1986). This conceptual basis is also aided by the notion of a conserved scalar, according to which the state of diffusion and the progress of an associated chemical reaction, in the limit of fast (diffusion-limited) chemical kinetics, is completely determinable by the local (Eulerian) state of the conserved scalar. See, for example, Bilger (1980) for a more complete description of this notion.

An important part of the proposed procedure is the normalization that will have to be imposed on the statistical weight (contribution) of each scale λ to the total amount of molecularly mixed fluid and associated chemical product. This is done via the expected interfacial surface per unit volume ratio that must be assigned to each scale λ. When totalled over all scales, these statistical weights must add up to unity.

The results are first obtained conditional on a uniform value of the dissipation rate ε. An attempt to incorporate and assess the effects of the fluctuations in the local dissipation rate $\varepsilon(\underline{x},t)$ will be made by folding the conditional results over a probability density function for ε.

In a similar vein, a refinement of the entrainment ratio idea is proposed, as noted earlier, in which it is recognized that the large scale spacing ℓ/x is a random variable and that therefore, by the force of equation 1.7a, the entrainment ratio is itself a random variable of the flow. Accordingly, the results will be obtained conditional on a given value of the entrainment ratio E, and will subsequently be folded over the expected distribution of values of E about its average value $\bar{E}$.

In the calculations that follow, it is assumed that the molecular diffusivities for all relevant species are equal to each other, but not necessarily equal to the kinematic viscosity. Heat release effects and temperature dependence effects of the molecular transport coefficients are also ignored. This is appropriate for the liquid phase measurements of Koochesfahani & Dimotakis (1986), and may be adequate for the description of the gas phase measurements of Mungal & Dimotakis (1984) and the Reynolds number study of Mungal et al (1985). The issue of heat release effects on the flow was specifically addressed elsewhere (see Hermanson et al 1987). In computing the temperature corresponding to the heat released in the reaction, equal heat capacities are also assumed for the two fluids brought together within the mixing zone. While some of these assumptions are not necessary for the proposed formulation outlined below, they allow calculations to be performed in closed form permitting, in turn, the examination of the dependence of the results on the various dimensionless parameters of the problem.

The proposed procedure assumes that the relevant statistics of the velocity field are known (or can be estimated) and computes the behavior of the passive scalar process in response to that velocity field. Finally, the procedure is "closed" in that it yields the (absolute) chemical product volume fraction δ_P/δ in the shear layer at x, with no adjustable parameters.

2.1 Turbulent diffusion of an entrained conserved scalar

Consider the shear layer as it entrains fluid from each of the two free streams and is interlaced with the resulting interfaces formed between the interdiffusing free stream fluids into a "vanilla-chocolate cake jelly roll" like structure. In describing the ensuing interdiffusion process it is useful to consider the scalar concentration field of, say, the high speed fluid mixture fraction $\xi(\underline{x},t)$, where $\xi = 0$ represents pure low speed fluid and $\xi = 1$ represents pure high speed fluid. A space curve intersecting the interface of the two interdiffusing fluids everywhere normal to this interface, i.e. in the direction of the local gradient of $\xi(\underline{x},t)$, would see at an instant in time a concentration field $\xi(s,t)$, where s is the arc length along the space curve. See figure 12. Note that, for an entrainment ratio E of high speed fluid relative to low speed fluid which is greater than unity, we would expect that the intervals along the space curve for which $\xi \sim 1$, labeled "a" in figure 12, would be longer, on average, than the intervals labeled "b", for which $\xi \sim 0$. In fact, the ratio of the expected $\xi > \xi_E$ time "a" to total time "a + b" for adjacent layers would be given by

$$\frac{\langle a \rangle}{\langle a+b \rangle} \approx \frac{E}{E+1} . \qquad (2.1)$$

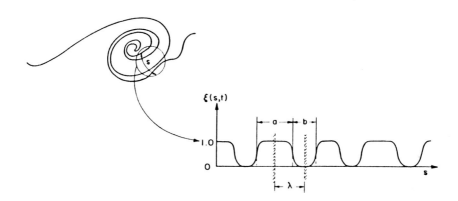

FIGURE 12. Shear layer mixing interface. Inset curve depicts values of the conserved scalar ξ(s,t) at fixed time, as a function of the arc length s on a line in the direction of ∇ξ .

Correspondingly, for portions of the segment that may have captured "jelly-roll" layers, which have been diffusing into each other for some time, we would expect that the high speed fluid fraction, in the resulting molecularly mixed fluid, would tend to homogenize to a local composition value ξ_E determined by the entrainment ratio E, i.e.

$$\xi \to \xi_E \approx \frac{\langle a \rangle}{\langle a+b \rangle} \approx \frac{E}{E+1} . \qquad (2.2)$$

In this context, ξ_E is equal to the long term (local) value of the scalar ξ , resulting from the interdiffusion of several successive ξ ∼ 1 and ξ ∼ 0 layers into each other, in a manner that preserves the (conserved) scalar ξ . This special role of the value of the scalar $\xi = \xi_E$ allows us to be more precise with the definition of the interfacial surface between the two entrained fluids, which we will define below as the three dimensional surface on which $\xi(\underline{x},t) = \xi_E$.

The evolution of the composition ξ(s,t) from the initial stages, which bring together adjacent layers of newly entrained high speed fluid (ξ ∼ 1) and low speed fluid (ξ ∼ 0), to the (local) completion of the molecular mixing ($\xi \to \xi_E$), is an unsteady diffusion problem that proceeds under the important influence of the straining field, imposed on the diffusion process by the turbulent velocity field. For the purposes of the present discussion, we will idealize this unsteady diffusion process as taking place in cells of length

$$\lambda = \frac{a+b}{2},$$ (2.3)

extending from the zero $\nabla\xi$ point in the $\xi \approx 1$ ("a") interval on one side of the interface to the zero $\nabla\xi$ point in the $\xi \approx 0$ ("b") interval on the other. See figure 12.

Using the scale λ, it is convenient to define a dimensionless space variable $\eta = s/\lambda$, for each cell of extent λ, where

$$0 \leq \eta \leq 1,$$ (2.4)

and a dimensionless time $\tau(\lambda)$, corresponding to the cell scale λ, given by

$$\tau(\lambda) = \frac{D\,t}{\lambda^2},$$ (2.5)

where D is the scalar species molecular diffusivity. The initial conditions for this problem are given by

$$\xi(\eta,0) = \begin{cases} 1, & \text{for } 0 \leq \eta < \frac{E}{E+1} \\ 0, & \text{for } \frac{E}{E+1} < \eta \leq 1, \end{cases}$$ (2.6a)

with adiabatic boundary conditions at the edges, i.e.

$$\frac{\partial}{\partial \eta}\xi(\eta,t) = 0, \quad \text{at } \eta = 0, 1.$$ (2.6b)

2.2 Strain-balanced diffusion

It is important to appreciate the role of the strain imposed on the interface, in the vicinity of some Lagrangian point of interest, in this unsteady diffusion process.

Imagine a point on the $\xi(\underline{x},t) = \xi_E$ surface associated with an arc interval λ between the two zero gradient points on either side of the interface. Imagine also a Taylor expansion of the velocity field component in the direction of the local $\nabla\xi$, in a frame convecting with that point. If we denote by s the arc length measured from the $\xi(\underline{x},t) = \xi_E$ surface and along the space curve in the direction of $\nabla\xi$, we expect the local expression

$$\underline{u} \cdot \frac{\partial \xi}{\partial \underline{x}} = u_s \frac{\partial \xi}{\partial s} \approx -\sigma(\lambda) s \frac{\partial \xi}{\partial s} ,$$

to be an adequate approximation for this scalar product, over the transverse extent of the diffusing layer on either side of the interface. The quantity $\sigma(\lambda)$ represents the expected value of the local strain rate, which we should be able to approximate as

$$\sigma(\lambda) \approx -\frac{1}{\lambda} \frac{d\lambda}{dt} . \qquad (2.7)$$

We note that $\sigma(\lambda)$ is not necessarily identified here with $-\sigma_3$, the local maximum contraction strain rate eigenvalue, where $\sigma_1 \geq \sigma_2 \geq \sigma_3$ are the local strain rate tensor eigenvalues and where $\sigma_1 + \sigma_2 + \sigma_3 = \nabla \cdot \underline{u} = 0$. We do expect that identification to represent an improving approximation as the viscous scales are approached, however, in as much as we expect the scalar interfaces to orient themselves normal to the direction of the local maximum contraction strain rate eigenvector in the limit of small scales, and the approximate relation of equation 2.7 to become exact in that limit. This was assumed by Batchelor (1959) in his discussion of the scalar spectrum at high wavenumbers, and recently corroborated by the analysis by Ashurst et al (1987) of the Rogers et al (1986) shear flow direct turbulence simulation data.

Returning to the unsteady diffusion problem, if the initial/boundary value problem has been proceeding in the cell of extent λ for a time $t(\lambda)$ that is large compared to the reciprocal of the imposed (contraction) strain rate $\sigma(\lambda)$ then the solution to the diffusion problem becomes independent of the time $t(\lambda)$ and a function of the strain rate $\sigma(\lambda)$ only. See figure 13. Specifically, for $\sigma(\lambda) \gg 1/t(\lambda)$, the appropriate dimensionless "time" for the problem is given by substituting $1/\sigma(\lambda)$ for t, in equation 2.5, or

$$\tau(\lambda) \to \frac{D}{\lambda^2 \sigma(\lambda)} , \quad \text{as } t \to \infty . \qquad (2.8)$$

This can be seen directly from the form of the diffusion equation, i.e.

$$\frac{\partial \xi}{\partial t} + \underline{u} \cdot \frac{\partial \xi}{\partial \underline{x}} = D \nabla^2 \xi ,$$

which can approximately be expressed in the local Lagrangian frame as

$$\frac{\partial \xi}{\partial t} - \sigma s \frac{\partial \xi}{\partial s} = D \frac{\partial^2 \xi}{\partial s^2} . \qquad (2.9)$$

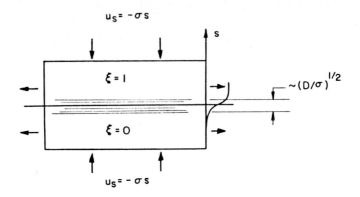

FIGURE 13. Strain-balanced diffusion process. Shaded region indicates thickness of equilibrium diffusion layer.

Physically, as the aspect ratio of the volume containing the strained interface changes, we can see that for long times the dominant species transport mechanism towards the interface becomes the convection owing to the strain field velocity normal to the interface. At equilibrium, the diffusive thickening of the mixed fluid layer is balanced by the steepening caused by the strain field, in a manner that tends to a time-independent concentration gradient and diffusive flux per unit area of interface. It can be ascertained, by solving the diffusion equation 2.9 for $\partial/\partial t \to 0$, that the resulting equilibrium flux corresponds to its value for the unsteady, time-dependent, zero strain problem, at the time $t = 1/\sigma(\lambda)$, hence, equation 2.8. It can be argued that, for $\lambda \ll \delta$, which will prove to be the important regime for the problem, the Lagrangian cascade time $t(\lambda)$ required to reach the scale λ is long compared to $1/\sigma(\lambda)$, the reciprocal of the strain rate we will associate with the scale λ. Consequently, we are encouraged to consider the additional simplification of the diffusion process, as it proceeds down the turbulent cascade of scales, as evolving in quasi-equilibrium with the associated strain rate $\sigma(\lambda)$, corresponding to the scale λ.

The unsteady diffusion problem in the normalized unit cell can be handled numerically in a straightforward manner. Nevertheless, it is worth noting that, for small $\tau(\lambda)$, the thickness of the diffusion layer will be small compared to its distance from either of the two cell edges. Consequently, the composition field can be approximated by the infinite domain solution to the problem, i.e.

$$\xi(z) = \frac{1}{2}\left[1 - \mathrm{erf}(z)\right], \qquad (2.10a)$$

where, corresponding to the boundary conditions of the problem (equation 2.6a),

$$z = \frac{1}{2\sqrt{\tau}} \left(\eta - \frac{E}{E+1} \right), \qquad (2.10b)$$

and erf(z) is the error function (equation 1.11b). This result should be valid for times τ that are short such that ξ is not appreciably different from 1 and 0 at the boundaries $\eta = 0, 1$ respectively, in which case the approximation that the imposition of the boundary conditions at a finite distance from the interface has not been felt as yet in the interior of the cell is a good one.

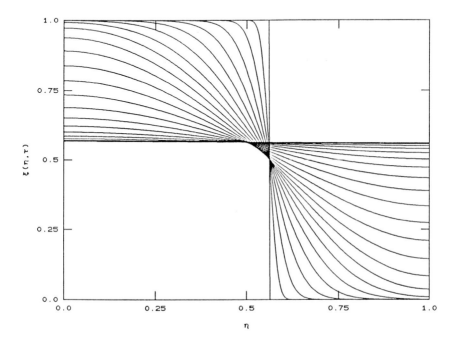

FIGURE 14. Numerical solution sequence $\xi(\eta,\tau)$ for unsteady diffusion of the conserved scalar in an adiabatic cell, corresponding to $\xi_E = E/(E+1)$ for $E = 1.3$. Curves computed for dimensionless times $\tau_{n+1} = \tau_n + n^2 \tau_0$, where $\tau_0 = 1.6 \times 10^{-4}$.

All this can, of course, be verified by the exact numerical solution to the problem. In particular, a numerical solution sequence, for a value of $E = 1.3$, is depicted in figure 14, for a sequence of values of the dimensionless time τ given by

$$\tau_{n+1} = \tau_n + n^2 \tau_0, \quad n = 1, 2, \ldots$$

where $\tau_0 = 1.6 \times 10^{-4}$. Note that, consistent with the area-preserving diffusion process, guaranteed in this case by the adiabatic boundary conditions, the composition field in the cell tends, for long times, to the value $\xi_E = E/(E+1)$, corresponding to the conserved value of

$$\langle \xi(\eta,\tau) \rangle = \int_0^1 \xi(\eta,\tau)\, d\eta = \xi_E \tag{2.11}$$

(recall also equation 1.8 and the related discussion).

2.3 Diffusion of chemically reacting species

Consider now a fast chemical reaction, with negligible heat release, between the two interdiffusing species. By fast here we mean that the thickness of the overlap region required to sustain a reaction rate, per unit area of interface, that can consume the diffusive flux of reactant towards the interface, is small compared to the diffusion layer thickness. In this fast reaction regime, commonly referred to as a "diffusion-limited" chemical reaction regime, the rate of production is dictated by the diffusive flux per unit area towards the interface and not by the reaction kinetics. More importantly, however, as a result of the inter-diffusion process, wherever the conserved scalar ξ is different from 0 or 1, the amount of chemical product formed will be equal to that corresponding to the complete local consumption of the lean reactant in the mixed fluid.

As noted earlier, we can use temperature rise (heat release) as a means of labeling the formation of chemical product. In that case, the fast chemistry assumption allows us to compute the amount of product (temperature rise) as a function of ξ, by assuming complete consumption of the lean reactant. Specifically, as was argued in the case of equation 1.9b,

$$\theta(\xi,\xi_\phi) = \frac{\Delta T(\xi,\phi)}{\Delta T_{flm}(\phi)} = \begin{cases} \dfrac{\xi}{\xi_\phi}, & \text{for } 0 \le \xi \le \xi_\phi \\ \\ \dfrac{1-\xi}{1-\xi_\phi}, & \text{for } \xi_\phi \le \xi \le 1, \end{cases} \tag{2.12a}$$

where ξ_ϕ is given by equation 1.6, i.e. $\xi_\phi = \phi/(\phi+1)$, and where

$$\Delta T_{flm}(\phi) = \frac{2\phi}{\phi+1}\, \Delta T_{flm}(1) = 2\,\xi_\phi\, \Delta T_{flm}(1) \tag{2.12b}$$

is the adiabatic flame temperature rise corresponding to the stoichiometric mixture ratio ϕ. See figure 15 and discussion following equation 1.4.

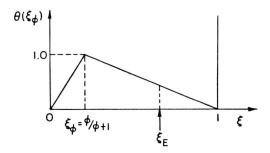

FIGURE 15. Normalized product function (temperature rise) as a function of the stoichiometric mixture fraction ξ, for a given free stream stoichiometric mixture fraction $\xi_\phi = \phi/(\phi+1)$.

Using equation 2.12 and the solution sequence depicted in figure 14 we can compute the amount of chemical product, or temperature (heat release) along η as a function of τ. Again, it is useful to consider the result for small "times" τ. In particular, we have for the total normalized temperature rise (chemical product) in the cell,

$$\Theta(\xi_\phi,\tau) = <\theta[\xi(\eta,\tau),\xi_\phi]> = \frac{1}{z_2 - z_1} \int_{z_1}^{z_2} \theta[\xi(z),\xi_\phi] dz , \qquad (2.13a)$$

where z_1 and z_2 are the values of the similarity coordinate (equation 2.10b) at the cell edges (at "time" τ), i.e.

$$z_1 = -\frac{1}{2\sqrt{\tau}} \left(\frac{E}{E+1}\right), \qquad z_2 = \frac{1}{2\sqrt{\tau}} \left(\frac{1}{E+1}\right). \qquad (2.13b)$$

Note that

$$z_2 - z_1 = \frac{1}{2\sqrt{\tau}} \qquad (2.14)$$

(independently of E), and therefore, for small τ,

$$\Theta(\xi_\phi,\tau) = \sqrt{\tau} \left\{ \frac{1}{\xi_\phi} \int_{-\infty}^{z_\phi} [1 - \mathrm{erf}(z)] dz + \frac{1}{1-\xi_\phi} \int_{z_\phi}^{\infty} [1 + \mathrm{erf}(z)] dz \right\} \qquad (2.15a)$$

where z_ϕ is the value of the similarity coordinate z at which the stoichiometric composition is attained, i.e.

$$\mathrm{erf}(z_\phi) = 2\xi_\phi - 1 = \frac{\phi - 1}{\phi + 1} . \qquad (2.15b)$$

Note also that, consistently with the small τ (boundary layer) approximation, the limits of integration have been taken to infinity. The indefinite integrals in equation 2.15 can be computed in closed form and we have, after a little algebra,

$$\Theta(\xi_\phi, \tau) = \sqrt{\tau}\, F(\xi_\phi)\,, \tag{2.16}$$

where $F(\xi_\phi)$ is the Marble strained flame sheet function (Marble & Broadwell 1977) plotted in figure 8, obtained here by different arguments. We note that the (weak) divergence of $F(\xi_\phi)$, as $\xi_\phi \to 0$ and $\xi_\phi \to 1$, is traceable to the "boundary layer" approximation and the additional approximation of taking z_1 and z_2 in the integral limits of equation 2.13a to infinity and can be lifted by folding the numerical solution sequence in equation 2.13 instead of the approximate closed form solution of equation 2.10.

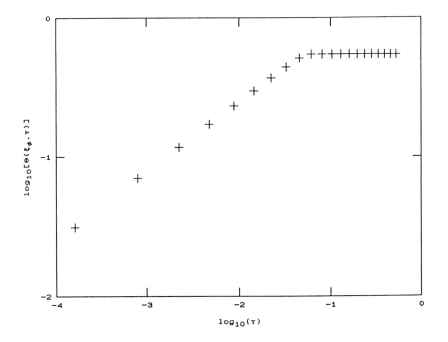

Figure 16(a)

Recall now that the $\sqrt{\tau}$ increase in the average temperature in the cell with "time", as indicated in equation 2.16, is expected to be valid for small τ only. For large τ, we know that the average temperature in the cell cannot exceed the temperature (total chemical product) resulting from the complete consumption of the lean reactant. Equivalently, if we <u>first</u> homogenize the reactants in the cell, to produce a composition

ξ_E, and <u>then</u> allow them to react, we will reach an average (total) temperature rise $\Theta_H(\xi_\phi)$ that cannot be exceeded by the transient diffusion problem. In other words,

$$\theta(\xi_\phi,\tau) \to \Theta_H(\xi_\phi) = \theta(\xi_E,\xi_\phi) , \quad \text{as } \tau \to \infty , \tag{2.17}$$

where $\theta(\xi,\xi_\phi)$ is given by equation 2.12. See also equation 1.9b and related discussion.

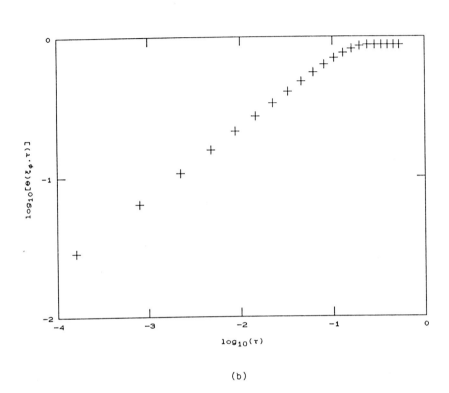

(b)

FIGURE 16. Normalized total chemical product (temperature rise) in λ-cell as a function of the (dimensionless) time τ, computed from exact numerical solution sequence in figure 14. (a) $\xi_\phi = 0.2$ ($\phi = 1/4$). (b) $\xi_\phi = 0.5$ ($\phi = 1$).

Sample exact numerical calculations of the solution to the original problem (without recourse to the "boundary layer" approximation) are plotted in figure 16 for $E = 1.3$ and for two values of ξ_ϕ, using the computed diffusion equation solutions depicted in figure 14. The results suggest that the "boundary layer" approximation can be used almost until the "time" τ the homogeneous composition temperature is attained, i.e.

$$\Theta(\xi_\phi, \tau) \approx \begin{cases} \sqrt{\tau}\, F(\xi_\phi), & \text{for } \tau < \tau_H \\ \Theta_H(\xi_\phi), & \text{for } \tau \geq \tau_H. \end{cases} \qquad (2.18a)$$

τ_H, in this approximation, is the dimensionless "time" when the homogeneous mixture total temperature rise (completion of the reaction) is attained by the boundary layer solution. In particular, matching the two regimes at $\tau = \tau_H$, we have

$$\sqrt{\tau_H} = \frac{\Theta_H(\xi_\phi)}{F(\xi_\phi)}. \qquad (2.18b)$$

See figure 17.

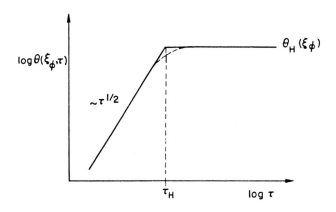

FIGURE 17. Proposed model scaled chemical product (temperature rise) function dependence on dimensionless time τ.

2.4 The scale diffusion "time" $\tau(\lambda)$

To proceed further, we need an estimate for $\sigma(\lambda)$, the strain rate associated with the scale λ. In particular, if $u(\lambda)$ is the expected velocity difference across a scale λ, we have the Kolmogorov (1941) relation for the self-similar inviscid inertial range,

$$u^2(\lambda) \sim \varepsilon^{2/3} \lambda^{2/3}, \qquad (2.19)$$

where ε is the local dissipation rate. Consequently, for diffusion interfaces spaced by distances λ in the inertial range, the associated strain rate $\sigma(\lambda)$ imposed on these interfaces should be scaled by

$$\sigma(\lambda) \sim \frac{u(\lambda)}{\lambda} \sim \varepsilon^{1/3} \lambda^{-2/3}. \qquad (2.20)$$

We note that the highest strain rates are associated with the smallest scales.

These power laws should hold for scales λ smaller than δ, where δ is identified here with the transverse extent of the vorticity-bearing region (δ_{vis} of Brown & Roshko 1974), but larger than the viscous dissipation (Kolmogorov 1941) scale λ_K, given by

$$\lambda_K = (\nu^3/\varepsilon)^{1/4}, \qquad (2.21)$$

where ε is the kinetic energy dissipation rate (per unit mass) in the shear layer turbulent region. Accepting ε as an integral quantity averaged over the extent δ of the turbulent region, and scaling with the outer flow variables, we can write

$$\varepsilon = \alpha \frac{(\Delta U)^3}{\delta}, \qquad (2.22)$$

where α is a dimensionless factor. This yields a relationship between λ_K and the outer variables given by

$$\frac{\lambda_K}{\delta} = (\alpha^{1/3} Re)^{-3/4} = \alpha^{-1/4} Re^{-3/4}. \qquad (2.23)$$

where,

$$Re = \frac{\Delta U \delta}{\nu} \qquad (2.24)$$

is the local Reynolds number for the shear layer. We note here that the dependence of λ_K/δ on the scaled dissipation rate α is weak.

In the opposite limit of small scales, corresponding to the viscous flow $\lambda \lesssim \lambda_K$ regime, the associated velocity gradients are imposed onto the small scales λ by the aggregate effect of the larger scales in the flow. In this case,

$$u(\lambda) \sim \lambda, \quad \text{for } \lambda \lesssim \lambda_K,$$

and therefore,

$$\sigma(\lambda) \approx \text{constant} = \sigma_c, \quad \text{for } \lambda \lesssim \lambda_K,$$

where σ_c is the expected contraction strain rate in the viscous regime. Consequently, we see that, in the inertial range, the strain rate increases as λ decreases, in accord with equation 2.20, until a maximum value is reached, corresponding to a scale λ_c. Below this scale, the associated expected strain rate can be taken to be a constant.

The assumption of a scale-independent expected strain in the viscous range was first proposed by Townsend (1951), who suggested (to quote Batchelor 1959), that "the action of the whole flow field on small-scale variations of any quantity ... is primarily to impose a uniform persistent straining motion". This idea was used by Batchelor (1959) to derive the k^{-1} conserved temperature fluctuation spectrum in a high Prandtl number ($Pr = \nu/\kappa$) fluid.

Gibson (1968) has argued that the estimate for σ_c can be bracketed by the inequality

$$\sqrt{3} < \frac{1}{\sigma_c t_K} \leq 2\sqrt{3}, \qquad (2.25a)$$

where $t_K = \sqrt{\nu/\varepsilon}$ is the Kolmogorov dissipation time scale, but notes that if fluctuations in the local dissipation rate ε are taken into account these bounds must be increased (see also Novikov 1961 and discussion in Monin & Yaglom **II** 1975, end of section 22.3). Defining

$$\beta = \frac{1}{\sigma_c t_K} \qquad (2.25b)$$

and in view of the bounds in the inequality 2.25a we shall accept an estimate for β of 3. Gibson's caveat, with respect to the effect of fluctuations in ε, will be dealt with below, as the effect of fluctuations in the local dissipation rate ε will be considered explicitly.

Matching to the $\lambda^{-2/3}$ behavior of $\sigma(\lambda)$ in the inertial range (equation 2.20), we now have for the expected value of $\sigma(\lambda)$, over the complete range of scales,

$$\sigma(\lambda) = \begin{cases} \sigma_c \left(\dfrac{\lambda_c}{\lambda}\right)^{2/3}, & \text{for } \lambda > \lambda_c \\ \\ \sigma_c, & \text{for } \lambda < \lambda_c, \end{cases} \qquad (2.26a)$$

where λ_c is a cut-off scale where the two regimes match. In other words,

$$\lambda_c = \beta^{3/2} \lambda_K \;\rightarrow\; \lambda_c \approx 5.2 \lambda_K, \quad \text{for } \beta \sim 3. \qquad (2.26b)$$

This yields for the maximum expected contraction strain rate,

$$\sigma_c \approx \frac{\nu}{\beta \lambda_K^2} = \frac{\beta^2 \nu}{\lambda_c^2} \sim \frac{9\nu}{\lambda_c^2}, \qquad (2.26c)$$

and where the numerical estimate is again for $\beta \sim 3$. A sketch of $\sigma(\lambda)$, versus λ, is depicted in figure 18.

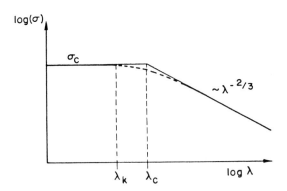

FIGURE 18. Proposed model contraction strain rate dependence on scale λ.

Using these relations for the strain rate $\sigma(\lambda)$ associated with the scale λ, we can now, in turn, associate the "time" $\tau(\lambda)$ to the scale λ, as required by the proposed approximate solution to the transient diffusion problem. In particular, combining equations 2.8 for $\tau(\lambda)$ and 2.23 for $\sigma(\lambda)$, we have

$$\tau(\lambda) = \begin{cases} \dfrac{1}{\beta^2 \, Sc} \left(\dfrac{\lambda}{\lambda_c}\right)^{-4/3}, & \text{for } \lambda > \lambda_c \\[2ex] \dfrac{1}{\beta^2 \, Sc} \left(\dfrac{\lambda}{\lambda_c}\right)^{-2}, & \text{for } \lambda < \lambda_c, \end{cases} \qquad (2.27)$$

where $Sc = \nu/D$ is the Schmidt number. See figure 19.

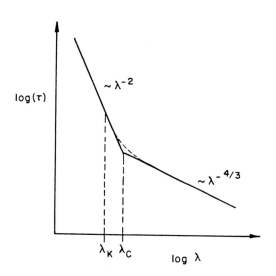

FIGURE 19. Proposed dimensionless "time" τ dependence on scale λ.

The implicit picture here is one in which the energy dissipation is more or less uniform in intermediate sized regions in the flow, of extent smaller that the outer scale δ of the flow but larger than the molecular diffusion scales. In the spirit of the earlier similarity hypotheses of Kolmogorov and Oboukhov, we would expect that the dynamics within these regions would be described in terms of their now local dissipation rate ε, which must be allowed to vary from one region to another, however, as formulated, for example, in the revised similarity hypotheses put forth by Kolmogorov (1962) and Oboukhov (1962). In the context of the model, the progress of the unsteady diffusion process is computed separately for each of these regions, conditional on their local value of the dissipation rate ε, and the total mixing is subsequently estimated as an ensemble average over regions whose dissipation rate can be treated as a random variable.

2.5 The reaction completion scale

The idealizations permitting the association of the unsteady diffusion "time" τ with a definite scale λ, through equation 2.27, and the "time" τ_H at which the homogeneous mixture temperature $\Theta_H(\xi_\phi)$ is attained (equation 2.18b), allow us to define, in turn, a reaction completion scale λ_H, at which the lean reactant in the cell has been consumed and the homogeneous temperature has been reached. Substituting in these equations, we find that the ratio λ_H/λ_c is determined, in turn, by the function

$$G(\xi_\phi) = \frac{F(\xi_\phi)}{\beta \sqrt{Sc}\, \Theta_H(\xi_\phi)} , \qquad (2.28)$$

where $F(\xi_\phi)$ is the flame sheet function of equation 2.16b. In particular, we have

$$\frac{\lambda_H}{\lambda_c} = \begin{cases} [G(\xi_\phi)]^{3/2} , & \text{for } G(\xi_\phi) > 1 \\ \\ G(\xi_\phi) , & \text{for } G(\xi_\phi) < 1 . \end{cases} \qquad (2.29)$$

We note here that the controlling function $G(\xi_\phi)$ can be made large or small, other things held constant, by manipulating the value of the Schmidt number. Accordingly, corresponding to the two cases of equation 2.29 dictated by the magnitude of $G(\xi_\phi)$, we will recognize two reacting flow regimes:

1. <u>gas-like</u>, for which the reaction is completed before λ_c is reached, i.e. $\lambda_H > \lambda_c$ [$G(\xi_\phi) > 1$, low Sc fluid],

and

2. <u>liquid-like</u>, for which the reaction is completed at scales smaller than λ_c, i.e. $\lambda_H < \lambda_c$ [$G(\xi_\phi) < 1$, large Sc fluid].

Combining these results with the expressions for the chemical product associated with a particular diffusion "time" $\tau(\lambda)$, see equations 2.18 and 2.27, we obtain for these two diffusion-reaction regimes,

<u>Gas-like</u>: $x_H = \lambda_H/\lambda_c = G^{3/2}(\xi_\phi) > 1$

$$\frac{\theta(x)}{\theta_H} = \begin{cases} 1, & x < 1 < x_H \\ 1, & 1 < x < x_H \\ \dfrac{G}{x^{2/3}}, & 1 < x_H < x, \end{cases} \quad (2.30a)$$

and

<u>Liquid-like</u>: $x_H = \lambda_H/\lambda_c = G(\xi_\phi) < 1$

$$\frac{\theta(x)}{\theta_H} = \begin{cases} 1, & x < x_H < 1 \\ \dfrac{G}{x}, & x_H < x < 1 \\ \dfrac{G}{x^{2/3}}, & x_H < 1 < x, \end{cases} \quad (2.30b)$$

where

$$x = \frac{\lambda}{\lambda_c} \quad (2.30c)$$

is the dimensionless interface scale λ, normalized by the strain rate cross-over scale λ_c.

2.6 The statistical weight of interface scales. Normalization.

The preceding approximations yield an estimate for the contribution to the total chemical product in the shear layer associated with each scale λ of the reactant interface, per unit surface area associated with λ. To compute the total product per unit volume of shear layer fluid, however, we need to estimate the statistical weight $W(\lambda)$ for the scale λ, in the range $d\lambda$, as the expected surface per unit volume of interface associated with the scale λ. Evidently, the resulting statistical weight (associated expected total surface-to-volume ratio) over the range of scales λ must be normalizable to the unit volume, i.e.

$$\int_0^\delta W(\lambda)\, d\lambda = 1. \quad (2.31)$$

Recall that, for every differential surface element dS of the $\xi(\underline{x},t) = \xi_E$ interface, the associated scale λ was defined as the arc length along $\nabla \xi$ between the two points on either side of the interface where $\nabla \xi$ has decreased to zero, or, operationally, to some nominally small fraction γ (say, $\gamma \approx 10^{-3}$) of its value at $\xi = \xi_E$. This operation is to be imagined as performed for every element dS on the interface, with $W(\lambda)$ the resulting probability density function of λ.

It will be convenient to first consider the interface that would be formed between the two entrained fluids in the absence of any scalar diffusion, i.e. in the limit of $Sc \to \infty$, or surface tension. It will also be convenient for the discussion below to factor $W(\lambda)$ into the surface to volume ratio of a scale λ and the probability $p(\lambda)$ of finding that scale λ in our Eulerian slice. This yields the relation

$$W(\lambda) d\lambda \sim \frac{1}{\lambda} p(\lambda) d\lambda = \tilde{p}[\ln(\lambda)] d \ln(\lambda), \qquad (2.32)$$

within a normalization constant.

If we may regard the self-similar inertial range ($\lambda_c \ll \lambda \ll \delta$) as not possessing an intrinsic characteristic scale, we must accept that the (dimensionless) product $W(\lambda) d\lambda$ can only depend on the scale λ itself. Accordingly, again within a normalization constant, we must have

$$W(\lambda) d\lambda \sim \frac{d\lambda}{\lambda}, \qquad (2.33)$$

as the only dimensionless, scale-invariant group that can be formed between $d\lambda$ and λ. It can be seen that, in this range of scales, $\tilde{p}[\ln(\lambda)]$, and therefore also $p(\lambda)$, must be uniform (independent of λ), as perhaps one might have argued a priori.

It is reasonable to assume that interface scales below the strain rate cross-over scale λ_c are primarily generated within regions of extent λ_c or smaller. We can imagine a Taylor expansion of the velocity field about the center of one such sub-λ_c region and a (non-inertial) frame of reference that convects with the velocity field at the point of expansion, rotating about the local vorticity axis at a rate that cancels the local value of the (nearly uniform) vorticity in that region. It has been a common assumption to regard the direction of the principal strain rate axes as also fixed in that frame (Townsend 1951, Batchelor 1959, Novikov 1961), at least for a time interval of the order of $t_K = (\nu/\epsilon)^{1/2}$. We shall accept this same approximation here, and also assume that within each of these sub-λ_c regions the local normal to the scalar interface has already been aligned with the principal contraction axis. As mentioned earlier, this latter was also assumed by Batchelor (1959) and recently corroborated for shear flows by the analysis of Ashurst et al (1987). We should note, however, that the time that the axes need to stay fixed in the rotating frame is scaled by the time t_{DK} to diffuse across λ_K, which at high Schmidt numbers can be longer than t_K†.

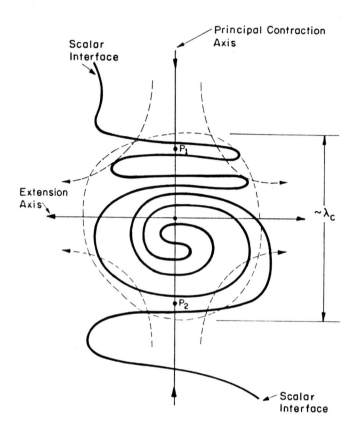

FIGURE 20. Schematic of scalar interface in the interior of a fluid element of extent λ_c.

Imagine now any two points P_1 and P_2 that remain on the principal contraction axis in this frame and, in view of our assumptions, can be regarded as moving with the fluid. It can be seen that the number of intersections of the interface and the principal

---

† Considering the diffusion geometry in figure 13 for $\lambda \approx \lambda_K$ and $Sc > 1$, this time can be estimated to be of the order of $t_{DK} \approx t_K \ln(\sqrt{Sc})$. Batchelor (1959) was aware of this time scaling, but the issue is ignored in the implicit assumption made about the contraction axis alignment during the diffusion process. We should also note, however, that if the diffusion geometry is one of sheets rolled up around vortex filaments, as assumed by Lundgren (1982, 1985), then the vorticity axis is normal to the maximum contraction axis and the scalar gradient, and the assumption (and the k^{-1} spectrum) remains valid. In that case, however, the scalar gradient would be at 45° to the maximum contraction axis (corroborated for the Kerr 1985 isotropic flow data analysis by Ashurst et al 1987) and not aligned with it, as appears to be the case in the Rogers et al (1986) shear flow data (Ashurst et al 1987) and as assumed here.

contraction axis between the points P_1 and P_2 will be constant, as the interface geometry is strained continuously reducing the normal spacings λ of the intersections of the interface with the principal contraction axis. This conclusion is the same regardless of whether the interface crosses the principal contraction axis with a zig-zag sheet topology, or as a rolled-up sheet, or a combination of these two possibilities. See figure 20. Moreover, the subsequent reduction of the normal spacings λ of these crossings along the contraction axis will proceed in accord with equation 2.7, which we may accept as exact for this flow regime and which we shall rewrite as

$$\frac{d}{dt} \ln(\lambda) = -\sigma_c . \qquad (2.34)$$

Imagine now that we are tracking a group of crossing spacings on the $\ln(\lambda)$ axis as they evolve, transformed in time by the strain field within the sub-λ_c region, initially between the limits, say, $\lambda_1 < \lambda < \lambda_2 < \lambda_c$, and described by a probability density function $\tilde{p}[\ln(\lambda) - \ln(\lambda_1)] = \tilde{p}[\ln(\lambda/\lambda_1)]$. Since they all "move" in Lagrangian time as a packet with a common (and constant) group velocity along the $\ln(\lambda)$ axis, we would find that their probability density function $\tilde{p}[\ln(\lambda/\lambda_1)]$ will be preserved, even as the spacings $\lambda(t)$ and $\lambda_1(t)$ themselves decrease (exponentially) with time, as dictated by equation 2.34. See figure 21.

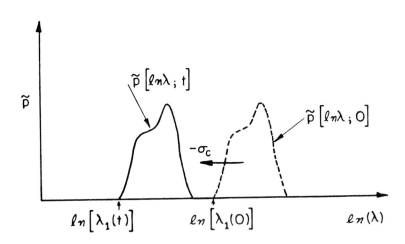

FIGURE 21. Scale packet evolution in the direction of the $-\ln(\lambda)$ axis under the action of a uniform and constant contraction strain rate σ_c.

We conclude that, in this sense, the straining field in the sub-λ_c regions does not alter the probability density function $\tilde{p}[\ln(\lambda)]$ of the larger scales that cascade to these regions from the inertial range.

These arguments suggest that $\tilde{p}[\ln(\lambda)]$, and therefore also $p(\lambda)$, must be constant not only within the inertial range but also in the viscous range and therefore throughout the spectrum of the interface scales. Consequently, for a self-similar surface, equation 2.33 may be accepted as a uniformly valid description of the expected surface to volume ratio of the interface as a function of λ, in the limit of $Sc \to \infty$.

To investigate the effect of a finite Schmidt number on the associated expected surface to volume ratio of a scale λ, we first consider the following model problem. Imagine that we are sliding the center of a ball of diameter d_b on the $Sc \to \infty$ interface and we wish to estimate the volume swept by this ball, per unit volume of flow, as its center scans the whole surface. It can be seen that for portions of the curve for which the local scale λ is large compared with the ball radius, the volume swept will be well approximated by the product of the ball diameter and the associated interface surface to volume ratio. Consequently, the volume swept by the ball as the interface scale decreases, per unit volume of fluid, will continue to increase in accord with equation 2.33, until we reach the scale $\lambda \sim d_b$, below which the contributions per unit interface surface area can be no larger that those at the scale $\lambda \sim d_b$. This picture suggests an estimate for the statistical weight of a scale λ, at finite values of the Schmidt number, given by

$$W(\lambda)\, d\lambda = \frac{1}{N(Sc, Re)} \begin{cases} \dfrac{d\lambda}{\lambda_D}, & \text{for } \lambda < \lambda_D \\[2mm] \dfrac{d\lambda}{\lambda}, & \text{for } \lambda > \lambda_D, \end{cases} \qquad (2.35)$$

where λ_D is an appropriate diffusion scale and $N(Sc, Re)$ is the normalization function, as required by equation 2.31. In particular, integrating over the range of scales, we have

$$N = \left\{ \int_0^{\lambda_D} + \int_{\lambda_D}^{\delta} \right\} W(\lambda)\, d\lambda ,$$

or

$$N = 1 + \ln\left(\frac{\delta}{\lambda_D}\right) = 1 + \ln\left(\frac{\lambda_C}{\lambda_D}\right) + \ln\left(\frac{\delta}{\lambda_C}\right) . \qquad (2.36)$$

To proceed further, we need an estimate for λ_D/λ_C, the ratio of the appropriate diffusion scale to the strain rate cut-off scale λ_C.

For high Schmidt number fluids ($Sc > 1$), we will base our estimate on the Batchelor (1959) scale. In particular,

$$\lambda_D = C_B \left(\frac{\beta}{Sc}\right)^{1/2} \lambda_K,$$

where λ_K is the Kolmogorov scale, $\beta \sim 3$ (recall equation 2.25 and related discussion), and C_B is a dimensionless constant of order unity. To assign a numerical value to C_B, we use the Batchelor (1959) estimates for the scalar space correlation function

$$D_{\xi\xi}(r) = \langle \xi(\underline{x}) \xi(\underline{x}+\underline{r}) \rangle_{\underline{x}},$$

which he expresses in terms of a double integral over $r' < r$. He finds that for distances r small compared with the diffusion scale, $D_{\xi\xi}(r) \sim \zeta/6$ asymptotically, whereas for distances large compared to the diffusion scale, but small compared with the Kolmogorov scale, $D_{\xi\xi}(r) \sim \ln(\zeta)$, where

$$\zeta = \frac{\beta}{Sc} \left(\frac{r}{\lambda_K}\right)^2.$$

Monin & Yaglom II (1975, section 22.4) express these results in terms of a differential equation for $D_{\xi\xi}(r)$, given by

$$4\zeta h''(\zeta) + (6+\zeta) h'(\zeta) = 1, \quad h(0) = 0,$$

where $h = h(\zeta)$ is the scaled (dimensionless) $D_{\xi\xi}(r)$, and which can be estimated by numerical integration of the differential equation. The resulting solution transitions from the linear behavior to the logarithmic behavior rather smoothly over the interval $1 \leq \zeta \leq 4$. Accepting the mid-point $\zeta_c = 2.5$ of this interval as the cross over between the linear (diffusive) behavior and the logarithmic (convective) behavior, we obtain the estimate $C_B = \sqrt{\zeta_c} \approx 1.6$. Finally, expressing the diffusion scale λ_D in terms of the strain rate cross-over scale λ_c, as required by the normalization function, we have

$$\frac{\lambda_D}{\lambda_c} \sim \frac{C_B}{\beta \, Sc^{1/2}}, \quad \text{for } Sc > 1.$$

For $Sc < 1$, Batchelor, Howells & Townsend (1959) find that $\lambda_D/\lambda_K \sim Sc^{-3/4}$. As we are not interested in Schmidt numbers that are much smaller than unity, and requiring continuity at $Sc = 1$, we will accept the estimate

$$x_D = \frac{\lambda_D}{\lambda_c} = \begin{cases} \dfrac{C_B}{\beta\, Sc^{1/2}}, & \text{for } Sc > 1, \\[2ex] \dfrac{C_B}{\beta\, Sc^{3/4}}, & \text{for } Sc < 1, \end{cases} \qquad (2.37)$$

with $C_B \sim 1.6$. Substituting for δ and λ_D in equation 2.36, we then obtain the expression for the normalization function,

$$N(Sc, Re) = 1 + \ln\left(\frac{\beta\, Sc^q}{C_B}\right) + \frac{3}{4}\ln\left(\frac{\alpha^{1/3} Re}{\beta^2}\right), \qquad (2.38)$$

where $q = 1/2$ for $Sc > 1$, $3/4$ for $Sc \leq 1$, $C_B \sim 1.6$, $\beta \sim 3$ and α is the dimensionless ratio of the dissipation rate ε and $(\Delta U)^3/\delta$ (recall equation 2.22 and related discussion).

These considerations suggest that the problem is characterized by <u>four</u> length scales, namely:

$\delta = (\alpha^{1/3} Re / \beta^2)^{3/4} \lambda_c$: large scale of the flow,

λ_H : reaction completion scale (equation 2.29)

λ_c : strain cross-over scale,

and,

λ_D : the species diffusion scale (equation 2.37)

All four scales have been referenced to λ_c, the strain cross-over scale, related, in turn, to the Kolmogorov scale through the constant β (see equation 2.26b). These scales define the arena in which the species diffusion and chemical reaction proceeds, bounded by δ as the large scale limit, on the one hand, and λ_D as the smallest scale at which it makes sense to attempt to track the species diffusion and chemical reaction interface.

The preceding arguments also lend credence to the conjecture that the preponderant fraction of molecularly mixed fluid, and hence chemical product, resides on interfaces associated with the smallest scales. This is owing to the fact that not only is the statistical weight of the smaller scales larger (equation 2.35) but also that the chemical product (molecularly mixed fluid) per unit surface area increases monotonically as the scale λ decreases (equation 2.30). The combination of these two effects renders the overall description of the mixed fluid and chemical product fortuitously forgiving to the treatment of the large scales in the flow.

2.7 The effect of dissipation rate fluctuations

The results thus far have been obtained conditional on a fixed value of the dissipation rate ε (corresponding to the particular station x). In particular, scaling with the outer variables of the flow we wrote for the dissipation rate per unit mass (equation 2.33),

$$\varepsilon = \alpha \frac{(\Delta U)^3}{\delta},$$

where α is a dimensionless factor. As Landau noted soon after Kolmogorov and Oboukhov formulated their initial similarity hypotheses, however, the local dissipation rate ε (and therefore α) cannot be treated as a constant in the turbulent region, but must be regarded as a strongly intermittent field. This objection was addressed in the revised similarity hypotheses of Kolmogorov (1962), Oboukhov (1962), and Gurvich & Yaglom (1967), which will be adopted here as yielding an adequate description for the purposes of assessing the effects of the dissipation rate fluctuations.

We can cast the revised similarity hypothesis in our notation by normalizing the fluctuating dimensionless factor α with its mean value $\bar{\alpha}$, i.e. $\alpha' = \alpha/\bar{\alpha}$, such that $\overline{\alpha'} = 1$. This yields a log-normal distribution for the values of the (scaled) dissipation rate α', averaged over a region of extent r_ε. In particular,

$$p(\alpha') d\alpha' = \frac{1}{\sqrt{2\pi}\, \Sigma\, \alpha'} \exp\left\{-\frac{1}{2}\left(\frac{\ln(\alpha')}{\Sigma} + \frac{\Sigma}{2}\right)^2\right\} d\alpha', \qquad (2.39a)$$

where $\Sigma^2 = \Sigma^2(r_\varepsilon)$, is the variance of the distribution, in this model given by

$$\Sigma^2(r_\varepsilon) = A + \mu \ln\left(\frac{\delta}{r_\varepsilon}\right). \qquad (2.39b)$$

The term A in this expression may depend on the large scales of the flow and μ is taken as a constant.

Monin & Yaglom II (1975, section 25) reviewed this hypothesis, and found that it represents a good approximation to measurements of the local dissipation rate. Additionally, they concluded on the basis of comparisons with data that the constant μ should be taken in the range of $0.2 \leq \mu \leq 0.5$. More recently, Van Atta & Antonia (1980) considered the consequences of this proposal on the dependence of velocity derivative moments on Reynolds number and concluded, if r_ε is taken all the way to the Kolmogorov scale λ_K, that μ should be taken as $\mu \sim 0.25$. Ashurst et al (1987) have also estimated the value of μ on the basis of direct turbulence simulation computation data (Kerr 1985, Rogers et al 1986) and concluded that $\mu \approx 0.3$. Considering all the evidence, this latter value will be accepted as representative and as our estimate for μ.

Estimates for $\bar{\alpha}$, corresponding to the mean value of the dissipation $\bar{\epsilon}$, are difficult to obtain for turbulent shear layers. There is enough information, however, in the data of Wygnaski & Fiedler (1970) to permit an estimate of $\bar{\alpha} \gtrsim 0.02$. An estimate can be made from the data of Spencer & Jones (1971), which also leads to the same value. It is difficult to assess the probable error of these estimates, not to speak of the possibility intimated by Saffman (1968) that $\bar{\alpha}$ may not necessarily be a constant, i.e. independent of the Reynolds number. Nevertheless, considering the nature of the experimental difficulties, the assumptions made about isotropy, and in view of the realizable spatial and temporal resolution relative to λ_K and λ_K / U_c, respectively, where U_c is the local flow convection velocity, we can say that this estimate is probably low, even though perhaps not by more than a factor of 2 to 3. Consequently, one would argue for a plausible range of values for $\bar{\alpha}$ of

$$0.02 \leq \bar{\alpha} \leq 0.06 . \tag{2.40}$$

In applying these results to the present discussion, we will take r_ϵ down to a (fixed) viscous scale λ_0 that is a function of the (local) Reynolds number and equal to the strain rate field cut-off scale λ_c corresponding to the mean value of the dissipation, i.e. (see equation 2.31)

$$\frac{\delta}{\lambda_0} \sim \left(\frac{\bar{\alpha}^{-1/3} \mathrm{Re}}{\beta^2} \right)^{3/4} . \tag{2.41}$$

This yields an expression for the variance of α given by

$$\Sigma_\alpha^2(\mathrm{Re}) = A + \frac{3\mu}{4} \ln \left(\frac{\bar{\alpha}^{-1/3} \mathrm{Re}}{\beta^2} \right) . \tag{2.42}$$

Finally, an estimate can be made for the constant A with the aid of the following argument. As the local Reynolds number is increased from very small values, the flow is initially essentially steady with no fluctuations in the dissipation rate field. At some minimum value of the Reynolds number, however, the flow will evolve into a fluctuating field with a spatial scale of the order of δ and an associated variance in the local dissipation rate fluctuations. At that critical value of the Reynolds number we must have $\Sigma_\alpha^2(\mathrm{Re}_{cr}) \sim 0$. This fixes the flow-specific constant A, and also removes the unpleasant dependence of the variance on the particular choice of the reference scale λ_0, and we have

$$\Sigma_\alpha^2(\mathrm{Re}) \approx \frac{3\mu}{4} \ln \left(\frac{\mathrm{Re}}{\mathrm{Re}_{cr}} \right) . \tag{2.43}$$

While we recognize that, strictly speaking, a free shear layer does not possess a critical Reynolds number, one can conclude from the linear stability analysis for viscous (but parallel) flow of a hyperbolic tangent profile (Betchov & Szzewczyk 1963) that an unstable mode with a spatial extent of order δ requires a minimum Reynolds number of the order of $15 \lesssim Re \lesssim 50$, which we will accept as bounds for Re_{cr}. Note that Re here is based on δ, the total width of the sheared region, and not on the (smaller) hyperbolic tangent maximum slope thickness. See also discussion by Betchov (1977). It is interesting that this estimate for a critical Reynolds number is not too different from the one made by Saffman (1968), who explored the idea that the structure of the flow in the dissipation range was essentially that resulting from the Taylor-Görtler instability of curved vortex sheets.

To compute the probability density function of the ratio λ_c/λ_0, we note that since

$$\lambda_c = \lambda_0 (\alpha')^{-1/4} , \qquad (2.44a)$$

we must have

$$p(y) \, dy = \frac{1}{\sqrt{2\pi}} \exp\left\{ -\frac{1}{2} \left(y - \Sigma_a/2 \right)^2 \right\} dy , \qquad (2.44b)$$

where

$$y = \frac{4}{\Sigma_a} \ln\left(\frac{\lambda_c}{\lambda_0} \right) . \qquad (2.44c)$$

This is correct to within a (near unity) normalization constant, as we wish to restrict λ_c to the range $0 \leq \lambda_c \leq \delta$.

To compute the effect of the dissipation fluctuations on the expected value of the scales normalization function $N(Sc, Re)$ discussed in the preceding section, we also note that if we assume that the ratio λ_c / λ_D is independent of the dissipation rate ε (a function of Sc only), we have

$$\langle N(Sc, Re) \rangle_\varepsilon = 1 + \ln\left(\frac{\lambda_c}{\lambda_D} \right) + \ln\left(\frac{\delta}{\lambda_0} \right) - \langle \ln\left(\frac{\lambda_c}{\lambda_0} \right) \rangle_\varepsilon .$$

Then since

$$\langle \ln\left(\frac{\lambda_c}{\lambda_0} \right) \rangle_\varepsilon = \frac{\Sigma_\alpha}{4} \langle y \rangle ,$$

where

$$\langle f(y) \rangle = \frac{\int_{-\infty}^{y_\delta} f(y) \, p(y) \, dy}{\int_{-\infty}^{y_\delta} p(y) \, dy} \quad ; \quad y_\delta = \frac{4}{\Sigma_a} \ln\left(\frac{\delta}{\lambda_0} \right) , \qquad (2.46)$$

and since, in particular (at high Reynolds numbers)

$$\langle y \rangle = \frac{\Sigma_\alpha}{2} - \frac{\exp\{-\frac{1}{2}(y_\delta - \Sigma_\alpha/2)^2\}}{\sqrt{2\pi}\,[1 - \frac{1}{2}\mathrm{erfc}(\frac{y_\delta - \Sigma_\alpha/2}{\sqrt{2}})]} \approx \frac{\Sigma_\alpha}{2}, \qquad (2.47)$$

we have (see equations 2.38 and 2.43)

$$\langle N(Sc, Re) \rangle_\varepsilon = 1 + \ln\left(\frac{\beta\, Sc^q}{C_B}\right) + \frac{3}{4}\left[\ln\left(\frac{\bar{\alpha}^{1/3} Re}{\beta^2}\right) - \frac{\mu}{8}\ln\left(\frac{Re}{Re_{cr}}\right)\right]. \qquad (2.48)$$

It is useful to rewrite this expression by defining a constant Γ through the relation

$$\Gamma = \frac{\bar{\alpha}^{1/3}}{\beta^2}\, Re_{cr}, \qquad (2.49a)$$

and where we note, at least on the basis of our numerical estimates for these quantities, that $\Gamma \sim 1$. In terms of Γ, we then have

$$\langle N(Sc, Re) \rangle_\varepsilon \approx 1 + \ln\left(\frac{\beta\, Sc^q}{C_B}\right) + \frac{3}{4}\left[(1 - \frac{\mu}{8})\ln\left(\frac{Re}{Re_{cr}}\right) + \ln(\Gamma)\right]. \qquad (2.49b)$$

Returning to the derivation of the quantity in the brackets and recognizing the role it plays in the normalization of the range of turbulent scales, however, we can argue that it should vanish as $Re \to Re_{cr}$ and that, therefore, we must have $\Gamma \approx 1$. This provides us with a consistency estimate and plausible bounds for Re_{cr} (see equation 2.49a and inequality 2.40) given by

$$23 \leq Re_{cr} \approx \frac{\beta^2}{\bar{\alpha}^{1/3}} \leq 33 \qquad (2.49c)$$

(recall $\beta^2 \sim 9$). This we can use to rewrite the expression for the ratio δ/λ_0 (equation 2.41), i.e.

$$\frac{\delta}{\lambda_0} = \left(\frac{Re}{Re_{cr}}\right)^{3/4}. \qquad (2.49d)$$

Finally, for high Reynolds numbers, we may certainly ignore the $\ln(\Gamma)$ term in favor of $\ln(Re/Re_{cr})$ and we have,

$$\langle N(Sc, Re) \rangle_\varepsilon \approx 1 + \ln\left(\frac{\beta\, Sc^q}{C_B}\right) + \frac{3}{4}(1 - \frac{\mu}{8})\ln\left(\frac{Re}{Re_{cr}}\right), \qquad (2.50)$$

where $\beta \sim 3$, $q = 1/2$ for $Sc > 1$, and $3/4$ for $Sc \leq 1$, $C_B \sim 1.6$, $\mu \sim 0.3$ and Re_{cr} is bounded by the limits in equation 2.49c. It may be worth noting that the resulting estimate for the normalization function is quite robust, as the various uncertainties in the constants appear as arguments of logarithms.

We conclude that the effects of the dissipation fluctuations on the expected value of the normalization function are small, being confined to the contribution to the final result owing to a non-zero value for μ.

2.8 The total product in the mixing layer

The total chemical product can now be computed as the weighted average of the contribution of each scale, i.e.

$$\Theta_T = \int_0^\delta \theta(\lambda/\lambda_c) W(\lambda) d\lambda = \int_0^{x_\delta} \theta(x) w(x) dx, \qquad (2.51)$$

where $x = \lambda/\lambda_c$, $x_\delta = \delta/\lambda_c$, $\theta(x)$ is given by equations 2.30, and

$$w(x) dx = \frac{1}{\langle N \rangle_\varepsilon} \begin{cases} \dfrac{dx}{x_D}, & \text{for } x < x_D \\[2ex] \dfrac{dx}{x}, & \text{for } x_D < x \end{cases} \qquad (2.52)$$

(see equation 2.36), with $x_D = \lambda_D/\lambda_c$ (see equation 2.37). We will first perform the computations conditional on a fixed value of the dissipation rate, and therefore λ_c, and then compute the total as the expectation value over the distribution of values of λ_c.

For gas-like flow-diffusion regimes (see equation 2.30 and related discussion), we have the relation

$$x_H = \frac{\lambda_H}{\lambda_c} = G^{3/2} > 1,$$

where $G = G(\xi_\phi, \xi_E, Sc)$ is given by equation 2.28, and we need to distinguish between two cases depending on the relative values of x_D and x_H. The first case corresponds to $x_D < x_H$, for which

$$\langle N \rangle_\varepsilon \frac{\Theta_T}{\Theta_H} = \frac{1}{x_D} \int_0^{x_D} dx + \int_{x_D}^{x_H} x^{-1} dx + G \int_{x_H}^{x_\delta} x^{-5/3} dx$$

or,

$$\langle N \rangle_\varepsilon \frac{\Theta_T}{\Theta_H} = \frac{5}{2} - \ln(x_D) + \frac{3}{2} \ln(G) - \frac{3}{2} G x_\delta^{-3/2} . \qquad (2.53a)$$

In the second case, we have $x_H < x_D$ and therefore

$$\langle N \rangle_\varepsilon \frac{\Theta_T}{\Theta_H} = \frac{1}{x_D} \int_0^{x_H} dx + \frac{G}{x_D} \int_{x_H}^{x_D} x^{-2/3} dx + G \int_{x_D}^{x_\delta} x^{-5/3} dx$$

or

$$\langle N \rangle_\varepsilon \frac{\Theta_T}{\Theta_H} = \frac{1}{x_D} \left(\frac{9}{2} G x_D^{1/3} - 2 G^{2/3} \right) - \frac{3}{2} G x_\delta^{-3/2} . \qquad (2.53b)$$

Of these two cases, the first [$x_D < x_H$] would typically be applicable for $Sc \lesssim 1$, if $\beta \sim 3$ and we have reasonable values for the stoichiometric mixture ratio ϕ and entrainment ratio E.

For <u>liquid-like</u> flow-diffusion regimes we have the relation

$$x_H = \frac{\lambda_H}{\lambda_c} = G < 1 ,$$

and, in principle, we need to distinguish between three cases. The first liquid-like case corresponds to the inequalities $x_D < x_H < 1$ and the integrals

$$\langle N \rangle_\varepsilon \frac{\Theta_T}{\Theta_H} = \frac{1}{x_D} \int_0^{x_D} dx + \int_{x_D}^{x_H} x^{-1} dx + G \int_{x_H}^{1} x^{-2} dx + G \int_1^{x_\delta} x^{-5/3} dx ,$$

which yields

$$\langle N \rangle_\varepsilon \frac{\Theta_T}{\Theta_H} = 2 - \ln(x_D) + \frac{G}{2} + \ln(G) - \frac{3}{2} G x_\delta^{-3/2} . \qquad (2.54a)$$

For the second liquid-like case, we have $x_H < x_D < 1$, and

$$\langle N \rangle_\varepsilon \frac{\Theta_T}{\Theta_H} = \frac{1}{x_D} \int_0^{x_H} dx + \frac{G}{x_D} \int_{x_H}^{x_D} x^{-1} dx + G \int_{x_D}^{1} x^{-2} dx + G \int_1^{x_\delta} x^{-5/3} dx ,$$

or

$$\langle N \rangle_\varepsilon \frac{\Theta_T}{\Theta_H} = G \left[\frac{1}{2} + \frac{1}{x_D} \left(2 + \ln(x_D) - \ln(G) \right) \right] - \frac{3}{2} G x_\delta^{-3/2} . \qquad (2.54b)$$

Finally, we may also have $x_H < 1 < x_D$, in which case

$$\langle N \rangle_\varepsilon \frac{\Theta_T}{\Theta_H} = \frac{1}{x_D} \int_0^{x_H} dx + \frac{G}{x_D} \left[\int_{x_H}^1 x^{-1} dx + \int_1^{x_D} x^{-2/3} dx \right] + G \int_1^{x_\delta} x^{-5/3} dx ,$$

which yields

$$\langle N \rangle_\varepsilon \frac{\Theta_T}{\Theta_H} = \frac{G}{x_D} \left(\frac{9}{2} x_D^{1/3} - 2 - \ln(G) \right) - \frac{3}{2} G x_\delta^{-3/2} . \qquad (2.54c)$$

Of the three cases for the liquid-like regime, the first one [$x_D < x_H < 1$] would typically be applicable for Sc $\gg$ 1.

If treating the variable β of equation 2.25 as a constant represents an adequate approximation, it can be seen that only the last term in each of these expressions will be modified by the fluctuations in the dissipation rate. With that proviso, since $x_\delta^{-2/3} = (\lambda_c/\delta)^{2/3}$, the contribution of the last term is small at large Reynolds numbers. In any event, expressing λ_c/δ in terms of the corresponding distribution variable y (equation 2.44), we find for Re $\gg$ Re$_{cr}$

$$\langle x_\delta^{-2/3} \rangle_\varepsilon = \left(\frac{\text{Re}}{\text{Re}_{cr}} \right)^{-(1-7\mu/48)/2} . \qquad (2.55)$$

Substituting in the results for the two typical cases (equations 2.53a and 2.54a), for example, we obtain for the expected value of the gas-like (G > 1), product volume fraction in the layer,

$$\langle \Theta_T \rangle_\varepsilon = \left\{ \frac{\frac{5}{2} - \ln(x_D) + \frac{3}{2} \ln(G) - \frac{3}{2} G \langle x_\delta^{-2/3} \rangle_\varepsilon}{1 - \ln(x_D) + \frac{3}{4}(1 - \frac{\mu}{8}) \ln\left(\frac{\text{Re}}{\text{Re}_{cr}} \right)} \right\} \Theta_H , \qquad (2.56)$$

and for the typical (Sc $\gg$ 1) liquid-like (G < 1) product fraction

$$\langle \Theta_T \rangle_\varepsilon = \left\{ \frac{2 - \ln(x_D) + \frac{G}{2} + \ln(G) - \frac{3}{2} G \langle x_\delta^{-2/3} \rangle_\varepsilon}{1 - \ln(x_D) + \frac{3}{4}(1 - \frac{\mu}{8}) \ln\left(\frac{\text{Re}}{\text{Re}_{cr}} \right)} \right\} \Theta_H , \qquad (2.57)$$

where x_D is given by equation 2.37 and $\langle x_\delta^{-2/3} \rangle_\varepsilon$ by equation 2.55.

2.9 The effect of entrainment ratio fluctuations

We should note, at this point, that in the discussions thus far we have treated the entrainment ratio E, and the resulting homogeneous mixture fraction ξ_E (equation 1.8), as single-valued. We should recognize, however, that ξ_E is a variable that we would expect to be distributed according to some probability density function $p(\xi_E)$. This can then be used to obtain the estimate for the expected product thickness, as would be measured in the laboratory, by weighting $\Theta_T(\xi_\phi, \xi_E)$, for each value of ξ_E, with $p(\xi_E)$. An estimate for $p(\xi_E)$ can be obtained with the aid of a refined model for the entrainment ratio E of the layer, which is outlined below.

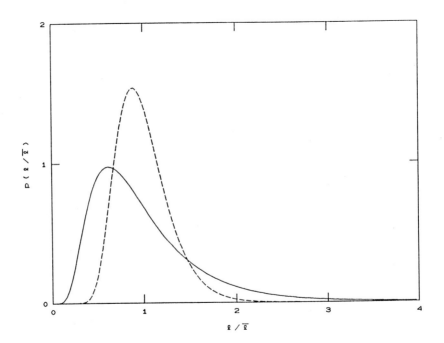

FIGURE 22. Probability density function $p(\ell/\bar{\ell})$ for normalized large scale structure spacings. Dashed curve for $\Sigma_\ell = 0.28$. Solid curve for $\Sigma_\ell = 0.56$.

If equation 1.7 can be used to estimate the entrainment ratio, then, even though the mean entrainment ratio would be given by

$$\bar{E} = \left(\frac{\rho_2}{\rho_1}\right)^{1/2} (1 + \bar{\ell}/x), \qquad (2.58)$$

we must allow for a distribution of possible values of E about $\bar{E}$, since, experimentally, one finds that the large structure spacing to position ratio ℓ/x is rather broadly distributed about its mean value $\bar{\ell}/x$. In particular, Bernal (1981) has presented data and theoretical arguments in support of a log-normal distribution, which we can write, following the notation in section 2.7, as

$$p_\ell(\ell') \, d\ell' = \frac{1}{\sqrt{2\pi}\, \Sigma_\ell} \exp\left\{ -\frac{1}{2}\left(\frac{\ln(\ell')}{\Sigma_\ell} + \frac{\Sigma_\ell}{2} \right)^2 \right\} \frac{d\ell'}{\ell'} , \qquad (2.59a)$$

where

$$\ell' = \ell/\bar{\ell} . \qquad (2.59b)$$

This is plotted in figure 22, where the dashed line is computed from equation 2.59, using the value recommended by Bernal of $\Sigma_\ell = 0.28$. See Roshko (1976, figure 5 and related discussion) as well as Bernal (1981, figure I.8) for a comparison with experimental data.

Equations 1.7 and 2.58 can be combined to yield an estimate for the expected distribution of the values of the entrainment ratio E. The picture to be borne in mind is one in which the entrainment ratio E corresponding to a particular large structure can be treated as more or less fixed, but that the value of E varies from structure to structure in accordance with the range of values of ℓ/x (as well as the history of ℓ/x of the structures that have amalgamated to form the ones passing through the station x). In particular, since

$$E = \left(\frac{\rho_2}{\rho_1} \right)^{1/2} \left[1 + (\bar{\ell}/x)\, \ell' \right] , \qquad (2.60)$$

we have, for equal free stream densities,

$$p_E(E) \, dE = p_\ell[\ell'(E)] \, \frac{dE}{\bar{\ell}/x} , \qquad \text{where} \quad \ell'(E) = \frac{E-1}{\bar{\ell}/x} ,$$

and therefore

$$p_E(E) \, dE = \frac{1}{\sqrt{2\pi}\, \Sigma_\ell} \exp\left\{ -\frac{1}{2}\left[\frac{1}{\Sigma_\ell} \ln\left(\frac{E-1}{\bar{\ell}/x} \right) + \frac{\Sigma_\ell}{2} \right]^2 \right\} \frac{dE}{E-1} . \qquad (2.61)$$

Using similar arguments, one also obtains the distribution $p(\xi_E)$ of the corresponding values of the mixture fraction ξ_E in terms of $p_E(E)$, in particular

$$p(\xi_E) \, d\xi_E = p_E[E(\xi_E)] \, \frac{d\xi_E}{(1-\xi_E)^2} , \qquad (2.62a)$$

where, inverting equation 1.8, we have

$$E(\xi_E) = \frac{\xi_E}{1 - \xi_E} \quad . \tag{2.62b}$$

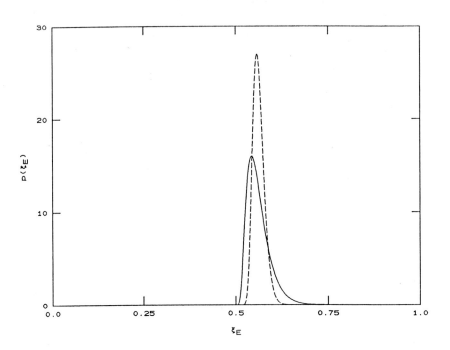

FIGURE 23. Probability density function for entrainment ratio mixture fraction $\xi_E = E/(E+1)$. Dashed curve for $\Sigma_\ell = 0.28$. Solid curve for $\Sigma_\ell = 0.56$.

The dashed line in figure 23 is the resulting probability density function $p(\xi_E)$, corresponding to equal free stream densities, a velocity ratio of $U_2/U_1 = 0.38$ and the Bernal value of $\Sigma_\ell = 0.28$. Note that, in spite of the width in the $\ell/\bar{\ell}$ distribution, the values of ξ_E are relatively narrowly distributed about the value of $\xi_{\bar{E}} = 0.567$, corresponding to a mean entrainment ratio of $\bar{E} = 1.305$, given by equation 2.36 for $\rho_2/\rho_1 = 1$ and $U_2/U_1 = 0.38$.

It should be noted that the experimental determination by Bernal (1981) of the histogram of values of $\ell/\bar{\ell}$ involved the identification of the intersection of the "braids" of each structure with the line corresponding to the trajectory of their centers. Consequently, structures in the process of tearing or coalescence, or at any other phase or configuration during which they could not be easily identified, were not included in his sample population. In other words, the distribution of spacings contributing to Bernal's experimental histogram and the resulting fit of the log-normal distribution width

Σ_ℓ was based on structures that were more or less clear of their neighbors and of interactions with them. Evidently, a full accounting of the possible large structure spacings will contribute values of ℓ/x, which if included in the population, would tend to broaden its width. Moreover, the expression for the entrainment ratio as given by equation 2.60 and as discussed elsewhere (Dimotakis 1986) refers to the entrainment flux ratio into a single large scale structure. The composition ratio of a given large scale structure, however, is the one resulting from the amalgamation of several structures, each of which was characterized by an entrainment ratio as dictated by its local ℓ/x and its fluctuations. While this consideration does not shift the mean value of E, it can be seen that it increases the variance of E, relative to its value referenced to the fluctuations of the local ℓ/x. Accordingly, in estimating the distribution of values of the entrainment ratio E, and the resulting homogeneous composition values ξ_E, one should accept a broader distribution of values of E, which we will approximate by accepting a larger value of Σ_ℓ.

The curves depicted with the solid lines in figures 22 and 23 were computed by doubling the Bernal value of the log-normal distribution width, i.e. $\Sigma_\ell = 0.56$, as representing a reasonable estimate for that quantity in view of the preceding discussion, and are plotted for comparison. It can be seen, however, that even this factor of two increase in the width Σ_ℓ does not significantly alter the resulting probability density function width for the distribution of values of ξ_E.

Using the computed probability density function for the values of ξ_E, the problem is finally closed and we can now estimate the expected product thickness δ_P/δ in the mixing layer, i.e.

$$\frac{\delta_P}{\delta} = \langle \Theta_T \rangle_{\varepsilon,E} = \int_0^1 \langle \Theta_T(\xi_E) \rangle_\varepsilon \, p(\xi_E) \, d\xi_E \,, \qquad (2.63)$$

where $\langle \Theta_T(\xi_E) \rangle_\varepsilon$ is the expected value of the normalized chemical product, averaged over the dissipation rate fluctuations, conditional on the fixed value of ξ_E, as discussed in the preceding section.

The dependence of the resulting estimates for δ_P/δ on the possible range of values of the variance Σ_ℓ of $\ell/\bar{\ell}$ is small and confined to values of the stoichiometric mixture fraction ξ_ϕ in the vicinity of $\xi_{\bar{E}}$. It will be discussed below in the context of the comparison of the theoretical values with the data.

3.0 RESULTS & DISCUSSION

Using the preceding formalism, one can estimate the expected volume (or mass) fraction of chemical product and molecularly mixed fluid generated within the two-dimensional turbulent shear layer wedge boundaries.

We recall here that the proposed model applies to incompressible flow, i.e. in the limit of zero Mach number. In the case of gas phase reactions, the heat release is assumed small and, in the case where the (small) temperature rise is used to label the chemical product, the heat capacities of the two free stream fluids are assumed matched. Differential diffusion effects have been ignored, i.e. all scalar species are assumed to diffuse with the same diffusivity. Also the chemical kinetics have been assumed fast. Finally, the Reynolds number has been assumed high enough for the shear layer to be in a fully developed three-dimensional turbulent state, i.e. $Re > 1.6 \times 10^4 - 2 \times 10^4$.

In evaluating the theoretical estimates, the following values will be used for the dimensionless parameters:

1. The expected value of the entrainment ratio $\bar{E}$ will be computed using equation 2.58.

2. The fluctuations of $E/\bar{E}$ will be estimated using equation 2.61 with a variance twice the Bernal value, i.e. $\Sigma_\ell = 0.56$, as discussed in section 2.9.

3. The value of the expected maximum contraction rate σ_c, at or below the Kolmogorov scale, will be estimated using the expression $\sigma_c t_K = 1/\beta$ with $\beta \sim 3$ (see equation 2.25 and related discussion).

4. Re_{cr}, the critical Reynolds number, will be estimated via equation 2.49c, using the mid-point of the expected dimensionless dissipation $\bar{\alpha}$ bounds (i.e. $\bar{\alpha} \approx 0.04$). This leads to the value of $Re_{cr} \approx 26$.

5. The Kolmogorov/Oboukhov coefficient μ in the variance of $\ln(\varepsilon)$ will be taken as 0.3 (equation 2.43).

6. The ratio x_D of the diffusion scale λ_D to the strain rate cut-off scale λ_c will be computed using equation 2.37, with a value of $C_B \sim 1.6$, as discussed in section 2.6.

and, finally,

7. The product volume fraction δ_P/δ will be computed using equation 2.63, with $\langle \Theta_T(\xi_E) \rangle_\varepsilon$ given by equations 2.56 or 2.57 (or 2.53, 2.54), as appropriate.

We note that the results are only weakly sensitive to these choices, appearing in the final expressions, by and large, as arguments of logarithms.

3.1 Comparison with chemically reacting flow data

The proposed model predictions for the chemical product volume fraction $\delta_{P1}/\delta = \xi_\phi \, \delta_P/\delta$, versus the stoichiometric mixture ratio ϕ, for the hydrogen-fluorine gas phase data (Mungal & Dimotakis 1984), for which $Sc \approx 0.8$, $U_2/U_1 \approx 0.40$, $\rho_2/\rho_1 \approx 1$, and $Re \approx 6.6 \times 10^4$, are plotted in figure 24. The predicted values are in good agreement with the gas phase chemical product volume fraction data. The essentially correct prediction of the absolute amount of product may perhaps be regarded as fortuitous but is nevertheless noteworthy.

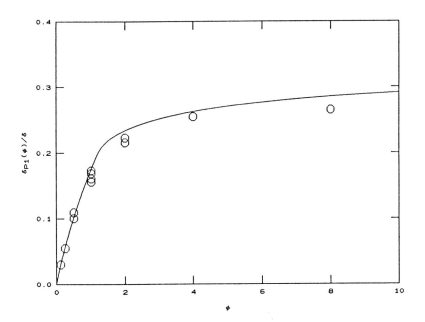

FIGURE 24. Model predictions for $\delta_{P1}(\phi)/\delta$ product thickness. Data legend as in figure 4.

A plot of the δ_P/δ predicted chemical product volume fraction, versus ξ_ϕ, appears in figure 25. The top solid curve and data points (circles) are transformed from figure 24. The corresponding predictions are also plotted for the liquid data ($Sc \approx 600$) of

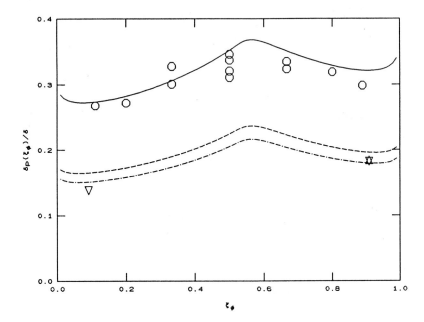

FIGURE 25. Proposed model predictions for δ_P/δ vs. ξ_ϕ data. Solid line for gas phase data (circles; $Sc \approx 0.8$, $Re \approx 6.6 \times 10^4$, Mungal & Dimotakis 1984). Dashed line for liquid phase ($Sc \approx 600$, Koochesfahani & Dimotakis 1986) data (inverted triangles; $Re \approx 2.3 \times 10^4$). Dot-dashed line for higher Reynolds number point (upright triangle; $Re \approx 7.8 \times 10^4$).

As can be seen, the Schmidt number dependence of the chemical product volume fraction, at comparable Reynolds numbers, appears also to be predicted essentially correctly. The prediction for the lower Reynolds number data is a little high. As mentioned earlier, however, it may be that the Reynolds number for those data may not be high enough.

Figure 26 depicts the model predictions (solid line, $Sc \approx 0.8$) for the dependence of the product thickness δ_P/δ on Reynolds number, as compared to the gas phase data of Mungal et al (1985) and the liquid phase (dashed line, $Sc \approx 600$) data of Koochesfahani & Dimotakis (1986). As can be seen, the experimentally observed drop in the chemical product volume fraction of approximately 6% per factor of two in Reynolds number in the gas phase data, appears correctly accounted for by the proposed model. We note again that the lower Reynolds number liquid phase data point of Koochesfahani & Dimotakis may be too close to the mixing transition Reynolds number regime to be considered representative of the asymptotic behavior at large Reynolds numbers.

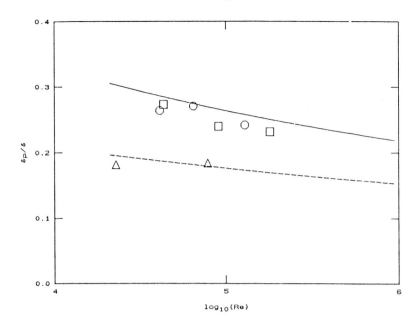

FIGURE 26. Model predictions for δp/δ chemical product volume fraction dependence on Reynolds number. Data as in figure 6. Solid line for gas phase data. Dashed line for liquid phase data. Data legend as in figure 6.

In figure 27, we investigate the sensitivity of the predictions on the value of the log-normal distribution width Σ_ℓ. The top cusped curve is computed for a single-valued entrainment ratio of $E = \overline{E}$, where $\overline{E}$ is given by equation 2.58, i.e. a Dirac delta function probability density function $p(\xi_E) = \delta[\xi_E - \overline{E}/(\overline{E}+1)]$, corresponding to a value for the variance of $\Sigma_\ell = 0$. The curve below the cusped curve corresponds to the Bernal value of $\Sigma_\ell = 0.28$. Finally, the bottom line corresponds to the value accepted here as representative of the entrainment ratio fluctuations as reflected in the composition within a single structure, i.e. double the Bernal value, or $\Sigma_\ell = 0.56$, and the probability density functions plotted as dashed lines in figures 22 and 23. As could have been anticipated, the effect of incorporating the expected distribution of values of the entrainment ratio is very slight and confined to the neighborhood of $\xi_\phi \sim \xi_{\overline{E}} = \overline{E}/(\overline{E}+1)$, corresponding to the mean entrainment ratio $\overline{E}$, and resulting in the removal of the cusp in the product thickness at $\xi_\phi = \xi_{\overline{E}}$.

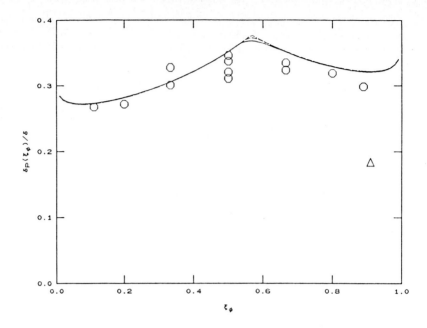

FIGURE 27. Model sensitivity to value of variance Σ_ℓ used in entrainment mixture fraction PDF. Corresponding predictions for $\delta_P(\xi_\phi)/\delta$ chemical product volume fraction. Top (cusped) curve for $\Sigma_\ell = 0$. Middle dashed curve for $\Sigma_\ell = 0.28$. Bottom (solid) curve for $\Sigma_\ell = 0.56$ used in the model.

As discussed earlier, the sensitivity of the computed values of the product volume fraction to the various choices of the flow constants is weak. By way of example, the smooth curves in figures 4 and 5, which do not differ substantially from those in figures 24 and 25, were computed using a value of $\beta = 5$[†], leaving all other constants at their nominal values.

Finally, we note that the model predictions for the chemical product volume fraction tend to be a little high (see figure 25 and 26). One could argue, considering what is being attempted here with a rather simple model and no adjustable parameters, that the agreement with experiment is more than satisfactory. We also note, however, that two assumptions that were made in the analysis may not be adequately justified in the case of the hydrogen-fluorine gas phase shear layer data. One, as Broadwell & Mungal (1986) have suggested, the kinetics may not be sufficiently fast. If this is indeed the case, the chemical product would be lower than would be observed in the case of infinite kinetics. Two, we recognize that the assumption of equal diffusivities for all the reactant species

---

[†] Such a value is, of course, inadmissible by virtue of the bounds in inequality 2.25

is also not justified, hydrogen possessing a diffusivity roughly four times higher than the other reactants/diluents in those experiments. It is difficult to give an a priori assessment of the effects of unequal diffusivities, possibly as coupled with the effects of finite kinetics, at this time.

3.2 The mixed fluid fraction

An important quantity in turbulent mixing is the mixed fluid fraction within the turbulent zone. It is to the mixing scalar (or scalar dissipation) field what intermittency is to the turbulent velocity (or energy dissipation) field. Operationally, it can be defined through the probability density function (PDF) of the conserved scalar, i.e. $p(\xi)$, integrated across the shear layer width. In particular the quantity

$$\frac{\delta_m(Sc,Re)}{\delta} \approx \int_{\xi_1}^{1-\xi_1} p(\xi,Sc,Re)\, d\xi \, , \quad (3.1)$$

for some small value of ξ_1 which excludes the unmixed fluid contributions from the neighborhood of $\xi \sim 0$ and $\xi \sim 1$, represents the volume fraction occupied by molecularly mixed fluid within the transverse extent of the turbulent shear layer. This quantity can be expected to be a function of the fluid Schmidt number and the flow Reynolds number (and potentially also of the free stream density ratio and velocity ratio). In particular, we would expect that an increase in the Schmidt number, at fixed Reynolds number, should result in a decrease of δ_m/δ, which should vanish in the limit of infinite Schmidt numbers. An a priori assessment of the behavior of the mixed fluid fraction at fixed Schmidt number in the limit of very large Reynolds numbers cannot be made as readily and will be discussed separately below.

While the integral indicated in equation 3.1 can, in principle, be estimated by direct measurement of the scalar field $\xi(\underline{x},t)$, and therefore also its PDF $p(\xi)$, it was pointed out by Breidenthal (1981) and Koochesfahani & Dimotakis (1986) that, as a consequence of the inevitable experimental finite resolution difficulties at high Reynolds numbers, such measurements will generally overestimate this quantity. It was also pointed out in Koochesfahani & Dimotakis, however, that reliable estimates are possible using the results of chemically reacting experiments, namely the chemical product fractions $\delta_P(\xi_0)/\delta$ and $\delta_P(1-\xi_0)/\delta$ from a "flip" experiment conducted at ϕ_0 and $1/\phi_0$, for small values of ϕ_0, corresponding to a $\xi_0 = \phi_0/(1+\phi_0) \ll 1$. In particular, the mixed fluid fraction can be estimated in terms of the chemically reacting flow results by means of the relation

$$\frac{\delta_m}{\delta} \approx \frac{1-\xi_0}{\delta} \left[\delta_P(\xi_0) + \delta_P(1-\xi_0) \right] . \quad (3.2)$$

This is very close to the expression in equation 3.1 and equivalent to computing the integral of the product of the probability density function with a "mixed fluid" function $\theta_m(\xi)$ given by,

$$\theta_m(\xi) = \begin{cases} \dfrac{\xi}{\xi_0}, & \text{for } 0 \leq \xi < \xi_0, \\ 1, & \text{for } \xi_0 < \xi < 1 - \xi_0, \\ \dfrac{1-\xi}{\xi_0}, & \text{for } 1 - \xi_0 < \xi < 1 \end{cases} \quad (3.3a)$$

(see figure 28), i.e.

$$\frac{\delta_m}{\delta} \approx \int_0^1 \theta_m(\xi)\, p(\xi)\, d\xi. \quad (3.3b)$$

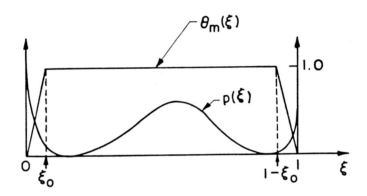

FIGURE 28. "Mixed fluid" normalized function $\theta_m(\xi)$. See equation 3.3.

We note that if the curvature in $p(\xi)$ in the edge regions $0 < \xi < \xi_0$ and $1 - \xi_0 < \xi < 1$ can be ignored, this expression reproduces the result of equation 3.1 for $\xi_1 = \xi_0/2$.

Gas phase (Sc ~ 0.8) "flip" experiments ($\phi_0 = 1/8$) are available from Mungal & Dimotakis (1984) at a Reynolds number of 6.6×10^4 (see figure 25). The liquid phase (Sc ~ 600) "flip" experiments ($\phi_0 = 1/10$) of Koochesfahani & Dimotakis (1986) were at a lower Reynolds number (Re ~ 2.3×10^4). The value of δ_p/δ for their higher Reynolds number data at $\phi = 10$, however, is so close to their corresponding lower Reynolds number value (see figure 26) that the Reynolds number dependence of the liquid data can probably be ignored as a first approximation in comparing the gas phase and liquid phase results to assess the Schmidt number dependence of δ_m/δ.

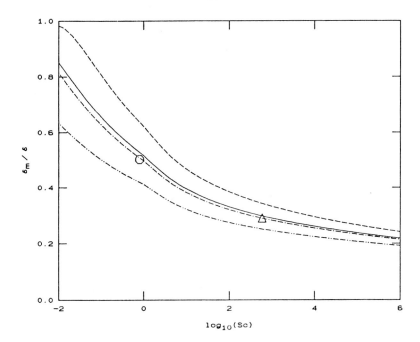

FIGURE 29. Proposed model predictions for mixed fluid volume fraction δ_m/δ as a function of Schmidt number and Reynolds number. Circle derived from Mungal & Dimotakis (1984) data. Triangle from Koochesfahani & Dimotakis (1986) data (see text). Solid curve for $Re = 6.6 \times 10^4$. Dashed curves, in order of decreasing mixed fluid volume fraction, for $Re = 10^4$, 10^5 and 10^6.

A plot of the model estimate for the mixed fluid volume fraction $\delta_m(Sc,Re)/\delta$, using a value for ϕ_0 of 1/8 corresponding to the gas phase data, is depicted in figure 29 as a function of Schmidt number. For the purposes of illustration of the qualitative behavior, the plot ranges from a value of the Schmidt number of 0.01, as would be appropriate in estimating the fluid at an intermediate temperature in a two-temperature free stream shear layer in mercury, for example, to a Schmidt number of 10^6, as would be appropriate for mixing of a particulate cloud that diffuses via Brownian motion. The solid line in that figure is for $Re = 6.6 \times 10^4$ corresponding to the gas phase data. The circle represents the gas phase experimental value while the triangle represents the liquid phase data. The other dashed lines are for $Re = 10^4$, 10^5 and 10^6, respectively, in order of decreasing values of δ_m/δ. The corresponding estimates using the Broadwell-Breidenthal model, with the values of the coefficients c_H and c_F in that model derived by fitting the gas/liquid difference (at low ϕ) of these data (equation 1.13), are plotted in figure 30 for comparison.

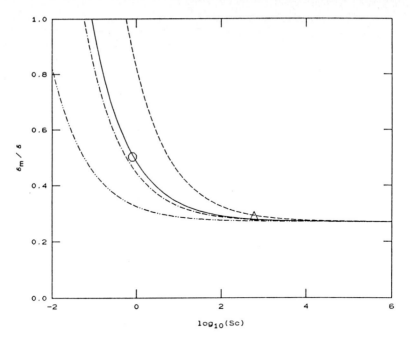

FIGURE 30. Broadwell-Breidenthal model predictions for mixed fluid volume fraction δ_m/δ as a function of Schmidt number and Reynolds number. Note Reynolds number dependence at fixed Schmidt number and asymptotic dependence for large Schmidt numbers. Data legend as in figure 29.

3.3 Discussion and conclusions

Several features of the predictions of the proposed theory for the expected mixed fluid or chemical product volume fraction, within the transverse extent of the shear layer, perhaps merit discussion.

The absolute amount of molecular mixing appears to be estimated essentially correctly, as a function of the Schmidt number and Reynolds number of the flow, with no adjustable parameters. In this context, recall that the various experimental values for the parameters used in the theory pertain to the statistics of the turbulent velocity (and dissipation) fields, which are assumed given. In particular, they are not derived from the results of mixing or chemically reacting experiments, which the theory attempts to describe. Moreover, the theory is relatively robust in that variations within the admissible range of these parameters do not have a significant effect on the predictions. The usually difficult question of an a priori estimate of intermittency, or, in the present context, of the volume fraction in the flow occupied by unmixed (unreacted) fluid, is addressed through the normalization of the volume-filling spectrum of scales. Finally, except for switching (matched) analytical expressions, depending on the relative magnitudes of the various (inner) scales of the problem, i.e. λ_D, λ_H, λ_C, liquids and gases are treated in a unified way through the explicit dependence of the results on Schmidt number.

The theory also predicts that, at sufficiently high Reynolds numbers, the amount of mixed fluid or chemical product in a gas phase shear layer would be less than what would be observed in a liquid layer at sufficiently low Reynolds numbers (recall figure 29). In fact, the interesting and controversial prediction is that the volume fraction of the mixed fluid tends to zero with increasing Reynolds number, albeit slowly, possessing no Reynolds number independent asymptotic (non-zero) value. One might intuitively argue that as the Reynolds number increases, the interfacial surface area available for mixing must increase (under the action of the higher sustainable strain rates in the flow) and therefore also the mixing rate. This argument, however, is incomplete in that it fails to recognize that the <u>thickness</u> of the mixed fluid layer straddling the interface must be decreasing, in fact, approximately inversely as the interface area is increasing, by the force of the same arguments. Consequently, these two effects must approximately cancel each other. In particular, the model predicts that as the Reynolds number increases, the associated diffused layer thickness must be decreasing at a slightly faster rate, since the flow volume fraction occupied by the mixed fluid is decreasing (slowly) with increasing Reynolds number. This behavior is corroborated by the limited data available, which indicate a monotonically decreasing volume fraction of chemical product with increasing Reynolds number, in good quantitative agreement with the predicted Reynolds number dependence.

It should be noted that the prediction is not that mixing ceases in the limit of infinite Reynolds numbers. If one were to increase the Reynolds number by increasing the downstream coordinate x, for example, the integrated mixed fluid thickness $\delta_m(x)$ would increase <u>almost</u> proportionally to the shear layer thickness $\delta(x)$, specifically

$$\delta_m(x) \sim \frac{b_1(Sc)}{b_0(Sc) + \ln(x/x_t)} \delta(x) . \qquad (3.4a)$$

Consequently, however, the mixed fluid <u>volume fraction</u> would decrease logarithmically, or in terms of the Reynolds number,

$$\frac{\delta_m}{\delta} \sim \frac{B_1(Sc)}{B_0(Sc) + \frac{3}{4}\left(1 - \frac{\mu}{8}\right)\ln\left(\frac{Re}{Re_{cr}}\right)} . \qquad (3.4b)$$

This result is a direct consequence of the assumed statistical weight and normalization over the range of scales of the problem. In particular, accepting the $W(\lambda)\,d\lambda \sim d\lambda/\lambda$ statistical weight for the moment, the dependence on the local Reynolds number enters through the ratio of the outer large scale δ to the inner scale $\lambda_c = \beta^{3/2}\lambda_K$, where $\beta \approx 3$ and λ_K is the Kolmogorov scale, i.e.

$$\frac{\delta_m}{\delta} \sim \frac{B_1(Sc)}{B_0(Sc) + \langle \ln(\frac{\delta}{\lambda_c}) \rangle_\varepsilon}, \qquad (3.5)$$

and where the subscripted angle brackets denote the ensemble average over the fluctuations in the local dissipation rate ε. We note that, again accepting the $W(\lambda)\,d\lambda \sim d\lambda/\lambda$ statistical weight, a non-zero asymptotic value for δ_m/δ at high Reynolds numbers would be the prediction only if $\langle \ln(\delta/\lambda_K) \rangle_\varepsilon \to$ constant, as $Re \to \infty$.

We have examples of such behavior in high Reynolds number turbulent flows. In particular, the skin friction coefficient for a turbulent boundary layer over a (smooth) flat plate at high Reynolds number appears to decrease approximately logarithmically with Reynolds number. For similar reasons, the pressure gradient coefficient for turbulent (smooth wall) pipe flow also decreases logarithmically with Reynolds number. It is interesting to note, however, that in these examples the behavior will asymptote to a Reynolds number independent regime, if the wall cannot be considered smooth compared to the smallest scales the turbulence can sustain and interferes with their participation in the dynamics, i.e. if a Reynolds number independent minimum scale is imposed on the dynamics of the flow. Analogously, in my opinion, the assignment of an Eulerian, Reynolds number independent volume fraction occupied by the homogeneously mixed fluid in the Broadwell-Breidenthal model leads perforce to a Reynolds number independent mixed fluid (and chemical product) volume fraction in the limit of large Reynolds numbers. In a free shear layer, however, in which the turbulence does not have to contend with any intruding rough walls or externally imposed minimum scales, the flow will generate its minimum (dissipation/diffusion) scales, of ever decreasing size as the Reynolds number increases, and which will participate in the mixing and dissipation dynamics unimpeded.

We recognize, however, that the Broadwell-Breidenthal argument is not Eulerian. These authors integrated the cascade time scale associated with each scale λ and concluded that the Lagrangian time to cascade to the Kolmogorov scale becomes independent of the Reynolds number at high Reynolds numbers. This is a central idea in the Broadwell-Breidenthal model. If one accepts it, one must also accept that fluid entrained at an upstream station x_1 cascades to the diffusion (Kolmogorov) scale by a station x_K, such that x_K/x_1 is independent of the Reynolds number. The argument is important and, if correct, difficult to reconcile with the proposed predicted shear layer mixing behavior at high Reynolds numbers.

We shall examine the Broadwell-Breidenthal argument by inverting equation 2.7 to yield the scaling for the differential time required to cascade from λ to $\lambda + d\lambda$. In particular, we have

$$-dt \sim \frac{1}{\sigma(\lambda)} \frac{d\lambda}{\lambda} \sim \frac{1}{\epsilon^{1/3}} \frac{d\lambda}{\lambda^{1/3}},$$

where $\epsilon = \alpha (\Delta U)^3/\delta$ is the dissipation rate. We note here that in the Broadwell-Breidenthal analysis, the dissipation rate was treated as a constant during the cascade. It can be argued that this is not a valid approximation for two reasons. First, because the distance to cascade is not small, corresponding to a non-negligible change in $\delta = \delta(x)$, and therefore $\bar{\epsilon}$ in the process, and second, because ϵ must be considered as a random variable with a Reynolds number dependent variance. It is possible to respond to these objections, however, by a proper separation of the problem variables, i.e.

$$-\epsilon^{1/3}(t) \, dt \sim \lambda^{-1/3} d\lambda .$$

Substituting for the dissipation rate and transforming to δ as the independent variable, we then obtain

$$-\alpha^{1/3}(\delta) \frac{d\delta}{\delta^{1/3}} \sim \lambda^{-1/3} d\lambda .$$

This can be integrated from a thickness $\delta_1 = \delta(x_1)$ to a thickness $\delta_K = \delta(x_K)$ to yield

$$\int_{\delta_1}^{\delta_K} \alpha^{1/3}(\delta) \frac{d\delta}{\delta^{1/3}} \sim \delta_1^{2/3} ,$$

where we have used that $\lambda_K \ll \delta_1$. To estimate the effect of the fluctuations in α we compute the expectation value of the left hand side, which for the purposes of the scaling estimate we will commute with the integration to write

$$\int_{\delta_1}^{\delta_K} \langle \alpha^{1/3} \rangle_\epsilon \frac{d\delta}{\delta^{1/3}} \sim \delta_1^{2/3} .$$

As before, the subscripted angle brackets denote the expectation value over the distribution of values of the dissipation rate. This can be estimated using the methods outlined in section 2.7 and we obtain

$$\langle \alpha^{1/3} \rangle_\epsilon \sim (\bar{\alpha})^{1/3} \left(\frac{Re}{Re_{cr}} \right)^{-\mu/12}$$

(recall $\mu = 0.3$). Substituting in our previous expression yields the desired result, i.e.

$$\frac{x_K}{x_1} \sim \frac{1}{\bar{\alpha}^{1/2}} \left(\frac{Re_1}{Re_{cr}} \right)^{\mu/8}, \qquad (3.6)$$

where Re_1 is the Reynolds number at $x = x_1$. While the preceding argument is not without its own shortcomings, if we accept $\bar{\alpha}$ as a non-increasing function of the Reynolds number, we note that the possibility that the distance to cascade is not Reynolds number independent must be entertained.

This is an interesting observation, bearing also on Saffman's (1968) concern that the available arguments in support of the assumption that $\bar{\alpha}$ itself is Reynolds number independent may not be sufficiently compelling. We note, however, that the conclusions of the present model would survive in the event (which has not been disallowed here). In particular, a weak dependence of $\bar{\alpha}$ on Reynolds number, say, $\bar{\alpha} \sim \bar{\alpha}_0 (Re/Re_0)^{-p}$, where presumably $0 < p \ll 1$, would produce only minor changes in the results (see equations 2.48 and 2.49). A weaker possible dependence, e.g. logarithmic, need not even be incorporated as a correction for any Reynolds number range of practical interest.

Finally, we return to the observation that the predicted asymptotic behavior in the limit of infinite Reynolds numbers is traceable to the assumed statistical weight distribution of scales in the inertial range, i.e. $W(\lambda) d\lambda \sim d\lambda/\lambda$, as discussed in section 2.6. A very small deviation from this expression, for example

$$W(\lambda) d\lambda \sim \left(\frac{\lambda}{\delta} \right)^r \frac{d\lambda}{\lambda},$$

where $|r| \ll 1$[†], would produce only minor differences in the range of Reynolds numbers of practical interest, but would alter the conclusions in the limit. In particular, the mixed fluid fraction δ_m/δ would tend to a (small, order r) non-zero asymptotic value in the limit of large Reynolds numbers, or to zero with a weak power dependence on Re, depending on whether r can be taken as positive or negative, and the (possible) dependence of the scaled mean dissipation rate $\bar{\alpha}$ on the Reynolds number.

We conclude by observing that, from an engineering vantage point, the volume fraction of mixed fluid within the shear layer, i.e. δ_m/δ, is expected to possess a (broad) maximum at a Reynolds number in the range of 2×10^4 to 3×10^4 (based on the local thickness δ and velocity difference ΔU). This corresponds to the region shortly after the flow has emerged from its "mixing" transition (Bernal et al 1979) to a fully three dimensional, turbulent state.

[†] This is admissible under the revised similarity hypotheses of Kolmogorov (1962) and Oboukhov (1962), which (even if weakly) impose the outer scale δ throughout the inertial range, or the fractal iddeas put forth by Mandelbrot (e.g. 1976). On the other hand, if a power law is appropriate, the exponent r is likely to be small, since the argument of no characteristic length scale in the inertial range (leading to the $d\lambda/\lambda$ distribution) must very nearly be right.

ACKNOWLEDGEMENTS

I would like to acknowledge the many discussions within the GALCIT community, which directly or indirectly have contributed to this paper. Without wishing to imply endorsement, I would specifically like to recognize the discussions with Dr. J. Broadwell, Prof. R. Narasimha and Prof. P. Saffman. Additionally, the critical comments by Prof. A. Leonard and Mr. P. Miller contributed to many important improvements and clarifications in the final form of the text. Finally, I would like to thank Mr. C. Frieler for performing the numerical integration of the differential equation for the velocity correlation function $h(\zeta)$ in section 2.6.

This work is part of a larger effort to investigate mixing and combustion in turbulent shear flows, sponsored by the Air Force Office of Scientific Research Contract No. F49620-79-C-0159 and Grant No. AFOSR-83-0213, whose support is gratefully acknowledged.

REFERENCES

ASHURST, W. T., KERSTEIN, A. R., KERR, R. M. and GIBSON, C. H. [1987] "Alignment of Vorticity and Scalar Gradient with Strain Rate in Simulated Navier-Stokes Turbulence", Phys. Fluids 30(8), 2343-2353.

BATCHELOR, G. K. [1959] "Small-scale variation of convected quantities like temperature in turbulent fluid. Part I. General discussion and the case of small conductivity", J. Fluid Mech. 5, 113-133.

BATCHELOR, G. K., HOWELLS, I. D. and TOWNSEND, A. A. [1959] "Small-scale variation of convected quantities like temperature in turbulent fluid. Part 2. The case of large conductivity", J. Fluid Mech. 5, 134-139.

BERNAL, L. P. [1981] The Coherent Structure of Turbulent Mixing Layers. I. Similarity of the Primary Vortex Structure, II. Secondary Streamwise Vortex Structure, Ph.D. Thesis, California Institute of Technology.

BERNAL, L. P., BREIDENTHAL, R. E., BROWN, G. L., KONRAD, J. H. and ROSHKO, A. [1979] "On the Development of Three Dimensional Small Scales in Turbulent Mixing Layers.", 2nd Symposium on Turbulent Shear Flows, 2-4 July 1979, Imperial College, England.

BETCHOV, R. [1977] "Transition", Handbook of Turbulence I (Ed. W. Frost and T. H. Moulden, Plenum Press), 147-164.

BETCHOV, R. and SZEWCZYK, A. [1963] "Stability of a shear layer between parallel streams", Phys. Fluids 6, 1391-96.

BILGER, R. W. [1980] "Turbulent Flows with Nonpremixed Reactants", Turbulent Reacting Flows (Springer-Verlag, Topics in Applied Physics 44, 1980, Ed. P. A. Libby, F. A. Williams), 65-113.

BREIDENTHAL, R. E. [1981] "Structure in Turbulent Mixing Layers and Wakes Using a Chemical Reaction", J. Fluid Mech. 109, 1-24.

BROADWELL, J. E. and BREIDENTHAL, R. E. [1982] "A Simple Model of Mixing and Chemical Reaction in a Turbulent Shear Layer", J. Fluid Mech. 125, 397-410.

BROADWELL, J. E. and DIMOTAKIS, P. E. [1986] "Implications of Recent Experimental Results for Modeling Reactions in Turbulent Flows", AIAA J. 24(6), 885-889.

BROADWELL, J. E. and MUNGAL, M. G. [1986] "The effects of Damköhler number in a turbulent shear layer", GALCIT Report FM86-01.

BROWN, G. L. and REBOLLO, M. R. [1972] "A Small, Fast-Response Probe to Measure Composition of a Binary Gas Mixture", AIAA J. 10(5), 649-652.

BROWN, G. L. and ROSHKO, A. [1974] "On Density Effects and Large Structure in Turbulent Mixing Layers", J. Fluid Mech. 64(4), 775-816.

DIMOTAKIS, P. E. [1986] "Two-Dimensional Shear-Layer Entrainment", AIAA J. 24(11), 1791-1796.

DIMOTAKIS, P. E. and BROWN, G. L. [1976] "The Mixing Layer at High Reynolds Number: Large-Structure Dynamics and Entrainment", J. Fluid Mech. 78(3), 535-560 + 2 plates.

FIEDLER, H. E. [1975] "On Turbulence Structure and Mixing Mechanism in Free Turbulent Shear Flows", from: (S. N. B. Murthy, Ed.) Turbulent Mixing in Non-Reactive and Reactive Flows (Plenum Press), 381-409.

GIBSON, C. H. [1968] "Fine Structure of Scalar Fields Mixed by Turbulence. II. Spectral Theory", Phys. Fluids 11(11), 2316-2327.

GURVICH, A. S. and YAGLOM, A. M. [1967] "Breakdown of Eddies and Probability Distributions for Small Scale Turbulence", Phys. Fluids (1967 Sup.), 59-65.

HERMANSON, J. C., MUNGAL, M. G. and DIMOTAKIS, P. E. [1987] "Heat Release Effects on Shear Layer Growth and Entrainment", AIAA J. 25(4), 578-583.

KERR, R. M. [1985] "Higher-order derivative correlations and the alignement of small-scale structures in isotropic numerical turbulence", J. Fluid Mech. 153, 31-58.

KOLLMAN, W. [1984] "Prediction of intermittency factor for turbulent shear flows", AIAA J. 22(4), 486-492.

KOLLMANN, W. and JANICKA, J. [1982] "The Probability Density Function of a Passive Scalar in Turbulent Shear Flows", Phys. Fluids 25, 1755-1769.

KOLMOGOROV, A. N. [1941] "Local structure of turbulence in an incompressible viscous fluid at very high Reynolds numbers", Dokl. Akad. Nauk SSSR 30, 299, reprinted in Usp. Fiz. Nauk 93, 476-481 (1967), transl. in Sov. Phys. Usp. 10(6), 734-736 (1968).

KOLMOGOROV, A. N. [1962] "A refinement of previous hypotheses concerning the local structure of turbulence in a viscous incompressible fluid at high Reynolds number", J. Fluid Mech 13, 82-85.

KONRAD, J. H. [1976] An Experimental Investigation of Mixing in Two-Dimensional Turbulent Shear Flows with Applications to Diffusion-Limited Chemical Reactions, Ph.D. Thesis, California Institute of Technology, and Project SQUID Technical Report CIT-8-PU (December 1976).

KOOCHESFAHANI, M. M. and DIMOTAKIS, P. E. [1986] "Mixing and chemical reactions in a turbulent liquid mixing layer", J. Fluid Mech. 170, 83-112.

LUNDGREN, T. S. [1982] "Strained Spiral Vortex Model for Turbulent Fine Structure", Phys. Fluids 25(12), 2193-2203.

LUNDGREN, T. S. [1985] "The concentration spectrum of the product of a fast bimolecular reaction", Chem. Eng. Sc. 40(9), 1641-1652.

MANDELBROT, B. [1976] "Intermittent turbulence and fractal dimension: kurtosis and the spectral exponent 5/3 + B", Turbulence and the Navier Stokes Equations (Conf. Proc., U. Paris-Sud, Orsay, 12-13 June 1975, Publ: Lecture Notes in Mathematics 565, Ed. A. Dold, B. Eckmann, Springer-Verlag), 121-145.

MARBLE, F. E. and BROADWELL, J. E. [1977] "The coherent flame model for turbulent chemical reactions", Project SQUID Technical Report TRW-9-PU.

MONIN, A. S. and YAGLOM, A. M. [1975] Statistical Fluid Mechanics: Mechanics of Turbulence 2 (English translation of the 1965 Russian text. Ed. J. L. Lumley, MIT Press).

MUNGAL, M. G. and DIMOTAKIS, P. E. [1984] "Mixing and combustion with low heat release in a turbulent mixing layer", J. Fluid Mech. 148, 349-382.

MUNGAL, M. G, HERMANSON, J. C. and DIMOTAKIS, P. E. [1985] "Reynolds Number Effects on Mixing and Combustion in a Reacting Shear Layer", AIAA J. 23(9), 1418-1423.

NOVIKOV, E. A. [1961] "Energy Spectrum of an Incompressible Fluid in Turbulent Flow", Reports Ac. Sc. USSR 139(2), 331-334, translated in Sov. Phys. Doklady 6(7), 571-573 (1962).

OBOUKHOV, A. M. [1962] "Some specific features of atmospheric turbulence", J. Fluid Mech. 13, 77-81.

POPE, S. B. [1985] "PDF Methods for Turbulent Reactive Flows", Prog. Energy Comb. Sc. 11, 119-192.

ROGERS, M. M., MOIN, P. and REYNOLDS, W. C. [1986] "The Structure and Modeling of the Hydrodynamic and Passive Scalar Fields in Homogeneous Turbulent Shear Flow", Stanford University Report TF-25.

ROSHKO, A. [1976] "Structure of Turbulent Shear Flows: A New Look", AIAA J. 14, 1349-1357, and 15, 768.

SAFFMAN, P. G. [1968] Topics in Non-Linear Physics (Ed. N. Zabusky, Springer-Verlag, Berlin), 485-614.

SPENCER, B. W. and JONES, B. G. [1971] "Statistical Investigation of Pressure and Velocity Fields in the Turbulent Two Stream Mixing Layer", AIAA Paper No. 71-613.

SREENIVASAN, K. R., TAVOULARIS, S. and CORRSIN, S. [1981] "A Test of Gradient Transport and its Generalizations", 3rd International Symposium on Turbulent Shear Flows, U. C. Davis, 9-11 September 1981 (Published by Springer-Verlag 1982, Eds. Bradbury, L. J. S., Durst, F., Launder, B. E., Schmidt, F. W. and Whitelaw, J. H. Turbulent Shear Flows 3), 96-112.

TENNEKES, H. and LUMLEY, J. L. [1972] A First Course in Turbulence (MIT Press).

TOWNSEND, A. A. [1951] "On the fine structure of turbulence" Proc. Roy. Soc. A 208, 534-542.

VAN ATTA, C. W. and ANTONIA, R. A. [1980] "Reynolds number dependence of skewness and flatness factors of velocity derivatives", Phys. Fluids 23(2), 252-257.

WYGNANSKI, I. and FIEDLER, H. E. [1970] "The Two Dimensional Mixing Region", J. Fluid Mech. 41(2), 327-361.

THE USE OF DIRECT NUMERICAL SIMULATION IN THE STUDY OF TURBULENT, CHEMICALLY-REACTING FLOWS

J. J. Riley
University of Washington
Seattle, Washington 98195

and

P. A. McMurtry
Sandia National Laboratories
Livermore, CA 94550

Abstract

At the present time the role of direct numerical simulation as applied to turbulent, chemically-reacting flows is twofold: to understand the physical processes involved, and to develop and test theories. In this paper we present and example of the former. We employ full turbulence simulations to study the effects of chemical heat release on the large-scale structures in turbulent mixing layers. This work not only aids in understanding this phenomena, but also gives insight into the strengths and limitations of the methodology.

We find, in agreement with previous laboratory experiments, the heat release is observed to lower the rate at which the mixing layer grows and to reduce the rate at which chemical products are formed. The baroclinic torque and thermal expansion in the mixing layer are shown to produce changes in the flame vortex structure that act to produce more diffuse vortices than in the constant density case, resulting in lower rotation rates of the large scale structures. Previously unexplained anomalies observed in the mean velocity profiles of reacting jets and mixing layers are shown to result from vorticity generation by baroclinic torques. The density reductions also lower the generation rates of turbulent kinetic energy and the turbulent shear stresses, resulting in less turbulent mixing of fluid elements.

Calculations of the energy in the various wave numbers shows that the heat release has a stabilizing effect on the growth rates of individual modes. A linear stability analysis of a simplified model problem confirms this, showing that low density fluid in the mixing region will result in a shift in the frequency of the unstable modes to lower wave numbers (longer wavelengths). The growth rates of the unstable modes decrease, contributing to the slower growth of the mixing layer.

Finally, we find that this methodology can be confidently applied for Reynolds numbers less than several hundred and for Damköhler numbers less than about ten. With some modification it is possible to treat the infinte Damköhler number case using a conserved scalar.

1. Background

The interactions between chemical heat release and fluid dynamics are one of the least understood aspects of chemically reacting flows. To increase our predictive capability of reacting flows and develop more reliable and efficient combustion models there is a need to understand more fully the fundamental physical processes occurring in such flows. In the work presented here, these processes are studied in a temporally growing mixing layer by the technique of direct numerical simulation.

Many chemically reacting flows are turbulent in nature and are characterized by large amounts of energy release, resulting in large density changes. If the Damköhler number of the flow is large (fast chemistry), the reaction rate will be controlled by molecular diffusion and fluid dynamical mixing. Properly treating the turbulent behavior is then clearly an essential part of any predictive method for this type of flow. In a diffusion flame in a turbulent mixing layer, for example, product formation is augmented by the stretching and wrinkling of the reaction zone. A highly strained flow field develops, which increases the area of the reaction surface and produces an increased diffusion of reactants to this zone. For reactions that are accompanied by large amounts of heat release, the fluid dynamics will be coupled to the chemistry through the nonhomogeneous density field that results. The main objective of this work is to investigate the effects of exothermic chemical reactions on the turbulent flow field and to explain these effects through an examination of the basic physical mechanisms that are involved.

To achieve this purpose, direct numerical simulations of two- and three-dimensional turbulent mixing layers have been performed. By the term direct numerical simulation we mean a numerical calculation which solves for the time development of the detailed, unsteady structure in a turbulent flow field. There are mainly two implementations of direct numerical simulation, namely, full turbulent simulations and large eddy simulations. Full turbulence simulations are calculations in which all of the dynamically significant length and time scales are resolved. In large eddy simulations, the governing equations are filtered to eliminate motions at the unresolved scales, and the effects of sub-grid scale motions are modeled.

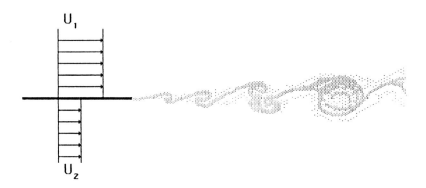

Figure 1. A mixing layer.

In the present study full turbulence simulations of chemically-reacting mixing layers have been carried out. An exothermic chemical reaction between initially unmixed chemical species is included. A mixing layer is formed when two initially separated parallel flowing fluids of different velocities come into contact and mix (figure 1). The somewhat idealized free shear flow without heat release has been extensively studied and has proven very useful in understanding the nature of turbulent flows. Laboratory experiments have shown that, at least in its early stages, the turbulent mixing layer is dominated by large-scale, quasi-two-dimensional vortices (Brown and Roshko, 1974), with the growth of the mixing layer dominated by the pairing of these vortices (Winant and Browand, 1974). Two-dimensional simulations have been shown to accurately portray the characteristic large scale-rollup and vortex-pairing processes (e.g., Riley and Metcalfe, 1980b; Ashurst and Barr, 1980; Aref and Siggia, 1980; Patnaik, Sherman and Corcos, 1976; Acton, 1976). Implications of heat releasing chemistry on this process can thus be meaningfully addressed with two-dimensional simulations. Since all turbulent flows are inherently three-dimensional, however, some potentially important physics will be lost by restricting the simulations to two spatial dimensions. For example, secondary instabilities can develop into streamwise vortices or "ribs" (Bernal, 1981), which can increase mixing and enhance product formation. Three-dimensional simulations are necessary to study this type of behavior and to realistically treat the turbulent behavior of the flow.

A large amount of work in the area of turbulent reacting flows has been directed at attempting to understand the effects of fluid dynamics on mixing and on reaction rates. A number of experiments have been performed on chemically-reacting constant density mixing layers to study the process of mixing and entrainment (Konrad, 1976; Breidenthal, 1981; Koochesfahani, 1984; Mungal and Dimotakis, 1984; Masutani, 1985). In these experiments the chemistry had no influence on the development of the velocity field due to the small amount of heat released. The results of these studies consistently showed that chemical reaction products were concentrated in large, spanwise-coherent structures that, at least initially, dominate the turbulent transport. In addition, three-dimensional effects were present and found to be important. Breidenthal noted a significant increase in product formation rate coinciding with the development of three-dimensional motions in the flow.

Previous direct numerical simulations of a temporally evolving, incompressible reacting mixing layer (Riley et al., 1986) focused on how the turbulent flow field affects the transport of chemical species. The simplified problem of a temperature independent, single-step, chemical reaction without heat release was utilized. This work has given physical insight into how vortex rollup and three-dimensional turbulence affect the chemical reaction. Despite the approximations used in the simulations, comparisons of these results with laboratory data have led to an increased confidence in the application of the direct numerical simulation methodology to chemically reacting flows.

Effects of heat release in turbulent reacting mixing layers have been performed by Wallace (1981) and Hermanson (1985). Wallace utilized a nitric oxide-ozone reaction to attain temperature rises of $400°K$. Hermanson used a hydrogen-fluorine reaction to produce temperature rises to $940°$ K. Large-scale structures were observed to persist over these temperature ranges. The results that were obtained in these experiments all indicated that the heat release resulted in a slower growth rate of the layer and a decrease in the total amount of mass entrained into the layer. Hermanson suggested that the decrease in mass entrainment could be attributed to a reduction in the turbulent shear stresses that was observed in the high heat release experiments. In both sets of experiments the streamwise pressure gradients were small. Hermanson performed additional experiments while imposing a favorable streamwise pressure gradient. These experiments showed an additional thinning of the layer.

Higher heat release experiments have been performed by Keller and Daily (1985), although under different conditions than the experiments of Wallace and Hermanson. Cold premixed reactants carried in a high velocity stream were ignited by hot, low density combustion products which were carried in a low speed stream. Results were reported for a range of equivalence ratios. In these experiments a large, favorable pressure gradient existed in the streamwise direction. The mixing layer thickness was observed to increase with increasing heat release, a result different from what Hermanson and Wallace reported in their experiments with a diffusion-limited reaction and uniform upstream density. The thickening of the layer with heat release was attributed to the downstream acceleration of the low speed, low density fluid. Large scale, two-dimensional vortices were again observed to persist over all heat release ranges.

Visual studies of diffusion flames in jets (Whol, Gazley, and Kapp, 1949; Hottel and Hawthorne, 1953; Takeno and Kotani, 1975) have shown that the existence of flames (heat release) delays the transition to turbulence. More recent turbulence measurements by Takagi et al. (1980) and Yule et al. (1981) show lower turbulence levels near the exit in jets with flames and an increase in turbulence downstream. Velocity profiles are steeper in the heat release runs and Yule reports humps (i.e., more than one inflection point) in the profiles that are not seen in the profiles for cold jets. Also, the frequencies of the most energetic instabilities (vortices at the fuel-air interface) were observed to decrease as heat release increased.

Some theoretical work performed by Marble and Broadwell (1977) and Karagozian (1982) is helpful in interpreting the results of our simulations. Marble and Broadwell studied the deformation of a constant density diffusion flame by turbulent motions. Karagozian examined the deformation of laminar diffusion flames in the flow field of two- and three-dimensional vortex structures, and also studied the effects of heat release on a laminar diffusion flame interacting with an inviscid vortex. The presence of density changes in the core caused the entire flow field to be shifted radially outward. The braids, or "flame arms" of the vortex were thus translated to a region of lower total straining (further from the point where the vortex was imposed), reducing the rate of which reactants were consumed by the flame. The effect of the reduction of the density in the core of the vortex structures was thus shown to reduce the rate at which reactants are consumed by the flame.

These previous investigations have demonstrated that there can be a very significant influence of combustion on the velocity field in reacting flows. The objective of the work presented in this paper is to use the simulation results of full three-dimensional chemically reacting mixing layers to better understand the mechanisms that produce these observed heat release effects.

The Temporally Developing Mixing Layer

The simulation results discussed here are of a temporally growing mixing layer. This is not the same flow as the spatially developing layer that is usually studied experimentally, but is an approximation if one follows the flow at the mean velocity. By studying a temporally growing layer, the requirement of specifying inflow-outflow boundary conditions, which are difficult to correctly implement for the spatial layer, is avoided. Furthermore, the amount of computer resources needed to obtain equivalent resolution is greater for the spatially growing layer. There are many dynamical features common to the two mixing layers (Metcalfe et al., 1987), so that a study of a temporal mixing layer can reveal important information about the spatial layer. Some of the differences between the two have been pointed out by Corcos and Sherman (1984). In the spatially developing layer, events that occur downstream can induce changes in the flow field upstream, whereas in the temporally developing layer, obviously no event can effect the flow at previous times. Also, the spatially developing layer has no symmetries around any spanwise axis so that entrainment rates of fluid into the layer from the two streams will not necessarily be the same. These points must be kept in mind when comparing simulation results to experiments.

In the following section a summary of the numerical approach used is discussed. Selected results that illustrate some of the effects of heat release on the flow are presented in Section 3. These effects are explained in section 4 by studying different aspects of the flow including the turbulent shear stress and turbulent kinetic energy distribution, the vorticity dynamics, and stability considerations. Section 5 summarizes the results and includes a discussion of how the different aspects of the flow presented in section 4 are related.

2. Methodology

Performing numerical studies specifically directed at investigating the basic interactions between the chemistry and the fluid dynamics calls for an approach in which the fundamental physical processes are an inherent part of the methodology. The approach used in this work is termed direct numerical simulation and involves solving the three-dimensional, time-dependent governing equations for the detailed development of the flow field (including the chemistry). The use of this method to study the nature of turbulent flows was developed by Orszag and Patterson (1972) and first applied to homogeneous, isotropic turbulence. This technique uses no closure modeling so that no assumptions are made pertaining to the turbulent behavior of the fluid.

Direct numerical simulations can supplement laboratory experiments well and have the advantage that they can give much more detailed information about the flow, since the entire flow field is known at every time step. In particular, properties that are difficult or impossible to measure experimentally can often be obtained and studied in the simulations. In addition, initial conditions are easily controlled, and flow field parameters can readily be varied to study their effect on the flow. Unfortunately, computer time and storage requirements limit the range of temporal and spatial scales that can be resolved, restricting full simulation to flows with moderate Reynolds and Damköhler numbers (although the latter limitation can be removed using a conserved scalar in the infinite reaction rate limit). At the present time direct numerical simulations are limited to fairly simple geometries and are used only in research applications. These applications consist of studying the fundamental physics of simple flows with the idea that the information obtained can be applied to predicting the behavior of more complicated flows.

Direct numerical simulations have been used successfully in many fluid mechanical applications including homogenous turbulence (Orszag and Patterson, 1972; Mansour et al., 1979; Rogallo, 1981), turbulent channel flow (Moin and Kim, 1982), mixing layers (Riley and Metcalfe, 1981), turbulent wakes (Orszag and Pao, 1974; Riley and Metcalfe, 1981) and turbulent boundary layers (Patera and Orszag, 1981). Extending the use of direct numerical simulation to reacting flows with heat release has been justified by two-dimensional simulations performed by McMurtry et al., 1986. In the following, results of three-dimensional simulations are presented to study the mechanisms by which heat release affects the flow.

Low Mach Number Approximation

To solve for the development of the flow field, a set of approximate equations valid for low Mach number flows is utilized. It has previously been presented, in various forms, by Rehm and Baum (1978), Sivashinsky (1979), Buckmaster (1985), Majda and Sethian (1985), and McMurtry et al., (1986). The basic idea behind this approximation is that for low Mach number flows, the acoustic waves generated by the turbulence and the combustion process have a much higher frequency and much faster propagation velocity than the motions characterizing convection processes. In addition, the energy in the acoustic waves is small compared both to the energy of the fluid convection velocity and to the thermodynamic internal energy. In the asymptotic limit as the Mach number goes to zero, the speed of sound becomes infinite and any disturbances in thermodynamic pressure will be felt instantaneously throughout the fluid. This approximation allows one to study effects of density changes due to heat release while filtering out the acoustic waves and hence avoiding the numerical stability problems and resolution difficulties associated with computing acoustic wave propagation.

Starting with the exact equations of combustion gas dynamics, then expanding each of the dependent variables in power series in the Mach number squared and equating coefficients of each power of the Mach number to zero, a set of ordered approximate equations is obtained. In this work, the equations solved include conservation equations for mass, momentum, energy, chemical species, and an equation of state. In the limit of low Mach number flows, these equations take the following nondimensional form (McMurtry et al., 1986):

$$\frac{\partial \rho^{(0)}}{\partial t} + \nabla \cdot \rho^{(0)} \mathbf{v}^{(0)} = 0 \tag{1a}$$

$$\nabla p^{(0)} = 0 \tag{1b}$$

$$\rho^{(0)} \frac{D^{(0)}}{Dt} T^{(0)} = -(\gamma - 1) p^{(0)} \nabla \cdot \mathbf{v}^{(0)} + \frac{1}{PrRe} \nabla \cdot \mathbf{q}^{(0)} + DaCeR_p \tag{1c}$$

$$\frac{\partial C_i^{(0)}}{\partial t} + \nabla \cdot C_i^{(0)} \mathbf{v}^{(0)} = \mathrm{Da} R_i + \frac{1}{\mathrm{Pe}} \nabla^2 C_i^{(0)} \tag{1d}$$

$$p^{(0)} = \rho^{(0)} T^{(0)}. \tag{1e}$$

The momentum equation, to the lowest order (eq. 1b) simply imposes the spatial uniformity of the thermodynamic pressure $p^{(0)}$. To complete the description of the velocity field the first order momentum equation must also be retained, and is given by

$$\rho^{(0)} \frac{D^{(0)}}{Dt} \mathbf{v}^{(0)} = -\nabla p^{(1)} + \frac{1}{Re} \nabla \cdot \tilde{\tau}^{(0)} . \tag{1f}$$

The nondimensional parameters appearing in these equations are the Damköhler number, $Da = k_o C_o L_o / U_o$, Peclet number, $Pe = L_o C_o / D_o$, and Reynolds number, $Re = U_o L_o / \nu$. The parameter Ce, a nondimensional number characterizing the amount of heat release, is given by $Ce = C_o \Delta H / \rho_o C_v T_o$.

D_o, ν, C_o, ρ_o, and T_o are the free stream molecular diffusivity, kinematic viscosity, chemical species concentration per unit volume, density, and temperature. U_o is the velocity difference across the layer, ΔH is the heat of reaction, k_o is the reaction rate frequency, and C_v is the specific heat at constant volume. For computational convenience, the length scale L_o is chosen such that the nondimensional wavelength of the most unstable mode (see section 4) is 2π ($L_o = \lambda/2\pi$, where λ is the dimensional wavelength). Time is nondimensionalized by L_o/u_o. Other variables appearing in equations $1a - 1f$ include R_i, the reaction rate of the ith species, R_p, the rate of generation of product, and $\mathbf{q}$, the heat flux vector.

The distinction between the two pressures that appear, $p^{(0)}$ and $p^{(1)}$, is essential both from a theoretical point of view and in the numerical solution procedure. The thermodynamic pressure, $p^{(0)}$, is constant in space, but may vary in time due to heat addition. $p^{(1)}$ is the dynamic pressure associated with the fluid motion and does not participate directly in thermodynamic processes.

These equations have previously been successfully applied to a two-dimensional problem (McMurtry et al., 1986). Comparisons of simulations using the exact equations and the low Mach number approximation were performed for a flow with a Mach number of 0.2 and confirmed the validity of the approximation in studying this type of flow. Details of the derivation of the equations and the numerical solution procedure are also discussed elsewhere (McMurtry, 1987).

Boundary Conditions and Numerical Methods

In the temporally developing mixing layer studied here, periodic boundary conditions can be applied in the streamwise (x) and spanwise (z) directions. In the transverse (y) direction, a free-slip, adiabatic boundary condition is applied. With these boundary conditions, very accurate pseudo-spectral (collocation) numerical methods can be efficiently implemented [see Gottlieb and Orszag (1977) for a description of these methods]. All dependent variables are expanded in Fourier series in the streamwise and spanwise directions. In the transverse direction a Fourier cosine series is used for all dependent variables except for the transverse velocity component, v, which is expanded in a Fourier sine series. Spatial derivatives are then computed by differentiating the series term by term. A second order accurate Adams-Bashforth time-stepping scheme is used to advance the equations in time.

The pseudo-spectral technique as implemented here exhibits very small phase errors and numerical diffusion compared to finite difference techniques. The numerical errors are truncation errors due to the finite wave number cutoff and time stepping errors. Care has been taken to minimize these errors by using small time steps and adjusting transport coefficients (viscosity, molecular diffusion coefficients) to be large enough to ensure accurate resolution. The lack of phase and diffusion errors is a very desirable property in simulations of reacting flows, where steep gradients of reacting species can develop and where reaction rates are controlled by molecular diffusion. Truncation errors can become significant, however, if gradients become too steep. A series of numerical experiments performed to assess the accuracy of the pseudo-spectral method in treating reacting flows is described by Riley, Metcalfe and Orszag (1986).

Flow Field Initialization

The computational domain for the simulations is chosen to be large enough to contain the most unstable mode and the subharmonic of the temporally evolving mixing layer (as determined from linear stability theory for an incompressible flow (Michalke, 1964)). The velocity field is initialized by adding a hyperbolic tangent velocity profile (figure 2) to a low amplitude, three-dimensional, spectrally broad-banded background perturbation. To specify the background perturbation we use a method similar to that introduced by Orszag and Pao (1984) and discussed in detail in Riley and Metcalfe (1980a). Energy is assigned to each wave number component as specified by a selected three-dimensional energy spectrum. A random phase is then given to each component which results in a random velocity field with a specified energy spectrum. The actual shape of the initial spectrum is found not to be critical as long as energy is contained in a fairly wide range of wave numbers. The spectrum used to initialize the turbulence in these simulations is given by

$$E_u(k) = \varepsilon_{3d}\Lambda \frac{k^4\Lambda^4}{(1+k^2\Lambda^2)^3}$$

This is a fairly broad banded, isotropic spectrum. Λ is the integral length scale, k is the magnitude of the wave number, and ε_{3d} is a coefficient that determines the level of the perturbation. The velocity field which then has this spectrum is multiplied by a "form function" in physical space to give a turbulence intensity profile characteristic of mixing layers (Riley and Metcalfe, 1980b). For simulations that start with a low enough initial amplitude, the most unstable frequencies will be selectively amplified.

The chemical reactant fields are initially one-dimensional and are given by the following functional relationships:

$$C_1(\mathbf{x},0) = \frac{1}{y_0\sqrt{\pi}} \int_{-\infty}^{y} \exp(-\varsigma^2/y_0^2)d\varsigma$$

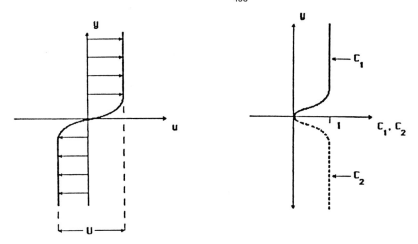

Figure 2. Initial mean velocity and chemical species field used in the simulation. 2a. Mean Velocity. 2b. Mean Species.

$$C_2(x,y,z,0) = 1 - C_1(x, y+\delta, z, 0)$$

δ is a value by which the two species fields have been offset so that there is initially no overlap between the two (figure 2). These functions are easily resolvable using spectral methods and are not unlike those measured experimentally. The particular chemical reaction used is the single step, irreversible reaction:

$$C_1 + C_2 \rightarrow \text{Products} + \text{Heat}$$

The reaction is a function only of the reactant concentrations and does not depend on temperature. Although some important features of real reacting flows are lost with this simple reaction, many important features of the effects of energy release on the dynamics of the flow can still be studied.

Summary of Approximations

The following is a summary of the major simplifications and approximations used in the numerical simulations.

A. Low Mach number approximation. To ease stability requirements in the numerical calculations a low Mach number approximation is used. This allows acoustic waves to be filtered out of the equations. The validity of the approximation has been illustrated for these simulations by McMurtry et al. (1986) by comparing results obtained using the complete set of fully compressible flow equations with the results from the equations obtained from the low Mach number approximation.

B. Temporally developing mixing layer. The simulations are of a mixing layer that develops in time and is periodic in the streamwise and spanwise directions. As pointed out earlier, this is not the same flow as the spatially growing layer that is usually encountered in practice. However, many dynamic similarities exist between the two so the temporally growing case is of relevance. Also, a more efficient and accurate computer code can be written for the temporally growing mixing layer for a given amount of computer resources.

C. Simple chemistry. A single step, irreversible reaction of the type $A + B \rightarrow$ Products + Heat is used. The reaction is taken to be a function only of the reactant concentrations and does not depend on temperature. This chemical reaction is highly idealized, but allows for effects of energy release in the flow to be studied.

D. Constant transport coefficients. The viscosity, thermal and molecular diffusivities, and the specific heats are taken to be temperature independent constants. Again, this is not a realistic feature of general combusting flows, but in addition to simplifying matters from a numerical point of view, it allows other effects on the flow to be isolated and studied in a simpler environment.

E. Moderate heat release. This restriction is necessary to ensure that velocities will not be generated by the combustion which are large enough to violate the low Mach number approximation. The condition that must be satisfied is $DaCe \ll 1/M$.

F. Finite wavenumber cutoff. The computations to be discussed in the following section were performed on a computational grid consisting of 64 x 65 x 64 modes. In the transformation to the wavenumber space this corresponds to computing the interactions of frequencies consisting of wavenumbers k_x, k_y, k_z = -32 to 32, where k_x, k_y, k_z are the wavenumber vector components in the three coordinate directions.

3. Simulation Results

Results are reported for two different values of the heat release parameter Ce: one with no heat release (Ce=0), and the other giving a maximum density decrease of approximately ρ/ρ_o = 0.5 (Ce=5). Runs 1 (Ce=0) and 2 (Ce=5) were initialized with random perturbations only. Runs 3 (Ce=0) and 4 (Ce=5) included, in addition to the random perturbations, a two-dimensional perturbation corresponding to the most unstable wavelength and its subharmonic. In the following, runs 3 and 4 will be referred to as the forced cases. These runs are analagous to laboratory experiments in which well defined harmonic perturbations are applied to the flow at selected frequencies. The parameters used in these runs are given in Table 1. Additional simulations have been run using different random initializations. The results presented here are representative of all these runs.

TABLE 1
THREE-DIMENSIONAL SIMULATION PARAMETERS

Run No.	Domain	Re	Da	$A_{1,0}$	$A_{1/2,0}$	ε_{3D}	Ce	Λ
1	4π	500	2	0.0	0.0	0.00003	0	2
2	4π	500	2	0.0	0.0	0.00003	5	2
3	4π	500	2	0.01	0.01	0.00003	0	2
4	4π	500	2	0.01	0.01	0.00003	5	2

$A_{1,0}$ and $A_{1/2,0}$ are the amplitude of most unstable mode and its subharmonic as determined from a linear stability analysis of the hyperbolic tangent velocity profile.[50]

All simulations were performed on the CRAY X-MP computer at NASA Lewis Research Center. The three-dimensional computations were performed on a 64 x 65 x 64 uniform mesh and required 12 seconds of cpu time per time step. Each simulation presented here required between 600 and 1200 time steps to complete.

Higher heat release cases than are presented here can be run, but, since the boundary conditions used here imply constant volume combustion, the background pressure rises with time. Test cases with a computational domain twice as large in the transverse (y) direction were performed to verify that pressure rise effects due to the presence of the boundaries were not significant for the values of the parameters studied here.

Some aspects of the nature of the flow can be seen in a three-dimensional perspective plot of the total vorticity. In figure 3 surfaces having a value of 50% of the maximum of the sum of the absolute values of all three vorticity components are plotted ($|\omega_x|+|\omega_y|+|\omega_z|$ = constant). This is a result from run 1 (initial random velocity perturbations, no heat release). Clearly evident is the approximately two-dimensional spanwise vorticity structure associated with the rolling up of the mixing layer. These structures produce a stretching and wrinkling of the reaction interface that can lead to enhanced product formation (Riley et al., 1986; McMurtry et al., 1986). In addition, tubes aligned in the streamwise direction are also apparent. These latter structures have been recognized as counter-rotating vortices (Bernal, 1981) and have been modeled by Lin and Corcos (1984) and Pierrehumbert and Widnall (1982), and computed in constant density mixing layers by Metcalfe et al. (1987). The effect of these structures on the flow field is to induce a velocity field that acts to pump fluid between the two layers, thereby further convoluting the reaction interface. This tends to increase mixing between the two streams and enhance chemical reactions. One purpose of this work was to examine the influence of heat release on these processes and study the influence this has on the product formation rate.

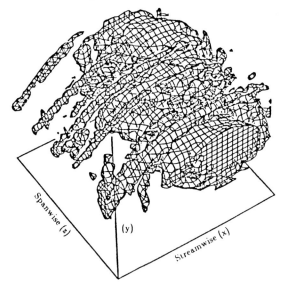

Figure 3. Magnitude of total vorticity, run 1, constant density calculation. Surface plotted is 50% of the maximum vorticity level.

The effect of heat release on the chemical product formation for runs 1 and 2 is shown in figure 4, where the total product (integrated over the entire computational domain) is plotted as a function of time. In agreement with previous two-dimensional results (McMurtry et al., 1986), the total product is seen to be lower for the case with heat release. Since the local instantaneous reaction rate is proportional to the product of the local species concentrations, this indicates that

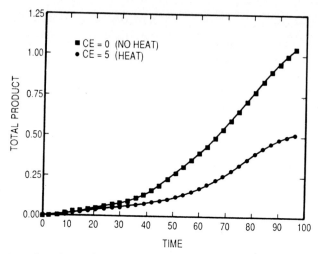

Figure 4. Total product formation as a function of time, run 1 (no heat release), run 2 (heat release).

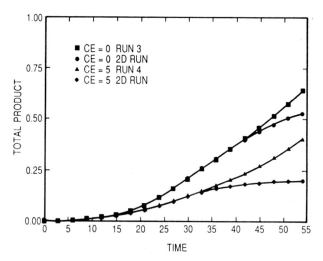

Figure 5. Total product formation as a function of time, run 3 (no heat release), run 3 (heat release). Also plotted are results from a purely two-dimensional calculation.

one effect of the heat release has been to reduce the amount of mixing between the two streams. An increase in the rate of product formation is seen to occur at about a time of 40 in the constant density case (run 1) and at a time of 60 in the heat release case. In both cases, this is due to the rollup of the subharmonic mode (i.e., vortex pairing), which engulfs large amounts of fluid into the vortex structures. Figure 5 shows the amount of product formed when the layer is forced at its most unstable wavelength and subharmonic in addition to the random three-dimensional purturbations. The overall rate of product formation is greater when the layer is forced, but again the decreases in the product with heat release is apparent.

The total product from a purely two-dimensional simulation with the same level of forcing as runs 3 and 4 is also plotted in figure 5. During the rollup and pairing process, the three dimensional growth is inhibited, giving nearly identical results for the two- and three- dimensional simulations. Other studies (Metcalfe et al., 1988) have shown rollup and pairing has a stabilizing effect on the three-dimensional growth when the amplitudes of the three dimensional modes are small, while absence of pairing can enhance the growth of the three-dimensional modes. After saturation of the two-dimensional modes (at a time of about t=35) the product formation in the three-dimensional runs continues to grow rapidly due to enhanced three-dimensional mixing, while in the two-dimensional simulations the product formation rate drops off rapidly.

One measure of the layer growth is the vorticity thickness, defined as the ratio of the maximum velocity difference across the layer to the maximum slope of the mean velocity profile. The mean velocity profiles for runs 1 and 2 (figure 6) at a time of 72 and runs 3 and 4 (figure 7) at a time of t=36 show a steeper profile for the heat release runs, indicating a slower rate of the layer growth. (Mean quantities are obtained by averaging over horizontal planes.) Note also the slight overshoot in the velocity profile that occurs in the heat release run. These observations are similar to those obtained from previous two-dimensional calculations by McMurtry et al., which were initialized by adding the most unstable mode and its subharmonic to the mean flow. The overshoot in the velocity profile is not as pronounced in the three-dimensional simulations as in the two-dimensional simulations. This is because of greater spanwise variation and lack of coherence that exists in the three-dimensional simulations. These overshoots seen in the velocity profile are similar to those observed in the jet flame experiments of Yule et al. (1981) and have also been observed in reacting mixing layer experiments (Mungal, 1986; Hermanson, 1986).

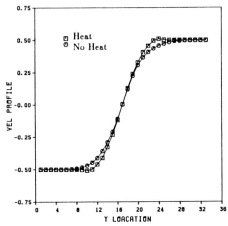

Figure 6. Velocity profile, run 1 (no heat release), run 2 (heat release) t=72.

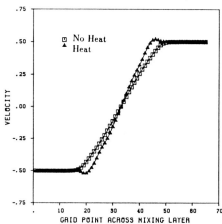

Figure 7. Velocity profile, run 3 (no heat release), run 4 (heat release), t=72.

Plots of the vorticity thickness and velocity half-width (the lateral distance from the centerline of the mixing layer to the location where the average streamwise velocity component attains one half of its free stream value) as a function of time reveal another interesting feature (figures 8-11). Initially, the growth of the layer given by these two measures is slightly greater in the heat release case; this is true up to a time of about 30 in the unforced runs (runs 1 and 2, figures 8 and 9) and up to t=15 in the forced runs (runs 3 and 4, figures 10 and 11). At later times the layer thickness is less in the heat release runs. The initial increase in the layer growth rate is a result of thermal expansion, which shifts the whole flowfield outward. Explanations for the smaller thickness seen in the heat release runs at the later times are given in the next section.

To summarize, the most obvious macroscopic effects of heat release on the mixing layer dynamics revealed in these numerical simulations are a decrease in the product formation and,

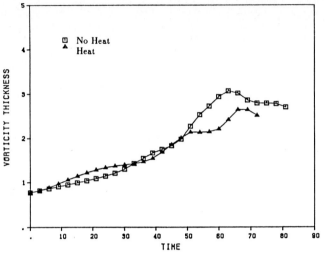

Figure 8. Vorticity thickness, run 1 (no heat release), run 2 (heat release).

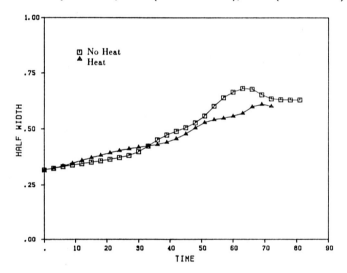

Figure 9. Velocity half width, run 1 (no heat release), run 2 (heat release).

after a small initial increase, a decrease in the layer growth rate. These results are qualitatively similar to those obtained experimentally by Wallace (1981) and Hermanson (1985), and are consequences of lower rates of fluid entrainment into the mixing region when exothermic chemical reactions occur. In the following sections physical mechanisms responsible for these observations are suggested and described.

4. Effects of Heat Release

In this section, the evolution of the flow field and the influence of heat release is discussed in terms of the shear stress distribution, the turbulent kinetic energy balance, vorticity dynamics,

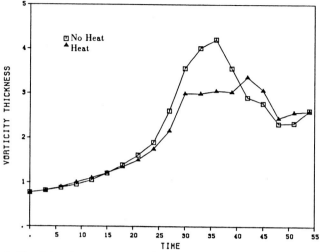

Figure 10. Vorticity thickness, run 3 (no heat release), run 4 (heat release).

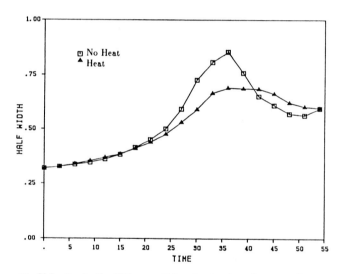

Figure 11. Velocity half width, run 3 (no heat release), run 4 (heat release).

and stability considerations. Since each of these aspects of the flow must be consistent with the others, they are not interpreted as separate mechanisms, but instead they provide viewpoints from which the effects of heat release can be explained.

Information related to the turbulence statistics and vorticity dynamics was computed for all simulations. Heat release affects were found to be qualitatively similar for the runs with and without the coherent two-dimensional perturbations added to the initial velocity field. Therefore, to clarify the discussion, only runs 1 and 2 (no coherent perturbations added to initial flow field) will be referenced in the remaining of this article.

Turbulent Stresses

It has been pointed out (Hermanson, 1985) that most of the observed heat release effects can be attributed to the decrease in the turbulent shear stresses. Velocity and density profiles obtained experimentally as well as the measured layer growth rate were used to compute the turbulent shear stress profiles in his experiments. In the numerical simulations performed here, it is possible to compute the turbulent shear stresses directly.

The computed Favre (density weighted) averaged turbulent shear stress profiles ($\widetilde{\bar{\rho} u_i'' u_j''}$) for runs with and without heat release are shown in figure 12. The wavy overbar denotes Favre averaged quantities, which are defined as

$$\widetilde{U} = \frac{\overline{\rho u}}{\bar{\rho}}$$

The fluctuating quantities are then given by

$$u'' = u - \widetilde{U}.$$

The straight overbar refers to conventional Reynolds averaging. In the temporal calculations this is an average over a horizontal plane. It is well known that the turbulent stresses, $\widetilde{\bar{\rho} u_k'' u_i''}$, represent a transport of mean momentum (per unit volume). In their Favre averaged form these stresses include momentum exchange mechanisms due to turbulent transport, interactions between the mean flow and volumetric changes, and fluctuation-fluctuation interactions $(\overline{\rho' u_i' u_j'})$ (Libby and Williams, 1981).

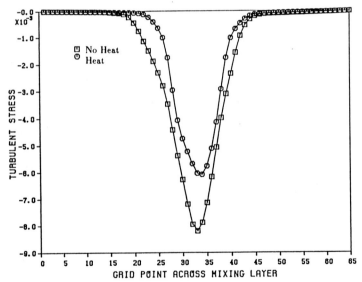

Figure 12. Favre average turbulent shear stress profile, run 1 (no heat release), run 2 (heat release), t=48.

The suppression of the shear stress in the heat release runs is clearly indicated. The lower turbulent stresses in the heat release case imply a lower transport of momentum by the turbulence. This also indicates, as will be shown in the next section, that less energy is being transferred from the mean flow to the turbulence. In addition, the turbulent shear stresses can be directly related to the stability characteristics of the mixing layer, providing explanations for the lower growth rates of the mixing layer.

Turbulent Kinetic Energy

Closely related to the turbulent shear stresses is the turbulent kinetic energy. Examination of the behavior of the turbulent kinetic energy can provide a useful way to interpret the effects of heat release on the turbulent motions and account for some of the observations in the heat release runs. Furthermore, the turbulent kinetic energy and its production, redistribution, and viscous dissipation are important aspects of the flow that must be treated in many of the models currently used to describe turbulent flows.

The turbulent kinetic energy profile $(\widetilde{\bar{\rho}u_i''u_k''})$ for runs with and without heat release is shown in figure 13 at a nondimensional time of t=48. This corresponds to a time after the rollup of the most unstable mode and before the rollup of its subharmonic is complete. From this figure it can be seen that the total turbulent kinetic energy is less for the heat release run. This is consistent with the earlier observations of less product formation and lower growth rates, since lower turbulence levels imply lower turbulent transport rates, resulting in a lower exchange of mass, momentum, and energy among fluid elements.

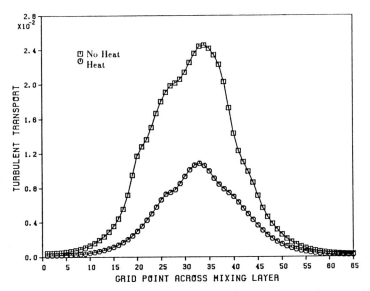

Figure 13. Turbulent kinetic energy profile, run 1 (no heat release), run 2 (heat release) t=48.

To understand how a lower turbulent kinetic energy profile results, it is useful to examine its transport equation. Written in Favre average form, the transport equation for the turbulent kinetic energy is given by

$$\frac{\partial \bar{\rho} q}{\partial t} + \frac{\partial \widetilde{U}_j \bar{\rho} q}{\partial x_j} = -\widetilde{\bar{\rho}u_k''u_i''}\frac{\partial \widetilde{U}_i}{\partial x_k} - \frac{\partial \widetilde{\bar{\rho}u_i''u_i''u_k''}}{\partial x_k} - \overline{u_i''\frac{\partial p'}{\partial x_i}} - \overline{\tau_{ki}\frac{\partial u_i''}{\partial x_k}}$$

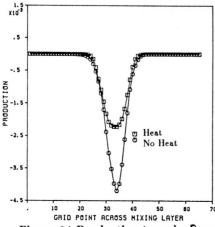
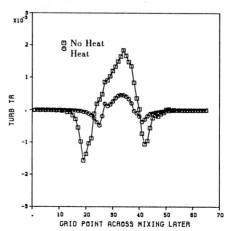

Figure 14 Production term in Favre averaged turbulent kinetic energy equation, run 1 (no heat release), run 2 (heat release), t=48.

Figure 15. Turbulent transport term in kinetic energy equation, run 1 (no heat release), run 2 (heat release), t=48.

where $q = \frac{1}{2}\widetilde{u_i'' u_i''}$ is the turbulent kinetic energy per unit mass.

The first term on the right hand side describes the exchange of energy between the mean flow and the fluctuating motion. This production of kinetic energy is seen to be equal to the product of the mean velocity gradient and the turbulent stresses. In the mixing layer simulated here, the only nonzero contribution of this term is for i=1 and k=2. The profile of the turbulent production is shown in figure 14 at t=48. Production is greatest at the center of the mixing layer where the velocity gradient and turbulent stresses are the largest. (In the following plots, negative values indicate a production of turbulent energy.) Although it was shown that the velocity gradient is steeper in the cases with heat release, the total turbulent kinetic energy production is less with heat release due to the smaller turbulent stress term as shown in the previous section (figure 12). The slope of the velocity profile profile is plotted in figure 6.

The second term on the right is a conservative term and describes the redistibution of kinetic energy by the fluctuating velocity field. Figure 15 shows that turbulent transport tends to convect energy from the center of the mixing layer, where the turbulent intensity is highest, to the outer regions. With the lower turbulence intensities that are seen in the heat release runs, this term is of a lower magnitude in this case.

The effects of pressure fluctuations on the turbulent kinetic energy are described by the term $\overline{u_i'' \partial p'/\partial x_i}$. The value of this term is plotted in figure 16 at t=48. This term is of the same order of magnitude as the production term, but has an opposite sign and does not appear to be as greatly affected by the heat release. The physical interpretation of this contribution can be clarified somewhat by writing it as

$$\overline{u_i'' \frac{\partial p'}{\partial x_i}} = \overline{\frac{\partial u_i'' p'}{\partial x_i}} - \overline{p' \frac{\partial u_i''}{\partial x_i}}$$

The first term on the right hand side is a conservative term describing the redistribution of kinetic energy by pressure fluctuations. The second term, which is zero for constant density flows, is a source of kinetic energy resulting completely from the combustion. The contribution of these two terms for the heat release run is shown in figure 17. The transport due to the pressure

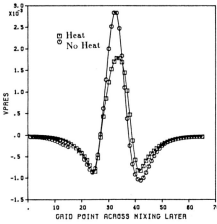
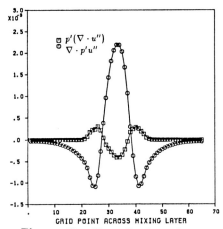

Figure 16. Velocity-pressure gradient correlation, run 1 (no heat release), run 2 (heat release), t=48.

Figure 17. Contribution of expansion and pressure fluctuations to the velocity-pressure gradient correlation.

fluctuations is not much changed between the runs with and without heat release, although the peak magnitude is less in the heat release runs. In both cases the pressure fluctuations act to transport energy away from the center of the layer to the outer regions. The contribution due to the velocity divergence leads to an increase in kinetic energy at the center of the vortices. In this respect the heat release acts to increase the turbulence level, although this effect is overshadowed by the decrease in the mean flow production term.

From the results presented here, it is seen that the most significant effects of heat release on the turbulent kinetic energy balance are felt through the reduction in the mean flow production of turbulent kinetic energy. A smaller contribution acting to increase the turbulence intensity is felt through the expansion part of the pressure gradient - velocity correlation. This production is, however, small in these simulations compared with the decrease in the mean flow production term, yielding an overall lower turbulent kinetic energy profile. For higher rates of heat release, the term $\partial u_i''/\partial x_i$ would increase, resulting in more internal energy being converted to kinetic energy, and possibly, an overall increase in the turbulent kinetic energy.

Vorticity Dynamics

Another approach to studying the flow and interpreting the effects of heat release on the flow field is in terms of vorticity dynamics. This allows a description of the flow directly in terms of the dynamics of the large-scale structures. In a two-dimensional flow without heat release or viscosity, the vorticity of a fluid particle is conserved following fluid particles, and the vorticity field can be used to directly visualize the flow field. In a three-dimensional flow, or a flow with density variations or expansion, this property is no longer valid and also the vorticity no longer serves as a reliable fluid marker. However, the dynamics of the flow field can still be understood and interpreted by studying the vorticity field.

In a general three-dimensional flow, the vorticity equation can be written in the following form:

$$\frac{D\omega}{Dt} = (\omega \cdot \nabla)\mathbf{v} - \omega(\nabla \cdot \mathbf{v}) + (\nabla \rho \times \nabla p)/\rho^2 + \frac{1}{Re}(\nabla^2 \omega)$$

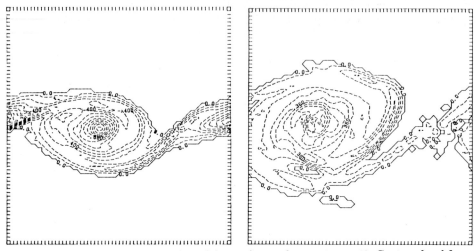

Figure 18. Spanwise component of vorticity, no heat release. 18a. t=48, Contour level from -1.3 to 0.0. 18b. t=72, Contour level from -.9 to 0.0

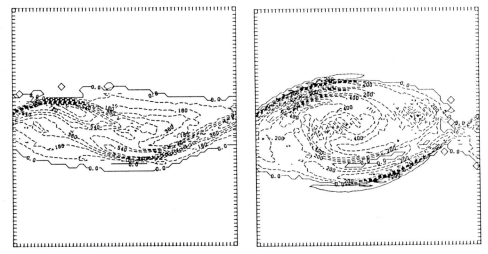

Figure 19. Spanwise component of vorticity, heat release. 19a. t=48, Contour level from -.8 to 0.0 19b. t=72, Contour level from -.7 to 0.2

Four different mechanisms can be identified that alter the vorticity: Vortex tube stretching and turning $(\omega \cdot \nabla)\mathbf{v}$, thermal expansion, $\omega(\nabla \cdot \mathbf{v})$, baroclinic torques, $(\nabla \rho \times \nabla p)/\rho^2$, and viscous diffusion. For a low Mach number flow without heat release, the expansion and baroclinic torque terms will be very small. When density changes due to heat release do occur, these two mechanism can be important in the vorticity dynamics.

The instantaneous spanwise component of vorticity (ω_3) at times of 48 and 72 is shown in figures 18 and 19 for runs 1 and 2. (All contour plots that follow are taken from one spanwise plane selected from the three-dimensional data field.) In the following figures dashed contour lines indicate negative vorticity (local rotation of fluid elements in the clockwise direction) and solid lines indicate positive vorticity. One of the most apparent differences between these two figures is that, for the simulations with heat release, the maximum amplitude of the vorticity, which occurs in the vortex cores, has decreased substantially. Furthermore, the vorticity is not as concentrated in the center of the large structures as in the constant density case. Also note the regions of positively signed vorticity that appear at the outer edges of the vortex structures for the case with heat release.

As mentioned above, the mechanisms that produce the observed changes in the vorticity field in the heat release runs are the baroclinic torques and thermal expansion. In an expanding flow, $\nabla \cdot \mathbf{v}$ is positive; therefore, the effect of thermal expansion results in a decrease in the magnitude of vorticity. This can be understood by angular momentum considerations, since as a fluid element expands due to heat release, the magnitude of its local rotation rate, and hence its vorticity, must decrease to conserve angular momentum.

The instantaneous value of the spanwise component of the expansion term at one spanwise location is shown in figure 20 at two different times. Thermal expansion occurs as heat is released by the chemical reaction so that this term also gives a good indication of the reaction zone. The rate of reaction is highest where the inflow of reactants in the highest. This occurs in the regions of highest strain, which are located in the braids. The result of the expansion is a decrease in the magnitude of vorticity, which in part explains the changes in the vorticity field that result when exothermic chemical reactions occur. [For a temperature dependent reaction, the high dissipation rates in the braids can cause local quenching of the flame so that the reaction rate is not necessarily highest in the regions where the inflow of reactants is the highest (Peters, 1983; Givi et al., 1986). This effect is not addressed here.]

The baroclinic torque, $(\nabla \rho \times \nabla p)/\rho^2$ describes differential fluid accelerations resulting from nonaligned pressure gradients and density gradients. Plots of the spanwise component of this term at specific spanwise locations show an alternating sign across the reaction surface and no contribution in the vortex cores (figure 21). This can be understood by recognizing that the density gradient changes sign across the reaction surface, and that the pressure gradient vector points approximately radially outward from the vortex cores. In the vortex cores the pressure and density gradients are small and are also approximately aligned, so there is no contribution of this term here. As the layer rolls up and winds around itself, the baroclinic torque takes on the complicated structure shown in figure 21b. Comparisons with the results of two-dimensional simulations show that these effects are dominated by two-dimensional dynamics.

Comparing figures for the spanwise vorticity component (figure 19) and the instantaneous baroclinic torque (figure 21) shows that the regions of positively signed vorticity at the outer regions of the vortices, and also the appearance of multiple extrema in the vorticity field, are a result of the baroclinic torque. The overshoot in the velocity profile seen in the previous section (figure 6) reflects the generation of positive vorticity in this region by the baroclinic torque. This is also possibly the mechanism that produces the previously unexplained inflection points seen in the velocity profile at the outer edges of jet diffusion flames by Yule, and in mixing layers with chemical heat release (Mungal, 1986).

In an attempt to understand the relative importance of the expansion and baroclinic torque on the development of the flow, average values of these two terms are plotted as a function of the height across the mixing layer for a sequence of times (figure 22). (Negative values of the expansion term indicate a decrease of vorticity while negative values of the baroclinic torque indicate an increase.) The magnitude of the expansion term is seen to initially be larger than the baroclinic torque term. At later times, the amplitude of the two terms are comparable. Note that the expansion term consistently produces a net decrease in vorticity. The baroclinic torque tends to decrease the vorticity near the upper and lower limits of the dynamically active region, at least

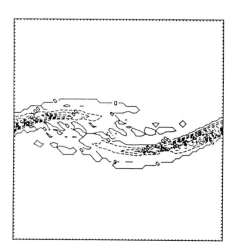 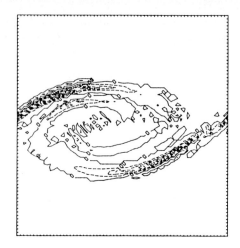

Figure 20. Instantaneous value of spanwise component of expansion term in vorticity equation, run 2, spanwise plane. $z=0$. 20a. $t=48$, Contour level from -.031 to .008. 20b. $t=72$, Contour level from -.028 to .012.

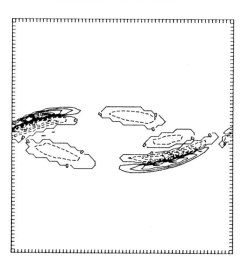 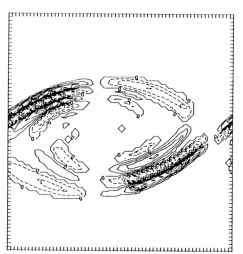

Figure 21. Instantaneous value of spanwise component of baroclinic torque, spanwise plane $z=0$. 21a. $t=48$, Contour level from -.04 to .06. 21b. $t=72$, Contour level from -.05 to .05.

during the initial stages when the two-dimensional modes dominate the flow. This is consistent with the generation of regions of positive vorticity at the edges of the cores (figure 21). Note also that the baroclinic vorticity generation on the centerline is occurring in the braids rather than the cores.

The weaker and more diffuse vortex structures which result from the baroclinic torque and thermal expansion lead to a slower rollup of the layer, thus reducing the straining of the reaction interface and decreasing the mass entrainment into the layer. This accounts for much of the decrease seen in the overall product formation and the changes in the layer growth rate. In section 3 it was shown (figures 8-11) that with heat release, the layer growth rate initially increases, and

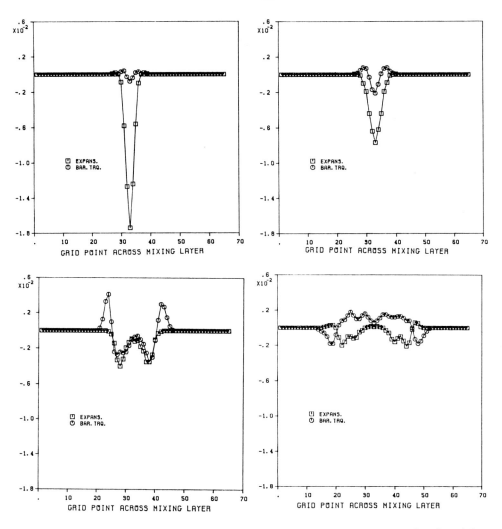

Figure 22. Average value of baroclinic torque and expansion terms transversing the mixing layer. 22a. t=12. 22b. t=24. 22c. t=48. 22d. t=72.

then decreases compared with the constant density case. The initial increase was explained to be due to thermal expansion, which tends to shift the whole layer outward. Later in the development of the mixing layer the growth is dominated by the large scale rollup and pairing processes. Since these processes are inhibited by heat release, the constant density mixing layer will then grow faster.

Stability Considerations

The question of how the stability characteristics of the mixing layer are affected by heat release is important from both physical and numerical viewpoints. If, as a result of density changes, the unstable modes shift to other wavelengths, or if the growth rates change, this certainly will affect the growth of the mixing layer and the rate at which chemical products are formed. From a computational point of view, it is not possible to look at a continuous distribution of wavelengths. Because of the periodic boundary conditions employed here, only disturbances of wavelengths that divide exactly into the computational domain are allowed (i.e., λ = domain length/n, where n is a positive integer less than the number of modes retained in the simulations). Therefore, if the most unstable modes shift to different wavelengths, dynamically important effects may be overlooked.

To address this question, a linear stability analysis of a simplified model problem was carried out in conjunction with the simulations. This analysis involved a shear layer with a piece-wise linear velocity profile and a low density in the velocity transition region (figure 23). The velocity and density in the free stream were constant, and the density in the transition zone was also constant and given by $\rho(1-\beta), 0 < \beta < 1$). Although this does not exactly describe the conditions in the simulations reported here, the basic characteristics of the two are similar, so that the general trends indicated by the analysis are expected to be true of the simulations. Details of the analysis have been given by McMurtry (1987).

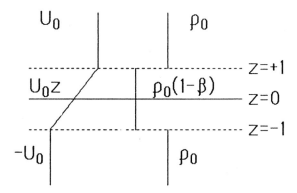

Figure 23. Configuration for stability analysis.

The results of this analysis are plotted in figure 24, which show the growth rate of any individual unstable mode (specified by its wave number) for a given value of β. As the density decreases (increasing β), the most unstable mode, represented as the maximum value on each curve, shifts to a lower wave number (longer wavelength). The growth rates of the unstable modes are also seen to decrease as the density is lowered, a result consistent with the lower growth rates discussed in section 3.

The energy $E(k_x, k_z)$ contained in the two-dimensional modes $k_x = 1, k_z = 0$ (the most unstable two-dimensional mode in the constant density mixing layer with a hyperbolic mean velocity profile) and $k_x = 1/2, k_z = 0$ (the subharmonic of the most unstable mode) as a function of time is shown in figures 25 and 26 for runs 1 and 2. This energy is defined as

$$E(k_x, k_z) = \int_{-\infty}^{\infty} |\hat{v}(k_x, k_z, y)|^2 \, dy$$

where

$$\hat{v}(k_x, k_z, y) = \sum_{k_y=0}^{N} \hat{u}(k_x, k_z, k_y, t) f(k_y, y)$$

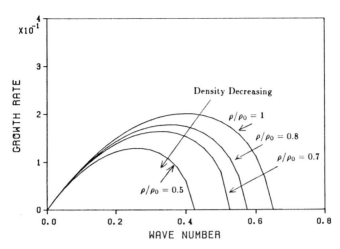

Figure 24. Stability analysis results for various density changes in the mixing region.

Here $\hat{u}(k,t)$ are the Fourier components of the velocity and $f(k_y, y)$ is either $\sin(k_y y)$ or $\cos(k_y y)$, depending on the component of the velocity under consideration. In the following discussion, we analyze the temporal behavior of the energies in these Fourier modes.

In figure 25, the energy contained in the fundamental mode $(E_{1,0})$ is compared for runs 1 and 2. For the constant density case this is the most unstable frequency and is seen to grow until a time of 40, at which point it reaches a quasi-equilibrium, saturated state (Riley and Metcalfe, 1980). The behavior of this mode changes significantly when heat release occurs. The growth drops off at a level well below the incompressible saturation level and at a much earlier time (figure 25), although the initial growth before much density change occurs (up to about t=7) is the same in the two cases. The energy in the subharmonic $(E_{1/2,0})$ grows until it reaches its saturation level at a time of 65 in the constant density case (figure 26). The subharmonic remains unstable with the addition of heat to the flow, but grows at a slower rate than in the constant density case. The energy in the three-dimensional modes (the sum of $E(k_x, k_z)$ for all k_x and $k_z \neq 0$) is also seen to be less for the heat release run as shown in figure 27.

The results of the simplified linear analysis for the linear profiles compare favorably with the simulation results. The density changes occurring in run 2 correspond to a β of 0.5. With this amount of heat release, the most unstable mode in the constant density case is predicted to be stable, which is consistent with the model behavior in the simulations (figure 25). The actual most unstable mode shifts to a wave number of the subharmonic of the most unstable mode in the constant density case (0.22). From figure 26, the growth rate $(E^{-1} \partial E / \partial t)$ of the subharmonic is shown to reduce from 0.18 to 0.13, while in the simulations the computed growth rate of the subharmonic decreased from 0.12 to 0.09 in the heat release case. In figure 26 this lower growth rate of the subharmonic is indicated by the smaller slope.

The laboratory experiments on mixing layers, which clearly show lower growth rates of the mixing layer as heat release is increased, do not lead to conclusions regarding the supression of modes or a shift in the wavelengths of the most unstable modes. Although the experiments performed by Hermanson (1985) appear to indicate there is little change in the spacing (wavelengths) of the vortex cores when heat release occurs, this observation does not dismiss the possibility that the frequencies of the unstable modes are affected by heat release. In particular, if the heat

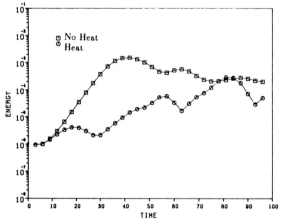

Figure 25. Comparison of energy in fundamental mode, runs 1 and 2 with and without heat release.

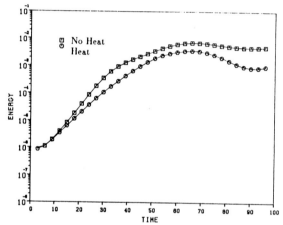

Figure 26. Comparison of energy in subharmonic, runs 1 and 2, with and without heat release.

release acts to eliminate only the highest wavenumber components as suggested in the previous analysis (Figure 24), this would only be seen in the near field of the splitter plate. A change in the frequencies has been observed, however, in experiments on jets (Yule et al., 1981) which showed the value of the most energetic frequencies decreased as heat release increased, a result consistent with the simulations and analysis presented here.

5. Summary

The simulation results discussed here and related laboratory experiments have shown that, when heat release accompanies the chemical reaction, the mixing layer grows at a lower rate and the amount of product formed decreases. In addition, an overshoot in the velocity profile appears. In the previous section, four different aspects of the flow were studied to explain the observed effects of heat release: The turbulent shear stresses, the turbulent kinetic energy, vorticity dynamics, and stability considerations. These different considerations must be consistent with one another and , therefore, only provide different approaches from which the flow can be studied.

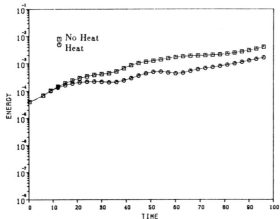

Figure 27. Comparison of energy in three-dimensional modes with and without heat release.

The study of the flow in terms of vorticity dynamics explored two mechanisms that do not act in constant density flows: the baroclinic torque and thermal expansion. The action of these mechanisms was shown to result in more diffuse and weaker vortices when heat release accompanies the chemical reaction. At the largest scales (which dominate the dynamics of the flow and account for most of the turbulent transport), the altered vorticity distribution resulted in slower rollup of the most unstable modes, giving lower growth rates and less entrainment of unmixed fluid. This was indicated by the lower energies and growth rates of the unstable modes computed in the simulations and confirmed by a stability analysis of a simplified model problem similar to the mixing layer flow simulated here. The appearance of "humps" in the velocity profile were shown to be the result of vorticity generation in the outer regions of the vortices by baroclinic torques.

The turbulent kinetic energy and the turbulent stresses are closely related, since the turbulent stresses are a direct indication of the kinetic energy transfer from the mean flow to or from the turbulence. Lower values of both the turbulent kinetic energy and the turbulent stresses were observed in the heat release runs. This can be related directly to the vorticity dynamics by realizing that in turbulent flows, the largest eddies (vortices) are responsible for most of the transport of momentum and scalar variables. The weaker large-scale vortices that result when heat release occurs transport less momentum, which is exactly what the lower values of the turbulent shear stresses indicate. The transport rates of the chemical reactants will also result in less product formation, as observed in the simulations and laboratory experiments.

Finally, some comments on the use of of direct numerical simulations in the study of turbulent, chemically-reacting flows are in order. (See Jou and Riley, 1987 for a more detailed discussion of this topic.) There are two implementations of direct numerical simulation: full turbulence simulations (FTS), in which all of the dynamically significant space and time scales are resolved; and large eddy simulation (LES), in which the governing equations are filtered at the numerical grid scales, and the sub-grid scale motions are modelled. Because of limited numerical resolution FTS is restricted to moderate Reynolds numbers (less than several hundred) and Damköhler numbers (less than about 10, although the conserved scalar approach can for certain cases eliminate the latter restriction), whereas LES can be applied, at least in theory, to high Reynolds number, high Damköhler number flows. The potential advantages of direct numerical simulation, when the method can be applied, are the following: the flow can be examined in detail, since all quantities are known at each point in space and time; parameters can be easily varied and experimental conditions are easily controlled; large-scale structures are directly addressed; and results are less sensitive to the models used, since only, at most, the smaller scales are modeled. The main disadvantages are the limitations in the spatial and temporal resolution available, and the large amount of computer resources required, both of which can be severe. Technical applications at the present are similar to those of laboratory experiments, i. e., understanding particular physical processes, and developing and testing models.

ACKNOWLEDGEMENTS

This study has benefited from numerous suggestions and ideas generously provided by Dr. W.-H. Jou. This work was supported in part by the NASA Lewis Research Center under contract NAS3-24229 to Flow Industries, Inc., by ONR Contract No. N00014-87-K-0174 to the University of Washington, and by the United States Office of Basic Energy Sciences, division of Chemical Sciences. All computer simulations were performed on the CRAY X-MP at the NASA Lewis Research Center.

REFERENCES

Acton, E. (1976) "The Modelling of Large Eddies in a Two-Dimensional Shear Layer," **J. Fluid Mech.**, Vol.76, pp. 561-592.

Aref, A. and Siggia, E.D. "Vortex Dynamics of the Two-Dimensional Turbulent Shear Layer," **J. Fluid Mech.**, Vol. 100, pp. 53-95.

Bernal, L. P. (1981) "The Coherent Structure of Turbulent Mixing Layers. I. Similarity of the Primary Vortex Structures. II. Secondary Streamwise Vortex Structure," Ph.D. Thesis, Calif. Inst. of Technology, Pasadena.

Bilger, R. W. (1980) in **Turbulent Reacting Flows**, P. A. Libby and F. A. Williams, ed., Springer-Verlag, New York.

Breidenthal, R. E. (1981) "Structure in Turbulent Mixing Layers and Wakes Using a Chemical Reaction," **J. Fluid Mech.**, Vol. 109, pp. 1-24.

Brown, G. L. and Roshko, A. (1974) "On Density Effects and Large Structure in Turbulent Mixing Layers", **J. Fluid Mech.**, Vol. 91 pp. 319-335.

Buckmaster, J. D., (1985) "An Introduction to Combustion Theory," in **The Mathematics of Combustion**, J. Buckmaster, ed. SIAM.

Corcos, G. M. and Sherman, F. S. (1984) "The Mixing Layer: Deterministic Models of a Turbulent Flow. Part 1. Introduction and the Two-dimensional Flow", **J. Fluid Mech.**, Vol. 139, pp. 29-65.

Givi, P., Jou, W.-H., and Metcalfe, R. W. (1986) "Flame Extinction in a Temporally Developing Mixing Layer," Twentyfirst Symposium (International) on Combustion, the Combustion Institute.

Gottlieb, D. and Orszag, S. A. (1977) **Numerical Analysis of Spectral Methods**, SIAM, Philadelphia.

Hermanson, J. Personal communication and unpublished data.

Hermanson, J. C. (1985) "Heat Release Effects in a Turbulent Shear Layer," Ph.D. Thesis, California Institute of Technology.

Hottel, H. C., and Hawthorne, W. R., (1953) Third Symposium (International) on Combustion, Williams and Wilkens, p. 254.

Hsiao, C. C., Oppenheim, A. K., Ghoniem, A. F. and Chorin, A. J. (1984) "Numerical Simulation of a Turbulent Flame Stabilized Behind a Rearward-Facing Step." Twentieth Symposium (International) on Combustion, The Combustion Institute.

Jou, W.-H. and Riley, J.J. (1987) "On Direct Numerical Simulations of Turbulent Reacting Flows," AIAA 19th Fluid Dynamics, Plasma Dynamics, and Lasers Conference, paper no. AIAA-87-1324.

Karagozian, A. (1982) "An Analytical Study of Diffusion Flames in Vortex Structures," Ph.D. Thesis, Calif. Inst. of Technology, Pasadena.

Keller, J. O. and Daily, J. W. (1985) "The Effects of Highly Exothermic Chemical Reaction on a Two-Dimensional Mixing Layer," **AIAA Journal**, Vol 23 (12).

Konrad, J. H. (1976) "An Experimental Investigation of Mixing in Two-Dimensional Turbulent Shear Flows with Application to Diffusion Limited Chemical Reactions," Ph.D. Thesis, Calif. Inst. of Technology, Pasadena.

Koochesfahani, M. M. (1984) "Experiments on Turbulent Mixing and Chemical Reactions in a Liquid Mixing Layer," Ph.D. Thesis, Calif. inst. of Technology, Pasadena.

Libby, P. A. and Williams, F. A. (1980) "Fundamental Aspects of Turbulent Reacting Flows," in **Turbulent Reacting Flows**, Libby and Williams, Eds. Springer Verlag.

Lin, S. J. and Corcos, G. M. (1984) "The Mixing Layer Deterministic Models of a Turbulent Flow. Part 3. The Effect of Plain Strain on the Dynamics of Streamwise Vortices," **J. Fluid Mech.**, Vol. 141, pp. 139-178.

Majda, A. and Sethian, J. (1985) "The Derivation and Numerical Solution of the Equations for Zero Mach Number Combustion," **Combust. Sci. and Tech.**, Vol. 42, pp. 185-205.

Marble, F. E. and Broadwell, J. E. (1977) "The Coherent Flame Model Turbulent Chemical Reactions," Project SQUID Technical Report, TRW-9-PU.

Masutani, S. M. (1985) "An Experimental Investigation of Mixing and Chemical Reaction in a Plane Mixing Layer," Ph.D. Thesis, Stanford University, California.

McMurtry, P. A. (1987) "Direct Numerical Simulations of a Reacting Mixing Layer with Chemical Heat Release," Ph.D. Thesis, University of Washington, Seattle, WA.

McMurtry, P. A., Jou, W.-H., Riley, J. J. and Metcalfe, R. W. (1986) "Direct Numerical Simulations of a Reacting Mixing Layer with Chemical Heat Release," **AIAA Journal**, Vol. 24 (6).

Metcalfe, R. W., Orszag, S. A., Brachet, M. E., Menon, S., and Riley, J. J., (1987) "Secondary Instability of a Temporally Growing Mixing Layer," **J. Fluid Mech.**, in press.

Mungal, M. G. and Dimotakis, P. E., (1984) "Mixing and Combustion with Low Heat Release in a Turbulent Shear Layer," **J. Fluid Mech.**, Vol. 148, pp. 349-382.

Mungal, M. G., (1987) Personal Communication. Orszag, S. A. and Pao, Y. (1974) "Numerical Computation of Turbulent Shear Flows," **Advances in Geophysics**, Vol. 18, pp. 225-236.

Peters, N. (1983) "Local Quenching Due to Flame Stretching and Non-Premixed Turbulent Combustion," **Combustion Science and Tech.**, Vol. 30 pp. 1-17.

Pitz, R. W., and Daily, J. W., (1983) "Combustion in a Turbulent Mixing Layer Formed at A Rearward-Facing Step," **AIAA Journal**, Vol. 21, pp. 1565-1570.

Rehm, R. G. and Baum, H. R., (1978) "The Equations of Motion for Thermally Driven, Buoyant Flows," **Journal of Research**, National Bureau of Standards, Vol. 83, pp. 297-308.

Riley, J. J. and Metcalfe, R. W., (1980a) "Direct Numerical Simulations of the Turbulent Wake of an Axisymmetric Body," **Turbulent Shear Flows II**, Bradbury, L.S., Durst, F., Launder, B.E., Schmidt, F.W., and Whitelaw, J.H., Eds. Springer-Verlag, pp. 78-93.

Riley, J. J. and Metcalfe, R. W., (1980b) "Direct Numerical Simulations of a Perturbed, Turbulent Mixing Layer," AIAA Paper No. 80-0274.

Riley, J. J. and Metcalfe, R. W., and Orszag, S. A. (1986), "Direct Numerical Simulations of Chemically Reacting Mixing Layers," **The Physics of Fluids**, Vol. 29, 2 pp. 406-422.

Rogallo, R. W., and Moin, P. (1984) "Numerical Simulation of Turbulent Flows," **Ann. Rev. Fluid Mech.** Vol. 16, pp. 99-137.

Sivashinsky, G. o. (1979) "Hydrodynamic Theory of Flame Propagation in an Enclosed Volume," **Acta Astronautica**, Vol. 6, pp. 631-645.

Takeno, T., and Kotani, Y., (1975) **Acta Astronautica**, Vol. 2, p. 999.

Takagi, T., Shin, H.-D., and Ishio, A. (1980) "Local Laminarization in Turbulent Diffusion Flames," **Combustion and Flame**, Vol. 37 pp. 163-170.

Wallace, A. K. (1981) "Experimental Investigation on the Effects of Chemical Heat Release on Shear Layer Growth and Entrainment," Ph.D. Thesis, University of Adalaide, Australia.

Whol, K., Gazley, C., and Kapp, A. J., (1949) "Diffusion Flames," Third Symposium on Combustion Flame and Combustion Phenomena, Combustion Institute, Pittsburgh, Pa.

Winant, C. D. and Browand, F. K., (1974) "Vortex Pairing, The Mechanism of Turbulent Mixing-Layer Growth at Moderate Reynolds Number," **J. Fluid Mech.,** Vol. 63, pp. 237-255.

Yule, A. J., Chigier, N. A., Ralph, S., Boulderstone, R., and Ventura, J. (1981) "Combustion-Transition Interaction in a Jet Flame," **AIAA Journal,** Vol. 19, 6, pp. 752-760.

Direct Numerical Simulation and Simple Closure Theory for a Chemical Reaction in Homogeneous Turbulence

Andy D. Leonard and James C. Hill

Department of Chemical Engineering, Iowa State University, Ames, IA 50011 USA

Abstract

The direct numerical simulation of turbulent flows serves as a useful test of simple closure theories, since one can examine the dynamics of the concentration and velocity fields in more detail than in laboratory experiments and learn how the interaction of turbulent motion and molecular diffusion affects the overall reaction rate. A brief review of the most popular methods available for full turbulence simulations is presented, and a demonstration of the usefulness of direct numerical simulation is given for simple single-point closure theories (*viz.*, those of Toor and of Patterson) applied to the irreversible, second-order chemical reaction of initially unmixed reactants.

Introduction

The direct computation of turbulent flows is a useful supplement to laboratory experiments, because it allows one to calculate quantities that cannot be measured easily in the laboratory. Furthermore, numerical experiments allow for complete reproducibility and much greater control than in laboratory experiments. Because the simulations involve actual solution of the unsteady Navier-Stokes and mass conservation equations, they are especially useful for testing theories that are based on approximations to the equations. In the present study, results of direct numerical simulations are used to evaluate two simple closure theories based on moment formulations. Before describing the numerical simulations, we first briefly review the methods available to carry out these simulations and comment in particular on the pseudospectral method, which we have used here.

Methods for the full turbulence simulation of reacting flows

In a full turbulence simulation (FTS) of a reacting flow, one must solve conservation equations for mass, momentum, energy, and the density of each reacting species, without modeling any effects of the turbulence. For the purposes of this paper we classify the methods used for FTS's into three groups: (1) finite-difference methods, (2) integral methods (including finite-element and spectral methods), and (3) vortex methods. Some hybridization of methods from different groups has also been used. A brief description of these methods used for the FTS of reacting flows and examples of such simulations are noted in this section. A more complete review of these methods is given in the book by Oran and Boris (1987), and some applications to turbulent reacting flows are described in an excellent paper by Jou and Riley (1987).

Finite-difference methods are the most commonly used numerical techniques for general fluid flow problems. A set of discrete points, $\{x_1, \ldots, x_N\}$, is used to approximate a continuous variable, $f(x)$, with a set of values, $\{f_1, \ldots, f_N\}$, known at the grid points. Taylor expansions are used to approximate the derivative at each grid point, using the discrete representation of the function. For example, one possible form for the first derivative of $f(x)$ (with equally spaced grid points) is

$$\left(\frac{df}{dx}\right)_{x_i} = \frac{f_{i+1} - f_{i-1}}{2\Delta x} + O(\Delta x^2). \tag{1}$$

Limitations on accuracy and stability determine the form of the derivatives. Finite-difference methods for turbulence simulations typically have high dispersive and diffusive errors, which causes poor resolution of sharp gradients. Artificial diffusivities and viscosities are often included in the methods to ensure stability. The grid points can be chosen to suit complex geometries, and adaptive grid techniques can be used to improve the efficiency of the method. Finite-difference algorithms are generally simpler to form than those for other methods. The text by Anderson *et al.* (1984) covers the fundamental concepts of finite-difference methods applied to fluid dynamics.

There are very few applications of finite-difference techniques to what we would call FTS's of reacting flows. Finite-difference simulations of propagating two-dimensional detonations have been performed (*e.g.*, Oran *et al.* 1982, Taki and Fujiwara 1981) using a flux corrected transport (FCT) algorithm (Boris and Book 1976) to treat steep gradients correctly. The calculations show the development of cellular structures in the detonation. A calculation of the transition to turbulence in a spatially-developing, two-dimensional mixing layer by Fujiwara *et al.* (1986) is a first step in the study of combustion in a rocket engine with finite-difference methods.

The second category, that of integral methods, includes both finite-element and spectral methods. In these methods the fundamental variables are approximated by a finite series of orthogonal basis functions,

$$f(x) = \sum_{k=1}^{N} a_k B_k(x). \tag{2}$$

Ordinary differential equations for the expansion coefficients are obtained by the method of weighted residuals. In finite-element methods one divides the computational domain into elements and then defines basis functions on each element, while in spectral methods one uses global basis functions.

With finite-element methods, computational elements can be concentrated in regions where gradients are large, and the elements can be arbitrarily shaped and used for complex geometries. The authors are not aware of any studies in which finite-element methods have been used in the FTS of reacting flows.

The methods most commonly associated with FTS are of the spectral type. The advantages of the spectral method far outweigh the disadvantages for the study of homogeneous turbulence, especially if it is implemented in pseudospectral form. The method is briefly described here, for the purpose of comparison with the other methods, and a more critical evaluation of pseudospectral methods is made in the next section. The monograph by Gotlieb and Orzsag (1977) gives a thorough treatment of spectral methods, and a review by Hussaini and Zang (1987) covers the application of spectral methods to fluid dynamics.

Spatial derivatives are evaluated easily and accurately with spectral methods if the function to be approximated is smooth enough and if the basis functions are chosen properly. If the basis functions are complex exponentials, for example, the expansion gives for the spatial derivative

$$df/dx = \sum_{k=1}^{N} ika_k e^{ikx}, \tag{3}$$

which is evaluated locally in the transform space. The error in the approximation of the derivative goes to zero faster than any power of $1/N$ as N goes to infinity. This is referred to as infinite-order or spectral accuracy. Test problems (Orszag 1972) show that a spectral method with N degrees of freedom in each direction is at least as accurate as a finite-difference method with $2N$ degrees of freedom in each direction. The choice of complex exponentials for basis functions requires the function $f(x)$ to have periodic boundary conditions on an interval of length 2π to avoid the appearance of boundary conditions in the transformed dynamical equations. Basis functions are generally chosen that meet the boundary conditions trivially and have derivatives that can be related to the basis functions themselves in the form (Peyret and Taylor 1983):

$$B'_k(x) = \sum_{k=1}^{N} \beta_{kj} B_j(x). \qquad (4)$$

The matrix β_{jk} is given by $ik\delta_{jk}$ for complex exponentials and by the recursion relationship $2B_k(x) = B'_{k+1}/(k+1) - B'_{k-1}/(k-1)$ for Chebyshev polynomials. Spectral tau methods can be used when the basis functions do not meet the boundary conditions, but these methods have not been used in the FTS of reacting flows.

The pseudospectral or spectral collocation method, rather than the full spectral or Galerkin spectral method, is usually used for the FTS of flows with suitable boundary conditions. Nonlinear terms are evaluated by transforming variables to physical space in order to evaluate the product—hence avoiding the computation of convolutions in transform space—and then transforming the product back to transform space so that derivatives are evaluated there to take advantage of spectral accuracy. Pseudospectral techniques set residuals to zero only at collocation points, instead of over the entire domain, and this introduces aliasing errors. These errors can be reduced by shifting the collocation points (Rogallo 1981) and by truncating the Fourier coefficients outside of a sphere (Kerr 1985), and a combination of the two eliminates aliasing errors arising from quadratic nonlinearities (Orszag 1972).

Pseudo-spectral methods have been used for the FTS of reacting mixing layers in order to to show the effects of fluid dynamics on the rate of chemical reaction in a constant density flow (Riley et al. 1986) and the effects of heat release in a variable density flow in the limit of low Mach number (McMurtry et al. 1986). A single-species reaction in homogeneous turbulence, with a simple kinetic mechanism to model a temperature-dependent reaction, was studied with a pseudospectral method by Picart et al. (1987), who also evaluated the joint pdf of the reacting species and its gradient. The present authors (Leonard and Hill 1986, 1987a,b) have used pseudo-spectral techniques to evaluate closure theories for an isothermal reaction of unmixed species and have examined the influence of the fluid motion on the reaction zone. In the next section we comment on the suitability of the pseudospectral method for the FTS of more general flows with chemical reaction.

In random vortex methods, the third category, one solves for the turbulent motion by calculating the velocity field induced by a discrete number of vortex elements. The method is Lagrangian, since the vortex elements are not restricted to a computational grid but are convected by the velocity field they induce. The new location of the elements is used to update the velocity field on each iteration. Random vortex methods have been used in the study of two-dimensional turbulent premixed flames by Ashurst et al. (1986) and the study of two-dimensional turbulent diffusion flames by Ghoniem and Givi (1987). While the dynamics of the velocity field are calculated in a Lagrangian frame, the method for calculating the scalar field may be either Eulerian or Lagrangian. Ghoniem and Givi used a Lagrangian method and represented scalar fields by discrete scalar elements. Ashurst et al. calculated scalar variables on an adaptive Eulerian grid. In both cases the heat release from the reaction was considered negligible and did not affect the flow. The main difficulties with random vortex methods lie in the modeling of viscous and diffusive effects. In the present authors' opinions, the method has potential for high Reynolds and Damköhler numbers, where viscous and diffusive effects are confined to small portions of the computational domain, and so all points in the flow need not be computed. A recent review by Leonard (1985) discusses the use of random vortex methods for three-dimensional flows.

Comments on pseudospectral methods

Since we make use of pseudospectral methods in the present study, it is important to keep their general usefulness in perspective. The main advantage of pseudospectral methods for incompressible flows is that the amount of machine computing time appears to be much less than for other methods to achieve comparable accuracy in a simulation, primarily because a Poisson equation does not need to be solved at each time step. Pseudospectral methods are currently the best choice for the full simulation of incompressible, homogeneous turbulent flows. There are difficulties, however, associated with their application to reacting flows or to flows of any practical consequence.

Some areas that cause problems with pseudospectral methods, or problems in dealing with the results of pseudospectral simulations, are: (1) realistic boundary conditions, (2) flows with variable physical properties, (3) compressible flows, (4) resolution of discontinuities, (5) storage and interpretation of large data sets, and (6) statistical sample size. The last two problem areas are common to all FTS methods. We discuss each of these points in turn.

1. **Realistic boundary conditions.** The problem concerning boundary conditions was introduced in the previous section. Full turbulence simulations with spectral methods have so far been limited to flows with periodic or no-slip boundary conditions, because fast transform methods can be used with Fourier and Chebyshev expansions. As a consequence, temporally developing flows are studied instead of spatially developing flows, although all numerical methods have difficulties with turbulent inflow and outflow conditions. In the simulation of an open domain,

information is being transferred between a finite computational domain and the part of the universe that is not being calculated. Some approaches for realistic boundary conditions that have been used in non-reacting flows are such hybrid methods as the spectral element method (Korczak and Patera 1986) and a combined finite-difference/spectral method (Lowery et al. 1987).

2. **Variable physical properties.** As an example of the problem with variable physical properties, consider the viscous terms in the momentum conservation equation. The divergence of the viscous part of the stress tensor,

$$\nabla \cdot \tau = \nabla \cdot \left[\mu \left(\nabla \mathbf{u} + \nabla \mathbf{u}^T \right) \right] + \nabla \left[(\kappa - \frac{2}{3}\mu) \nabla \cdot \mathbf{u} \right], \tag{5}$$

becomes $\mu \nabla^2 \mathbf{u}$ for an isochoric motion in a fluid with constant viscosity. This term is evaluated in transform space and can be integrated integrated exactly when a Fourier expansion is used. If the coefficients of viscosity are functions of position through temperature dependence, and if a pseudospectral method is used, then 9 additional transforms to physical space (one each for μ, κ, and $\nabla \cdot \mathbf{u}$ and 6 for the symmetric tensor $\nabla \mathbf{u} + \nabla \mathbf{u}^T$) and 7 additional transforms back to Fourier space are needed in order to evaluate the nonlinear viscous terms.

3. **Compressible flow.** Compressible flows usually require different techniques of solution than incompressible flows, because the nature of the governing equations is different. The compressible Navier-Stokes equations are a set of hyperbolic-parabolic equations, while the incompressible Navier-Stokes equations are a set of elliptic-parabolic equations (Anderson et al. 1984). The problem with spectral solutions for compressible flows is that disturbances are felt instantaneously throughout the computational domain when variables are expanded in global basis functions, whereas a finite rate of propagation is required for the solution to match physical phenomena. The speed of sound is infinite in an incompressible fluid, so the global basis functions are a natural representation for this case. Spectral methods have been used, however, in the calculation of some compressible flows (*e.g.*, Hussaini et al. 1985, Feiereisen et al. 1981).

4. **Discontinuities.** The use of global basis functions also limits the resolution of discontinuities and steep gradients. Gibbs ringing contaminates the solution of sharp gradients if the resolution is inadequate. Sharp concentration gradients are produced if the reaction rate coefficients are large, with the limiting case of an infinitely fast reaction giving a discontinuous surface separating the reactant species in the case of nonpremixed flows. Riley et al. (1986) have studied the accuracy of a pseudospectral method for a test case including a chemical reaction. The simulation retained spectral accuracy, provided that the temporal and spatial

resolution are both adequate. The need for a FTS to resolve all scales of both the velocity and scalar fields restricts both the turbulent Reynolds number and the Damköhler number. Jou and Riley (1987) estimate an upper limit on the Damköhler number of about 30 for FTS's.

5. **Storage and interpretation.** Pseudospectral simulations, like all simulation methods, yield a solution of the conservation equations at every grid point. The sizable amount of information from a direct simulation must be interpreted graphically or be reduced by averaging. In addition, the visualization of three-dimensional vector and tensor fields is not a trivial problem. The data fields are best viewed interactively, but data storage for interactive visualization after the completion of a simulation can be expensive, especially if one is interested in the time-dependent behavior of the data (Leonard and Hill 1987b). The collection of statistical information from the simulation data is normally done during the simulation, and the full data sets are rarely archived. It is usually cheaper to duplicate a simulation if different statistics are studied than it is to store enough data to analyze at later times. A review by Hesselink (1988) covers the techniques available for the interpretation of both laboratory and numerical data. One of the more novel techniques presented is the use of holography to display data.

6. **Sample size.** Despite the large amount of data that is generated by FTS's, each simulation is only one member of an ensemble of numerical experiments. Dynamical systems theory tells us that the solutions to the Navier-Stokes equations are sensitive to initial conditions. Statistical averages should be made over an ensemble of experiments but, because a sufficient number of experiments would be excessively expensive, a quasi-ergodic assumption is made, and averages are taken over space (when homogeneous) or time (when stationary). In the case of spatial averaging, the size of the largest structures in the flow must be a small fraction of the computational domain, if the statistics are to be meaningful. The smallest scales of motion must still be resolved, and so the range of scales—and hence the Reynolds number—is limited by the resolution. In the case of time averaging, the duration of the simulation must be long enough to make a large number of measurements, but then the sensitivity of the differential equations to initial conditions casts doubt on the degree to which the computation is the "desired" solution. Also, in the case of decaying turbulence, there are time limitations imposed by the growth of spatial scales beyond the size of the domain.

Despite the fact that practical combustion computations are not currently possible with spectral methods due to the difficulties outlined above, these methods are very useful for studying isolated phenomena in reacting flows, for serving as complements to laboratory experiments, and for the evaluation of turbulence closure theories. An example of the application of a pseudospectral simulation to a reacting turbulent flow is discussed in more detail in the following section. We are able to use the results of a FTS to test simple single-point closure theories for the reaction term in the dynamical equations of a simple chemical reaction and also to examine the effect of the reaction rate coefficient on important statistical quantities, such as the length scale for the dissipation of concentration fluctuations.

An application of full turbulence simulation to a reacting flow

An example of the application of a FTS to chemically reacting flows is the study of an isothermal, irreversible, second-order, two-species reaction of the form

$$A + B \to \text{products}.$$

The decaying velocity field is spatially homogeneous with zero mean. The reactant species are segregated initially; hence this application can be thought of as a highly simplified diffusion flame. Another way of looking at this initial value problem, is that of a turbulent multi-jet plug flow reactor with a reference frame moving at the mean velocity (see Figure 1, Hill 1976). The objective of the present study is to describe the effects of turbulent motion on the mean reaction rate, and attention is focused on the statistical nonlinearity arising from the reaction term.

The velocity field is governed by the Navier-Stokes equation and the incompressibility condition

$$\frac{\partial \mathbf{u}}{\partial t} + \mathbf{u} \cdot \nabla \mathbf{u} = -\frac{\nabla p}{\rho} + \nu \nabla^2 \mathbf{u}, \tag{6}$$

$$\nabla \cdot \mathbf{u} = 0. \tag{7}$$

The concentration of species A is governed by the mass conservation equation

$$\frac{\partial A}{\partial t} + \mathbf{u} \cdot \nabla A = D \nabla^2 A - R_A, \tag{8}$$

where R_A is the rate of consumption due to reaction

$$R_A = k_R A B. \tag{9}$$

An equivalent equation governs the decay of species B. It is assumed that the molecular diffusion coefficients D of species A and B are the same constant values. The rate constant k_R is also assumed to be constant.

Table 1. Conditions for the simulations.

Run:	A	B	C	D
Parameters:				
ν	0.04	0.04	0.04	0.025
D	0.1	0.1	0.1	0.036
k_R	10	5	1	10
Dimensionless groups:				
R_λ	19	19	19	30
Sc	0.4	0.4	0.4	0.7
Da_I	10	5	1	10
Da_{II}	60	30	6	170
Initial values common to all runs:				
u'	L_f	$\bar{A}_0, \bar{B}_0$	$\overline{a_0^2}, \overline{b_0^2}$	$\overline{ab}_0$
1.0	1.0	1.0	0.86	−0.86

Note: $Da_I \equiv k_R \bar{A}_0 L_f / u'$ and $Da_{II} \equiv k_R \bar{A}_0 \lambda^2 / D$, where u' is the initial turbulence intensity and where L_f and λ are the initial longitudinal integral scale and Taylor microscale, respectively. R_λ and Sc are defined to be $u'\lambda/\nu$ and ν/D, respectively.

Numerical method

A pseudo-spectral method based on a code developed by Kerr (1985) was used to integrate the conservation equations for the velocity and three scalars (*i.e.*, the concentration of reactants A and B and a nonreacting species that has the same initial conditions as A) using expansions in Fourier series and third-order Runge-Kutta timestepping in Fourier space. The spatial resolution of the calculations was 32^3 Fourier modes. A constant timestep of 0.033, in arbitrary units consistent with those in Table 1, was used to integrate the equations. The initial Courant number was about 0.4 for these simulations. The spatial domain is a cube with sides of length 2π, thus giving integer values for wavenumber k. The velocity field decayed freely from an initially Gaussian energy spectrum, defined with a peak wavenumber k_0 of 2.50, while the initial concentrations of the reactant species were defined as

$$A(\mathbf{k}, 0) = \begin{cases} 1 & \text{if } \mathbf{k} = \mathbf{0}; \\ A_0(k_1) & \text{if } k_1 \text{ odd and } k_2, k_3 = 0; \\ 0 & \text{otherwise}; \end{cases} \quad (10a)$$

and

$$B(\mathbf{k}, 0) = \begin{cases} 1 & \text{if } \mathbf{k} = \mathbf{0}; \\ -A_0(k_1) & \text{if } k_1 \text{ odd and } k_2, k_3 = 0; \\ 0 & \text{otherwise}; \end{cases} \quad (10b)$$

where

$$A_0(k_1) = (-1)^{(k_1-1)/2} 2 e^{-k_1^2 D_A t^*} / \pi k_1. \quad (10c)$$

This initial scalar field corresponds to spatially segregated slabs of reactants parallel to the x_1 plane, such that B occurs in the center half of the domain and A lies to the sides, as indicated in Figure 1. Diffusion is allowed to act alone for a small time t^* before the simulation begins, in order to smooth out the Gibbs ringing in the initial condition. The concentrations are not required to be nonnegative, as in the simulations by Riley et al. (1986), but the concentrations are initially out of phase and the diffusivities of the species are equal, and therefore computational instabilities do not occur.

Initial conditions and parameters for each run are given in Table 1. The initial velocity and scalar fields were identical for each run, only the parameters were different for each run. The values of Damköhler numbers, based on both convective and diffusive time scales, are given in the table. Any reference to Damköhler number in this paper is to Da_I, unless noted otherwise. Each simulation was carried out for 60 steps, to a dimensionless time of 2, based on the initial conditions. Each calculation used about 80 minutes of machine time on a NAS 9160 computer. The routines used for the Fast Fourier transform were not vectorized on this machine. This amount of time is not excessive, since the entire study could performed with a Cray X-MP in 30 minutes of cpu time.

Statistical equations

The traditional approach to the solution to turbulent flow problems is to use averaging techniques and then to model the effects of the turbulence in the averaged dynamical equations. Modeling methods for reacting turbulent flows are not well developed, and the results of FTS's can be used to evaluate existing models and to aid in the development of future models. The statistical equations for the particular reaction kinetics used in the simulation are presented below.

The moment equations for the mean and variance of reactant A and the covariance of reactants A and B can be obtained by averaging the conservation equations and imposing the homogeneity condition, giving

$$\frac{d\overline{A}}{dt} = -k_R \left(\overline{A}\,\overline{B} + \overline{ab} \right) \tag{11}$$

$$\frac{d\overline{a^2}}{dt} = -2D\overline{|\nabla a|^2} - 2k_R \left(\overline{a^2}\,\overline{B} + \overline{ab}\,\overline{A} + \overline{a^2 b} \right) \tag{12}$$

$$\frac{d\overline{ab}}{dt} = -2D\overline{\nabla a \cdot \nabla b} - k_R \left(\overline{a^2}\,\overline{B} + \overline{ab}\,\overline{A} + \overline{a^2 b} + \overline{b^2}\,\overline{A} + \overline{ab}\,\overline{B} + \overline{ab^2} \right) \tag{13}$$

where $A = \bar{A} + a$, and where the overbar denotes volume-averaging. Equations for the mean and variance of reactant B can be obtained by interchanging the variables A and B, and a and b. The set of first- and second-order moment equations contains five differential equations, if those for $\overline{B}$ and $\overline{b^2}$ are included, and ten unknown terms, and is therefore not a closed system.

The closure problem due to the reaction term can be avoided by considering the joint pdf of the reacting species at a point. The equation for the joint pdf of reactants A and B then contains unknown terms due to molecular diffusion, and so one type of stochastic nonlinearity has been traded for another. One significant problem with the pdf method is the high dimensionality of the formulation, especially for multi-point densities. Cumulative probability distributions can be measured from FTS data by averaging Heaviside step functions (Pope 1985); for the single point joint distribution of species A and B, for example, this is given by the equation

$$F(\alpha,\beta) = \frac{1}{N^3} \sum H(\alpha - A)H(\beta - B), \qquad (14)$$

where the sum is taken over the N^3 spatial points at which A and B are evaluated. The evaluation of n-point distributions is expensive when $n > 1$, because the summation in Equation (14) is over all combinations of n points in the numerical data (Leonard and Hill 1987b), and the number of calculations thus increases as N^{3n}. Consequently, if a two-point joint distribution were to be estimated, there are N^6 points to evaluate using Equation (14), and this is a formidable computational problem.

A hybrid model proposed by Eswaran and O'Brien, (1987) has advantages of both the moment and pdf methods. This model uses the pdf method to treat the reaction exactly, uses EDQNM theory to describe the dynamics of the turbulence and scalar mixing, and is less expensive to evaluate than full multi-point pdf methods. We are currently studying this model but will not discuss it further in this paper. Further discussion of the applications of statistical theory to turbulent reacting flow can be found in the reviews by Hill (1976), Pope (1985), and O'Brien (1986).

Closure methods

The first of two closure methods that are tested here with data from the numerical simulations was proposed by Toor in 1969. The covariance of the reactant concentration fluctuations for the case of two unmixed streams can be related to the intensity of segregation of a non-reacting scalar in the limits of very high and very low Damköhler numbers. The theory is restricted to stoichiometric mixtures of reactants with equal mass diffusivities, and it assumes that the pdf of the conserved scalar, $A - B$, is Gaussian in shape.

For very low Damköhler numbers the covariance ratio, $\psi^2 = \overline{ab}(t)/\overline{ab}_0$, is equal to the variance ratio, $\sigma^2 = \overline{a^2}(t)/\overline{a_0^2}$, of either reactant or of the conserved scalar, $A - B$. For very high reaction rate both the mean concentration and—because the reactants remain segregated—the covariance can be related to the variance of a nonreacting scalar. The relationship is the same for both extremes and led Toor to propose that the equation $\psi^2 = \sigma_m^2$ holds for any intermediate reaction rate. The subscript m indicates the variance ratio for a non-reacting tracer species. Kosály (1987) has shown that this relationship is only correct if the pdf of the conserved scalar is initially Gaussian. (Actually, a simple mathematical argument shows that it is only necessary that the pdf of the conserved scalar be of the same form at the initial and later times, and that it scale with the variance of the conserved scalar.) The pdf of the conserved scalar approaches a Gaussian form asymptotically (Eswaran and Pope 1987, Givi and McMurtry 1987), and corrections must be made if the initial form is other than Gaussian. When the initial pdf is a bimodal delta function, the mean concentration is overestimated by a factor of 1.25 in the fast reaction limit, and Toor's relationship becomes $\psi^2 = 2\sigma_m^2/\pi$.

An alternative closure method is due to Patterson (1981). Simple concentration profiles or, equivalently, a simple degenerate form of the joint single-point pdf of reactant concentrations, are assumed. The knowledge of the profiles or pdf allows any single-point moments of the reactant concentration to be calculated in terms of the two means and two variances.

Table 2. Summary of closure theories for terms due to chemical reaction.

unknown term	closure	author
$\overline{ab}$	$\overline{ab}_0(\overline{a^2}/\overline{a_0^2})_m$	Toor
$\overline{ab}$	$-\overline{a^2}\,\overline{b^2}/(\bar{A}\bar{B})$	Patterson
$\overline{a^2 b}$	$-\left(\overline{a^2}^2/\bar{A}\right)\left(1 - \overline{a^2}/\bar{A}^2\right)$	Patterson

Using this method, the unknown term, $\overline{ab}$, in Equation (11) for the mean concentrations can be replaced with a form shown in Table 2. When Patterson's method is used, the equations for the variance of the concentrations, including Equation (12), must be solved in addition to the equation for the mean. The additional unknown term due to reaction, $\overline{a^2 b}$, can be closed using the assumed pdf with the form given in Table 2, but Patterson recommends using $\overline{a^2 b} = 0$ to improve agreement with experiment. We have used $\overline{a^2 b} = 0$ when testing the theory by integrating the closed forms of Equations (11) and (12). The dissipative term in the variance equation is still not closed. Corrsin's model (1964) for the decay of the variance in an isotropic mixer with no reaction is recommended by Patterson for this term. This model uses simple forms of the spectra for energy and concentration fluctuations to approximate the concentration microscale. The forms assumed for $Sc \leq 1$ give

$$\lambda_A^2 = \frac{24D}{(3-Sc^2)}\left(\frac{5L_A}{\pi\sqrt{\varepsilon}}\right)^{2/3}, \tag{15}$$

where $\overline{|\nabla a|^2} = 6\overline{a^2}/\lambda_A^2$, ε is the rate of dissipation of kinetic energy per unit mass, and L_A is an integral scale for the concentration field. Instead of using Corrsin's equation, we chose to model the dissipation term with a constant scalar microscale. This seems to fit the results of the simulations fairly well, and in any case, any theory based on a fully-developed isotropic turbulent scalar field would not be expected to correlate the data from the simulation, because of the high degree of anisotropy associated with the initial scalar field.

Results of the simulations

Before we present the results of tests of the closure theories, we show some "snapshots" of the concentration fields as they develop for one of the runs, and show the effect of Da on the behavior of statistics of the concentration fields. The statistical quantities reported here are evaluated using volume averaging, except in Figure 5, despite the fact that the concentration fields are inhomogeneous in the x_1 direction. The net transport in the x_1 direction is zero, however, because of the periodic boundary conditions. The closure theories that are tested in this study attempt to predict the average reaction rate, and so volume averaging is an appropriate treatment.

Instantaneous values of the concentrations of each reactant in a cross section of the physical domain are shown in perspective plots at several times in Figure 1 for Run A. The time is made dimensionless with an eddy turnover time based on initial values of u' and L_f. The unaveraged rate of molecular dissipation of the concentration fluctuations of species A, $\varepsilon_A = 2D|\nabla a|^2$, and the rate of consumption of reactant A due to chemical reaction, $R_A = k_R A B$, are shown in Figure 2 for the same values of x_3 and time as for the concentrations in Figure 1. The reacting species remain highly segregated during the simulation because of the high Damköhler number. The reaction occurs in the two fronts, or reaction zones, that exist between the regions containing each reactant. (An even number of zones is needed because of periodic boundary conditions.) These reaction zones can be seen as the regions where $k_R A B$ is nonzero in the perspective plot of the local reaction rate.

The dissipation rate of the fluctuations of concentration of species A are nonzero in the regions next to the reaction zone, on the side containing species A. The scalar dissipation rate is highly correlated with the reaction rate for this case, where the

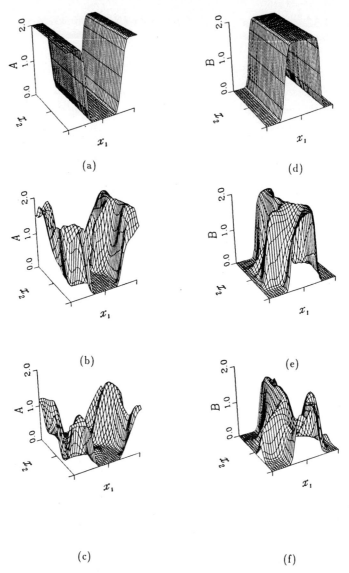

Figure 1. Evolution of reactant concentration in the plane $x_3 = 0$ for $Da = 10$ and $R_\lambda = 19$ (Run A). (a)-(c), Reactant A at times $t = 0$, $t = 1$, $t = 2$, respectively; (d)-(f), Reactant B at times $t = 0$, $t = 1$, $t = 2$, respectively.

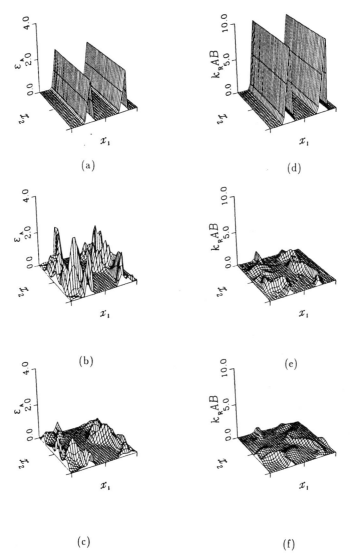

Figure 2. Evolution of unaveraged scalar dissipation rate and reaction rate in the plane $x_3 = 0$ for $Da = 10$ and $R_\lambda = 19$ (Run A). (a)-(c), Unaveraged scalar dissipation rate at times $t = 0$, $t = 1$, $t = 2$, respectively; (d)-(f), Rate of reactant consumption at times $t = 0$, $t = 1$, $t = 2$, respectively.

reactants are not mixed, because the concentration values change significantly only in the fairly thin reaction zones. The dissipation rate for reactant B is not shown, but it has a similar structure to that for reactant A, but with the nonzero regions located on the side of the reaction zone opposite A. The magnitudes of the gradients of reactant concentration are increased, and the molecular diffusion is thereby enhanced, by the stretching and folding action characteristic of turbulence. The convolution of the reaction zones by the turbulent velocity field can be seen, to some extent, in the perspective plots in Figures 1 and 2. The use of three-dimensional contour surfaces of reactant species in another study (Leonard and Hill 1987b) shows that structures of the velocity field resembling vortex rings or horseshoes may be responsible for the structure of the reaction zone. The present study does not examine how different velocity fields affect the reaction rate; rather, different reaction rate coefficients are used here for identical flows, and the effect of varying the reaction rate coefficient on the statistics of the concentration fields is examined.

The instantaneous values of the concentration and the reaction rate for species A (Figure 3) in Run D, which has the same initial value of Da_I but a different initial value of Da_{II}, illustrate the previously addressed problem of the resolution of steep gradients. The same initial velocity field is used in each run, but the viscosity and molecular diffusivity are different in Runs A and D. The largest scales of motion are those least affected by viscous dissipation, and, therefore, the structure of the concentration field looks similar. The simulation is not able to capture the smallest scales of the concentration field, however, because the spatial resolution is inadequate for the value of Da_{II} that was used in this run. The Gibbs error in the representation of the concentration field causes spurious negative values for the concentrations of the reactants and, therefore, negative values of the reaction rate. The data for the chemical reaction rate in the higher Reynolds number simulation are therefore suspect, and so the data for only the lower Reynolds number simulations are used here to test the closure theories.

The reaction rate is very high within the reaction zone at the beginning of the simulation (Figure 2), and the initial stage of the reaction is therefore dominated by the initial conditions for the concentration fields. The evolution of the mean concentration of reactant species is not influenced by the velocity field in this stage. The segregation coefficient, $\zeta = \overline{ab}/\bar{A}\bar{B}$, for the reactant species decreases rapidly in a few time steps from $\zeta = -0.85$ to $\zeta = -0.95$ and then remains nearly constant for the case with the highest Damköhler number. The reaction is limited by diffusion after the initial stage for this high reaction rate coefficient, but not to such an extent that a conserved scalar approach (Bilger 1980) can be applied.

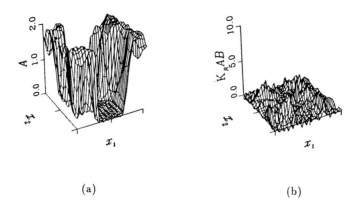

Figure 3. (a) Reactant A concentration and (b) reaction rate for $Da_I = 10$ and $R_\lambda = 30$ (Run D), in the plane $x_3 = 0$ at time $t = 1$.

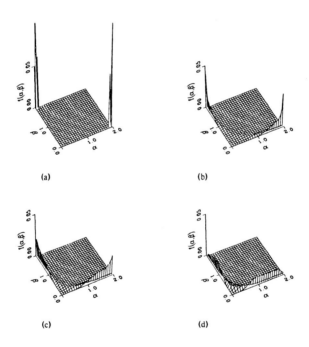

Figure 4. Evolution of joint single-point pdf for reactant concentrations at times $t = 0, 0.5, 1.0$, and 1.5, respectively, for a simulation with $Da = 10$.

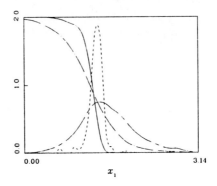

Figure 5. Local values and planar averages of the concentration of reactant A and the reaction rate. ———, $A(x_1,0,0)$; ——·——, $\bar{A}(x_1)$; — — —, $k_R ab(x_1,0,0)$; — — —, $k_R \overline{ab}(x_1)$.

The single-point joint pdfs of reactant concentrations were calculated at various times from simulation data for $Da = 10$ (Figure 4). The shape of the pdf is not the idealized form used in Patterson's closure, which consists of three delta functions located at the points $(\alpha, \beta) = (A^*, 0)$, $(0, B^*)$, and (A^*, B^*) in probability space. The closure implies that the concentration of a species in the reaction zone is the same value as the concentration in the region containing only that species. The results of the simulation show that the concentrations of the reacting species are much less in the reaction zone than are assumed in Patterson's closure, and the average reaction rate is therefore overestimated by the closure, because the delta function at (A^*, B^*) in the joint pdf does not actually exist. This shortcoming of the closure theory is reflected in the tests discussed below.

Local values of reactant concentrations along the line x_1 at $x_2, x_3 = 0$, are compared to averages over x_2, x_3 planes in Figure 5, for a 64^3 calculation with conditions similar to those used in Run D. This demonstrates the loss of information that occurs when statistical methods are used. The magnitude of the mean scalar gradient is less than the maximum local value of the gradient, and the reaction zone is much thinner than the average reaction rate values would indicate. Overestimation of the reaction concentrations in the reaction zone by Patterson's closure method is also apparent here. The Gibbs errors in the reaction rate term are obviously less pronounced than those in Figure 3.

The time dependence of the mean and variance of the concentration of reactant A in Figure 6 shows the effect of the reaction rate coefficient on the progress of the reaction and the amount of mixing. The small change in the mean concentration

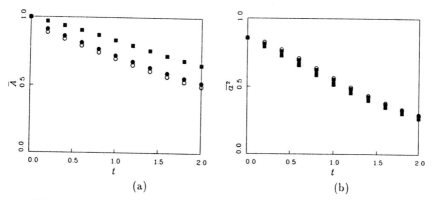

Figure 6. (a) Mean and (b) variance of the concentration of reactant A as a function of time. The mean concentration does not change for $Da = 0$, and, therefore the results have not been plotted. All four data points in (b) at each time have been plotted, but it is hard to distinguish between the points. □, $Da = 0$; ■, $Da = 1$; ○, $Da = 5$; ●, $Da = 10$.

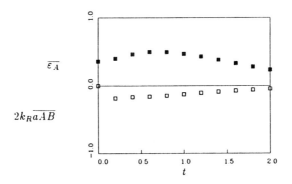

Figure 7. Contributions to the rates of change of concentration variance due to dissipation (■) and to reaction (□) for Run A.

when the reaction rate coefficient is doubled illustrates the fact that the reaction has approached a diffusion-limited case. The reaction rate coefficient has little net effect on $\overline{a^2}$. The relative contributions to the rate of change of the variance due to diffusion and to reaction are shown for Run A in Figure 7. The scalar dissipation term dominates the reaction term in Equation (12). The reaction acts to increase the concentration variance by maintaining the segregation between the reactants, while the dissipation of

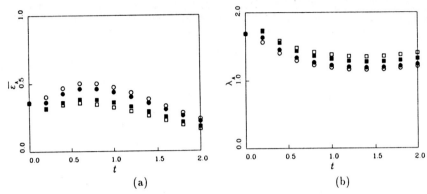

Figure 8. (a) Scalar dissipation rate and (b) microscale for the dissipation of fluctuations of reactant A. □, $Da = 0$; ■, $Da = 1$; ○, $Da = 5$; ●, $Da = 10$.

concentration fluctuations is increased by the larger magnitudes of the concentration gradients resulting from the segregation.

The evolution of the volume-averaged dissipation rate of the concentration fluctuations and the length scale for scalar dissipation are shown in Figure 8 for each reaction rate constant. The Damköhler number has a small effect on the dissipation rate, but the microscale for scalar dissipation is nearly insensitive to the reaction rate coefficient, except in the early chemical kinetics-controlled stage of the reaction.

Tests of the closure theories

The numerical experiments were used to test the single-point moment closures of Toor and Patterson in two ways. First, the results of the simulation for the means and variances were used to directly compare the closure forms given in Table 2 with the measured values for the unknown terms. Second, the initial conditions for the simulation were used with Equations (11) and (12) and the closure forms in Table 2 to integrate the moment equations, in order to compare predictions of the closures with results from the numerical experiments.

The covariance of the reactant concentration fluctuations is better evaluated by Toor's method than by Patterson's. Figure 9 shows the ratio of the value of the covariance evaluated with FTS data substituted directly into the forms from the closure theories, as given in Table 2, to the value from the simulations. The correction for the shape of the pdf of the conserved scalar (Kosály 1987) has not been included in Toor's theory, as it can only be determined when the shape of the pdf of the conserved scalar

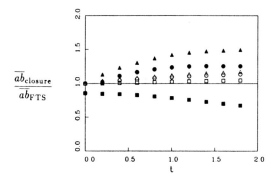

Figure 9. Ratio of the approximated form of the covariance, for each closure method, to the measured data. ■, •, ▲ Patterson's form (i.e., $-\overline{a^2}\,\overline{b^2}/(\bar{A}\bar{B}\overline{ab})$) for $Da = 1$, 5, and 10, respectively. □, ○, △ Toor's form (i.e., $(\overline{ab}_0/\overline{ab})(\overline{a^2}/\overline{a_0^2})_m$) for $Da = 1$, 5, and 10, respectively.

data is, therefore, expected, especially at higher values of Da. Patterson's method should be especially accurate at high values of Da because of the segregation of reactant species, but the predictions are not as accurate as those of Toor. The accuracy of the method does not improve as the reaction rate increases, which is not expected. This is a disappointment because Patterson's theory corresponds to a somewhat plausible physical model, whereas Toor's theory does not, at least for finite values of Da.

The closure theories were also used to integrate the moment equations for the concentrations of the reacting species, with the same the initial conditions that were used in the simulation. Toor's method was used to close Equation (11) and Patterson's method was used to close Equations (11) and (12). A constant microscale was used to close the dissipation term in Equation (12) and an equivalent equation for the variance of a nonreacting scalar. The results are compared with results of the simulation (Figure 10) for Run A. The agreement is much better for Toor's closure at these conditions. The mean concentration is overestimated by Toor's method because of the change in the shape of the pdf of the conserved scalar. The mean concentration is underestimated by Patterson's model because the mean reaction rate is overestimated.

Predictions of Patterson's closure theory for the contribution of reaction to the decay of the concentration variance were compared to the actual values. The terms in Equation 12 that involve k_R were evaluated with the data for the means and variances

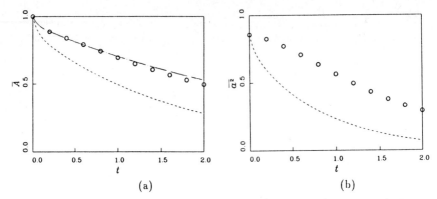

Figure 10. Evaluation of the (a) mean and (b) variance of reactant A by integration of the moment equations for the conditions of Run A. The unknown terms due to reaction were closed using the methods of Toor (———) and Patterson (- - -). The results are compared to FTS data (o).

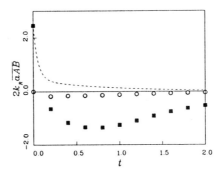

Figure 11. Contribution of the chemical reaction term to the rate of change of the variance of reactant A (o, FTS data). The term is evaluated with Patterson's closure in two ways. First, the means and variances from integration of the moment equations for the conditions of Run A are used in the closure forms of Table 2 (- - -). Second, the FTS data were used in the closure forms (■).

from both the integrated moment equations and the FTS solution. The terms due to reaction are not modeled accurately in Patterson's method (Figure 11). The predicted rate of change due to reaction is initially incorrect. The actual rate of change from the terms that Patterson models is initially zero and acts to increase the concentration variance throughout the rest of the simulation. The terms modeled by Patterson predict an accelerated rate of mixing when the integrated values of the mean and variance are used and, thus, a corresponding accelerated rate of consumption of reactant species. The use of FTS data in the modeled terms shows that Patterson's simple closure does not hold at any time during the simulation. The predictions of the mean and variance were not improved when the value of the scalar microscale from the simulation, rather than a constant value, was used to close the dissipation term.

While the errors in the estimation of the unknown term in Equation (11) are relatively small for moderate Da (Figure 9), the effect of the chemical reaction on the concentration variance is poorly modeled. The need to solve simultaneous equations for the means and variances results in poor prediction of the mean when the approximations in the rate of change of the variance are in error. Accurate modeling of behavior due to chemical reaction and molecular diffusion is seen to be crucial to correctly predicting the behavior of reacting turbulent flows.

Conclusions

Methods that are being used in the full turbulence simulations (FTS's) of nonreacting flows, and the applications of these methods to reacting flows, are briefly reviewed in this paper. While the pseudospectral method is the most popular, the flows that can be solved with it are limited, partly because of restrictions on boundary conditions, and partly because of the difficulty of dealing with variable-property flows. Spectral element methods can potentially be applied to more general flows than can the pseudospectral method. Finite-difference and finite-element methods are not generally used for FTS: these methods are more commonly used for large-eddy simulations and other modeled equations. Random vortex methods may be useful in the future for high Reynolds number reacting flows.

The pseudospectral method can be used for FTS's if one is aware of the limitations. The simulation of flows of practical interest is presently beyond the reach of researchers; however, full turbulence simulation is valuable as a complement to laboratory experimentation and for examining isolated phenomena.

The results of FTS's have been used in this study to test two single-point closure methods for unknown terms in the set of moment equations for a second-order, two-species reaction. The method proposed by Toor is more accurate than the method proposed by Patterson. Although Patterson's method is based on a physical model, this model is an inaccurate description of the reaction zone, and the predictions of the means and variances of reactant concentrations are, therefore, in error. Toor's theory is only valid in the limits of very low and very high reaction rates, and the restrictions for Toor's theory to hold for very fast reactions were not met by the initial conditions of this study. Nevertheless, the method is better than Patterson's for the conditions that we have used here.

The results of the simulations show that the microscale for the rate of scalar dissipation is only a weak function of Da. This is somewhat surprising, since the reaction rate and the scalar dissipation rate are strongly correlated; this fact may be important in future modeling of reacting flows.

Limitations on the Damköhler number were noted in this study, and it was found that the value of Da_{II} proved to be crucial in determining whether or not steep gradients could be resolved adequately. The simulation was able to accurately solve the detailed equations for cases up to $Da_I = 10$ and $Da_{II} = 60$.

Acknowledgements

The support of the Iowa State University Computation Center, the National Science Foundation (Grant ECS 8515047), the Minnesota Supercomputer Center, and the Iowa State Engineering Research Institute are appreciated, as is the fellowship support provided A. D. Leonard by the Phillips Petroleum Company.

References

Anderson, D. A., Tannehill, J. C. and Pletcher, R. H. (1984). *Computational Fluid Mechanics and Heat Transfer*. McGraw-Hill: New York.

Ashurst, W. T., Peters, N. and Smooke, M. D. (1986). Numerical simulation of turbulent flame structure with non-unity Lewis number, *Combust. Sci. Tech.* **53**, 339–375.

Bilger, R. W. (1980). Turbulent flows with nonpremixed reactants. In Libby, P. A. and Williams, F. A. (eds.) *Turbulent Reacting Flows*, Topics in Applied Physics, **44**, Springer-Verlag: Berlin, pp. 65–113.

Boris, J. P. and Book, D. L. (1976). Solution of the continuity equation by the method of flux-corrected transport, *Methods in Computational Physics* **16**, 85–129.

Corrsin, S. (1964). The isotropic turbulent mixer: Part II. Arbitrary Schmidt number, *AIChE J.* **10**, 870–877.

Eswaran, V. and O'Brien, E. E. (1987). Chemically reacting scalars in homogeneous, isotropic, turbulent flows. Submitted for publication.

Eswaran, V. and Pope, S. B. (1987). Direct numerical simulations of the turbulent mixing of a passive scalar. *Fluid Dynamics and Aerodynamics Report* FDA-87-12. Cornell University: Ithaca, New York.

Feiereisen, W. J., Reynolds, W. C. and Ferziger, J. H. (1981). Numerical simulation of a compressible, homogeneous turbulent shear flow. *Dept. Mech. Engng. Rep.* TF-13, Stanford University: Stanford, California.

Fujiwara, T., Taki, S. and Arashi, K. (1986). Numerical analysis of a reacting flow in H_2/O_2 rocket combustor. Part 1: Analysis of turbulent shear flow, AIAA Paper 86-0528.

Ghoniem, A. F. and Givi, P. (1987). Vortex-scalar element calculations of a diffusion flame, AIAA Paper 87-0225.

Givi, P. and McMurtry, P. A. (1987). Non-premixed simple reaction in homogeneous turbulence: Direct numerical simulation, *AIChE J.* To be pubished.

Gotlieb, D. and Orzsag, S. A. (1977). *Numerical Analysis of Spectral Methods: Theory and Applications.* SIAM: Philadelphia.

Hesselink, L. (1988). *Ann. Rev. Fluid Mech.*, to appear.

Hill, J. C. (1976). Homogeneous turbulent mixing with chemical reaction, *Ann. Rev. Fluid Mech.* **8**, 135–161.

Hussaini, M. Y., Salas, M. D. and Zang, T. A. (1985). Spectral methods for inviscid, compressible flows. In Habashi, W. G. (ed.) *Advances in Computational Transonics.* Pineridge: Swansea, Wales. pp. 875–912.

Hussaini, M. Y. and Zang, T. A. (1987). Spectral methods in fluid dynamics, *Ann. Rev. Fluid Mech.* **19**, 339–367.

Jou, W.-H. and Riley, J. J. (1987). On direct numerical simulations of turbulent reacting flows, AIAA Paper 87-1324.

Kerr, R.M. (1985). Higher-order derivative correlations and the alignment of small-scale structures in isotropic numerical turbulence, *J. Fluid Mech.* **153**, 31–58.

Korczak, K. Z. and Patera, A. T. (1986). An isoparametric spectral-element method for solution of the Navier-Stokes equations in complex geometry, *J. Comp. Phys.* **62**, 361–382.

Kosály, G. (1987). Non-premixed simple reaction in homogeneous turbulence, *AIChE J.* **33**, 1998–2002.

Leonard, A. (1985). Computing three-dimensional incompressible flows, *Ann. Rev. Fluid Mech.* **17**, 523–559.

Leonard, A. D. and Hill, J. C. (1986). Direct simulation of turbulent mixing with irreversible chemical reaction, *Proc. World Congress III of Chem. Eng.*, **4**, 177-80.

Leonard, A. D. and Hill, J. C. (1987a). A simple chemical reaction in numerically simulated homogeneous turbulence, AIAA Paper 87-0134.

Leonard, A. D. and Hill, J. C. (1987b). Direct numerical simulation of turbulent flows with chemical reaction. Second Nobeyama Workshop on Fluid Dynamics and Supercomputers. Nobeyama, Japan.

Lowery, P. S., Reynolds, W. C. and Mansour, N. N. (1987). Passive scalar entrainment and mixing in a forced, spatially-developing mixing layer, AIAA Paper 87-0132.

McMurtry, P. A., Jou, W.-H., Riley, J. J. and Metcalfe, R. W. (1986). Direct numerical simulations of a reacting mixing layer with chemical heat release, *AIAA J.* **24** 962–970.

O'Brien, E. E. (1986). Recent contributions to the statistical theory of chemical reactions in turbulent flows, *PhysicoChem. Hydrodyn.* **7**, 1–15.

Oran, E. S. and Boris, J. P. (1987). *Numerical simulation of reactive flow.* Elsevier, New York.

Oran, E., Young, T. R., Boris, J. P., Picone, J. M. and Edwards, D. H. (1982). A study of detonation structure: The formation of unreacted gas pockets, *Nineteenth Symp. (Int.) Combust.*, 573–582.

Orszag, S. A. (1972). Comparison of pseudospectral and spectral approximations, *Studies Appl. Math.* **51**, 253–259.

Patterson, G. K. (1981). Application of turbulence fundamentals to reactor modelling and scaleup, *Chem. Eng. Commun.* **8**, 25–52.

Peyret, R. and Taylor, T. D. (1983). *Computational Methods for Fluid Flow.* Springer-Verlag: New York.

Picart, J., Borghi, R. and Chollet, J. P. (1987). Numerical simulation of turbulent reactive flows. Sixth Symposium on Turbulent Shear Flows, Toulouse, France.

Pope, S. B. (1985). Pdf methods for turbulent reactive flows, *Prog. Energy Combust. Sci.* **11**, 119–192.

Riley, J. J., Metcalfe, R. W. and Orszag, S. A. (1986). Direct numerical simulations of chemically reacting turbulent mixing layers, *Phys. Fluids* **29**, 406–422.

Rogallo, R. S. (1981). Numerical experiments in homogeneous turbulence. *NASA TM* 81315.

Taki, S. and Fujiwara, T. (1981). Numerical simulation of triple shock behavior of gaseous detonation, *Eighteenth Symp. (Int.) on Combust.*, 1671–1681.

Toor, H. L. (1969). Turbulent mixing of two species with and without chemical reactions, *Ind. Eng. Chem. Fundam.* **8**, 655–659.

THE INTERACTION BETWEEN TURBULENCE AND CHEMISTRY
IN PREMIXED TURBULENT FLAMES

K. N. C. Bray*, M. Champion[†], Paul A. Libby**

* University Engineering Department,
Trumpington Street,
Cambridge, CB2 1PZ, England.

[†] ENSMA, Université de Poitiers,
Rue Guillaume VII,
86034 Poitiers; France.

** Department of Applied Mechanics and Engineering Sciences,
University of California San Diego,
La Jolla,
California, 92093, U.S.A.

Abstract

Recent studies of turbulent premixed combustion by the authors and their coworkers are reviewed with emphasis on two topics, related to the interaction between turbulence and chemistry. The first is the use of conditional averaging techniques to describe terms in the transport equations for Reynolds stress and flux components. Secondly a laminar flamelet description of the mean reaction rate terms is presented. Some problems which remain to be solved are identified.

Introduction

This paper reviews some recent work by the authors and their coworkers in the field of premixed turbulent combustion. Theories of premixed combustion find important applications for example in spark ignition engine design and in the assessment of explosion hazards. Also the premixed flame provides a convenient vehicle for testing general theories relating to turbulent reacting flows.

Various regimes of premixed turbulent combustion may be defined [1 - 3] in terms of the values of a turbulence Reynolds number and a Damköhler number, the latter being defined here as the ratio of a turbulent mixing time to a chemical time. In most practical problems the Reynolds and Damköhler numbers are both large in comparison with unity. A turbulent flame is then found to consist of thin reacting interfaces, which are distorted and convected by the turbulent flow, and which form the boundaries between regions of unburned reactants and fully burned products. Two dimensional images formed by laser tomography [4] graphically illustrate the validity of this description both for Bunsen type burners [5] and for spark ignition engines [6].

In this thin-flame burning regime the probability density function (p d f) of a progress variable, c, which is defined to be zero in the unburned reactants and unity in fully burned products, may be written

$$P(c; \underline{x}) = \alpha(\underline{x}) \delta(c) + \beta(\underline{x}) \delta(c-1) + \gamma(\underline{x}) f(c; \underline{x}) \quad (1)$$

Here α, β and γ are the probabilities of observing reactants, products and burning interfaces, respectively, at location $\underline{x}$; δ is the Dirac delta function, and $f(c; \underline{x})$ the p d f for the internal burning mode. In this notation the thin-flame burning regime is characterised by $\gamma \ll 1$. Numerous experiments [7, 8] illustrate this behaviour.

In an important portion of the thin-flame burning regime an additional constraint is met, namely, that the laminar flame thickness is smaller than the Kolmogorov microscale length of the turbulence. This is the laminar flamelet burning regime [9] in which the reacting interfaces separating reactant and product regions consist of strained laminar flames. The interior mode p d f $f(c; \underline{x})$ is then determined [9] from the internal structure of the laminar flame. Here we are mainly concerned with laminar flamelet combustion.

For large Damköhler number, D_a, the quantity γ is of order D_a^{-1} [1]. To the lowest order in a series expansion, we may set $\gamma \approx 0$ in all terms in the transport equations except where γ appears in combination with D_a. The product γD_a appearing in chemical reaction terms is of order unity and these terms must of course be retained.

Three aspects of the problem of describing the structure and properties of premixed turbulent flames may now be identified and addressed. The first is to describe the fluid mechanics of a medium consisting of regions or pockets of high and low density separated by thin interfaces. For this purpose we assume the interfaces to be of infinitesimal thickness so $\gamma \approx 0$. The formalism of conditional averaging is helpful here and this topic is reviewed in the following section. The second aspect of the general problem is to describe the statistical geometry of the wrinkled interface between reactant and product regions in order either to model the interface area or to obtain equivalent information. Again for this purpose the interface can usually be regarded as possessing negligible thickness, unless the local radius of curvature becomes comparable with the interface thickness. The third and final aspect is a description of the interface structure, and it is only here that finite γ effects appear.

A strained laminar flamelet approach [10, 11] is described. Because the flamelets are thin the time within which they adjust to flow field variations is assumed to be small. Thus only steadily strained laminar flames are considered and transient effects such as ignition and extinction are excluded. A "library" of strained laminar flame solutions is called for, including appropriately detailed descriptions of chemical kinetics and molecular transport. The turbulent flow field model employs information stored in the strained flame "library".

The present paper reviews progress in these various developments and identifies some problems which remain to be solved. These problems are listed at the end of the paper.

Turbulent Flow with Heat Release: Use of Conditional Averaging.

In the work reviewed here a second-order Reynolds stress, Reynolds flux model [12] is selected in which individual components of the Favre mean Reynolds stress tensor $\overline{\rho u_i'' u_j''}$ and flux vector $\overline{\rho u_i'' c''}$ are characterised by their own transport equations. A simplified thermochemical description is chosen [13] in which the instantaneous value of the progress variable $c(\underline{x}, t)$ determines the instantaneous density and temperature:

$$\frac{\rho_r}{\rho} = \frac{T}{T_r} = 1 + \tau c$$

where τ is a heat release parameter, defined from $1 + \tau = T_p/T_r$, and subscripts r and p denote reactants and products, respectively.

The full set of second-order transport equations may be found elsewhere: see, for example [12]. Here we list only two of these equations in order to illustrate the types of term which must be modelled. Terms whose representation will be discussed here are displayed on the right hand side of these equations. The equation for the mean progress variable $\tilde{c}$ is

$$\frac{\partial}{\partial x_k} (\overline{\rho} \tilde{u}_k \tilde{c} + \overline{\rho u_k'' c''}) = \overline{w} \qquad (2)$$

where the Reynolds number is assumed to be sufficiently large for molecular transport to be neglected and where $\overline{w}$ is the mean chemical source term for $\tilde{c}$. In a second-order closure the Reynolds flux $\overline{\rho u_i'' c''}$ is determined from the following transport equation:

$$\frac{\partial}{\partial x_k}(\bar{\rho}\,\tilde{u}_k\,\frac{\overline{\rho u_i" c"}}{\bar{\rho}}) + \overline{\rho u_k" u_i"}\,\frac{\partial \tilde{c}}{\partial x_k} + \overline{\rho u_k" c"}\,\frac{\partial \tilde{u}_i}{\partial x_k} + \overline{c"\,\frac{\partial \bar{p}}{\partial x_i}}$$

$$= -\frac{\partial}{\partial x_k}(\overline{\rho u_k" u_i" c"}) - \overline{c"\,\frac{\partial p'}{\partial x_i}} + (-\overline{u_i"\,\frac{\partial j_k}{\partial x_k}} + \overline{c"\,\frac{\partial \tau_{ik}}{\partial x_k}}) + \overline{u_i" w} \quad (3)$$

where τ_{ij} is the molecular stress tensor and j_i is the molecular flux vector of the progress variable. Favre averaging is employed in these equations. Thus $\tilde{c} = (\overline{\rho c})/\bar{\rho}$ and $c" = c - \tilde{c}$.

A second-order, Reynolds stress and flux model is selected because terms, representing the interaction between density fluctuations and mean pressure gradients, and leading to phenomena of non-gradient transport [14] and flame-generated turbulence production [15], then appear explicitly. For example, the term $\overline{c"\,\partial \bar{p}/\partial x_i}$ in (3) leads [14, 15] to non-gradient transport driven by the mean pressure gradient.

In this review we stress the value of conditional averaging for reactant and product regions in a turbulent flame. Thus for example we shall define the conditional mean velocity and turbulence intensity, $\bar{u}_r$ and $\overline{u'^2_r}$ respectively, in reactant regions where $c = 0$, together with similar quantities, $\bar{u}_p$ and $\overline{u'^2_p}$, in product regions where $c = 1$. These mean quantities are all functions of position $\underline{x}$.

Other than the chemical heat release terms, $\bar{w}$ and $\overline{u_i" w}$, which will be considered later, the full set of second-order Reynolds stress and flux transport equations contain three types of term for which closure assumptions must be made [12]:

a) third-order terms: $\overline{\rho u_i" u_j" c"}$ and $\overline{\rho u_i" u_j" u_m"}$;

b) molecular dissipation terms: $\overline{u_i"\,\partial \tau_{jk}/\partial x_k}$, $\overline{u_i"\,\partial j_k/\partial x_k}$,

$\overline{c"\,\partial \tau_{ik}/\partial x_k}$;

c) Pressure fluctuation terms: $\overline{c"\,\partial p'/\partial x_i}$, $\overline{u_i"\,\partial p'/\partial x_j}$;

In each case it is possible to make use of conditional variables in formulating the closure assumptions.

We commence with the conditional velocities and their use in modelling the third-order terms. With $\gamma \ll 1$, the p d f expression (1) may be generalised [9] to read

$$P(u_i, c; \underline{x}) = \alpha(\underline{x}) \, P_0(u_i, 0; \underline{x}) \, \delta(c)$$
$$+ \beta(\underline{x}) \, P_1(u_i, 1; \underline{x}) \, \delta(c-1) + 0(\gamma) \quad (4)$$

where $P_0(u_i, 0)$ and $P_1(u_i, 1)$ are the conditional joint pdf's for reactant and product modes respectively. Equation (4) states that the unconditional joint pdf $P(u_i, c)$ is dominated by these entries at $c = 0, 1$. The unconditional Favre mean velocity $\tilde{u}$ can now be written

$$\bar{\rho} \tilde{u}_i = \int_{-\infty}^{\infty} du_i \, u_i \int_0^1 dc \, \rho(c) \, P(u_i, c)$$
$$= \bar{\rho}(1-\tilde{c}) \, \bar{u}_{ir} + \bar{\rho} \tilde{c} \, \bar{u}_{ip} + 0(\gamma) \quad (5)$$

where

$$\bar{u}_{ir} = \int_{-\infty}^{\infty} du_i \, u_i \, P_0(u_i, 0)$$

$$\bar{u}_{ip} = \int_{-\infty}^{\infty} du_i \, u_i \, P_1(u_i, 1)$$

It is then found [9] that the mean flux of c in direction x_i is

$$\overline{\rho u_i'' c''} = \bar{\rho} \tilde{c}(1 - \tilde{c}) \, (\bar{u}_{ip} - \bar{u}_{ir}) + 0(\gamma) \quad (6)$$

while the Favre mean turbulence intensity may be written

$$\overline{\rho u_i''^2}/\bar{\rho} = (1 - \tilde{c}) \, \overline{u_{ir}'^2} + \tilde{c} \, \overline{u_{ip}'^2} + \tilde{c}(1 - \tilde{c}) (\bar{u}_{ip} - \bar{u}_{ir})^2 + 0(\gamma) \quad (7)$$

As may be seen from (7), and is confirmed by experiment [16], the mean intensity contains an important contribution from the mean slip velocity between products and reactants, in addition to the weighted average of intensities in burned and unburned regions.

The required third-order quantities are [12]:

$$\overline{\rho u_i'' u_j'' c''} = \bar{\rho} \tilde{c}(1 - \tilde{c}) [(1 - 2\tilde{c}) (\bar{u}_{ip} - \bar{u}_{ir}) (\bar{u}_{jp} - \bar{u}_{jr})$$
$$+ \overline{(u_i' u_j')}_p - \overline{(u_i' u_j')}_r] + 0(\gamma) \quad (8)$$

and

$$\overline{\rho u_i'' u_j'' u_m''} = \bar{\rho}\tilde{c}(1-c)[(1-2\tilde{c})(\bar{u}_{ip}-\bar{u}_{ir})(\bar{u}_{jp}-\bar{u}_{jr})(\bar{u}_{mp}-\bar{u}_{mr})$$
$$+ \{(\overline{u_i' u_j'})_p - (\overline{u_i' u_j'})_r\}(\bar{u}_{mp}-\bar{u}_{mr})$$
$$+ \{(\overline{u_j' u_m'})_p - (\overline{u_j' u_m'})_r\}(\bar{u}_{ip}-\bar{u}_{ir})$$
$$+ \{(\overline{u_i' u_m'})_p - (\overline{u_i' u_m'})_r\}(\bar{u}_{jp}-\bar{u}_{jr})]$$
$$+ \bar{\rho}(1-\tilde{c})(\overline{u_i' u_j' u_m'})_r + \bar{\rho}\tilde{c}(\overline{u_i' u_j' u_m'})_p + O(\gamma) \quad (9)$$

The slip velocities $(\bar{u}_{ip} - \bar{u}_{ir})$ etc. are related to second-order fluxes via (6) and so are determined from transport equations. Closure at second-order level therefore depends on expressions for the second- and third-order conditional moments appearing in (8) and (9). The most general expressions so far proposed [12] are

$$(\overline{u_i' u_j'})_p - (\overline{u_i' u_j'})_r$$
$$= [(1 - K_{ij_1})\tilde{c} + (K_{ij_0} - 1)(1 - \tilde{c})] \overline{\rho u_i'' u_j''} / \bar{\rho} \quad (10)$$

and

$$(1 - \tilde{c})(\overline{u_i' u_j' u_m'})_r + \tilde{c}(\overline{u_i' u_j' u_m'})_p$$
$$= - c_s \frac{\tilde{k}}{\tilde{\varepsilon}} \frac{\overline{\rho u_k'' u_m''}}{\bar{\rho}} \frac{\partial}{\partial x_k} \left[\frac{\overline{\rho u_i'' u_j''}}{\bar{\rho}} \right] \quad (11)$$

where K_{ij_1}, K_{ij_0} and c_s are constants. Substitution of these equations into (8) and (9) leads [12] to the required expressions for third-order quantities in terms of known second-order quantities.

A consequence of Equation (11) is that, when $\tilde{c}$ is zero or unity, the third-order velocity correlation $\overline{u_i' u_j' u_m'}$ is described by a conventional gradient transport expression. Therefore the turbulent flows upstream or downstream of the reaction zone can be described by normal second-order Reynolds stress models. On the other hand, if the turbulent flows upstream and downstream of the flame are assumed to be homogeneous, a simpler model than Equation (11) can be employed [14, 15, 17] namely

$$(\overline{u_i' u_j' u_m'})_r = (\overline{u_i' u_j' u_m'})_p = 0 \quad (12)$$

In planar turbulent flames the expression

$$\overline{u_p'^2} - \overline{u_r'^2} = K(\overline{u}_p - \overline{u}_r)^2 \qquad (13)$$

was initially used [14, 15] and gave satisfactory agreement with experiment if the constant K was suitably chosen. More recently Libby [17] has repeated these one-dimensional flame calculations using Equation (10) in place of (13) and reports very similar results to those obtained previously [14, 15]. Although the calculations are thus found to be relatively insensitive to these alternative closure assumptions, further work is still required. This should include comparisons with conditional and unconditional velocity measurements to explore the validity of these and other approximations.

The above analysis can be extended [18] to problems where either heat transfer or non-uniformity of the mixture leads to variations in enthalpy. We assume that thin-flame burning occurs, that unburned reactant regions have a constant specific enthalpy, h_o, and that fully burned product regions can possess a range of specific enthalpies from h_o to h_e. A normalised enthalpy variable is then defined as

$$\zeta = \frac{h - h_o}{h_e - h_o} \qquad (14)$$

and the joint p d f $P(c, \zeta; \underline{x})$ takes the form (see Figure 1):

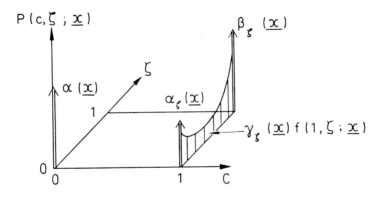

Figure 1: Joint p d f $P(c, \zeta; \underline{x})$

$$P(c, \zeta; \underline{x}) = \alpha(\underline{x}) \delta(c) \delta(\zeta)$$
$$+ \delta(c-1)[\alpha_\zeta(\underline{x})\delta(\zeta) + \beta_\zeta(\underline{x})\delta(\zeta-1) + \gamma_\zeta(\underline{x})f(1,\zeta;\underline{x})]$$
$$+ 0(\gamma) \qquad (15)$$

The distribution of product enthalpies in square brackets in Equation (15) consists of a delta function of strength α_ζ at $\zeta = 0$, a delta function of strength β_ζ at $\zeta = 1$ and a distribution $\gamma_\zeta f(1, \zeta; \underline{x})$, where $\alpha(\underline{x}) + [\alpha_\zeta(\underline{x}) + \beta_\zeta(\underline{x}) + \gamma_\zeta(\underline{x})] = 1$. Retaining this notation, the joint p d f of progress variable, c, normalised enthalpy, ζ, and velocity component, u_i, is

$$P(c, \zeta, u_i; \underline{x}) = \alpha(\underline{x}) \delta(c) P(0, 0, u_i; \underline{x})$$
$$+ \delta(c-1)[\alpha_\zeta(\underline{x}) \delta(\zeta) P(1, 0, u_i; \underline{x})$$
$$+ \beta_\zeta(\underline{x}) \delta(\zeta-1) P(1, 1, u_i; \underline{x}) + \gamma_\zeta(\underline{x}) f(1, \zeta, u_i; \underline{x})]$$
$$+ 0(\gamma) \qquad (16)$$

Despite its complexity this expression leads [18, 19] to conditional velocity equations of the same form as (3 - 9). However, because the product density is now a variable conditional Reynolds means must be replaced by conditional Favre means. For example, (6) becomes

$$\overline{\rho u_i'' c''} = \bar{\rho} \tilde{c}(1-\tilde{c})(\tilde{u}_{ip} - \bar{u}_{ir}) \qquad (17)$$

where $\tilde{u}_{ip} = \overline{(\rho u_i)}_p / \bar{\rho}$. Consequently the requirements for third-moment closure are formally unchanged.

We now turn to the molecular dissipation terms appearing in the second-order Reynolds stress and flux transport equations. These terms are

$$\bar{\rho}\tilde{\varepsilon}_{u_i u_j} = -\left[\overline{u_i'' \frac{\partial \tau_{ik}}{\partial x_k}} + \overline{u_j'' \frac{\partial \tau_{ik}}{\partial x_k}}\right] \qquad (18)$$

$$\bar{\rho}\tilde{\varepsilon}_{u_i c} = -\left[\overline{c'' \frac{\partial \tau_{ik}}{\partial x_k}} - \overline{u_i'' \frac{\partial j_k}{\partial x_k}}\right] \qquad (19)$$

As shown by Libby and Bray [9] conditional averaging can once more be helpful to the description of these terms. Thus we may write

$$\overline{\rho \varepsilon u_i u_j} = \alpha \rho_r (\overline{\varepsilon u_i u_j})_r + \beta \rho_p (\overline{\varepsilon u_i u_j})_p + \gamma \overline{(\rho \varepsilon u_i u_j)}_f \qquad (20)$$

where subscripts r, p and f refer to reactants, products and flamelets, respectively. The corresponding statement for Equation (19) is simply

$$\overline{\rho \varepsilon u_i c} = \gamma \overline{(\rho \varepsilon u_i c)}_f \qquad (21)$$

because the contributions from reactant and product regions are usually considered to be negligible, on the grounds that only small scales contribute significantly. At small scales the turbulence is isotropic so the correlation between scalar and velocity gradients is zero.

It might be thought that the flamelet terms in (20) and (21) would themselves become negligibly small in the thin flamelet regime as $\gamma \to 0$. However, this is not necessarily the case [9]. Thinner flamelets not only reduce the value of γ; they also lead to steeper gradients within the flamelet structure, and these gradients are always highly correlated with each other. Consequently the flamelet contributions to Equations (20) and (21) remain finite as $\gamma \to 0$. Flamelets are also the cause of heat release so it is not surprising to find [9] that the flamelet contributions to the molecular dissipation terms are proportional to the mean chemical source term $\overline{w}$ which appears in (2). The following expressions have been proposed [9, 12]:

$$\gamma \overline{(\rho \varepsilon u_i u_j)}_f = K_2 \frac{\overline{\rho u_i'' c''} \; \overline{\rho u_j'' c''}}{[\overline{\rho} \tilde{c}(1 - \tilde{c})]^2} \overline{w} \qquad (22)$$

$$\gamma \overline{(\rho \varepsilon u_i c)}_f = K_1 \frac{\overline{\rho u_i'' c''}}{\overline{\rho} \tilde{c}(1 - \tilde{c})} \overline{w} \qquad (23)$$

where K_1 and K_2 are constants.

The most detailed theoretical analyses of the reactant and product contributions to Equation (20) is due to Champion and Libby [20]. In their study of a reacting boundary layer, they relate these terms to the dissipation of turbulence kinetic energy which they write

$$\bar{\rho}\tilde{\varepsilon}_q = \alpha\rho_r\bar{\varepsilon}_{qr} + \beta\rho_p\bar{\varepsilon}_{qp} + \bar{\rho}\tilde{\phi}_F \qquad (24)$$

where $\bar{\rho}\tilde{\phi}_F$ represents the flamelet contribution. They choose an algebraic length scale model of the form

$$\bar{\varepsilon}_{qa} = c_\mu \bar{q}_a^{3/2} / \ell \qquad a = r, p \qquad (25)$$

to represent both $\bar{\varepsilon}_{qr}$ and $\bar{\varepsilon}_{qp}$ where $\bar{q}_a$ ($a = r, p$) is the conditional mean turbulence kinetic energy in reactants and products, respectively, c_μ is a constant, and ℓ is the length scale for dissipation which is taken to be the same in reactants and in products. However, taking into account the effect of temperature on the kinematic viscosity and the relative strain in reactant and product, they deduce that

$$\bar{\varepsilon}_{qr} = \bar{\varepsilon}_{qp} / (1 + \tau)^4 \qquad (26)$$

Now the heat release parameter τ is typically in the range 4 to 7 so Equation (26) implies $\bar{\varepsilon}_{qp} \gg \bar{\varepsilon}_{qr}$. Evaluating α and β, Equation (24) then becomes

$$\tilde{\varepsilon}_q = \tilde{q}^{3/2} h(\tilde{c}) / \ell + \tilde{\phi}_F \qquad (27)$$

where

$$h(c) = c_\mu \left\{ \frac{\tilde{c} + (1-\tilde{c})/(1+\tau)^4}{[\tilde{c} + (1-\tilde{c})/(1+\tau)^{8/3}]^{3/2}} \right\}$$

Finally they assume the sum of the reactant and product terms in Equation (20) to be proportional to the combined reactant-plus-product term in (27) and the result using (22) is

$$\bar{\rho}\tilde{\varepsilon}_{u_i u_j} = \frac{\overline{\rho u_i'' u_j''} \tilde{q}^{3/2} h(\tilde{c})}{\tilde{q} \quad \ell}$$

$$+ K_2 \frac{\overline{\rho u_i'' c''} \; \overline{\rho u_j'' c''}}{[\bar{\rho}\tilde{c}(1-\tilde{c})]^2} \bar{w} \qquad (28)$$

A similar analysis is employed by Bray et al [19] in the case of a non-adiabatic flow. In both of these reacting boundary layer problems the form of the similarity solution subsequently shows that dissipation due to flamelets is unimportant in comparison with dissipation within constant density reactant and product regions of the flow.

While an algebraic turbulence length scale expression such as that employed in (25) and (28) can be satisfactory in a boundary layer analysis it is less likely to be successful in more complex flows. Consequently Bray et al [12] have proposed that, for such flows, the first term on the right hand side of (28) should be replaced by a term proportional to $\tilde{\varepsilon}$, where, despite its known shortcomings, $\tilde{\varepsilon}$ is obtained from the $\tilde{\varepsilon}$ transport equation.

Insufficient experimetnal data is available at present to make a confident test of the accuracy of the conditional dissipation models which have been described above. Gokalp et al [21] report conditional velocity spectra obrained in vee flames. A length scale corresponding to the mean size of the most energetic eddies is calculated and is found to be larger in products than in reactants by a factor of almost two. This observation suggests that the ratio of conditional dissipation rates may be even larger than indicated in Equation (26) since the derivation of this expression assumed equality of the two conditional length scales, see Equation (25).

The question of conditional dissipation rates is relevant to the rates of strain to which laminar flamelets are subjected. We shall therefore return to this topic later.

Effects of pressure fluctuations appear as

$\overline{c'' \partial p'/\partial x_i}$ and $\overline{u_i'' \partial p'/\partial x_j}$

in the second-moment equations. These terms are a special problem in reacting flows since heat release provides a distinct source of pressure fluctuations. Strahle [22] reviews modelling strategies. Bray et al [15] neglect the terms completely, in studies of planar turbulent flames, on the grounds that they represent processes which are too slow to influence the turbulence within the flame. This assumption is not applicable to combustion in a boundary layer. For this case Champion and Libby [20] and Bray et al [19] follow current practice in ignoring combustion induced pressure fluctuations and assuming that a conventional non-reacting flow model may be applied. They use the models given by Launder et al [23] and Hanjalic and Launder [24] as interpreted in terms of Favre averaged quantities by Jones [25]. Bray et al [12] also recommend these expressions.

On the other hand Masuya [26] includes only a combustion contribution to pressure fluctuation effects and neglects contributions from turbulence phenomena. Employing a conditional averaging approach, he assumes that the change in pressure across the laminar flamelets is the dominant source of pressure fluctuations. He writes the equations of motion for a stationary unstrained laminar flamelet in the form

$$d(\rho u_\eta)/d\eta = 0$$
$$\rho u_\eta dc/d\eta = w + dj_\eta/d\eta$$
$$\rho u_\eta du_\eta/d\eta = d(-p + \tau_{\eta\eta})/d\eta$$

where the η coordinate is perpendicular to the flamelet and fixed in it. He shows that flamelet pressure changes appear only in combination with flamelet viscous and molecular diffusion terms and that these terms can be combined together and represented through the flamelet model. For example he shows that the combined viscous dissipation and pressure fluctuation term is

$$\overline{\gamma\{u_i'' \partial(-\delta_{jk}p' + \tau_{jk})/\partial x_k\}}$$
$$= \overline{w} \iota u_\ell^o \{(\overline{u}_{ia} - \tilde{u}_i)\overline{o}_j + \overline{u_{ia}' o_j'} + \iota u_\ell^o \overline{o_i o_j}/2\} \quad (29)$$

where u_ℓ^o is the unstrained laminar flame burning velocity, o_i is the direction cosine specifying the instantaneous direction of the flamelet, and $\overline{u}_{ia}$ is the conditional mean value of u_i in the flamelet mode with $c \to 0$. Thus the statistical geometry of the flamelet surface enters the description and must be modelled as must the conditional velocity at the edge of the flamelet.

A rational generalisation is to include both conventional non-reacting flow terms [23, 24] and Masuya's flamelet contribution [26]. See [27] for experimental data from flames.

Laminar Flamelet Description of Turbulent Mean Reaction Rates

The reaction rate terms to be modelled in a second order description based on the progress variable, c, are $\bar{w}$ in Equation (2) and $\overline{u_i''w}$ in Equation (3). The second of these is usually expressed in terms of $\bar{w}$ through an equation of the form [28]

$$\overline{u_i''w} = (\phi_n - \tilde{c}) \frac{\overline{\rho u_i''c''}}{\bar{\rho}\tilde{c}(1-\tilde{c})} \bar{w} \tag{30}$$

where ϕ_n is a constant. An alternative formulation is provided by Masuya [26] who combines $\overline{u_i''w}$ with viscous and molecular diffusion terms all of which are represented together via unstrained laminar flamelets. In either case $\overline{u_i''w}$ is linked to the mean reaction rate $\bar{w}$ which is therefore the main focus of modelling effort.

In the laminar flamelet burning regime chemical reaction and gradients in c are both confined to the interior structure of the laminar flamelets. It follows [9] that $\bar{w}$ is proportional to the scalar dissipation function for c

$$\chi_c = 2D \left(\frac{\partial c}{\partial x_k}\right)^2$$

If it is naively assumed that $\overline{\chi_c}$ can be modelled as in a cold nonreacting flow then Spalding's eddy break-up model [29] is recovered

$$\bar{w} = K_3 \bar{\rho} \tilde{\varepsilon} \tilde{c}(1 - \tilde{c})/\tilde{q} \tag{31}$$

where $\tilde{q}$ and $\tilde{\varepsilon}$ are the turbulence kinetic energy and its dissipation rate and K_3 is a constant. Although the assumption underlying Equation (31), that the mean rate is more influenced by turbulent mixing than by chemical kinetics, is often qualitatively correct, experiments show that chemical composition is also important. An accurate description of $\bar{w}$ must therefore include realistic chemical kinetic mechanisms and rates. One important advantage of laminar flamelet models is that laminar flame calculations with appropriate kinetic and molecular transport descriptions are performed separately from turbulent flow field calculations.

Perhaps the most obvious way to describe $\bar{w}$ in the flamelet regime is in terms of the flamelet area per unit volume, Σ, say. Then [30, 31]

$$\bar{w} = \rho_r \bar{u}_\ell \Sigma \qquad (32)$$

where ρ_r is the reactant density and $\bar{u}_\ell$ is the mean laminar burning velocity at the average strain rate in the turbulent flow. We find an alternative formulation in terms of crossing frequencies to be attractive because it deals in quantities many of which can easily be measured. The crossing frequency expression can be written either in terms of a time series [32] or in terms of a spatial variation [5, 11].

In terms of a time series the mean reaction rate is written [32]

$$\bar{w} = \bar{w}_F \, \nu \qquad (33)$$

where ν is the number of times per second that a flamelet crosses the location $\underline{x}$ at which $\bar{w}$ is to be evaluated and

$$\bar{w}_F = \langle \int w \, dt \rangle \qquad (34)$$

is the mean reaction per crossing, where the integration is over the time interval during which a flamelet passes over location $\underline{x}$ and angled brackets denote an average. Thus

$$\bar{w}_F = \langle \frac{1}{|\bar{V}_n|} \int w \, d\eta \rangle \approx \bar{W}/|\bar{V}_n| \qquad (35)$$

where $|\bar{V}_n|$ is the mean speed, normal to itself, at which flamelets pass location $\underline{x}$ and

$$\bar{W} = \langle \int w \, d\eta \rangle \qquad (36)$$

in which the average is over flamelets of different strain rates, η being a coordinate perpendicular to the flamelet and fixed in it. The simplest procedure is to evaluate Equation (36) at the mean strain rate. The consequences of assuming a distribution of strain rates will be considered later. Note that in order to obtain $w(\eta)$ from a laminar flame calculation it is necessary to define the progress variable c in the flamelet. This may be done in terms of the normalised temperature or product concentration.

The mean crossing frequency ν is modelled [32, 33] noting that the progress variable $c(\underline{x}, t)$ at a fixed location $\underline{x}$ is a square wave time series. If the durations of both the $c = 0$ and $c = 1$ intervals in this time series have a given distribution, modelled as an exponential [32] or a more realistic gamm-two p d f [33], see also [34], then

$$\nu = \frac{g \bar{c}(1 - \bar{c})}{\hat{T}} \qquad (37)$$

where $\hat{T}$ is the integral time scale of the square wave time series and $g = 2$ and 1 respectively for the exponential and gamma-two p d fs.

An alternative but equivalent formulation [5, 10, 11] expresses $\bar{w}$, in terms of spatial rather than temporal variations, as

$$\bar{w} = \bar{w}_y \, n_y \qquad (38)$$

where the crossing frequency n_y is now the average number of flame crossings of a specified line per unit distance y along this line, see Fig. 2. Choosing the line y as a contour along which $\bar{c}$ is

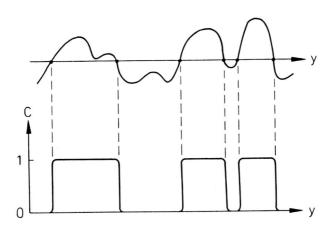

Figure 2: Variation of c along a contour y.

constant then, at any instant in time, a graph of c against y is again a square wave function. If the burned and unburned interval lengths along this line are assumed to be distributed according to either an exponential or a gamma-two p d f then

$$n_y = \frac{g\bar{c}(1-\bar{c})}{\hat{L}_y} \tag{39}$$

where as before $g = 2$ for the exponential pdf and $g = 1$ for the gamma-two pdf, while $\hat{L}_y$ is the integral length scale of the 0 - 1 square wave function $c(y)$. The mean chemical reaction per flamelet crossing is now

$$\bar{w}_y = \langle w\,dy \rangle \approx \bar{W}/|\bar{\sigma}_y| \tag{40}$$

where $|\bar{\sigma}_y|$ is the mean magnitude of the direction cosine of the flamelet surface in direction y and $\bar{W}$ is defined in Equation (36).

The two crossing frequency formulations presented above are clearly related to each other. Comparing Equation (33) with (38) and Equation (35) with (40) we find

$$\frac{\nu}{n_y} = \frac{\bar{w}_y}{\bar{w}_F} = \frac{\hat{L}_y}{\hat{T}} = \frac{|\bar{V}_n|}{|\bar{\sigma}_y|} \tag{41}$$

where it has been assumed that the factor g takes the same value in Equations (37) and (39). Another informative set of relationships may be derived in terms of the flamelet area per unit volume Σ. Considering the mean flame area contained within an elementary volume we find

$$\Sigma = \frac{n_y}{|\bar{\sigma}_y|} = \frac{g\bar{c}(1-\bar{c})}{|\bar{\sigma}_y|\,\hat{L}_y} \tag{42}$$

or alternatively from (37, 41)

$$\Sigma = \frac{\nu}{|\bar{V}_n|} = \frac{g\bar{c}(1-\bar{c})}{|\bar{V}_n|\,\hat{T}} \tag{43}$$

These relationships allow us to interlink the various formulations presented above and also to convert experimental data from one form into antoher. However, unfortunately, they do not remove the basic modelling problems, which are to predict a characteristic length or time scale of the wrinkled flame ($\hat{T}$, $\hat{L}_y$ or Σ), and to predict a second property $|\bar{\sigma}_y|$ or $|\bar{V}_n|$ related to the statistical geometry or dynamics. It is fortunate that many of the quantities appearing in the equations presented above can be studied experimentally.

Experimental data reviewed by Bray et al [35] indicates that $\hat{T}$ is almost independent of position in a variety of flames. Less comprehensive spatial measurements [5] in a turbulent bunsen flame show that $\hat{L}_y$ also does not vary strongly with position and is similar in magnitude to the integral length scale, ℓ, of the turbulence in the unburned gas. Cine films of oblique turbulent flames suggest that the wrinkles travel along the flame with a velocity similar to the resultant mean flow velocity $\tilde{u}_R$. Equation (41), with $|\overline{V}_n|$ replaced by $\tilde{u}_R$ and $|\overline{\sigma}_y|$ close to 0.5 [5], then shows that $\hat{T} \sim \ell/\tilde{u}_R$ as suggested previously [35]. In other circumstances it may be more appropriate to assume $\hat{T} \sim \ell/u_o'$ [35] where u_o' is the turbulence intensity in the unburned mixture.

For turbulence diffusion flames Marble and Broadwell [30, 31] have proposed a phenomenological transport equation for the quantity, Σ, with terms representing convection, turbulent, transport, production and destruction of flame surface. Premixed flame calculations currently assume either that $\hat{T}$ or $\hat{L}_y$ is constant, as indicated above, or [11] that $\tilde{L}_y = \tilde{q}^{3/2}/\tilde{\varepsilon}$.

Equations (36, 38, 40) assume that all flamelets are characterised by the local mean rate of strain. Laminar flame studies [36] show that increased strain rates lead to reduced burning rates and perhaps to extinction. Consequently this mean strain rate model will lead to a strong reduction in $\overline{w}$ as the limiting strain rate is approached. In reality however turbulent flows involve a nonuniform spatial distribution of strain rates. Consequently some flamelets remain only weakly strained even when others are strained close to or beyond extinction. This important feature should be included in a detailed description of flamelet combustion.

The local strain rate may be characterised by a dimensionless Karlovitz number

$$\kappa = \frac{\ell_\ell^o}{u_\ell^o} \left(\frac{\varepsilon}{\nu_v}\right)^{\frac{1}{2}} \tag{44}$$

where ℓ_ℓ^o and u_ℓ^o are the thickness and burning velocity, respectively, of an unstrained laminar flame, ν_v is the molecular kinematic viscosity and ε is the instantaneous value of the viscous dissipation. Given that either κ or ε can be used to label a particular strained laminar flame solution we may denote the Karlovitz number at which a laminar flame is just quenched by strain as κ_q. Then

$$\varepsilon_q = \nu_v [\kappa_q \frac{u_\ell^o{}^2}{\ell_\ell^o}].$$

is the viscous dissipation at quenching. Equation (40) may then be replaced [5] by

$$\bar{w}_y = \frac{\rho_r u_\ell^o}{|\bar{\sigma}_y|} \int_0^{\varepsilon_q} \frac{W(\varepsilon)}{W(o)} P(\varepsilon) d\varepsilon \qquad (45)$$

where

$$W(\varepsilon) = \int_{-\infty}^{\infty} w(\eta; \varepsilon) d\eta \qquad (46)$$

$W(o) = \rho_r u_\ell^o$ and $P(\varepsilon)$ is the pdf of viscous dissipation rates at the flamelet location. This conditional pdf is assumed to be represented by the log-normal distribution of Kolmogorov [37]

$$P(\varepsilon) = \frac{1}{\sqrt{2\pi} \sigma \varepsilon} \exp\left[-\frac{1}{2\sigma^2}(\ln \varepsilon - \mu)^2\right] \qquad (47)$$

in which $\mu = \ln \varepsilon + \frac{1}{2}\sigma^2$ and σ is the standard deviation.

We must now recall our earlier discussion of the viscous dissipation function suggesting that the conditional dissipation in reactants is much smaller than that in products, see Equation (26). If, as is conventional, the strained flame "library" is characterised in terms of reactant properties, then ε in Equations (45, 46, 47) should be the reactant quantity.

In an approach making full use of a strained laminar flame library, values of u_ℓ^o and $W(\varepsilon)/W(o)$ are computed and stored, to be interpolated to match the conditions required in a turbulent flow calculation. A simplification, demonstrated by Cant and Bray [11], assumes a linear variation of $W(\varepsilon)/W(o)$, namely

$$\begin{aligned}\frac{W(\varepsilon)}{W(o)} &= 1 - \frac{\varepsilon}{\varepsilon_q} \quad ; \quad \varepsilon \leq \varepsilon_q \\ &= 0 \quad\quad ; \quad \varepsilon > \varepsilon_q\end{aligned} \qquad (48)$$

so that integration of Equation (45) can be carried through analytically to yield

$$\bar{w}_y = \frac{\rho_r u_\ell^o}{|\bar{\sigma}_y|} I_o(\frac{\tilde{\varepsilon}}{\varepsilon_q}, \sigma) \tag{49}$$

where

$$I_o = \frac{1}{2} \left[\text{erfc}\left\{ \frac{\ln(\frac{\tilde{\varepsilon}}{\varepsilon_q}) - \frac{1}{2}\sigma^2}{\sqrt{2}\sigma} \right\} - \frac{\tilde{\varepsilon}}{\varepsilon_q} \text{erfc}\left\{ \frac{\ln(\frac{\tilde{\varepsilon}}{\varepsilon_q}) + \frac{1}{2}\sigma^2}{\sqrt{2}\sigma} \right\} \right] \tag{50}$$

Assembling Equations (38, 39, 49) the final mean rate expression becomes [11]

$$\bar{w} = \frac{g\bar{\rho} u_\ell^o I_o(\frac{\varepsilon}{\varepsilon_q}, \sigma)(1 + \tau)\tilde{c}(1 - \tilde{c})}{|\bar{\sigma}_y| \hat{L}_y (1 + \tau\tilde{c})} \tag{51}$$

Note that this equation has been written in terms of $\tilde{c}$, rather than $\bar{c}$, and that no allowance has been made for differences between conditional dissipation values in reactants and products.

Under conditions where most flamelets are close to extinction, i.e. when $\tilde{\varepsilon}/\varepsilon_q$ is not much larger than unity, the standard deviation σ emerges as an important parameter in Equations (50, 51). Smaller values of σ lead to a sharper reduction in $\bar{w}$ as $\tilde{\varepsilon}/\varepsilon_q$ approaches unity.

Concluding Remarks

Considerable progress has been made, in understanding both the physics and the chemistry of premixed turbulent combustion, and also in devising ways of incorporating this increased understanding into predictive models.

Nevertheless several areas have been identified in this review where current knowledge is inadequate and where further experimental or numerical studies are required. These are as follows:

i. The Reynolds stress and Reynolds flux transport equations contain turbulent transport terms which are closed through assumed relationships between second-order and third-order conditional turbulence quantities in reactants and products. Equations (10-13) present two alternative sets of assumed relationships, neither of which has yet been confirmed experimentally.

ii. The description of the pressure fluctuation terms presents a particular problem since heat release provides a separate mechanism for the production of pressure fluctuations. Neither the use of conventional non-reacting flow models [23, 24] nor the assumption that flamelet pressure fluctuations are dominant [26] can easily be justified.

iii. Perhaps the most challenging area where further work is required is the statistical geometry of the wrinkled flame interface. We have defined various different measures of a characteristic length or time scale for this interface ($\hat{L}_y$, $\hat{T}$, Σ) and have explored relationships between them. However, the interaction between these scales and the characteristic scales of the velocity field is not well understood. Furthermore we have identified separate conditional dissipation functions for reactants and products. The relationships proposed for these, e.g. Equation (25), imply that at least some characteristic velocity scales will have different magnitudes in reactants and in products. Further work is required on the prediction of relevant scales. Another relevant property of the statistical geometry of the flame surface is its direction cosine. This appears at various points in the analysis, see for example Equations (29, 40, 52).

iv. In a strained laminar flamelet description of turbulent reaction rates it is necessary to assume, see for example Equation (45), that the flamelet location is uncorrelated with the local instantaneous viscous dissipation rate. This assumption is questionable as is the treatment of the standard deviation σ of the assumed log-normal p d f $P(\varepsilon)$ as an assigned parameter.

Two powerful techniques which have not yet been fully exploited in the study of these questions are firstly the various laser sheet flame visualisations and secondly direct numerical simulations of turbulence with thin-flame burning.

Acknowledgements

The research performed at UCSD is supported by the Department of Energy, Office of Basic Energy Sciences, Division of Chemical Sciences under Contract DE-FG03-86ER13527, that at Cambridge by grants from the Science and Engineering Research Council and the European Community and that at ENSMA by the European Community. Our collaboration is facilitated by an International Exchange Grant from the National Science Foundation.

References

1. Bray, K. N. C., Turbulent flows with premixed reactants, in Turbulent Reacting Flows, ed. by P. A. Libby and F. A. Williams, Springer-Verlag, Berlin, 1980.

2. Borghi, R., On the structure and morphology of turbulent premixed flames in Recent advances in the aerospace sciences, C. Casci, ed., Plenum Pub. p.117, 1985.

3. Peters, N., Laminar flamelet concepts in turbulent combustion. Twenty First Symp. (Int.) on Combustion, Combustion Institute, 1986.

4. Boyer, L., Laser tomographic method for flame front movement studies. Combustion and Flame 39, 321, 1980.

5. Chew, T. C., Britter, R. E., Bray, K. N. C. Laser tomography of turbulent premixed bunsen flames. Combustion and Flame, to appear, 1988.

6. Zur Loye, A. R., Bracco, F. V., Two dimensional visualisation of ignition kernels in an I.C. engine. Combustion and Flame 69 59, 1987.

7. Gouldin, F. C., Halthore, R. N., Rayleigh scattering for density measurements in premixed flames. Experiments in Fluids 4, 269, 1986.

8. Namazian, M., Talbot, L., Robben, F., Density fluctuations in premixed turbulent flames. Twentieth Symp. (Int.) on Combustion, pp. 411-419, Combustion Institute, 1984.

9. Libby, Paul A. and Bray, K. N. C., Implications of the laminar flamelet model in premixed turbulent combustion. Combustion and Flame 39, 33, 1980.

10. Bray, K. N. C., Scales and burning rates in premixed turbulent flames. Proc. Ninth Australasian Fluid Mechanics Conf. pp. 230-233, University of Auckland, New Zealand, 1986.

11. Cant, R. S. and Bray, K. N. C., Strained laminar flamelet calculations of premixed turbulent combustion in a closed vessel. Submitted for publication, 1988.

12. Bray, K. N. C., Libby, Paul A., Moss, J. B. Unified modelling approach for premixed turbulent combustion - Part I: General formulation. Combustion and Flame 61, 87, 1985.

13. Bray, K.N. C., Moss, J. B. A unified statistical model of the premixed turbulent flmae. Acta Astronautics 4, 291, 1977.

14. Libby, Paul A., Bray, K. N. C., Countergradient diffusion in premixed turbulent flames, AIAA J. 19 205, 1981.

15. Bray, K. N. C., Libby, Paul A., Masuya, G., Moss, J. B. Turbulence production in premixed turbulent flames, Combustion Science and Tech. 25, 127, 1981.

16. Cheng, R. K., Conditional sampling of turbulence intensities and Reynolds stress in premixed flames. Combustion Science and Tech. 25, 109, 1984.

17. Libby, Paul A., Theory of normal premixed turbulent flames revisited. Prog. Energy Combustion Sci. 11, 83, 1985.

18. Bray, K. N. C., Champion, M., Dave, N., Libby, Paul A., On the thermochemistry of premixed turbulent combustion in variable enthalpy systems. Combustion Science and Tech., 46 31, 1986.

19. Bray, K. N. C., Champion, M., Libby, Paul A., The turbulent premixed boundary layer with variable enthalpy. Combustion Science and Tech. 55, 139, 1987.

20. Champion, M., Libby, Paul A., Turbulent premixed combustion in a boundary layer, Combustion Science and Tech. 38, 267, 1984.

21. Gokalp, I., Shepherd, I. G., Cheng, R. K., Spectral behaviour of velocity fluctuations in premixed turbulent flames. Private communication, 1987.

22. Strahle, W. C., Duality, dilatation, diffusion and dissipation in reacting turbulent flows. Nineteenth Symp. (Int.) on Combustion, pp. 337-347, Combustion Inst. 1982.

23. Launder, B. E., Reece, G. J., Rodi, W., Progress in the development of a Reynolds stress turbulence closure. J. Fluid Mech. 68, 538, 1975.

24. Hanjalic, K., Launder, B. E., A Reynolds stress model of turbulence and its application to thin shear flows. J. Fluid Mech. 52, 609, 1972.

25. Jones, W. P., Models for turbulent flows with variable density and combustion, in Prediction Methods for Turbulent Flows, Ed. W. Kollmann, Hemisphere Pub. Corp., Washington, New York, London, 1980.

26. Masuya, G., Influence of laminar flame speed on turbulent premixed combustion, Combustion and Flame, 64, 353, 1986.

27. Chandran, S. B. S., Komerath, N. M., Strahle, W. C., Scalar-velocity correlations in a turbulent premixed flame. Twentieth Symp. (Int.) on Combustion, pp. 429-435, Combustion Inst. 1984.

28. Libby, Paul A., Bray, K. N. C., Moss, J. B., Effects of finite reaction rate and molecular transport in premixed turbulent combustion. Combustion and Flame, 34, 285, 1979.

29. Mason, H. B., Spalding, D. B., Prediction of reaction rates in turbulent premixed boundary layer flows, in Combustion Institute European Symposium, ed. F. J. Weinberg, pp. 601-606, Academic Press, London, 1973.

30. Marble, F. E. and Broadwell, J. E., The coherent flame model for turbulent chemical reactions, Project Squid Report TRW-9-PU, 1977.

31. Marble, F. E. and Broadwell, J. E., A theoretical analysis of nitric oxide production in a methane-air turbulent diffusion flame, EPA Technical Report, 1979.

32. Bray, K. N. C., Libby, Paul A., Moss, J. B., Flamelet crossing frequencies and mean reaction rates in premixed turbulent combustion. Combustion Science and Tech. 41, 143, 1984.

33. Bray, K. N. C., Libby, Paul A., Passage times and flamelet crossing frequencies in premixed turbulent combustion. Combustion Science and Tech. 47, 253, 1986.

34. Bray, K. N. C., Champion, M., Libby, Paul A., The correlation functions for flamelet crossings in premixed turbulent flames. Combustion Science and Tech., to appear, 1988.

35. Bray, K. N. C., Champion, M., Libby, Paul A., Mean reaction rates in premixed turbulent flames. Submitted, 1988.

36. Rogg, B., Response and flamelet structure of stretched premixed methane-air flames. Combustion and Flame, to appear, 1988.

37. Kolmogorov, A. N., A refinement of previous hypotheses concerning the local structure of turbulence in a viscous incompressible fluid at high Reynolds number. J. Fluid Mech. 13, 82, 1962.

ON THE PROBLEM OF MODELLING TIME OR LENGTH SCALES IN TURBULENT COMBUSTION

R. BORGHI, A. PICART, M. GONZALEZ
Faculté des Sciences de Rouen - U.A. C.N.R.S. N° 230
B.P. 118   76134 Mont-Saint-Aignan Cedex (France)

I. INTRODUCTION

All the existing methods for turbulent reaction rates modelling need the knowledge of time or length scales characterizing the reacting turbulent medium. The old (but usefull) Eddy Break Up model assumes that a characteristic time for the destruction of the fluctuation of the progress variable is proportional to the integral Eddy Turn over time of the turbulence [1] ; a similar assumption has been used with the presumed pdf approach, when the chemistry is not infinitely fast [2]. The conserved scalar approach for diffusion flames needs an assumption relating χ, the mean scalar dissipation rate, to ε, the mean dissipation rate for the turbulence kinetic energy [3]. A similar hypothesis has to be involved also in approaches using modelled equations for the joint probability function of the velocity and species mass fraction [4]. The coherent flame model of [5] involves a different quantity : the mean flame surface by unit of volume ; but this quantity is nothing but a length scale of the turbulent medium. The flamelets approaches [6] are something more detailed: they need a full distribution of stretch rates, that means a range of turbulent times scales, instead of a single one.

Several questions appear about such time or length scales

i) what is their proper connection with the turbulent velocity field scales ?

ii) what is their relation with the chemical time scales in the case where these later are not infinitely small ?

iii) in what conditions is it necessary to take into account a full range of time scales ?

Our purpose is to discuss here these important questions in order to put into evidence what can be done or what is not to be done.

These time or length scales are central for turbulent combustion modelling; they have been discussed or studied by several researchers, explicitely or implicitely. We will synthetize first these previous studies. In a second part we will present a numerical investigation of these scales by means of a full direct simulation of a reactive, premixed, turbulent homogeneous medium ; of course, the simulation has been restricted to low values of Reynolds and Peclet numbers.

Finally, in a third part, we will show that taking into account a full range of turbulent time is important when the implied reactions are sudden (that is with high activation energy) ; we will present a model specially built for that situation, as do the flamelets models, but taking into account interactions of flamelets as well as their extinctions and reignitions. The results of such a model for the case of a diffusion flame will be discussed.

II. FIRST PART : A CRITICAL REVIEW ABOUT THE TIME OR LENGTH SCALES IN TURBULENT COMBUSTION MODELLING

II.1 When the reactions are very fast

II.1.1 When the reactions are very fast, turbulent <u>diffusion flames</u> can be fully described if we are able to compute the probability density function of one inert tracer released in the turbulent flow (at least when the Lewis numbers are close to unity). This finding is due to H.L. Toor, and is discussed at length by R.W. Bilger [7]. In such a case, the problem is then a purely turbulent one.

The prediction of the inert pdf, however, is not a solved problem. Either with a presumed pdf approach, or with a modelled equation for the pdf, a closure assumption concerning $\widetilde{\chi} = d \, \overline{\dfrac{\partial \xi}{\partial x_\ell} \cdot \dfrac{\partial \xi}{\partial x_\ell}}$ (where ξ is the inert tracer mass fraction) is needed. The presumed pdf approaches use explicitly (1): $\widetilde{\chi} = C_\chi \cdot \widetilde{\xi^{\prime 2}} \cdot \dfrac{\varepsilon}{k}$, where ε is the dissipation rate of the turbulence and k its kinetic energy, C_χ being a modelling constant. The pdf approach of Pope [4] or Kollmann and Jones [8] use implicitly the same closure assumption : they define an exchange frequency and take it proportionally to k/ε ; but if we extract from the pdf modelled equation a balance equation for $\widetilde{\xi^{\prime 2}}$, we obtain again the relation (1).

The equation (1) states that the scalar time $\tau_\xi = \widetilde{\xi^{\prime 2}}/\widetilde{\chi}$, and the kinetic time k/ε are proportional. Such a relation has been discussed from long time. Experimental verifications [9], as well as theoretical ones [10], have been performed, from its first statement, due to S. Corrsin. Even in the very idealistic case of homogeneous isotropic turbulence, it seems from [10] that such a relation can be obtained only for a very large turbulence Reynolds number. In addition, for more complicated non homogeneous turbulence, with intermittency phenomena, that relation has yet to be ascertained. The only possible way to overcome this problem now in the framework of a one point closure is the use of a balance equation for $\widetilde{\chi}$, as in Zeman and Lumley [11] ; but a more general approach would be a two point closure theory.

II.1.2 Concerning <u>premixed turbulent flames</u>, a similar problem appears, for the destruction rate $\widetilde{\chi_Y}$ of the progress variable (reactive) Y : $\widetilde{\chi_Y} = d_Y \overline{\frac{\partial Y'}{\partial x_\alpha} \cdot \frac{\partial Y'}{\partial x_\alpha}}$. Indeed it has been shown by K. Bray and B. Moss [12] as early as 1974, that the mean reaction rate for $\widetilde{Y}$ in the limit of very fast chemistry was proportional to $\widetilde{\chi_Y}$. Using again a closure assumption similar to (1) but in the framework of the Prandtl mixing length theory of turbulence : $\widetilde{\chi_Y}/\widetilde{Y'^2} = C_\chi \left|\frac{\partial \tilde{u}}{\partial y}\right|$ (2), it demonstrates the proposal of D.B. Spalding in 1970 [1].

There is now an additional reason to suspect the validity of (1) or (2) in this case : the variable Y is here a reactive scalar, and the reactions could probably well modify the instantaneous gradient $\partial Y'/\partial x_\alpha$ and, consequently, $\widetilde{\chi_Y}$. That question has been put into evidence and discussed in a number of papers : [14], [15], [16], [17], [18].

For reasons of dimensions, the characteristics time $\tau_Y = \widetilde{Y'^2}/\widetilde{\chi_Y}$ can be a function of $\tau_t = k/\varepsilon$ or $|\partial \tilde{u}/\partial y|^{-1}$, τ_c the chemical time, d_Y the diffusion coefficient for Y (assumed equal to ν), and k, the turbulence kinetic energy ; in an homogeneous turbulent reacting medium, no additional variable occurs. In the case of very fast reactions, τ_c/τ_t, going to zero, has to disappear ; if we assume also that the turbulence Reynolds number $Re_T = k^2/\nu\varepsilon$ is very large, it remains that τ_Y/τ_t has to be a function of $U_L/k^{\frac{1}{2}}$ only, when this quantity is finite, or a constant, if $k^{\frac{1}{2}} \gg U_L$ (U_L is the laminar flame speed, proportional to $(d_Y/\tau_c)^{\frac{1}{2}}$). That seems to confirm very simply the relations (1) or (2), but in the case where $k^{\frac{1}{2}} \gg U_L$ only ; this finding does not contradict the discussions of the previous paragraph II.1, because we have kept here the large Reynolds number assumption. In addition, the case $U_L/k^{\frac{1}{2}}$ finite is fully compatible with both $Re_T \gg 1$ and $\tau_c/\tau_t \ll 1$.

A something similar result has been found in [15], but apparently without the restriction of $U_L \ll k^{\frac{1}{2}}$!

The demonstration begins with picture of a wrinkled turbulent flame, with non stretched flamelets, and expresses the contribution $\widetilde{\chi_Y}^F$ of these flamelets to $\widetilde{\chi_Y}$ in terms of $\overline{\Sigma}$, the mean flame surface by unit of volume ; assuming similarity properties in the Y profiles within the flamelets, for any orientation or position of these flamelets, one obtains :

(3) $\quad \widetilde{W} \propto \widetilde{\chi_Y}^F \quad$ similarly to Bray and Moss [12] , $\quad \widetilde{\chi_Y}^F \propto d_Y \frac{\overline{\Sigma}}{\delta_L} \quad$ (3')

($\widetilde{W}$ is the mean reaction rate)

where δ_L is the laminar flame thickness : $\delta_L \propto (d_Y \tau_c)^{\frac{1}{2}}$.

(4) $\quad \widetilde{Y(1-Y)} \propto \delta_L \overline{\Sigma} \quad , \quad \widetilde{Y\dot{w}} \propto \dfrac{\delta_L \overline{\Sigma}}{\tau_c} \quad$ (4')

It follows then from (4) that $\widetilde{\chi}_Y^F \propto \dfrac{\widetilde{Y(1-Y)}}{\tau_c}$ (5) and $\widetilde{Y\dot{w}} \propto \widetilde{\chi}_Y^F$ (5').

The formula (5) is undeterminated when $\tau_c/\tau_t \to 0$, as $\widetilde{Y(1-Y)} \to 0$ also because $\widetilde{Y'^2} \to \widetilde{Y}(1-\widetilde{Y})$, and cannot be used directly.

But, if we look now at the balance equation for $\widetilde{Y(1-Y)}$ itself, which can be obtained from the primitive equation for Y in a similar manner as $\widetilde{Y'^2}$, and if we introduce the asymptotic expansion : $\widetilde{Y(1-Y)} = a_1 \dfrac{\tau_c}{\tau_t} + a_2 \left(\dfrac{\tau_c}{\tau_t}\right)^2 + \ldots$, we end up finally with

(6) $\quad \widetilde{\chi}_Y^F \propto \widetilde{\chi}_Y^i$

where $\widetilde{\chi}_Y^i$ is the contribution from outside the flamelets to $\widetilde{\chi}_Y$; the magnitude of gradients are probably higher within the flamelets than outside, but, when $\tau_t/\tau_c \to 0$ the flamelets thickness δ_L goes to zero, and then the percentage of volume occupied by the flamelets is probably small ; it seems then not convenient to neglect $\widetilde{\chi}_Y^i$, and, following [14], it is the only way to avoid the unrealistic result of zero dissipation rate $\widetilde{\chi}_Y$.

It remains to assume that $\widetilde{\chi}_Y^i \propto \widetilde{Y'^2}/\tau_t$, as in the non reacting case, to find (1) ; but in fact $\widetilde{\chi}_Y^i$ is the dissipation rate conditionned at $Y \cong 0$, and not $\widetilde{\chi}_Y$ itself...

There could be a simple physical explanation for that result, a little bit surprising, that the flamelets do not modify the dissipation rate. Indeed, if we follow the idea of the Kolmogorov cascade, $\widetilde{\chi}_Y$ is not governed by small scale phenomena, but at the contrary by large scale ones. In that case, when the reactions are very fast and the flamelets very thin, they will modify only the smallest scales of the scalar spectrum, so without consequence for $\widetilde{\chi}_Y$. However, it remains to explain if, when U_L is of the same order than k, the influence of flamelets could be sensible also at larger scales.

Libby, Bray and Moss, in [14], have adressed also this problem. Using also a picture of the reacting medium with flamelets, they have found also the undeterminated form of $\widetilde{\chi}_Y$. But they have proposed to overcome it with :

(7) $\quad \widetilde{w} \propto \dfrac{u_L}{l_t} \widetilde{Y}(1-\widetilde{Y})\quad$, where l_t is the integral length scale of the turbulence.

That leads simply to :

(8) $\quad\quad\quad\quad \tau_Y \propto \dfrac{l_t}{u_L} \quad\quad ;\ (\ l_t = k^{\frac{1}{2}}\tau_t\ )$

which departs from (1) or (2) and put into evidence the right number $U_L/k^{\frac{1}{2}}$.

In a more recent paper, the same authors propose [17] :

(9) $\quad \widetilde{w} \propto \dfrac{\widetilde{Y}(1-\widetilde{Y})}{\widehat{T}} \dot{w}_F$, with $\quad \dot{w}_F \propto \dfrac{u_L}{\widetilde{u}}$

$\widehat{T}$ is the integral time scale obtained from the one point time spectrum of Y, and $\widetilde{U}$ the mean velocity in which the turbulent flame is stabilized. With the theory of [17], $\widehat{T}$ is a constant all over the flame brush, and so could be taken proportional to the same quantity in a non reacting case, i.e. to τ_t, with a large enough Re_T. On the other hand, $\widetilde{U}$ is nothing but U_T, the so called "turbulent burning velocity" ; taking as tentative $U_T \propto k$, one finds again a formula identical to (7).

One could think, then, that $\tau_Y \propto \tau_t$ is probably justified (at large Reynolds number and for homogeneous turbulence) when $U_L \ll k^{\frac{1}{2}}$. But do we replace that relation by (8) when $U_L/k^{\frac{1}{2}}$ is of order one ? In fact, it is somewhat surprising that the formula (9) in [17] is justified again with $U_L/k^{\frac{1}{2}} \ll 1...$.

The problem of the influence of flamelets on the closure assumption for $\widetilde{\chi}_Y$ has been also tackled by Pope and Anand [16], in the framework of the modelled pdf approach. Here the problem appears when closing the small scale mixing term for the joint pdf equation ; in the case of so called distributed reaction (that is when the reaction is not very fast and flamelets are not present) this term is modelled in such a way that the balance equation for $\widetilde{Y'^2}$ involves a destruction term $\widetilde{\chi}_Y$ equal to $C_Y \widetilde{Y'^2}/\tau_t$. In the case of flamelets combustion, Pope and Anand mention also that there are two contributions, inside and outside the flamelets, which have to be modelled separately ; they model the part from outside the flamelets as in the distributed reaction case, and the first one is modelled together with the reaction term refering to the profile of Y inside steady laminar flamelets (see [16] for details). They have been able to compute the turbulent flame velocity U_T, as a function of τ_t/τ_c, the Damköhler number ; the results show an increase of $U_T/k^{\frac{1}{2}}$ with τ_t/τ_c, very rapidly with a saturation phenomenon for the flamelet combustion, much more slowly for the distributed combustion. In no case it appears a dependance of U_T on U_L.

From the pdf equation used in [16] for the case of flamelets combustion, we can extract a new equation for $\widetilde{Y'^2}$; because the new closure assumption of [16] involves a term proportional to $1/\tau_c$, and because of the linear shape of the $h^*(\phi)$ in the non reacting part of the curve (see fig. 1 of [16]) we will end up with a new closure assumption of the type :

(10) $$\widetilde{\chi}_Y = \widetilde{Y'^2}\left(c_\phi \frac{1}{\tau_t} + c_F \frac{1}{\tau_c} \right)$$

Then, it appears clearly that the dependence of the results with τ_t/τ_c will be modified, but that no influence of $U_L/k^{\frac{1}{2}}$ can be found. In fact, this study deals again with $k^{\frac{1}{2}} \gg U_L \ldots$.

What conclusions could we extract from these apparently contradictory studies? Our tentation is to consider that each study has been able to handle only a part of the reality. The contribution inside the flamelets should be, as found by Bray, Libby, Moss, proportional to $U_L/k^{\frac{1}{2}}$; but, as pointed out in [15] and [16], a contribution from outside the flamelets exists, may be similar to the case of a non reacting medium. Although the convenient weighting of each contribution does not appear very clearly at the moment, it would lead in any case to :

(11) $$\widetilde{\chi}_Y = \frac{\widetilde{Y'^2}}{\tau_t} \cdot \left(1 + \alpha \frac{U_L}{k^{\frac{1}{2}}} \right)$$

which represent, at least, a first order approximation of the reality... .

A recent numerical simulation, by means of cellular automata, has been proposed in [18] for an idealized premixed flame propagation at high Damköhler number. Although the cellular automaton algorithm has been adopted on purely intuitive basis, the result has been obtained that $\widetilde{W}$ is depending on $U_L/k^{\frac{1}{2}}$ with a simular linear relation.

II.2 Finite Damköhler number

II.2.1 Let us consider first, again, _diffusion flames_. Finite values of Damköhler number can be taken into account in predictions either through a model using a balance equation for a multidimensional pdf of reactive species as proposed first by O'Brien (see [19]), or through a "flamelet model", following Liew and Bray [20] or Peters [6] ; a third possibility, less known, is the "coherent flame model" of Marble and Broadwell [5]. A big problem in all cases is the chemical scheme; the reactive pdf approach is limited, up to now, to a few steps mechanism, while the flamelets or coherent flame approaches seem more able to take into account quite complex schemes. But the range of validity of the approaches are different...

Concerning the time scales, the situation is quite clear for the "flamelets model"; it is needed first $\tau_s = \widetilde{\xi'^2}/\widetilde{\chi}$, again, in order to compute the pdf of the inert scalar ξ; the discussion of § II.1.1 is always valid. Secondly, in order to take into account the stretch effect on the flamelets, it is needed in addition a distribution of stretch rates, again a pure turbulent characteristic. The "coherent flame model" uses a modelled equation for $\overline{\Sigma}$, the flame surface by unit of volume; that quantity is nothing but a length scale (or the inverse of a length scale) for the reacting turbulent medium. To find an equation for $\overline{\Sigma}$, including the effects of turbulence, small scale diffusion, local extinctions, and even small scale interaction of flamelets is not easy; the proposal of Marble and Broadwell represents only a first step toward that. The situation appears here far more complicated.... For the "modelled reactive pdf" approach, the problem is more clearly set up; several time scales like $\tau_Y = \widetilde{Y'^2}/\widetilde{\chi_Y}$ are needed, and, up to now, they are modelled using again (2) or (1), that is neglecting any influence of chemical reactions on it. In principle, that assumption can be justified only if the shape of the spectrum of the fluctuations of the reactive species Y is not significantly modified by the reactions, and we do not know practically nothing about such a spectrum.... However, a fortunate result is that the predictions using that assumption did not show any pathological behaviors (see [21] or [22]); they look like quite realistic, on the contrary.

II.2.2 In the case of <u>premixed flames</u>, the situation is similar. In fact, the "modelled reactive pdf" approach can handle premixed flames as well as diffusion flames. However, for a premixed flame, if the chemical scheme can be reduced to a single step mechanism, and if other, restrictive, assumptions are made, it is possible to reduce the dimensionality of the pdf to one, and to use, even, a "presumed pdf" approach (see [23]).

In such a case, the problem to find a modelled equation for $\widetilde{\chi_Y}$ itself can be attacked; such a way has been attempted in [13] and it has been found a clear influence of the reaction on λ_Y, that is on τ_Y itself. But the final effect on the transverse profiles of $\widetilde{Y}$, was of the same order of the uncertainties about the chemical constants....

The study of [16], previously described, represents also an attempt for a more convenient closure of $\widetilde{\chi_Y}$; it is simpler than [13]; results are quite plausible, but its assessment remains to be made.

Another possibility to improve the pdf approach is to search for a modelled equation for the joint pdf of Y and $\vec{\nabla}Y$; that has been proposed by Meyers and O'Brien [24]. This approach is very appealling, because it can solve (approximately, of course) three problems at the same time : i) the problem of τ_t/τ_s in a non reactive medium, ii) the problem of the influence of reaction on τ_t/τ_Y, iii) the problem of using a distribution of time scales, as in the flamelets approach. But there are very few experimental results, at this time, for our guidance into the modelling of such a very complicated equation ; what is the way, then, to test the equation proposed by Meyers and O'Brien and to improve it, if necessary ?

II.2.3 What are the right parameters to be included in the models, in order to take into account finite rate chemistry in both cases of premixed or non premixed flames ? Let us assume here for simplicity that an unique τ_c can characterize the chemistry.

The "flamelets model" will involve as key parameter τ_c/τ_K, where τ_K is the Kolmogorov time, as τ_K^{-1} is the mean stretch rate. On the contrary the "modelled pdf" approaches [8], [16], or the "presumed pdf" approach [13] will involve τ_c/τ_t, where τ_t is the integral turbulent time. That is not surprising, because the flamelets model is valid in the flamelets regime of turbulent flames, bounded by the $\tau_K = \tau_c$ line (see [25]) while the others have been built for the distributed combustion regime, bounded by the $\tau_t = \tau_c$ line.

On the basis of the Meyers and O'Brien model for the joint pdf of Y and $\vec{\nabla}Y$, we would obtain τ_t/τ_c as well as τ_K/τ_c as interesting parameters because the turbulence Reynolds number $Re_T = (\tau_t/\tau_K)^2$ appears explicitly in it (see [26]).

More than just a small improvement of a "modelled reactive pdf method", it seems therefore that, this type of approach could lead to a model valid for both regimes. Although for the moment the physical implications of the present form of the model are not clear ; it can surely be a base for further studies.

III. PART TWO : NUMERICAL EXPERIMENTS IN ORDER TO STUDY TIME AND LENGTH SCALES IN AN HOMOGENEOUS TURBULENT PREMIXED REACTIVE MEDIUM

III.1

Experimental results concerning fluctuations of reactive species are difficult to obtain, specially when we are interested on two points or two times measurements. But numerical experiments with a full direct simulation of the exact balance equations can lead to results of this type. We have chosen, for the first case, to study a turbulent single species reactive homogeneous medium, without mean velocity, without heat release, and with a non linear reaction rate

$$\dot{w}(Y) = \frac{7}{\tau_c} Y(1-Y)^5 \tag{12}$$

Then the flow field is governed by :

$$\frac{\partial \vec{u}}{\partial t} + (\vec{u} \cdot \vec{\nabla}) \times \vec{u} = -\frac{1}{\rho} \vec{\nabla} p + \nu \Delta \vec{u} \tag{13}$$

and the concentration field Y by :

$$\frac{\partial Y}{\partial t} + (\vec{u} \cdot \vec{\nabla}) Y = d \Delta Y - \dot{w}(Y) \tag{14}$$

From a given initial field of $(\vec{u}, Y)$, randomly determined but homogeneous and isotropic, the direct numerical integration of (12), (13), (14) can be performed with a spectral method in space and finite differences in time, provided that the Reynolds number is not too large ; that need, of course a CRAY-1 computer ; the details are given in [28]. We can vary τ_t^o/τ_c (where τ_t^o is the integral time of the turbulence at $t = 0$), l_c^o/l_t^o, the ratio of length scales of the Y field and velocity field, and, too lesser extend, ν/d and Re_T. As time is increasing, the turbulence is decaying and the reaction is consuming Y ; as we will see, the rate of decrease of Y depends strongly on turbulence, if τ_c and τ_{t_o} are of the same order.

III.2 Some results about the length scales

The figures 1 and 2 give the results of time evolution of $\widetilde{Y}$, $\widetilde{\dot{w}}$, and also the length scales l_t, l_c and λ_t, λ_c respectively for $\tau_c = 0.5$ and 2.

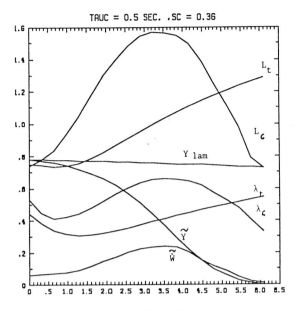

Figure 1 :

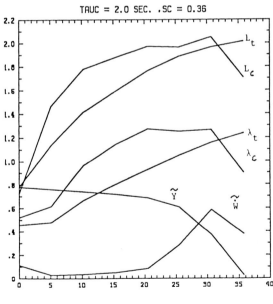

Figure 2 :

In fig. 1 is plotted also Y_1, which is simply the result of $d/dt(Y_1) = -(7/\tau_c) \cdot Y_1 \cdot (1-Y_1)^5$; one can notice the strong difference between Y_1 and $\widetilde{Y}$. The time for consumption of fuel is much larger with $\tau_c = 2$ than 0.5, but $\widetilde{Y}$ evolution are very different, and in no way similar, in both cases ; the consumption is neither chemically controlled, nor controlled by the turbulence.

The integral length scales l_t and l_c can be compared after a short period where the influence of initial spectra is too strong ; l_t is increasing, as expected in decaying turbulence, but l_c first increases, and then decreases, the decrease of l_c occurs at the same time as $|\widetilde{W}|$, approximately, for all the cases.

The integral time scales, in fact, are directly related to the Taylor microscales λ_t and λ_c, as $\varepsilon/\nu k \propto (\lambda_t^2)^{-1}$ and $\widetilde{\chi}/(d\widetilde{Y}^2) \propto (\lambda_c^2)^{-1}$. A similar behavior of λ_t with respect to λ_c is seen after the initial period; λ_t increases monotonically, but λ_c increases until $|\widetilde{W}|$ attains its maximum, and then decreases. The same behavior occurs for τ_t and τ_Y.

These results show clearly that, in no way, neither τ_Y/τ_t nor l_c/l_t could be approximated by a constant value. A closure assumption like (1) or (2) is therefore not valid here. Similarly a formula like (8) desagrees with the numerical simulation, because U_L ($\propto (d/\tau_c)^{\frac{1}{2}}$) is here a constant in time, for each case. Equivalently, the tentative formula of Pope and Anand : the formula (10) cannot explain the non monotonic behavior of τ_Y/τ_t.

III.3 Some results about the spectra and pdf's

The figures 3 and 4 show the spectra for the velocity fluctuations and scalar Y, respectively, at different times. One see fig. 3 the velocity spectrum attaining an equilibrium shape after few time steps, and then decreasing with keeping this shape. On the contrary, for the reactive scalar, the total $\widetilde{Y'^2}$ increases at the beginning and decreases further ; the spectral distribution of the fluctuations is much more flat : the reaction plays, in a complicated manner, at large scales as well at small ones ; a self similarity of the spectra at different times is not clear.

It is possible also to extract from the numerical experiments the joint pdf of Y and $(\nabla Y)^2$; the figures 5.a) to e) show them for different times, in the case $\tau_c = 0.5$. The large gradients at $Y = 1$, for $t = 0$, are due to the initial fields, in which we have simply avoided values larger than 1 by truncation. One see, as the time proceeds, more and more events at small Y, and, at a sufficient time, zero

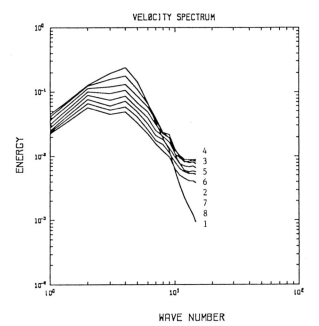

Figure 3: Evolution from 0 to 1 Sec.

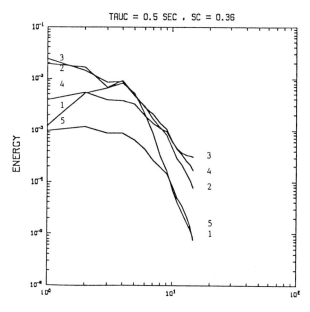

Figure 4: Scalar spectrum from 0 to 2 Sec.

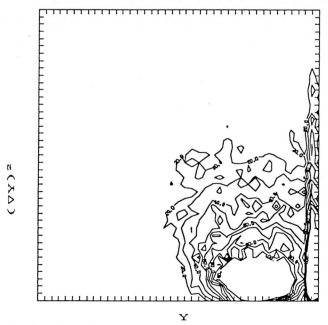

Figure 5 :a) at time t=0

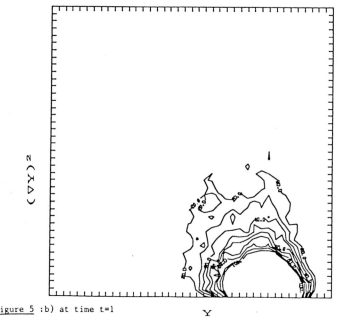

Figure 5 :b) at time t=1

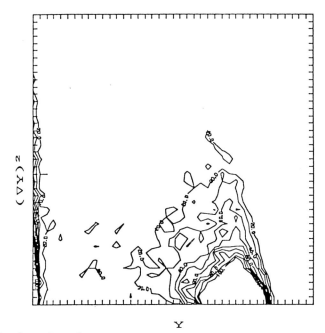

Figure5 :c) at time t=2

Figure5 : d) at time t=3

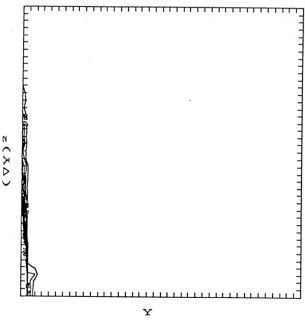

Figure 5 :e) at time t=4
(Y varies from 0 to 1 but the $(\nabla Y)^2$ scale varies from a figure to the other)

values appear ; note that the scale for $(\nabla Y)^2$ changes for each time : the maximum value of $(\nabla Y)^2$, due to that, increases and then decreases when the times goes on. It is quite understantable that there is a low pdf value around $Y = 0.2$, because it is the location of higher instantaneous reaction rate ; it is more surprising that quite large gradients occurs at $Y \cong 0$; that corresponds probably to strong curvatures of instantaneous Y profiles close to zero.

We have seen that τ_Y and l_c evolve during the combustion differently from τ_t and l_t. A closure assumption, even in this very simple case of homogeneous and isotropic turbulent medium, is not easy to find. A detailed study of spectra and joint pdf's could be of great interest for that.

IV. PART THREE : A NEW MODEL WITH INTERMITTENT COMBUSTION

IV.1 The principles

From the discussions of the first part, the conclusion has arosen of the need of a model taking into account the influence of the chemical reaction on τ_Y ; in addition the best way to attack this problem with the possibility to cover the distributed combustion regime as well as the flamelets regime has appeared to be using explicitely a distribution of turbulent characteristic times ; finally the model would be based on more physical ground than the one of Meyers and O'Brien [24]. A first step toward that direction has been done in [26] ; further improvements and an attempt of application to a diffusion flame have been performed in [29], that we will describe now.

The model is based on a lagrangian approximation of the balance equations for any species Y_i, called the I.E.M. model (Interaction by Exchange with the Mean); details about such a model, and other lagrangian models, can be found in [30] and [31]. The lagrangian equation describing the model is

(15) $$\frac{d Y_i}{dt} = \frac{\widetilde{Y_i} - Y_i}{\tau_{ex}} + \dot{w}_i(Y_1, \ldots Y_n, T)$$

It must be noticed that this model leads to the same eulerian, balance equations for $\widetilde{Y_i}$ and $\widetilde{Y_i'^2}$ than used previously in "presumed pdf" models, and the same one also that the ones given by "modelled pdf" approaches.

But we will extend now the model assuming that there is a distribution of exchange times τ_{ex}, instead of a single one. In addition we will assume now that the combustion is very sudden, that is the global activation energy is very large ; that simplify the question of the integration of (15), and on the other hand allows a clearer definition of τ_{ex}, as we will see later.

Let us consider now a non premixed flame ; the flow enters the burner through two entrances : A and B. If we consider Y_o as an inert tracer and Y_1 as the fuel, and if we look at the plane (Y_o, Y_1), the solutions of (15) can be represented by trajectories within this plane ; some of them are beginning at A, others at B.

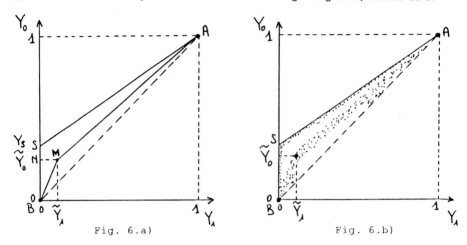

Fig. 6.a)   Fig. 6.b)

Following equation (15), with a very large activation energy and in an homogeneous turbulent medium (where $\widetilde{Y}_o$ and $\widetilde{Y}_1$ are constant everywhere), it can be demonstrated that these trajectories are made with segments (see [31]) : for instance for a particle entering the flow in A a path along AM followed by a jump on AS, a path along AS and SN ; for a particle issued from B a similar evolution occurs on BM and BN. As indicated in fig. 6.a), the result of that is that every fluid particle can be found on the lines AS, SB, BM and MA, but not elsewhere, because the jumps are very fast due to the sudden combustion ; so the joint pdf $P(Y_o, Y_1)$ is supported only by BS, SA, AM and MB. Due to the fact that the IEM model is not exact, and that we are interested in non homogeneous flows, the joint pdf is not actually as simple ; instead it looks like the figure 6.b) ; indeed such a picture has been found experimentally by Masri and Bilger [32], but in the plane : Nitrogen, temperature.

IV.2 The formulae from the intermittent combustion model

The IEM model, associated with fig. 6b), allows us directly to find a mean reaction rate expression for Y_1, even with random τ_{ex} ; indeed the combustion takes places only on the AS and BS lines, and, following (15)

(16)
$$\begin{cases} \dot{w}_1 = - \tilde{Y}_1/\tau_{ex} & \text{on SB} \\ \dot{w}_1 = - \dfrac{1}{\tau_{ex}} \left(\tilde{Y}_1 - \dfrac{\tilde{Y}_o - Y_s}{1 - Y_s} \right) & \text{on AS} \end{cases}$$

(see [21] for details).

On the other hand, any jump can be physically interpreted as an ignition. It can be assumed, then, that the probability of jump at each value of Y_o in the vicinity of the AM and MB lines is just the quantity α as

(17) $\quad \alpha(Y_o) = \displaystyle\int_{\tau_{ig}(Y_o)}^{\infty} P_\tau(\tau_{ex}) \, d\tau_{ex} \quad$ where P_τ is distribution of τ_{ex}

That means that the fluid particles with a frequency of exchanges $1/\tau_{ex}$ which is to high with respect to a reciprocal ignition time τ_{ig}^{-1} cannot be ignited.

It is possible now to compute the mean reaction rate $\tilde{\dot{w}}_1$:

(18) $\quad \tilde{\dot{w}}_1 = \displaystyle\int_0^1 \tilde{P}(Y_o) \int_{\tau_{ig}(Y_o)}^{\infty} \dot{w}_1 \, P_\tau(\tau_{ex}) \, d\tau_{ex} \, dY_o$

where $\dot{w}_1$ is given by the formulae (16).

The formulae (18) can be used directly in an eulerian balance equation for Y_1 ; the other species can be calculated also owing to similar formulae (for oxydizer or products) or even simpler (see [26]) ; but the assumption of sudden chemistry precludes the computation of many species.

The use of (18) demands three entries :

i) the knowledge of $P(Y_o)$; it is the pdf of an inert scalar ; the problem is classical ;

ii) the knowledge of $P_\tau(\tau_{ex})$; it is a new entry. With our assumption of sudden combustion, τ_{ex} is just an exchange time due to the turbulence before ignition; no influence of chemistry is to be taken into account, a distribution, as a function of τ_t and τ_K, deduced from theoretical or experimental basis ;

iii) the knowledge of $\tau_{ig}(Y_o)$; that is a pure chemical quantity. It is not τ_c, the classical chemical time, used for instance in a premixed flame for $U_L \alpha (d/\tau_c)^{\frac{1}{2}}$. It is an ignition time, taken roughly along the AM and MB lines ; therefore it depends on Y_o and $\widetilde{Y}_o$, $\widetilde{Y}_1$, $\widetilde{Y}_3$ ... $\widetilde{Y}_n$. It could be calculated knowing a full mechanism, or deduced from experiments... .

IV.3 An application

In order to apply the previous model, we have chosen as a first attemps $P_\tau(\tau_{ex})$ and τ_{ig} as follows :

a) P_τ has been assumed a uniform distribution between τ_t and τ_K ; τ_t and τ_K are calculated by means of a k-1_t model.

b) τ_{ig} has been computed from an Arrhenius law $W_1(Y_1, Y_{ox}, T)$ with Y_{ox} related to Y_1 and Y_o (inert scalar) assuming a single step reaction, and T related to Y_o and Y_1 assuming that enthalpy behaves like an inert scalar (see [23]). We adopted

$$\tau_{ig}^{-1}(Y_o) = \int_{Y_1^{(1)}(Y_o)}^{Y_1^{(2)}(Y_o)} \dot{w}_1(Y_o, Y_1) dY_1$$; for a given value of Y_o, $Y_1^{(2)}$ and $Y_1^{(1)}$

are respectively the values of Y_1 on the AMB line or on the AB line ; the purpose of that formula is to take into account, at some extent, the scattering around the AMB line, as shown fig. 6b).

We have used this model to compute the diffusion flame in a model burner shown fig. 7, a little bit more complicated than a jet flame. The model described is able to represent unsteady regimes, and particularly the ignition process ; to do that, it is necessary only to prescribe a consumption of Y_1 at some place within the burner (thus simulating an ignition device) and to wait until the ignition lead to a steady regime (if successfull). The figure 7 shows also the boundary of the reaction zone at different times during such an evolution.

The profiles of species and temperature, in this steady regime, are shown fig. 8a), b), c) and 9.

V. CONCLUDING REMARKS

We have now to list the conclusions from the previous discussion, in order to determine the future works :

a) The problem of τ_t/τ_s relation is yet unsolved ; it will be probably solved with a two point theory, based on extensive experimental results. Very interesting results of this type are the ones of [27]. Our tentation is to leave this problem to fluid dynamicists... .

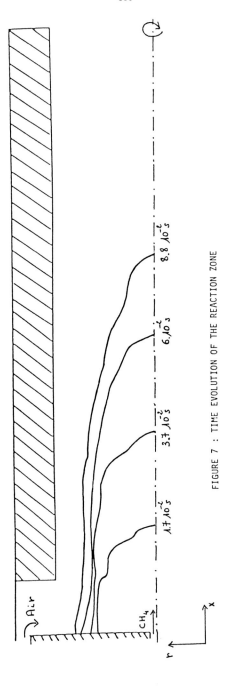

FIGURE 7 : TIME EVOLUTION OF THE REACTION ZONE

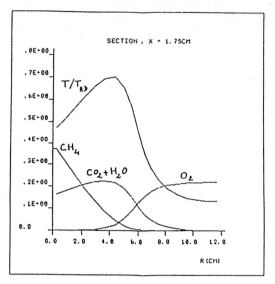

FIGURE 8.a)

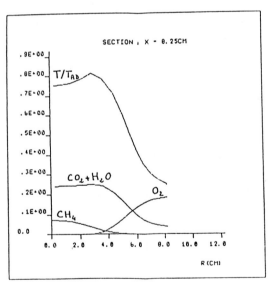

FIGURE 8.b)

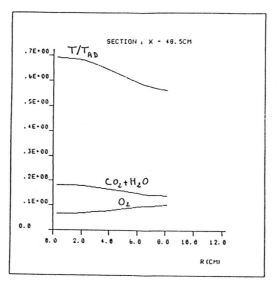

FIGURE 8.c)

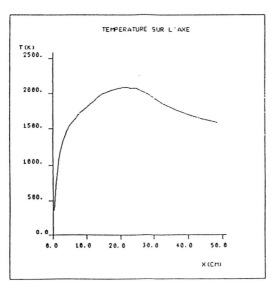

FIGURE 9

b) In premixed flames with very fast reactions, the problem of the dependence of τ_Y/τ_t on $U_L/k^{\frac{1}{2}}$ is open. A formula like (11) has to be verified or improved. It is necessary to note, in addition, that we have assumed a large turbulence Reynolds number : what are the modifications for small Reynolds numbers ?

c) For distributed combustion (in the case of both premixed or non premixed flames), the existing models with the very simplistic relation $\tau_Y \propto \tau_t$ lead to results that are, apparently, not unrealistic, compared to the experiments. But, at this time, more precise and detailed measurements are needed. Numerical experiments appear very valuable for that ; unfortunately the results of our simulations do not agree with anyone of the simple existing closure formulae...

d) Although for flamelets combustion, there are apparently no problems other than a), the difficulties arise when one is interested in a model that has to cover also distributed combustion. A model with a distribution of turbulence time has to be done in order to cover both regimes. Our Lagrangian Intermittent Model (MIL) is an attempt toward that way, which needs to be studied in details and compared with experiments.

ACKNOWLEDGEMENT

We thank Dr. Jean-Pierre Chollet of "Institut de Mécanique de Grenoble" for his very helpful contribution to numerical simulation by pseudo-spectral method in homogeneous and isotropic turbulence.

REFERENCES

[1] D.B. SPALDING - "Mixing and chemical reaction in steady confined turbulent flames" - 13th Symp. on Combustion - The Comb. Inst. Pittsburg - p. 649 (1971).

[2] R. BORGHI, P. MOREAU - "Turbulent combustion in a premixed flow" - Acta Astronautica - vol. 4, pp. 321-341 (1977).

[3] F.C. LOCKWOOD, A.S. NAGUIB - "A unified method for the prediction of turbulent diffusion and premixed flames" - Proceedings of the Second Eur. Symposium of combustion - p. 502 - Orléans (sept. 1975).

[4] S.B. POPE - "Transport equation for the joint probability density function of velocity and scalars in turbulent flow" - Phys. Fluid - Vol. 24, n° 4, pp. 588-596 (1981).

[5] F.E. MARBLE, J.E. BROADWELL - "The coherent flame model for turbulent chemical reactions" - Project squid tech. Rep. - TRW-9-PU - Purdue University, West Lafayette, Indiana (1977).

[6] N. PETERS - "Laminar flamelets concepts in turbulent combustion" - Munich (aug. 1986).

[7] BILGER R.W. - "Turbulent diffusion flames", in : Turbulent Reacting Flows - Ed. P.A. Libby, F.A. Williams - Topics in Applied Physics - vol. 44, Springer-Verlag Heidelberg (1980).

[8] W.P. JONES, W. KOLLMANN - "Multi Scarlar pdf transport equations for turbulent diffusion flames" - 5th Turbulent Shear Flows Conference - Cornell University (aug. 1985).

[9] C. BEGUIER, I. DEKEYSER, B.E. LAUNDER - "Ratio of Scalar velocity dissipation time scales in shear turbulence" - Phys. Fluids 26, pp. 1222-1227 (1978)

[10] M. LARCHEVEQUE, J.P. CHOLLET, J.R. HERRING, M. LESIEUR, G.R. NEWMAN, D. SCHERTZER - "Two point closure applied to a passive scalar in Decaying Isotropic turbulence" - pp. 50-66 - Turbulent shear flows 2, Springer-Verlag, (1980).

[11] O. ZEMAN, J.L. LUMLEY - "Modeling Buoyancy driven mixed Layers" - Journal of Atmospheric Sciences - Vol. 33, p. 1974 (1976).

[12] K.N.C. BRAY, J.B. MOSS - "A unified statistical model of premixed turbulent flame"- Acta Astronautica - Vol. n° 3-4, pp. 291-319 (1977).

[13] R. BORGHI, D. DUTOYA - "On the scales of the fluctuations in turbulent combustion" - 17th Symp. (Int.) on Combustion - The Comb. Inst., Pittsburgh - pp. 235-244 (1978)

[14] P.A. LIBBY, K.N.C. BRAY - "Implications of the laminar flamelet model in premixed turbulent combustion" - Comb. and Flame - Vol. 39, pp. 33-41 (1980).

[15] R. BORGHI - "Turbulent premixed combustion : further discussion on the scales of fluctuation" - 9th Int. Coll. Dyn. Expl. React. Syst. - Poitiers (4-8 July 1983) - to be published in Comb. Flame.

[16] S.B. POPE, M.S. ANAND - "Flamelet and distributed combustion in premixed turbulent flames" - 20th Symp. (Int.) on Combustion - Ann Arbor (aug. 1984).

[17] K.N.C. BRAY, P.A. LIBBY, J.B. MOSS - "Flamelet crossing frequencies and mean reaction rate in premixed turbulent combustion" - to be published in Combustion Sciences and Technology.

[18] R. BORGHI, R. SAID - "Using a simulation by "cellular automata" as a help for Turbulent Combustion Modelling" - 9th Australasian Fluid Mechanics Conference - Auckland (9-12 déc. 1986).

[19] E.E. O'BRIEN - "The probability density function approach to reacting turbulent flows" in : Turbulent Reacting Flows - Ed. P.A. Libby, F.A. Williams - Topics in Applied Physics - Vol. 44, Springer-Verlag - Heidelberg (1980)

[20] S.K. LIEW, K.N.C. BRAY, J.B. MOSS - "A flamelet model of turbulent non premixed combustion" - Comb. Sci. Tech. - vol. 27, pp. 69-73 (1981).

[21] C. BONNIOT, R. BORGHI - "Joint probability density function in turbulent combustion" - Acta Astronautica - Vol. 6, pp. 309-327 (1979).

[22] J. JANICKA, W. KOLLMANN - "A two-variables formalism for the treatment of chemical reactions in turbulent Ar_2 - A_{12} diffusion flame" - 17th Symp. (Int) on combustion - pp. 421-430 - The Comb. Inst., Pittsburgh (1979).

[23] R. BORGHI - "Models of Turbulent Combustion for Numerical Predictions". In : Prediction methods for turbulent flows - Ed. W. Kollmann - A von Karman Institute Book, Hemisphere Pub. Corp. (1980).

[24] R.E. MEYERS, E.E. O'BRIEN - "The Joint pdf of a scalar and its gradient at a point in a turbulent fluid" - Combustion Science and Technology - Vol. 26, pp. 123-124 (1981).

[25] R. BORGHI - "On the structure and morphology of turbulent premixed flames". In : Recent advances in the aerospace sciences - Ed. Corrado Casci (Plenum Publishing Corporation, 1985).

[26] R. BORGHI, M. GONZALEZ - "Further application of Lagrangian Models to turbulent combustion - Combustion and Flame - Vol. 63, pp. 239-250 (1986).

[27] F. ANSELMET, R.A. ANTONIA - "Joint statistics between temperature and its dissipation in a turbulent jet" - Phys. Fluid. - vol. 28, n° 4, pp. 1048-1054 (1985).

[28] A. PICART, R. BORGHI, J.P. CHOLLET - "Numerical simulation of turbulent reactive flows : new results and discussion - 6th Symp. on Turbulent Shear Flows, Toulouse (sept. 1987).

[29] M. GONZALEZ - "Contributions à la simulation numérique d'écoulements avec combustion". Thèse de Physique - Université de Rouen (1986).

[30] R. BORGHI, E. POURBAIX - "Lagrangian models for turbulent combustion" - 4th Symp. on Turbulent Shear Flows, Karlsruhe (sept. 1983).

[31] R. BORGHI - "Turbulent combustion - on its structure and its modelling" - In : Combustion and non linear phenomena - Ed. P. Clavin, B. Larrouturou, P. Pelce - Les éditions de Physique (1986).

[32] A.R. MASRI, R.W. DIBBLE, R.W. BILGER - "Turbulent non-premixed flames of Methane near extinction : mean structure from Raman measurements - 21th International Symposium on Combustion - Munich (August 3-8, 1986).

STATISTICAL MODELLING OF TURBULENT REACTIVE FLOWS

by

S. B. Pope

Sibley School of Mechanical and Aerospace Engineering
Cornell University
Ithaca, NY 14853

ABSTRACT

Turbulent combustion occurs in many engineering applications: spark-ignition engines, gas-turbine combustors, and furnaces, for example. In each of the applications cited the design process is lengthy and expensive. The industries involved are attempting to improve their design procedures by using computer models of turbulent reacting flows.

Since the fundamental governing partial differential equations are known, the direct approach is to solve them numerically. However, because of the wide range of length and time scales involved, this direct approach is computationally impracticable, now and in the foreseeable future. The alternative is to use a statistical approach. Such approaches face a formidable challenge: superimposed on the difficulties of calculating inert, constant-density turbulent flows, are those associated with reaction and heat release. The reaction rates are typically highly nonlinear functions of the composition variables, which are subject to large turbulent fluctuations. Often reaction takes place in laminar flamelets that are thin compared to turbulent scales. Due to heat release the specific volume of the mixture can increase by a factor of ten which, as may be expected, is found to have a large effect on the turbulence. For example, due to heat release, the turbulence energy can increase by an order of magnitude, and new transport process become dominant and can lead to countergradient diffusion.

REFERENCES

Pope, S.B. 1985. "PDF"Methods for Turbulent Reactive Flows". *Prog. Energy Combust. Sci.* **11**, 119-92.

Pope, S.B., Cheng, W.K. 1986. "Statistical Calculations of Spherical Turbulent Flames". *Twenty-First Sym. (Int'l) on Combustion.*

Pope, S.B., Correa, S.M. 1986. "Joint PDF Calculations of a Non-Equilibrium Turbulent Diffusion Flame". *Twenty-First Sym. (Int'l) on Combustion.*

Anand, M.S., Pope, S.B. 1987. "Calculations of Premixed Turbulent Flames by PDF Methods". *Combust. Flame* **67**, 127-42.

Pope, S.B. 1987. "Turbulent Premixed Flames". *Ann. Rev. Fluid Mech.* **19**, 237-70.

COHERENT FLAME DESCRIPTION OF TURBULENT PREMIXED DUCTED FLAMES

N. Darabiha, V. Giovangigli, A. Trouvé, S.M. Candel and E. Esposito

E.M2.C. Laboratory, CNRS and Ecole Centrale des Arts et Manufactures
92295 Châtenay-Malabry, FRANCE

Abstract

This paper describes some aspects of our effort to analyze turbulent combustion on the basis of an extension of the coherent flame model initially proposed by Marble and Broadwell.

At this stage the model comprises a local description (flamelets) and a global representation of the turbulent flow-field including a balance equation for the mean flame area per unit volume.

The flamelets are non-adiabatic premixed strained flames, a model suggested by Libby, Linan and Williams. Complex chemistry calculations have been carried out for a large number of propane-air flames and a large data-base of flamelets is being constructed. These calculations provide consumption rates, extinction and ignition characteristics which are used in the global turbulent calculation to model the mean reaction terms. Numerical results obtained for turbulent premixed flames stabilized in a duct are discussed.

Experiments performed on a model combustor provide distributions of the mean heat release rate. These distributions are compared with those determined numerically. This comparison indicates that the coherent flame description accounts for important features found in the experiment.

1. INTRODUCTION

Many recent studies of turbulent combustion are based on flamelet models. In these models the turbulent reaction zone is viewed as a collection of laminar flame elements (flamelets) embedded in the turbulent stream (Williams 1975, Carrier et al. 1975, Marble and Broadwell 1977, 1979, Peters 1984, 1986, Bray 1986, Spalding 1978, Libby and Bray 1984, Clavin and Williams 1982).

An advantage of this concept is that it essentially decouples complex chemistry calculations from the turbulent flow description. Chemical kinetics and multi-component transport properties may be treated separately and the results of the local flamelet analysis are then included in the calculation of the turbulent flow field. In practical applications a flamelet library (data base) is first constructed and provides specific properties such as the consumption rates per unit flame area, ignition and extinction conditions that are required in the computation of the turbulent flow field. Because chemical kinetics are explicitly treated it is expected that this approach will be able to describe its influence on the turbulent flame structure.

Flamelet models are believed to provide a viable description of turbulent combustion in the range of large Damkohler numbers and for flows characterized by typical length scales which are much larger than the typical flame thickness (Peters 1986, Bray 1986, Borghi 1985, Williams 1985). Another limit of the flamelet regime may result from quenching mechanisms due to the turbulent fluctuations. According to some authors (Borghi 1985, Williams 1985, Peters 1986) the turbulent flame structure is modified if the Kolmogorov length scale ℓ_k becomes larger than the flame thickness. It is indicated that in that situation the smallest eddies enter into the inner flame structure and that the rates of strain associated with these eddies locally quench the flamelets. This is a regime of distributed reaction in which the flamelets have lost their identifiable structure. While some authors believe that this regime cannot be described by flamelet models, others suggest that flame quenching and the subsequent mechanisms and modifications of the reactive flow may be treated with flamelet descriptions. The question is not settled because very little is actually known on the structure of the turbulent small scales and on their interaction with flames. This point may be examined with detailed numerical simulations but the current understanding remains essentially intuitive and speculative. Relevant contributions dealing with these aspects are due to Williams (1975, 1985), Broadwell and Breidenthal (1982), Peters (1984, 1986), Borghi (1985), Bray (1986). Further comments on the problem of flame quenching and on its modelling are given in Section 2 of this paper.

While the domain of application of flamelet models cannot be precisely defined at this time one may say that such models are generally valid in the range of large Damkohler numbers. The typical turbulent scale must also be much larger than the flame thickness. These two conditions are certainly satisfied in many practical situations and flamelet models are relevant at least in a portion of the domain of operation of IC engines and of continuous flow combustors. Leaving aside further discussions of this point let us consider some of the problems encountered in the construction of flamelet models. These problems are related to the basic ingredients of such models :
(1) The laminar flamelet submodel (or submodels) providing the local structure and properties of the reactive elements
(2) The description of the turbulent flow based on dynamic equations for the flow variables and the main species and the relevant closure rules.
(3) The rule or rules or additional balance equations which are flamelet submodels into the turbulent flow description.
(4) Any additional submodel accounting for chemical reactions taking place outside the flamelets.

There are many alternative ways of specifying these elements. Some methods are reviewed in Peters (1984, 1986) and in Bray (1986). Our own work is based on the coherent flame model initially proposed by Marble and Broadwell (1977) for the analysis of nonpremixed combustion. Fundamental ideas of this model are also contained in an earlier study due to Carrier, Fendell and Marble (1975). These investigations suggest that turbulent combustion of unmixed reactants is controlled in the early stages by a competition between <u>straining</u> of the flame elements and mutual annihilation of flame area due to the interaction of neighbouring reactive elements. The coherent flame model recognizes some other important features of turbulent combustion such as the production of flame area by stretching and its destruction by flame shortening (mutual annihilation). Because these mechanisms are important in the large scale coherent motions found in turbulent shear flows, the model accounts in some sense for these organized fluctuations.

Our own effort has been to extend the coherent flame description to premixed flow configurations (Candel et al. 1982, Darabiha 1984, Darabiha et al. 1986, Darabiha et al. 1987) and to explore its potential in nonpremixed situations (Veynante et al. 1986, Lacas et al. 1987). The present paper reports on our progress in this entreprise. The current state of the model is described and comparisons between calculations and experiments are provided in the case of a turbulent flame stabilized in a duct. From the many experiments performed in this configuration we will only examine the flow structure obtained from schlieren visualizations and spatial distributions of the light emission from C_2 radicals. These distributions may be interpreted as giving a qualitative mapping of the local mean heat release in the turbulent flame and they are compared with distributions of the mean source terms obtained from the model. This comparison has obvious imperfections but it also constitutes an original test of turbulent combustion models. It is of special interest because it deals with the mean source term appearing in the right hand side of the energy balance equation. This differs from the more common tests performed in the literature on velocity, temperature and species mass fractions all of which are only indirectly related to the modeled source terms. Because these variables are the outcome of an integration process they are less sensitive to the modeling assumptions and do not allow a direct evaluation of the combustion models and do not give access to the source terms. Accurate representations of the mean source terms are however essential if one wishes to describe the effects of chemical kinetics or Reynolds number on the structure of the turbulent flame.

To provide the proper background to this study the basic elements of the coherent flame model are reviewed in Section 2. Special attention is devoted to the balance equation describing the transport of flame area. This equation put forward by Marble and Broadwell is derived and adapted to the premixed flow case. Production of flame area by stretch and destruction by mutual annihilation are considered and illustrated with recent numerical calculations. The effect of flame quenching on the balance of flame area is discussed and a model is proposed for this mechanism.

The strained flamelet submodel is considered in Section 3. Important features of strained premixed flames with complex chemistry are examined. Calculations of strained propane-air flamelets under adiabatic and non adiabatic conditions are discussed.

Experimental results for premixed ducted flames are presented in Section 4. Results of calculations for the same configuration are presented in Section 5 and compared to the measurements contained in Section 4.

2. BASIC ELEMENTS OF THE COHERENT FLAME MODEL

The coherent flame model adopts the view that the turbulent reaction field may be described as a collection of laminar flamelets. The flamelets are transported and distorted by the turbulent stream but each flamelet retains an identifiable structure. In this sense the flamelets remain "coherent" (i.e. organized). A rough sketch of this situation is given in Fig. 1. A flamelet embedded in this flowfield is essentially affected by the local strain rate. The flamelet structure is modified and the local reaction rate per unit flame area may be obtained from an analysis of strained laminar flames. The simplest geometry allowing this analysis is shown in Fig. 1c.

The turbulent motion and the associated field of strain rates also has the effect of increasing the available density of flame surface. The production of flame area is balanced by various destruction mechanisms such as flame shortening (mutual annihilation of adjacent flamelets) and flame quenching.

The basic elements of the coherent flame description are now reviewed. We will focus on aspects which are most specific and apparently less well known such as the balance of flame area.

2.1 Turbulent flow description

The description of the turbulent reacting flow employs standard mass-averaged balance equations. Using standard notations to designate the density, velocity components, pressure enthalpy and species mass fractions these equations may be cast in the form :

Overall mass

$$\partial \bar{\rho}/\partial t + \partial \bar{\rho}\widetilde{V}_k/\partial x_k = 0 \quad (1)$$

Momentum

$$\partial \bar{\rho}\widetilde{V}_j/\partial t + \partial \bar{\rho}\widetilde{V}_k\widetilde{V}_j/\partial x_k = -\partial \overline{P}/\partial x_j + \partial(-\overline{\rho V_k''V_j''})/\partial x_k \quad (2)$$

Energy

$$\partial \bar{\rho}\widetilde{h}_t/\partial t + \partial \bar{\rho}\widetilde{V}_k\widetilde{h}_t/\partial x_k = \partial \overline{P}/\partial t + \partial(-\overline{\rho V_k''h''})/\partial x_k - \sum_i h^\circ_{f\,i}\overline{\dot{w}}_i \quad (3)$$

Species

$$\partial \bar{\rho}\widetilde{Y}_i/\partial t + \partial \bar{\rho}\widetilde{V}_k\widetilde{Y}_i/\partial x_k = \partial(-\overline{\rho V_k''Y_i''})/\partial x_k + \overline{\dot{w}}_i \quad (4)$$

In writing these equations for high Reynolds number flows it has been assumed that molecular fluxes could be neglected with respect to the turbulent fluxes.

Two distinct types of closure models are required to obtain a complete set of balance equations. The first closure model concerns the momentum, enthalpy and species transport terms ($-\overline{\rho V_k''V_j''}$, $-\overline{\rho V_k''h''}$, $-\overline{\rho V_k''h''}$). A conventional second order closure based on transport equations for the turbulent kinetic energy and the dissipation is used in this study. These equations have well known imperfections but they are provisionally adopted because they provide an acceptable description of turbulent transport in simple flow configurations.

The second closure model concerns the mean reaction terms $\overline{\dot{w}}_i$. The corresponding closure problem has been the subject of many studies and research in this area is reviewed in the monography of Libby and Williams (1980) and in articles due to Borghi (1979), Bray (1980, 1986), Peters (1984, 1986) Jones and Whitelaw (1982). An up to date discussion of the problem is also contained in Williams (1985).

Standard models for the mean source terms are based on probability density functions (PDF). A pdf shape is presumed or calculated. In certain formulations the pdf describes a single parameter like a progress variable : $p(c, x)$. The pdf then gives the distribution of this variable at each point of the flow and the mean source terms are evaluated as integrals of the source terms weighted by the pdf

$$\overline{\dot{w}}_i = \int \dot{w}_i \, p(C\,;\,x)\,dC \quad (5)$$

The coherent flame description provides an alternate solution to the closure problem. The mean reaction rates are deduced from a consideration of the mean flame area per unit volume Σ_f. The consumption rate per unit flame area are also calculated and the mean source terms are obtained by combining these two quantities. These elements are now considered in more detail.

2.2 Balance equation for the flame area per unit volume

The balance equation for the density of flame area plays an essential role in the coherent flame description. This equation is now established in the case of premixed combustion. To begin with let us consider a *material* surface element with a vector area represented by δs. The time rate of change of this element is given by Batchelor (1967, p. 132)

$$d(\delta s_i)/dt = \delta s_i\, \partial v_j/\partial x_j - \delta s_j\, \partial v_j/\partial x_i \qquad \text{as } \delta s \to 0 \qquad (6)$$

Taking the scalar product of this relation by δs_i and dividing by $(\delta s)^2 = \delta s_i\, \delta s_i$ one obtains

$$1/\delta s\; d(\delta s_i)/dt = \partial v_j/\partial x_j - n_i\, n_j\, \partial v_j/\partial x_i \qquad \text{as } \delta s \to 0 \qquad (7)$$

where $n_i = \delta s_i/\delta s$ designates the unit vector normal to the area element δs (this vector is parallel to δs). Consider now an elementary material volume δv. The time rate of change of this volume is given by

$$1/\delta v\; d(\delta v)/dt = \partial v_j/\partial x_j \qquad \text{as } \delta v \to 0 \qquad (8)$$

The material surface density may be defined by

$$\Sigma = \lim_{\delta v \to 0} \partial s/\partial v \qquad (9)$$

and the fractional variation of this quantity is obtained by combining expressions (7) and (8). It is found that

$$1/\Sigma\; d(\Sigma)/dt = -n_i\, n_j\, \partial v_i/\partial x_j = -n_i\, n_j\, S_{ij} \qquad (10)$$

where $S_{ij} = 1/2\, (\partial V_i/\partial x_j + \partial V_j/\partial x_i)$ are the components of the strain rate tensor. Equation (10) may also be written in the form

$$\partial \bar{\rho}\Sigma/\partial t + \partial \rho V_k \Sigma/\partial x_k = -\rho\, n_i n_j S_{ij} \Sigma \qquad (11)$$

This equation describes the instantaneous balance of material surface density. In this equation $n_i n_j S_{ij} \Sigma$ represents the material surface rate of change associated with the strain rates existing in the flow. Taking the mass average of this equation one obtains

$$\partial \bar{\rho}\widetilde{\Sigma}/\partial t + \partial \bar{\rho}\widetilde{V_k \Sigma}/\partial x_k = \partial(-\overline{\rho V_k'' \Sigma''})/\partial x_k - \bar{\rho}\,\widetilde{n_i n_j S_{ij}}\,\Sigma \qquad (12)$$
$$\quad 1 \qquad\qquad 2 \qquad\qquad\qquad 3 \qquad\qquad\qquad 4$$

The turbulent flux of the surface density which appears in the third term of this expression may be modeled like the other turbulent fluxes by making use of the turbulent viscosity μ_T and the mean gradient of $\widetilde{\Sigma}$

$$-\overline{\rho V_k'' \Sigma''} = \mu_T/\sigma_\Sigma\, \partial\widetilde{\Sigma}/\partial x_k \qquad (13)$$

where σ_Σ is a Schmidt number for the surface density.

The last term in equation (12) is more unusual. It represents the mean variations of surface density produced by the local strain rates. Following Marble and Broadwell (1977) this term may be modeled by

$$-\bar{\rho}\,\widetilde{n_i n_j S_{ij}}\,\Sigma = \bar{\rho}\,\varepsilon_s \widetilde{\Sigma} \qquad (14)$$

where ε_s designates a mean strain rate. According to this model the surface density Σ increases linearly with the mean strain rate. In simple shear flows (mixing layers, jets, wakes) the strain rate may be related to the mean velocity gradient in the transverse direction

$$\varepsilon_s = \alpha\, |\partial u/\partial y| \qquad (15)$$

where α is a constant. In more complex flows it is more appropriate to deduce the strain rate ε_s from the turbulent kinetic energy and from the dissipation rate

$$\varepsilon_s = C_s\, \varepsilon/k \qquad (16)$$

where C_s is a constant. An alternate method for prescribing this strain rate consists of estimating the scalar dissipation rate χ and assuming that ε_s is proportional to this quantity. The mean scalar dissipation rate may be obtained from (see for example Peters 1984)

$$\chi = C_\chi \, \varepsilon / k \, \widetilde{z''^2} \qquad (17)$$

where $C_\chi = 2$, and $\widetilde{z''^2}$ designates the mass average of the square of the fluctuations of a conserved scalar. This last quantity is usually determined by solving an additional transport equation.

The mean scalar dissipation χ may also be derived from a modeled transport equation as given for example by Borghi and Dutoya (1979).

Leaving aside any further discussion of this aspect, one finds that the transport equation for the mean surface density $\widetilde{\Sigma}$ may be cast in the form

$$\partial \bar{\rho}\widetilde{\Sigma}/\partial t + \partial \bar{\rho}\widetilde{V}_k \widetilde{\Sigma}/\partial x_k = \partial(\mu_T/\sigma_\Sigma \, \partial \widetilde{\Sigma}/\partial x_k)/\partial x_k + \bar{\rho}\, \varepsilon_s \, \widetilde{\Sigma} \qquad (18)$$

In fact this equation does not describe all the mechanisms which determine the balance of flame area. Indeed the elements of flame area cannot be identified with material surfaces because they are in relative motion with respect to the local flow. One may however neglect this displacement because the flame velocities are usually small when compared to the flow velocities.

Furthermore it is indicated by Marble and Broadwell that flame area disappears when two adjacent flame elements interact and consume the intervening reactants. Finally in certain situations the flame element is located in a field of high strain rates and it may be extinguished. This process causes the dissipation of flame area. To represent these mechanisms one has to include additional terms in the transport equation describing $\widetilde{\Sigma}$. The balance equation for the density of flame area $\widetilde{\Sigma}_f$ will take the general form

$$\left\{\begin{array}{c}\text{Rate of change}\\ \text{of } \widetilde{\Sigma}_f\end{array}\right\} = \left\{\begin{array}{c}\text{Turbulent diffusion}\\ \text{of } \widetilde{\Sigma}_f\end{array}\right\} + \left\{\begin{array}{c}\text{Production of}\\ \widetilde{\Sigma}_f \text{ by strain}\end{array}\right\}$$

(1) (2) (3)

$$- \left\{\begin{array}{c}\text{Dissipation of } \Sigma_f\\ \text{caused by extinction}\end{array}\right\} - \left\{\begin{array}{c}\text{Dissipation}\\ \text{of } \widetilde{\Sigma}_f \text{ by mutual}\\ \text{annihilation}\end{array}\right\}$$

(4) (5)

The first two terms are similar to those obtained for the material surface density.

The third term has to be modified because local conditions of temperature mass fractions and strain rate influence the production of flame area. Whereas a material surface submitted to strain is augmented, this may not be the case for the flame surface. If for example the strain rate is too high extinction will occur, no flame area will be produced and in some cases the flame surface density will be decreased. To illustrate this point we have extracted some data from a study of vortex driven combustion instabilities (Poinsot et al. 1987). In this experiment large scale vortices are periodically formed behind a backward facing step (Fig. 2). The vortices of fresh gases are convected downstream and simultaneously grow in size. At the vortex cap the fresh mixture is in contact with burned products and in a moving frame of reference the situation is practically that of a stagnation point flow. A rough estimate of the strain rate at the cap indicates that it reaches values of about $10^4 s^{-1}$. This value corresponds to partial extinction of a strained laminar propane-air flame. Looking now at C_2-emission measurements (Fig. 3) obtained in the same experiments and corresponding to the same instant in the cycle, no light emission is found at the cap indicating the absence of reaction in that region. Combustion takes place in the vortex skirts where the strain rate is lower.

Another process which may limit the production of flame area is the absence of one of the reactants. For instance in the recirculation region the gases are near chemical equilibrium and no heat production occurs.

Other mechanisms may limit the production of flame area but those mentioned are among the most important. The production term $\bar{\rho}\varepsilon_s \tilde{\Sigma}$ must be moderated to account for these processes. In the general case, for adequate values of the strain rate, mass fractions and temperature the production term keeps its standard form $\bar{\rho}\varepsilon_s \tilde{\Sigma}_f$. When these values are outside the domain in which a flame may develop, the production term will vanish. The domain Ω of possible flames is derived from the local flamelet analysis. If $f_\Omega(\varepsilon_s, Y_i, T_b)$ designates a function which vanishes outside Ω and is equal to one inside Ω then the production of flame area may be cast in the form $\bar{\rho}\varepsilon_s f_\Omega \tilde{\Sigma}_f$. According to this expression flame area is produced if the flame existence is assured.

If the local strain rate is very high the local flamelets are quenched and the strain rate does not create flame area. In that situation the strain rate will actually participate in the destruction of flame surface. This mechanism is expected to occur when the strain rate ε_s exceeds a certain critical value ε_{se} which depends on the local conditions and is determined from the local flamelet analysis. In that circumstance the flame area will disappear in proportion to the strain rate itself. A model for this destruction term is of the form

$$\gamma_e \bar{\rho} \varepsilon_s h(\varepsilon_s - \varepsilon_{se})\tilde{\Sigma}_f$$

where h is the Heaviside function and γ_e is a constant. One may also argue that the rate of destruction of flame area is proportionl to the difference between the strain rate and the critical strain rate. In that case the destruction term becomes

$$\gamma_e \bar{\rho} (\varepsilon_s - \varepsilon_{se}) h(\varepsilon_s - \varepsilon_{se}) \tilde{\Sigma}_f$$

In both cases the net result of the production and destruction terms (3) and (4) appearing in eq. (19) will be a decrease of the flame surface density when ε_s exceeds ε_{se}.

It is not difficult to take into account the fact that the strain rates found at a certain point in the turbulent stream are distributed in a finite range around the mean value. This is achieved with a probability density function like that proposed by Peters (1984) or Bray (1986). When such a pdf is introduced the transition between production and dissipation of flame area is more progressive.

Dissipation of flame area is also due to the mutual annihilation of adjacent flame elements. Flamelets coming at a close distance consume the intervening reactant and this shortens the available flame area.
This mechanism is well illustrated by the theoretical work of Marble (1985), Karagozian and Marble (1986) and by some recent calculations of Laverdant and Candel (1987a, 1987b). Let us first consider the case of a diffusion flame interacting with a vortex. The flame is initially plane. A viscous core vortex is suddenly turned on and the flame is wound up in the vortex flow field. Flame area is created by the vortex motion (the local production of flame surface is proportional to the strain rate induced by the vortex). The flame elements formed near the vortex core interact and the intervening reactant is consumed (Fig. 4). A core of burned products appears and most of the flame area initially generated near the vortex center is annihilated in the mutual interaction of the flamelets.

The flame shortening mechanism is also found in the premixed flame case (see Fig. 5 extracted from Laverdant and Candel 1987). The pattern formed in the vortex interaction differs from that obtained in the diffusion flame case but the basic features are quite similar. Flame area is created by the vortex motion and it is dissipated by mutual annihilation.

A model for this process may be established on the basis of a simple argument due to Marble and Broadwell. Their derivation is easily extended to the premixed case. Consider now two flamelets separated by combustion products. The distance between these two elements will tend to increase and mutual annihilation will not occur. On the other hand if fresh reactants separate the two flamelets the reaction fronts progress towards each other, the reactants are consumed and mutual annihilation takes place. Now consider the mean distance between the flamelets. This distance is of the order of $1/\widetilde{\Sigma}_f$. If $\widetilde{X}_i$ designates the volume fraction (the molar fraction) of one of the main species involved in the reaction, then the effective distance occupied by this species is $l_i = \widetilde{X}_i / \widetilde{\Sigma}_f$. Let v_{Di} designate the volume rate of consumption of the i-th species per unit flame area (this quantity has the dimensions of a velocity). This species is consumed at the rate v_{Di} and it disappears in a characteristic time

$$\tau_{Di} \sim l_i / v_{Di} = \widetilde{X}_i / (v_{Di} \widetilde{\Sigma}_f)$$

The rate of annihilation of flame area associated with the consumption of the i-th species is then proportional to

$$\widetilde{\Sigma}_f / \tau_{Di} = v_{Di} \widetilde{\Sigma}_f^2 / \widetilde{X}_i$$

The rate of flame annihilation associated with the consumption of the main species involved in the reaction is then of the form

$$\sum_i (v_{Di} / \widetilde{X}_i) \widetilde{\Sigma}_f^2 \qquad (20)$$

When the main species are a fuel F and an oxidizer O, the flame shortening term is proportional to

$$(v_{Do} / \widetilde{X}_O + v_{Df} / \widetilde{X}_F) \widetilde{\Sigma}_f^2 \qquad (21)$$

This term is evaluated per unit volume. The flame shortening term is proportional to the square of the flame surface density and to a weighted sum of the consumption velocities of the main species.

It is now possible to write a complete balance equation for the flame surface density :

$$\partial\bar{\rho}\tilde{\Sigma}_f/\partial t + \partial\bar{\rho}\tilde{V}_k\tilde{\Sigma}_f/\partial x_k = \partial(\mu_T/\sigma_\Sigma \, \partial\tilde{\Sigma}_f/\partial x_k)/\partial x_k + \bar{\rho}\,\varepsilon_s\, f_\Omega\, \tilde{\Sigma}_f$$
$$- \gamma_e\, \bar{\rho}\, (\varepsilon_s - \varepsilon_{se})h(\varepsilon_s - \varepsilon_{se})\tilde{\Sigma}_f$$
$$- \beta\, \bar{\rho}\, (v_{Do}/\tilde{X}_O + v_{Df}/\tilde{X}_F)\tilde{\Sigma}_f^2$$
(22)

where ε_s may be estimated as

$$\varepsilon_s = \alpha\, (S_{ij}\, S_{ij})^{1/2} \qquad \text{or} \qquad \varepsilon_s = C_s\, \varepsilon/k$$

In these expressions $\alpha, \beta, \gamma_e, C_s$ are constants and σ_Σ is a Schmidt number for the flame surface density.

The volume rate of consumption per unit flame appearing in this expression are obtained from the local flamelet analysis.

2.3 Determination of the mean reaction rates

Knowing the mean flame surface density Σ_f and the consumption rates per unit flame surface one may easily determine the mean consumption terms per unit volume :

$$\overline{\dot{w}_f} = -\bar{\rho}\, v_{Df}\, \tilde{\Sigma}_f$$

$$\overline{\dot{w}_o} = -\bar{\rho}\, v_{Do}\, \tilde{\Sigma}_f$$
(23)

These simple expressions may be refined if one takes into account the fact that the local flamelets correspond to a distribution of strain rates. The probability density function describing the distribution of the strain rates $p_S(\varepsilon_s)$ may be presumed and the mean consumption terms per unit volume take the general form

$$\overline{\dot{w}_i} = -\bar{\rho}\,\tilde{\Sigma}_f \int v_{Di}(\varepsilon_s)\, p_S(\varepsilon_s)\, d\varepsilon_s$$
(24)

A slightly more complex expression is also obtained if one assumes that the burnt gas temperature which arises in the local flamelet analysis is also distributed around its mean value.

3. THE LOCAL FLAMELET STRUCTURE

3.1 The flamelet model

An idea shared by all flamelet models is that the strain rate imposed by the turbulent flow on the laminar flame elements plays an essential role. The flamelets are strained by the turbulent motion, their inner structure is modified and the consumption of reactants and the heat release rate per unit flame area are affected. Large strain rates may lead to partial or total extinction of the flamelets. The simplest example illustrating the effect of strain rate is that of a diffusion flame in the fast chemistry limit. It is easily shown that in this limit the rate of reaction is augmented when the strain rate is increased (see Carrier et al. 1975). It is actually found that the consumption rates per unit flame area increase like $(D \varepsilon_s)^{1/2}$ where D is a diffusion coefficient. The strain rate augments the flow of gases to the flame sheet and the reaction rate is enhanced. This process is limited by finite chemical rates of reaction and for high strain rates the flame may be cooled by the rapid convection of fresh material and it may be extinguished.

Notable analyses of strained diffusion flames are due to Carrier et al. (1975), Linan (1974), Tsuji (1982), Law (1984). The case of premixed strained flames is considered by Klimov (1963), Buckmaster and Ludford (1983), Libby and Williams (1981, 1983, 1984), Libby et al. (1983).

Many studies of premixed strained flames are based on activation energy asymptotics. Numerical calculations are due, among others, to Smooke et al. (1986), Giovangigli and Candel (1986), Darabiha et al. (1986), Giovangigli and Smooke (1985). Detailed chemistry calculations are now being performed by many groups. Dixon-D Lewis et al. (1984) compare calculations of methane-air diffusion flames. Complex chemistry is also used in the studies of Marble and Brodwell (1977, 1979), Liew et al. (1984), Rogg et al. (1986) and Giovangigli and Smooke (1986). The structure and properties of strained propane-air flames are examined by Darabiha et al. (1987).

The basic flamelet model and the main results of this study are reviewed in this section. The presentation will essentially focus on the effect of strain and hot stream temperature on the flamelet structure and extinction/ignition characteristics.

3.2 Problem formulation

We consider a strained premixed laminar flame in a counterflow of fresh reactants and burnt gases, as shown in Fig. 6. We model our system by employing a boundary layer approximation. For a planar geometry, the governing equations along the stagnation streamline can be written in the form :

$$\partial V/\partial y + \varepsilon_s \rho U = 0 \qquad (25)$$

$$\partial (\mu \, \partial U/\partial y)/\partial y - V \, \partial U/\partial y + \varepsilon_s (\rho_f(\phi) - \rho U^2) = 0 \qquad (26)$$

$$-\partial (\rho Y_k V_{ky})/\partial y - V \, \partial Y_k/\partial y + W_k \dot{\omega}_k = 0, \qquad k = 1, \ldots, K \qquad (27)$$

$$\partial (\lambda \, \partial T/\partial y)/\partial y - c_p V \, \partial T/\partial y - \sum_k \rho Y_k V_{ky} c_{pk} \, \partial T/\partial y$$

$$- \sum_k h_k W_k \dot{\omega}_k = 0 \qquad (28)$$

$$\rho = pW/RT \qquad (29)$$

where y denotes the spatial coordinate normal to the stagnation plane (see Fig. 6) ; V, the density times the velocity v in the normal direction ; ρ, the mass density ; U, a similarity function related to the velocity u in the transverse direction ; μ, the mixture viscosity ; ρ_f, the mass density in the fresh reactant stream ; Y_k, the mass fraction of the kth species ; K, the number of species ; V_{ky}, the diffusion velocity of the kth species in the normal direction ; W_k, the molecular weight of the kth species ; ω_k, the molar rate of production rate of the kth species ; T, the temperature ; λ, the thermal conductivity of the mixture ; c_p, the constant pressure heat capacity of the mixture ; c_{pk}, the constant pressure heat capacity of the kth species ; h_k, the specific enthalpy of the kth species ; p, the pressure ; W, the mean molecular weight of the mixture and R, the universal gas constant.

These equations have to be completed by formulas expressing the transport coefficients λ and μ, the diffusion velocities V_k, the thermodynamic properties c_p, c_{pk} and h_k and the chemical production rates $\dot{\omega}_k$ in terms of the state variables T, p, Y_k, k = 1,..., K and their gradients. These expressions, which involve detailed transport nd complex chemical kinetics, can be found in Kee et al. (1986) and Kee et al. (1983).

The velocity components u and v in the transverse (x) and normal (y) direction, respectively, may be obtained from the solution vector (V, U, Y_1,... Y_K, T) by forming

$$u = \varepsilon_s \, xU \qquad (30)$$

$$v = V/\rho \qquad (31)$$

Complete specification of the problem requires that boundary conditions be imposed at each end of the computational domain and the origine must be fixed. As $y \longrightarrow +\infty$ we have

$$U = 1 \qquad (32)$$

$$Y_k = Y_{kf}(\phi) \qquad k= 1,........, K \qquad (33)$$

$$T = T_f \qquad (34)$$

at y = 0 we have

$$V = 0 \qquad (35)$$

and as $y \longrightarrow -\infty$ we have

$$U = (\, \rho_f(\phi) \, / \rho_b(\phi, T_b) \,)^{1/2} \qquad (36)$$

$$Y_k = Y_{kb}(\phi, T_b) \qquad k= 1,........, K \qquad (37)$$

$$T = T_b \qquad (38)$$

where T_f is the specified temperature in the fresh reactant stream ; Y_{kf}, the specified mass fractions in the fresh reactant stream ; T_b, the specified temperature in the burnt gases stream and Y_{kb}, the equilibrium mass fractions of the reactant mixture at temperature T_b and pressure p. Note that the origin is located at the stagnation plane.

We observe that our model depends on three parameters ε_s , T_b and ϕ . The quantity ε_s denotes the strain rate, T_b represents the temperature in the burnt gas stream and ϕ is the equivalence ratio of the incoming fresh reactant stream. The strain rate is a measure of the "stretch" in the flame due to the imposed flow, ϕ is a measure of the relative proportion of fuel and air in the reactant stream and T_b represents how much the incoming burnt products have gained or lost energy subsequent to their creation, by a mechanism inoperative within the flamelet. Whenever these products have neither gained nor lost any energy, the temperature T_b is the so called adiabatic flame temperature T_{ad}. We will refer to this case as the "adiabatic" case, whereas the situations $T_b > T_{ad}$ and $T_b < T_{ad}$ are referenced as "superadiabatic" and "subadiabatic". These terms characterize the history of the burnt gases and not the flamelets which do not exchange heat with the surroundings.

Our goal is now to study the dependence of the solution on the parameters ε_s and T_b for propane-air mixtures.

3.3 Numerical method

The solution method combines a phase-space, pseudo-arclength continuation method, Newton-like iterations and global adaptive gridding.

Once the set of equations (25) - (38) is discretized, the problem reduces to solving systems of the form

$$F(X,\lambda) = 0 \qquad (39)$$

where X is the discrete solution vector and where λ denotes any parameter which influences the solution, e.g., T_b or ε_s. Following standard continuation procedures, the solution path $(X(\lambda), \lambda)$ is reparametrized into $(X(s), \lambda(s))$ where s is a new independent continuation parameter, while λ becomes an eigenvalue. This procedure suppresses the Jacobian matrix singularities at simple turning points.

The dependence of s on the augmented solution (X, λ) is specified by an extra scalar equation

$$N(X(s), \lambda(s), s) = 0 \qquad (40)$$

chosen such that s approximates the arclength of the solution branch in a given phase space. More precisely, in the neighborhood of a previously obtained solution $(X(s_0), \lambda(s_0))$, N is taken to be

$$N = \int_{-\infty}^{+\infty} \{\partial X/\partial s(s_0)\}^T \{X(s) - X(s_0)\} dy + \partial\lambda/\partial s(s_0)\{X(s) - X(s_0)\} - (s - s_0) \qquad (41)$$

in which case s locally approximates the arclength of the solution arc (X, λ) with a discrete l_2 norm for X.

Finally the coupled system of equations (39)-(40) is solved by using a first order Euler predictor and a corrector step involving Newton-like iterations and adaptive gridding. A global adaptive gridding procedure equidistributing a positive weight function is used. We refer to Giovangigli and Smooke (1987a, 1987b, 1987c) and Smooke (1982) for more details. Lastly, vectorized versions of the CHEMKIN and TRANSPORT packages are used to evaluate transport coefficients and thermo-chemical properties (Giovangigli and Darabiha 1987).

3.4 Numerical results

The flamelet data base may be constructed by performing sequence of continuation with respect to the strain rate for different values of the burnt gas temperature. It is possible in this way to investigate the dependence of the propane-air flame on ε_s and T_b. Typical calculations done for an equivalence ratio of $\phi = 0.75$, a fresh reactant stream temperature of $T_f = 300°K$ and a pressure p of one atmosphere are now described. The reaction mechanism for the propane-air system is taken from Warnatz (1983, 1985).

Based on the theoretical work of Libby, Linan and Williams (1983) it is anticipated that the burnt gas temperature T_b will play an important role in the extinction process. In particular, depending on the value of T_b, a flame subject to varying strain rates may undergo abrupt ignition or extinction transitions involving altered locations of the reaction zone and altered creation of product.

Figure 7 illustrates the dependence of the temperature profiles on the hot stream temperature T_b for a strain rate of $\varepsilon_s = 340$ sec^{-1}. For these low values of ε_s many features of the flame are essentially independent of T_b, as already pointed out in Libby and Williams (1983), Libby et al. (1983). For instance it is observed in Figure 7 that the reaction zone location remains unchanged and the peak temperature is still close to T_{ad}.

For higher values of ε_s however, the variations of T_b significantly affect the flamelet since the reaction zone itself is modified. This is illustrated in Figure 8 where temperature profiles for different T_b and for a strain rate of $\varepsilon_s = 1000$ sec^{-1} are plotted. In Figure 8 a decrease in T_b gives rise to a shift of the flame front towards the hot gas stream. Note also that the peak temperature is exactly T_b.

Figure 9 shows the dependence of the temperature profiles on the strain rate for a hot stream temperature of $T_b = 1700°$ K. It is found that as ε_s increases, the flame front moves continuously towards the hot stream. Moreover the peak temperature is reached within the flame front for low values of ε_s whereas the peak temperature is T_b and reached in the burnt gases for large values ε_s.

For lower values of T_b, however, multiple solutions appear. This is illustrated in Figure 10 where the temperature profiles corresponding to a hot stream temperature of $T_b = 1700$ K and to different strain rates are plotted. Indeed, for the intermediate value $\varepsilon_s = 640$ sec^{-1}, three solutions are obtained, corresponding to the curves b, c and d. The profiles b and d are standard and extinguished solutions while the curve c is an unstable solution.

To investigate this solution multiplicity, the heat release rate h per unit flame area

$$\dot{h} = \int_{-\infty}^{+\infty} \left(\sum_k h_k W_k \dot{\omega}_k \right) dy \qquad (42)$$

is plotted in Figure 11 as a function of the strain rate ε_s for different temperatures T_b. These curves show that for flames with product stream having elevated enthalpies, i.e., a sufficiently large T_b, the heat release rate varies continuously with ε_s. On the other hand, for flamelets with sufficiently cooled product streams, i.e., a sufficiently low T_b, there are abrupt ignition and extinction transitions and the characteristic response curve (ε_s, $\dot{h}$) is S-shaped. The upper branch corresponds to standard flame structures, the lower branch to extinguished solutions and the intermediate branch to unstable solutions. The corresponding flame front locations are presented in Figure 12. These plots show that during ignition or extinction transitions there are abrupt flame front location and peak temperature changes. These results are in excellent qualitative agreement with the theoretical work of Libby, Linan and Williams (1983).

Lastly, for the kinetic scheme involving 33 species and 126 reactions used in these calculations it is found that the first extinction point is reached at about $T_b = 1500$ K and for $\varepsilon_s = 800$ sec^{-1}.

4. EXPERIMENTS ON TURBULENT DUCTED FLAMES

4.1 Experimental configuration

The experimental set up used in this study is shown in Fig. 13. A mixture of air and propane is injected through a long duct into a rectangular combustor. The height, depth and length of this element are respectively 50, 100 and 300 mm. The inlet plane comprises a V-gutter flame holder placed at the duct center and producing a 50 % blockage. The upper and lower combustor walls are made of thick ceramic material while the lateral walls are transparent artificial quartz windows. A stainless steel structure holds the various parts together. Ignition of the premixed stream is obtained with a spark plug. The plug may be lowered down behind the flameholder, it is activated and removed from the chamber after ignition. Combustion is then stabilized by hot gases recirculating behind the V-gutter.

The stabilizer may be set into a pitching motion with an electromechanical shaker as shown in Figure 13 but this device is not used in the present study.

4.2 C_2-radical light emission measurements

While exact measurements of the local energy release are not available many observations indicate that the light radiated by the combustion zone is related to the reaction intensity and hence to the heat-release process. Certain radicals like C_2, CH, OH appear almost exclusively in reactive zones and their concentration is always small. Hence the self absorption of the light emitted by these radicals is not important and the radiated light intensity is directly related to the reaction rate or equivalently to the heat relese rate (see Poinsot et al. 1987 for further discussion of this point and an application to the analysis of the nonsteady heat release under unstable combustion).

While a linear relation between the heat release source term and the light emission from free radicals has been proved in some special situations it may be safely stated that a monotonic relation exists between these two quantities. The measurement of the mean heat release source term relies on this assumption. The emission band used to this purpose is the (0,0) C_2 band at a wavelength λ = 5165 A. This wavelength is isolated in the emission spectrum with a narrowband interferential filter ($\Delta\lambda \sim$ 50 A). The light emitted by the flame is collected by an f = 54 mm convex lens located at 100 mm from the combustor center plane. The light beam is then filtered and the filtered light intensity is detected by a photomultiplier through a 1 mm diameter pinhole to provide a local measurement : the emitting area actually seen by the detector has a diameter of about 1 mm. The detector output is amplified, low pass filtered to prevent aliasing and transmitted to a PDP 11/23 computer through an 8 channel 12 bit A/D converter. This computer also controls the optical system displacement in the vertical and horizontal directions. The photomultiplier scans a grid comprising 3000 points. At each point $\underset{\sim}{x}$ of this grid, 1000 data samples are acquired and averaged. This process yields the spatial distribution of the mean light emission $I(\underset{\sim}{x})$.

4.3 Schlieren visualizations

It is intructive to first examine the flame structure using spark Schlieren photographs.

For the whole range of Reynolds numbers considered in this paper $Re_H < 10^5$ (the Reynolds number is based on the stabilizer height, on the velocity at the lipe and on the cold flow viscosity), the flow in the flame zone is dominated by large scale coherent vortices (Fig. 14a to f). The small fluctuations observed in the recirculation zone is due to a small leak between the flameholder and the window leading to some spurious burning of the mixture near the wall. These vortices resemble those found in studies of non reactive two dimensional shear layers (for example Brown and Roshko 1974). There are however some distinctive features. It is observed in particular that the initial small scale vortices formed near the stabilizer lips are perfectly rolled up in a succession of "cat's eye" patterns (Fig. 14c). However at a greater distance from the stabilizer the inner boundary of the vortex patterns disappears and a core of relatively uniform gases is formed on the hot side of the flame. This behavior is a direct consequence of the competition taking place in the vortices between the production of flame area by stretch and its destruction by mutual annhihilation. Another specific feature is the sudden growth of the vortex structures as they are convected downstream (Fig. 14a, b). This growth may be caused by vortex pairing. It is more probably a result of the gas expansion produced by the sudden reaction of fresh material entrained by the vortices. This behavior is not always observed. It is found mainly for small values of the equivalence ratio near the lean extinction limit of the combustor (Fig. 14a,c).

When the Reynolds number is increased (compare Fig. 14c, d) the typical scale of the structures in the downstream part of the flow is diminished and the zone of active reaction comes closer to the flame holder.

For a fixed Reynolds number when the equivalence ratio is augmented (compare Fig. 14d and 14e). the reaction zone is longer in the axial direction and the burning vortex patterns are more evenly distributed. The size of these reacting vortices also grows more uniformly.

These observations obtained from spark Schlieren photographs are confirmed by the measurements of C_2-radical light emission discussed below.

4.4 C_2-radical light emission measurements

As already indicated, the measurement of the light emission from radicals like C_2, CH, OH provides information on the local rate of heat release.

Spatial distributions of the light emitted by C_2 radicals are displayed in Figures 15a to h. The first four maps (Fig. 15a to d) correspond to the same Reynolds number Re_H = 46000 but the equivalence ratio takes four different values (0.6, 0.7, 0.8 and 0.9). These data indicate that the equivalence ratio notably influences the mean flame structure. At low values of ϕ, concentrated regions of reaction are observed in the vicinity of the flame holder. For ϕ = 0.6 these regions form at a close distance and a merged reaction zone is observed. For = 0.7 the concentrated regions of reaction are located further apart. Some reaction still takes place between the two main flames. For ϕ = 0.8 the inital regions of concentrated reaction have disappeared and two flames develop from the lips of the flameholder. The light emission reaches a maximum in the vicinity of the two sidewalls. This peculiar phenomenon is observed because the flame now touches the boundaries which are in the present case nearly adiabatic. At the points where the flame reaches the walls, the flow velocity in the boundary layer becomes less than that existing in the main flow and the flame angle increases (this angle is measured with respect to the axial direction). The wall temperature takes large values and the wall region acts like a secondary stabilization zone for the incoming stream of fresh mixture. In this region combustion is activated and the C_2 light emission is enhanced. For ϕ = 0.90 (Fig. 15d) the two flames are separated by a region of little reaction. In each flame region the C_2 light emission intensity increases progressively and reaches its maximum in the vicinity of the combustor walls.

Consider now the set of light emission maps obtained by varying the mass flow rate for a fixed value of the equivalence ratio (Fig. 15e to h). At low values of the mass flow rate or the Reynolds number the two flame regions are distinct and the maximum rate of heat release is found near the duct walls where the flames interact with the boundaries (Fig. 15e and f). As the mass flow rate is increased the flame angle decreases slightly and the maximum of the heat release rate is shifted upstream (Fig. 15g and h).

These data clearly show that the equivalence ratio and the Reynolds number have a notable influence on the flame structure. The modifications associated with these two parameters are not extensively documented in the previous literature (see Libby et al. 1987 for a review of the available data, Zukoski and Marble 1955 for an early analysis of the influence of the Reynolds number on the wake transition behind a flame stabilizer).

5. NUMERICAL RESULTS AND DISCUSSION

We are now going to discuss numerical results obtained for the combustor geometry described in the previous section. These results may be compared to the experimental data presented in Section 4. All calculations correspond to a single set of constants (standard values are adopted for the $k-\varepsilon$ model ($C_\mu = 0.09$, $c_1 = 1.44$, $c_2 = 1.92$, $\sigma_k = 1$, $\sigma_\varepsilon = 1.3$, $\sigma_T = 0.7$) and the constants involved in the balance of flame area are :

$$\sigma_\Sigma = 1 \quad , \quad C_s = 1 \quad , \quad \gamma_e = 1 \quad , \quad \beta = 1$$

No effort was expanded to adjust these parameters but some improvements are possible. The calculations are carried with a time dependent finite difference implicit code developed at ONERA and described by Dupoirieux and Scherrer (1985).

The coherent flame description was introduced in the code in place of the preexisting "Eddy-Break Up" model.

Taking into account the mean flow symetry the computation is performed on a domain extending from the combustor axis to the wall. The finite difference grid is nonuniform in the axial direction and comprises 21 vertical and 63 horizontal nodes.

The reacting flow calculations are initiated by introducing a temperature spot with a Gaussian shape and a localized distribution of flame surface density in the recirculation zone of the cold flow. It turns out that the temperature spot is sufficient to start the reaction, the initial distribution of flame surface may be chosen arbitrarily.

The inlet temperature is 300 K, and the ambiant pressure at the combustor exhaust is atmospheric. Results presented correspond to a stationary flow configuration which is obtained after 2000 time step iterations after ignition. The typical CFL number used in the calculations is 7.

To compare the numerical results with the experimental results, we have displayed in Fig. 16a the fuel consumption rate distribution in the combustor using a scale of grey symbols. This figure corresponds to an inlet velocity of 22 m/s, a mass flow rate of m_a = 100 g/s and a fresh mixture equivalence ratio of ϕ = 0.75. The fuel consumption is essentially concentrated in a region located at about 50 mm from the flame holder edge. The maximum value is 2.14 kg/m^3/s. Note that for clarity the geometry of the plot is distorted, the vertical scale differs from the horizontal one. The ratio of the combustor height to the combustor length is 0.357 in the figure while it is 0.166 in reality. However the numerical results may be directly compared with the experimental distributions (Fig. 15g).

The fuel consumption rate diminishes progressively in the downstream direction. There is no consumption of fuel near the horizontal walls. This result is in agreement with the experimentl observations of the C_2 radiall light emission (Fig. 15g). The main difference between these two results is that the calculation predicts heat release in the recirculation zone. This is at variance with the experiment and also with common knowledge. This difference may be explained in various ways. One may argue for example that the turbulent diffusion of flame area is overestimated and that the Schmidt number σ_I should be increased. A detailed examination of the distribution of fuel indicates that a certain amont of this species penetrates into the recirculation region and burns in that region. This indicates an important weakness of the gradient diffusion modeling used in the calculation.

Figure 16b displays the temperature distribution in the combustor for the same conditions (v_I = 22 m/s, $\dot{m}_a$ = 100 g/s, ϕ = 0.75). The hot recirculation zone has the maximum temperature slightly above 2000 K. The temperature then diminishes on the combustor axis. In the middle of the chamber and on this axis the temperature is about 1300 K. Near the combustor exhaust the temperature increases and reaches 1500 K K. Near the walls, the temperature slowly increases in the axial direction and the temperature profile at the exhaust has not reached a uniform state.

Another calculation is performed for an equivalence ratio of ϕ = 0.6 and an inlet velocity of 15 m/s (corresponding to a mass flow rate $\dot{m}_a$ = 69 g/s). Figure 17a displays the distribution of the mean consumption rate of fuel. Comparing this figure with the previous calculation it is found that the length of the reaction region diminishes but its location is almost the same. This result agrees with the experimental observations (see Fig. 15a). We note that the maximum fuel consumption rate reaches a value of 1.59 kg/m^3/s which is less than that of the previous calculation. This change is due in part to the lower value of the equivalence ratio and it also corresponds to a change in the spatial distribution of the mean reaction rate term.

Figure 17b represents the distribution of the temperature for this calculation (Da = 0.6, $\dot{m}_a$ = 69 g/s). The temperature in the recirculation zone reaches 1700 K instead of 2000 K in the previous calculation. On the jet axis, the temperature diminishes more rapidly to about 1000 K. It increases in the downstream direction. In this region the temperature varies from about 450 K to 1400 K.

The agreement between calculated and measured mean heat release terms is not perfect but qualitative features and trends are retrieved. In view of these results one may ask if the flamelet description is indeed applicable to the ducted flame configuration considered in this paper.

As indicated in the introduction a precise answer to this question cannot be given but some information may be obtained from an examination of the Barrère-Borghi-Peters phase diagram. This diagram defines the different regimes in premixed turbulent combustion in terms of two dimensionless parameters :

(1) the ratio of the local turbulent length scale to a typical flame thickness ℓ_T/ℓ_F)
(2) the ratio of the local velocity fluctuation to the laminar flame speed v'/v_F.

A simplified phase diagram adapted from Peters (1986) is shown in Fig. 18. In this plot lines of constant Damkohler number are parallel and make a 45° angle with the horizontal axis. The line Da = 1 separates the region of homogeneous combustion corresponding to well stirred reactors (Da < 1) from the region of corrugated flamelets and distributed reaction zones (Da > 1).

Conditions which exist at each point of the grid representing the combustor in a typical calculation are indicated in the diagram. The domain defined in this way is located below the Da = 1 line and, it therefore belongs to the region of corrugated flamelets and distributed reaction zones. The domain covers regions where the flamelets are well defined and other regions where extinction takes place.

From this discussion it may be concluded that the flamelet model will be applicable if it includes a proper description of flame quenching and subsequent mechanisms. The present model accounts for this aspect at least in some sense but it certainly requires some further refinements.

6. CONCLUSION

A description of premixed turbulent ducted flames is proposed on the basis of the coherent flame model. This model initially proposed by Marble and Broadwell for turbulent diffusion flames is extended to the premixed situation.

The balance equation which fixes the amount of flame area per unit volume is derived from basic principles and a consideration of important mechanisms which participate in the production and destruction of flame surface. These mechanisms (flame stretch, extinction, flame shortening) are illustrated by recent experimental observations and numerical calculations. The local flamelet structure of strained premixed flames with complex chemistry is discussed and extinction/ignition properties of these flames are examined.

Numerical calculations based on the coherent flame description are carried for turbulent flames stabilized in a duct. The results are compared with experiments performed on a model combustor. It is shown that the spatial distributions of the mean heat release computed in the model are similar to the distributions of the light intensity emitted by C_2 radical. This indicates that the model, in its present form, already provides a suitable description of turbulent ducted flames. Imperfections are also made evident but the structure of the model allows refinements and improvements at various levels.

ACKNOWLEDGMENTS

This work was supported in part by a contract from SNECMA. The schlieren views included in the paper were obtained with the help of Frank Bourienne. The authors wish to acknowledge many discussions with François Lacas, Drs Denis Veynante and Thierry Poinsot. We would also like to thank E. Djavdan and E. Maistret for their help in the computation.

References

Batchelor, G.K. (1967). An introduction to fluid mechanics. Cambridge V. Press., Cambridge.

Borghi, R., and Dutoya, D. (1979). On the scales of the fluctuations in turbulent combustion. 17th Symposium (International) on Combustion, Combustion Institute, 235-244.

Borghi, R. (1979). Models of turbulent combustion for numerical predictions. Prediction Methods for Turbulent Flows. Lecture Series 1979-2, Von Karman Institute, Rhode St Genese, Belgium.

Borghi, R. (1985). On the structure and morphology of turbulent premixed flames. Recent Advances in the Aerospace Sciences. C. Cashi, ed. Plenum Press, New York, 117-138.

Bray, K.N.C. (1980). Turbulent flows with premixed reactants. Tubulent Reacting Flows. Topics in Applied Physics Vol. 44, P.A. Libby and F.A. Williams ed. Springer Verlag, New York.

Bray, K.N.C, Libby, P.A., and Moss, J.B. (1984). Flamelet crossing frequencies and mean reaction rates in premixed turbulent combustion. Comb. Sci. and Tech. 41, 143-172.

Bray, K.N.C. (1986). Methods of including realistic chemical reaction mechanisms in turbulent combustion models. Second workshop on modelling of chemical reaction systems. Heidelberg, Aug. 1986.

Broadwell, J.E., and Breidenthal, R.E. (1982). A simple model of mixing and chemical reaction in a turbulent shear layer. J. of Fluid Mech. 125, 397-410.

Buckmaster, J.D., and Mikolaitis, D. (1982). The premixed flame in a counterflow. Comb. Flame 47, 191-204.

Buckmaster, J.D., and Ludford, G.S.S. (1983). The theory of laminar flames. Cambridge University Press, Cambridge.

Candel, S.M., Darabiha, N., and Esposito E. (1982). Models for a turbulent premixed dump combustor. AIAA Paper 82-1261, AIAA, New York.

Carrier, G.F., Fendell, F.E., and Marble, F.E. (1975). The effect of strain rate on diffusion flames. SIAM J. of Appl. Math. 28, 463-500.

Clavin, P., and Williams, F.A. (1982). Effects of molecular diffusion and thermal expansion on the structure and dynamics of premixed flames in turbulent flows of large scale and low intensity. J. Fluid Mech. 116, 215

Darabiha, N. (1984). Un modèle de flamme cohérente pour la combustion prémélangée: Analyse d'un foyer turbulent à élargissement brusque. Doctoral Thesis. Ecole Centrale des Arts et Manufactures, Châtenay-Malabry, France.

Darabiha N., Candel, S.M., and Marble, F.E. (1986). The effect of strain rate on a premixed laminar flame. Comb. Flame 64, 203-217.

Darabiha, N., Giovangigli, V., Candel, S.M., and Smooke, M.D. (1987). Extinction of strained premixed propane-air flames with complex chemistry. Submitted to Comb. Sci. and Tech.

Dixon-Lewis, G., David, T., Gaskell, P.H., Fukutani, S., Jinno, H., Miller, J.A., Kee, R.J., Smooke, M.D., Peters, N., Effelsberg, E., Warnatz, J., and Behrendt, F. (1984). Calculation of the structure and extinction limit of a methane-air counterflow diffusion flame in the forward stagnation region of a porous cylinder. Twentieth Symposium (International) on Combustion, The Combustion Institute, Pittsburgh, 1893.

Dupoirieux, F., and Scherrer, D. (1985). Methodes numériques à convergence rapide utilisées pour le calcul des écoulements reactifs. Conference on Simulation of the Combustion Phenomena, INRIA, Sophia Antipolis, France.

Giovangigli, V., and Candel, S.M. (1986). Extinction limits of premixed catalysed flames in stagnation point flows. Comb. Sci. and Tech..

Giovangigli, V., and Smooke, M.D. (1985). Calculation of extinction limits for premixed laminar flames in stagnation point flow. Yale Univ. Rep. ME-105-85. Submitted to J. Computat. Phys.

Giovangigli, V., and Smooke, M.D. (1986). Calculation of critical points in flames. Second workshop on modelling of chemical reaction systems, Heidelberg, Aug. 1986.

Giovangigli, V., and Darabiha, N. (1987). Vector computers and complex chemisry. Presented at SIAM Meeting on Numerical Combustion. San Francisco, March 1987.

Giovangigli, V., and Smooke, M. (1987). Extinction limits for premixed laminar flames in a stagnation point flow. J. Comp. Phys. 68, 327-345.

Giovangigli, V., and Smooke, M. (1987). Extinction limits of strained premixed laminar flames with complex chemistry. Comb. Sci. and Tech., in press.

Giovangigli, V., and Smooke, M. (1987). Adaptive continuation algorithms with application to combustion problems. Submitted to Appl. Numerical Methods.

Jones, W.P. and Whitelaw, J.H. (1982). Calculation methods for reacting turbulent flows: a review. Comb. Flame 48, 1-26.

Karagozian, A., and Marble, F.E. (1986). Study of a diffusion flame in a stretched vortex. Comb. Sci. and Tech. 45, 65.

Kautsky, J., and Nichols, N.K. (1980). Equidistributing meshes with constraints. SIAM J. Sci. Stat. Comput. 1, 499

Kee, R.J., Miller, J.A., and Jefferson, T.H. (1980). CHEMKIN : A general - purpose problem - independent, transportable Fortran chemical kinetics code package.SANDIA Rep. SAND 80-8003, Sandia Lab., Livermore.

Kee, R.J., Warnatz, J., and Miller, J.A. (1983). A Fortran computer code package for the evaluation of gas-phase viscosities, conductivities, and diffusion coefficients. SANDIA Nat. Lab. Report, SAND 83-8209.

Lacas, F., Zikikout, S., and Candel, S. (1987). A comparaison between calculated and experimental mean source terms in non premixed turbulent combustion. AIAA Paper 87-1782.

Laverdant, A., and Candel, S.M. (1987). Numerical calculations of a diffusion flame vortex interaction. Submitted to Comb. Sci. and Tech..

Laverdant, A., and Candel, S.M. (1987). Computation of diffusion and premixed flames rolled-up in vortex structures. AIAA Paper 87-1779.

Law, C.K. (1984). Heat and mass transfer in combustion : fundamental concepts and analytical techniques. Progr. in Energy and Comb. Sci. 10, 295-318.

Libby P.A., Sivasegaram, S., and Whitelaw J.H. (1986). Premixed Combustion. Progr in. Energy and Comb. Sci. 12, 393-405.

Libby, P.A., and Williams, F.A. (1980). Fundamental aspects in turbulent reacting flows. Topics in Applied Physics, Vol. 44, P.A. Libby and F.A. Williams ed. Springer Verlag, New York.

Libby, P.A., and Williams, F.A. (1981). Structure of laminar flamelets in premixed turbulent flames. Comb. Flame 44, 287

Libby, P.A., and Williams, F.A. (1983). Strained premixed laminar flames under nonadiabatic conditions. Comb. Sci. and Tech. 31, 1

Libby, P.A., and Williams, F.A. (1984). Strained premixed laminar flames with two reaction zones. Comb. Sci. and Tech. 37, 221

Libby, P.A., Linan, A., and Williams, F.A. (1983). Strained premixed laminar flames with nonunity Lewis numbers. Comb. Sci. and Tech. 34, 257-293.

Liew, S.K., Bray, K.N.C., and Moss, J.B. (1984). A stretched laminar flamelet model of turbulent non-premixed combustion. Comb. Flame 56, 199-213.

Linan, A. (1974). The asymptotic structure of counterflow diffusion flames for large activation energies. Acta Astronautica 1, 1007.

Marble, F.E. (1985). Growth of a diffusion flame in the field of a vortex. Advances in Aerospace Science, C. Cashi, Ed. Plenum Press, New York, 395-413.

Marble, F.E., and Broadwell, J.E. (1977). The coherent flame model for turbulent chemical reactions. Project Squid Rep. TRW-9-PU.
Marble, F.E., and Broadwell, J.E. (1979). A theoretical analysis of nitric oxide production in a methane-air turbulent diffusion flame. EPA Tech. Rep.

Norton O.P. (1983). The effects of a vortex field on flame with finite reaction rates. PhD Thesis, California Inst. of Technology, Pasadena.

Peters, N. (1984). Laminar diffusion flamelets model in non-premixed turbulent combustion. Progr. in Energy and Comb. Sci. 10, 319-339.

Peters, N. (1986). Laminar flamelet concepts in turbulent combustion. 21st Symposium (International) on Combustion.

Poinsot, T.J., Trouvé, A.C., Veynante, D.P., Candel, S.M., and Esposito, E. (1987). Vortex driven acoustically coupled combustion instabilities. J. Fluid Mech. 177, 265-292.

Rogg, B., Behrendt, F., and Warnatz J. (1986). Turbulent non-premixed combustion in partially premixed diffusion flamelets with detailed chemistry. 21st International Symposium on Combustion, Munich, Aug. 3-8, 1986.

Smooke, M.D. (1982). Solution of burner stabilized premixed laminar flames by boundary value methods. J. Computat. Phys. 48, 72.

Smooke, M.D., Puri, I.K., and Seshadri, K. (1986). A comparison between numerical calculations and experimental measurements of the structure of a counterflow diffusion flame burning diluted methane in diluted air. Yale Univ. Rep. ME-101-86.

Spalding, D.B. (1978). The influence of laminar transport and chemical kinetics on the time mean reaction rate in a turbulent flame. Seventeenth Symposium (International) on Combustion. The Combustion Institute, Pittsburgh, 431

Tsuji, H. (1982). Progr. in Energy and Comb. Sci. 8, 93.

Veynante, D., Candel, S.M., and Martin, J.P. (1986). Coherent flame modelling of chemical reaction in a turbulent mixing layer. Second workshop on modelling of chemical reaction systems, Heidelberg, Aug. 1986.

Warnatz, J. (1983). The mechanism of high temperature combustion of propane and butane. Comb. Sci. and Tech. 34 177-200.

Warnatz, J. (1985). Private communication.

Williams, F.A. (1975). A review of some theoretical combustions of turbulent flame structure AGARD Conf. Proc. 164, p. II 1.1.

Williams, F.A. (1985). Combustion theory. 2nd ed. Benjamin/Cummings, Menlo Park.

Zukoski, E.E., and Marble F.E. (1955). The role of wake transition in the process of flame stabilization on bluff bodies. Comb. Research and Rev., Butterworths Scientific Publications, London.

Figure captions

FIG. 1. Schematic representation of physical processes descxribed in the coherent flame model.
(a) Production of flame area by the field of strain rates
(b) Mututal annihilation of adjacent flamelets
(c) Strained laminar flamelet model.

FIG. 2. Spark-Schlieren photograph of large scale vortices formed in a dump combustor under unstable combustion. The vortices appear at a frequency of f = 530 Hz. The vortex cap separates the fresh mixture of air and propane and hot combustion products. In this region the strain rate is of the order of $10^4 s^{-1}$ (photograph and data obtained from Poinsot et al. 1987).

FIG. 3. Instantaneous map of the C_2-emission unsteady component corresponding to the flow situation visualized in Fig. 2. The light emission in the regions of high strain rates such as the vortex cap (data obtained from Poinsot et al. 1987).

FIG. 4. Diffusion flame interacting with a vortex. An initially planar flame is wound-up in the field of a viscous-core vortex. Countours of constant product mass fraction are plotted at two instants in time. Flame area is created by the vortex motion. The flame elements formed near the vortex center interact and the intervening reactant is consumed. A core of burned products is formed. The flame shortening mechanism is made apparent in this calculation (further details may be found in Laverdant and Candel 1987)
(a) $t_* = 0.005$ (b) $t_* = 0.013$
$Re = 50$, $Sc = 1$, $\phi = 1$.

FIG. 5. Premixed flame interacting with a vortex. An initially planar laminar flame is wound up in the field of a viscous-core vortex. Contours of constant progress variable are plotted at two instants in time. The constant density assumption is used in the calculation. Flame area is created by the vortex motion. Flame area is dissipated near the vortex center due to mutual interactions of neighbouring elements (further details may be found in Laverdant and Candel 1987).
(a) $t_* = 0.0003$ (b) $t_* = 0.0022$ (c) $t_* = 0.0022$, 3D plot of the progress variable
$Da = 2.5 \cdot 10^9$, $T_A = 20000$ K, $Re = 50$, $Sc = 1$.

FIG. 6. Strained flame geometry

FIG. 7. Temperature distributions for different hot stream conditions. The strain rate is fixed $\epsilon_s = 340$ s^{-1}.

FIG. 8. Temperature distributions for different hot stream conditions. The strain rate is fixed $\epsilon_s = 1000$ s^{-1}.

FIG. 9. Temperature distributions for different strain rates. The hot stream temperature $T_b = 1700$ K is below the adiabatic flame temperature, but the temperature difference is moderate $T_{af} - T_b = 265$ K.

FIG. 10. Temperature distributions for different strain rates. The hot stream temperature is $T_b = 1450$ K. In this situation ignition and extinction transitions are observed
(a) and (b) distributions correspond to standard flames belonging to the upper branch of the extinction curve (c) is an unstable solution belonging to the middle branch (d) and (e) are extinguished solutions belonging to the lower branch

FIG. 11. Heat release rate per unit flame area plotted as a function of the strain rate for different hot stream temperatures. The curves are monotonic when T_b exceeds 1530 K, they take an S - shape below that value. The vertical scale is divided by 10^8 J/m^3/s.

FIG. 12. Reaction zone location plotted as a function of the strain rate for different hot stream temperatures. The curves are monotonic when T_b exceeds 1530 K, they take an S - shape below that value.

FIG. 13. Experimental configuration. A mixture of air and propane is injected through a long duct into a rectangular combustor inlet plane comprises a V-gutter flameholder placed at the duct center and producing a 50 % blockage.

FIG. 14. Spark schlieren photographs of the wake of the flameholder.
(a), (b) $\dot{m}_a = 33$ g/s, $\dot{m}_F = 1.19$ g/s, $\phi = 0.56$, $Re_H = 22000$
The equivalence ratio is close to the lean extinction limit of the combustor. Large burning structures are observed in this case.
(c) $\dot{m}_a = 41$ g/s, $\dot{m}_F = 1.65$ g/s, $\phi = 0.63$, $Re_H = 28000$
This photograph shows a well developed row of small scale vortices formed in the initial wake.
(d) $\dot{m}_a = 66.7$ g/s, $\dot{m}_F = 2.59$ g/s, $\phi = 0.61$, $Re_H = 46000$
The Reynolds number is higher than in case (c) and the reaction zone is shifted towards the flameholder.

(e) $\dot{m}_a$ = 67,5 g/s, $\dot{m}_F$ = 3.4 g/s, ϕ = 0.79, Re_H = 46500
The Reynolds number is close to that of case (d) but the equivalence ratio is higher than in (b). Reaction takes place over a broader spatial domain and the burning structures form a regular pattern and their size grows more uniformly.
(f) Same conditions as in case (d). A single reacting shear layer is displayed.

FIG. 15. Spatial distributions of the light emitted by C_2 radicals. The distributions are plotted on a scale of grey levels. The darkest symbol corresponds to the maximum light emission, the lightest symbol represents the minimum light intensity.
(a) Re_H = 46000, ϕ = 0.6, $\dot{m}_a$ = 67 g/s
(b) Re_H = 46000, ϕ = 0.7, $\dot{m}_a$ = 67 g/s
(c) Re_H = 46000, ϕ = 0.8, $\dot{m}_a$ = 67 g/s
(d) Re_H = 46000, ϕ = 0.9, $\dot{m}_a$ = 67 g/s
(e) Re_H = 21000, ϕ = 0.75, $\dot{m}_a$ = 31 g/s
(f) Re_H = 34000, ϕ = 0.75, $\dot{m}_a$ = 50 g/s
(g) Re_H = 68000, ϕ = 0.75, $\dot{m}_a$ = 100 g/s
(h) Re_H = 83000, ϕ = 0.75, $\dot{m}_a$ = 120 g/s

FIG. 16. Spatial distributions obtained by numerical calculation of (a) the fuel consumption rate and (b) the temperature for $\dot{m}_a$ = 100 g/s and equivalence ratio of ϕ = 0.75
Part "a" of this figure may be compared to Fig. 15g.

FIG. 17. Spatial distributions obtained by numerical calculation of (a) the fuel consumption rate and (b) the temperature for $\dot{m}_a$ = 69 g/s and equivalence ratio of ϕ = 0.75
Part "a" of this figure may be compared to Fig. 15a.

FIG. 18. Simplified version of the Barrère-Borghi-Peters diagram showing the different regimes of premixed turbulent combustion. Conditions existing at each point of the computational grid are plotted in this diagram.

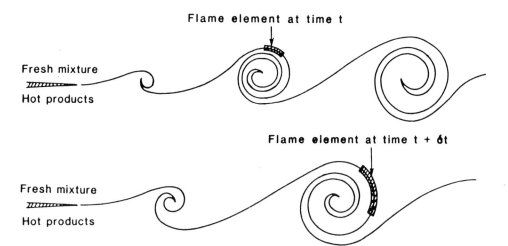

a) Production of flame surface by the strain rate

b) Flame annihilation process

c) Local model of the strained premixed flame element

Figure 1

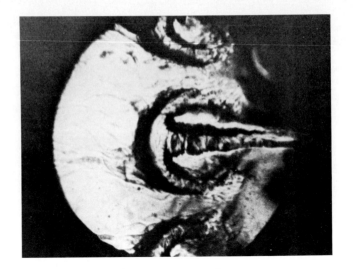

Figure 2

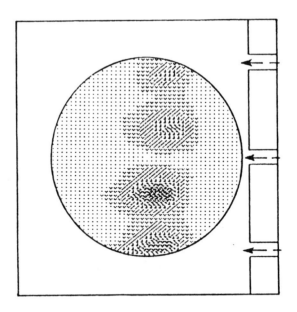

Figure 3

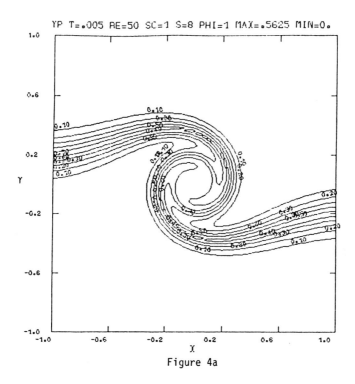

Figure 4a

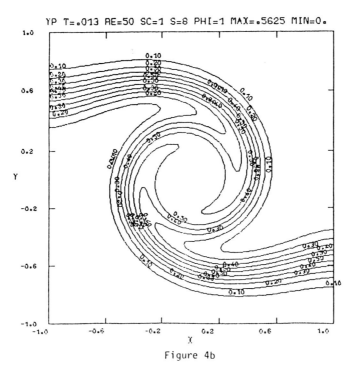

Figure 4b

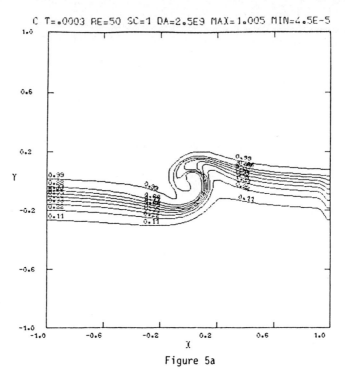

Figure 5a

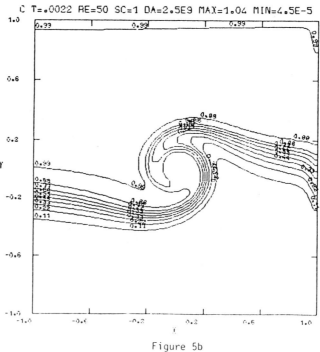

Figure 5b

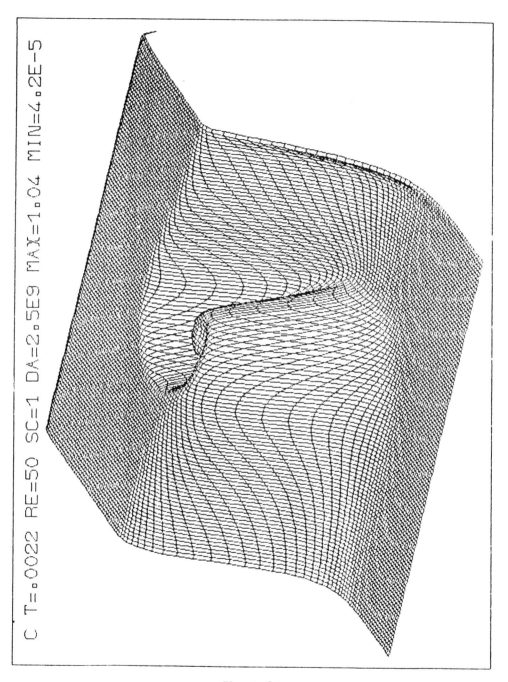

Figure 5c

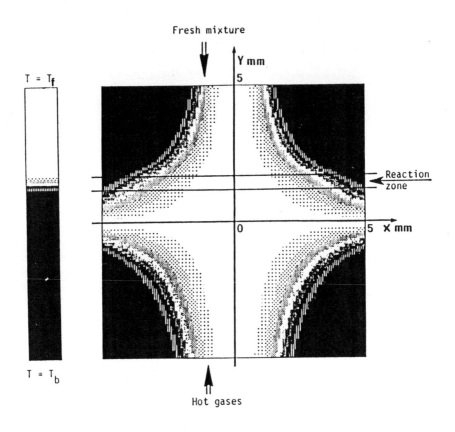

Figure 6

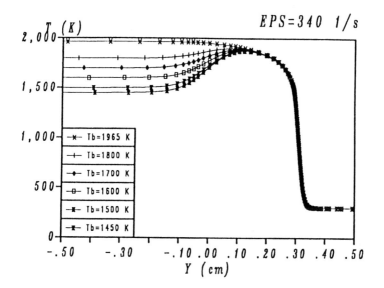

Figure 7

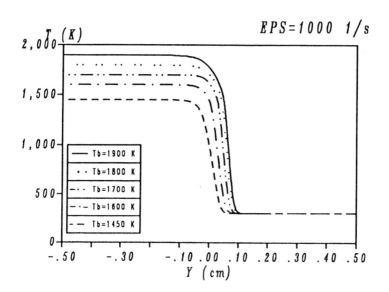

Figure 8

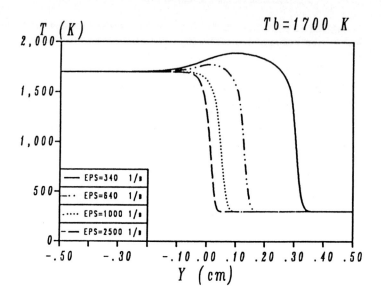

Figure 9

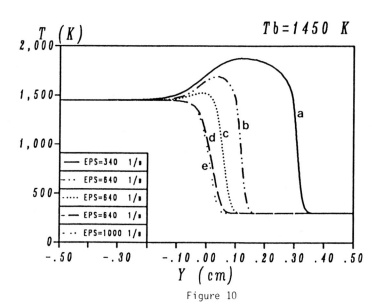

Figure 10

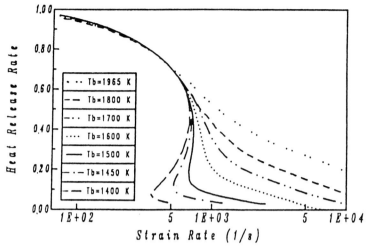

Figure 11

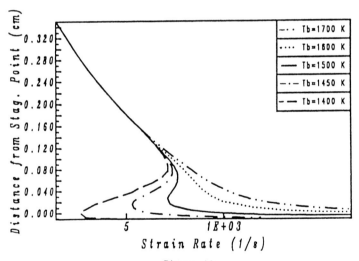

Figure 12

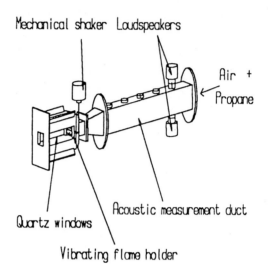

Figure 13

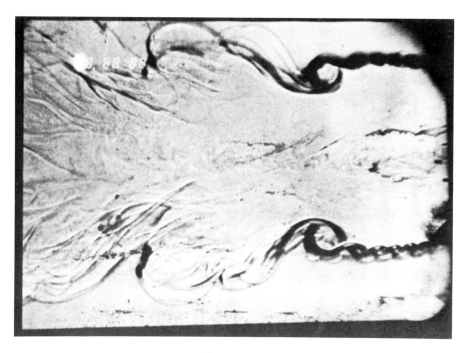

Figure 14a

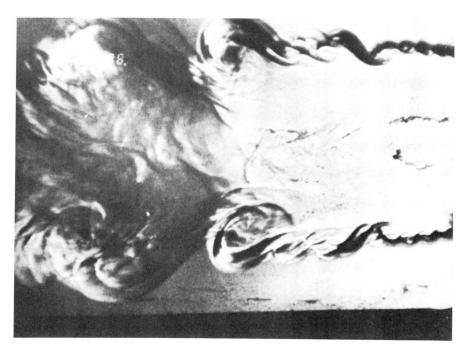

Figure 14b

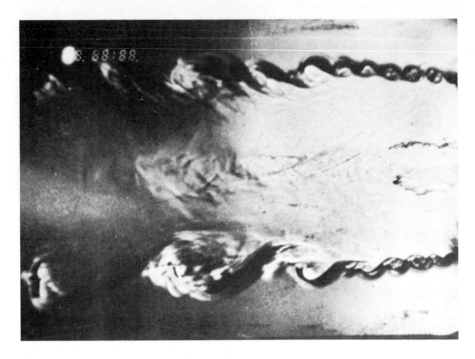

Figure 14c

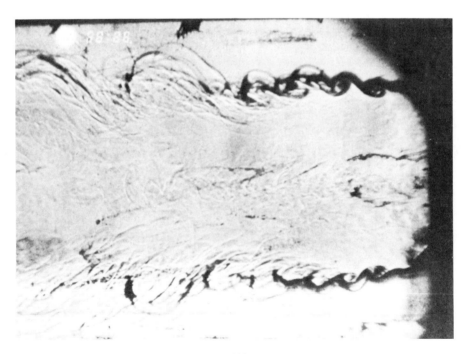

Figure 14d

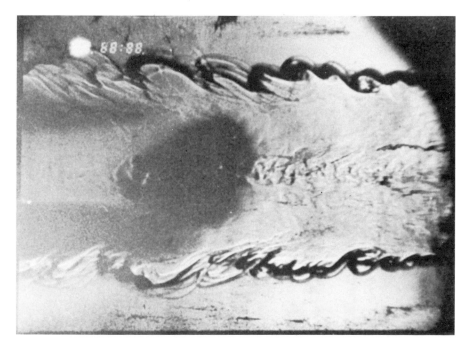

Figure 14e

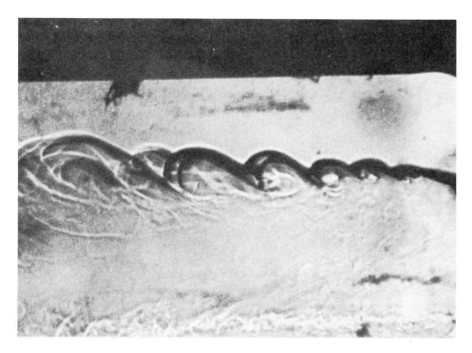

Figure 14f

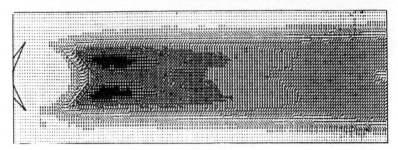

Figure 15a

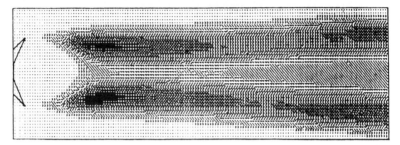

Figure 15b

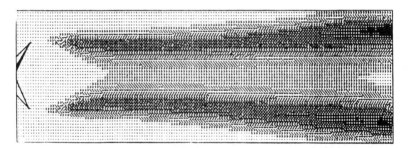

Figure 15c

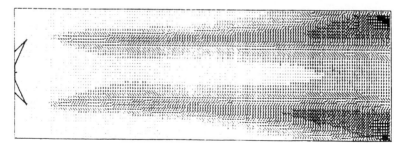

Figure 15d

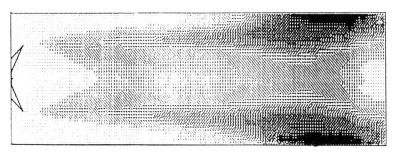

Figure 15e

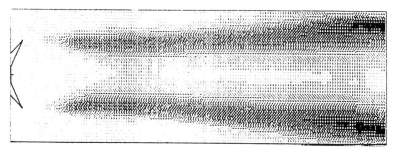

Figure 15f

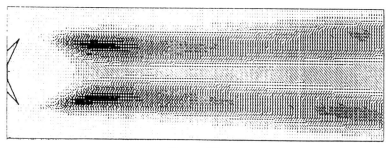

Figure 15g

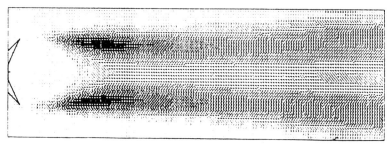

Figure 15h

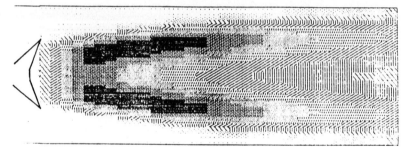

Figure 16a

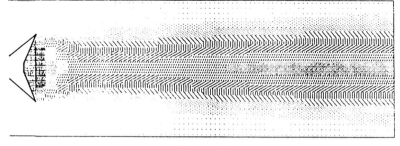

Figure 16b

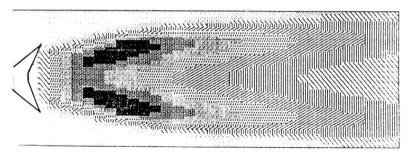

Figure 17a

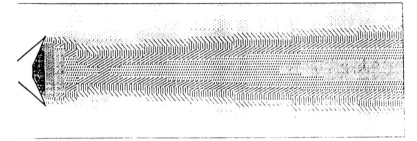

Figure 17b

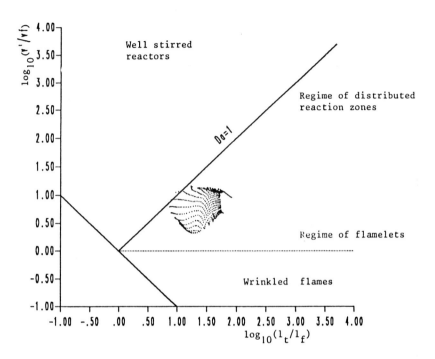

Figure 18

TURBULENCE-COMBUSTION INTERACTIONS IN A REACTING SHEAR LAYER

Ahmed F. Ghoniem, Ghassem Heidarinejad, and Anantha Krishnan

Department of Mechanical Engineering
Massachusetts Institute of Technology
Cambridge, MA 02139

ABSTRACT

Turbulence-combustion interactions are analyzed using results of a numerical simulation of a reacting shear layer. Premixed combustion at finite activation energy, moderate chemical kinetic rates and finite diffusivities is considered. The transport element method, a numerical scheme based on the accurate discretization of the vorticity and the scalar gradient fields into Lagrangian finite elements, is used to perform the numerical simulation. Processes that lead to burning enhancement, flame deceleration or possible extinction are analyzed. We find that the rollup of the shear layer accelerates burning by stretching the reaction surface. However, by comparing the local burning velocities within the shear layer to that of a laminar flame, we find that stretch, which accompanies the rollup, decelerates the rate of burning per unit area. This is due to the local cooling effects associated with the enhanced heat flux out and mass flux into the reaction zone. Both phenomena are strong functions of the turbulence field and the Damkohler number.

The preliminary version of this article was presented at the AIAA/SAE/ASME/ASEE 23rd Joint Propulsion Meeting, La Jolla, CA, June 1, 1987, AIAA-87-1718.

I. INTRODUCTION

Turbulent combustion is governed by the complex interaction between convective and diffusive transport processes on the one hand, chemical reaction and heat release on the other hand. Chemical reactions are strong nonlinear functions of temperature and species concentrations, and hence their rates are critically dependent on transport fluxes which determine these variables. Meanwhile, rates and magnitudes of heat release, associated with the chemical oxidation of practical fuels, are large enough to affect the dynamics of the flow, and hence the transport phenomena. Understanding the outcome of these interactions is an important leap on the way to achieve better control of burning processes in combustion systems. It is the objective of this work to: (1) develop numerical models capable of predicting turbulent combustion processes; (2) identify the most important modes of turbulence-combustion interactions; and (3) elucidate the subtle outcome of some of these interactions. We confine our attention to shear layers since they are relatively simple to analyze, and since they represent a generic model for many reacting flows.

Turbulent combustion has been the subject of extensive experimental, theoretical and numerical investigations over the years[1]. However, many of its fundamental mechanisms remain unclear.[2] Progress in phenomenological turbulent combustion models, based on the closure of a system of averaged transport equations which describe the statistical behavior of the aerothermodynamic variables, has made it possible to produce results that agree with experimental measurements. However, since most of the interesting dynamics of turbulence-combustion interactions are hypothesized a priori in these models, solutions do not provide a better understanding of the phenomena and are limited by the modelling assumptions.

Two problems have been identified as most challenging in the study of turbulence in reacting flow: the origin of the statistical correlations between fluctuating quantities; and the nature of the source terms in the energy and species conservation equations. In turbulent shear layers, the first problem is complicated by the presence of large scale structures that cannot be modelled by gradient diffusion terms. The second problem stems from the fact that chemical reactions are strongly affected by fluctuations in local variables in a nonlinear way, e.g., the Arrhenius form. In the following, the two issues are discussed in more detail.

Time-resolved flow visualization and instantaneous point measurements in nonreacting and reacting shear layers have revealed the existence of large scale periodic turbulent structures for a long distance downstream the separation point.[3,4,5,6] It has been shown experimentally, and supported by numerical studies, that these structures appear via the Kelvin-Helmholtz instability of the vorticity layer which forms between two initially-separate streams[7]. By a different mechanism, the subharmonic instability, these structures persist through successive pairings, thus maintaining the periodicity downstream though at different frequencies. Qualitatively, it is known that the role played by these structures in the mixing process is to engulf, then stretch layers of fluids to scales where molecular diffusion is most effective. The two processes, which have been called entrainment, may create a bottleneck impeding mixing if the rate of molecular diffusion is high due to small scale turbulence[8]. However, analytical results that support, or reproduce these effects are not yet available.

The importance of these results in the context of investigating turbulence-combustion interactions is how the existence of different scales can be properly represented. Even though gradient diffusion models, which assume that only local conditions can affect turbulence, may provide an adequate description of the small diffusive scales, they do not contain enough information to identify the large convective scales. The latter is a feature of the unsteady flow field and depends strongly on the initial and boundary conditions. Thus, it must be resolved by solving the unsteady unaveraged equations using accurate schemes, while the effect of the small scales, for the sake of economy, may be modelled based on an understanding of the "substructural phenomena."

Since chemical reaction depends on the rate of molecular mixing which, as previously described, is a consequence of entrainment by the large scales and diffusion at the small scales, it is important that these two processes be represented accurately. Overestimating the rate of mixing by turbulence models, which do not account for the effect of the large scales on the entrainment process, results in erroneous predictions of combustion.[8,9,10] The solution algorithm must, therefore, be able to resolve large scale convective structures as well as small scale diffusive eddies. It should also give careful consideration to their continuous interactions. Since the typical size of a large scale eddy is on the order of magnitude of the thin, but finite vorticity layers, they can only be resolved if the unsteady unaveraged equations are integrated using accurate numerical methods. These methods must be non-diffusive, i.e. they should not dissipate the flow energy by distributing it on large cells.

Resolving the unsteadiness of the flows is particularly important in combustion modelling since chemical source terms are strongly nonlinear functions of the instantaneous values of the temperature and species concentrations, as exemplified by the Arrhenius form. The magnitude of the source term depends on the fluctuations of the aerothermodynamic variables and, to a lesser extent, on the mean values. Moreover, since a chemical reaction, which is a time-dependent process, occurs on the molecular level, using averaged modelled equations in which mixing is described by mean fluctuations is not expected to properly describe it. An important aspect of combustion is that the chemistry is a Lagrangian process which proceeds as fluid elements move. Averaging this intrinsically unsteady process removes information that cannot be recovered using few moments.

A better description of chemical reaction in an unsteady flow field may be based on the Lagrangian formulation of the conservation equations. A numerical scheme, which employs the Lagrangian description of the flow field, should then lead to more accurate results if fine resolution is achieved. In this work, we describe such numerical simulation algorithms. Mathematical theories constructed to address issues of accuracy and convergence are exemplified in [11,12,13].

Numerical simulation of turbulent combustion attempts to minimize the use of phenomenological modelling. Thus, their results can be used to investigate some of the mechanisms of turbulence-chemistry interactions. Furthermore, since the instantaneous behavior of the variables are known at all points and at all times, accurate simulations offer a method of probing the flow when experimental techniques are not available. Ultimately, after validating and verifying the results against experimental measurements, ab initio predictions will be possible. Finite difference methods[14,15], spectral methods[16], and vortex methods[17] have been utilized in numerical simulation of nonreacting shear layers. Some have also been extended to reacting shear layers[18,19]. The first two methods are based on the Eulerian description of the flow field, using grids to discretize derivatives of the aerothermodynamic variables, or to expand these variables in harmonic functions, respectively. Vortex methods are grid free, Lagrangian schemes, which have been used to obtain solutions at high Reynolds number.

Vortex methods optimize the computational efforts by distributing computational elements around regions of high vorticity.[20,24] However, five factors have limited their utilization to study combustion problems: (1) Eulerian methods, which were used to solve the energy and species conservation equations in thin flame sheet models, seemed, in some sense, to defeat the purpose of using vortex methods to simulate the hydrodynamic field;[25] (2) the limit of fast chemistry, which was used in thin flame sheet models, did not allow realistic finite rate chemical kinetics to be part of the model;[26] (3) vortex methods, while maintaining reasonable accuracy in the majority of the field, lost resolution within the part of the field where the strain field is very strong;[27] (4) vortex methods were limited to handling incompressible flows, thus the models neglected the distributed expansion and the baroclinic torque generated during combustion; and (5) three-dimensional effects were only included in specific cases,[28,29] and no attempt has been made yet to represent small scale dissipation in two-dimensional methods.

In this article, we introduce the transport element method. When applied to obtain a solution of the vorticity transport equation, the method becomes the vortex element method in which particles are treated as finite elements that accurately discretize the vorticity field and change their shape, configuration or distribution to accommodate distortions of the vorticity field caused by the development of strong strain fields. The transport element method, moreover, extends the concepts of the vortex element method to obtain solutions of the scalar conservation equations, which govern reacting flow, in terms of moving Lagrangian gradients. Both schemes are formulated to preserve the effect of compressibility at low Mach number. The transport element method is applied to study the evolution of combustion in a reacting shear layer in premixed gases. Results are used to investigate different modes of turbulence-combustion interactions in a shear layer, and to study the outcome of these interactions in different regimes of the governing parameters. Processes of burning enhancement and flame deceleration or complete extinction via the effect of stretch which develops within the rolling shear layer are analyzed in light of the numerical results.

In Section II, the formulation of the model equations governing a reacting shear layer at low Mach number is described. In Section III, we give a summary of the vortex element method, and a comprehensive development of the transport element method. Emphasis is placed on the latter since it clarifies some subtle issues regarding the effect of the strain field on the local scalar gradients. In Section IV, results of the application of these methods to a spatially developing, thermally stratified mixing layer are described. In section V, results for a reacting, temporally developing mixing layer are analyzed. In Section VI, conclusions and future work are summarized.

II. FORMULATION

The non-dimensional form of the conservation equations governing a two dimensional, unsteady reacting flow is summarized is Table I. We assume that initially, and at all times at the inlet section, a premixed reactant R and a product P are present at given concentrations c_{Ro} and c_{Po} in the top high speed and bottom low speed streams, U_1 and U_2, respectively. For computational simplicity, chemistry is assumed to be governed by a single-step, irreversible, Arrhenius reaction of order n. Adding more steps to the chemical kinetics scheme will require integrating more species conservation equations. The Mach number is assumed to be small, which leads to the following simplifications in the governing equations: (1) pressure variation due to the flow field is small compared with the total pressure, and hence neglected in the equation of state; and (2) spatial variations of pressure and energy dissipation due to viscosity are neglected in the energy equation; and (3) acoustic interactions are removed. This isobaric approximation allows partial decoupling of the continuity, momentum and energy equations so that they can be integrated sequentially instead of simultaneously[30,31]. We assume that the reactants and products behave as perfect gases with equal molecular weights and specific heats, and that the thermal and mass diffusivities are constants, but not necessarily equal. The Reynolds number is high and the effect of viscosity is neglected.

The definition of symbols is as follows: $d/dt = \partial/\partial t + u \cdot \nabla$ is the Lagrangian derivative along a particle path. $u = (u,v)$ is velocity, $x = (x,y)$ where x and y are the streamwise and the cross-stream directions, respectively. t is time, ϕ is a velocity potential, $\psi = \psi e_z$ is a stream function defined such that $u_\omega = \nabla \times \psi = (\partial\psi/\partial y, -\partial\psi/\partial x)$, and $\omega = \nabla \times u$ is vorticity. e_z is the unit vector normal to the x-y plane. u_p is a potential velocity, $\nabla \cdot u_p = 0$, added to satisfy the normal boundary condition across the boundary of the domain. c is the concentration per unit mass, T is temperature. ∇ and ∇^2 are the gradient and Laplacian operators, respectively. A_f W is the rate of formation of products per unit mass per unit time, $W = c_R^n \exp(-T_a/T)$, and n is the reaction order. Variables are non-dimensionalized with respect to the appropriate combination of the velocity of the high speed stream U_1, the channel height H, the free stream concentration of R, c_{Ro}, and the free stream temperature of the reactants at $x = 0$, T_o. T_a is the activation energy, non-dimensionalized with respect to $(R_g T_o)$, R_g being the gas constant. Q is the enthalpy of reaction, non-dimensionalized with respect to $C_p T_o$, where C_p is the specific heat at constant pressure. $P_e = U_1 H/\alpha$ is the Peclet number, where $\alpha = k/(\rho C_p)$ is the thermal diffusivity, taken as a constant. $A_f = A H/U_1$ is the non-dimensional frequency factor of the chemical reaction rate constant. The Damkohler number $D_a = A_f W(c_m, T_m)$, c_m and T_m corresponding to conditions of maximum reaction rate, is the ratio of flow time to chemical time. $L_e = \alpha/D$ is the Lewis number, and D is the mass diffusivity.

Equation (2) is the decomposition of the velocity field into irrotational, solenoidal and potential components. Equation (3) is obtained by substituting u into the continuity equation and using $p = \rho T =$ constant since the flow is unconfined and is at low Mach number. Equation (5) is obtained by taking the curl of the momentum equation of an inviscid flow and using $\nabla p = - \rho \, du/dt$ to substitute for the pressure gradient in the baroclinic torque term to allow the integration of the equations without explicitly computing the pressure distribution. Equations (6-8) are the conservation of energy and species, respectively, for a reacting mixture at finite heat and mass diffusivities. Equation (9) is the equation of state at low Mach number in an unconfined flow.

The equations form a five-parameter system: P_e, L_e, D_a, Q and T_a. The properties of the solution and the characteristics of the interaction between the flow field and the chemical reaction depend on the values, or the combination of values, of the individual parameters. If the system is not adiabatic, i.e., $T_R = T_o$ while $T_P - T_R \neq Q$, one more parameter, such as T_P/T_R, must be specified in the formulation. The equations identify four different processes of turbulence-combustion interactions: (1) the generation of an irrotational velocity due to volumetric expansion as the temperature rises during heat release, $\nabla \phi$, is Eqs. (2,3); (2) the generation of baroclinic vorticity due to pressure gradient-density gradient interactions during heat release, $\nabla \rho \times \nabla p$ in Eq. (5); (3) the advection and straining of the flame structure in Eqs. (6,7 and 8); and, (4) the inhomogeniety in the diffusive fluxes due to non-unity Lewis number in Eqs. (6) and (7).

TABLE I GOVERNING EQUATIONS

REACTION	$R \xrightarrow{k} P$	(1)
VELOCITY	$u = \nabla \phi + \nabla \times \psi + u_p$	(2)
EXPANSION	$\nabla^2 \phi = \frac{1}{T} \frac{dT}{dt}$	(3)
ROTATION	$\nabla^2 \psi = - \omega(x,t)$	(4)
VORTICITY	$\frac{d}{dt}(\frac{\omega}{\rho}) = - \frac{1}{2} \nabla \rho \times (\frac{du}{dt})$	(5)

ENERGY $\quad \dfrac{dT}{dt} = \dfrac{1}{P_e} \nabla^2 T + A_f Q W$ (6)

REACTANTS $\quad \dfrac{dc_R}{dt} = \dfrac{1}{P_e L_e} \nabla^2 c_R - A_f W$ (7)

PRODUCTS $\quad \dfrac{dc_P}{dt} = \dfrac{1}{P_e L_e} \nabla^2 c_P + A_f W$ (8)

STATE $\quad \rho T = $ constant (9)

III. NUMERICAL METHOD

III.1. THE VORTEX ELEMENT METHOD

An important step in improving the accuracy and extending the application of vortex schemes to flow fields that develop large strain rates, such as shear layers, is the formulation of the vortex element method.[27] In this method, the vorticity field is accurately discretized among finite elements that move along particle paths, or particles that transport finite elements of vorticity. The strain field is used to redistribute the vorticity among the computational elements as time progresses so that small scales generated by planar stretch can be captured. This allows accurate long-time computations of the vorticity field after the strain field has developed. Capturing the strain field accurately is very important in computing turbulent flames since: (1) it governs the mixing process, which occurs after the original fluid layers have been stretched to very small scales, since it defines the diffusive flux; and (2) it may lead to flame quenching, or to burning enhancement, due to the generation of strong gradients as will be shown later. Below, we summarize the method and show how it can be extended to compute a compressible non-barotropic flow at low Mach number.

The vorticity field is initially discretized among vortex elements of finite structure. The distribution of vorticity associated with each element is described by a radially symmetric function, f_δ, with a characteristic radius, δ, such that most, or all of the vorticity is concentrated within $|x-X_i| < \delta$. X_i denotes the center of a vortex element at time $t = 0$. Vortex elements are initially distributed in the area where $|\omega| > 0$ such that the distance between neighboring elements is h in the two principal directions. The accuracy of the discretization depends on the choice of f_δ, the value of h, and the ratio δ/h. The strength of the vortex element located at X_i, denoted by ω_i, is obtained from the solution of the system of equations:

$$\omega(X_i, 0) = \sum_{j=1}^{N} \omega_j h^2 f_\delta(X_i - X_j) \quad (10)$$

where $\omega(\mathbf{x},0)$ is the vorticity distribution at $t = 0$. It can be shown that $f_\delta(r) = (1/\pi\,\delta^2)\exp(-r^2/\delta^2)$ leads to a second-order discretization. We found that for accurate representation of the vorticity distribution, δ must be slightly larger than h, i.e., $\delta/h \sim 1.1\text{--}1.3$, and that h must be varied until $\|\Gamma - \Sigma\Gamma_i\| < \varepsilon$ and $\|\omega(\mathbf{x},0) - \Sigma\,\omega_i\,h^2\,f_\delta(\mathbf{x}-\mathbf{X}_i)\| < \varepsilon$, where Γ is the total circulation of the vorticity field, $\Gamma(\mathbf{x},t) = \Sigma\,\Gamma_j\,K(\mathbf{x}-\mathbf{X}_j)$, $K(r) = \int_0^r r'f(r')dr'$ and $r = |\mathbf{x}|$. $\Gamma_i = \omega_i h^2$ is the total circulation of each individual vortex element. $\|\cdot\|$ denotes the second norm and ε is a small number which determines the accuracy.

For an incompressible flow, Eq. (5) leads to the Helmholtz theorem, which states that vorticity is constant along particle paths, i.e.

$$\omega(\mathbf{x},t) = \sum_{i=1}^{N} \Gamma_i\, f_\delta(\mathbf{x}-\mathbf{X}_i(\mathbf{X}_i,t)) \qquad (11)$$

and

$$\frac{d\mathbf{X}_i}{dt} = \mathbf{u}(\mathbf{X}_i(\mathbf{X}_i,t),t) \qquad (12)$$

where $\mathbf{X}_i$ is the particle path $\mathbf{X}_i(\mathbf{X}_i,0) = \mathbf{X}_i$. To obtain the velocity field of a collection of vortex elements in the form of Eq. (11), we note that the stream function of a single vortex element is obtained by integrating Eq. (4). Using polar coordinates to integrate this equation for a vortex element placed at x=0, we get $\partial\psi_\delta/\partial r = -K(r/\delta)/r$. The velocity field of a single element is thus radially symmetric since $u_\theta = -\partial\psi_\delta/\partial r$. The velocity field induced by a distribution of finite-core vortex elements, of shape f_δ and strength Γ_i located at $\mathbf{X}_i(\mathbf{X}_i,t)$ is:

$$\mathbf{u}_\omega(\mathbf{x},t) = \sum_{i=1}^{N} \Gamma_i\, \mathbf{K}_\delta(\mathbf{x}-\mathbf{X}_i(\mathbf{X}_i,t)) \qquad (13)$$

where

$$\mathbf{K}_\delta(\mathbf{x}) = -\frac{(y,-x)}{r^2}\, K\!\left(\frac{r}{\delta}\right) \qquad (14)$$

Vortex elements move at the local velocity computed at their centers. As time progresses, the distance between neighboring elements increases in the direction of maximum strain rate such that $\Delta\chi > h$, where $\Delta\chi$ is the distance in the direction of maximum strain defined as $\Delta\chi = \Delta\mathbf{x}\cdot\Delta\mathbf{u}/|\Delta\mathbf{u}|$ and Δ is the difference operator. This leads to a deterioration of the discretization accuracy, which requires that $\delta > \Delta\chi$. Thus, an algorithm must be used such that when $\Delta\chi > \beta h$, where $\beta \sim 1.5$, a computational element is inserted at the midpoint between the original elements and $\Delta\chi' = \Delta\chi/2$. The circulation of the new element, and that of the original two neighboring elements, is one third the sum of the circulation of the original two elements.[27]

For compressible barotropic flow, Eq. (5) shows that $d(\omega/\rho)/dt = 0$. Moreover, $\Gamma = \int \omega dA$, where A is the area, while $\int \rho dA = $ constant. Thus, the circulation is constant along a particle path -- Kelvin theorem -- and Eqs. (11-14) can be used to compute the evolution of the vorticity and velocity field provided that Eq. (3) is used to compute the irrotational component of the velocity due to volumetric expansion, as will be shown in the next section. When $\nabla\rho \times \nabla p \neq 0$, the circulation of each vortex element must be updated each time step. Using the definition of the circulation in Eq. (5), we get:

$$\frac{d\Gamma}{dt} = - \int \frac{\nabla \rho}{\rho} \times \left(\frac{du}{dt}\right) dx \tag{15}$$

Since $\Gamma = \Sigma \Gamma_i \kappa(x-X_i)$, $\nabla \rho = \Sigma \Delta \rho_i f_\delta(x-X_i)$, and $\Delta \rho_i = \nabla \rho_i h^2$ as will be shown in the next section, Eq. (15) can be written as:

$$\frac{d\Gamma_i}{dt} = - \frac{\Delta T_i}{T_i} \times \left(\frac{du}{dt}\right)_i \tag{16}$$

where, according to the low Mach number approximation, $\nabla \rho / \rho = - \nabla T / T$, while $\nabla T = \Sigma \Delta T_i f_\delta(x-X_i)$ and $\Delta T_i = \nabla T_i h_i^2$. In the next section, we will show how to compute $\nabla \rho$, ρ, ∇T and T. Moreover, $(du/dt)_i$ is computed by numerically differentiating the velocity of the vortex element using a high-order formula. Equations (12) and (16) are integrated using a fourth order Runge-Kutta-Merson method with variable time step for error control.

III.2. THE TRANSPORT ELEMENT METHOD

Another important development in the application of particle methods to reacting flows is the formulation of the transport element method to compute the temperature and species concentration distributions in a Lagrangian form[27]. In this scheme, the gradient of the scalar field is discretized into a number of finite elements using Eq. (10) with ω replaced by $g = \nabla s$, where s is a generalized scalar, being either T or c. Like vortex elements, transport elements are distributed where $|\nabla s| > 0$ and are moved with the local velocity field with time. Particles are used to transport scalar gradients, however, contrary to vorticity, scalar gradients are not conserved along particle path, and should be modified according to the local straining and tilting of the material elements. The extension of this method to reacting flow will require changing the gradient transported by each element according to the reaction source term in Eqs. (6,7,8) in a way similar to changing the circulation with the baroclinic torque. Thus, the evolution of the chemical reaction with time will be computed in a Lagrangian frame of reference as the interacting species flow. In the following, we describe the conservative form of the transport element scheme and its extension to solve Eqs. (6,7,8).

Initially, the scalar gradient g is discretized on a square mesh hxh according to

$$g(X_j,0) = \sum_{i=1}^{N} g_i h^2 f_\delta(X_j - X_i) \tag{17}$$

where f_δ, δ and h have been defined as before, and should be chosen to satisfy the same requirements. Note that the values of g_i depend on the choice of h and δ, and are obtained by solving the system of linear algebraic equations formed by applying

Eq. (17) to all mesh points. To see how to transport the scalar gradient in a Lagrangian form, we start by the incompresssible, non-diffusive, non-reactive case. If s is a passive, non-diffusive scalar, the conservation equations for s and g = ∇s are:

$$\frac{ds}{dt} = 0 \qquad (18)$$

and

$$\frac{dg}{dt} = - g \cdot \nabla u - g \times \omega \qquad (19)$$

where $\omega = \omega\, e_z$. Thus, s remains constant along a particle path, while g changes due to the straining and rotation of the material line by the local strain field and vorticity. If the material is exposed to a strong strain in the direction normal to the gradient, the value of g must increase by the same amount as the stretch in the material element. This can be seen by expanding g in terms of n and g, where g = |g| and n = g/g, and noting that $dn/dt = - n \times \omega/2 - (1.(n.\nabla u^s))1$, where ∇u^s is the symmetric part of the strain tensor ∇u and 1 is the unit vector normal to n (see Batchelor[32]):

$$\frac{dg}{dt} n = - g\, (n \cdot \nabla u + \frac{1}{2}\, n \times \omega - (1.(n.\nabla u^s))1\,) \qquad (20)$$

Moreover, g = (ds/dn) n ~ (δs/δn) n, where δs is the variation of s across a small material line δn. The variation of a material vector element δl is given by $d(\delta l)/dt = \delta l \cdot \nabla u$, where $\delta l = \delta l\, 1$.[32] Furthermore, for an incompressible flow, δl δn = constant along a particle path defined by $dX/dt = u(X(X,t),t)$. From these kinematic relations, the variation of the material line δl along a particle path can be written as:

$$\frac{d\delta l}{dt} n = - \delta l\, (n \cdot \nabla u + \frac{1}{2}\, n \times \omega - (1.(n.\nabla u^s))1\,) \qquad (21i)$$

From Eqs. (20) and (21i), it follows that g/δl = constant along a particle path. For a graphical representation of this concept, see Fig. 1. Thus, the flux initialized by Eq. (17) evolves according to:

$$g(x,t) = \sum_{i=1}^{N} g_i(t)\, h^2\, f_\delta(x - X_i(X_i,t)) \qquad (22i)$$

where

$$g_i(t) = \frac{\delta s_i\, \delta l_i(t)}{h^2}\, n_i(t) \qquad (23i)$$

while

$$\frac{dX_i}{dt} = u(X_i(X_i,t),t)$$

where $X_i(X_i,0)=X_i$. δl_i is updated using Eq. (21i) and $n_i \cdot 1_i = 0$. While using Eq. (23) is equivalent to updating $g_i(t)$ according to Eq. (19), applying the expression in Eq. (23i) guarantees the conservation of δs_i. Moreover, instead of integrating Eq. (21i) to update δl_i, one can save computational effort by recalling that $\delta l_i(t) = (X_{i+1} - X_{i-1})/2$. Thus, it suffices to move the centers of the transport elements, while remembering the near neighbors at $t = 0$ to compute the scalar flux. When an element is inserted between two neighboring elements, the values of δl_i are redistributed between the three elements and h^2 in Eqs. (22) and (23) is changed to h_i^2 so that the total material area is conserved. In this case, Eq. (22) becomes

$$g(x,t) = \sum_{i=1}^{N} g_i(t) h_i^2 f_\delta(x - X_i(X_i,t)) \qquad (22s)$$

where

$$g_i(t) = \frac{\delta s_i \, \delta l_i(t)}{h_i^2} n_i(t) \qquad (23s)$$

For a compressible flow, the above analysis should be modified to reflect the fact that $\rho \, \delta l \, \delta n = $ constant along a particle path. Using the kinematic relations listed above, Eq. (21i) becomes:

$$\frac{d}{dt}(\rho \, \delta l) \, n = - \rho \, \delta l \, (\, n \cdot \nabla u + \frac{1}{2} n \times \omega - (1.(n.\nabla u^s))1 \,) \qquad (21c)$$

Thus, $g/(\rho \, \delta l) = $ constant along a particle path. In terms of the variational change in s, δs_i, across a material element δl_i, Eq. (23i) is modified as follows:

$$g_i(t) = \frac{\delta s_i \, \delta l_i(t) \, \rho_i(t)}{\rho_i(0) \, h_i^2(0)} n_i(t) \qquad (23c)$$

The value of ρ is computed using the relation $\rho T = $ constant, in accordance with the low Mach number approximation. Note that the area of the material element is expanding such that $\rho_i(t) h_i^2(t) = $ constant. Thus,

$$g_i(t) = \frac{\delta s_i \, \delta l_i(t)}{h_i^2(t)} n_i(t) \qquad (23c)$$

and

$$g(x,t) = \sum_{i=1}^{N} g_i(t) h_i^2(t) f_\delta(x - X_i(X_i,t)) \qquad (22c)$$

Given the location and strength of the transport elements, the scalar concentration can be computed as follows. By taking the gradient of s, $\nabla s = g$, we get $\nabla^2 s = \nabla \cdot g$. The solution of this equation in an infinite domain can be written as: $s = - \int \nabla \cdot g \, G \, dx$, where $G = -1/2\pi \ln r$ is the Green function of the Poisson equation. This last equation shows that the transport elements act as sources of strength equal to the divergence of the scalar flux, $\nabla \cdot g$. Integrating by parts, one gets $s = \int g \cdot \nabla G \, dx$. Using Eqs. (22) and (23) for g, we get:

$$s(x,t) = \sum_{i=1}^{N} g_i(t) h_i^2(t) \cdot \nabla G_\delta(x - X_i(X_i,t)) \qquad (24)$$

where

$$\nabla G_\delta(x) = \frac{(x,y)}{r^2} \kappa(\frac{r}{\delta}) \tag{25}$$

where $\kappa(r) = \int r' f(r') dr'$, as defined before. If the distance between neighboring elements in the direction of principal strain exceeds a maximum distance βh, one element is inserted halfway between the two elements and the value of δl_i and h_i^2 are adjusted for the three elements. A recombination procedure can also be implemented to curb the growth in the number of computational elements. The need for this insertion-recombination procedure is more apparent here since the magnitude of the gradient increases where the strain field is high; and to maintain accuracy, more elements must be used to transport this gradient.

With finite diffusivity, the first term on the right hand side of Eqs. (6-8) should be simulated in the solution. In gradient form, the conservation equation can be written as:

$$\frac{dg}{dt} = -g \cdot \nabla u - g \times \omega + \alpha \nabla^2 g \tag{26}$$

where α is the molecular diffusivity, or the inverse of the Peclet number. At high speed, this is typically 10^2-10^5. To solve Eq. (26) using the scheme that we have developed so far, each element g_i must be updated according to the diffusion equation:

$$\frac{\partial g_i}{\partial t} = \alpha \nabla^2 g_i \tag{27}$$

without changing the shape of the core function or the value of g_i. Taking $\delta = \delta(t)$, and substituting Eq. (17) into Eq. (27), we obtain $d\delta^2/dt = 4\alpha$. Thus, to simulate the effect of diffusion, the core radius must grow according to:

$$\delta^2 = \delta_o^2 + 4\alpha t \tag{28}$$

where δ_o is the core radius at $t = 0$. If the diffusivities of momentum, heat and mass are different, the core of the vortex elements and of different scalar transport elements become different as time progresses. At high diffusivities, or small Peclet numbers, the cores of the elements will experience rapid growth and $\delta >> \beta h$. In this case, transport elements must be subdivided into elements with smaller cores while preserving their total strength. However, this will not be used here since we are interested in cases where the Peclet number is large. Values typical to this study are: $\delta_o = h = 0.3$, $t_{max} = 20$, and $\alpha = 0.001$, then $\delta_{max} = 0.41$.

If the chemical source term is non-zero, then Eq. (26) is modified to become:

$$\frac{dg}{dt} = -g \cdot \nabla u - g \times \omega + \alpha \nabla^2 g + \sum_{j=1}^{k+1} \frac{dW}{ds_j} g_j \tag{29}$$

where k is the number of reacting species. Using the definitions of g, the gradient transported by each element must be modified according to:

$$\frac{d}{dt} \delta s_i = \sum_{j=1}^{k+1} \frac{dW}{ds_j} \delta s_j \qquad (30)$$

In this case, the element strength should be modified as:

$$g_i(t) = \frac{\delta s_i(t) \, \delta l_i(t)}{h_i^2(t)} n_i(t) \qquad (23r)$$

while all the kinematic relations, and Eq. (22c) hold as before.

Recognizing the fact that $h_i^2(t)$ appears in the numerators of Eqs. (22i), (22s) and (22c); while it appears in the denominator of Eqs. (23i), (23is), (23c), and (23r), we will define a new quantity $\Delta g_i = g_i h_i^2$ and rewrite these equations as: [6~

$$g(x,t) = \sum_{i=1}^{N} \Delta g_i(t) \, f_\delta(x - X_i(X_i,t)) \qquad (31)$$

$$\Delta g_i(t) = \delta s_i(t) \, \delta l_i(t) \, n_i(t) \qquad (32)$$

$$s(x,t) = \sum_{i=1}^{N} \Delta g_i(t) \cdot \nabla G_\delta(x - X_i(X_i,t)) \qquad (33)$$

Equations (31), (32) and (33) apply for the most general case. The transport elements generate an expansion field as their temperatures change, according to Eq. (3). The velocity field associated with this expansion within each element at the low Mach number limit can be written as:

$$\nabla^2 \phi_i = \frac{1}{T_i} \left(\frac{dT}{dt}\right)_i \qquad (34)$$

The total velocity produced by the expansion field is:

$$\nabla \phi(x,t) = \sum_{i=1}^{N} \frac{1}{T_i} \left(\frac{dT}{dt}\right)_i h_i^2(t) \, \nabla G_\delta(x - X_i(X_i,t)) \qquad (35)$$

where h_i^2 is the area of the material element which is divided every time one element is inserted due to stretch and is varied according to mass conservation, and $\rho_i(t) h_i^2(t)$ = constant.

The algorithm of the transport element method proceeds as follows: (1) update the locations of the elements X_i according to the velocity at their centers using Eq. (12); (2) update the values of δl_i and n_i either according to the integration of Eq. (21) or by keeping track of the neighboring elements; (3) update the core radii of different elements according to the corresponding Peclet number using Eq. (28); and (4) compute the concentrations of all the scalars using Eq. (24); and (5) update the value of δs_i according to Eq. (30). In most cases, it is possible to use the same set of particles to transport elements of different scalars, as well as the vortex elements, resulting in substantial savings in the transport step.

IV. THE SPATIALLY-DEVELOPING, NON-REACTING SHEAR LAYER

The vortex element and the transport element methods are applied to simulate the initial stages of development of a spatially-developing, thermally-stratified, two-stream shear layer. On the left boundary of the domain, it is assumed that the wake region behind the splitter plate, where the two incoming boundary layers merge to form the shear layer, is negligibly small. Thus, at $x = 0$; for $y > \Delta_s$: $u \rightarrow U_1 = 1$, $T \rightarrow T_1^* = 1$, and for $y < -\Delta_s$, $u \rightarrow U_2 = 0.333$, and $T \rightarrow T_2^* = 0$, where $\rightarrow$ means "approaches asymptotically." $\Delta_s^2 = 2 \sigma^2$, where σ is the standard deviation of the Gaussian distribution that describes the vorticity and the scalar gradients and $2\Delta_s$ is the nominal shear layer thickness at $x = 0$. The normalized temperature is defined as $T^* = (T-T_2)/(T_1-T_2)$. For the results in Fig. 2, $\Delta_s = 1/26.4$. The corresponding most unstable wavelength, as predicted by the linear theory, is 0.5. Within the shear layer, the velocity and temperature distributions are represented by error functions.

The rate at which vorticity is convected into the upstream side of the computational domain, at $x = 0$, is $d\Gamma/dt = \Delta U \cdot U_m$, where $U_m = (U_1+U_2)/2$. At each time step, five elements arranged vertically, are used to discretize this vorticity according to Eq. (10). The potential velocity component, u_p, is computed by adding two source flows at $x = -\infty$ and $y = +0$ and $y = -0$ to the velocity field in Eq. (2) to satisfy the boundary condition at $x = 0$. The no-flow boundary condition across the solid walls is implemented by using conformal mapping and image vortices with the opposite sign of vorticity in the transformed plane[7].

In the solution of the energy equation, the walls are considered insulated, i.e., $dT/dn = 0$ where n is the unit vector normal to the wall. To satisfy this boundary condition, the images of the temperature transport elements in the transformed plane must have the opposite of the signs of the elements. Energy sources are utilized to impose the boundary condition at $x = 0$. At the downstream side of the computational window, $x = 5$, vortex and transport elements are deleted. This induces a perturbation which ensures that the rollup and first pairing will always take place within the computational window. Since this perturbation is not applied in an organized manner, the resulting shear layer will be considered as an unforced layer.

Figure 2 shows the location and velocity of all vortex elements used in the computations for four different time steps. The time step of the computations is $\Delta t = 0.15$. The plots exhibit a very clear and accurate portrait of the rollup. During rollup, the vorticity within the shear layer is attracted towards the center of a large eddy entraining fluid from both sides and forming what appears to be a moving focal point of a spiral. Between neighboring large eddies, a zone of strong strain is developing where the vorticity is depleted and the gradients are growing. This "braid" zone can be described as a moving saddle point where locally the fluid flow experiences a separation into two streams; one moving towards the left and the other moving towards the right with respect to the saddle stagnation point. Downstream, the process of rollup continues until a stronger perturbation forces two neighboring eddies to interact in a pairing process. It is important to stress that the algorithm of inserting elements as the strain field develops is responsible for maintaining the organization of the calculation for a long time.

The natural frequency of shedding can be defined as $f_n = U_m/\lambda$, where λ is the wavelength of the large eddy. The corresponding Strouhal number, as computed from the computational results, is $S_t = 1/f_n = 0.033$. This is the same value as the frequency of the most unstable mode computed from the linear stability theory of a spatially developing shear layer under the conditions described above. Results for the growth rate, average velocity and turbulent statistics were presented in the study of Ghoniem and Ng[7] for the forced shear layer. Comparison with the corresponding analytical and experimental data were also performed in the same reference.

If the layer is forced at a frequency close to the most unstable mode by oscillating the incoming vorticity layer according to $\Delta y = a_f \sin(2\pi \Omega_f t)$, where Δy is displacement of the center of the vortex element due to forcing and a_f and Ω_f are the amplitude and frequency of forcing, the evolution is expected to be more organized.[7] In Fig. 3, we plot the results of such a case with $a_f = 0.025$ and $\Omega_f = \lambda_f/U_m = 1.33$, where λ_f is the wavelength of forcing. The evolution of the eddy which has the forcing frequency through the various stages of rollup is shown clearly at each time step when moving downstream, or with time when observed from the same location.

The effect of rollup on the temperature distribution within the eddy is shown in Fig. 4. Here, we plot the temperature distribution across several sections downstream, superimposed on the distribution of vortex elements at the same location. In these plots, we assume that the thermal diffusivity is negligibly small, and we concentrate on the effect of the convection field on the entrainment of hot and cold fluid within the large eddies. Note that the temperature profiles become more ragged as the core spins further, and that the temperature distribution is not symmetric around the midsection of the eddy.

The high resolution of the transport element method demands the use of a large number of transport elements. Moreover, the number of elements grows rapidly with time due to the severe stretch produced in the shear layer. This makes the computation of a wide window, which contains a number of successive eddies, rather expensive. In the next section, we direct our attention towards a model of this problem that requires less effort computationally while essentially preserving all the physical processes involved in the spatially developing layer. This is the temporal shear layer model in which a computational window that moves at the average speed of the flow is imposed on a single wavelength while the eddy is growing.

V. TEMPORALLY-DEVELOPING, REACTING SHEAR LAYER

Computational results showing the evolution of a large eddy in a temporal shear layer are presented in Fig. 5. In this case, the boundary conditions are periodic, i.e., $\omega(x,y,t) = \omega(x+\lambda,y,t)$ and $u(x,y,t) = u(x+\lambda,y,t)$, where λ in the wavelength of the perturbation. Since detailed analysis of the evolution of the temporal, thermally stratified shear layer was presented in Ghoniem et al.[27], it will not be repeated here. The qualitative resemblance between the development of large eddies in a spatial and a temporal shear layer is clearly seen by comparing Figures 3 and 5. Moreover, the shedding frequency, i.e. the frequency of the most amplified mode, is almost the same in both cases. However, the growth rate of the perturbation is different since it depends on the velocity ratio across the layer; a parameter that does not appear in the analysis of the temporal layer. Moreover, the asymmetric growth of the eddies, which is observed in the spatially-growing case, Fig. 3, is not present in the temporally-developing layer results, Fig. 5.

In the computation of the temporal layer, the window is limited to one wavelength and one can afford to use more elements within the domain to improve the resolution. One can also conduct, inexpensively, parametric studies on the effect of various physical parameters that appear in the model, Eqs.(1-9). Thus, the temporal layer will be used as a model for the spatial layer to study turbulence-combustion interactions in shear flow. Since the flow in unconfined, the shear layer thickness Δ_s is used instead of H to non-dimensionalize the length.

The temperature profile across the midsection of the eddy is exhibited in Figure 6. The rollup brings fluid from one side to the opposite side, while stretch increases the gradient across each layer. Thus, the rollup of the shear layer is the mechanism of entrainment that leads to strong mixing enhancement as the two fluids diffuse across the stretched interface. The temperature profiles show that after the relaxation of the first rollup, a secondary instability develops which forces the core through another turn, creating a more ragged temperature distribution. It is also noticed, by comparing Figs. 4 and 6, that the asymmetric growth of the spatially-developing layer is responsible for creating asymmetric temperature profiles across the midsection of the eddies. The relationship between these temperature profiles and the asymmetric entrainment observed in experimental measurements [33,34] will be explored in detail in future studies. [35]

Since rollup is associated with strong stretch that reduces the thickness of the material layers, it increases the gradients across these intertwining layers, thus enhancing the diffusion fluxes. Quantitatively, the rate of mixing can be expressed as $\dot{M} = \int q \cdot n \, da$, where q is the diffusion flux, n is the unit vector normal to the material surface, and da is the surface area element. Moreover, for two dimensional flow, $da = dl$, and since $q/\delta l$ = constant, then $\dot{M}$ is proportional to $(\delta l)^2$. The net result is that stretch by a factor ζ enhances mixing by a factor ζ^2. The quadratic rise in mixing during rollup will have a significant effect on the rate of reaction.

In the reacting layer calculations, the full system of equations is integrated using particles that transport vortex elements, temperature gradient elements, and reactant and product gradients elements. At time t = 0, the vorticity layer and the flame front coincide, and the thickness of the vorticity layer as well as the flame thickness are equal. A small sinusoidal perturbation with amplitude $\varepsilon = 0.05 \lambda$ is imposed on both distributions. The first case to be computed corresponds to the following set of parameters: $P_e = 200$, $L_e = 1$, $A_f = 1$, $Q = 4$, and $T_a = 10$ and $n = 1$. The corresponding Damkohler number, measured at the conditions of maximum reaction rate, is around 0.02, and the temperature ratio across the layer is $T_p/T_R = 5$.[1]

Figure 7 shows the reacting shear layer as rollup and the chemical reaction proceed simultaneously. At the early stages, the reacting eddy strongly resembles the nonreacting eddy shown in Fig. 5. However, as rollup starts, the following is observed: (1) a swelling, due to the increase in the rate of heat release, continues as more reactants are entrained into the burning core; (2) the growth of the instability, as measured by the angle between the major axis of the elliptical structure and the main stream direction, is encumbered because the volumetric expansion causes the vorticity intensity to decrease and the eddy to become weaker and less coherent; and (3) the eddy loses its symmetry and becomes eccentric due to the asymmetric expansion and due to the generation of baroclinic torque associated with density gradients. As more of the initial core is burnt, the fluid inside the eddy ceases to spin, contrary to the nonreacting case in which the secondary instabilities force the core to continue its spinning. Meanwhile reactants move through the side to enter the reaction region.

These results agree qualitatively with the experimental results of Keller and Daily[5] on the reacting premixed shear layer at intermediate values of the equivalence ratio. The Schlieren photographs of the experiment show that as the equivalence ratio is increased, the rate of growth of both the individual eddies as well as the entire shear layer increases due to heat release. In the meantime, the rollup of individual eddies slows down, leading to the formation of elliptical eddies. The major axes of the eddies remain at a finite angle with respect to the streamwise direction. Moreover, at low equivalence ratios, most burning occurred within the cores of the eddies, and the flame did not leave the shear layer.

On the same figure, a solid line is plotted through points of maximum reaction rate. The line indicates where the flame front, or the maximum heat release rate, is within the shear layer. Below this line, the product concentration approaches unity and the temperature reaches T_p. During the early stages of rollup, the line of maximum reaction rate follows one of the material lines closely, i.e., the growth of perturbation merely changes the topology of the flame front. At later stages,

---

1. In the following results, the value of h in Eq. (23c) was taken as a constant for all elements and for all times. In more recent computations, when we varied h with stretch and expansion, while all the trends were the same, the rate of reaction was found to be less than what was obtained with constant h. Thus, the results will only be interpreted qualitatively.

this line, while staying close to another material line, forms a boundary of products across which the reactants are entrained into the burning core. Below this line, where products form, the core almost ceases to rotate. At the last stage of burning of the eddy, the two sides of the flame within the core burn to close this entry way, and the flame moves out of the eddy and becomes an ordinary laminar flame.

The effect of heat release on the structure of the eddy, which is generated by the rollup of the shear layer, can be seen from the temperature profiles across the midsection of the wavelength shown in Fig. 8. Since the Lewis number is one, $c_R = 1-(T-T_R)/Q$. As reactants are entrained into the core of the growing eddy from the right side, a Z-shaped flame is formed. At the initial stages when the rate of entrainment is faster than the rate of burning, the flame extends deep into the lower stream. As the reactants within this zone burn, heat is released within the core of the rotating eddy causing the eddy to swell while maintaining its elliptical shape. The baroclinic vorticity generated around this zone causes the observed eccentricity of the large eddy. The temperature profiles show that the higher order instabilities observed in the nonreacting case are suppressed by the heat release, and that the core of the eddy stops its rotation. As the reactants within the eddy burn, the flame leaves the structure and moves into the reactants stream. This results in the formation of a temperature profile which is very similar to the temperature profile at t=0.

Figure 7 also shows the effect of rollup on the shape of the flame front, which, as will be shown in the next paragraph, has a strong effect on the overall rate of burning and the local burning velocity. In the early stages, and until $t \sim 7$, the flame front maintains its sinusoidal shape and its length is approximately the same as the flame length at $t = 0$. In the second stage, as the eddy starts to roll up, the flame front forms a fold within the eddy. Within this fold, reactants are trapped, and a situation in which two flames are burning towards each other is created. Rollup increases the length of flame front and exposes the flame to a strong strain. The extent of the fold within the eddy is limited by the consumption of the reactants trapped between the two sides of the flame front. It is also limited by the fact that burning inhibits the spinning of the core. The consumption of reactants and the continuous stretch of the flame reduces the distance between the two folds around $t \sim 15$, and the two flames become much closer to each other than before.

To study the effect of the shear layer on the chemical reaction, we plot the total mass of products, M_p, formed since the rollup starts at $t = 0$ in Fig. 9. At the early stages, when the flame stretch is negligibly small, the rate of burning is linear and identical to that of a laminar flame. As the layer starts to roll up, the area of the reaction surface increases and the flame is convoluted around the growing eddy. The increase in the flame area, or length in a two-dimensional sense, L_f, due to its folding within the eddy is shown in Fig. 10. The rate of product formation, $\dot{M}_p$, which is the slope of the curve in Fig. 9, can be approximated by the product of the flame length times the average burning velocity along the flame, S_u.

Since $\dot{M}_p$ is almost constant in the second stage, then the value of S_u must be decreasing with increasing L. Thus, as the flame stretches, its burning velocity decreases. This is in accordance with the previous results on stretched laminar flames at high strain rates.[36,37] In both studies, a drop in the flame burning velocity and partial extinction was observed as the strain rate increased.

The drop of the local burning velocity when a strain rate develops along the flame can be explained as follows. As the strain along the flame front becomes finite and positive, the local gradients normal to the front increase, enhancing the diffusion fluxes of heat from and of reactants into the flame. This can lead to flame cooling if the chemical time scale is relatively large, i.e., if the reaction is not fast enough to produce heat that could balance the cooling effect of the diffusion fluxes. Moreover, cooler flames burn slower than adiabatic flames. Thus, strong strain may lead to slower flames at moderate values of the reaction rate.

In Fig. 11, we plot the temperature, T, the strain rate, $\dot{s}$, and the rate of expansion, $\dot{e}$, along one particular layer of fluid within the reacting eddy. The rate of expansion is an indication of the rate of temperature rise due to the combined effect of diffusion and chemical reaction, as seen from Eq. (6). The layer along which these parameters are plotted is shown in Fig. 12. Figure 11 shows that within the fold of the flame, the temperature is very close to the temperature of maximum reaction rate, indicating that most of the burning occurs within the eddy core. This is in agreement with the experimental results.[5] On the other hand, the temperature at the side of the eddy which is exposed to the reactants is relatively low, and burning is not expected to proceed at an appreciable rate there.

When we decrease the frequency factor to $A_f = 0.5$, which reduces the Damkohler number by the same ratio, we see a stronger effect of stretch and a better distinction between the different stages of development. For this case, the large eddy is shown in Fig. 13 at t = 17.57, while the total mass of products is shown in Fig. 14. The swelling of the eddy is reduced since the rate of chemical reaction is one half of its value in the first case. Figure 14 shows that at the early stages, the reaction proceeds in the same way as before: a laminar flame followed by a stretched laminar flame. Around time t = 14, the slope of the curve of M_p vs. t, i.e. $\dot{M}_p = S_u L_f$, increases. While the value of S_u is still decreasing as the flame length increases, its value is somewhat higher than before. A possible explanation for this phenomenon can be found by observing that the two sides of the flame fold become much closer around t = 14, as seen in Fig. 7. As the two sides of the flame fold approach each other, the temperature of the reactants trapped between the two sides rises. This leads to an increase in the burning velocity.

At t = 17.57, the reaction slows down approaching a state of total extinction, [6~ shown by the total mass of products exhibited in Fig. 14. To explain what happens around extinction, we refer to Fig. 15. In this figure, plots of T, $\dot{s}$, and $\dot{e}$ are shown along one particular layer at t = 17.57. The geometry of the same layer is shown in Fig. 16. Plots for $\dot{e}$ show that the temperature is falling and the layer is experiencing cooling. This is in spite of the fact that T corresponds to maximum reaction rate along most of the layer. Moreover, the values of T and $\dot{s}$ are now negatively correlated, i.e. temperature maxima correspond to minima in $\dot{s}$, and $\dot{s}$ and $\dot{e}$ are also negatively correlated. The strain thus cools the flame leading to its eventual extinction.

When the frequency factor is lowered further to $A_f = 0.25$, extinction occurs earlier at around t = 10, as shown in Fig. 17. A laminar flame at the same condition shows the expected linear rise in M_p. These results confirm that the local burning velocity decreases with stretch, and that this phenomena is caused by the imbalance between the rates of diffusion and the rate of the chemical reaction.

VI. CONCLUSIONS

Numerical methods enable one to: (1) integrate elaborate and detailed models, which cannot be done analytically, so that complex mechanisms may be revealed and analyzed; and (2) provide detailed information about the flow field which may not be possible using traditional experimental techniques. Computer output, rich in data, offers a challenge of extracting and presenting valuable information about the phenomena under investigation. Finding the appropriate diagnostics to probe computational results represents half the journey to reaching the conclusions.

In this article, we have introduced the transport element method; a Lagrangian particle scheme based on the discretization of the vorticity and the gradients of the scalars into finite elements. The particles move along material lines, in accordance with their transport equations. As strong strains develop in the dynamic field, the finite elements may change their shape or configuration to accommodate the distortion which is produced by these strain fields. In case of chemical reaction: (1) the strength of the elements, i.e. the source strength, changes according to the rate of reaction; and (2) the chemical heat release induces volumetric expansion and baroclinic vorticity into the dynamic field.

The simplest model which can be proposed to study turbulence-combustion interactions contains five parameters: (1) the Peclet number which defines the ratio between the rate of convective and diffusive heating; (2) the Lewis number which represents the ratio between the rate of heat and mass diffusion; (3) the frequency factor which defines the ratio between the rate of chemical reaction and mass convection; (4) the activation energy of the reaction; and (5) the enthalpy of reaction. The outcome of these interactions can, thus, be presented on a five dimensional space on which one can identify several subdomains for burning enhancement, flame extinction, flame oscillations, etc. To accomplish this goal, computations must be performed for a matrix of parameters. The compiled data can then be plotted on this space. Under the idealization of high activation energy and thin flame structure, results of the asymptotic analysis can be used to fill some parts of this space and show the limiting trends[39,40,41].

In this article, we presented results for the effect of changing the frequency factor, which leads to changing the Damkohler number, at fixed values of the rest of the parameters. We showed that for $P_e = 200$, $L_e = 1$, $T_a = 10$ and $Q = 4$, at $A_f = 1.0$, the stretch associated with the rollup of large eddies in the mixing layer enhances the rate of reaction by extending the flame surface area within the large eddies, in spite of the fact that the local burning velocity decreases as the flame surface is stretched. At lower values of A_f, combustion is interrupted under strong stretch, and the lower the values of A_f become, the earlier the flame is extinguished. This is due to the fact that the rise in the mass flux into the reaction zone and the heat flux out of the flame is not balanced by an increase in heat release by chemical reaction within this zone. The reaction zone is thus cooled, followed by the extinction of the flame. Work is underway to vary the rest of the controlling parameters and study their effect on flame stability.

ACKNOWLEDGEMENT

This work was supported by the Air Force Office of Scientific Research Grant AFOSR 84-0356, the U.S. Department of Energy, Office of Energy Utilization, Energy Conservation and Utilization Technologies Program Contract DE-AC04-86AL16310, the National Science Foundation Grant CPE-8404811, and the Edgerton Professorship at M.I.T.

REFERENCES

1. Libby, P.A. and Williams, F.A., eds., Turbulent Reacting Flows, Springer-Verlag, Berlin, 1980, xiii + 243 p.

2. Chigier, N.A., ed., Progress in Energy and Combustion Science, special issue on Turbulent Reacting Flows, 12, (1986).

3. Brown, G.L. and Roshko, A., J. Fluid Mech., 64, 775 (1974).

4. Ho, C.-H., and Huerre, P., Ann. Rev. Fluid Mech., 16, 365, (1985).

5. Keller, J.O. and Daily, J.W., AIAA J., 23, 1937, (1985).

6. Mungal, M.G., and Dimotakis, P.E., J. Fluid Mech., 148, 349, (1984).

7. Ghoniem, A.F. and Ng, K.K., Phys. Fluids, 30, 706, (1987).

8. Broadwell, J.E. and Breidenthal, R.E., J. Fluid Mech., 125, 397, (1982).

9. Driscoll, J.F., Tangirala, V. and Chen, R.H., Combust. Sci. Tech., 51, 75, (1986).

10. Kelly, J. private communications.

11. Hald, O., SIAM J. Num. Anal., 16, 726, (1979).

12. Beale, J.T. & Majda, A., Math. Comp., 39, 28, (1982).

13. Anderson, C., J. Comput. Phys., 61, 417, (1985).

14. Corcos, G.M. and Sherman, F.S., J. Fluid Mech., 139, 29, (1984).

15. Grinstein, F.F., Oran, E.S. and Boris, J.P., J. Fluid Mech., 165, 201, (1986).

16. Riley, J.J., and Metcalfe, R.W., AIAA paper 80-0274.

17. Ashurst, W.T., in Turbulent Shear Flows, ed. Durst et al. (Springer-Verlag, Berlin, 1979), p. 402.

18. McMurtry, P.A., Jou, W.A., Riley, J.J., and Metcalfe, R.W., AIAA Journal, 24, 962 (1986).

19. Ghoniem, A.F. and Givi, P., AIAA paper 87-0225.

20. Chorin, A.J., J. Fluid Mech., 57, 785 (1973).

21. Chorin, A.J., "Vortex models and boundary layer instability." SIAM J. Sci. Stat. Comput., 1, 1980, pp. 1-24.

22. Leonard, A., J. Comput. Phys., 37, 289 (1980).

23. Ghoniem, A.F. and Gagnon, Y., J. Comput. Phys., 68, 342, (1987).

24. Sethian, J.A. and Ghoniem, A.F., "Validation of the vortex method," J. Comput. Phys., to appear.

25. Ashurst, W.T. and Barr, P.K., "Lagrangian-Eulerian calculation of turbulent diffusion flame propagation," Sandia Report SAND80-9950, Sandia National Laboratories, 1982.

26. Ghoniem, A.F., Chorin, A.J. and Oppenheim, A.K., Phil. Trans. Roy. Soc. Lond., A304, 303, (1982).

27. Ghoniem, A.F., Heidarinejad, G. and Krishnan, A. "Vortex element simulation of the rollup and mixing in a thermally stratified shear layer," J. Comput. Phys., submitted for publication.

28. Leonard, A. Ann. Rev. Fluid Mech., 17, 523, (1985).

29. Ghoniem, A.F., Knio, O.M., and Aly, H.F., AIAA Paper 87-0379.

30. Majda, A., and Sethian, J.A., Combust. Sci. Tech., 42, 185, (1987).

31. Ghoniem, A.F., Lectures in Applied Mathematics, 24, ed. by A. Ludford, 199, (Amer. Math. Soc., 1986), p. 199.

32. Batchelor, G.K., An Introduction to Fluid Dynamics, Cambridge University Press, 1967, xviii + 615 p.

33. Koochesfahani, M.M, Dimotakis, P.,E. and Broadwell, J.E., AIAA Journal, 23, 1985, pp. 1191-1194.

34. Dimotakis, P.E., AIAA Journal, 24, 1986, pp. 1791-1796.

35. Krishnan, A., and Ghoniem, A.F., "Baroclinic effect of a density stratified shear layer," for presentation at the 1st National Congress on Fluid Dynamics, Cincinati, OH, July 1988.

36. Darabiha, N., Candel, S.N. and Marble, F.E., Combust. Flame, 64, 1986, pp. 203-217.

37. Giovangigli, V. and Smooke, M.D., J. Comput. Phys., 68, 1987, pp. 327-345.

38. Rogg, B., "Response and flamelet structure of stretched premixed methane-air flames," Combust Flame, 1987, in print.

39. Clavin, P., Prof. Energy Combust. Sci., 11, 1 (1985).

40. Williams, F.A., Combustion Theory, 2nd ed., Benjamin/Cummings, 1985, xxiii + 680 p.

41. Buckmaster, J.D. and Ludford, G.S.S., Lectures on Mathematical Combustion, SIAM, 1983, V + 126 p.

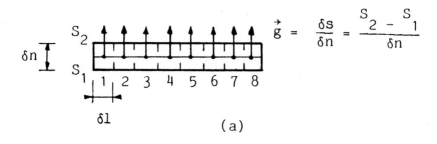

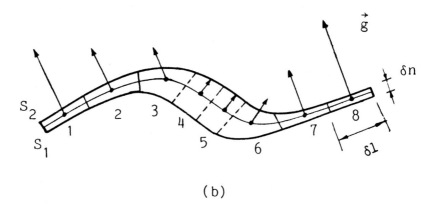

Figure 1. Schematic sketch showing the evolution of a material layer separating two values of the conserved scalar, s, and the associated scalar gradient, g, under the effect of stretch.

Figure 2. The development of large scale vortex structures in an unforced, spatial shear layer. Each point in the figure represents a vortex element and the line attached to it is the velocity vector. The velocity ratio across the layer is 3:1.

Figure 3. The development of large scale vortex structures in a forced, spatial shear layer with the same velocity ratio as in Fig. 2. The forcing frequency is 1.333.

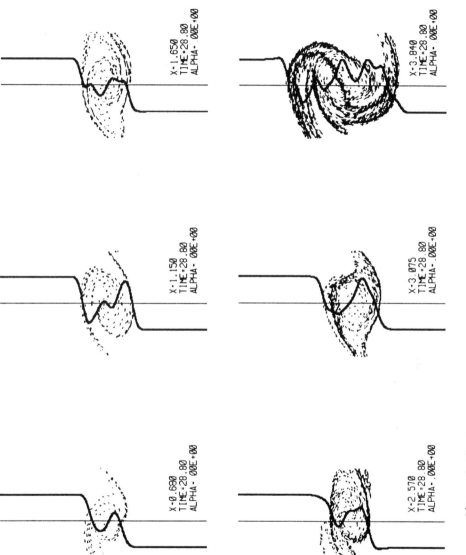

Figure 4. The temperature distribution across selected cross sections for the layer in Fig. 3.

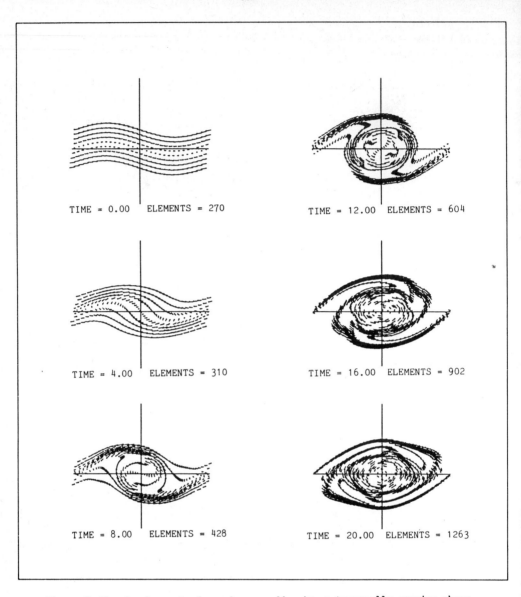

Figure 5. The development of a large eddy in a temporally growing shear layer.

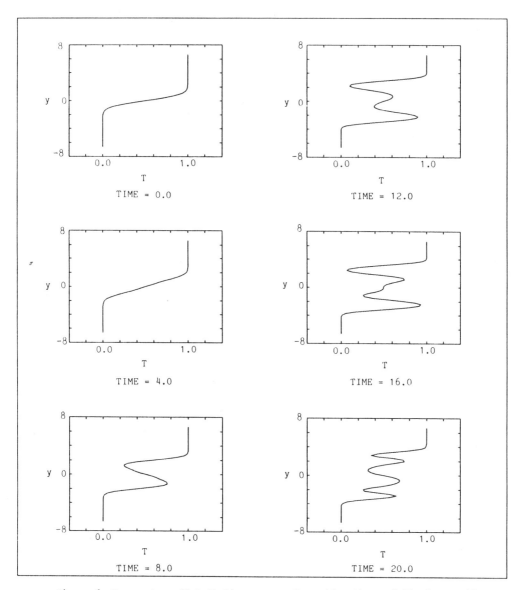

Figure 6. Temperature distribution across the midsection of the large eddy shown in Fig. 5.

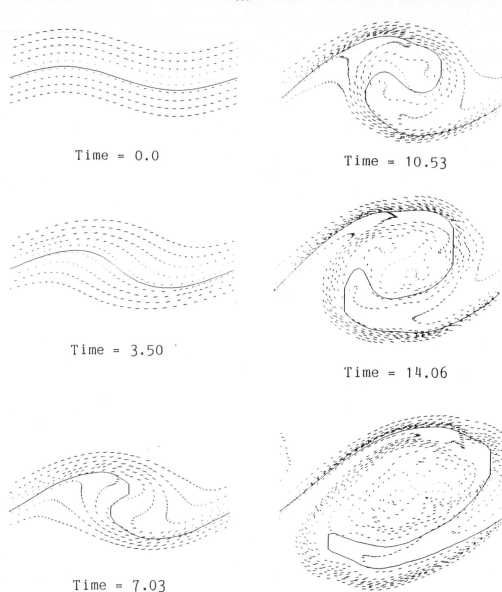

Figure 7. The development of a large eddy in a reacting temporal shear layer at the same conditions as in Fig. 5. The solid line defines the flame front.

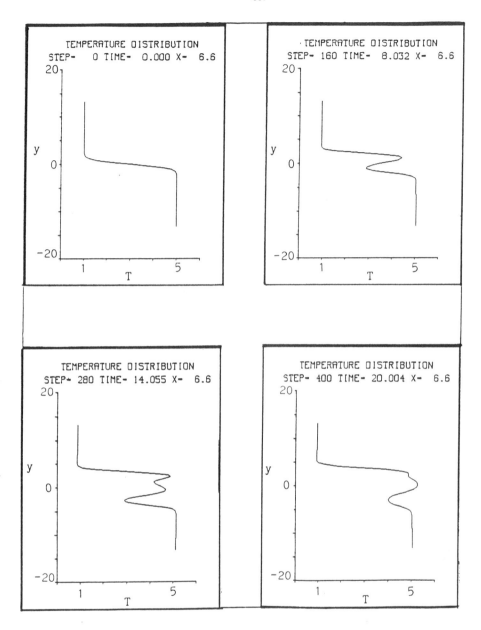

Figure 8. Temperature distribution across the midsection of the reacting large eddy shown in Fig. 7.

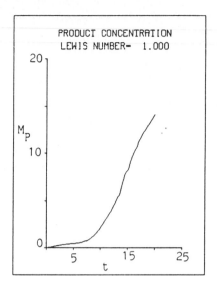

Figure 9. Total mass of products M_p formed since $t = 0$, in the reacting mixing layer with $A_f = 1.0$.

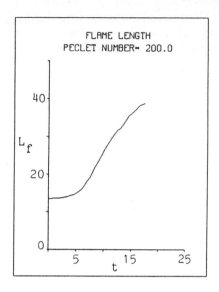

Figure 10. The total length of the flame in the reacting shear layer of figure 7.

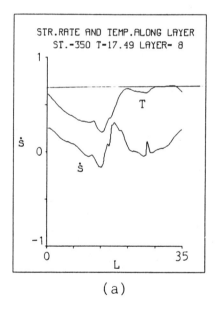

(a)

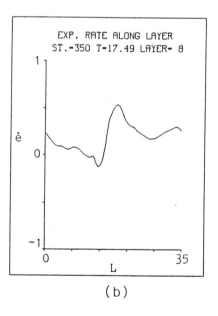

(b)

Figure 11. (a) the temperature T and the strain rate $\dot{s}$, (b) the expansion rate $\dot{e}$ along layer 8 in the reacting eddy shown in Fig. 7.

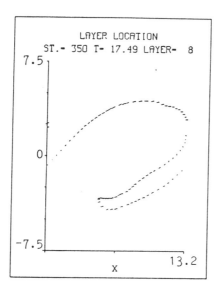

Figure 12. Layer 8 in the reacting eddy of Fig. 7.

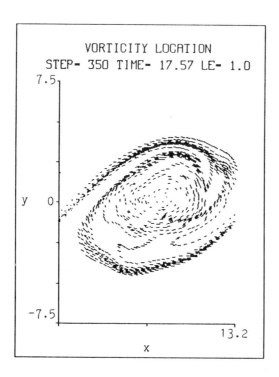

Figure 13. The large eddy in a reacting temporal shear layer at the same conditions as in Fig. 7 but with $A_f = 0.5$ at $t = 17.57$.

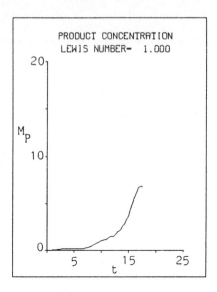

Figure 14. The total mass of products M_p formed since t = 0 for a reacting shear layer with $A_f = 0.5$.

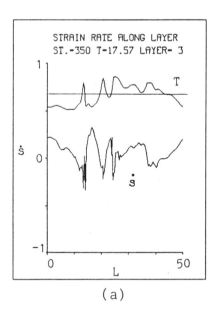
(a)

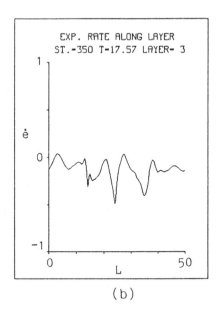
(b)

Figure 15. (a) the temperature T and the strain rate $\dot{s}$, and (b) the expansion rate $\dot{e}$ along layer 3 in the reacting eddy of Fig. 13.

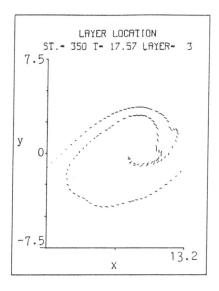

Figure 16. Layer 3 in the reacting eddy of Fig. 13.

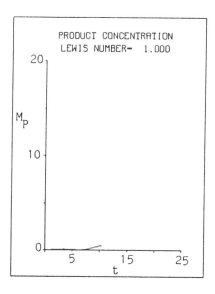

Figure 17. The total mass of products M_p formed since $t = 0$ for a reacting shear layer with $A_f = 0.25$.

A PDF METHOD FOR TURBULENT RECIRCULATING FLOWS

M. S. Anand*, S. B. Pope**, and H. C. Mongia*

* Allison Gas Turbine Division
General Motors Corporation
P.O. Box 420
Indianapolis, Indiana 46206

**Sibley School of Mechanical and Aerospace Engineering
Upson Hall, Cornell University
Ithaca, New York 14853

ABSTRACT

A novel approach for the application of probability density function (PDF) methods to multidimensional turbulent recirculating flows is presented. The method is applicable to turbulent recirculating and reacting flows such as in gas turbine combustors. The method is based on a judicious combination of the conventional finite-volume technique for the solution of the Reynolds-averaged equations and the Monte Carlo technique for the solution of the transport equation for the velocity-scalar joint PDF. An important aspect of the approach is that the use of conventional turbulence closure models is avoided. The method is applied to the flow over a backward-facing step investigated experimentally by Pronchick and Kline [1]. The results predicted using the present approach are in good agreement with data.

I. INTRODUCTION

The flow in most practical combustion devices, such as in gas turbine combustors, is characterized by recirculation zones, swirl, and strong interaction between turbulence and reaction. Predictions of turbulent flows, reacting and nonreacting, require the use of turbulence models to overcome the turbulence closure problem [2]. For reacting flows, additional closures are needed due to the nonlinear reaction rates and large density variations. The turbulence/chemistry interaction—the effect of turbulence on mean reaction rates and the effect of heat release due to reaction on the turbulence—is one of the least understood aspects of turbulent reacting flows [3].

The turbulence closures used currently are the mean-flow closure (e.g., k-ε model [4], Reynolds stress--algebraic (RSA) [5, 6, 7] models) and the second-order closures (e.g., Reynolds stress--differential (RSD) [8, 9]). In mean-flow closures, the Reynolds stresses (which transport momentum) are modeled, and in second-order closures, the triple velocity correlations (which transport Reynolds stresses) are modeled. These transport processes are usually modeled by gradient diffusion. The same ideas are used in modeling scalar transport as well. The conventional models can treat reactions only under special circumstances--when the reaction rate is linear or when it is very fast (e.g., [10]) or very slow compared with the turbulent time scale [11]. Also, sometimes these models involve an assumed form for the PDF of an appropriate scalar variable (e.g., [10]). However, a general nonlinear finite-rate reaction eludes treatment by these conventional models.

Since the first extensive application of analytical models as a design aid [12], based primarily on the k-ε model, a number of model improvements and combustor analytical design procedures have been proposed, but none of these has shown a consistently significant improvement over the basic approach [13]. Although the utility of the current models should not be overlooked, there is clearly a need to develop radically new and alternative methods for calculating turbulent recirculating and reacting flows. A most promising alternative is the PDF transport approach [14].

The joint probability density function (JPDF or PDF for short) transport approach [14] involves the solution of the evolution equation of the velocity-composition joint PDF. The joint PDF, $f(\underline{V}, \psi; \underline{x}, t)$, is the probability density of the simultaneous event $\underline{U}(\underline{x}, t) = \underline{V}$ and $\underline{\phi}(\underline{x}, t) = \underline{\psi}$, where $\underline{U}$ is the velocity vector, $\underline{\phi}$ is a set of scalars, and $\underline{V}$ and $\underline{\psi}$ are independent variables in the velocity-scalar space.

The PDF transport approach overcomes most of the serious closure problems previously mentioned. The terms involving convection, reaction, body forces, and the mean pressure gradient (including the variable-density effects in those terms) appear in closed form. Hence, the use of conventional turbulence models (and consequently the gradient transport assumptions) is avoided. Also, arbitrarily complex reactions can be treated without approximations. It should be noted that the PDF transport method is different from the assumed PDF method in that the form of the PDF of the scalar is not assumed. Rather, the joint PDF of velocities and scalars is obtained from the solution of the transport equation for the PDF. An additional advantage of the PDF transport approach is that the joint PDF contains a rather complete representation of the flow field, and any single-point correlation can be determined. This contrasts with conventional methods in which only a limited number of moments (usually the first and the second) of the PDF are determined.

The PDF f is a function of a large number of variables, 7 + σ--three velocity components, three space variables, time, and σ scalars. Due to the large dimensionality of the PDF, standard techniques (such as finite differences to solve the PDF transport equation) are prohibitive [15]. Monte Carlo methods are computationally efficient in solving the transport equation and have been used for calculating many flows, some of which are listed in the next paragraph.

Some of the flows studied using the PDF transport method include plane jets [16], thermal wakes [17], evolution of homogeneous turbulence [18], mixing layers between turbulent streams of different length scales [19], self-similar turbulent free shear flows [20], premixed turbulent flames [21, 22, 23], and syngas diffusion flames [24]. These studies have clearly demonstrated the advantages offered by this method over conventional methods. However, the flows studied can be classified as simple flows--primarily time-dependent one-dimensional or two-dimensional boundary-layer type flows. It is necessary to apply the PDF transport method to more complex flows, such as confined, recirculating, swirling, and reacting flows, for it to be useful as a design tool for practical combustors.

The PDF transport method is described briefly in Section II. Although there are no theoretical restrictions for applying the method to multidimensional flows, there are certain limitations for making routine calculations of multidimensional flows, due to computer resource requirements. A novel approach, which combines the simplicity of conventional finite-volume solution algorithms for mean-flow equations and the advantages of the solution of the PDF transport equation, is presented in Section III. The approach is used to calculate the flow over a backward-facing step. The setup, results, and discussion are presented in Section IV. Section V presents the conclusions and directions for future work.

II. THE PDF TRANSPORT METHOD

The reader is referred to the review paper by Pope [14] for the details of the derivation and the solution of the PDF transport equation, and to the work of Anand [25] for additional details on computations for turbulent reacting flows. A brief review is presented in the following paragraphs.

The evolution of the reactive flow field is governed by the conservation equations of mass (continuity), momentum, chemical species, and enthalpy. They are:

$$\frac{\partial \rho}{\partial t} + \frac{\partial}{\partial x_i}(\rho U_i) = 0, \tag{1}$$

$$\rho \frac{DU_j}{Dt} = \frac{\partial \tau_{ij}}{\partial x_i} - \frac{\partial p}{\partial x_j} + \rho g_j, \tag{2}$$

and

$$\rho \frac{D\phi_\alpha}{Dt} = -\frac{\partial J_i^\alpha}{\partial x_i} + \rho S_\alpha, \quad \alpha = 1,2,\ldots\sigma, \tag{3}$$

where the material derivative (the rate of change following the fluid) is

$$\frac{D}{Dt} = \frac{\partial}{\partial t} + U_i \frac{\partial}{\partial x_i}.$$

In these equations, g_j is the body force (per unit mass) in the x_j-direction; τ_{ij} is the sum of the viscous and viscous-diffusive stress tensors, S_α is the source for the scalar ϕ_α due to reaction, or in the case of the scalar being enthalpy, it is the source for specific enthalpy due to compressibility, viscous dissipation, and radiation; $\underline{J}^\alpha$ is the diffusive mass flux vector of species α (or the specific energy flux vector due to molecular transport for the scalar being enthalpy). A set of σ-scalars is considered, where there is no specific limitation on σ. For a given reference pressure p_0 (assumed constant), the density and sources depend only on the set of scalars $\underline{\phi}$:

$$\rho = \rho(\underline{\phi}), \tag{4}$$

$$S_\alpha = S_\alpha(\underline{\phi}). \tag{5}$$

The set of σ scalars, $\underline{\phi}$, provides a complete description of the thermochemical properties of the reactive mixture. Many combustion problems involve a large number of species and consequently σ is large. However, in idealized premixed and diffusion flames, the gas composition (by assumption) is determined by a single scalar. For constant-density, nonreacting flows, like the one considered in the present study, ρ is a constant and S_α is zero.

The transport equation for the joint PDF can be derived from the set of conservation equations (1-3) and the definition of "expectation" or mean (see [14] for details). The resulting PDF transport equation is:

$$\rho(\underline{\psi})\frac{\partial f}{\partial t} + \rho(\underline{\psi})V_j\frac{\partial f}{\partial x_j} + \left(\rho(\underline{\psi})g_j - \frac{\partial \langle p \rangle}{\partial x_j}\right)\frac{\partial f}{\partial V_j} + \frac{\partial}{\partial \psi_\alpha}[\rho(\underline{\psi})S_\alpha(\underline{\psi})f]$$

$$= \frac{\partial}{\partial V_j}\left[\left\langle -\frac{\partial \tau_{ij}}{\partial x_i} + \frac{\partial p'}{\partial x_j}\middle| \underline{V}, \underline{\psi} \right\rangle f\right] + \frac{\partial}{\partial \psi_\alpha}\left[\left\langle \frac{\partial J_i^\alpha}{\partial x_i}\middle| \underline{V}, \underline{\psi} \right\rangle f\right]. \quad (6)$$

The terms on the right side involve conditional expectations. The conditional expectation $\langle Q | \underline{V}, \underline{\psi}, \underline{x}, t \rangle$ is the expectation of the quantity Q at position $\underline{x}$ and time t on the condition that the velocity and the scalar values are $\underline{U}(\underline{x}, t) = \underline{V}$ and $\underline{\phi}(\underline{x}, t) = \underline{\psi}$. If the conditional expectations appearing on the right side were known, the transport equation (6) could be solved for f (ρ and $\underline{S}$ are known functions of $\underline{\psi}$ and the mean pressure $\langle p \rangle$ can be determined from f). However, the terms on the left-hand side of equation (6) are entirely "known" in terms of f and the independent variables $\underline{x}$, t, $\underline{V}$, and $\underline{\psi}$. Thus, all the processes represented by those terms are accounted for exactly, without any approximation. Those processes are: variation with time; transport in physical space (convection); effects of body forces, mean pressure gradient, and reaction, respectively. It is remarkable that in a variable-density flow with arbitrarily complicated reactions, all these processes can be treated without approximation. The terms on the right side of equation (6) represent transport in velocity space by the viscous stresses and by the fluctuating pressure gradient; and transport in composition space by the molecular fluxes, respectively. Before the equation can be solved for f, the conditional expectations appearing in those terms must be modeled.

The conditional expectation due to viscous stresses and fluctuating pressure gradients is modeled by the simplified Langevin model [17, 20]. The conditional expectation due to the molecular fluxes of scalars is modeled by the improved mixing model [26].

These models are usually prescribed as functions of the joint PDF and a specified turbulent time scale $\tau(\underline{x}, t)$. The modeled PDF transport equation is solved by a Monte Carlo method. As mentioned in Section I, this method has been successfully applied to many simple flows.

Theoretically, there is no restriction for extending the method to two-dimensional and three-dimensional flows; however, there are practical limitations. First, although the Monte Carlo method is vastly more efficient than finite-difference techniques for solving the PDF equations, it is still computer intensive for practical multidimensional flows. To reduce the statistical errors, a large number of stochastic (or computational) particles are needed in the Monte Carlo solution. Second, the mean pressure gradient (which appears in the transport equation for the PDF [equation 6]) can be calculated from the particle properties as a part of the solution by solving an elliptic equation for the mean pressure [14]. However, the accuracy to which some of the terms in the elliptic equation need to be determined further emphasizes the need for a large number of stochastic particles, especially for variable-density flows. The solutions may require several hundred thousand particles. The computational time increases nearly linearly with the number of particles, although the calculation of simple flows noted in Section I have typically required only a few minutes of CPU time on an IBM mainframe such as a 3081, but using only tens of thousands of particles. In view of these facts, a straightforward extension of the method described in this section will require computer resources which are not readily available to all but a few for making routine calculations of multidimensional flows. Further improvements in the method and the solution algorithm are expected to make such an extension of the PDF method much more affordable. In the present study, however, a simpler approach is presented as a first step to demonstrate the feasibility of the PDF method for multidimensional flows.

III. PRESENT APPROACH

An approach which exploits both the simplicity of the SIMPLE-based finite-volume (FV) algorithm [27] for the solution of the Reynolds-averaged flow equations and the modeling advantages of the PDF transport approach is presented in this section as a viable tool for analyzing complex multidimensional flows.

The basic idea behind the approach is to take advantage of the PDF method's capability for calculating higher order correlations (such as Reynolds stresses) rather than expend considerable computational effort in computing the mean pressure field. The mean pressure field is more easily calculated using the SIMPLE-based finite-volume algorithm. This results in a very significant reduction in the computer resources needed and makes the method feasible. However, the basic idea has to be carefully incorporated to achieve consistent results.

The ultimate objective of this approach is to develop a Monte Carlo code to solve a modeled joint PDF equation for multidimensional elliptic turbulent reacting flows. The immediate objective is to couple the Monte Carlo method with the finite-volume (FV) algorithm to solve for the joint PDF. The theme of the approach is to start with the FV code and to superimpose the Monte Carlo method.

The Monte Carlo and the finite-volume codes are linked in the following manner. The FV code supplies the Monte Carlo code with the mean velocity, the turbulent time scale (τ), and the mean pressure fields. The turbulent time scale, τ (= k/ε, where k is the turbulent kinetic energy and ε is its rate of dissipation), is needed for the Langevin and the stochastic mixing models mentioned in Section II. The need to transfer the mean pressure field has already been discussed. Given the mean pressure field and the turbulent time scale, the Monte Carlo code can compute the mean velocities and all the other statistics. However, the mean velocity field is still transferred and the mean velocity field computed by the Monte Carlo code is periodically adjusted to correspond to that provided by the FV code. This is done to keep the mean velocity consistent with the mean pressure field.

The Monte Carlo code in turn supplies the Reynolds stresses and k (computed by the Monte Carlo code) to the FV code. The Reynolds stresses are incorporated in the momentum equations in the FV code, thus eliminating the need for modeling the stresses through the k-ε or the Reynolds stress models. The kinetic energy field supplied by the Monte Carlo code is held fixed in the FV code while ε is computed.

The sequence of solution is as follows. A converged solution of the flow field is obtained with the FV code using one of the conventional models such as k-ε or Reynolds stress--algebraic (RSA) models. The required mean fields are transferred to the Monte Carlo code as initial conditions. After a steady-state solution is obtained with the Monte Carlo code, the necessary quantities are transferred back to the FV code. Now the velocity and pressure fields are perturbed due to the new stresses. The entire process is repeated until the fields are no longer perturbed beyond a specified limit upon return from the Monte Carlo code.

For the case of variable density flows (reacting and non-reacting), the mean density field is also shared by the two codes.

Before the approach just described can be used to compute confined flows, the treatment of solid boundaries (walls) in the Monte Carlo code has to be addressed. The flows studied previously using the Monte Carlo method did not have solid boundaries to be treated in the solution domain. Pope [28] recently developed the treatment for walls and applied it successfully for calculating the developing turbulent flow through a pipe. The essence of the treatment is presented here while the details are omitted. The treatment ensures that the stochastic particles in the Monte Carlo solution lose the required amount of momentum at the wall so that the resulting shear stress and mean velocity profiles near the wall are consistent with the log-law of the wall. The wall treatment has been incorporated in the present study for calculating confined recirculating flows.

IV. RESULTS AND DISCUSSION

The approach described in Section III was applied to the planar flow over a backward-facing step investigated experimentally by Pronchick and Kline [1]. The schematic of flow is shown in Figure 1. The flow enters through the inlet channel of height $(W_1 + W_2)$ and flows over a step of height H into an expanded channel of height W_3 (= $W_1 + W_2 + H$). A recirculation zone results just behind the step, and the distance of the reattachment point from the step, defined as the reattachment length, is denoted by x_R. The problem shown in Figure 1 is a slight variant of the problem studied by Pronchick and Kline. In their case, the inlet is a single passage of height $(W_1 + W_2)$ and a single fluid (water) flows through the inlet at a Reynolds number of 12,000 based on the step height H and the centerline velocity in the inlet channel (U_{ref}). In the present study, a scalar (say fuel) is introduced through part of the inlet channel over the height W_2 as shown in Figure 1. The fuel is chosen, for simplicity, to be of nearly the same density as the fluid through the other part of the inlet channel (the primary fluid). For example, with air as the primary fluid, the choice of fuel is CO. The Reynolds number is maintained at 12,000 so that the fluid dynamics of the two problems is identical. The width over which the fuel, say CO, is introduced was determined such that the overall equivalence ratio is equal to 1.0. Thus, defining the mass fraction of CO to be the scalar ϕ, the value of ϕ is unity in the upper stream and zero in the lower stream at the inlet.

The computational domain extended from the step (inlet) to 15 step-heights downstream of the step, bounded by the top and bottom walls. Fully developed profiles for mean velocity, kinetic energy, and dissipation were used at the inlet. The profiles were based on the full inlet channel height ($W_1 + W_2$), assuming the wall separating the scalar stream to be a full-slip wall, to be consistent with Pronchick and Kline's experiments.

The results are presented in Figures 2 through 11. The transverse profiles of various quantities are plotted at different streamwise locations. The normalized streamwise coordinate x^* is defined by

$$x^* = \frac{(x - x_R)}{x_R} \qquad (7)$$

where x is the distance from the step (Figure 1) and x_R is the reattachment length. This choice of normalized streamwise coordinate for comparing profiles is suggested, among others, by Westphal et al. [29] and Pronchick and Kline [1].

The streamwise stations shown in the figures consist of two stations in the recirculation region, the third near the reattachment point, and the fourth in the redevelopment region downstream of the reattachment point. The PDF results are also contrasted with the results from the k-ε model in Figures 1 through 3. The reattachment length obtained are (as multiples of the step-height):

o k-ε--4.8H
o PDF--5.3H
o experiment--6.75H

Traditional finite-volume calculations generally tend to predict a lower value for the reattachment length than that observed experimentally. It has been shown that a part of the reason for this is the numerical diffusion introduced by the hybrid differencing scheme used here in the finite-volume calculations. The use of higher order differencing schemes results in improved predictions for the reattachment length [30, 31]. The second reason for the lower reattachment length is the turbulence model used. In this respect, it is encouraging to note that the PDF calculation shows considerable improvement over the k-ε model. However, this result could be fortuitous, and further investigation is necessary.

Figures 2 and 3, respectively, show the mean streamwise velocity ($<U>$) and mean transverse velocity ($<V>$) profiles normalized by U_{ref}. The streamwise velocity profiles are uniformly in good agreement with data. Also, the profiles calculated by the k-ε model and those by the present PDF approach are nearly the same although some differences are evident. The agreement of the calculated profiles for transverse mean velocity with data is moderate. The calculated profiles do display the qualitative features of the experimental data. It should be recognized that the transverse velocities are an order of magnitude lower than the streamwise velocities.

Figure 4 shows the kinetic energy (k) profile, and Figure 5 shows the streamwise velocity variance ($<u'^2>$) profiles. The profiles are normalized by U_{ref}^2. No experimental data are available for the kinetic energy. However, the kinetic energy is of the same order as $<u'^2>$ in these flows. The k-ε model predicts higher values of the kinetic energy than the PDF calculations. Given the agreement between the PDF calculations and data for $<u'^2>$ seen in Figure 5, and for $<v'^2>$ seen in Figure 6, it is reasonable to assume that the kinetic energy calculated from the PDF is closer to the actual experiment and that the k-ε model overpredicts the kinetic energy.

Figure 5 shows that the calculated PDF accurately predicts the locations of the peaks in the streamwise velocity variance profiles at all stations though the magnitudes of the peaks do not match the data exactly. The same observation is true for the transverse velocity variance profiles shown in Figure 6. Nevertheless, the calculated variances are generally in good agreement with data.

The turbulent shear stress ($<u'v'>$) profiles calculated from the PDF are compared against data in Figure 7. The profiles are normalized by U_{ref}^2. Again, the peak locations are well predicted while there are some differences in magnitude between predictions and data. The agreement is still quite satisfactory. Further, the agreement of the computed and measured shear stresses near the wall lends support to the wall treatment used in the Monte Carlo calculations.

Figures 8 through 11 show various third-order velocity correlations normalized by U_{ref}^3. The values calculated from the joint PDF compare well with data. The agreement further implies that the shapes of the PDFs from calculations and experiments are in good agreement at least up to the third moments. The agreement between the calculated profiles and the data for the third order correlations is especially noteworthy since the third-order correlations are one to two orders of magnitude smaller than the second-order correlations, and three to four orders of magnitude smaller than the mean velocities.

We now present the results of the scalar calculations. Though data is not available for comparison, the calculations are presented to demonstrate the capability for scalar calculations using the present approach. In fact, the scalar calculations with the present approach are no different from the fully Monte Carlo based calculations used for the flows referred to in Section II. For the present problem, the scalar calculations were performed entirely in the Monte Carlo part of the code.

Figures 12 through 15 present the calculated profiles for the mean, variance, streamwise flux, and transverse flux of the scalar (fuel mass-fraction), respectively. The figures show the expected trends. The peaks of the mean and variance (Figures 12 and 13) occur at the expected locations, consistent with the inlet conditions and the computed velocity field. The scalar fluxes in Figures 14 and 15 are consistent with the mean scalar and velocity fields.

Some information on the computational times is given in this paragraph only to give the reader an idea of the CPU times involved and to demonstrate that the method is indeed affordable. A set of finite-volume calculations followed by the Monte Carlo calculations is termed a cycle. The Monte Carlo code takes approximately 45 seconds of CPU time for 50 time-steps with 100,000 particles on a CRAY-XMP, which is typical of the runs made for each cycle. This time reflects a moderate effort to vectorize the code as well as improve the Monte Carlo algorithm, resulting in substantial reductions in the CPU time required since the initial formulation of the approach. The finite-volume calculations were performed on an IBM 3084. Typically, about 200 iterations are performed with the FV code for each cycle, requiring approximately 2 minutes of CPU time. Taking into account the difference in execution speeds between the CRAY and the IBM, it is seen that the computational times for the Monte Carlo and the finite-volume codes are of the same order of magnitude. Usually two or three cycles of calculations are sufficient for obtaining a converged solution, i.e., one that is not perturbed significantly upon return from the Monte Carlo calculations. A systematic study of the optimum number of particles required or the optimum number of time-steps or iterations required during each cycle has not been performed.

V. CONCLUSIONS

A PDF method for calculating turbulent recirculating flows has been proposed, and for the first time joint PDF calculations for such flows have been performed. The results of the calculations for the flow over a backward-facing step show good agreement with experimental data of Pronchick and Kline [1]. The proposed method is applicable to turbulent recirculating and reacting flows such as in gas turbine combustors and other practical combustion devices.

Several developmental tasks for the method can be identified. The developments can be classified into two areas--improvements to existing models or development of new models, and improvements to the solution algorithm. One of the objectives of the present approach is to progressively eliminate the dependence on the finite-volume code and solve the joint PDF equation entirely by the Monte Carlo method. Two important aspects (rather, limitations) of the method have to be addressed to achieve this goal. They are the need to supply a turbulent time scale externally and the determination of the mean pressure gradient.

There has been some preliminary work on determining the rate of viscous dissipation (ε) within the Monte Carlo solution [19]. This is done by treating ε as one of the scalar variables of the PDF so that the mean dissipation rate can be determined from the PDF (i.e., from the properties of the stochastic particles in the solution). This particle-dependent dissipation model needs to be further developed and tested. With the advances in the capability of the current computers in regard to available storage and speed, and with further improvements to the Monte Carlo algorithm for determining the mean pressure field, it will be possible to eliminate the need to supply the mean pressure field from the finite-volume code. Some progress has already been made in this respect. Preliminary results, obtained by the present authors, from improvements to the solution algorithm have shown that the mean pressure field can be computed within the Monte Carlo solution with moderate computational effort.

There are several advantages offered by the PDF method for variable-density and reacting flows. The present approach has to be developed to incorporate swirl, variable density, and reactions in the calculations. In this regard, it will be advantageous to eliminate the link between the Monte Carlo and the finite-volume codes.

Thus, the theme of the present approach is to superimpose the Monte Carlo and the finite-volume codes for calculating turbulent recirculating and reacting flows, but as the Monte Carlo parts of the code develop, the finite-volume parts of the code are progressively eliminated.

REFERENCES

1. S. W. Pronchick and S. J. Kline, "An Experimental Investigation of the Structure of Turbulent Reattaching Flow Behind a Backward-Facing Step," Report MD-42, Stanford University, Stanford, California, 1983.
2. W. Rodi, "Turbulence Models and Their Application in Hydraulics," book publication of International Association for Hydraulic Research, Delft, The Netherlands, 1980.
3. P. A. Libby and F. A. Williams, "Turbulent Reacting Flows," Topics in Applied Physics 44, Springer-Verlag, New York, 1980.
4. B. E. Launder and D. B. Spalding, Mathematical Models of Turbulence, Academic Press, 1972.
5. B. E. Launder, "On the Effects of a Gravitational Field on the Turbulent Transport of Heat and Momentum," Journal of Fluid Mechanics, 67, 1975, pp 569-581.
6. W. Rodi, "A New Algebraic Relation for Calculating the Reynolds Stresses," ZAMM, 56, 1976, pp 219-221.
7. G. L. Mellor and T. Yamada, "Development of a Turbulence Closure Model for Geophysical Problems," Reviews of Geophysics and Space Physics, 20, 1982, pp 851-876.
8. B. E. Launder, E. J. Reece, and W. Rodi, "Progress in the Development of a Reynolds-Stress Turbulent Closure," Journal of Fluid Mechanics, 68, 1975, pp 537-566.
9. J. L. Lumley, "Computational Modeling of Turbulent Flows," in Advances in Applied Mechanics, Vol 18 (ed C. S. Yih), Academic Press, New York, 1978.
10. W. P. Jones, "Models for Turbulent Flows with Variable Density and Combustion," Prediction Methods for Turbulent Flows (ed. W. Kollman), 1980.

11. D. B. Spalding, "Development of the Eddy Break-up Model of Turbulent Combustion," <u>Sixteenth Symposium (International) on Combustion</u>, The Combustion Institute, Pittsburg, Pennsylvania, 1976, pp 1657-1663.
12. H. C. Mongia and K. F. Smith, "An Empirical/Analytical Design Methodology for Gas Turbine Combustors," AIAA 78-998.
13. H. C. Mongia, R. S. Reynolds, and R. Srinivasan," Multidimensional Gas Turbine Combustion Modeling: Applications and Limitations," <u>AIAA Journal</u>, 24, No. 6, 1986, pp 890-904.
14. S. B. Pope, "PDF Methods for Turbulent Reactive Flows," <u>Progress in Energy and Combustion Science</u>, 11, 1985, pp 119-192.
15. S. B. Pope, "A Monte Carlo Method for the PDF Equations of Turbulent Reactive Flow," <u>Combustion Science and Technology</u>, 25, 1981, pp 159-174.
16. S. B. Pope, "Calculations of a Plane Turbulent Jet," <u>AIAA Journal</u>, 22, 1984, pp 896-904.
17. M. S. Anand and S. B. Pope, "Diffusion Behind a Line Source in Grid Turbulence," <u>Turbulent Shear Flows 4</u> (eds. L. J. S. Bradbury et al.) Springer-Verlag, New York, 1985.
18. D. C. Haworth and S. B. Pope, "A Generalized Langevin Model for Turbulent Flows," <u>Physics of Fluids</u>, Vol 29, 1986, pp 387-405.
19. S. B. Pope and D. C Haworth, "The Mixing Layer Between Turbulent Fields of Different Scales," <u>Turbulent Shear Flows 5</u> (eds. L. J. S. Bradbury et al.) Springer-Verlag, Berlin, Heidelberg, 1987.
20. D. C. Haworth and S. B. Pope, "A PDF Modeling Study of Self-Similar Turbulent Free Shear Flows," <u>Physics of Fluids</u>, 20, 1987, pp 1026-1044.
21. M. S. Anand and S. B. Pope, "Calculations of Premixed Turbulent Flames by PDF Methods," <u>Combustion and Flame</u>, Vol 67, No. 2, 1987, pp 127-142.
22. S. B. Pope and M. S. Anand, "Flamelet and Distributed Combustion in Premixed Turbulent Flames," <u>Twentieth Symposium (International) on Combustion</u>, The Combustion Institute, Pittsburg, Pennsylvania, 1984, pp 403-410.
23. S. B. Pope and W. K. Cheng, "Statistical Calculation of Spherical Turbulent Flames," <u>Twenty-First Symposium (International) on Combustion</u>, The Combustion Institute, Pittsburg, Pennsylvania, 1986, in press.
24. S. B. Pope and S. M. Correa, "Joint PDF Calculations of a Non-Equilibrium Turbulent Diffusion Flame," <u>Twenty-First Symposium (International) on Combustion</u>, The Combustion Institute, Pittsburg, Pennsylvania, 1986, in press.
25. M. S. Anand, "PDF Calculations for Premixed Turbulent Flames," Ph.D. Thesis, Cornell University, Ithaca, New York, 1986.
26. S. B. Pope, "An Improved Mixing Model," <u>Combustion Science and Technology</u>, 28, 1982, pp 131-145.

27. S. V. Patankar, <u>Numerical Heat Transfer and Fluid Flow</u>, Hemisphere, 1980.
28. S. B. Pope, Private communication, Cornell University, Ithaca, New York, 1987.
29. R. V. Westphal, J. P. Johnston, and J. K. Eaton, "Experimental Study of Flow Reattachment in a Single-Sided Sudden Expansion," NASA CR-3765, January 1984.
30. S. V. Patankar, K. C. Karki, and H. C. Mongia, "Development and Evaluation of Improved Numerical Schemes for Recirculating Flows," AIAA 87-0061.
31. S. Syed and L. Chiappetta, "Finite Difference Methods for Reducing Numerical Diffusion in TEACH-Type Calculations," AIAA 85-0057.

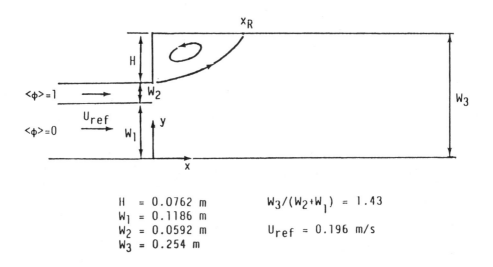

$H = 0.0762$ m
$W_1 = 0.1186$ m
$W_2 = 0.0592$ m
$W_3 = 0.254$ m

$W_3/(W_2+W_1) = 1.43$

$U_{ref} = 0.196$ m/s

Figure 1: Schematic of the backward-facing step with a fuel stream.

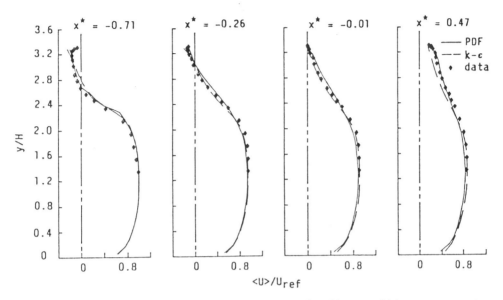

Figure 2: Predicted mean streamwise velocity profiles compared against data [1].

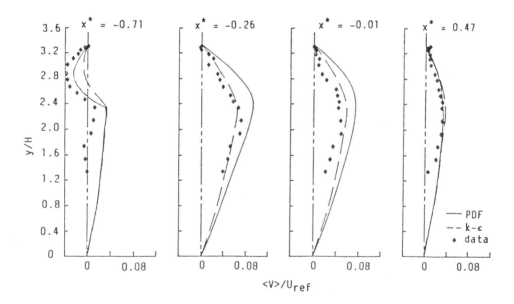

Figure 3: Predicted mean transverse velocity profiles compared against data [1].

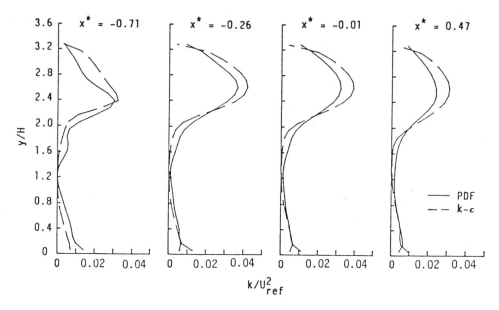

Figure 4: Predicted kinetic energy profiles.

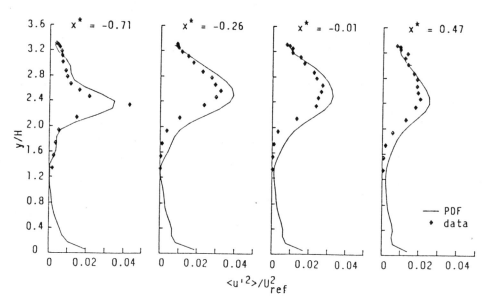

Figure 5: Predicted streamwise velocity variance profiles compared against data [1].

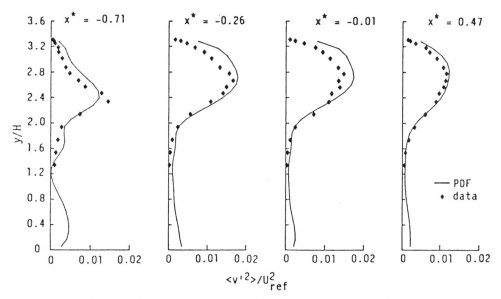

Figure 6: Predicted transverse velocity variance profiles compared against data [1].

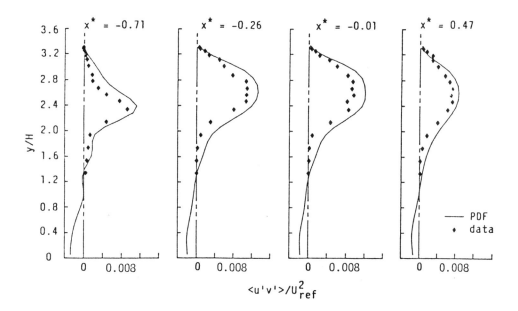

Figure 7: Predicted shear stress profiles compared against data [1].

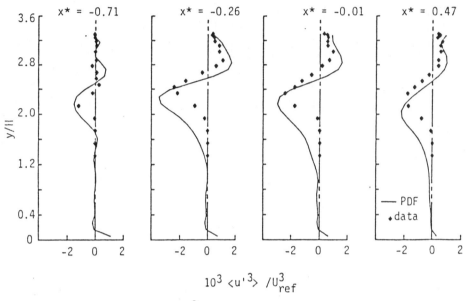

Figure 8: Predicted $<u'^3>$ profiles compared against data [1].

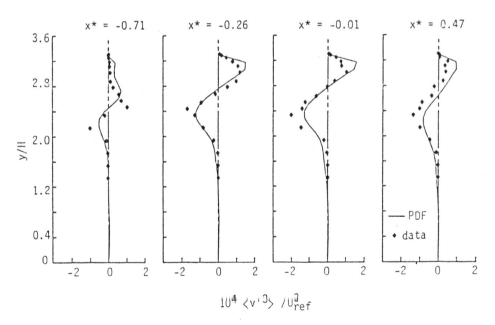

Figure 9: Predicted $<v'^3>$ profiles compared against data [1].

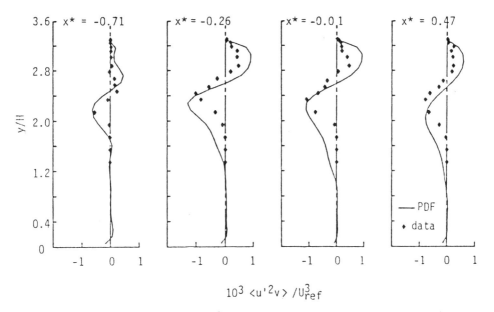

Figure 10: Predicted $\langle u'^2 v'\rangle$ profiles compared against data [1].

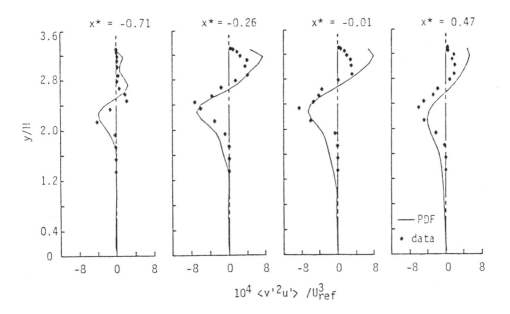

Figure 11: Predicted $\langle v'^2 u'\rangle$ profiles compared against data [1].

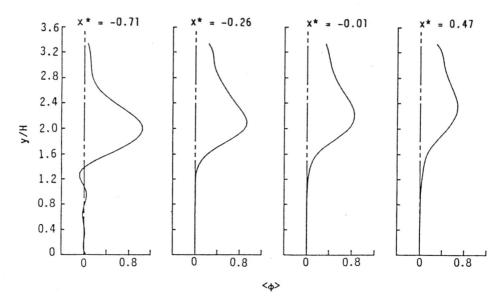

Figure 12: Predicted profiles of mean fuel mass-fraction.

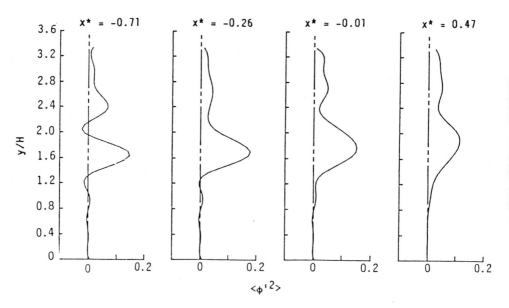

Figure 13: Predicted profiles of the variance of fuel mass-fraction.

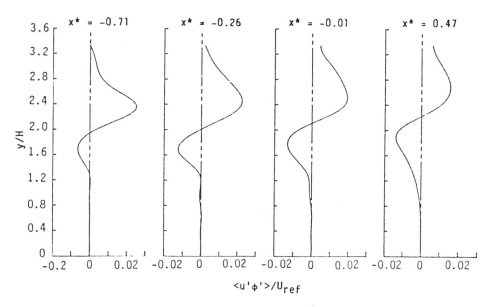

Figure 14: Predicted profiles of streamwise flux of fuel mass-fraction.

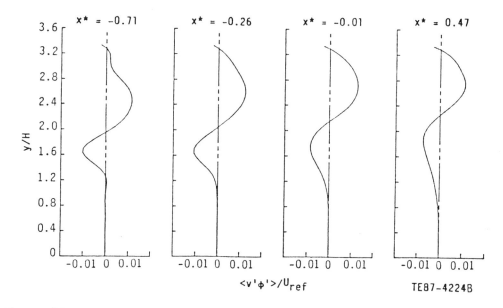

Figure 15: Predicted profiles of transverse flux of fuel mass-fraction.

DYANMICS OF COLD AND REACTING FLOWS ON BACKWARD FACING STEP GEOMETRY

A. GIOVANNINI

ONERA-CERT/DERMES

BP 4025   31055 TOULOUSE, FRANCE

INTRODUCTION

Aerodynamic of recirculating flow is one of the most difficult problem encountered in fluid mechanics. Particulary, classical turbulent closure of equations and finite difference treatment are handicaped by several drawbacks and does not work so well in view to prediction on these flows.

<u>Unsteadiness and multi-scale vortical structures</u> ; who interact from the smallest one due to Kelvin-Helmoltz instabilities and so at the scale of initial displacement thickness, to the largest one corresponding to reverse aera size ; are to be modelised. Modeling difficulties are enforced when adding a combustion reaction because its stiffness on time (fast chemical kinetics) and space (thin fronts), but also non linear Arrhenius term. Physical process identified as aerodynamic – combustion interaction is still unperfectly known, and we present here a contribution based on numerical Lagrangian time dependent calculations using a vortex code RVM coupled with SLIC and SLIC VOF flame calculations. We finally compare some results on backward facing step geometry with experimental data from Berkeley.

I - NON REACTING FLOW

1°) Presentation

The more general governing equation in terms of vorticity vector is :

$$(1) \quad \frac{D(\xi/\rho)}{Dt} = \frac{\xi}{\rho}\,\text{grad}\,U + \frac{\mu}{\rho^2}\nabla^2\xi - \frac{\text{grad}\,p \wedge \text{grad}\,\rho}{\rho^3}$$

advection   stretching   diffusion   source

Under assumptions of incompressible, 2D plane flow, general equation (1) adimensionalised reduces to :

$$(2) \quad \frac{D\xi}{Dt} = \frac{1}{Re}\nabla^2\xi$$

As shown figure 1, velocity field is decomposed in two parts, potential one U_p and solenoïdal U_ξ satisfying separately slip-condition at the walls, the total velocity U fitting the no-slip one. Potential solution is obtained by conformal mapping (Schwartz-Christoffel) turning physical domain into complex half plane. Elements of vorticity are created at the wall (sheets with constant repartition of circulation) and turned into vortex blobs, as they leave the wall, inducing velocity on the whole field calculated by Biot-Savart law modified by a core filter of size r_o. The general procedure is based on operator splitting for the solution of (2). Elements are displaced by convection, and random walks whose standard deviation is related to Reynolds Number simulate the diffusion step.

2°) Calculations

Characteristic parameters through suggestions of A. Ghoniem and Y. Gagnon, and numerical experiments are : time step $\Delta t = 0,2$ (Scheme Runge Kutta order 2), elementary circulation $\Delta\Gamma = 0,04$, and length scale for vortex blobs $r_o = 0,08$. Publications of O.H. Hald, J.T. Beale and A. Majda, C. Marchioro and M. Pulvirenti give many proofs upon convergence and stability for an infinite domain. But, here, as we deal with internal flows, we should notice that as we have not a set of complete proofs, numerical experiments have to be carried out prudently in order to give credit to that method. Accuracy is order 2 in time and space, and operator splitting give a convergence rate as $\Delta t/re$ who enforces efficiency for Large Reynolds Number flows.

3°) Results

On figure 2, we can notice during flow establishement, a stable growing of main circulation. After $t = 10$, destabilisation is mainly due to 2 effects :

- amplifications to oscillations of shear layer and consequently break of main vortice in 2 parts (1) and (2), one of them being convected by main flow.

- breathing of counterotating vortice (3) at the step corner who eject main vortice on direct flow (5) and so favor engulfment of fluid from the detachment point through low velocity dead zone (4).

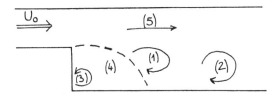

When flow is established, typical results are presented figure 3 by plotting a sequence of streamlines, enlighting time evolution and interaction of large vortical structures. Generally, main vortice reach a critical size, is cutted in two parts, and at this time as one of them is convected downstream, it occurs large fluctuations of reattachment point x_R ($\pm$ 3 step height). As noticed previously by S.W. Prondchick and S.J. Kline, phenomenum time scale is very large compared to K.H. instability (ratio 10). We don't find a perfect periodocity, and that scatters explain that in corresponding experiments, Fourier velocity analysis should exhibit a broad band spectra near the corresponding frequency. We present figure 4 mean axial velocities and stream lines obtained with a sample of 60 successive time steps and compared to data J.R. Pitz and J. Daily. In spite of a good agreement we have to notice that Pitz results use a sample of 1024 values and for a step aspect ratio of 6.9, flow structure is surely 3D.

II - REACTING FLOW

1°) Introduction

Most of industrial flames burnt in turbulent condition, and very often as in step geometry process is stabilized by a reverse flow inducting lower velocities. Going through classification of turbulent premixed flames, we are here in the case of integral scale ℓ_T larger than flame thickness e_T. More over laminar burning velocity U_L is also smaller than turbulent fluctuations $K^{1/2}$. We are in wrinkled flames, with or without pockets, regimes. Generally turbulence increase the rate of burning $\frac{dV\,b}{dt}$, as contact surface between reactants and products is more important, and also because local stretching of fluid particles near the flame enhances turbulent mixing. One quantifies usually, this effect by introducing a pseudo turbulent burning velocity U_T, different from U_L by the ratio of local wrinkled flame A_f to mean flame $\overline{A}$ areas.

$$(3) \qquad \frac{U_T}{U_L} = \frac{A_f}{\overline{A}}$$

Corresponding correlations, reproduced in figure 5 exhibit large scattering as conditions for turbulent Reynolds Numbers and length scales are different, but ratio $\frac{U_T}{U_L}$ is still larger than 1.

Numerical calculations on that regime are difficult because :

- turbulence model classically introduced is mono-scale,

- turbulent combustion model treats Arrhenius term which is strongly non linear versus species and temperature,

- as we have large gradients of velocity and density, difficulties arise from Eulerian grid, introducing numerical diffusion ans instabilities. Solutions using coarse meshes, flux corrected algorithm or adaptative grid are more efficient.

Model presented here based on SLIC alogorithm and main assumptions are the following :

- flowing substance consists in two incompressible media : unburnt (u) and burnt (b) separated by a flame front interface propagating with normal combustion velocity U_N and advected by main flow.

- fresh mixture crossing flame front experiences strong dilatation due to exothermicity of reaction, and a volume source term appears in the continuity equation corresponding to a transfer of a fixed mass of fluid Λ m with density ρ_u to ρ_b.

- chemical kinetics are very fast compared to all turbulent time scales.

- compressibility effects are neglected, so sound waves travel infinitely fast, and tangential pressure-normal density gradients interaction corresponding to vorticity creation at the flame front is not accounted. Modeling of that term in infinite medium has been done by M. Pindera, but its order in our case is difficult to appreciate.

2°) Physical model

a) Interface tracking

Tracking a Lagrangian front is difficult because discontinuities, cusps, foldings, arising when two fronts collide. Method used here gathers works well known as SLIC, SLIC-VOF due to W. Noh and P. Woodward, A. Chorin, C.W. Hirt and B.D. Nichols, A. Ghoniem. Physical domain is divided in square cells of size h_c, and position and shape of flame front is recognized by checking burnt fraction $f_{i,j}$ on neighbouring cells (figure 6).

b) Interface dynamics

Flame position r_f is governed by propagation equation

$$(4) \quad \frac{d \vec{r}_f}{dt} = \vec{u} + U_N \cdot \vec{n}_f$$

and two following operators are successively applied :

- advection

on a cell (i, j), balance on burnt fraction $f_{i,j}$ is written as :

$$(5) \quad h_c^2 \frac{d}{dt}(f_{i,j}) + (u_f h_c)\bigg|_{i-1/2,j}^{i+1/2,j} + (v_f h_c)\bigg|_{i,j-1/2}^{i,j+1/2} = 0$$

The two last terms represent fluxes of f through cells faces and velocities u are evaluated on a staggered grid (i ± 1/2, j ± 1/2). During a time step Δt_c, satisfying courant stability critera based on h_c, interface is moved in two fractional steps (x, y) and pattern recognition is done.

- combustion

Flame is moved according to fresh mixture consumption rate with a normal velocity U_N. For unsteady flows, P. Clavin and G. Joulin have shown that an unique scalar quantity, the flame strectch, controls flame propagation. The stretch can be splitted in two parts : one accounting for front geometry (curvature), and the other one for flow non-uniformity near the flame (rate of strain tensor). Corresponding normal velocity law can be expressed as :

$$(6) \quad \frac{U_N}{U_L} = 1 + \frac{L}{U_L} \left(\frac{U_L}{R} - \frac{1}{\delta} \frac{d\delta}{dt} \right)$$

Curvature $1/R$ is evaluated, following B.D. Nichols, in SOLA-VOF at the nearest mesh point on fresh mixture size from burnt fractions values at 8 neighbouring cells. Local stretch $\frac{1}{\delta} \frac{d\delta}{dt}$ is computed, on 2D plane configuration, always on fresh mixture size from velocities interpolations.

c) exothermicity

Fluid particles acceleration, with an expansion ration $\nu = \frac{\rho_u}{\rho_b}$, as they cross the flame, is modelised by volumic sources flow characterized by a velocity $u_s = \frac{U_N}{2} (\nu - 1)$ and intensity $\Delta_{i,j}$ related to burn fraction variation during combustion step.

$$(7) \quad \Delta_{i,j} = \frac{h_c^2}{2} (\nu - 1) \frac{\Delta f_{i,j/b}}{\Delta tc}$$

Sources are discretized, and placed just behind front on products size. In figure 1, a third component of velocity u_ϕ appears and is evaluated from source position and intensity treated as blobs singularities.

3°) Results

Combustion calculation presented now are carried out from lean propane-air mixture of equivalence ratio $\emptyset = 0{,}57$ corresponding expansion ratio $\nu = 5$ and reduced laminar burning velocity $\frac{UL}{U\infty} = 0{,}015$ (Markstein length $L = 0{,}005$). Calculations start from a non reacting time step, assuming $f_{i,j} = 1$ in recirculating area. Figure 7 compares time evolution of burnt rate for constant (U_L) and varying (equation 6) normal velocity U_N. After establishement of combustion process, we can compare to non-reacting corresponding case, velocity fields and flame position for the two laws Figure 8.1 and 8.2. On the figure 8.3. we can see that flame front position is corrugated, engulfing from time to time a pocket of reactants who disappear later very quickly. Clavin and Joulin law gives flame front more wrinkled and consequently burnt rate more important. For last case, we find a ration $\frac{UT}{UL}$ equal to 2.9 whose value fits well to Clavin and Willliams law of figure 5, evaluated for a turbulence ratio around the mean flame position of 9%. Near channel exit, velocities are increased, and resulting pressure gradient imposed over reattachment area is smaller than adverse one in the non reacting case. So as shown by D.M. Khuen experiments (figure 9) surimposed pressure gradient by exothermicity, reduces by that way recirculating length.

Moreover, all litterature data plot reduced reattachment length xR/h versus inlet cold flow Reynolds number. In fact growth of the shear layer proceed at a temperature between reactant and products ones. In particular, if we plot Pitz and Daily results at a Reynolds number evaluated at a bulk temperature we find a shift on the corresponding coordinate by a ratio $(\frac{\nu}{2})^{1.5} \simeq 4$; and a decrease between $Re = 10^3$ to 10^4 of reattachment length. That effect seems not to be more important than the first one due to exothermicity.

Attempt is made finally figure 11 to compare influence of a reaction to mean velocities obtained from 60 successive time steps results : mean recirculation is lowered from $\frac{xR}{h} = 7,15$ to $\frac{xR}{h} = 4,7$ by combustion effects and corresponding maximum recirculating volumic flow-rate increased from 13,5% to 17,5%.

CONCLUSION

Effect of large vortical structure on flame front wrinkles can be studied by time dependent and Lagrangian calculations of this type, coupled to very efficient pattern recognition algorithm. Variable propagation law of Clavin an Joulin, seems, more performing as we find length and flame position, velocities, reverse flow extension more realistic.

In the future, more detailed comparisons have to be carried out using more time steps on computation for turbulent statistics, improving 2D approach for experiments, and instantaneous flame front visualisations by laser tomographies.

This work is partly supported by DRET (Direction des Recherches et Etudes Techniques) of the French Ministry of Defence under Contract n° 86002.

REFERENCES

A.J. CHORIN "J. Fluid Mech. 57", n° 4 (1973) p. 785.
A.J. CHORIN "J. Comp. Phys. 27", (1978) p. 423.
A. GHONIEM and Y. GAGNON "AIAA paper 860370".
O.H. HALD "SIAM J. Num. Anal 16" (1979) p. 726.
J.T. BEALE and A. MAJDA "Math Comp 37" n° 156 (1981) p. 243.
C. MARCHIORO and M. PULVIRENTI "Comm. Math. Phys 84" (1982) p. 483.
SW PRONDCHICK and S.J. KLINE "Stanford University Report" (1983) MD 42.
J.R. PITZ and J. DAILY "NASA C.R. 165427" (August 1981).
R. BORGHI "Journal de Physique Chimie" (1984) 81 n° 6 p. 361.
M. PINDERA "Ph. D Thesis University of California" Berkeley (1986).
A. CHORIN "J. Comp. Phys. 35" (1980) p. 1.
C.W. HIRT and B.D. NICHILS "J. Comp. Phys. 39" (1981) p. 201.
A. GHONIEM "Combustion and Flame" Vol. 64 (1986) p. 321.
P. CLAVIN and G. JOULIN "Le Journal de Physique - lettres 1" (1983) p. 1.
P. CLAVIN and F. WILLIAMS "J. Fluid Mech. 116" (1982) p. 251
D.M. KHUEN "AIAA Journal 18 n° 3" (1980).

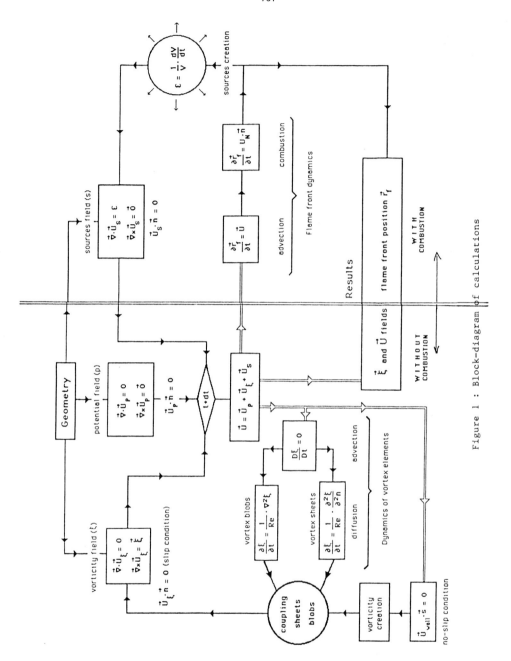

Figure 1 : Block-diagram of calculations

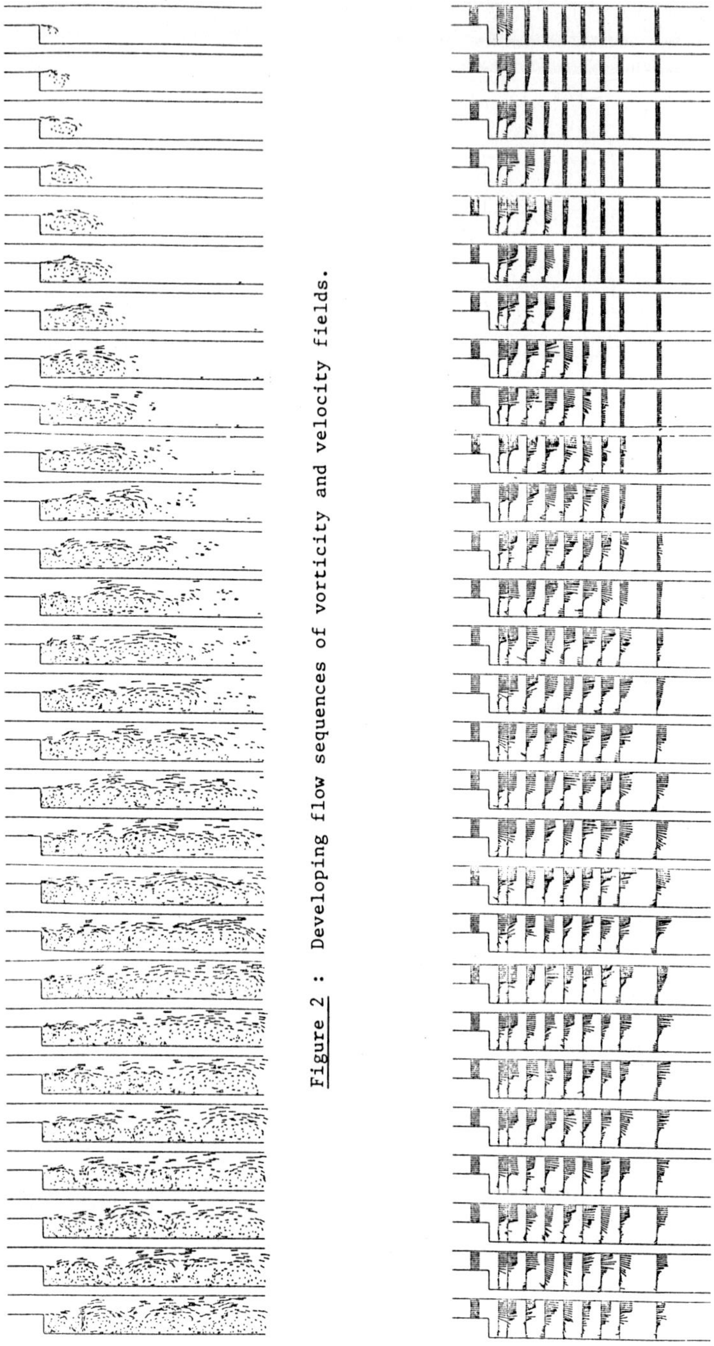

Figure 2 : Developing flow sequences of vorticity and velocity fields.

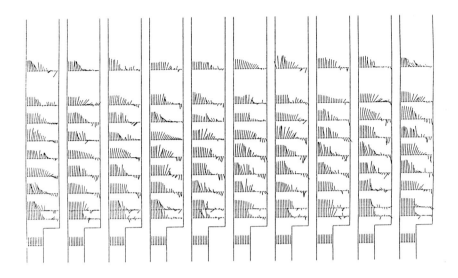

Figure 3.1. : Established flow : sequence of vorticity and velocity fields

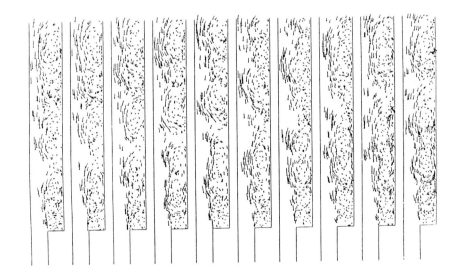

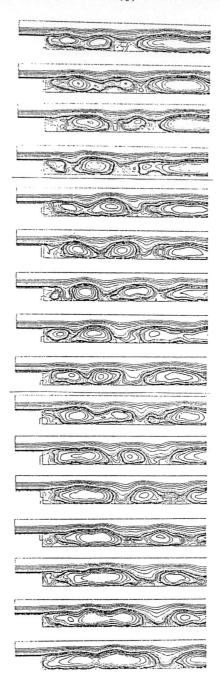

Figure 3.2. :

Established flow : sequence of streamlines

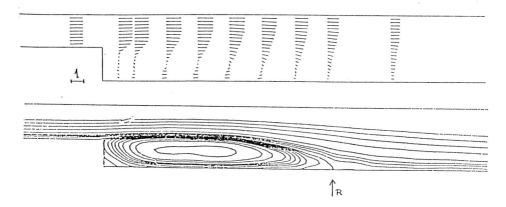

Figure 4 : Mean velocity and streamlines
Comparison with PITZ and DAILY experiments

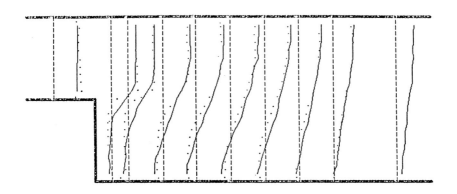

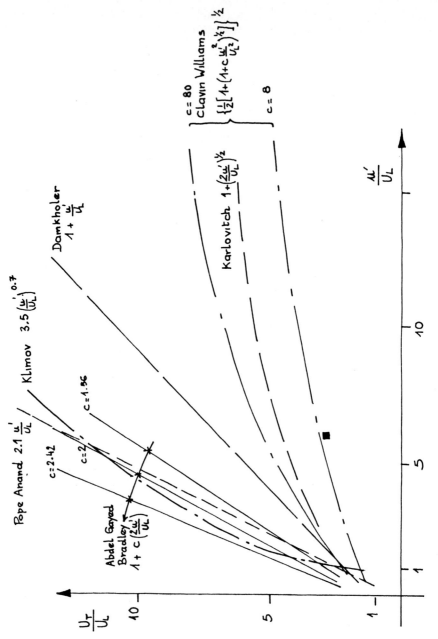

Figure 5 : Turbulent burning velocity

Figure 6 : Flame front pattern recognition

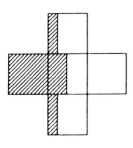
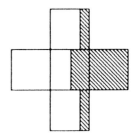

a) Vertical interface : products on left b) Vertical interface : products on right

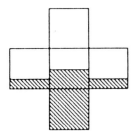
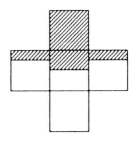

c) Horizontal interface : products on bottom d) Horizontal interface : products on top

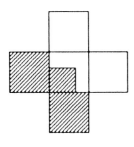
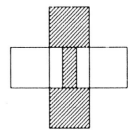

e) corner f) strip

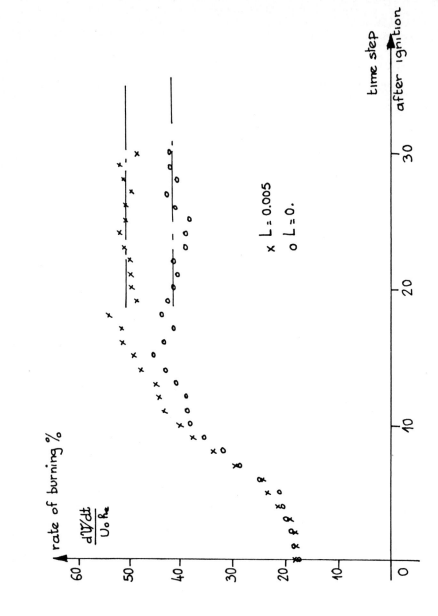

Figure 7 : Time evolution of the rate of burning

Figure 8.1. : Sequence of velocity fields with and without reaction law $U_N = U_L$

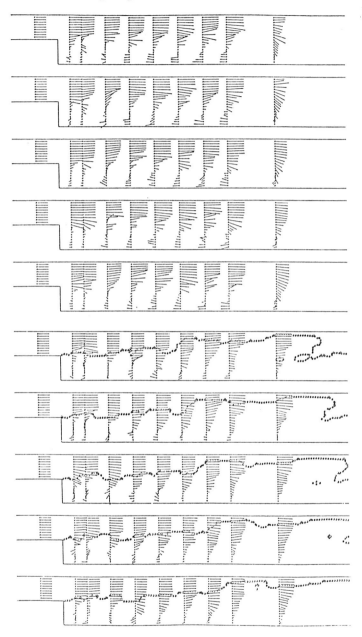

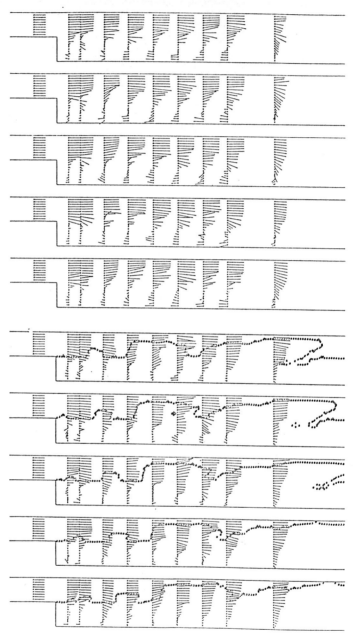

Figure 8.2. : Sequence of velocity fields with and without reaction law U_N from Clavin and Joulin

Figure 8.3. : Sequence of flame front position and velocity fields $\Delta t = 0.2$

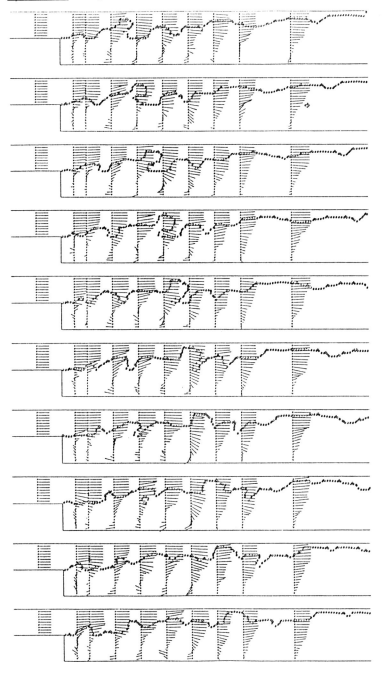

Figure 9 : Influence of external pressure gradient on reattachment

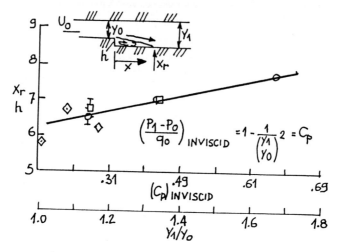

Effect of adverse pressure gradient on reattachment distance for parallel-walled channels.

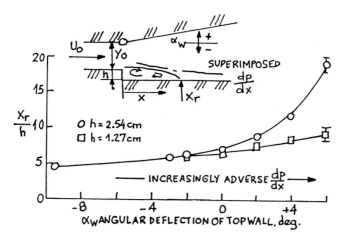

Reattachment location where a pressure gradient is superimposed on the flow over a rearward-facing step; $U_0 \sim 45$ m/s.

FROM D.M. KUEHN
AIAA JL Vol.18 N°3 March 80

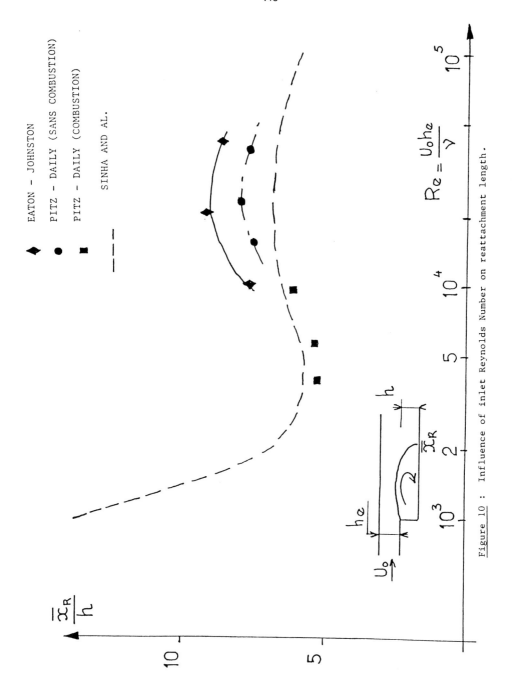

Figure 10 : Influence of inlet Reynolds Number on reattachment length.

Figure 11 : Mean velocities and recirculation rate evolution with and without combustion

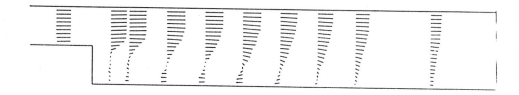

without

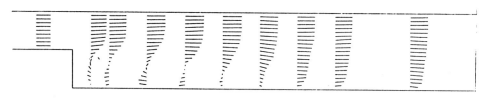

with

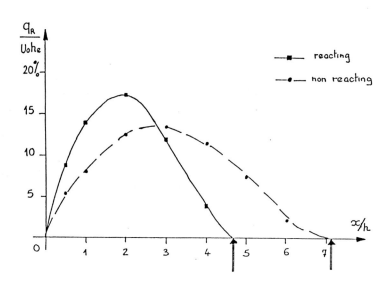

PDF - TRANSPORT EQUATIONS FOR CHEMICALLY REACTING FLOWS

W. Kollmann
Department of Mechanical Engineering
University of California
Davis, CA 95616

1. Introduction

The analysis and the computation of turbulent flows with chemical reactions can be carried out by means of probability density functions (pdf's) or characteristic functions. Pdf's and characteristic functions offer several advantages over moment methods: Finite-rate chemistry can be dealt with rigorously as well as turbulent diffusion. The transport equations for pdf and characteristic functions require closure assumptions for the effects of viscosity and pressure on the pdf. this closure problem is the central part of the paper. It will be analyzed in terms of pdf's and characteristic functions. First the properties of the linear and closed equations for the characteristic functional for Eulerian and Lagrangian variables will be established. Then the closure problem for the finite-dimensional case will be discussed for pdf and characteristic function. It will be shown for instance that the closure for the scalar dissipation term in the pdf equation developed by Dopazo [18] and Kollmann et al. [15] results in terms of characteristic functions in a single integral in contrast to the pdf, where double integration is required. Finally some recent results using pdf methods, which were obtained for turbulent flows with combustion including effects of chemical non-equilibrium, will be discussed.

2. Probability density functions and characteristic functions

2.1 The conservation laws for reacting flow

The flow of a mixture of Newtonian fluids in the gaseous phase is governed by the balances for mass, momentum, energy. These balance equations must be complemented with thermodynamic state relations in order to arrive at a closed system of equations. For these state relations ideal gas behavior is assumed. The balance equations can be set up in various ways. The Lagrangian and Eulerian frames are the most convenient systems of variables for the present purpose of analyzing turbulent reacting flows. The Lagrangian frame is advantageous for the investigation of the dynamics of material properties such as turbulent diffusion of particles [1] and numerical solution procedures based on stochastic simulation techniques [2], [3]. The Eulerian frame equations are less complicated than their Lagrangian counterparts and appropriate for the investigation of properties at an arbitrary observer position. The independent variables in the Eulerian frame are the observer position $\underline{x}$ and time t. In the Lagrangian frame are label $\underline{a}$ and time t the independent variables. The position of a material point identified by $\underline{a}$ at time t is in the Lagrangian frame a dependent variable and serves as transformation between Eulerian and Lagrangian frames

$$\underline{x} = \underline{X}(\underline{a},t) \quad , \quad \underline{a} = \underline{X}^{-1}(\underline{x},t)$$

where $\underline{X}^{-1}$ denotes the inverse mapping of $\underline{X}(\underline{a},t)$. If both mappings $\underline{X}$ and $\underline{X}^{-1}$ are twice continuously differentiable, then the partial derivatives in the Eulerian frame can be transformed into the Lagrangian frame (Euler relations, [19]). It follows, that

$$\frac{\partial}{\partial x_{\beta}} = \frac{1}{2J} \varepsilon_{\beta\delta\eta} \varepsilon_{\varphi\omega\varphi} \frac{\partial X_{\delta}}{\partial a_{\omega}} \frac{\partial X_{\eta}}{\partial a_{\varphi}} \frac{\partial}{\partial a_{\gamma}} \tag{1}$$

and

$$\frac{\partial^2}{\partial x_{\beta} \partial x_{\zeta}} = \frac{1}{2J} \varepsilon_{\beta\delta\eta} \varepsilon_{\varphi\omega\varphi} \frac{\partial X_{\delta}}{\partial a_{\omega}} \frac{\partial X_{\zeta}}{\partial a_{\varphi}} \frac{\partial}{\partial a_{\gamma}} \left(\frac{1}{J} \frac{\partial X_{\delta}}{\partial a_{\gamma}} \frac{\partial X_{\zeta}}{\partial a_{\eta}} \frac{\partial}{\partial a_{\beta}} \right) \tag{2}$$

hold, where J denotes the Jacobian determinant.

$$J = \frac{1}{6} \varepsilon_{\beta\delta\eta} \varepsilon_{\varphi\omega\varphi} \frac{\partial X_{\beta}}{\partial a_{\beta}} \frac{\partial X_{\omega}}{\partial a_{\delta}} \frac{\partial X_{\varphi}}{\partial a_{\eta}} \tag{3}$$

The label variable $\underline{a}$ is defined as usual by

$$\underline{a} = \underline{X}(\underline{a}, t_o)$$

where t_o is a fixed reference time. The time derivative in the Lagrangian frame transforms according to

$$\left(\frac{\partial}{\partial t}\right)_{\underline{a}} = \left(\frac{\partial}{\partial t}\right)_{\underline{x}} + \underline{v}(t,\underline{x}) \cdot \nabla_{\underline{x}} = \frac{D}{Dt} \tag{4}$$

and is called substantial or Stokes derivative in the Eulerian frame. The basic laws governing the flow of a mixture of ideal gases can now be set up in both frames.

Mass conservation appears in the Lagrangian frame as

$$\frac{R(\underline{a},t_o)}{R(\underline{a},t)} = J \tag{5}$$

and in the Eulerian frame as

$$\frac{\partial \rho}{\partial t} + \frac{\partial}{\partial x_{\alpha}} (\rho v_{\alpha}) = 0 \tag{6}$$

where $\rho(\underline{x},t) = R(\underline{a},t)$, $\underline{x} = \underline{X}(\underline{a},t)$ denote density in Eulerian and Lagrangian frames. For the balance of mass of a component in the mixture in terms of the mass fraction $Y_i(\underline{a},t)$ we obtain in the Lagrangian frame

$$\frac{\partial Y_i}{\partial t} = Q_i(Y_1, \ldots, Y_n)$$

$$+ \frac{1}{2R_o} \varepsilon_{\beta\delta\eta} \varepsilon_{\varphi\omega\varphi} \frac{\partial X_{\beta}}{\partial a_{\omega}} \frac{\partial X_{\zeta}}{\partial a_{\varphi}} \frac{\partial}{\partial a_{\gamma}} \left(\frac{R^2 \Gamma_i}{R_o} \frac{\partial X_{\beta}}{\partial a_{\gamma}} \frac{\partial X_{\zeta}}{\partial a_{\eta}} \frac{\partial Y_i}{\partial a_{\beta}} \right) \tag{7}$$

and in the Eulerian frame

$$\frac{\partial}{\partial t}(\rho Y_i) + \frac{\partial}{\partial x_{\alpha}}(\rho v_{\alpha} Y_i) = \frac{\partial}{\partial x_{\alpha}} \left(\rho \Gamma_i \frac{\partial Y_i}{\partial x_{\alpha}} \right) + \rho Q_i \tag{8}$$

where $R_o = R(\underline{a},t_o)$ and Q_i denotes the kinetic source term and the diffusive flux is given by a Fick-type relation with diffusivity Γ_i.

The momentum balance can be deduced from Newton's second law and appears in the Lagrangian frame as [4], [5].

$$\frac{\partial V_\alpha}{\partial t} = -\frac{1}{2R_o} \varepsilon_{\alpha\sigma\eta} \varepsilon_{\gamma\omega\varphi} \frac{\partial X_\sigma}{\partial a_\omega} \frac{\partial X_\eta}{\partial a_\varphi} \frac{\partial P}{\partial a_\gamma} +$$

$$+ \frac{1}{2R_o} \varepsilon_{\alpha\sigma\eta} \varepsilon_{\gamma\omega\varphi} \frac{\partial X_\sigma}{\partial a_\omega} \frac{\partial X_\eta}{\partial a_\varphi} \frac{\partial \tau_{\alpha\beta}}{\partial a_\gamma} + F_\alpha \tag{9}$$

and in the Eulerian frame as

$$\frac{\partial}{\partial t}(\rho v_\alpha) + \frac{\partial}{\partial x_\beta}(\rho v_\alpha v_\beta) = -\frac{\partial p}{\partial x_\alpha} + \frac{\partial \tau_{\alpha\beta}}{\partial x_\beta} + \rho f_\alpha \tag{10}$$

where $p(x,t) = P(a,t)$ denotes the pressure and $f_\alpha(x,t) = F_\alpha(a,t)$ the external force per unit mass. The viscous stress tensor $\tau_{\alpha\beta}$ for a Newtonian fluid is given by

$$\tau_{\alpha\beta} = \mu\left(\frac{\partial v_\alpha}{\partial x_\beta} + \frac{\partial v_\beta}{\partial x_\alpha} - \frac{2}{3}\delta_{\alpha\beta}\frac{\partial v_\gamma}{\partial x_\gamma}\right)$$

The first law of thermodynamics leads to the energy equation, which can be established in terms of the specific enthalpy of the mixture $h(x,t) = H(a,t)$. In the Lagrangian frame the equation

$$\frac{\partial H}{\partial t} = \frac{1}{R}\frac{\partial P}{\partial t} + \frac{1}{R}\Phi -$$

$$- \frac{1}{2R_o}\varepsilon_{\alpha\sigma\eta}\varepsilon_{\gamma\omega\varphi}\frac{\partial X_\sigma}{\partial a_\omega}\frac{\partial X_\eta}{\partial a_\varphi}\frac{\partial Q_\alpha}{\partial a_\gamma} \tag{11}$$

and in the Eulerian frame

$$\frac{\partial}{\partial t}(\rho h) + \frac{\partial}{\partial x_\alpha}(\rho v_\alpha h) = \frac{Dp}{Dt} + \Phi - \frac{\partial q_\alpha}{\partial x_\alpha} \tag{12}$$

is obtained. The dissipation function Φ is given by

$$\Phi = \tau_{\alpha\beta}\frac{\partial v_\beta}{\partial x_\alpha} \tag{13}$$

or

$$\frac{1}{R}\Phi = \frac{1}{2R_o}\varepsilon_{\alpha\sigma\eta}\varepsilon_{\gamma\omega\varphi}\tau_{\alpha\beta}\frac{\partial X_\sigma}{\partial a_\omega}\frac{\partial X_\eta}{\partial a_\varphi}\frac{\partial V_\beta}{\partial a_\gamma} \tag{14}$$

and $q_\alpha(x,t) = Q_\alpha(a,t)$ denotes the energy flux vector. The constitutive relation for the energy flux incorporates Fourier's law of heat conduction, the energy flux due to diffusion and the radiative flux. It is given in the Eulerian frame by

$$q_\alpha(x,t) = -k\frac{\partial T}{\partial x_\alpha} + \sum_{i=1}^{n}\frac{\bar{h}_i}{M_i}j_\alpha^i + q_\alpha^R \tag{15}$$

where k denotes the thermal conductivity, $\bar{h}_i$ the molar-specific enthalpy of component i,

$$\bar{h}_i = \Delta\bar{h}_i^o + \int_{T_o}^{T}dT'\bar{C}_p(T') \tag{16}$$

which is composed of enthalpy of formation $\Delta\bar{h}_i^o$ and the change of sensible enthalpy, and q_α^R denotes the radiative energy flux. The diffusive flux j_α^i is given by the Fick-type relation

$$j_\alpha^i = -\rho\Gamma_i\frac{\partial Y_i}{\partial x_\alpha} \tag{17}$$

in the Eulerian frame. The equation for ideal gases

$$p_i = \rho_i R_i T \quad , \quad i = 1,2,\cdots,n \tag{18}$$

and the chemical source terms Q_i close the system of equations describing the flow of a reacting mixture of gases. For the present purpose it is sufficient to know that Q_i is a local and differentiable function of the thermodynamic variables density, temperature and composition.

Species balances and energy equation are written more compactly in common form. In the Eulerian frame this system is given by

$$\frac{\partial}{\partial t}(\rho \psi_i) + \frac{\partial}{\partial x_\alpha}(\rho v_\alpha \psi_i) = \frac{\partial}{\partial x_\alpha}\left(\rho \Gamma_i \frac{\partial \psi_i}{\partial x_\alpha}\right) + \rho Q_i(\psi_1, \cdots \psi_n) \tag{19}$$

and in the Lagrangian frame by

$$\frac{\partial \psi_i}{\partial t} = Q_i(\psi_1, \cdots, \psi_n) +$$

$$+ \frac{1}{2R_0}\varepsilon_{\beta\delta\eta}\varepsilon_{\tau\omega\gamma}\frac{\partial X_\tau}{\partial a_\omega}\frac{\partial X_\delta}{\partial a_\gamma}\frac{\partial}{\partial a_\tau}\left(\frac{R^2\Gamma_i}{R_0}\frac{\partial X_\tau}{\partial a_\beta}\frac{\partial X_\varepsilon}{\partial a_\eta}\frac{\partial \psi_i}{\partial a_\beta}\right) \tag{20}$$

where the variables $\psi_1, \cdots, \psi_n$ are an appropriate combination of Y_i, ρ, T, h such, that the local thermodynamic state is determined by this set. The source terms Q_i may be strongly nonlinear functions of the $\psi_1, \cdots, \psi_n$ at the same point (x,t) or (a,t) as the other terms in the equation.

2.2. Characteristic functionals

The turbulent flow of a chemically reacting mixture of gases is viewed as a stochastic phenomenon. Hence, are probabilistic tools appropriate for its description. The characteristic functional [7] - [10]

$$m[d, v, \varphi_1, \cdots, \varphi_n] \equiv \langle \exp i\{\int_0^T d\tau(\rho, d) + \int_0^T d\tau(v, \underline{v}) + \sum_{i=1}^n \int_0^T d\tau(\psi_i, \varphi_i)\}\rangle \tag{21}$$

for Eulerian variables or

$$M[d, \underline{X}, \varphi_1, \cdots, \varphi_n] \equiv \langle \exp i\{\int_0^T d\tau(d, R) + \int_0^T d\tau(\underline{X}, \underline{x}) + \sum_{i=1}^n \int_0^T d\tau(\psi_i, \varphi_i)\}\rangle \tag{22}$$

for Lagrangian variables contains all the statistical information. The set of scalar variables $\psi_1, \cdots, \psi_n$ is chosen such, that the values those variables at a given point (x,t) or (a,t) determine uniquely the local thermodynamic state. The right hand side of (21) and (22) implies the existence of a probability measure on the space of weak solutions [6] of the Navier-Stokes system. This measure exists for flows in bounded domains [7] and incompressible flows. In the following it is assumed to exist for reacting flows also (the method of proof given in ref. [7] can be expected to work for the reacting case as well). The functional derivatives of the characteristic functionals defined by (21) and (22) are governed by linear transport equations. First the Eulerian functional $m[d, v, \varphi_1, \cdots, \varphi_n]$ is considered. Mass conservation (6) leads to the transport equation

$$\frac{\partial}{\partial t}\frac{\delta m}{\delta d(x,t)} = i\frac{\partial}{\partial x_\alpha}\frac{\delta^2 m}{\delta d(x,t)\delta v_\alpha(x,t)} \tag{23}$$

where $\delta/\delta d(x,t)$ etc. denote functional differentiation [11]. Momentum balance (10) leads to

$$\frac{\partial}{\partial t}\frac{\delta^2 m}{\delta d(x,t)\delta v_\alpha(x,t)} = i\frac{\partial}{\partial x_\beta}\frac{\delta^3 m}{\delta d(x,t)\delta v_\alpha(x,t)\delta v_\beta(x,t)} + \frac{\partial \Pi}{\partial x_\alpha} - \frac{\partial T_{\alpha\beta}}{\partial x_\beta} + if_\alpha \frac{\delta m}{\delta d(x,t)} \tag{24}$$

where Π denotes the pressure functional [8]

$$\Pi[d, v, \varphi_1, \cdots, \varphi_n; x, t] \equiv \langle p(x,t) \exp i\{\int_0^T d\tau(\rho, d) + \cdots \}\rangle \tag{25}$$

and $T_{\alpha\beta}$ the stress functional

$$T_{\alpha\beta}[d,\underline{v},\varphi_1;\cdots,\varphi_n;\underline{x},t] \equiv \langle \tau_{\alpha\beta}(\underline{x},t) \exp i\{\int_0^T d\tau(\rho,d) + \cdots\}\rangle \qquad (26)$$

The external force $f_\alpha(\underline{x},t)$ does not fluctuate (gravity) and is thus unaffected by averaging. The variables in the thermodynamic set satisfying (19) lead to

$$\frac{\partial}{\partial t}\frac{\delta^2 m}{\delta d(\underline{x},t)\delta\varphi_i(\underline{x},t)} = i\frac{\partial}{\partial x_\alpha}\frac{\delta^3 m}{\delta d(\underline{x},t)\delta v_\alpha(\underline{x},t)\delta\varphi_i(\underline{x},t)} + i\frac{\partial}{\partial x_\alpha}\left(\rho T_i \frac{\partial}{\partial x_\alpha}\frac{\delta m}{\delta\varphi_i(\underline{x},t)}\right)$$

$$+ iQ_i\left(\frac{\delta}{i\delta\varphi_1(\underline{x},t)},\cdots,\frac{\delta}{i\delta\varphi_n(\underline{x},t)}\right)\frac{\delta m}{\delta d(\underline{x},t)} \qquad (27)$$

Here it was assumed, that the transport coefficient ρT_i does not fluctuate in order to avoid unnecessary complications. The source term Q_i in (19) appears in (27) as pseudo-differential operator [12] on the functional level. If Q_i is a polynomial, then the pseudo-differential operator reduces to a standard differential operator. For instance [8]

$$Q_i(\varphi_1,\cdots,\varphi_n) = \sum_{\ell=1}^n a_{i\ell}\,\varphi_p^\ell \quad,\quad 1\leq p \leq n$$

appears in (27) as operator

$$Q_i\left(\frac{\delta}{i\delta\varphi_1(\underline{x},t)},\cdots,\frac{\delta}{i\delta\varphi_n(\underline{x},t)}\right) = \sum_{\ell=1}^n a_{i\ell}\,\frac{\delta^\ell}{i^\ell\delta\varphi_p^\ell}$$

of order n. Finally is the stress functional for Newtonian fluids with nonfluctuating dynamic viscosity given by

$$T_{\alpha\beta}[d,\underline{v},\varphi_1,\cdots,\varphi_n;\underline{x},t] = -i\mu\left(\frac{\partial}{\partial x_\beta}\frac{\delta m}{\delta v_\alpha(\underline{x},t)} + \frac{\partial}{\partial x_\alpha}\frac{\delta m}{\delta v_\beta(\underline{x},t)} - \frac{2}{3}\delta_{\alpha\beta}\frac{\partial}{\partial x_\gamma}\frac{\delta m}{\delta v_\gamma(\underline{x},t)}\right) \qquad (28)$$

The characteristics functional in the Lagrangian frame depends on the same number of arguments as its Eulerian counterpart. The main difference is mass conservation (5), which allows to express the current density in terms of the initial density. The basic laws lead now to the following system of functional equations, which was given first by Monin [13] for incompressible flows. First the mixed differential operator

$$\frac{\partial}{\partial a_\beta}\frac{\delta}{\delta v_\alpha(\underline{a},t)}\circ M \equiv \lim_{\substack{\underline{a}^*\to\underline{a}\\ t^*\to t}}\frac{\partial}{\partial a_\beta^*}\frac{\delta M}{\delta v_\alpha(\underline{a}^*,t^*)} \qquad (29)$$

is defined. Note that ordinary and functional differentiation do not commute. For multiple application is assumed, that all limits are carried out along different paths $(\underline{a}^*,t^*)$ towards the same point $(\underline{a},t)$. Mass conservation (5) leads now to the functional equation

$$\frac{\delta M}{\delta d(\underline{a},0)} = \frac{i}{6}\varepsilon_{\alpha\beta\gamma}\varepsilon_{\sigma\eta\omega}\frac{\partial}{\partial a_\alpha}\frac{\delta}{\delta x_\sigma(\underline{a},t)}\circ\frac{\partial}{\partial a_\beta}\frac{\delta}{\delta x_\eta(\underline{a},t)}\circ\frac{\partial}{\partial a_\gamma}\frac{\delta}{\delta x_\omega(\underline{a},t)}\circ\frac{\delta M}{\delta d(\underline{a},t)} \qquad (30)$$

and momentum balance (9) results in

$$\frac{\partial^2}{\partial t^2}\frac{\delta^2 M}{\delta d(0)\delta x_\kappa(t)} = -\frac{1}{2}\varepsilon_{\kappa\delta\eta}\varepsilon_{\gamma\omega\varphi}\frac{\partial}{\partial a_\omega}\frac{\delta}{\delta x_\gamma(t)}\circ\frac{\partial}{\partial a_\varphi}\frac{\delta}{\delta x_\eta(t)}\circ\frac{\partial\Pi}{\partial a_\gamma}$$

$$+ \frac{1}{2}\varepsilon_{\beta\delta\eta}\varepsilon_{\gamma\omega\varphi}\frac{\partial}{\partial a_\omega}\frac{\delta}{\delta x_\gamma(t)}\circ\frac{\partial}{\partial a_\varphi}\frac{\delta}{\delta x_\eta(t)}\circ\frac{\partial T_{\alpha\beta}}{\partial a_\gamma} + iF_\alpha\frac{\delta M}{\delta d(0)} \qquad (31)$$

where the dependence on the label variable $\underline{a}$ was omitted in the functional derivatives. The thermodynamic variables $\psi_i(\underline{a},t)$ determine the dynamics of the corresponding derivatives of the characteristic functional. The fluctuations of the molecular transport coefficient

$$\mathcal{D}_i \equiv \frac{R^2 T_i}{R_o}$$

are neglected and the equations for $\frac{\delta M}{\delta \varphi_i}$ are then given by

$$\frac{\partial}{\partial t} \frac{\delta^2 M}{\delta d(o) \delta \varphi_i(t)} = \frac{i}{2} \varepsilon_{\alpha\beta\gamma} \varepsilon_{\delta\eta\omega} \frac{\partial}{\partial a_\alpha} \frac{\delta}{\delta x_\xi(t)} \circ \frac{\partial}{\partial a_\omega} \frac{\delta}{\delta x_\zeta(t)} \circ \frac{\partial}{\partial a_\gamma}$$

$$\left(\mathcal{D}_i \frac{\partial}{\partial a_\beta} \frac{\delta}{\delta x_\xi(t)} \circ \frac{\partial}{\partial a_\gamma} \frac{\delta}{\delta x_\xi(t)} \circ \frac{\partial}{\partial a_\alpha} \frac{\delta M}{\delta \varphi_i(t)} \right) + i Q_i \left(\frac{\delta}{i \delta \varphi_i(t)}, \cdots, \frac{\delta}{i \delta \varphi_n(t)} \right) \frac{\delta M}{\delta d(o)} \quad (32)$$

for $i = 1, \cdots, n$.

The transport equations for the characteristic functionals in Eulerian and Lagrangian frames share two important properties. They are linear and the equations for the functional derivatives associated with thermodynamic variables do not have an integer order unless the source terms in (19) and (20) are polynomials in the thermodynamic variables. Chemical reactions with finite speed lead to Arrhenius-type source terms, which are transcendental functions of temperature, density and composition. Hence, there is no integer order associated with the functional derivatives in equations (27) and (32).

2.3 Finite-dimensional characteristic functions and pdf's

Finite-dimensional characteristic functions can be obtained, if the functionals $m[\hat{a}, \underline{\hat{v}}, \hat{\varphi}_i]$ and $M[\hat{a}, \underline{x}, \varphi_i]$ are considered at particular test functions (points in phase space). One-dimensional characteristic functions result from the choice [14]

$$d(\underline{x}', t') = \hat{a}\, \delta(\underline{x}' - \underline{x}) \delta(t' - t)$$

$$\underline{v}(\underline{x}', t') = \underline{\hat{v}}\, \delta(\underline{x}' - \underline{x}) \delta(t' - t) \quad (33)$$

$$\varphi_i(\underline{x}', t') = \hat{\varphi}_i\, \delta(\underline{x}' - \underline{x}) \delta(t' - t) \quad , \quad i = 1, \cdots, n$$

for the test functions $\hat{a}, \underline{\hat{v}}, \hat{\varphi}_i$. Functional differentiation reduces now to standard partial differentiation, as for instance in

$$\frac{\delta}{\delta d(\underline{x}, t)} m[\hat{a}\,\delta(\underline{x}' - \underline{x}) \delta(t' - t), \cdots] = \frac{\partial m_i}{\partial \hat{a}}$$

if $\underline{x}^o = \underline{x}$ and $t^o = t$ hold and where

$$m_i \equiv \langle \exp i [\hat{a} \rho(\underline{x}, t) + \underline{\hat{v}}_\alpha \underline{v}_\alpha(\underline{x}, t) + \sum_{j=1}^{n} \hat{\varphi}_j \psi_j(\underline{x}, t)] \rangle \quad (34)$$

denotes the single point characteristic function. The time derivative of m_i is related to the functional derivatives by

$$\frac{\partial m_i}{\partial t^o} =$$

$$= \lim_{\substack{\underline{x} \to \underline{x}^o \\ t \to t^o}} \left[\hat{a} \frac{\partial}{\partial t} \frac{\delta m}{\delta d(\underline{x}, t)} + \hat{v}_\alpha \frac{\partial}{\partial t} \frac{\delta m}{\delta v_\alpha(\underline{x}, t)} + \sum_{j=1}^{n} \hat{\varphi}_j \frac{\partial}{\partial t} \frac{\delta m}{\delta \varphi_j(\underline{x}, t)} \right] \quad (35)$$

where m is taken at the point given by the test functions (33). The transport equation for the single-point characteristic function m_i can be obtained from the functional equations taken at the point defined by (33) with the aid of (35) or directly by differentiating m_1 and using the dynamical equations. The resulting equation can be given in the form

$$\frac{\partial}{\partial t}\frac{\partial m_1}{i\partial d} + \frac{\partial}{\partial x_\alpha}\frac{\partial^2 m_1}{i\partial d\, i\partial v_\alpha} - \sum_{j=1}^{n} Q_j\left(\frac{\partial}{i\partial \varphi_1},\cdots,\frac{\partial}{i\partial \varphi_n}\right)\frac{\partial m_1}{i\partial d} =$$

$$= -id\int d\xi \int d\underline{w}\int d\underline{x}\, \langle \rho^2 \frac{\partial v_\beta}{\partial x_\beta}\hat{f}\rangle e^{i(\ )} - iv_\alpha \int d\xi \int d\underline{w}\int d\underline{x}\, \langle \frac{\partial p}{\partial x_\alpha}\hat{f}\rangle e^{i(\ )} \qquad (36)$$

$$+ iv_\alpha \int d\xi \int d\underline{w}\int d\underline{x}\, \langle \frac{\partial \tau_{\alpha\beta}}{\partial x_\beta}\hat{f}\rangle e^{i(\ )} + \sum_{j=1}^{n} i\varphi_j \int d\xi \int d\underline{w}\int d\underline{x}\, \langle \frac{\partial}{\partial x_\beta}(\rho \Gamma_j \frac{\partial \psi_j}{\partial x_\beta})\hat{f}\rangle e^{i(\ )}$$

where $(\ ) = (\xi d + \underline{w}\cdot\underline{v} + \sum_{j=1}^{n} x_j \varphi_j)$ and where $Q_j\left(\frac{\partial}{i\partial\varphi_1},\cdots,\frac{\partial}{i\partial\varphi_n}\right)$ denotes a pseudo-differential operator, which can be represented by

$$Q_j\left(\frac{\partial}{i\partial\varphi_1},\cdots,\frac{\partial}{i\partial\varphi_n}\right)\frac{\partial m_1}{i\partial d} = \int d\xi \int d\underline{w}\int d\underline{x}\, \xi\, Q_j(x_1,\cdots,x_n)\cdot$$

$$\cdot f_1(\xi,\underline{w},x_1,\cdots,x_n)\exp i(\xi d + \underline{w}\cdot\underline{v} + \sum_{j=1}^{n} x_j\varphi_j) \qquad (37)$$

The function f_1 denotes the Fourier transform of m_1

$$f_1(\xi,\underline{w},x_1,\cdots,x_n) =$$

$$= \frac{1}{(2\pi)^{4+n}}\int d\alpha \int d\underline{v}\int d\underline{\varphi}\, m_1 \exp i(\xi d + \underline{w}\cdot\underline{v} + \sum_{j=1}^{n} x_j\varphi_j) \qquad (38)$$

and is called pdf of density, velocity and thermodynamic scalars. Furthermore denotes $\hat{f}$ the pdf

$$\hat{f} \equiv \delta(\rho-\xi)\,\delta(\underline{v}-\underline{w})\prod_{j=1}^{n}\delta(\psi_j - x_j) \qquad (39)$$

The counterpart of the characteristic function m_1 is the pdf f_1. The transport equation for f_1 can be obtained as Fourier-transform of equation (36) or directly using well-known methods [2], [15]. The result can be stated as follows

$$\xi\frac{\partial f_1}{\partial t} + \xi w_\alpha \frac{\partial f_1}{\partial x_\alpha} = -\sum_{j=1}^{n}\frac{\partial}{\partial x_j}[\xi\, Q_j(x_1,\cdots,x_n)f_1] +$$

$$+ \frac{\partial}{\partial \xi}\langle \rho^2 \frac{\partial v_\beta}{\partial x_\beta}\hat{f}\rangle + \frac{\partial}{\partial w_\alpha}\langle \frac{\partial p}{\partial x_\alpha}\hat{f}\rangle - \frac{\partial}{\partial w_\alpha}\langle \frac{\partial \tau_{\alpha\beta}}{\partial x_\beta}\hat{f}\rangle -$$

$$- \sum_{j=1}^{n}\frac{\partial}{\partial x_j}\langle \frac{\partial}{\partial x_\beta}(\rho\Gamma_j\frac{\partial\psi_j}{\partial x_\beta})\hat{f}\rangle \qquad (40)$$

Characteristic function m_1 and pdf f_1 are equivalent descriptions of the single-point statistics of density, velocity and scalars. Both equations (36) and (40) contain expressions that cannot be represented in terms of single-point characteristic function m_1 or single-point pdf f_1. Hence are (36) and (40) indeterminate and closure models are required in order to obtain a solvable system.

Equations for multi-point characteristic functions can be derived without difficulty by modifying the test functions such as $d(\underline{x}',t')$ to

$$d(\underline{x}', t') = \sum_{\ell=1}^{N} \hat{d}^\ell \delta(\underline{x}' - \underline{x}^\ell) \delta(t' - t^\ell)$$

The transport equations for characteristic function and pdf for a finite number of points $(\underline{x}^\ell, t^\ell)$ are always indeterminate. They will not be discussed here in detail and the subsequent chapters will focus on single-point functions.

3. Closure problem

The transport equations for the characteristic function m, and the pdf f, at a single point contain three groups of terms, that are not closed. Each group represents a distinct physical process. The properties of all groups will be discussed briefly. A more detailed analysis is then devoted to the group representing molecular transport. The first group reflects the influence of the rate of volume expansion on m, and f, . The term for m, is given by

$$S_d \equiv -id \int d\xi \int d\underline{w} \int d\underline{x} \, \langle \rho^2 \frac{\partial v_s}{\partial x_s} \hat{f} \rangle \exp i(\xi d + \underline{w} \cdot \underline{v} + \sum_{j=1}^{n} x_j \varphi_j) \quad (41)$$

and for the pdf f, by

$$\hat{S}_d = \frac{\partial}{\partial \xi} \langle \rho^2 \frac{\partial v_s}{\partial x_s} \hat{f} \rangle \quad (42)$$

The effect of volume expansion on the pdf can be deduced from (42). Positive volume expansion acts as transport of pdf along the density axis towards lower density. This term can be expected to be important for combustion flows. Since the rate of volume expansion depends on the rate of heat release due to the chemical reactions, a closure expression would be proportional to the rate of heat release.

The second group represents the effect of the pressure gradient and is for the characteristic function m, given by

$$S_p \equiv -i v_\alpha \int d\xi \int d\underline{w} \int d\underline{x} \, \langle \frac{\partial p}{\partial x_\alpha} \hat{f} \rangle \exp i(\xi d + \underline{w} \cdot \underline{v} + \sum_{j=1}^{n} x_j \varphi_j) \quad (43)$$

and for the pdf f, by

$$\hat{S}_p = \frac{\partial}{\partial w_\alpha} \langle \frac{\partial p}{\partial x_\alpha} \hat{f} \rangle \quad (44)$$

The properties of this group are rather complex and can be split into three different terms [2], representing the effects of return to isotropy, fast response and boundaries on characteristic function and pdf. Several closure models have been suggested [2] for return to isotropy and fast response parts of $\hat{S}_p$.

The third group of non-closed terms reflects the effect of molecular transport on characteristic function and pdf. It is given by

$$S_m \equiv i v_\alpha \int d\xi \int d\underline{w} \int d\underline{x} \, \langle \frac{\partial \tau_{\alpha s}}{\partial x_s} \hat{f} \rangle \exp i(\xi d + \underline{w} \cdot \underline{v} + \sum_{j=1}^{n} x_j \varphi_j)$$

$$+ i \sum_{j=1}^{n} \varphi_j \int d\xi \int d\underline{w} \int d\underline{x} \, \langle \frac{\partial}{\partial x_\alpha} (\rho \Gamma_j \frac{\partial \psi_j}{\partial x_\alpha}) \hat{f} \rangle \exp i(\xi d + \underline{w} \cdot \underline{v} + \sum_{j=1}^{n} x_j \varphi_j) \quad (45)$$

for the characteristic function and by

$$\hat{S}_m \equiv -\frac{\partial}{\partial w_\alpha}\langle\frac{\partial \tau_{i\alpha}}{\partial x_i}f\rangle - \sum_{j=1}^{n}\frac{\partial}{\partial x_j}\langle\frac{\partial}{\partial x_i}(\rho\Gamma_j\frac{\partial \psi_j}{\partial x_i})f\rangle \qquad (46)$$

for the pdf. It is well known [2], [15], that molecular transport induces transport of the pdf in physical space and in scalar-velocity space. The transport in physical space is analogous to classical diffusion and is negligible for high Re-number turbulence and Prandtl/Schmidt numbers of under unity. The transport in scalar-velocity space is the dominant part and is fundamentally different from classical diffusion. The properties of this part of the molecular transport terms $\hat{S}_m$ can be clarified by considering its effect on statistical moments. It follows, that normalization and first order moments are unaffected by it. Higher order moments are reduced such that in the absence of any mechanism, that produces fluctuation, the Dirac-delta function is obtained as time goes to infinity. In order to illustrate the power of the dual approach using characteristic function and pdf, we consider the case of a single scalar variable and constant transport coefficients. S_m can be rearranged as follows

$$S_m(\varphi) = \Gamma\frac{\partial^2 m_i}{\partial x_s \partial x_s}$$
$$-\Gamma\int dx\, e^{ix\varphi}\frac{\partial^2}{\partial x^2}(\langle\nabla\psi\cdot\nabla\psi/\psi=x\rangle f_i(x)) \qquad (47)$$

and

$$\hat{S}_m(x) = \Gamma\frac{\partial^2 f_i}{\partial x_s \partial x_s} - \Gamma\frac{\partial^2}{\partial x^2}(\langle\nabla\psi\cdot\nabla\psi/\psi=x\rangle f_i(x)) \qquad (48)$$

The first term on the right hand side of both equations represents the transport in $\underline{x}$-space and the second term transport in scalar space. The decay of scalar fluctuations is governed by S_m and $\hat{S}_m$ respectively. The limit for the pdf f_i is then

$$\lim_{t\to\infty} f_i(x) = \delta(x) \qquad (49)$$

whereas for the characteristic function

$$\lim_{t\to\infty} m_i(\varphi) = 1 \qquad (50)$$

is obtained. This elementary consequence of Fourier transformation implies that $\hat{S}_m$ must decrease variance and higher order moments by reducing the width of f_i, whereas S_m does the opposite to m_i, to achieve the same effect on the moments. This leads to the conclusion, that $\hat{S}_m(x)$ cannot be modelled as diffusion in scalar space, because the diffusivity would be negative, whereas $S_m(\varphi)$ could be modelled as diffusion in the Fourier-transformed scalar space, because the width of the characteristic function has to increase in order to decrease its curvature at the origin. The second advantage of considering both characteristic function and pdf for the construction of closure models is the possibility of translating closure models for one quantity (such as $\hat{S}_m$) into closure models for the other quantity (S_m) by Fourier transformation. For instance is a pdf-closure for $\hat{S}_m(x)$ [15] given by

$$\hat{S}_m(x) \cong \Gamma\frac{\partial^2 f_i}{\partial x_\alpha \partial x_\alpha} +$$
$$+ 3\frac{C_\varepsilon}{\tau}[\int_{-\infty}^{\infty}dx'\int_{-\infty}^{\infty}dx''\, f_i(x')f_i(x'')\,T(x',x'';x) - f_i(x)] \qquad (51)$$

where C_F is a constant, τ a turbulent time scale and

$$T(x', x'', x) = \begin{cases} \frac{1}{|x'-x''|} & , x' \le x \le x'', \, x'' \le x \le x' \\ 0 & \text{otherwise} \end{cases} \tag{52}$$

Fourier transformation leads to the corresponding closure model for $S_m(\varphi)$

$$S_m(\varphi) \cong \Gamma \frac{\partial^2 m_i}{\partial x_\alpha \partial x_\alpha} + 3 \frac{C_F}{\tau} \left[\frac{1}{\varphi} \int_0^\varphi d\varphi' m_i(\varphi') m_i(\varphi - \varphi') - m_i(\varphi) \right] \tag{53}$$

It should be noted, that T transforms to

$$\hat{T}(\varphi', \varphi'', \varphi) = \frac{1}{\varphi} \delta(\varphi' + \varphi'' - \varphi) H(\varphi', \varphi'', \varphi) \tag{54}$$

where

$$H(\varphi', \varphi'', \varphi) = \begin{cases} 1 & , -\varphi \le \varphi'' - \varphi' \le \varphi \, , \, -\varphi \le \varphi' - \varphi'' \le \varphi \\ 0 & \text{otherwise} \end{cases}$$

Comparison of (51) with (53) shows that (53) has for numerical solution the distinct advantage of a single integral in contrast to the double integral in (51).

4. Illustrative example for the pdf-approach

The pdf approach to the calculation of turbulent reacting flows allows the treatment of chemical non-equilibrium. An illustrative example for this case is the calculation of a propane-air diffusion flame based on the method developed in ref. [16]. A simplified scheme for the combustion of C_3H_8 with air was constructed using the constrained equilibrium approach [17]. The thermo-chemical state is then determined by three scalar variables [16]: The mixture fraction f, the mole number of the mixture n and the mass fraction $Y_{C_3H_8}$ of the fuel. These three variables $(f, n, Y_{C_3H_8})$ satisfy instantaneously transport equations of the form (19) or (20). The mixture fraction $f(\underline{x}, t)$ is conserved, hence is $Q_f = 0$, but the equations for n and $Y_{C_3H_8}$ contain Arrhenius-type source terms. Density is here a local function of the three scalars and the closed pdf-transport equation appears now in the form

$$\langle \rho \rangle \frac{\partial \tilde{f}_i}{\partial t} + \langle \rho \rangle \tilde{u}_\alpha \frac{\partial \tilde{f}_i}{\partial x_\alpha} + \langle \rho \rangle \sum_{j=1}^{n} \frac{\partial}{\partial \varphi_j} (Q_j \tilde{f}_i) = \frac{\partial}{\partial x_\alpha} \left(\frac{\mu_t}{\sigma_\rho} \frac{\partial \tilde{f}_i}{\partial x_\alpha} \right) +$$

$$+ 3 \frac{C_F}{\tau} \left[\int \cdots \int d\underline{\varphi}' \int \cdots \int d\underline{\varphi}'' \, \tilde{f}_i(\underline{\varphi}') \tilde{f}_i(\underline{\varphi}'') T(\underline{\varphi}', \underline{\varphi}'', \underline{\varphi}) - \tilde{f}_i(\underline{\varphi}) \right] \tag{55}$$

where $\tau = \frac{\tilde{k}}{\tilde{\varepsilon}}$ and tilde denotes Favre-averaging

$$\tilde{f}_i = \frac{\rho(\varphi_1, \cdots, \varphi_n)}{\langle \rho \rangle} f_i \quad , \quad \varphi_1 = f, \, \varphi_2 = n, \, \varphi_3 = Y_{C_3H_8}$$

and T is given by (52). The pdf-equation is combined with a $\tilde{k} - \tilde{\varepsilon}$ closure for the velocity statistics [16]. The pdf equation was solved numerically with a Monte-Carlo procedure [2], [16] and the moment equations for mean velocity and $\tilde{k}$ and $\tilde{\varepsilon}$

with a standard finite-difference procedure. This method has several new features not present in prediction methods based on chemical equilibrium. It is necessary to ignite the flame by inserting near the jet pipe exit (between X/D = 0.01 and X/D = 0.2 at r/D = 0.5) hot products such that the local mean temperature exceeds the flammability limit of C_3H_8-air combustion. If no ignition takes place, the pure mixing flow of C_3H_8 with air can be calculated. Furthermore opens the inclusion of chemical non-equilibrium the road towards the treatment of extinction and blow-off phenomena.

A few selected results are presented in figs. 1-8 for a flame with a Reynolds number Re $\bar{=}$ 40,000 based on exit diameter D and exit bulk velocity $U_o = 32$ m/s (Round jet configuration). All results are taken at X/D = 25 and r/D = 1.52. The marginal pdf of mixture fraction in fig. 1 was smoothed using cubic splines. It shows considerable skewness but only a single maximum. The simultaneous pdf of mixture fraction f and CO - mass fraction in fig. 2 and in fig. 3 has a more complex structure. Chemical equilibrium would require, that the pdf be concentrated along a line $Y_{CO} = Y_{CO}(f)$. The calculation exhibit however at this radial location considerable spread of the pdf in an approximately triangular area of the $f - Y_{CO}$ plane indicating the presence of fluid elements out of equilibrium. The characteristic functions for mixture fraction f, total number of moles per unit volume n, mass fracture of fuel $Y_{C_3H_8}$ are shown in figs. 4-8 without smoothing. It should be noted, that characteristic functions can be calculated from the results of stochastic simulations without recourse to a grid. The graphs contain therefore the statistical error, which is zero at the origin since

$$Re\,(m_j) = 1 \quad \text{and} \quad Im\,(m_j) = 0$$

at the origin and increases with the distance from it. The appearance of sinosoidal oscillations (as in fig. 5 and fig. 6) indicates step-like variation of pdf's at the end of their domain of definition.

5. Conclusions

Turbulent flows with strongly exothermic reactions show significant variations of density and thus the complete Navier-Stokes system including species balances, energy equation and state relations are required for the description of the instantaneous dynamics. This fundamental set of equations was established in the Eulerian and Lagrangian frames. It was shown that the order of nonlinearity in the two frames is different, because the transformation between the frames is a nonlinear mapping. A closed set of equations for turbulent flows, if turbulence is regarded as stochastic phenomenon, can only be obtained on the functional level. The equations for the characteristic functionals for Eulerian and Lagrangian variable fields were established. It was found that the chemical source term in the species balances plays on the functional level the role of the symbol of a pseudo-differential operator. The equations are still closed and linear but do not have an integer order in the usual sense. The equations for characteristic function and pdf for finitely-many variables (corresponding to finitely-many points in the flow field) are not closed and require therefore closure models. The case of the single-point characteristic function and pdf is discussed in detail. It appears that the use of both characteristic function and pdf has several advantages as exemplified for the scalar dissipation model. Finally are selected results from the stochastic simulation of a propane-air diffusion flame discussed. The pdf-approach proves an appropriate tool to handle the effects of chemical non-equilibrium in this type of turbulent reacting system.

Acknowledgement

This research was supported by NASA-Lewis grant no. NAG 3-667. This paper was presented at the USA-France workshop on Turbulent Reacting Flows. Support for travel expenses by NSF is gratefully acknowledged.

References

[1] T. S. Lundgren, JFM 111 (1981), pp. 27.

[2] S. B. Pope, Progr. Energy Comb. Sci. 11 (1985), pp. 199.

[3] D. C. Haworth, S. B. Pope, Stoch. Anal. Appl. 4 (1986), pp. 151.

[4] S. Corrsin, in Mecan. de la Turbulence (A. Favre, ed.), CNRS (1962), pp. 27.

[5] A. S. Monin, A. M. Yaglom, "Statistical Fluid Mechanics", Vol. 1, MIT Press 1971.

[6] W. von Wahl, "The Equations of Navier-Stokes and Abstract Parabolic Equations", Aspects of Mathematics Vieweg, 1985.

[7] M. I. Vishik, A. I. Komech, A. V. Fursikov, Usp. Mat. Nauk 34 (1979), pp. 135.

[8] E. Hopf, J. Rat. Mech. Anal. 1 (1952), pp. 87.

[9] C. Dopazo, E. E. O'Brien, Phys. Fluids 17 (1974), pp. 1968.

[10] C. Foias, Usp. Mat. Nauk 29 (1974), pp. 282.

[11] V. I. Averbukh, O. G. Smolyanow, Usp. Mat. Nauk 22 (1967), pp. 201.

[12] M. Taylor, "Pseudo Differential Operators", Lect. Notes Math No. 416, Springer (1974).

[13] A. S. Monin, PMM 26 (1962), pp. 320.

[14] A. S. Monin, PMM 31 (1967), pp. 1057.

[15] W. Kollmann, J. Janicka, Phys. Fluids 25 (1982), pp. 1755.

[16] W. P. Jones, W. Kollmann, in Turbulent Shear Flows 5, (F. Durst et al. eds.), Springer V. (1987), pp.

[17] J. C. Keck, in Maximum Entropy Formalism, (R. D. Levine et al. eds.), MIT Press (1976), pp. 219.

[18] C. Dopazo, Phys. Fluids 22 (1979), pp. 20.

[19] C. Truesdell, The Kinematics of Vorticity, Indiana Univ. Press, 1956.

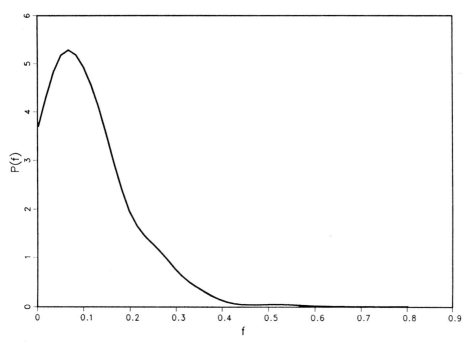

Fig. 1. Propane-air diffusion flame: Pdf of mixture fraction at X/D = 25, Y/D = 1.52.

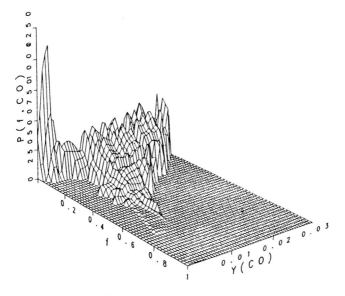

Fig. 2. Propane-air diffusion flame: Pdf of mixture fraction f and Y_{CO} at X/D = 25, Y/D = 1.52.

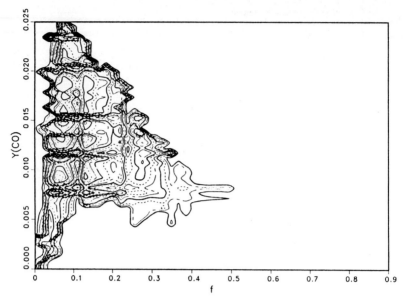

Fig. 3. Propane-air diffusion flame: Isoprobability lines for the pdf of mixture fraction f and Y_{CO} at X/D = 25, Y/D = 1.52.

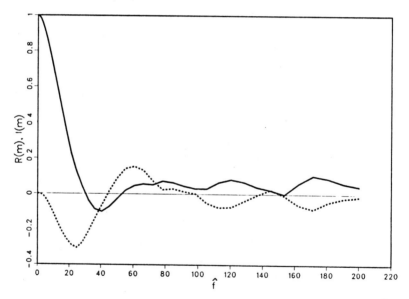

Fig. 4. Propane-air diffusion flame: Characteristic function on $m(\hat{f})$ of mixture fraction f at X/D = 25, Y/D = 1.52 (R(m): Real part, I(m): Imaginary part).

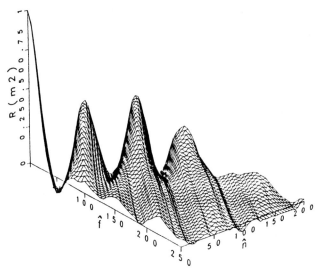

Fig. 5. Propane-air diffusion flame: Real part of the characteristic function $m_2(\hat{f},\hat{n})$ of mixture fraction f and mole number n at X/D = 25, Y/D = 1.52.

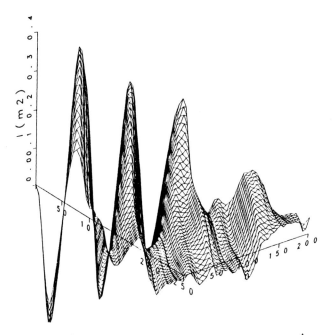

Fig. 6. Propane-air diffusion flame: Imaginary part of $m_2(\hat{f},\hat{n})$ at X/D = 25, Y/D = 1.52.

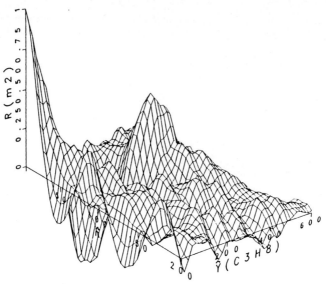

Fig. 7. Propane-air diffusion flame: Real part of the characteristic function $m_2(\hat{f}, \hat{Y}_{C_3H_8})$ of mixture fraction f and fuel mass fraction $Y_{C_3H_8}$ at X/D = 25, Y/D = 1.52.

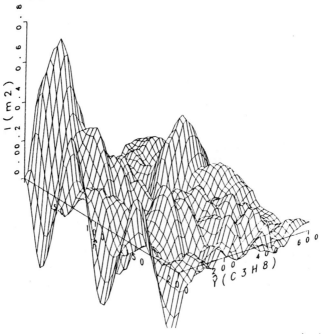

Fig. 8. Propane-air diffusion flame: Imaginary part of $m_2(\hat{f}, \hat{Y}_{C_3H_8})$ at X/D = 25, Y/D = 1.52.

MODELLING THE EFFECTS OF COMBUSTION
ON A PREMIXED TURBULENT FLOW : A REVIEW

M. Champion
Laboratoire d'Energétique et de Détonique
U.A. 193, E.N.S.M.A., 86034 Poitiers - France

1. INTRODUCTION

Combustion modifies the turbulent flow where it occurs through large density fluctuations due to heat release and through large variations in the molecular properties of the reactive fluid : viscosity and diffusivities. Interaction between these effects and pressure or velocity gradients leads to new or, at least, strongly modified mechanisms of production and destruction of turbulence. Following a modeller's point of view and using the moment approach of turbulent combustion, these interaction mechanisms are represented by various production and dissipation terms in the second order balance equations for mean quantities such as turbulent kinetic energy $\bar{k}$, Reynolds stress tensor components and mass and energy turbulent fluxes. General features of such an unclosed set of equations are given by Libby and Williams /1/ /2/ and Bray /3/. In the absence of a unified model for turbulent reactive flows the closure equations depend on the chemical and dynamical characteristic of the particular flow investigated.

In the present paper, we deal with reactive flows involving fast chemistry and intense turbulence, corresponding to large Damkolher and turbulent Reynolds numbers. Although such conditions are restrictive, they are relevant in many practical cases related to stationary regimes in industrial burners, gas turbines and jet engine chambers. To emphasize this point, it is interesting to consider the well known diagram, given by figure 1, where characteristic velocity and length scales of combustion and turbulence are compared. This diagram has been already discussed by Borghi /4/ /5/, Bray /3/ and recently by Pope /6/, who have shown that the various regimes of turbulent combustion can be described in terms of the ratio $\sqrt{\bar{k}}/u_L$ and the turbulent Reynolds number $Re = \bar{k}^{1/2} \ell / \nu$ where ℓ, u_L and ν are the macroscale of turbulence, the laminar burning velocity and the kinematic viscosity respectively. Then, with the assumption of a unity Schmidt number, the Damkolher number of the reactive flow is $D_A = \ell \, u_L^2 / \sqrt{\bar{k}} \nu$ and curves with constant D_A are straight lines. Data relative to practical turbulent reactive flows discussed by Andrews et al. /7/ are represented by segments whose extent corresponds to the usual range for laminar burning velocities of hydrocarbon-air mixtures (0.1 m/s $\leqslant u_L \leqslant$ 1 m/s). Clearly, both conditions $Re \gg 1$ and $D_A \gg 1$ hold, indicating that, in such conditions, turbulent combustion can be approximated as turbulent mixing controlled, except for mixtures possessing large laminar burning velocities which are characterized by values of the ratio $\sqrt{k}/u_L$ of order unity.

Starting with these restrictive but useful assumptions, we shall give a short account of the role of current statistical models (i.e. using a Probability Density Function either presumed or calculated) in dealing with the modification of a turbulent flow properties by combustion. In a first part, we review and discuss the existing models and the main results obtained in the case of turbulent planar flames. We emphasize the mechanisms related to the effect of combustion and heat release on the mass and energy turbulent fluxes and viscous dissipation rates. The second part is devoted to the modelling of these mechanisms in the case of bidimensional reactive shear flows.

Figure 1 :

Internal combustion engines /7/
Gas turbines /7/

Exp. Meunier et al. /33/
— Exp. Yanagi and Mimura /22/
o Exp. Moss /18/
◊ Exp. Yoshida /41/
□ Exp. Cheng /21/

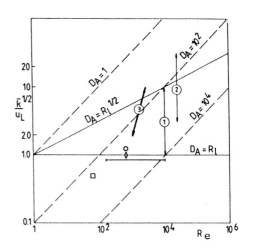

2. THE PLANAR FLAME

The planar flame problem, in the limit of large Damkolher and Reynolds numbers, stabilized in a homogeneous turbulent flow of a reactant mixture (r) has been primarily investigated by Bray, Moss, Libby and coworkers /8/ /9/ /10/ and more recently by Libby /11/, Anand and Pope /12/, Pope and Anand /13/ and Masuya /14/ /15/. Figure 2 gives a schematic view of this reactive turbulent flow.

With the simplest assumptions of a unity Lewis number and isenthalpic conditions, the combustion is represented by a single progress variable c . c $\equiv$ 0 in the unburnt reactant mixture and c $\equiv$ 1 in fully burnt products. In the case of a flame normal to the incoming flow of fresh reactant, balance equations for mean quantities, up to second order correlations can be written as :

$$\frac{d}{dx}(\bar{\rho}\,\tilde{u}) = 0 \qquad (1)$$

$$\frac{d}{dx}(\bar{\rho}\,\tilde{u}^2 + \overline{\rho u''^2}) = -\frac{d\bar{P}}{dx} - \frac{d\bar{\tau}_{xx}}{dx} \qquad (2)$$

$$\frac{d}{dx}(\bar{\rho}\,\tilde{u}\,\tilde{c} + \overline{\rho u''c''}) = \bar{\omega} + \frac{d}{dx}\overline{(\rho D \frac{\partial c}{\partial x})} \qquad (3)$$

$$\frac{d}{dx}(\bar{\rho}\,\tilde{u}\,\widetilde{u''^2} + \overline{\rho u''^3}) = -2\,\overline{\rho u''^2}\frac{d\tilde{u}}{dx} - 2\overline{u''\frac{\partial P}{\partial x}} - 2\,\overline{u''\frac{\partial \tau_{xk}}{\partial x_k}} \qquad (4)$$

$$\frac{d}{dx}(\bar{\rho}\,\tilde{u}\,\widetilde{v''^2} + \overline{\rho u'v''^2}) = -2\,\overline{v''\frac{\partial P}{\partial y}} - 2\,\overline{v''\frac{\partial \tau_{yk}}{\partial x_k}} \qquad (5)$$

$$\frac{d}{dx}(\bar{\rho}\,\tilde{u}\,\widetilde{u''c''} + \overline{\rho u''^2 c''}) = -\overline{\rho u''c''}\frac{d\tilde{u}}{dx} - \overline{\rho u''^2}\frac{d\tilde{c}}{dx} - \overline{c''\frac{\partial \tau_{xk}}{\partial x_k}}$$

$$+ \overline{u''\frac{\partial}{\partial x_k}(\rho D \frac{\partial c}{\partial x_k})} - \overline{c''\frac{\partial P}{\partial x}} + \overline{u''\omega} \qquad (6)$$

where τ_{xy} is the strain rate tensor. $g = \tilde{g} + g''$ defines the usual Favre averaging with $\tilde{g} = \overline{\rho g}/\bar{\rho}$.

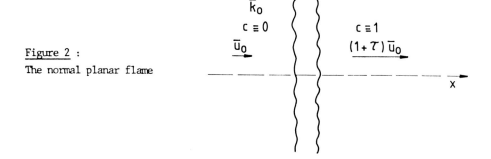

Figure 2 :
The normal planar flame

In the right hand sides of eq. (4)-(6) molecular terms can be written as sums of diffusion terms and dissipation functions :

$$\overline{u'' \frac{\partial}{\partial x_k}(\rho D \frac{\partial c}{\partial x_k})} - \overline{c'' \frac{\partial \tau_{xk}}{\partial x_k}} = \frac{d}{dx}\overline{(u'' \rho D \frac{\partial c}{\partial x} - \overline{c'' \tau_{xk}})} - \overline{\chi}$$

$$\overline{u'' \frac{\partial \tau_{xk}}{\partial x_k}} = \frac{d}{dx}\overline{(u'' \tau_{xk})} - \overline{\epsilon}_{xx}$$

where $\overline{\epsilon}_{xx}$ is the dissipation rate of streamwise velocity fluctuations and

$$\overline{\chi} = \overline{\rho D \frac{\partial c}{\partial x_k} \frac{\partial u''}{\partial x_k}} - \overline{\tau_{xk} \frac{\partial c''}{\partial x_k}}$$

is the scalar dissipation rate. In the limit of large Reynolds numbers, molecular diffusion terms are negligible /1/ /3/ compared to turbulent fluxes.

Using the equation of state $\rho(1 + \tau c) = \rho_r$, terms involving pressure gradients can be written as :

$$\overline{u'' \frac{\partial P}{\partial x}} = \tau \frac{\overline{\rho u'' c''}}{\rho_r} \frac{d\bar{P}}{dx} + \overline{u'' \frac{\partial P'}{\partial x}} \tag{8}$$

$$\overline{c'' \frac{\partial P}{\partial x}} = \tau \frac{\overline{\rho c''^2}}{\rho_r} \frac{d\bar{P}}{dx} + \overline{c'' \frac{\partial P'}{\partial x}} \tag{9}$$

where $\tau = (T_{ad} - T_r)/T_r$ is the heat release parameter.

The correlation involving the streamwise mean pressure gradient have been shown /9/ /10/ to be key quantities in the modelling of turbulent reactive flows as they provide an important mechanism for turbulence production through the flame : when τ is large enough (a typical value for a propane-air flame is 7), even a small drop of pressure across the flame leads to relatively large positive values of the production term $-\tau\overline{\rho c''^2}(d\bar{P}/dx)/\rho_r$ in Eq. (6) and $\overline{\rho u'' c''}$ is positive. The term $-\tau\overline{\rho u'' c''}(d\bar{P}/dx)/\rho_r$ in Eq. (4) also becomes positive which leads to a production of the streamwise part of the turbulent kinetic energy $\overline{\rho u''^2}$. Such an unconfined planar flame calculation has been carried out by Bray et al /10/ and Libby /11/ in the limiting case $D_A \rightarrow \infty$, when the pdf is dominated by two entries, i.e. c = 0 and c = 1. In both calculations molecular diffusion is neglected as well as terms involving pressure fluctuations. Countergradient diffusion ($\overline{\rho u'' c''} > 0$) and associated turbulence production ($\overline{\rho u''^2}$) is found to occur across the flame for values of τ larger than 3.

An important result, related to this mechanism, is given by :

$$\overline{\rho u'' c''} = \bar{\rho} \tilde{c}(1 - \tilde{c})(\bar{u}_p - \bar{u}_r) + O(1/D_A) \tag{10}$$

where $\bar{u}_r$ and $\bar{u}_p$ are conditional mean velocities within reactant and products respectively. Equation (10) shows clearly that $\overline{\rho u''c''}$ can be positive due to the velocity of burned products (light) packets being larger than the velocity of (heavy) reactant packets when $\bar{u}_p$ and $\bar{u}_r$ are defined as :

$$\bar{u}_p = \int_{-\infty}^{+\infty} u\, \mathbb{P}(u,\, c = 1)\,du$$

$$\bar{u}_r = \int_{-\infty}^{+\infty} u\, \mathbb{P}(u,\, c = 1)\,du$$

where $\mathbb{P}(c, u\,;\,x)$ is the joint pdf, bimodal in c, defined as :

$$\mathbb{P}(c,\, u\,;\, x) = \alpha(x)\, \mathbb{P}_0(u)\, \delta(c) + \beta(x)\, \mathbb{P}_1(u)\, \delta(1 - c) + \gamma\, f(c, u) \quad (11)$$

with $\quad \gamma = O(1/D_A) \ll 1$

Another important feature of the model is the introduction of a flamelet concept /17/ to close the velocity - chemical rate correlations and the viscous dissipation terms of Eqs. (4)-(6). The following form is assumed for the variation of u across a flamelet whose thickness is $\delta_L = O(1/D_A)$

$$u = (\bar{u}_p - \bar{u}_r)c + \bar{u}_r \quad (12)$$

Equation (12) is consistent with Eq. (10) and shows that $\partial u/\partial x_k \sim \partial c/\partial x_k \sim O(1/\delta_L) \sim O(D_A)$. Thus the viscous dissipation rate $\bar{\varepsilon}_{xx}$ has a contribution due to flamelet crossings of order unity. Calculations using Eqs. (11) and (12) give :

$$\bar{\varepsilon}_{xx} \simeq \left(\frac{\overline{\rho u''c''}}{\tilde{c}(1-\tilde{c})}\right)^2 \tilde{\chi}$$

which provides an evaluation of the dissipation rate of velocity fluctuations through the turbulent flame.

It must be pointed out that the local dynamics of the flame front (whose velocity is u_L) is neglected in Eq. (12). So that such a model is valid for intense turbulence only, i.e. for $u_L/k^{1/2} \ll 1$, and therefore Re $\gg D_A \gg 1$ (cf. fig. 1).

The general following comments about the BML model can be made :
- Mechanisms leading to production of turbulence through the flame as these described above are evidenced by using second order balance equations. The closure problem related to the third order correlations introduced by these equations is solved only partially by expressing them in terms of conditional

averages /10/ and using equations such as (10) as additional closure expressions.

- Quantitative conclusions regarding the order of magnitude of countergradient diffusion effects and associated production of turbulence have been obtained /10/ by using the following assumptions (cf. Eqs (8) - (9)) : the intensity $\overline{\rho c''^2}$ keeps its maximum value $\bar{\rho}\,\tilde{c}(1-\tilde{c})$ valid when the pdf is bimodal ($D_A \to \infty$) and effects of pressure fluctuations included in Eqs (8) and (9) have been neglected. This leads to an uncertainty about the exact order of magnitude of the complete terms $\overline{u''\partial P/\partial x}$ and $\overline{c''\partial P/\partial x}$. These points require comparison with experimental data and also with predictions from more sophisticated models.

- As pointed out by Anand and Pope /13/ such a configuration of 1-D planar flame with a relatively large turbulent Reynolds number cannot be stabilized in an homogeneous grid turbulence and is only neutrally stable in the case of a non decaying upstream turbulence. Therefore comparisons with experimental data can be carried out with attached flows only, such as turbulent bunsen burners /18/ /19/ /20/ /22/ /49/ or V-shape flames stabilized by a rod /21/ /23/. In some of these flows the production of turbulence by shear is present and must be measured. Moreover, the main characteristics of several of the flows quoted above indicate rather low levels of turbulence (cf. fig. 1), which implies that the basic assumption of the BML model i.e. Re $\gg D_A$ is not always fullfilled. However, taking these remarks into account, comparison between numerical results obtained from the BML model and experimental data on $\overline{\rho u''c''}$ and $\overline{\rho u''^2}$ provided by Moss /18/ does confirm the existence of a strong countergradient diffusion and the production of turbulence within the reactive region of the flow, at least for velocity fluctuations normal to the flame (Figs. 3, 4). The experiment of Sheperd and Moss /24/ relative to a reactive flow stabilized by a step has the advantage of being characterized by a larger value of Re but comparisons with an (oblique) planar flame calculation are more difficult. More recently Driscoll and Gulati /50/ have compared experimental data obtained from a flame stabilized on the edge of a rectangular burner with corresponding numerical results from the BML model. In particular, they have investigated experimentally the various terms of eq. (4) for $\overline{\rho u''^2}$. Results exhibit important increases of $\overline{\rho u''^2}$ across the flame, though slightly smaller than predicted by the model. Moreover, it is shown that, using both these measurements and the expression of $\overline{\rho u''^2}$ derived from the BML model :

$$\frac{\overline{\rho u''^2}}{\bar{\rho}} = \tilde{c}\,\overline{u'^2_p} + (1-\tilde{c})\,\overline{u'^2_r} + \tilde{c}(1-\tilde{c})(\bar{u}_p - \bar{u}_r) + O(1/D_A)$$

effects due solely to the intermittency of c can be identified, which leads to an evaluation of the "true" flame generated turbulence, i.e. $\tilde{c}\overline{u'^2_p} + (1-\tilde{c})\overline{u'^2_r}$.

Figures 3-4 :

Turbulent flux and streamwise velocity fluctuations across a planar flame (from Bray et al. /10/).

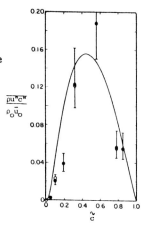
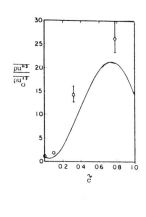

To deal with the more general case $Re > D_A \gg 1$ a significant progress has been made recently by Masuya /14/ /15/ who developed a flamelet model which includes effects due to the local dynamics of interfaces between unburnt and fully burnt regions (flamelets). The turbulent flame is considered as an ensemble of unstrained planar laminar flames whose burning velocity is u_L. Accordingly, equations of a flamelet are :

$$\rho_r u_L \frac{dc}{dn} = \omega + \frac{d}{dn}(\rho D \frac{dc}{dn})$$

$$\rho_r u_L \frac{du_n}{dn} = \rho_r u_L^2 \tau \frac{dc}{dn} = -\frac{dP}{dn} - \frac{d\tau_{nn}}{dn}$$
(13)

where n is the direction normal to the flamelet front.

The streamwise (x) velocity component u within a flamelet ($0<c<1$) is given by the conservation of mass and can be approximated as :

$$u \simeq (\bar{u}_p - \bar{u}_r)c + \bar{u}_r + \tau c u_L \phi_x \tag{14}$$

where $\phi_x = \underline{n} \cdot \underline{x}$ is the direction cosine of the flamelet.

Using Eqs. (13) and taking into account the bimodal form of the pdf given by Eq. (11), unclosed mean quantities of eqs. (4) - (6) related to pressure and molecular effects can be evaluated. For example, from Eq. (6) the following result is obtained:

$$\overline{u''(\omega + \frac{\partial}{\partial x_k}(\rho D \frac{\partial c}{\partial x_k}))} = \bar{\omega} \int_0^1 (u - \tilde{u}) \, \delta_L \, f(c) \frac{dc}{dn} \, dc \qquad (15)$$

where the product $\delta_L (dc/dn)$ is of order unity in D_A.

The integral in Eq. (15) can be evaluated by using Eq. (14) and presuming that the part $f(c)$ of the pdf relative to the burning state (Eq. 11) is of the form $(1/\delta_L)(dn/dc)$ /25/.

Relative to this extension of turbulent mixing controlled combustion towards wrinkled flames, three important comments can be made

- Equation (14) leads to the following expression for the turbulent mass flux:

$$\overline{\rho u'' c''} \approx \bar{\rho} \, \tilde{c}(1 - \tilde{c})(\bar{u}_p - \bar{u}_r + \tau \, u_L \, \bar{\phi}_x) \qquad (16)$$

which generalizes Eq. (10) deduced from the BML theory by taking into account the velocity of local flame fronts i.e. the laminar burning velocity u_L into account. If the fluctuation u'' is scaled by $\sqrt{k_r}$, it is clear that the contribution $\tau \, u_L \, \bar{\phi}_x$ will become negligible as soon as $u_L/\sqrt{k_r}$ becomes small. Moreover, in the limit of large fluctuations, the existence of corrugated flame fronts and pockets of reactant and products /16/ leads to $\bar{\phi}_x \approx 0$, if we consider that ϕ_x is approximately uniformly distributed between - 1 and + 1. Therefore, expression (10) is recovered. On the contrary low turbulence (associated to a simply connected flame front) is such that $\bar{\phi}_x > 0$. Thus this latter case corresponds to a countergradient diffusion effect and production of turbulence larger than the one predicted by the BML model.

The extension of this method to more complicated flows such as 2-D shear flows requires models for ϕ_x and ϕ_y.

- Such a flamelet model deals with molecular mechanisms as well as pressure fluctuations. In particular in the balance equation for the transverse velocity fluctuations $\overline{\rho v''^2}$ (Eq. (15)) the right hand side can be calculated as:

$$-\overline{(v'' \frac{\partial P'}{\partial y} + v'' \frac{\partial \tau_{yk}}{\partial x_k})} = \bar{\omega}(\tau \, u_L)^2 \, \bar{\phi}_y^2 > 0 \qquad (17)$$

which corresponds to a production of $\overline{\rho v''^2}$ across the flame. This mechanism, which is absent in the BML theory has been predicted by Clavin and Williams /26/ in the case of wrinkled flames. Kuznetsov obtains a similar result /27/ by using an alternative method ; no pdf is used but the left hand side of Eq. (17) is evaluated by averaging on a control volume around the flamelet i.e. integrating along the normal to the flamelet and then along its surface S. Numerical calculations relative to the planar flame /15/ suggest that the correlation $\overline{v''\partial P'/\partial y}$ becomes significant as soon as the ratio $u_L/k_0^{1/2}$ is of order unity. A closed form of the correlation $\overline{u''\partial P'/\partial x}$ has been obtained independently by Strahle /28/ which is found to be, at least for planar situations, closed to an expression derived by Jones /29/ in extending constant density results to variable density flows by using Favre averaging.

A similar flamelet treatment is used also by Anand and Pope /13/ and Pope and Anand /12/ to close the term corresponding to molecular diffusion in the pdf balance equation. The use of such a pdf evolution equation permits extension of flamelet models to finite values of the Damkolher number D_A as no special form of the pdf has to be presumed a priori and corresponds to range of Re and D_A such that Re$\gg D_A>1$. In fact, it is found /12/ that D_A is larger than 20, the pdf is nearly bimodal, which confirms the validity of the BML model for large (finite) values of D_A. Accordingly, comparisons between results, from pdf calculations and the BML model, on turbulent kinetic energy and turbulent fluxes profiles through the flame exhibit a good agreement /13/. It must be emphasized also that by introducing finite values of D_A, thus non zero values of the chemical time, the flamelet pdf model leads to the determination of a turbulent flame speed u_T, depending on the upstream turbulence, which is obtained from the stationary solution of the pdf equation.

3. MODELLING TWO-DIMENSIONAL TURBULENT REACTIVE FLOWS

The extension of models such as those described in section 2 of this paper to two dimensional planar or axisymmetric turbulent reactive flows is not straightforward for several reasons :

Considering the BML model first, it is clear that the increased number of unknown correlations introduced by the second order balance equations leads, when they are closed by using the bimodal pdf $\mathbb{P}(u_k,c;x_k)$, to an even larger number of conditional averages. Thus additional closure equations are required to express all the conditional averages in terms of primary unknown functions i.e. the corresponding Favre averages which are solutions of balance equations. The same remark holds for the flamelet model of Masuya. Moreover turbulent equations have to be solved not only in the flamelet regions ($0 < c < 1$) but also in region of pure reactants ($c=0$) and regions of burnt products ($c=1$). Considering now the pdf models /30/, as the structure of turbulence cannot be simplified in the general case, the equation for $P(c, x_k ; t)$ has to be solved in a multi-dimensional space. Finally many practical 2-D reactive flows as open flows or internal flows with heat or radiative losses or variable fuel-air ratios are not isenthalpic, which leads to the introduction of enthalpy h as a new independent variable and thus to a bidimensional pdf $P(c,h;x_k)$. The main lines of the first two of these problems are discussed by Bray et al /31/ in the case of a flame stabilized by a bluff-body /32/.

Nevertheless, some of these difficulties can be avoided by first considering the case of simple shear flows such as equilibrium boundary layers. Such a reactive turbulent boundary layer can be obtained /33/ /34/ by injecting fresh reactants through a porous plate into an incoming flow of hot products parallel to the plate (fig. 5).

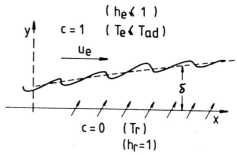

Figure 5 :

Turbulent reactive boundary layer with injection of reactant /33/, /34/.

In the limit of large Reynolds and Damkolher numbers the evolution of the Reynolds stress tensor components and the turbulent mass fluxes through this boundary layer is given by :

$$\bar{\rho} \frac{D}{Dt} (\overline{\rho u''v''}/\bar{\rho}) + \frac{\partial}{\partial y} (\overline{\rho u''v''^2}) = - \overline{\rho v''^2} \frac{\partial \tilde{u}}{\partial y} - \tau \left(\frac{\overline{\rho u''c''}}{\rho_r} \frac{\partial \bar{P}}{\partial y} + \frac{\overline{\rho v''c''}}{\rho_r} \frac{\partial \bar{P}}{\partial x} \right)$$

$$- \left(\overline{u'' \frac{\partial P'}{\partial y}} + \overline{v'' \frac{\partial P'}{\partial x}} \right) - \bar{\rho} \tilde{\epsilon}_{xy} \quad (18)$$

$$\bar{\rho} \frac{D}{Dt} (\overline{\rho u''^2}/\bar{\rho}) + \frac{\partial}{\partial y} (\overline{\rho u''^2 v''}) = - 2 \overline{\rho u''v''} \frac{\partial \tilde{u}}{\partial y} - 2 \frac{\tau}{\rho_r} \overline{\rho u''c''} \frac{\partial \bar{P}}{\partial x} - 2 \overline{u'' \frac{\partial P'}{\partial x}}$$

$$- \bar{\rho} \tilde{\epsilon}_{xx} \quad (19)$$

$$\bar{\rho} \frac{D}{Dt} (\overline{\rho v''^2}/\bar{\rho}) + \frac{\partial}{\partial y} (\overline{\rho v''^3}) = - 2 \frac{\tau}{\rho_r} \overline{\rho v''c''} \frac{\partial \bar{P}}{\partial y} - 2 \overline{v'' \frac{\partial P'}{\partial y}} - \bar{\rho} \tilde{\epsilon}_{yy} \quad (20)$$

$$\bar{\rho} \frac{D}{Dt} (\overline{\rho w'^2}/\bar{\rho}) + \frac{\partial}{\partial y} (\overline{\rho w'^2 v'}) = -2 \overline{w' \frac{\partial P}{\partial z}} - \bar{\rho} \tilde{\epsilon}_{zz} \qquad (21)$$

$$\bar{\rho} \frac{D}{Dt} (\overline{\rho v'c''}/\bar{\rho}) + \frac{\partial}{\partial y} (\overline{\rho v'^2 c''}) = -\overline{\rho v'^2} \frac{\partial \tilde{c}}{\partial y} - \frac{\tau}{\rho_r} \tilde{c}(1-\tilde{c}) \frac{\partial \bar{P}}{\partial y} - \overline{c'' \frac{\partial P'}{\partial y}}$$
$$- \overline{v' \omega} - \bar{\rho} \tilde{\epsilon}_{yc} \qquad (22)$$

$$\bar{\rho} \frac{D}{Dt} (\overline{\rho u''c''}/\bar{\rho}) + \frac{\partial}{\partial y} (\overline{\rho u''v'c''}) = -\overline{\rho u''v'} \frac{\partial \tilde{c}}{\partial y} - \frac{\tau}{\rho_r} \tilde{c}(1-\tilde{c}) \frac{\partial \bar{P}}{\partial x} - \overline{u''\omega}$$
$$- \bar{\rho} \tilde{\epsilon}_{xc} \qquad (23)$$

where $D/Dt = \tilde{u}\frac{\partial}{\partial x} + \tilde{v}\frac{\partial}{\partial y}$ is the convective operator and the $\tilde{\epsilon}_{ij}$ are the various dissipation rates.

Taking the streamwise derivative of the boundary layer thickness $d\delta/dx$ as a small parameter, Champion and Libby /34/ have shown that the second order equations, Eqs. (18) - (23), yield equilibrium between production and dissipation. Then, using the BML model for flamelet regions and classical constant density flows closures for reactant and burnt products regions, the following expression for the turbulent flux $\overline{\rho v'c''}$ is obtained :

$$\overline{\rho v'c''} = - \{\mu_t(\hat{k},\ell) \bar{\rho} \frac{\partial \tilde{c}}{\partial y} + K \frac{\tau}{\rho_r} \tilde{c}(1-\tilde{c}) \frac{\partial \bar{P}}{\partial y} / \frac{\partial \tilde{u}}{\partial y}\} \qquad (24)$$

where μ_t is a turbulent viscosity function and K a constant coefficient, both introduced by the model. Meanwhile a gradient diffusion form is recovered for the Reynolds stress $\overline{\rho u''v'}$ /34/.

The first term of the right hand side of Eq. (24) is the usual gradient assumption though the second one is a function of the heat release τ and the transverse mean pressure gradient. Depending on the sign of $\partial \bar{P}/\partial y$, this latter contribution to the turbulent flux can be either positive or negative, the second of these possibilities leading possibly to countergradient diffusion. However the analysis shows that, in such a situation of equilibrium boundary layers, the corresponding production terms in $\overline{\rho u''^2}$ and $\overline{\rho v'^2}$ equations are always dominated by production of turbulence by the mean shear $\partial \tilde{u}/\partial y$. Nevertheless it must be kept in mind that these results are obtained from an asymptotic analysis i.e. $d\delta/dx \ll 1$ corresponding to a highly parabolic flow.

Starting from the a priori assumption of an equilibrium between production and dissipation within the turbulent flux equation, Borghi and Dutoya /35/ and Borghi and Escudié /36/ have derived an expression for $\overline{\rho v''c''}$ which is comparable to Eq. (24) but exhibits a turbulent Schmidt number whose value depends on the mean chemical rate $\bar{\omega}$, although in Eq. (24) the Schmidt number is found to be unity. This discrepancy between the two findings is a consequence of the analysis of Champion and Libby /33/ which leads to a term $\overline{v''\omega}$ in the turbulent flux Eq. (22) found to be of the same order as the convective - diffusive term and thus to be negligible at first order in $d\delta/dx$. Then, it is suggested that in the case of a general 2-D flow both $\overline{v''\omega}$ and convection must be retained. However in both studies (/34/ and /37/) the contribution to $\overline{\rho v''c''}$ from the pressure gradient is found to be of prime importance.

Concerning now the dissipation rate of turbulence due to molecular effects, it is also found /33/ that contributions due to burnt products ($\bar{\epsilon}_p$) and fresh reactant ($\bar{\epsilon}_r$) are dominant. Accordingly,

$$\tilde{\epsilon}_{ij} \cong \tilde{c} (\overline{\epsilon_{ij}})_p + (1 - \tilde{c})(\overline{\epsilon_{ij}})_r \qquad (25)$$

In fact, when acceleration of gases through flamelets and change in viscosity due to temperature are taken into account, the contribution $(\overline{\epsilon_{ij}})_p$ from the fully burnt products regions can be evaluated as :

$$\bar{\epsilon}_p \cong (1 + \tau)^4 \bar{\epsilon}_r$$

which shows that dissipation of turbulence is much more important within the hot regions of the flow.

When the flow is not isenthalpic, because of convective or radiative heat losses due to boundary conditions, the BML thermochemistry can be extended by considering a pdf $\mathbb{P}(c, h ; x_k)$, which is bimodal in c only, and the corresponding mean balance equations for the enthaply $\tilde{h}$ and the enthalpy turbulent flux $\overline{\rho v''h''}$ /37/. It is of interest to notice also that such an assumption, i.e., defining c and h as independent variables, allows relaxation of the assumption of a unity Lewis number required by the basic BML model. This can be of importance in regions of the flow, near solid boundaries for example, where turbulent and molecular transport are of the same order of magnitude.

This extension has been used by Bray et al /38/ in the case of the boundary layer with injection described above, when the temperature T_e of the external flow is lower than the adiabatic temperature T_{ad} of the injected fresh mixture. The turbulent mass and energy fluxes are found to be modified by enthalpy variations according to :

$$\overline{\rho v''c''} = - (\mu_t(\tilde{k}, \ell) \ \bar{\rho} \ \frac{\partial \tilde{c}}{\partial y} + K \ \overline{c''} \ \frac{\partial \bar{P}}{\partial y} / \frac{\partial \tilde{u}}{\partial y})$$

$$\overline{\rho v''\zeta''} = - (\mu_t(\tilde{k}, \ell) \ \bar{\rho} \ \frac{\partial \tilde{\zeta}}{\partial y} + K \ \overline{\zeta''} \ \frac{\partial \bar{P}}{\partial y} / \frac{\partial \tilde{u}}{\partial y})$$

with

$$\overline{c''} = \frac{1}{\rho_e(h_e + \tau)} (\tau \ \overline{\rho c''^2} - (1 - h_e) \ \overline{\rho c''\zeta''})$$

$$\overline{\zeta''} = \frac{1}{\rho_e(h_e + \tau)} (\tau \ \overline{\rho c''\zeta''} - (1 - h_e) \ \overline{\rho \zeta''^2})$$

where ζ is the reduced enthalpy $(1 - h)/(1 - h_e)$ and h_e the external enthalpy ($h_e < 1$) whose minimum value is $1 - \tau$.

Numerical calculations show /38/ that $\overline{\rho v''c''}$ retains the same sign as in the isenthalpic case (eq. (24) but with a larger negative value. A different conclusion emerges from calculations of the corresponding correlation involving the reduced temperature field $\theta = (T-T_r)/(T_{ad}-T_r)$:

$$\overline{\rho v''\theta''} = \overline{\rho v''c''} - \frac{(1 - h_e)}{\tau} \ \overline{\rho v''\zeta''}$$

which appears to vary in the opposite direction i.e. towards the positive values, becoming positive when $|h_e|$ is large enough. This result, which is illustrated by figure 7, is of importance as numerical results on turbulent fluxes are generally compared /18/, /22/, /39/, with experimental data on the temperature field. Therefore, it is clear that countergradient diffusion effects and turbulence production can be underestimated by a model based on the assumption of isenthalpic flow. Extending this comment to other 2-D shear flows we deduce that these non isenthalpic effects should be taken into account in more cases, in particular in open flames /18/ /19/ mixing with the surrounding ambient which is generally a gas with different enthalpy.

Figure 6 :

Turbulent kinetic energy $\tilde{k}/(u_e^2 \frac{d\delta}{dx})$ across a reactive boundary layer with injection of reactant, for various values of the external enthalpy h_e /38/.

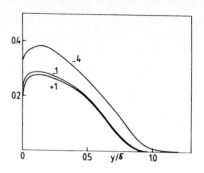

The main features of the non isenthalpic model discussed above have been used also by Davé and Kollmann /40/ to predict mean quantities profiles in the case of the open flame investigated experimentally by Yoshida /19/ /41/. In this modelling, the difficulty raised by the closure of third order correlations following the lines of the BML model is avoided by re-introducing gradient assumptions at this level.

Figure 7 :

Turbulent fluxes in a boundary layer /38/ for various values of the external enthalpy h_e. $\tau = 7$

—— - $\overline{\rho v'' c''}/(\bar{\rho} u_e \frac{d\delta}{dx})$ isenthalpic case (h = 1)

---- - $\overline{\rho v'' c''}/(\bar{\rho} u_e \frac{d\delta}{dx})$

—·— - $\overline{\rho v'' \theta''}/(\bar{\rho} u_e \frac{d\delta}{dx})$

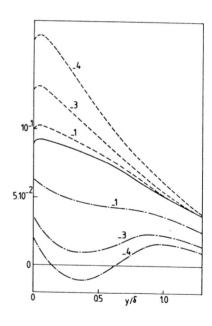

Moreover, they neglect dissipation through flamelets, which is in accordance with Eq.(25). A good general agreement with experimental data is obtained and the term related to the mean pressure axial gradient is found to be important. However the production of turbulence ($\overline{\rho u''^2}$, $\overline{\rho v''^2}$) through the flame zone following the radial distance is underestimated. This discrepancy could be explained by mechanisms related to the laminar flamelet structure and dynamics, since, as illustrated by figure 1, the level of turbulence corresponding to Yoshida experiment is relatively low. Therefore it is suggested that, the use of a flamelet model such as described by Masuya /15/ should be of interest.

The more practical case of a large Reynolds number elliptic flow obtained by stabilizing a premixed flame on a disc-baffle has been investigated by Heitor et al both numerically /42/ and experimentally /43/. Employing a model which follows the concepts put forward in the present paper, i.e. dealing with three different states (burnt, unburnt and reactive), these studies show that the interaction between mean pressure gradients and large density changes represent the main source term in the transport equations for the turbulent fluxes, although it is less important for the Reynolds stress. This last finding generalizes the conclusion drawn by Champion and Libby /34/ and Bray et al /38/ in the case of a simple shear flow.

4. CONCLUDING REMARKS

From the preceding discussion, it can be concluded that moment and pdf methods based on large Damkolher and turbulent Reynolds numbers assumptions and using second order modelling allow the numerical prediction of important effects related to the modification of turbulence by combustion. To this end, in the limit $Re \gg D_A \gg 1$ the BML model provides the basic mechanisms of turbulent transport and production of turbulence by interaction between pressure forces and density variations. Flamelet models correspond to an extended range of validity as they apply also in cases where the laminar burning velocity is no longer negligible when compared to the characteristic velocity of turbulence.

Moment models reviewed in this paper introduce conditionally averaged quantities within unburnt reactants and burnt products, which lead to new closure problems relative to turbulence properties within these constant density regions, thus, on both sides of a flamelet. These problems, in the more general case of multi-dimensional flows have not received a totally satisfactory treatment yet as additional empirical relations are still required. Nevertheless such a closure strategy proves to be useful for the following reasons :

- it provides a clear understanding of mechanisms in term of relative dynamics of packets of gas of different densities,
- it can be expected that local analysis of flamelet structure under various conditions of strain will provide useful informations on jump conditions for turbulent quantities across a flamelet /44/ and will improve closure approximations in the case of 2-D flows. The generality of these moment models can be checked by comparisons with pdf models (i.e. using a pdf equation) which are now becoming operative in 2-D flows calculations /45/.

Considering now comparisons with experimental data, it appears that many experimental devices such as open flames /23/, /41/ operate in ranges of relatively low Damkolher and turbulent Reynolds numbers, where there are mechanisms which modify turbulence that are not taken into account by models such as these discussed in the present review ; this raises the problem of extending these flamelet models towards both cases of finite Damkolher number flows and wrinkled flame regime. Relative to this remark two important points must be emphasized :

- In many cases the local rate of strain cannot be neglected and the flamelets structure cannot be considered as unstretched any longer. Therefore u_L depends on the local stretch rate and flow curvature induced by turbulence /46/ /53/. A priori this complicating effect can be introduced in expressions of u across flamelets such as the one given by eq. (14). However, this requires additional experimental and numerical data on the coefficients relating flame burning velocity to stretch effects for actual reactive mixtures /51/ /52/. Moreover the statistical distribution of this stretch rate has to depend on directions of flamelets relatively to the flow.
- When molecular and turbulent transport are of the same order of magnitude, laminar flame instability phenomena /47/ may become important. Various processes such as low frequency oscillations /48/, production or dissipation of turbulence by the local flame front can be observed experimentally /23/ /39/. Which one of the two latter mechanisms is effective depends on the stabilizing or destabilizing characteristic of the local flame fronts and the reactive mixture. Within a balance equation for turbulent kinetic energy $\tilde{k}$ an attempt has been made by Ballal /54/ to predict such a production rate quantitatively in the case of low frequency oscillations. Nonetheless, a general treatment of these unsteady and molecular mechanisms remains beyond the scope of methods described in this paper.

REFERENCES

1. Libby, P.A. and Williams, F.A., eds, Turbulent Reacting Flows, Springer-Verlag, Berlin, 1980.
2. Libby, P.A. and Williams, F.A., Some implications of recent theoretical studies in turbulent combustion, AIAA Journal, 19, 3, 1981.
3. Bray, K.N.C., Turbulent flows with premixed reactants, in Turbulent Reacting Flows, ed. by P.A. Libby and F.A. Williams, Springer-Verlag, Berlin, 1980.
4. Borghi, R., On the structure and morphology of turbulent premixed flames in Recent advances in the aerospace sciences C. Casci ed., Plenum Pub., p. 117, 1985.
5. Borghi, R., Mise au point sur la structure des flammes turbulentes, Journal de Chimie Physique, 81, 6, 1984.
6. Pope, S.B., Turbulent premixed flames, Ann. Rev. Fluid. Mech., 19, p. 237, 1987.
7. Andrews, G.R., Bradley, D. and Lwakabamba S.B., Turbulence and turbulent flame propagation. A critical appraisal, Comb. and Flame, 24, p. 285, 1975.
8. Bray, K.N.C., Libby, P.A., Interaction effects in turbulent premixed flames, Phys. of Fluids, 19, 11, p. 1687, 1976.
9. Libby, P.A. and Bray, K.N.C., Countergradient diffusion in premixed turbulent flames, AIAA J., 19, p. 205, 1981.
10. Bray, K.N.C., Libby, P.A., Masuya, G. and Moss, J.B., Turbulence production in premixed flames, Comb. Sci. and Tech., 25, p. 127, 1981.
11. Libby, P.A., Theory of normal premixed turbulent flames revisited, Prog. Energy Comb. Sci., 11, p. 83, 1985.
12. Pope, S.B., and Anand, M.S., Flamelet and distributed combustion in premixed turbulent flames, 20th Symp. (Int.) on Combustion, the Combustion Institute, p. 403, 1984.
13. Anand, M.S. and Pope, S.B., Calculations of premixed turbulent flames by pdf methods, Comb. and Flame, 67, p. 127, 1987.
14. Masuya, G., Comb. and Flame, 56, p. 123, 1984.
15. Masuya, G., Influence of laminar flame speed on turbulent premixed combustion, Comb. and Flame, 64, p. 353, 1986.
16. Peters, N., Length and time scales in turbulent combustion, presented at EUROMECH 220 on Turbulent Reactive Flows, Cambridge, March 1987.
17. Libby, P.A. and Bray, K.N.C., Implications of the laminar flamelet model in premixed turbulent combustion, Comb. and Flame, 39, p. 33, 1980.
18. Moss, J.B., Simultaneous measurements of concentration and velocity in an open premixed turbulent flame, Comb. Sci. and Tech., 22, p. 119, 1980.
19. Yoshida, A. and Gunther, R., An experimental study of structure and reaction rate in turbulent, Comb. Sci. and Tech., 26, p. 43, 1981.
20. Shepherd, I.G. and Moss, J.B., Measurement of conditioned velocities in a turbulent premixed flame, AIAA J., 20, p. 566, 1982.
21. Cheng, R.K., Conditional sampling of turbulence intensities and Reynolds stress in premixed turbulent flames, Comb. Sci. and Tech., 41, p. 109, 1984.
22. Yanagi, T., and Mimura, Y., Velocity - temperature correlation in premixed flame, 18th Symp. (Int.) on Comb., The Combustion Institute, 1981.
23. Cheng, R.K. and Ng, T.T., Velocity statistics in premixed turbulent flames, Comb. and Flame, 52, p. 185, 1983.
24. Shepherd, I.G., Moss, J.B. and Bray, K.N.C., Turbulent transports in a confined premixed flame, 19th Symp. on Comb., The Combustion Institute, p. 423, 1982.
25. Bray, K.N.C. and Moss, J.B., A unified statistical model of the premixed turbulent flame, Acta Astronautica, 4, p. 291, 1977.
26. Clavin P. and Williams, F.A., Effects of molecular diffusion and of thermal expansion on the structure and dynamics of premixed flames in turbulent flows of large scale and low intensity, J. Fluid Mech., 116, p. 251, 1982.
27. Kuznetsov, V.R., Estimate of the correlation between pressure fluctuations and the divergence of the velocity in subsonic flows of variable density, Translated from Izvestiya Akademii Nauk SSSR, Mekhanika Zhidkosti i Gaza, 3, p. 4, 1979.

28. Strahle, W.C., Estimation of some correlations in a premixed reactive turbulent flow, Comb. Sci. and Tech., 29, p. 243, 1982.
29. Jones, W.P., Models for turbulent flows with variable density in <u>Prediction methods for turbulent flows</u>, Hemisphere Pub., 1980.
30. Pope, S.B., Pdf methods for turbulent reactive flows, Prog. Energy Combust. Sci., 11, p. 119, 1985.
31. Bray, K.N.C., Libby, P.A. and Moss, J.B., Unified modeling approach for premixed turbulent combustion - Part I : General formulation, Comb. and Flame, 61, p. 87, 1985.
32. Wright, F.H. and Zukoski, E.E., 8th Symp. (Int.) on Comb., Williams and Wilkins, Baltimore, p. 933, 1962.
33. Meunier, S., Champion, M. and Bellet, J.C., Premixed combustion on a turbulent boundary layer with injection, Comb. and Flame, 50, p. 231, 1983.
34. Champion, M. and Libby, P.A., Turbulent premixed combustion in a boundary layer, Comb. Sci. and Tech., 38, p. 267, 1984.
35. Borghi, R. and Dutoya, D., On the scales of the fluctuations in turbulent combustion, 17th Symp. (Int.) on Comb., The Combustion Institute, p. 235, 1979.
36. Borghi, R. and Escudié, D., Assesment of theoretical modeling of turbulent combustion with a simple experiment, Comb. and Flame, 56, 149, 1984.
37. Bray, K.N.C., Champion, M., Davé, N. and Libby, P.A., On the thermochemistry of premixed turbulent combustion in variable enthalpy systems, Comb. Sci. and Tech., 46, p. 31, 1985.
38. Bray, K.N.C., Champion, M. and Libby, P.A., The turbulent premixed boundary layer with variable enthalpy, To appear in Comb. Sci. and Tech., 1987.
39. Tanaka, H. and Yanagi, T., Cross correlation of velocity and temperature in a premixed turbulent flame, Comb. and Flame, 51, p. 183, 1983.
40. Davé, N. and Kollmann, W., A second-order closure prediction of premixed turbulent combustion in jets, Phys. Fluids, 30, 2, p. 345, 1987.
41. Yoshida, A., Experimental study of wrinkled laminar flame, 18th Symp. (Int.) on Combustion, The Combustion Institute, p. 931, 1981.
42. Heitor, M.V., Taylor, A.M.K.P. and Whitelaw, J.H., Simultaneous velocity and temperature measurements in a premixed flame, Experiments in Fluids, 3, p. 323, 1985.
43. Heitor, M.V., Taylor, A.M.K.P. and Whitelaw, J.H., The interaction of turbulence and pressure gradients in a baffle-stabilised premixed flame, to appear in J.F.M., 1987.
44. Williams, F.A., Structures of flamelets in turbulent reacting flows and influences of combustion on turbulence fields, Workshop France - USA on turbulent reactive flows, Rouen, July 1987.
45. Anand, M.S., Pope, S.B. and Mongia, H.C., A pdf method for turbulent recirculating flows, Workshop France - USA on turbulent reactive flows, Rouen, July 1987.
46. Clavin, P., Dynamic behavior of premixed flame fronts in laminar and turbulent flows, Prog. Energy Combust. Sci., 11, p. 1, 1985.
47. Sivashinsky, G.I., Instabilities, pattern formation and turbulence in flames, Ann. Rev. Fluid Mech., 15, p. 179, 1983.
48. Matsumoto, R., Nakajima, T., Kimoto, K., Noda, S. and Maeda, S., An experimental study on low frequency oscillation and flame-generated turbulence in premixed/diffusion flame, Comb. Sci. and Tech., 27, p. 103, 1982.
49. Boukhalfa, A., Sarh, B., Debbich, M. and Gökalp, I., Spatial and temporal characteristics of the density field in turbulent premixed conical methane - air flames, Proceedings of the Joint meeting of the French and Italian sections of the Combustion Institute, Amalfi, June 1987.
50. Driscoll, J.F. and Gulati, A., Measurements of various terms in the turbulent kinetic energy balance within a flame and comparison with theory, submitted to Combustion and Flame, 1987.
51. Cambray, P., Deshaies, B., Experimental investigation of stretch and curvature effects on premixed flame burning velocities, Proceedings of the Joint meeting of the French and Italian sections of the Combustion Institute, Amalfi, June 1987.

52. Cambray, P., Deshaies, B. and Joulin, G., Gaseous premixed flames in non-uniform flows, Proceedings of the AGARD Conference on Combustion and Fuel in gas turbine engines, Chania, Crête, p. 36.1, 1987.
53. Clavin, P. and Joulin, G., Premixed flame in large scale and high intensity turbulent flow, Journal de Physique-Lettres, 44, 1, 1983.
54. Ballal, D.R., Studies of turbulent flow-flame interaction, AIAA Jal, 24, 7, p. 1148, 1985.

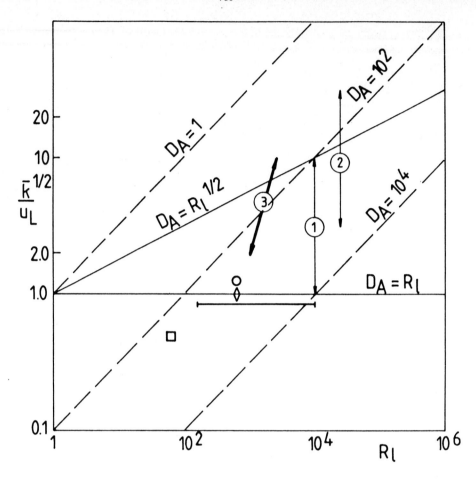

Figure 1

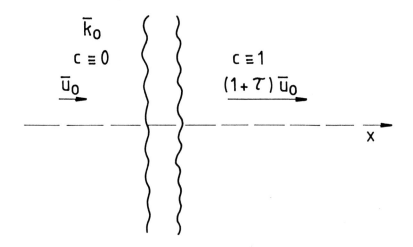

Figure 2

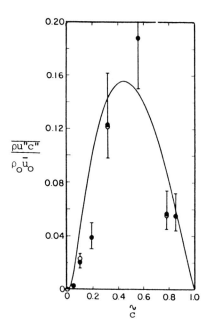

Figure 3

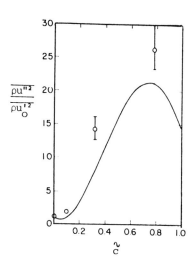

Figure 4

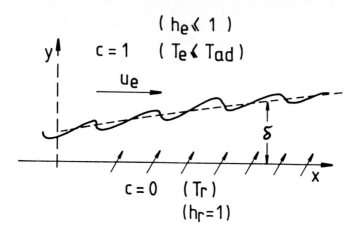

Figure 5

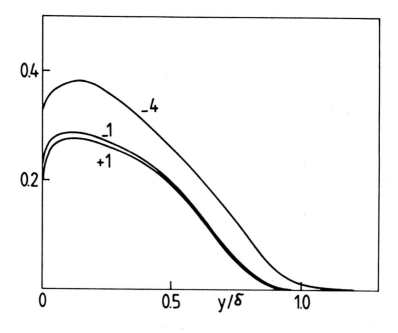

Figure 6

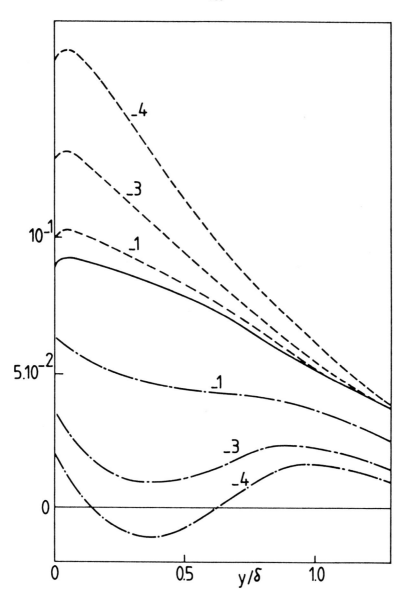

Figure 7

THE NUMERICAL SIMULATION OF COMPRESSIBLE AND REACTIVE TURBULENT STRUCTURES

Jay Boris, Elaine Oran, and Kazhikathra Kailasanath
Laboratory for Computational Physics and Fluid Dynamics
Code 4400, Naval Research Laboratory
Washington, D.C. 20375

ABSTRACT

Computational research on the simulation of compressible and reactive turbulent structures in the Laboratory for Computational Physics and Fluid Dynamics requires the development of algorithms to exploit parallel and vector processing, the application of these techniques to simulate compressible, turbulent reactive flow structures, and the initiation of laboratory experiments to calibrate and complement the simulations.

Our algorithmic research includes the development of explicit and implicit monotone methods for compressible convection, techniques for coupling disparate timescales which do not involve inverting matrices, variable, adaptive and unstructured gridding in multidimensions suitable for turbulent flows, realistic models for inflow and outflow boundary conditions in bounded domain problems, and the use of triangular grids in two dimensions and tetrahedronal grids in three dimensions to adapt the solution automatically to evolving flow structures. This paper will review recent important contributions in each of these areas and will discuss the practical limits attainable by finite resolution detailed modeling techniques.

Our applications of these advances in simulation technology include research into the turbulent structure, propagation and extinction of detonations and flames using detailed chemistry models and multispecies diffusion coefficients, strong shock and transonic flows in complex geometries, and the study of turbulence and boundary layer phenomena in subsonic and supersonic shear layers and jets.

A brief review will be given using illustrative simulations from the Naval Research Laboratory Cray and the Graphical and Array Processing System (GAPS), a multitasking, low cost hardware and software system assembled to provide a parallel processing capability to perform and analyze turbulent reactive flow simulations interactively. The GAPS approaches Cray performance on spatially evolving, compressible transition-to-turbulence simulations.

INTRODUCTION

We review recent research on techniques for simulating compressible and reactive flows performed in the Laboratory for Computational Physics and Fluid Dynamics at the Naval Research Laboratory. Colleagues who made major contributions to this work include Theodore Young, John Gardner, K. Kailasanath, Fernando Grinstein, Raafat Guirguis, David Fyfe, Gopal Patnaik, J. Michael Picone, and Rainald Löhner. In this paper, we emphasize high-speed reactive flows, in which characteristic flow velocities may be comparable to the sound speed. We conclude with discussion of a two-dimensional compressible simulation of the wake shed by a bluff body. This latter calculation was carried to one million timesteps on the experimental Graphical and Array Processing System (GAPS). Such long calculations are necessary to accumulate meaningful statistics about body drag and drag reduction and to determine the connection between fluid dynamics and chaotic dynamical systems.

There are three directions to our work. The first is the development of accurate, optimized numerical methods, which become extremely important when computer resources are limited. We want methods that can make noticeable improvements in what can be calculated. The second direction is the application of the these methods to specific physical problems. We want to be able understand important physical interactions and mechanisms that have eluded earlier research efforts. In our work, a set of applications has always motivated the development of numerical methods, which in turn have to be applicable to other types of problems. The third direction is the development and exploitation of new computer systems. We want to have the resources to compute the answers to problems that are sufficiently computer intensive that previous attempts have not been fully successful. Much of the material presented here is based on a review article and book by Oran and Boris[1,2].

REACTIVE FLOW CONSERVATION EQUATIONS

The equations used to model neutral gas-phase reactive flows are the continuum time-dependent equations for conservation of mass density ρ, individual chemical species number densities, $\{n_i\}$, momentum density $\rho\mathbf{v}$, and total energy E,

$$\frac{\partial \rho}{\partial t} = -\nabla \cdot (\rho \mathbf{v}), \tag{1}$$

$$\frac{\partial n_i}{\partial t} = -\nabla \cdot (n_i \mathbf{v}) - \nabla \cdot (n_i \mathbf{v}_{di}) + Q_i - L_i n_i, \tag{2}$$
$$i = 1, ... N_s,$$

$$\frac{\partial \rho \mathbf{v}}{\partial t} = -\nabla \cdot (\rho \mathbf{v}\mathbf{v}) - \nabla \cdot \mathbf{P} + \sum_i \rho_i \mathbf{a}_i, \tag{3}$$

and

$$\frac{\partial E}{\partial t} = -\nabla \cdot (E\mathbf{v}) - \nabla \cdot (\mathbf{v} \cdot \mathbf{P}) - \nabla \cdot (\mathbf{q}) + \mathbf{v} \cdot \sum_i m_i \mathbf{a}_i + \sum_i \mathbf{v}_{di} \cdot m_i \mathbf{a}_i . \tag{4}$$

The first term on the right hand side of each of these equations describes the convective fluid dynamics effects. The remaining terms contain the source, sink, coupling, external force, and diffusive transport terms that drive the fluid dynamics. Pressure is a tensor,

$$\mathbf{P} \equiv P(N,T)\mathbf{I} + \left(\frac{2}{3}\mu_m - \kappa\right)(\nabla \cdot \mathbf{v})\mathbf{I} - \mu_m[(\nabla \mathbf{v}) + (\nabla \mathbf{v})^T], \tag{5}$$

where μ_m and κ_m are the shear and bulk viscosity coefficients and the superscript T means matrix transpose. The heat flux, $\mathbf{q}$, total number density, N, and internal energy density, ϵ, can be found from

$$\mathbf{q}(N,T) \equiv -\lambda_m \nabla T + \sum_i n_i h_i \mathbf{v}_{di} + P \sum_i K_i^T \mathbf{v}_{di} + \mathbf{q}_r, \tag{6}$$

$$N \equiv \sum_i n_i, \tag{7}$$

and

$$E \equiv \frac{1}{2}\rho \mathbf{v} \cdot \mathbf{v} + \rho \epsilon. \tag{8}$$

Here K_i^T is the thermal diffusion coefficient of species i, λ_m is the mixture thermal conductivity, $\mathbf{v}_{di}$ is the diffusion velocity for species i, $\mathbf{q}_r$ is the heat flux due to radiation, and h_i is the specific enthalpy.

In addition, we need equations of state, for gases usually taken as

$$P = Nk_BT = \rho RT, \tag{9}$$

$$h_i = h_{io} + \int_{T^o}^{T} c_{pi}\, dT, \tag{10}$$

where h_{io} and c_{pi} are the heat of formation and specific heat of species i.

A lot has been hidden in the chemical production terms and loss rates, $\{Q_i\}$ and $\{L_i\}$, which can be complicated functions of of temperature, pressure, and the other species densities. The set of species diffusion velocities $\{\mathbf{v}_{di}\}$ are found by inverting the matrix equation

$$\mathbf{W}_i = \sum_k \frac{n_i n_k}{N^2 D_{ik}}(\mathbf{v}_{dk} - \mathbf{v}_{di}), \tag{11}$$

where the source terms $\{\mathbf{W}_i\}$ in Eq. (11) are defined as

$$\mathbf{W}_i \equiv \nabla\left(\frac{n_i}{N}\right) - \left(\frac{\rho_i}{\rho} - \frac{n_i}{N}\right)\frac{\nabla P}{P} - K_i^T \frac{\nabla T}{T}. \tag{12}$$

The diffusion velocities are subject to the constraint

$$\sum_i \rho_i \mathbf{v}_{di} = 0. \tag{13}$$

These equations, the physics and chemistry implicit in them, and their soluion in a number of important circumstances are described in some detail by Oran and Boris[2].

There are basically four types of physical processes in these continuum reactive flow equations, and these are summarized in Table 1. The first two, chemical

Table 1.   Terms in the Reactive Flow Equations

$\frac{\partial \rho}{\partial t} = \gamma \rho + S$	Local Processes:	Source, Sink, Coupling
		Chemical Reactions
$\frac{\partial \rho}{\partial t} = \nu \nabla^2 \rho$	Diffusive Processes:	Molecular Diffusion
		Thermal Conduction
		Thermal Diffusion
$\frac{\partial \rho}{\partial t} = \nabla \cdot (\rho \mathbf{v})$	Convective Processes:	Advection
		Compression
		Vorticity (Rotation)
$\frac{\partial^2 \rho}{\partial t^2} = c_w^2 \nabla^2 \rho$	Waves and Oscillations:	Sound Waves
		Gravity Waves
		Oscillations

reactions and diffusive transport, originate in the atomic and molecular nature of matter. The third and fourth, convection and wavelike phenomena, are collective, continuum properties.

Chemical reactions (or chemical kinetics), represented by the production and loss terms, Q_i and $L_i n_i$ in Eq. (2), are examples of "local" phenomena which do not depend on spatial gradients. Other examples of local phenomena include phase changes, external source terms such as laser or spark heating, and sinks such as optically thin radiation loss. The macroscopic models of these processes used in the continuum equations arise from averages over microscopic effects. The computational difficulties posed by these local processes, generally called "stiffness," come about when it is necessary to integrate the ordinary differential equations with computational timesteps appreciably longer than some of the physical timescales in the problem. Volumes have been written about stiff ordinary differential equations and how to solve them. Oran and Boris[2] review these techniques in the context of solving equations (1)–(4). Here we will concentrate on the nonlocal processes which depend on spatial variation and hence on spatial as well as temporal derivatives in the equations.

Diffusion (or diffusive transport), has the general form $\nabla \cdot Y$ where Y may be $n_i \mathbf{v}_{di}$, $\mu \nabla \mathbf{v}$, or $\mathbf{q}$. The last two terms in Eq. (5) describe the diffusive effects of viscosity. The last terms in Eq. (6) describe the change in energy due to molecular diffusion, chemical reactions, and radiation transport. When the diffusion coefficients are large, diffusion processes, like chemical kinetics, can be stiff. In such situations the solution everywhere in space is essentially coupled together on timescales shorter than we can afford to resolve in a simulation. The resulting parabolic equations can be solved by several expensive but relatively well understood numerical techniques. Further, strong diffusion generally means that the numerical solution is quite smooth and hence relatively easy to represent accurately once the intrinsic instability of the numerical time integration procedure has been overcome.

The third equation in Table 1 is a continuity equation describing convection. Convection effects are represented in the equations by fluxes of conserved quantities through volumes, e.g., $\nabla \cdot \mathbf{v} X$ where X is $\rho, n_i, E, \rho\mathbf{v}$ or P. Convection is a continuum concept, which assumes that quantities such as density, velocity, and energy are smoothly varying functions of position. The fluid variables are defined as averages over the actual particle distributions, so fewer degrees of freedom are necessary to describe the local state of the material. This process in the reactive flow equations is generally acknowledged as the most difficult to implement in an accurate simulation. The discipline which has grown up around integrating these convective transport processes is known as Computational Fluid Dynamics and is the most extensive of the fields of Computational Physics.

Wavelike and oscillatory behavior are described implicitly in the reactive flow equations by coupled continuity equations. To solve for nonlinear wave fields accurately often combines the difficulties of convection with those of reversible oscillatory phenomena. The main type of waves considered are shock waves, which move as discontinuities through the system, and sound waves, in which there are alternating compressions and rarefactions in density and pressure of the fluid. Other types of waves included in these reactive flow equations are gravity waves and chemical reaction waves.

This extremely rich set of equations, combined with appropriate initial and boundary conditions, describes flames, detonations, the evolution of coherent structures, reacting turbulence, and many other important reactive flow phenomena. The complete set of reactive flow equations is very difficult to solve so independent scientific communities flourish solving subsets of these equations for particular applications. Several things make it difficult to solve the equations: lack of knowledge of the input parameters such as chemical reaction rates, physical diffusion coefficients, or boundary conditions; inadequate numerical methods to resolve the physical phenomena, and inadequate computer time and memory.

NUMERICAL METHODS

Of the four different types of terms in Table 1, we have concentrated the most numerical development effort on the solution of continuity equations, on ordinary differential equations, and on methods for coupling algorithms for various types of processes. The guiding principle for ordinary differential equations is that the method must be particularly efficient when coupled to convection calculations. The guiding principle for solutions of the continuity equation is that the method must be accurate, efficient computationally, and allow variably spaced and moving grids.

The method we use for solving the ordinary differential equations of chemical reactions is the algorithm CHEMEQ[3,4]. This is a hybrid method combining a second-order Adams-Moulton method and an asymptotic method. The method used for each equation depends on whether the equation currently is stiff or not. The method is self starting, which is crucial for multiprocess numerical simulations, and fully optimized for vector and parallel processing with respect to species and computational cells. It is extremely efficient when coupled to algorithms for convection. Because the intrinsic algorithm is nonlinear, however, exact conservation of the atomic species numbers cannot be guaranteed. This lack of exact conservation means that the conserved species numbers can be used as a measure of accuracy and the resulting numerical algorithm is one to two orders of magnitude faster.

The Flux-Corrected Transport (FCT) method was developed for solving continuity equations and extended to the solution of coupled continuity equations[5-7]. FCT is a compressible, one-dimensional, fourth-order, nonlinear monotone convection algorithm that is particularly good at resolving and maintaining gradients. It is also straightforward to combine the FCT modules and construct multidimensional programs using directional-splitting methods. FCT was the first monotone method enforcing positivity in the solution. A history of its development from 1971 until recently is summarized in Table 2. Several of the monotone convection methods including Flux-Corrected Transport now approach optimal performance. As a consequence, recent numerical development efforts are focusing on adaptive gridding techniques that will improve spatial resolution without increasing the cost of a computation. By putting smaller computational cells only where they are needed to resolve the steeper gradients in a flow, a given number of cells in a mesh can resolve finer detail in the flow.

The final section of this paper describes our Graphical and Array Processing System and the highly optimized Reactive Flow Model using FCT which was developed for efficient simultaneous use of the GAP's multiple array processors. This two-dimensional model can be run with a variey of geometries, flow speeds, and reaction chemistry models and it is used to generate the examples of vortex shedding and coherent structure interactions described below.

COUPLING NUMERICAL MODELS OF PHYSICAL PROCESSES

Several generic approaches have evolved for solving sets of equations such as Eqs. (1)-(10) in which a number of different physical processes, represented by different types of mathematical terms, interact. The major methods are the

Table 2. History of Flux-Corrected Transport Algorithm Development

Boris[5], 1971	Basic nonlinear, monotone SHASTA algorithm
Boris & Book[6], 1976	FCT adapted to general finite-difference algorithms
Boris[7], 1976	Optimization for vector and parallel processing
Zalesak[8], 1979	Fully multidimensional FCT and limiters, generalized to arbitrary high- and low-order convection algorithms
Löhner[9], 1985	Finite-element FCT on unstructured grids
Patnaik[10], 1986	Implicit FCT for slow subsonic flows
LCP&FD staff, 1988	LCPFCT with generalized boundary conditions
To Be Done, 19??	Fully Lagrangian FCT

global-implicit method, also called the block-implicit method, and the fractional-step method, also called timestep splitting. Other approaches to coupling, such as the method-of-lines and general finite-element methods, are also used. The various approaches are often combined into "hybrid" algorithms.

The sets of coupled partial differential equations describing a reactive flow system can be written

$$\frac{d}{dt}\rho(\mathbf{x},t) = \mathbf{G}(\rho, \nabla\rho, \nabla\nabla\rho, \mathbf{x}, t) \qquad (14)$$

where ρ is a vector, each component of which is a function of the time t and the vector position $\mathbf{x}$. A global-implicit method solves this equation using implicit finite-difference formulas in which the right side of Eq. (14) is evaluated using information at the advanced time. To do so, the nonlinear terms in $\mathbf{G}$ have to be linearized locally about the known solutions at the previous timestep. For example, write Eq. (14) in the form

$$\Delta\rho \equiv \rho^n - \rho^{n-1} \approx \Delta t \, \mathbf{G}^n \, . \qquad (15)$$

If the changes in ρ are small during a timestep, it is consistent to assume that the changes in **G** are also small, so that $\mathbf{G}^n$ at the new time can be approximated by

$$\mathbf{G}^n = \mathbf{G}^{n-1} + \Delta\rho \frac{\partial \mathbf{G}^{n-1}}{\partial \rho}, \qquad (16)$$

where the Jacobian, $\partial \mathbf{G}^{n-1}/\partial \rho$, is a tensor when ρ is a vector of variables evaluated using the known values. We can combine Eqs. (15) and (16) to obtain a linearized global implicit approximation for the new values of ρ,

$$\Delta\rho^n = \Delta t\, \mathbf{G}^{n-1}\left[1 - \Delta t \frac{\partial \mathbf{G}^{n-1}}{\partial \rho}\right]^{-1}. \qquad (17)$$

An implicit matrix equation results from this procedure, and a large evolution matrix, $1 - (\Delta t)\partial \mathbf{G}^{n-1}/\partial \rho$, must generally be inverted. The cost for this matrix inversion is often very high.

This formalism requires linear finite-difference approximations to the convective derivatives and validates the use of more accurate, nonlinear methods. In one dimension the problem usually involves a block tridiagonal matrix with M physical variables at N_x grid points. In this case an $(MN_x \times MN_x)$ matrix must be inverted at each iteration of each timestep. The blocks on, or adjacent to, the matrix diagonal are $M \times M$ in size, so that the overall matrix is quite sparse. The matrices are $(MN_xN_y \times MN_xN_y)$ in two dimensions and $(MN_xN_yN_z \times MN_xN_yN_z)$ in three dimensions. In complex chemical kinetics problems with no spatial variation, the M fluid variables are the species number densities plus the temperature of the homogeneous volume of interest.

When only convection is involved, implicit methods can be relatively inexpensive. It is then possible to combine the equations to obtain an implicit elliptic equation, which is often much easier to solve than the general case given in Eq. (17). In these global implicit algorithms, first-order numerical damping (or diffusion) is required to maintain positivity. Because the equations are linearized, iterations are needed to include nonlinear terms which are evaluated implicitly.

The second coupling method is the fractional-step approach[11], also called timestep splitting, in which the individual processes are solved independently and the changes resulting from the separate partial calculations are coupled (added) together. The processes and the interactions among them may be treated by analytic, asymptotic, implicit, explicit, or other methods. Advantages of this approach are that it avoids many costly matrix operations and allows the best method to be used for each type of term. The exact way the processes are coupled depends on the individual properties of the different algorithms used[1,2] and on the specific conditions of the problem being solved. Different methods are needed for simulations of detonations and flames because the flow mach numbers differ so radically for example.

Consider writing Eq. (14) in a form suited to timestep splitting,

$$\frac{d}{dt}\rho(\mathbf{x},t) = \mathbf{G}_1 + \mathbf{G}_2 + \mathbf{G}_3 + \cdots + \mathbf{G}_M , \qquad (18)$$

where $\mathbf{G}$ has been broken into its constituent processes. Each of the functions $\{\mathbf{G}_i\}$ contributes a part of the overall change in ρ during the numerical timestep. Thus

$$\left. \begin{aligned} \Delta\rho_1^{n-1} &= \Delta t\,\mathbf{G}_1^{n-1} \\ \Delta\rho_2^{n-1} &= \Delta t\,\mathbf{G}_2^{n-1} \\ &\vdots \quad \vdots \\ \Delta\rho_M^{n-1} &= \Delta t\,\mathbf{G}_M^{n-1} \end{aligned} \right\} . \qquad (19)$$

For example, $\mathbf{G}_1^{n-1}$ might be the chemical kinetics terms, $\mathbf{G}_2^{n-1}$ the diffusion terms, $\mathbf{G}_3^{n-1}$ the thermal conduction terms, and so on.

The solution for the new values ρ^n is found by summing all of the partial contributions,

$$\rho^n = \rho^{n-1} + \sum_{m=1}^{M} \Delta\rho_m^{n-1} . \qquad (20)$$

If each of the processes is simulated individually, Eq. (20) gives a simple prescription for combining the results.

The method-of-lines approach focuses on separating the spatial and temporal parts of the problem regardless of the processes involved or nonlinearity. The full nonlinear set of equations is reduced to a set of stiff ordinary differential equations in time, which can be solved by a standard method. Finite-element methods (FEM) have yet another focus, though some of their features are common to both the method-of-lines and global-implicit methods. The finite-element approach finds the "best" solution by error minimization techniques. Thus FEMs are extremely appealing even though the computational expense is comparable to, or can exceed, global-implicit methods. One added feature of FEMs is that the locations of the grid points as well as the values of the dependent variables at these points can be made into degrees of freedom for the error minimization process.

Our method of choice has been timestep splitting. The reasons for this are its relatively straightforward applicability and its flexibility to incorporate the best available numerical algorithm for each particular physical process described in Table 1. For example, we have found that small errors in calculating convection can cause large errors in the overall reactive-flow simulation. This means that very accurate calculations of convection are generally required. However, the solution of ordinary differential equations representing chemical kinetics does not have to be as accurate when coupled to convection as it does when it is solved alone at constant volume or constant pressure. With timestep splitting, the order of the algorithms can be adjusted accordingly. Specific tricks for using timestep splitting for fast and slow reactive flows are discussed in Reference 2.

COMBINING PARALLEL PROCESSING WITH GRAPHICS

GAPS is an asynchronous, multi-tasking, high-performance, scientific parallel processing system developed in the Laboratory for Computational Physics and Fluid Dynamics specifically for reactive flow simulations in complex geometries. The GAPS consists of an APTEC 2400 I/O Computer with four powerful array processors connected to a VAX 11/780 computer. The objectives of the research project centered on the GAPS[19,20] are to develop expertise in multiprocessing systems, to develop an interactive, user-friendly system designed to accommodate interactive graphics for large data sets such as generated by multidimensional fluid dynamics simulations or experiments, and to provide inexpensive floating point computer cycles for a wide class of computationally intensive physical problems.

The VAX system in GAPS acts as host to the APTEC I/O Computer, compiling the Fortran program that runs on the array processors, controlling the file structure of the IBIS disk and mass memory on the APTEC, and providing user-friendly interactive control of one or more ongoing simulations in the GAPS. To make this composite system work efficiently for large computational fluid dynamics problems, our APTEC has been configured with 12 megabytes of additional fast memory and about 2 gigabytes of online disc storage. The GAPS also contains several major computational components, including a Floating Point Systems 5305 array processors (maximum performance of 12 Megaflops) and three Numerix MARS 432 array processors (maximum performance of 30 Megaflops each). The GAPS will be expanded in the next year to six MARS array processors, bringing the composite system performance to about that of a Cray X-MP processor.

The array processors can be programmed in Fortran and have an extensive library of vectorized routines and synchronization software. Our standard Flux-Corrected Transport routines have been optimized for the system and are augmented by fully optimized complex geometry and color graphics routines. Several high-resolution color graphics monitors, a Tectronix 4115B raster system, a Metheus Ω raster system, an Evans and Sutherland PS 300 vector system, and an IRIS 4D vector system, are connected to the GAPS. These graphics systems all have distributed processing capabilities such as zooming and rotating images.

The APTEC is a high-speed (24 megabyte/s) data bus used to move data among the VAX, the array processors, and other GAPS components through programmable data interchange adapters (DIAs). The ability of the VAX to control data flow is superb, but its ability to actually move the data is limited. In practice, a quarter megabyte/s on the VAX is the maximum information transfer rate it can systain, and this severely limits the overall system performance. A very high-speed data switchyard is necessary behind the VAX to display high-resolution fluid dynamics simulations at the rate at which they are calculated. The VAX reaches its skinny arm into the APTEC to turn large data faucets, this alleviating the data flow bottleneck which results when data hungry array processors are connected directly to the undersized VAX UNIBUS. The I/O 2400, working at its full 24 megabyte/s potential, can barely keep up with the 12 megabyte/s speed required to show 12 frames per second of a 1024 × 1024 pixel array.

The GAPS currently has 102 megaflops of array processors. The NRL Cray X-MP/12 is rated as a 210 megaflop system at full use. When the 30 megaflops array processors are processing a two-dimensional FCT benchmark code, simultaneously pixelating the data arrays as it computes them, and sending every third or fourth frame to one of the high resolution color graphics monitors, only about 30% of the APTEC data bus capacity is being used. Operating together these array processors give about 50% of the performance of the CRAY. The FPS array processor and extra graphics monitors permit "instant replay" of archived data sets and detailed analysis of previous simulations while the GAPS is cranking away on new simulations using the faster MARS array processors. The array processors can be used together on one computation or singly by separate users, thus allowing a flexible user environment.

An important part of our research has been to develop optimally efficient algorithms that can take advantage of an asynchronous multi-tasking parallel architecture such as the GAPS. We have begun with algorithms that can take advantage of more parallelism than the pipeline and vector register architectures used in current supercomputers. The Reactive Flow Model (RFM), currently implemented on the GAPS, can be run for very long times on relatively high-resolution fluid simulations. Because the GAPS runs in the background without affecting the operation of the VAX, days or weeks of computer time can be devoted to a problem. The GAPS effectively gives us an extra 60 hours of CRAY time every week.

Results from GAPS simulations using RFM can be displayed as they are calculated using VOYEUR and RTGS (Real Time Graphics System), high-bandwidth graphics packages that can select data from the VAX, from the GAPS array processors, or from the CRAY. Figures 1, 2, 3, 6, and 7 each show contour plots of several physical variables at one timestep in two-dimensional planar simulations performed on GAPS of the flow over a bluff body. These figures are all produced by the VOYEUR graphics system as the computation proceeds.

APPLICATIONS TO COMPRESSIBLE AND REACTIVE FLOWS

We now describe several simulations of compressible and reactive flows. These include the onset of symmetric and asymmetric vortex shedding in two-dimensional flow over a bluff body with a rectangular trailing edge which we call a "wide splitter plate" (WSP), further studies in this configuration using a thin transverse barrier to suppress vortex shedding and hence to reduce drag (carried to one million timesteps), and simulations of an axisymmetric reactive flow in an idealized ramjet combustion chamber[23,27]. Related basic research areas for these techniques include complex reacting shock dynamics, multidimensional detonations[21,22], and supersonic shear flow mixing[26]. Applications include the hypersonic aerospace plane, reactor safety calculations, and shock-generated turbulence.

ONSET OF VORTEX SHEDDING ON A BLUFF BODY

As an example of two-dimensional coherent vortex structures interacting in a compressible flow, consider the problem shown after 4173 timesteps in Figure 1. Standard temperature and pressure air is flowing from left to right at 100 m/s away from the base of the bluff body, here a rectangular "wide splitter plate" (WSP) which is 1.6 cm thick (high) as shown by the shaded rectangle at the left edge of the four panels in the figure. The boundary conditions and initial flow are symmetric above and below the midplane of the WSP. The initial density was constant everywhere and the initial vector velocity field (V_X, V_Y) was everywhere (100 m/s, 0 m/s). The sound speed of the incoming gas is a ≈330 m/s. After an initial transient lasting a couple of acoustic transit times, the solution rapidly relaxes into the symmetric quasi-steady recirculation shown in Figure 1. The fluid moves toward the midplane near the tip of the recirculation and moves away from the midplane along the bluff base of the wide splitter plate.

The full color spectrum used to represent (contour) the three primary fluid dynamic variables in these figures, the horizontal (X) velocity, vertical (Y) velocity and mass density, is reduced to a gray scale rendition but the different qualitative flow features can still be identified. The calculations are being performed on the GAPS and displayed using the Voyeur interactive graphics system described above. The various intensities of gray are in fact renditions of 16 of the 256 colors available on the system. The figures on the far left are, from top to bottom, the X-velocity, the Y-velocity, and the density. During the course of the calculation, we needed a blowup of the region near the body, and so created the inset panel on the right.

In the simulation, the flow field initially started out symmetric about the WSP midplane, and then became asymmetric as computer roundoff errors triggered the expected vortex shedding mode which eventually dominates the flow field. The finite-volume FCT grid has 600 columns of cells by 200 rows of cells in two dimensions. The cells are stretched away from the region of the recirculation to the right and away from the WSP midplane to the top and the bottom of the computational mesh. Figure 2 shows the beginning of this laminar (symmetric) to "turbulence" (asymmetric shedding) transition at timestep 13,800 in the simulation. Figure 3 is computed using a 300 x 100 grid, about the same resolution as our longest Cray production runs, and shows the "vortex street" after 99,700 timesteps have elapsed. This run develops the usual wake structure with essentially periodic vortex shedding.

The inset panel in Figure 1, entitled V_Y Blowup, is a three-to-one enlargement of data near the left end of the middle panel in the figure. At this time the remnants of the initial strong rarefaction can be seen as an oblong density contour in white around the edge of the lower panel. This rarefaction occurs at the base of the WSP as the moving fluid first pulls away and lasts until the entirely horizontal flow has curved effectively around the corner at the trailing edge of the WSP bluff body. The quasi-steady recirculation is clearly visible in the blowup and lasts for thousands of timesteps.

Figure 2 shows a continuation of the same very high-resolution simulation through 13,800 steps where asymmetric vortex shedding has transformed the initial steady recirculation of Figure 1 into the usual Kelvin vortex street solution. The density contours, shown in the blowup, clearly indicate that one weak and two strong vortices have been shed with a fourth vortex in the process of being shed. A number of inflow and outflow boundary conditions were tested for these very long runs. Figure 1 and Figure 2 both had ideal, free-slip reflecting top and bottom boundaries stretched 21 cm above and below the midplane by exponentially increasing cell sizes before row 30 and after row 70. The WSP is 16 cells thick ($\delta y = 0.1$ cm) and 40 cells long ($\delta x = 0.2$ cm) in the 300×100 resolution calculations. Cell sizes are simply halved in the higher resolution simulations.

The computation in Figure 3 shows the limiting behavior of such a vortex street in a fluid system where the top and the bottom boundary are treated as ideal smooth but impenetrable walls. Many vortices have been shed on this somewhat coarser 300 x 100 finite-difference grid. The long sequence of relatively regular alternating sign vortices are shed at a Strouhal frequency of .233 obtained using the flow speed, 100 m/s, the WSP width of 1.6 cm, and measuring the vortex shedding rate from the simulations. Detailed inspection of the solutions shows an essentially periodic behavior with a single dominant frequency, about 1500 Hz.

DRAG REDUCTION BY INTERFERING WITH VORTEX SHEDDING

In switching to a grid of 300 cells by 100 cells, as shown in Figure 3, much longer runs can be carried out to study the physical mechanisms governing several unexpected low-frequency (subharmonic) modes and resonances which appeared in our previous simulations of idealized ramjet chambers[23,27]. In these simulations we have seen that the initial relaxation of the flow, in the absence of any vortex shedding control barriers, to an approximately periodic vortex shedding phenomenon varies greatly depending on a range of seemingly reasonable boundary conditions. Using a number of simulations and tests at intermediate and high resolution, we determined that acoustic effects are of paramount importance, in particular how acoustic waves from downstream sources are reflected at the fluid inflow plane. The confined and unconfined behaviors were very different, the confined case in Figure 7 shows a musical instrument-like resonance.

A very long sequence of related tests were begun on GAPS as a single computation to investigate the bluff body vortex shedding computationally with a simulation encompassing hundreds of vortex sheddings rather than tens. The initial 400,000 steps of this computation were made with ideal reflecting top and bottom boundaries at 21 cm above and below the horizontal midplane of the computation. These distances, obtained by stretching the mesh near the top and bottom boundaries, are expected to be far enough from the recirculation region that mininal effects of the boundaries are felt. The remaining 600,000 steps of the simulation were performed with another boundary condition specified at +21 cm and −21 cm to model the free, unconfined but irrotational breathing of the flow at the boundary. Only small quantitative changes in the solutions are observed by changing the top and bottom boundary conditions as described.

Figure 6 shows typical contour plots of V_X, V_Y, and density after 642,000 timesteps of the 1,000,000 step run have elapsed. A vanishingly thin barrier of horizontal length L = 1.6 cm is located in the recirculation region from 1.6 cm to 3.2 cm behind the center of the WSP which is also 1.6 cm thick. The calculation is suggested by the experiments of Roshko[15], Bearman[16], and Rockwell[16] and analyses published by Simmons[17] and Griffin[18]. Figure 6 shows a very different flow pattern from that seen in Figure 3. The shedding frequency and amplitude are both reduced by a factor of about 2.5 over the case with no barrier to interfere with crossflows in the near wake.

When the barrier is retracted into the WSP or pulled downstream far enough, the usual rapid asymmetric shedding pattern returns. This generation of additional vorticity in the wake necessarily produces drag. With the barrier leading edge at D = 0.875×W, drag on the plate is a minimum and only a third of the Euler drag measured with unhindered shedding. Detailed studies show that an exchange of flow patterns occurs at the point of minimum drag where several different frequencies are apparent in Fourier analyses of the base pressures at the rear of the bluff body (WSP). In these regimes long-time Fourier diagnostics indicate that the vortex shedding appears chaotic with appreciable subharmonic activity.

REACTIVE FLOWS IN AXISYMMETRIC COMBUSTORS

The FCT algorithms described above have been used on variably spaced grids with appropriate inflow and outflow boundary conditions to calculate the properties of compressible axisymmetric jets and planar mixing layers[24-26]. These types of calculations have been performed on the NRL Cray, the AFWAL Cray, the NAS Cray-2, and in GAPS. Figure 4 shows a complicated axisymmetric flow in which a gas flows out of a long cylindrical inlet into a chamber of larger diameter[23,27]. The exit condition from this larger chamber is choked flow, in which the flow becomes sonic at the exit nozzle. The top portion of the figure shows a schematic of the chamber, and the bottom portion shows a series of instantaneous streamlines of the flow in the cross-hatched region. Lines originating at the centers of vortices have been drawn between the streamlines to show the evolution of the vortex structures.

The reactive flow aspects of this problem have been modeled by combining the FCT algorithm with a phenomenological model describing the combustion of hydrogen and oxygen[12]. The model is derived by solving numerically a set of equations representing about 50 chemical reactions among the species H_2, O_2, OH, H, O, HO_2, H_2O_2, H_2O, and a diluent that can be either Ar, N_2, or He, combined with appropriate thermochemical data[13,14].

Figure 4 shows a subsonic high-speed slightly preheated mixture of hydrogen and air entering into a combustor. The mixture evolved to a quasi-periodic pattern as vortices were shed, merged, and exited through the nozzle. Temperature contours before ignition are essentially uniform, but showed the effects of compressibility. There is a strong coupling between the vortex structures formed in the shear layer and the acoustics in a ramjet combustion chamber[23,27]. For example, the flowfield shown in Figure 4 has an overall repetition cycle of 6000 steps, corresponding to a low-frequency mode of $\approx$ 150 Hz. In addition, the first vortex rollup appears with a frequency of $\approx$ 450 Hz. The low frequency mode, corresponding to the quarter wave mode of the inlet, controls the overall merging pattern in the chamber. The higher frequency mode corresponds to the first longitudinal acoustic mode of the chamber.

Figure 5 shows another calculation similar to that shown in Figure 4 in which the mixture is "ignited" in a small region around the inlet[23]. Chemical reactions and heat release are allowed to occur according to an induction parameter model. The instantaneous streamlines and temperature contours show the flame front moving down the chamber and a quasi-steady pattern is eventually set up. The flame front is located on the temperatue contours both by the the high temperatures and the dark lines caused by closely spaced contours. In time, the roll-up in the shear layer causes the reaction front to curve downward and engulf the cold mixture which subsequently burns. As the reaction moves downstream, a new vortex forms near the step between timesteps 175,000 and 180,000. This mixes the burned gases with the incoming mixture and acts as an ignition source. The flow field undergoes a cycle of roughly 25,000 timesteps, or 3.463 ms. The low-frequency mode persists in the reacting flow, but its amplitude is higher.

SUMMARY

We have reviewed recent research on techniques for simulating compressible and reactive flows in the Laboratory for Computational Physics and Fluid Dynamics at the Naval Research Laboratory. In this paper, high-speed reactive flows, in which characteristic flow velocities may be comparable to the sound speed, were emphasized. There have been three directions to this work. The first is the development of accurate, optimized numerical methods. We want methods that can make noticeable improvements in what it was previously possible to calculate, even when computer resources are limited.

The second direction we discussed is the development and exploitation of new, parallel-processing computer systems to conduct more extensive simulations cost effectively. To have the resources to simulate problems that are sufficiently computer intensive for previous attempts to have been unsuccessful, we have developed the Graphical and Array Processing System (GAPS) as described above. Running now at about half a Cray X-MP processor in speed, the GAPS is being applied to a number of important fluid dynamic problems.

The third direction is the application of these new capabilities and methods to specific physical problems. We want to be able understand important physical interactions and mechanisms that have eluded earlier research efforts. In our work, a set of applications has always motivated the development of numerical methods, which in turn have to be applicable to other types of problems. Much of the material presented in this paper is based on a review article and book by Oran and Boris[1,2]

We concluded with presentation of extensive two-dimensional compressible simulations of the wake shed by a wide splitter plate and vortex-acoustic interactions in an idealized ramjet combustor. These calculations have been carried to as long as one million timesteps on the GAPS and the NRL Cray systems. The length of these calculations stems from the need to simulate the shedding of many vortices to study the connections between chaos, as derived from nonlinear system dynamical mappings, and realistic fluid dynamics with an infinite number of degrees of freedom. Such long calculations are also necessary to accumulate meaningful statistics about body drag and drag reduction through control of the vortex shedding process.

R.H. Guirguis and T.R. Young have recently conducted a series of simulations on the GAPS and the Cray designed to investigate techniques for enhancing the mixing in supersonic shear flows with and without bluff centerbodies and to hold the flame in the combustor at supersonic flow speeds. Related computations of detonation cells are also being pursued using RFM on the GAPS [21,22,26].

Simplified reactive flow models have been included in these simulations to release the correct amount of chemical energy at the right location and at the correct rate. Years of research has gone into calibrating these models[1,2,12,21,22]. Our current research also includes detailed flame simulations in two dimensions using the full hydrogen-oxygen reaction mechanism and the "Barely Implicit Correction" [10] to allow calculations for many thousands of acoustic transit times.

ACKNOWLEDGEMENTS

This work was sponsored by the Office of Naval Research and the Naval Research Laboratory. Algorithmic developments reported here were also developed with the sponsorship of DNA, DARPA, NASA and AFOSR.

REFERENCES

1. Oran, E.S. and J.P. Boris, 1981, Detailed Modeling of Combustion System, *Prog. Ener. Combust. Sci.* 7: 1–70.
2. Oran, E.S. and J.P. Boris, 1987, *Numerical Simulation of Reactive Flows*, Elsevier, New York.
3. Young, T.R., and J.P. Boris, 1977, A Numerical Technique for Solving Stiff Ordinary Differential Equations Associated with the Chemical Kinetics of Reactive-Flow Problems, J. Phys. Chem. 81, 2424-2427.
4. Young, T.R., 1979, CHEMEQ: Subroutine for Solving Stiff Ordinary Differential Equations, Naval Research Laboratory Memorandum Report 4091, Naval Research Laboratory, Wash., D.C., 20375, 1979.
5. Boris, J.P., 1971, A Fluid Transport Algorithm that Works, in *Computing as a Language of Physics*, pp. 171–189, International Atomic Energy Agency, Vienna.
6. Boris, J.P., and D.L. Book, 1976, Solution of the Continuity Equation by the Method of Flux-Corrected Transport, *Methods in Computational Physics*, vol. 16, Academic Press, New York, p. 85–129.

7. Boris, J.P., 1976, Flux-Corrected Transport Modules for Generalized Continuity Equation, NRL Memorandum Report 3237, Naval Research Laboratory, Washington, D.C.

8. Zalesak, S., 1979, Fully Multidimensional Flux-Corrected Transport Algorithms for Fluids, *J. Comp. Phys.* 31: 335–362.

9. Löhner, R., K. Morgan, M. Vahdati, J.P. Boris, and D.L. Book, 1986, FEM-FCT: Combining Unstructured Grids with High Resolution, submitted to *J. Comp. Phys.*.

10. Patnaik, G., R.H. Guirguis, J.P. Boris, and E.S. Oran, 1986, A Barely Implicit Correction for Flux-Corrected Transport, to appear in *J. Comp. Phys.*.

11. Yanenko, N.N., 1971, *The Method of Fractional Steps*, Springer-Verlag, New York.

12. Oran, E.S., T.R. Young, J.P. Boris, and A. Cohen, 1982a, Weak and Strong Ignition. I. Numerical Simulations of Shock Tube Experiments, *Combustion and Flame* 48, 135–148.

13. Stull, D.R., and H. Prophet, 1971, JANNAF Thermochemical Tables, 2nd edition, National Standard Reference Data Series, U.S. National Bureau of Standards, No. 37, Gaithersburg, Md.,

14. Gordon, S., and B.J. McBride, 1976, Computer Program for Calculation of Complex Chemical Equilibrium Compositions, Rocket Performance, Incident and Reflected Shocks, and Chapman-Jouguet Detonations, NASA SP-273, National Aeronautics and Space Administration, Wash., D.C.

15. Roshko, A., 1954, *On the Drag and Shedding Frequency of Two-Dimensional Bluff Bodies*, National Advisory Committee for Aeronautics, Technical Note 3169, Washington, July 1954 and On the Wake and Drag of Bluff Bodies, *Journal of the Aeronautical Sciences* 124–132 February 1955.

16. Bearman, P.W., 1965, Investigation of the Flow Behind a Two-Dimensional Model with Blunt Trailing Edge and Fitted With Splitter Plates, *Journal of Fluid Mechanics* 20: Pt 2, 241–255 , also Rockwell, D., 1986, private communications.

17. Simmons, J.E.L., 1977, Similarities Between Two-Dimensional and Axisymmetric Vortex Wakes, *The Aeronautical Quarterly* 26: Pt 1, 15–20, February 1977.
18. Griffin, O.M., 1981, Universal Similarity in the Wakes of Stationary and Vibrating Bluff Structures, *Journal of Fluids Engineering* 103: 52–58.
19. Boris, J.P., 1987, Supercomputing at the U.S. Naval Research Laboratory, *Proceedings of the 1986 SPIE Meeting on Optical Computing*.
20. Boris, J.P., E. Reusser, T.R. Young, and R.Guirguis, 1987, *The Graphical and Array Processing System (GAPS)*, to appear as NRL Memorandum Report, Washington, D.C.
21. Kailasanath, K., E.S. Oran, J.P. Boris, and T.R. Young, 1985, Determination of Detonation Cell Size and the Role of Transverse Waves in Two-Dimensional Detonations, *Comb. Flame* 61, 199–209.
22. Guirguis, R., E.S. Oran, and K. Kailasanath, 1986, Numerical Simulations of the Cellular Structure of Detonations in Liquid Nitromethane — Regularity of the Structure, *Comb. Flame* 61: 199–209.
23. Kailasanath, K., J.H. Gardner, E.S. Oran, and J.P. Boris, 1986, Numerical Simulation of Combustion Oscillations in Compact Ramjets, *Proceedings of the JANNAF Propulsion Meeting*, Chemical Propulsion Information Agency, Johns Hopkins University, Applied Physics Laboratory, 1986.
24. Grinstein, F.F., E.S. Oran and J.P. Boris, 1986, Numerical Simulations of Asymmetric Mixing in Planar Shear Flows, *J. Fluid Mech.* 165: 201–220, 1986.
25. Grinstein, F.F., E.S. Oran, and J.P. Boris, 1986, Direct Numerical Simulation of Axisymmetric Jets, to appear, *AIAA J.*.
26. Guirguis, R., K. Kailasanath, E.S. Oran, J.P. Boris, and T.R. Young, 1987, Mixing and Enhancement in Supersonic Shear Layers, AIAA 25th Aerospace Sciences Meeting, Paper No. AIAA-87-0373, AIAA, NY.
27. Kailasanath, K., J. Gardner, J. Boris and E. Oran, 1986, Numerical Simulations of Acoustic-Vortex Interactions in a Central-Dump Ramjet Combustor, to appear, *J. Prop. Power*.

WSP3 600X200 2/87

ISTEP = 4,173   TIME = 1050.27   DT = 3.500E-07

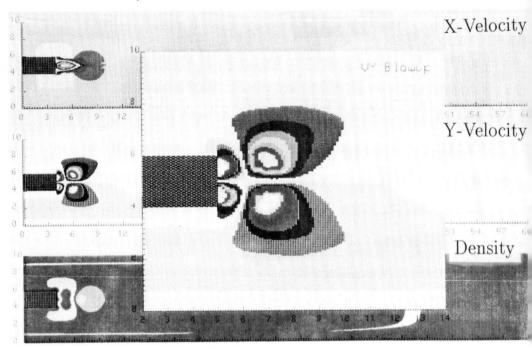

V_X min = 100 V_X max = 11000   NV = 200

V_Y min = 100 V_Y max = 4000   NH = 600

Rho min = 1.100E-03   Rho max = 1.300E-03

Figure 1. Gray scale contour plots of horizontal velocity (upper panel), vertical velocity (middle panel) and density (lower panel) for STP air flowing at a constant velocity of 100 m/s over a wide splitter plate (bluff body) which is 1.6 cm thick shown as the shaded rectangle to the left of the three panels. The inset figure is a three-to-one blowup of the vertical (V_y) velocity in the center panel. This solution is initially a symmetric, nearly steady recirculation pattern.

WSP3 600X200 2/87

ISTEP = 13,800   TIME = 4593.27   DT = 3.500E-07

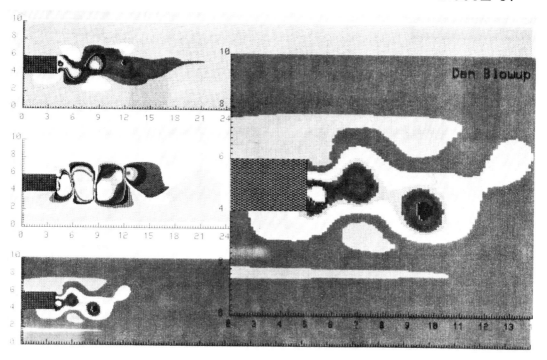

V_X min = 100 V_X max = 11000   NV = 200

V_Y min = 100 V_Y max = 4000   NH = 600

Rho min = 1.100E-03   Rho max = 1.300E-03

Figure 2. Gray scale contour plots of the wide splitter plate calculation at 13800 timesteps after unstable asymmetric vortex shedding has transformed the initial steady recirculation of Figure 1 into the usual Kelvin Vortex Street solution.

WSP3 300X100 4/87

ISTEP = 99,700   TIME = 62637.9   DT = 6.69E-07

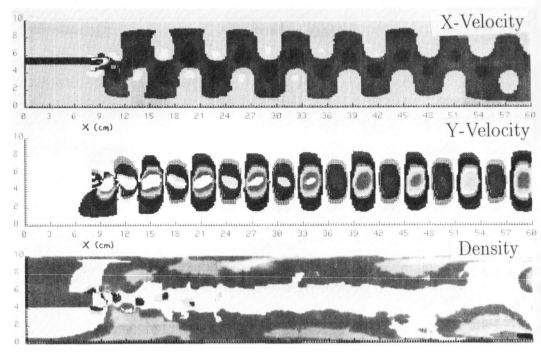

V_X min = 100 V_X max = 11000   NV = 100

V_Y min = 100 V_Y max = 4000   NH = 300

Rho min = 1.240E-03   Rho max = 1.310E-03

Figure 3. Gray scale contour plots of the WSP computations continued on a 300 x 100 finite difference grid. The long sequence of relatively regular alternating sign vortices are shed at a Strouhal frequency obtained using the flow speed, 100 m/s, the WSP width of 1.6 cm, and a Strouhal factor of 0.233. Detailed inspection of the solutions shows an essentially periodic nonlinear limit cycle behavior. The lower panels, showing localized minima in the contours of the density fluctuations, form a very good diagnostic for coherent structures.

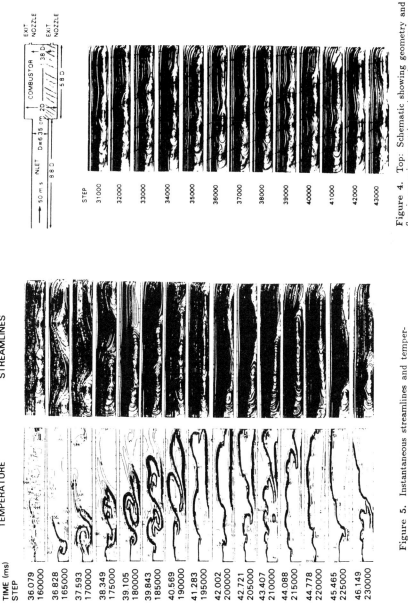

Figure 4. Top: Schematic showing geometry and flow in an axisymmetric ramjet combustor. Bottom: Calculated instantaneous streamlines in the hatched region of the schematic. Lines through the streamlines connect the centers of vortex structures.

Figure 5. Instantaneous streamlines and temperature contours for the continuation of calculation shown in Figure 4 after chemical reactions have been initiated.

WSP3 300X100 5/87

ISTEP = 642,000   TIME = 428701   DT = 6.69E-07

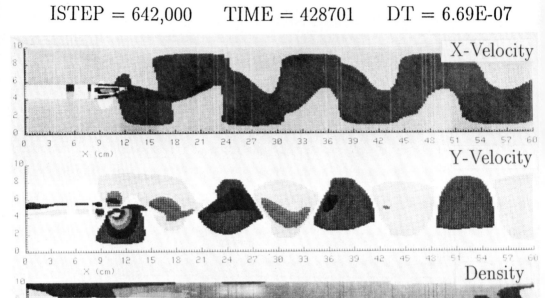

V_X min = 100 V_X max = 11000   NV = 100

V_Y min = 100 V_Y max = 4000   NH = 300

Rho min = 1.250E-03   Rho max = 1.300E-03

Figure 6. Gray scale contour plots of the same wide splitter plate (WSP) problem treated in Figures 1, 2, and 3. A thin barrier 1.6 cm long is included in the flow behind the center of the trailing edge of the WSP at a distance 1.6 cm downstream. This ideal barrier is edge on to the incident flow and so contributes no net drag. It interferes with the alternating asymmetric shedding pattern established in Figure 3, however, and thus reduces drag on the WSP by a factor of three.

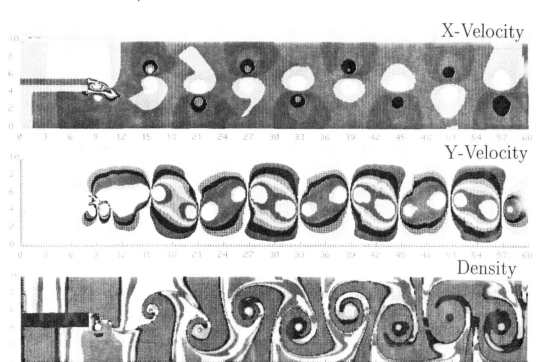

Figure 7. Horizontal and vertical components of velocity and density for the wide splitter plate calculation performed as above but in a bounded channel which reflects acoustic waves back in and allows much larger resonances to set up. The barrier is again absent. The progressing flow pattern appears again as a basically periodic nonlinear limit cycle solution with one dominant frequency.

TURBULENT MULTIPHASE FLOWS

G. M. Faeth
Department of Aerospace Engineering
The University of Michigan
Ann Arbor, MI 48109-2140

ABSTRACT

Recent measurements and predictions concerning turbulent multiphase flows are considered, emphasizing findings of the author and his associates. The properties of both dense sprays (comparable phase volume fractions) and dilute dispersed multiphase flows (dispersed-phase volume fractions less than 1-10 percent) are considered.

Results for dense sprays are limited to the near-injector region of noncombusting, combusting monopropellant, and combusting bipropellant sprays from pressure-atomizing injectors. The results suggest that these flows approximate locally-homogeneous flow properties in the atomization regime, but exhibit much slower mixing rates as the first wind-induced breakup regime is approached, in a manner which is not anticipated by predictions. Flow properties for the atomization regime are strongly influenced by the degree of flow development and turbulence levels at the injector exit. However, existing measurements of the structure of dense sprays are very limited: more work is required to assess the appropriate flow regimes and the effectiveness of locally homogeneous flow analysis for these flows.

Contemporary stochastic analysis of dilute multiphase flows has provided encouraging predictions of the mean structure and mixing properties (turbulent dispersion) of a variety of dilute dispersed flows. However, effects of turbulence modulation (the modification of turbulence properties by the dispersed phase) have been observed, which existing theoretical methods cannot treat effectively, due to inadequate consideration of the response of the dispersed phase to various wave numbers of the turbulence spectrum. Interphase transport phenomena associated with high relative turbulence intensities, virtual mass forces, Basset history forces, and the existence of envelope flames around drops, are also not sufficiently understood to provide reliable predictions of the properties of the dilute portions of combusting sprays.

INTRODUCTION

Liquid and solid fuels are widely used, and condensed phases often appear during combustion processes; therefore, turbulent reacting flows frequently involve multiphase flows. The objective of this paper is to discuss recent studies of multiphase flows in order to identify several research issues. Both noncombusting and combusting multiphase flows are considered: the former to examine the mixing and turbulence properties of multiphase flows without the added complexities of combustion. Attention is limited to turbulent flows, since few practical multiphase flows are laminar.

In the following, research of the author and his associates is emphasized. More complete summaries of the field are provided by several recent review of multiphase flow and combustion processes, and references cited therein, e.g., Bracco [1,2], Clift et al. [3], Crowe [4], Drew [5], Law [6], Sirignano [7], and Faeth [8-10].

Most combusting multiphase flows consist of two phases: either gas/solid, e.g., pulverized coal combustion; or gas/liquid, e.g., spray combustion. For two-phase flows, there are two main regimes of multiphase flow, as follows: (1) dense flows, where the volume fractions of both phases are comparable, e.g., the near-injector region of sprays; and (2) dilute dispersed flows, where volume fractions of the dispersed phase (drops, particles, etc.) are less than 1-10 percent. Dilute flows provide several simplifications for analysis: the topography is known, e.g., drops or particles with a given shape in a continuous gas phase; due to the small volume of the dispersed phase, it is common to integrate over phenomena on the scale of dispersed-phase elements, e.g., the details of the flow within or near dispersed-phase elements is rarely considered; and dispersed-phase elements only interact weakly with each other by influencing the local properties of the continuous phase, e.g., interphase transport rate coefficients are nearly independent of the dispersed-phase volume fraction, and collisions and coalescence of dispersed-phase elements can generally be ignored. In contrast, these simplifications are generally not appropriate for dense flows; where the topography of irregularly-shaped phase elements, the flow properties within each phase, effects of phase volume fractions on interphase transport, and the nature of collision and coalescence phenomena, are all matters of importance.

Aspects of both dense and dilute-dispersed multiphase flows will be considered in the following. Discussion of dense flows is limited to the near-injector region of pressure-atomized sprays injected into a still gaseous environment, since this is a representative configuration with few parameters to characterize the injection process and the flow field. (Pressure atomization involves injection of a pure liquid from a passage, in contrast to twin-fluid injection where both liquid and gas pass through the injector passages.) Experimental and theoretical results are considered for the following flows: nonevaporating large-scale sprays, involving water injection into air; premixed combusting sprays, involving liquid monopropellant injection into near-adiabatic combustion products at high pressures; and nonpremixed combusting sprays, involving fuel sprays burning in room-temperature air at high pressures. Due to their complexity, theory is not highly developed for dense sprays; therefore, analysis will be limited to simple mixing-controlled methods, based on the locally-homogeneous flow (LHF) approximation of multiphase flow theory. The main issues are the flow regimes and the operating conditions where LHF methods provide useful estimates of the structure and mixing properties of dense sprays.

Dilute dispersed flows are more tractable than dense flows, both theoretically and experimentally.

Earlier reviews of dilute-dispersed flows by the author [8-10], have considered the properties of particle-laden jets in gases, nonevaporating and evaporating sprays, and noncondensing and condensing bubbly jets. Simplified stochastic analysis has proven to be useful for these flows, providing a means of estimating some properties of turbulence/dispersed-phase interactions, e.g., aspects of nonlinear interphase transport and effects of turbulent dispersion. These considerations are extended in the following to particle-laden water jets and nonpremixed monodisperse combusting sprays. Particle-laden water jets are of interest since their phase densities are comparable, which provides a simplified simulation of high-pressure sprays, highlighting aspects of turbulence/dispersed-phase interactions that are not observed at other conditions. In an analogous fashion, consideration of monodisperse combusting sprays also highlights aspects of turbulence/drop combustion interactions in a simple manner. The main issues examined for dilute-dispersed flows are the enhancement of interphase transport by turbulence and the modification of continuous-phase turbulence properties by interphase transport – the last often being called turbulence modulation [11].

The paper begins with consideration of dense sprays, treating nonevaporating sprays, premixed combusting sprays, and nonpremixed combusting sprays in turn. Dilute dispersed flows are then discussed, treating noncombusting particle-laden jets and then combusting monodisperse sprays – the last more briefly. The paper concludes with a summary of research issues that merit particular attention at this time.

DENSE SPRAYS

Background

The properties of pressure-atomized sprays are influenced by the breakup regime of the flow. Ranz [12] and Reitz and Bracco [13] identify several breakup regimes for the pressure-atomized sprays; however, the atomization regime is the most important, in practice. For particular injector characteristics, liquid properties, and ambient conditions, the atomization regime is present for all injector Reynolds numbers higher than a fixed threshold value.

A sketch of spray structure for the atomization breakup regime appears in Figure 1. Breakup begins at the injector exit, initiallly forming a drop-containing shear layer around an all-liquid core (which is similar to the potential core of a single-phase jet). Due to difficulties in defining flow properties over any crossection containing the liquid core, dilute spray analysis is generally deferred to positions downstream of the end of the liquid core. Existing measurements of the length of the core are scattered and controversial [14-16], however, the core can extend substantial distances from the injector yielding a sizable dense-spray region, e.g., ca. 100 injector diameters at atmospheric pressure.

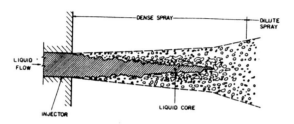

Figure 1. Sketch of the near-injector region of a pressure-atomized spray for atomization breakup.

Use of the locally-homogeneous-flow (LHF) approximation provides an attractive approach to initiate analysis of complex multiphase flows, like dense sprays, since LHF analysis involves minimal empiricism [1,2,8-10]. The LHF approximation implies infinitely-fast interphase transport rates, so that relative velocities between the phases are small in comparison to flow velocities and local thermodynamic equilibrium is maintained. The LHF approximation is well known in the multiphase flow literature [17], and was first applied to spray combustion more than thirty years ago [18]. However, the potential accuracy of the approach is controversial and criteria for its use have not been established. For example, Bracco and coworkers [1,2,19,20] suggest good results with LHF analysis for high pressure sprays, while others find that the LHF approach generally overestimates the rate of development of multiphase flows, even at high pressures [8-10,21-23].

Results from this laboratory concerning the structure of the dense-spray region for pressure atomized sprays, and the accuracy of LHF analysis for predicting flow properties, are considered in the following. Nonevaporating, premixed combusting monopropellant, and nonpremixed combusting (fuel burning in air) sprays are considered. The present description is brief; more details can be found in Ruff et al. [24,25], Lee et al. [26], and Mao et al. [23].

Nonevaporating Sprays

Experimental methods. Issues being considered relate to the dynamics of spray mixing, which are not thought to depend strongly on the injector diameter; therefore, large-scale (9.5 and 19.1 mm injector diameters) jets were studied to provide reasonable spatial resolution. Water was used as the test liquid, injecting vertically downward in still air. The injector consisted of a honeycomb flow straightener followed by a rounded contraction section with either injection at the end of the contraction section (slug flow) or at the end of a constant-area passage having a length of 41 injector diameters (fully-developed flow).

The flows were visualized using flash and laser-light sheet photography, each providing roughly 1 µs exposure times. Distributions of mean liquid volume fractions were obtained by gamma-ray absorption measurements which were deconvoluted to yield radial distributions of mean liquid volume fractions. Laser Doppler anemometry was used to measure jet exit conditions by submerging the injector exit in a water bath.

Three test conditions were used for each injector, yielding operation in the first wind-induced, second wind-induced and atomization breakup regimes. In all these regimes breakup is due to instabilities caused by the relative motion of the liquid and gas: breakup occurs far from the injector, yielding drop sizes comparable to the injector diameter, for the first wind-induced breakup regime; breakup occurs at some distance from the injector, yielding drops generally an order of magnitude smaller than the injector diameter, for the second wind-induced breakup regime; and breakup begins right at the injector exit, and has other properties similar to the second wind-induced breakup regime, for the atomization regime [12,13]. The Reynolds numbers of the experiments were quite high, in the range $5.2 \times 10^4 - 1.1 \times 10^6$.

Theoretical methods. Analysis of the flows followed methods used in the past with the LHF approximation [10]. Major assumptions are as follows: steady (in the mean) round boundary-layer flow with no swirl; negligible potential and kinetic energy; buoyancy/turbulence interactions ignored; equal exchange coefficients of all species and heat; and the LHF approximation, e.g., local kinematic and thermodynamic equilibrium between the phases.

These assumptions satisfy the approximations of the conserved-scalar formalism, which was used in conjunction with a Favre-averaged k-ε-g turbulence model. This approach requires state relationships, providing scalar properties as a function of mixture fraction. State relationships were found allowing for liquid vaporization to maintain local thermodynamic equilibrium, since the ambient air was not saturated; however, effects of vaporization were small. A clipped-Gaussian Favre-averaged probability density function of mixture fraction was used to compute time-averaged liquid volume fractions for comparison with the measurements. All empirical constants in the turbulence model were established earlier using measurements from constant and variable density, single-phase jets [27], and were not changed for predictions described in this paper. Values used, however, are not very different from an early proposal by Lockwood and Naguib [28].

Results and discussion. For the first and second wind-induced breakup regimes, the liquid surface can be observed near the injector exit; however, the drop-containing shear layer obscures the surface for atomization conditions. When there is fully-developed flow in the first wind-induced breakup regime, the liquid surface exhibits fine-grained roughness near the injector exit; which becomes smoother, with large-scale irregularities appearing, far from the injector. This appearance suggests a shift in the turbulence spectra near the surface, toward lower wave numbers, with increasing distance from the injector. This behavior is caused by reduced turbulence production in the liquid, once the retarding action of the passage wall is lost, similar to the decay of grid-generated turbulence. For slug flow, the region of fine-grained roughness is not observed, suggesting that the surface roughness is largely caused by liquid-phase turbulence for present test conditions.

Similar behavior is seen near the injector in the second- wind-induced breakup regime, before the surface is obscured by breakup and the formation of the drop-containing shear layer. For atomization breakup, the drop-containing shear layer has a wispy appearance, with drop-free regions penetrating progressively farther toward the axis with increasing distance from the injector, suggesting an all-liquid core extending some distance from the injector – similar to the other breakup regimes. The wispy appearance of the shear layer for atomization breakup, similar to tracer particles in a turbulent jet, is qualitatively supportive of LHF flow.

Predicted and measured time-averaged mean liquid volume fractions along the axis, are plotted in Figures 2 and 3, for fully-developed and slug flow, respectively. The effect of the breakup regime, and

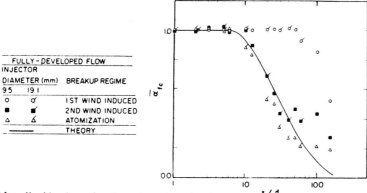

Figure 2. Mean liquid volume fractions along the axis of pressure-atomized water sprays in still air at atmospheric pressure (fully-developed injector flow). From Ruff, et al. [25].

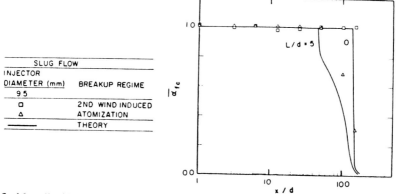

Figure 3. Mean liquid volume fractions along the axis of pressure-atomized water sprays in still air at atmospheric pressure (near-slug injector flow). From Ruff, et al. [25].

the degree of flow development at the injector exit, on the mixing properties of the flow is striking. For fully-developed flow, mixing is most rapid for the atomization breakup regime, slightly slower for the second wind-induced breakup regime, and much slower for the first wind-induced breakup regime. The same trends are observed for slug flow but mixing rates in all breakup regimes are vastly reduced, e.g., for the atomization breakup regime, comparable levels of mean liquid volume fractions occur roughly an order-of-magnitude farther from the injector for slug flow than for fully-developed flow.

Predictions using the LHF approximation exhibit encouraging agreement with measurements for the atomization regime and for fully -developed flow. For slug flow, computations indicate a significant influence of boundary layer buildup on the surfaces of the contraction section of the injector, cf., computations for passage length-to-diameter ratios of 0 and 5 in Figure 3. The thin wall layer for the present slug flow measurements could not be resolved with sufficient accuracy to adequately test predictions; however, the measurements for atomization conditions are bounded by the predictions at reasonable limits of the properties of the wall layer, which is encouraging.

Radial distributions of time-averaged mean liquid volume fractions are illustrated in Figure 4, for

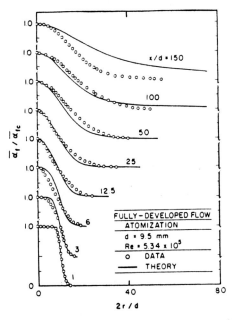

Figure 4. Mean liquid volume fraction distributions for pressure-atomized water sprays in still air at atmospheric pressure (fully-developed injector flow and atomization breakup). From Ruff, et al. [25].

atomization breakup and fully-developed flow at the injector exit. The comparison between predictions and measurements is good, well within experimental uncertainties. Results for the second wind-induced breakup regime were similar.

For the high Reynolds numbers of present experiments, LHF predictions show little effect of Reynolds number on mixing rates. Thus, the analysis provides no warning of the vastly reduced mixing rates in the first wind-induced breakup regime, or the transition to this state as breakup distances from the injector increase with reduced flow rates in the second wind-induced breakup regime. Present results suggest that the flow should be in the atomization regime for the LHF approximation to be effective, but a larger data base must be considered before this criterion can be firmly established.

Conclusions. Major conclusions, based on the present observations of nonevaporating sprays, are as follows:

1. The length of the all-liquid core, and the mixing properties of the sprays, are strongly influenced by the breakup regime and the degree of flow development at the injector exit; it is probable that inadequate specification of these properties (particularly the degree of flow development) is largely responsible for earlier controversy concerning liquid core lengths and the effectiveness of the LHF approximation [1,2,8-10].

2. Predictions using the LHF approximation yielded encouraging agreement with measurements for the atomization breakup regime, including effects of initial flow development. However, LHF analysis provides no warning of the vastly-reduced mixing rates as the first wind-induced breakup regime is approached – when its predictions are clearly deficient.

Present conclusions are based on large-scale sprays, which have appreciably lower rates of flow deceleration than practical injectors, favoring the LHF approximation. Drop-size distributions for present flows are not known and would vary for different fluids even for the same range of test conditions, potentially modifying present conclusions as well. Resolution of these issues will require measurements of drop properties in the shear layer as well as consideration of other injector sizes, fluids and ambient conditions.

Combusting Monopropellant Sprays

Experimental methods. Liquid monopropellant spray combustion is the multiphase counterpart of premixed gas combustion; thus, this flow is of some fundamental importance. Nearly adiabatic combustion of these sprays in their combustion products was considered, for pressure-atomized injection in the atomization breakup regime. A hydroxyl ammonium nitrate (HAN-based monopropellant was considered, consisting of a mixture of HAN, tri-ethanol ammonium nitrate (TEAN), and water. The fuel/oxidant mixture ratio of the propellant composition was stoichiometric, yielding a mixture of carbon dioxide (25%), nitrogen (21%) and water vapor (54%), at roughly 2000 K, after adiabatic combustion.

Details of the experimental apparatus are provided by Birk and Reeves [29]. Spray experiments were carried out in a windowed chamber (57 mm diameter and 406 mm long). The axis of the chamber was vertical, with the spray injected vertically upward along the axis from round 1 and 2 mm diameter passages. The interior of the injector passages had a complex geometry; however, the exit portions were constant-area passages having lengths of 1-2 injector diameters.

The propellant was injected into a hot precombusted gas mixture, to simulate near-adiabatic combustion, at pressures of 6-8 MPa. This yielded liquid/gas density ratios corresponding to adiabatic combustion of the propellant in its combustion products at pressures of 10-13 MPa. Injector Reynolds numbers were in the range $1\text{-}2 \times 10^{-4}$. The HAN-based propellants burn very cleanly, producing no particulates; therefore, the liquid-containing portion of the flow was readily visualized as motion picture shadowgraphs. The shadowgraphs were backlighted by flash sources synchronized with the camera. Flash durations were less than 2 µs, which effectively stopped the motion of the spray on the films.

The boundaries of the spray were very well defined on the films, suggesting thin reaction zones associated with the liquid surfaces of drops and irregular liquid elements near the all-liquid core. Measurements of strand burning rates by McBratney [30-31] similarly imply thin reaction zones, ca. 1 µm thick, near the liquid surface for the spray test conditions. Furthermore, phenomena associated with the thermodynamic critical point require much higher pressures (ca. 100 MPa) than present test conditions; thus the liquid surface was well defined. In view of this behavior, shadowgraphs could be analyzed to yield time-averaged mean and fluctuating liquid volume fractions, by assigning dark zones to unburned liquid reactant and light zones to the combustion products. Since the measurements were obtained from line-of-sight projections, they are biased downstream and radially outward from correct point measurements of these properties. The effect of this biasing was not quantified, but is probably not large in comparison to other experimental uncertainties – as will be seen subsequently.

Theoretical methods. Major assumptions of the analysis involved the use of LHF approximations and other assumptions of the noncombusting spray analysis. The details of the analysis appear in Lee et al. [26].

Effects of premixed reaction were treated using the approach proposed by Bray [32,33], at the thin-flame limit. In this case, flow properties are found by solving Favre-averaged governing equations for conservation of mass, momentum, reaction progress variable, turbulence kinetic energy, and the rate of dissipation of turbulence kinetic energy. The governing equation for mean reaction progress variable is similar to that of mixture fraction in the conserved-scalar formulation, except for a source term due to reaction which has the following form at the thin-flame limit:

$$C_r \bar{\rho} \tilde{c}(1-\tilde{c}) \varepsilon / k \qquad (1)$$

C_r is an empirical constant whose value is roughly equal to 2; it was taken to be 1.87 (the same value as the sink term constant in the ε equation) for present computations. All other empirical constants in the turbulence model were unchanged from the values used earlier [27]. The reaction term, Eq. (1), is independent of the chemistry of the combustion process under the thin-flame approximation – yielding a mixing controlled limit [32,33].

Under the thin-flame approximation, the process only has two scalar states: unburned reactant liquid, and adiabatic combustion product gases at thermodynamic equilibrium. Gas properties were computed using the Gordon and McBride [34] computer program. The thin-flame approximation only admits a double-delta probability density function for the reaction progress variable, yielding all scalar properties from a knowledge of $\tilde{c}$ alone [32,33]. In particular, the equation for time-averaged mean liquid volume fraction is as follows:

$$\bar{\alpha}_f = \bar{\rho}(1-\tilde{c}) / \rho_0 \qquad (2)$$

where ρ_0 is the density of the unreacted liquid.

Results and discussion. Predictions and measurements of time-averaged liquid volume fractions along the axis are plotted as a function of x/d in Figure 5. Measurements for both injector

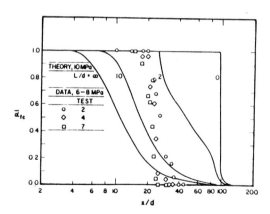

Figure 5. Predicted and measured mean liquid volume fractions along the axis of pressure-atomized combusting monopropellant sprays. From Lee, et al. [26].

diameters are very similar when plotted in this manner. By itself, this suggests a turbulent mixing-controlled process for the conditions of the experiments. Similarly, predictions are also insensitive to injector diameters and Reynolds numbers over the range of the tests; therefore, only a single prediction line is shown for all injector conditions. Computations at various pressures exhibited little influence of pressure on predictions of flow properties; thus predictions at 10 MPa (matching the density ratio of the test flow) serve equally well for pressures of 6-8 MPa.

In contrast to other variables, the degree of flow development at the injector exit (represented by the injector length-to-diameter ratio, L/d, and computed for flow development in a constant-area passage [26] has a strong influence on the predictions illustrated in Figure 5 – similar to earlier findings for nonevaporating sprays illustrated in Figures 2 and 3. Due to uncertainties concerning experimental injector exit conditions, this large effect precludes definitive assessment of predictions, including the effectiveness of the LHF approximation, or any attempt to optimize C_r. Nevertheless, estimates of the properties of the test injectors suggest an effective L/d in the range 2-10; thus, it is encouraging that predictions at these limits generally bound the measurements. However, the slopes of predictions and measurements are somewhat different in the region where $\bar{\alpha}_{fc}$ rapidly decreases; this may be due to line-of-sight biasing and difficulties in distinguishing low concentrations of gas or liquid on the photographs (a form of gradient-truncation). Further study is required to quantify these effects.

Some additional predictions for pressure atomized combusting monopropellant sprays with fully-developed injector exit conditions, are illustrated in Figure 6. The parameters $\tilde{c}$ and $\bar{\alpha}_{fc}$ along the axis of

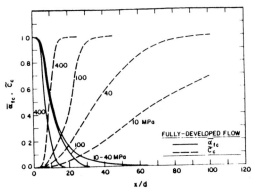

Figure 6. Predicted mean scalar properties along the axis of pressure-atomized combusting monopropellant sprays (fully-developed injector flow). From Lee, et al. [26].

the sprays are plotted as a function of normalized distance from the injector for various pressures. Pressure has a strong effect on the variation of $\tilde{c}_c$, which is a measure of the mass of propellant reacted. Reaction nears completion much closer to the injector at high pressures. This behavior is caused by higher jet entrainment rates at higher pressures, which increases reaction rates at the mixing-controlled limit of the present analysis. In contrast, $\bar{\alpha}_{fc}$ is essentially independent of pressure in the range 10-40 MPa, due to the compensation of increased entrainment rates by reduced phase-density ratios, as the pressure is increased.

Conclusions. Major conclusions based on present observations of premixed combusting sprays are as follows:

1. Use of the LHF and thin-flame approximations yielded encouraging agreement with measurements; however, uncertainties concerning injector exit conditions for the experiments precluded definitive assessment of predictions.

2. These flows exhibited strong sensitivity to the degree of flow development and turbulence levels at the injector exit, similar to findings for nonevaporating sprays at much higher Reynolds numbers; fully developed flows with enhanced turbulence intensities require significantly-smaller combustion volumes than slug flows with low initial turbulence intensities.

In a sense, present findings, using the LHF approximation, suggest that premixed spray flames, are analogous to premixed gas flames having boundary layer flow properties, reported by Bray [32,33]. However, improved measurements, over a wider range of test conditions, are needed to assess the limitations of this methodology; and to gain a better understanding of the structure and mixing properties of premixed spray flames.

Combusting Bipropellant Sprays

Experimental methods. The combustion properties of nonpremixed spray flames, and the performance of LHF analysis for the process, was considered by Mao et al. [23]. Test conditions involved combustion of n-pentane in room temperature air at pressures of 3-9 MPa. Pressure-atomized injection was used with a relatively-short injector passage, L/d ca. 4, and an injector exit diameter of 200 µm. The spray was ignited with a hot wire.

Motion-picture shadowgraphs were obtained for the combusting sprays. The liquid-containing region could be identified reasonably well at 3 MPa; however, this became increasingly difficult at higher pressures due to the large gas-phase density gradients present for nonadiabatic combustion. Another difficulty at high pressures involves the approach of the liquid surface to the thermodynamic critical point, which reduces density differences needed to distinguish the phases. The measured spray boundaries were taken as the average position of the edge of the spray determined from the shadowgraphs.

Theoretical methods. The analysis was similar to the approach described for noncombusting sprays, except that these earlier calculations used a Reynolds-averaged formulation [23]. This introduces differences in the computation of mixture-density and time-averaged scalar properties, from the Favre-averaged formulation; however, the net effect of the changes on scalar property predictions is not large [27].

The conserved-scalar formalism was used in conjunction with the LHF approximation. State relationships were computed assuming local thermodynamic equilibrium up to fuel-equivalence ratio of roughly 4, with adiabatic nonreactive mixing of this mixture and the injected liquid used to find properties at larger fuel-equivalence ratios. A more up-to-date approach would involve use of the laminar flamelet concept of Bilger [35] and Lieu et al. [36], where measured or predicted properties from analogous laminar flames are used to find state relationships [10]. Such corrections, however, do not have a large effect on predictions of the extent of the liquid-containing region, which generally involves mixture fractions which are much larger than the near-stoichiometric conditions where non-equilibrium effects are important.

For high-pressure conditions, liquid surface properties are influenced by real-gas effects associated with the liquid approaching thermodynamic-critical conditions. These effects were treated using the Redlich-Kwong equation of state, allowing for the solubility of combustion-product gases and nitrogen in the liquid.

A typical state relationship for the combusting spray, for a pressure of 6 MPa, is illustrated in Figure 7. Properties at low mixture fractions are qualitatively similar to gaseous diffusion flames, with

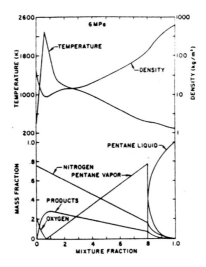

Figure 7. State relationships for combusting pressure-atomized n-pentane sprays burning in air at 6 MPa. From Mao, et al. [23].

temperatures reaching a maximum and densities a minimum near stoichiometric conditions. Properties are modified at high mixture fractions, however, due to the presence of liquid. As mixing proceeds from the pure liquid state, liquid vaporizes, causing the mass fraction of liquid to decrease and the mass fraction of fuel vapor to increase. The liquid is finally fully vaporized at a mixture fraction near 0.8, for the conditions of Figure 7; the mass fraction of fuel vapor reaches a maximum at the same mixture fraction. Effects of gas solubility in the liquid-containing region are evident from the nonlinear behavior of the mass fraction of gases for the portion of the state relationship where liquid is present.

Results and discussion. Predicted and measured spray boundaries at pressures of 3, 6 and 9 MPa are illustrated in Figure 8. Due to uncertainties in initial conditions, the difficulties of distinguishing liquid and gas at high pressures, and effects of biasing due to line-of-sight measurements; any absolute agreement between predictions and measurements is not very meaningful. However, it is encouraging that the LHF approach provides qualitative agreement concerning the effects of pressure on the boundaries and a general indication of the extent of the liquid-containing region.

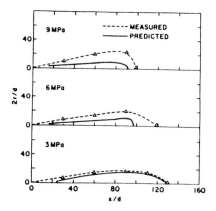

Figure 8. Spray boundaries for pressure-atomized n-pentane sprays burning in high-pressure air. From Mao, et al. [23].

Potential effects of finite interphase transport rates for these test conditions were also examined, by completing drop trajectory computations using the estimates of the LHF calculations to describe the structure of the flow [23]. The results indicate that only drops having initial diameters less than 10 μm had velocities and temperatures nearly equal to the flow and disappeared near the spray boundary estimated using the LHF approximation – a conclusion that is not very different from atmospheric pressure conditions [22]. These rather stringent requirements on drop size are due to the small diameter of the injector passage, which causes very large deceleration rates of the flow. Increasing pressures did not directly increase the allowable drop size range acceptable for use of the LHF approximation, since particle drag is not strongly influenced by gas density (e.g., drag is independent of gas density in the Stokes regime) and the flow itself develops more rapidly due to higher entrainment rates at high pressures (a trend already seen in Figure 6). However, higher pressures do imply less stability for large drops due to lower surface tension as the thermodynamic-critical point of the liquid surface is approached; thus, the potential effectiveness of the LHF approximation for high-pressure nonpremixed combusting sprays hinges on effects of secondary breakup in many instances, a phenomena that is not well understood at present [9,10].

Conclusions. Major conclusions, based on these observations of nonpremixed combusting sprays, are as follows:

1. Use of the LHF approximation and the conserved-scalar formalism yielded predictions that were in qualitative agreement with measurements in the atomization regime; however, difficulties in distinguishing gas and liquid at high pressures, effects of line-of-sight biasing, and uncertainties concerning the extent of flow development at the injector, preclude any definitive assessment of predictions.

2. At high pressures, sprays develop more rapidly due to increased entrainment rates, increasing deceleration rates of the flow, while the drag properties of drops are not a strong function of gas pressure; therefore, potential improvements of the LHF approximation at high pressures largely rests on improved atomization properties of the spray and effects of secondary breakup phenomena due to reduced surface tension as the drop liquid nears the thermodynamic critical point.

These conclusions are based on relatively early measurements where the importance of flow properties at injector exit were unrecognized and, thus, not well defined. Furthermore, measurements provided only limited information concerning the structure of the flow. Additional measurements of nonpremixed combusting sprays at high pressures are clearly needed to gain a better understanding of the process and more adequate assessment of the value of the LHF approximation for these conditions.

DILUTE DISPERSED FLOWS

Background

Dilute dispersed flows are more tractable for measurements and analysis than dense flows, and have attracted more attention [4,6-10]. A series of studies of dilute dispersed flows have been completed in this laboratory, including: monodispersed particle-laden gas jets in gases [37-40], monodisperse particle-laden liquid jets in liquids [41], noncondensing and condensing initially monodisperse bubbly jets in liquids [42,43], nonevaporating and evaporating polydisperse sprays [44,45], and initially monodisperse combusting sprays [46]. The findings of these studies will be reviewed in the following, emphasizing recent results for particle-laden water jets in water and combusting sprays, both of which raise interesting issues concerning turbulence/dispersed-phase and turbulence/combustion interactions.

In all these studies [37-46], three types of analysis were considered in order to help interpret the measurements and to gain insight concerning the analysis of dilute-dispersed flows, as follows: (1) locally-homogeneous flow (LHF), described earlier; (2) deterministic separated flow (DSF), where finite interphase transport rates are considered, but turbulence/dispersed-phase interactions are ignored; and (3) stochastic separated flow (SSF), where both finite interphase transport rates and turbulence/dispersed-phase interactions are considered, using random sampling for turbulence properties in conjunction with random-walk computations for particle trajectories and life histories. Exact numerical simulation of turbulence properties of the dilute-dispersed flows was not feasible, due to the large range of length scales in the test flows. Thus, continuous-phase turbulence properties were estimated using the Favre-averaged turbulence model discussed in conjunction with dense-spray phenomena, but including source terms to account for interphase transport. This approach involves averaging over phenomena on the scale of dispersed-phase elements, e.g., the use of semi-empirical interphase transport coefficients for drag, mass transfer, and heat transfer.

Comparison of predictions and measurements for dilute dispersed flows generally indicates that finite interphase transport rates and turbulence/dispersed-phase interactions are important for most dilute dispersed flows [37-46]. Thus, LHF and DSF analysis, the first ignoring finite interphase transport and the second ignoring turbulence/dispersed-phase interactions, provided poorer estimates of the measurements than the SSF approach [10]. One exception was that the LHF approach was somewhat more successful for treating near-injector properties where flow velocities are high in comparison to potential relative velocities between the phases – a property that has been exploited to treat dense-spray processes, as discussed earlier [39,41-43]. Three types of turbulence/dispersed-phase interactions have been identified, as follows: (1) turbulent dispersion of the dispersed phase; (2) transient and nonlinear interphase transport, i.e., the enhancement of interphase transport by turbulence; and (3) turbulence modulation, i.e., the direct modification of continuous-phase turbulence properties by interphase transport. The simplified SSF approach considered thus far has been most successful in treating turbulent dispersion and some aspects of nonlinear interphase transport (particularly the biasing of drag toward values higher than estimates based on mean properties), yielding encouraging agreement with measurements for conditions as disparate as particles in gases and bubbles in liquids [10].

Current SSF methods have been far less successful in treating effects of the enhancement of interphase transport rates due to high relative turbulence intensities, effects of turbulence modulation, and effects of combustion [10,41,46]. The problems are due to oversimplified theoretical methods and the difficulties of measuring these phenomena. In the following, evidence for these deficiencies is examined, based on results for particle-laden water jets in water [41] and combusting sprays [46].

Particle-Laden Water Jets

Experimental methods. The test flows consisted of a water jet containing glass particles (injector diameter 5.08 mm, passage length 350 mm, particle Sauter mean diameter 505 µm, and a particle density 2450 kg/m^3) injected vertically-downward in still water. The flow was observed within a windowed tank (410 × 530 × 910 mm high).

Measurements consisted of mean and fluctuating phase velocities and mean particle number fluxes. Mean and fluctuating phase velocities were measured with a phase-discriminating laser-Doppler anemometer (LDA), along the lines of the approach used by Modarress et al. [47]. The LDA was conventional, involving a dual-beam forward-scatter arrangement with frequency shifting to control directional bias and ambiguity. Amplitude discrimination provided particle velocities with no difficulty, since the amplitude of particle signals was much greater than signals from the natural seeding particles in the water. However, amplitude discrimination alone is not adequate for continuous-phase properties, since low-amplitude signals from grazing collisions of particles with the LDA measuring volume can be misinterpreted as coming from the continuous phase. This was handled by observing light scattered from a third laser beam (at 632.8 nm as opposed to the 514.5 nm line used for the LDA) whose measuring volume surrounded the LDA measuring volume. A signal pulse from the third beam indicated the presence of a particle near the LDA measuring volume, allowing measurements at such conditions to be deleted from the data used to find continuous-phase velocities.

Mie scattering was used to measure particle number fluxes in the streamwise direction. This involved observing a small section of a laser light sheet, with a pulse counter recording the passage of particles through the sheet. The area of observation was calibrated by collecting particles in a uniform flow.

Three test conditions were considered, a pure liquid jet (as a baseline) and two dilute particle-laden jets, having initial particle volume fractions of 2.4 and 4.8%. Jet Reynolds numbers were roughly 8500, yielding reasonably turbulent flow. Initial conditions of the flow, needed for separated-flow computations, were measured at x/d = 8, since high particle densities nearer the injector tended to block LDA signals.

Theoretical methods. Theoretical methods were similar to past methods used in this laboratory and will only be described briefly [37-46]. The continuous-phase analysis was similar to the LHF approach, discussed earlier, except for the appearance of source terms due to particles with the separated-flow methods.

The separated-flow methods involved use of the particle-source-in-cell approach described by Crowe [4]. The particle phase was treated by solving Lagrangian equations for the trajectories of a sample of individual particles (n groups defined by initial position, velocity, direction, and sample). These trajectories were based on the mean properties of the continuous phase for DSF calculations. SSF calculations treated the motion of the particles as they move through the flow and encounter a random distribution of turbulent eddies. Results of these computations were averaged over all particle groups to provide mean (for both DSF and SSF calculations) and fluctuating (SSF approach only) properties of the particles, as well as the particle source terms needed to solve the continuous-phase governing equations.

Several variations of SSF analysis were considered. The baseline version was the same as Shuen et al. [37-39], which is a modification and extension of an approach proposed by Gosman and Ioannides [48]. The stochastic simulation of the moments and Lagrangian correlations of continuous-phase turbulence properties are treated quite simply. Particles are assumed to interact with a succession of eddies, each having uniform properties, with eddy properties changing randomly from one eddy to the next. At the start of particle/eddy interaction, the velocity components of the eddy are found by making random selections from the velocity PDF's at the position of the particle. A particle is assumed to interact with an eddy for a time which is the minimum of either the eddy lifetime or the time required for the particle to traverse the eddy. Characteristic eddy sizes, L_e, and lifetimes, t_e, are estimated from the following expressions [37-39]:

$$L_e = C_\mu^{3/4} k^{3/2} / \varepsilon; \qquad t_e = L_e / (2k/3)^{1/2} \tag{3}$$

Particles and eddies are assumed to interact as long as the time of interaction and the relative displacement of the particle and eddy are both less than t_e and L_e. Following Gosman and Ioannides [48], these prescriptions were calibrated using results for the turbulent dispersion of particles in homogeneous turbulent flows [37]. In flows considered thus far, predictions of turbulent dispersion have not been very sensitive to the selection of L_e and t_e, within a range of 2:1 [37-46].

During baseline SSF analysis, the velocity PDF was taken to be Gaussian and isotropic, with mean values, $\bar{u}$, $\bar{v}$, $\bar{w}$ and standard deviations in each direction of $(2k/3)^{1/2}$. In jet flows, however, streamwise velocity fluctuations are larger than the other components, causing streamwise particle velocity fluctuations to be underestimated using the baseline approach. Thus, an extended version was considered, allowing for effects of continuous-phase anisotopy, based on measured levels of anisotropy in the flow.

When the densities of the phases are comparable, all terms in the BBO equation of particle motion must be considered: particle inertia, drag, virtual mass, the Basset history force, and the force of gravity. The formulation of Odar and Hamilton [49] as reviewed by Clift et al. [3], was used to treat these effects, with coefficients for virtual mass and the Basset history forces allowing for the particle acceleration. For baseline calculations, the standard drag correlation for solid spheres was used – calibrated by measuring terminal velocities of the test particles in a still liquid. Computations revealed, however, that turbulent fluctuations of the continuous phase were often relatively large in comparison to relative velocities of the particles; therefore, the extended version of the SSF approach accounted for this effect following Clift et al. [3] and Lopes and Dukler [50]. These empirical expressions were developed for unknown length-scale distributions of the turbulence in comparison to particle dimensions, and for relative turbulence intensity levels less than 50%, while values up to 150% were seen during present calculations; thus, the approach used is highly speculative for present test conditions.

All the separated-flow methods to be considered allow for interphase momentum transfer in the mean momentum equation. Comparable source terms in the k and ε equations, the turbulence modulation terms, were ignored for the DSF and baseline SSF methods. However, two approaches were examined to treat effects of turbulence modulation, denoted SSF-KMOD and SSF-EXT, similar to the limiting cases recently considered by Reitz and Diwakar [51]. Both versions adopt the following particle source term in the k equation:

$$\bar{S}_{pk} = \overline{u\,S_{pu}} - \bar{u}\,\bar{S}_{pu} \tag{4}$$

where S_{pu} is the momentum source term due to particles. This term appears directly in the derivation of the k equation and the stochastic simulations allow $\bar{S}_{pu}$ and $\bar{u}\,\bar{S}_{pu}$ to be computed without further approximation; therefore, $\bar{S}_{pk}$ is exact within the other limitations of single-scale analyses of turbulence represented by the k-ε turbulence model. A similar term in the ε equation is simply ignored for the SSF-KMOD version. For the SSF-EXT approach, $\bar{S}_{p\varepsilon}$ is modeled by assuming that it is proportional to the source term in the k equation, similar to treatment of this term by Reitz and Diwakar [51] for multiphase flows, and conventional turbulence models for single-phase flows, e.g., Lockwood and Naguib [28]. The term becomes

$$\bar{S}_{p\varepsilon} = C_{\varepsilon 3}\,\bar{S}_{pk}\varepsilon/k \tag{5}$$

The value of $C_{\varepsilon 3}$ was chosen by considering equilibrium requirements in a homogeneous stationary flow where turbulence is only generated by particle motion: this implies $C_{\varepsilon 3} = C_{\varepsilon 2}$.

Results and discussion. Complete theoretical and experimental results are reported by Parthasarathy and Faeth [41]. Comparison of predictions and measurements for the single-phase water jet was reasonably good, which is typical of past experience using the turbulence model for a variety of free jets [10]. Continuous-phase properties of the particle-laden jets were very similar to the single-phase jet (exceptions are noted later) since the flows were very dilute. Predictions of all the models, except the SSF-KMOD version, were also in good agreement with measurements.

The SSF-KMOD predictions substantially underestimated the rate of development of the flow, similar to the findings of Reitz and Diwakar [51] for a similar treatment of turbulence modulation limited to a source term in the governing equation for k. This term acts like a sink, reducing the turbulent kinetic energy of the flow, and thus, the rate of mixing. The problem is an effect of scale, where particles influence the high wave-number portions of the turbulence spectrum for present flows, rather than the large-scale motion that is primarily responsible for turbulent mixing. The SSF-EXT version avoids the problem, by empirically correcting the ε equation so that gross mixing levels are unchanged The baseline SSF approach achieves the same effect by neglecting turbulence modulation entirely, which is tantamount to assuming that effects of turbulence modulation are limited to small scales which don't influence the low wave-number range of the turbulence spectrum that is primarily responsible for mixing. Clearly, consideration of multiscale effects and the turbulence spectrum are required for a rational treatment of turbulence modulation. Initial attempts along these lines by Al Taweel and Landau [11], Elghobashi and Abou-Arab [52], merit further development.

Predicted and measured mean streamwise particle velocities for the two particle-laden jets are illustrated in Figure 9. Effects of finite-rate interphase transport rates and nonlinear interphase transport

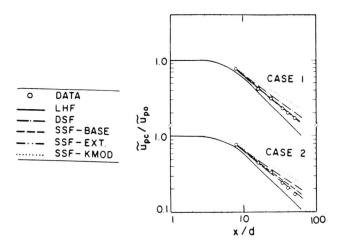

Figure 9. Mean particle velocities along the axis of particle-laden water jets. From Parthasarathy and Faeth [41].

can be seen by comparing predictions and measurements: LHF predictions underestimate particle velocities since finite interphase transport rates are ignored; DSF predictions overestimate particle velocities since particle drag is underestimated, by ignoring turbulence/particle interactions due to use of mean properties in a nonlinear drag expression; and the SSF-KMOD version yields poor results due to errors in continuous-phase properties discussed earlier. Only the SSF-BASE (the baseline version) and SSF-EXT methods yield good agreement with measurements of both continuous and dispersed-phase properties. However, the extensions of the SSF-EXT version do not have a great influence on predictions of mean particle properties.

Predictions and measurements of particle number fluxes along the axis are illustrated in Figure 10. This variable is a sensitive indicator of capabilities for predicting turbulent dispersion of particles; therefore, it is encouraging that the SSF-BASE and SSF-EXT methods yield reasonably-good predictions – extending to flows having comparable phase densities the reasonably successful use of this approach for large and small phase-density ratios [37-40,42,43]. The LHF method overestimates particle mixing rates, since neglecting particle inertia causes the particles to diffuse like fluid particles, which is too fast for present conditions. This is most problematical when relative velocities become large in comparison to liquid velocities, due to effects of gravity, far from the injector. The DSF approach underestimates particle mixing rates substantially, since turbulent dispersion is ignored. The SSF-KMOD predictions are even worse, due to poor predictions of liquid-phase properties.

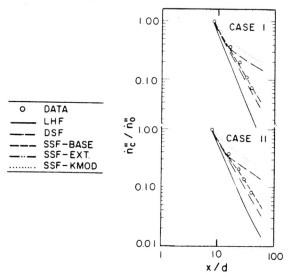

Figure 10. Mean particle number fluxes along the axis of particle laden water jets. From Parthasarathy and Faeth [41].

Predictions and measurements of liquid-phase properties in the most heavily-loaded particle-laden jet are illustrated in Figure 11. LHF and SSF-BASE predictions are shown; however, SSF-EXT and DSF predictions are essentially the same as the latter. Flow properties are generally similar to the single-phase jet, since the flow is dilute; and are predicted reasonably well, except for the high values of velocity fluctuations and k near the axis. This feature is not present for the single-phase flow, and the bulge in $\bar{u}'$, $\bar{v}'$ and k progressively increases with increasing particle loading: clearly, it is an effect of turbulence modulation. Increased fluctuation levels are probably due to particle wake disturbances, and are most evident near the axis, where effects of turbulence production by shear forces in the continuous phase are small. The effect does not influence turbulence mixing properties significantly, and is probably confined to relatively high wave numbers, but measurements of spectra are needed to resolve this issue. None of the theoretical methods can predict the phenomenon, which is probably theoretically inaccessible with simple turbulence models based on single-scale ideas, like the present k-ε model.

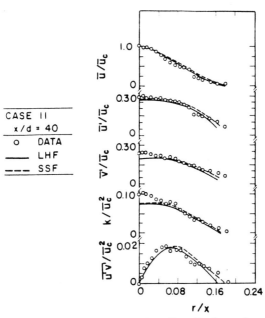

Figure 11. Liquid properties in particle-laden water jets. From Parthasarathy and Faeth [41].

Predictions and measurements of mean and fluctuating particle velocities are illustrated in Figure 12. Only LHF, SSF-BASE and SSF-EXT predictions are shown, since the deficiencies of the DSF and

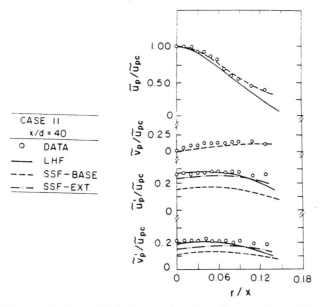

Figure 12. Particle properties in particle-laden water jets. From Parthasarathy and Faeth [41].

SSF-KMOD methods have already been discussed. LHF predictions are virtually identical to liquid properties, providing a baseline for relative velocity effects. The normalization used in Figure 12 tends to minimize the apparent errors of the LHF approach; recall that this method provides relatively poor predictions of particle velocities along the axis. Overall, the SSF-EXT version provides best agreement with measurements, serving as an indication of the importance of treating anisotropic velocity fluctuations and effects of high relative turbulence intensities for particle drag predictions. In particular, the SSF-EXT approach provides a good estimate of the anisotropy levels of particle velocity fluctuations, rectifying errors observed when gas velocity fluctuations are assumed to be isotropic [39,42,43].

Conclusions. Major conclusions, based on present observations, are as follows:

1. Effects of turbulence modulation were observed, evidenced by increased turbulence levels near the flow axis, but the phenomenon did not influence the mixing and turbulent dispersion properties of the flow. It appears that effects of turbulence modulation for present flows were confined to high wave numbers, rather than the energy-containing range which is largely responsible for mixing. Several proposals for treating turbulence modulation were examined; however, none was particularly successful since they did not incorporate effects of turbulence scale and the response of the dispersed phase to a range of turbulence scales. Additional measurements and analysis are needed to gain a better understanding of turbulence modulation phenomena.

2. Stochastic analysis of the process (particularly a version allowing for anisotropic velocity fluctuations of the continuous phase, and high relative turbulence intensities of particles on particle drag properties) yielded reasonably-good predictions. Further study is needed, however, to resolve effects of particle size, turbulence scales and particle inertial properties on drag enhancement; and to examine the performance of multistress turbulence models for treating anisotropy in flows of this type.

3. The performance of LHF analysis was better for comparable phase densities than for flows involving large and small phase density ratios – particularly near the jet exit.

Combusting Spray

Experimental methods. The combustion properties of dilute sprays were studied by Shuen et al. [46]. An ultra-dilute flow was considered, injecting drops from a monodisperse drop generator into the base of a turbulent jet diffusion flame whose properties were known from earlier work by Jeng and Faeth [27]. The turbulent jet diffusion flame was fueled by methane, burning in still air. Methanol sprays were studied, having initial drop diameters of 105 and 180 µm.

Measurements considered included phase velocities, drop sizes and drop number fluxes. The flows were very dilute; therefore, gas-phase properties were known from Jeng and Faeth [27]. Drop velocities were measured using LDA; and particle number fluxes were measured using Mie scattering, similar to the particle-laden jet study [41]. Although the particles were initially monodisperse, drop sizes varied even at a given point in the flows since all drops do not experience the same environment when reaching a given position, due to effects of turbulence. For convenience, however, both predictions and measurements were averaged over all drop sizes at each point.

Theoretical methods. DSF, and SSF-BASE computations were carried out for these flows. Effects of turbulence modulation and high relative turbulence intensities were small for these flows; thus, SSF-EXT predictions would be nearly the same as SSF-BASE predictions, except for effects of anisotropic continuous-phase velocity fluctuations.

Scalar properties of the continuous phase were found using the laminar flamelet technique of Bilger [35] and Liew et al. [36]. This involved correlating measurements of scalar properties in analogous diffusion flames as a function of mixture fraction, in order to obtain state relationships needed for the conserved-scalar method. These measurements exhibited nearly universal correlations of scalar properties as a function of mixture fraction, except near points of flame attachment, justifying use of the laminar-flamelet approximation. The correlation approximated thermodynamic equilibrium for lean conditions, but departed appreciably from equilibrium predictions at rich conditions [27].

A new issue for drop transport involves the presence or absence of envelope flames around the drops for lean conditions [10]. Reliable criteria for envelope flames are not available for present conditions; therefore, limiting cases of envelope flames present (in oxygen-containing surroundings) or always absent, were considered. Predictions of drop transport rates were calibrated using measurements of drop properties in a known flame environment – matching the Reynolds numbers of the drops in the spray flame.

Results and discussion. Complete results for this study are reported by Shuen et al. [46]. Earlier work by Jeng and Faeth [27] showed that the properties of the flames could be predicted reasonably well using present methods, as a baseline.

Predictions and measurements of drop number fluxes along the axis of the sprays are illustrated in Figure 13. These predictions are based on ignoring the presence of envelope flames around the drops. The stochastic approach shows reasonable capabilities for predicting effects of turbulent dispersion in these combusting flows – similar to the other cases considered. Notably, drop trajectories with and without envelope flames were not very different, since the drops spend most of their lifetime in the fuel-rich region of the flames and drop gasification rates for the two limits are not very different until fuel-equivalence ratios are somewhat below stoichiometric [10]. Similar conditions may not always be experienced, however, and methods for estimating the presence or absence of envelope flames are needed for a better understanding of combusting sprays.

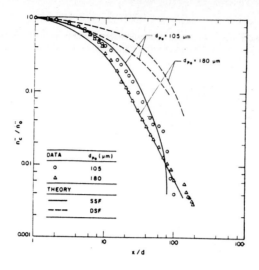

Figure 13. Mean drop number fluxes along the axis of ultra-dilute combusting sprays. From Shuen, et al. [46].

Conclusions. Major conclusions, based on present observations of ultra-dilute combusting sprays, are as follows:

1. The stochastic method, in conjunction with the laminar flamelet approximation for relating scalar properties to mixture fraction, was reasonably successful for predicting the properties of sprays under present test conditions.

2. The distinction between drop transport properties with envelope flames present or absent was not very important for present test conditions, since most drops evaporated before penetrating the flame zone. However, this is probably not always the case and reliable criteria for the existence of envelope flames for conditions in spray flames are needed.

CONCLUSIONS

Measurements and predictions of the structure of several multiphase flows have been considered, emphasizing recent findings of the author and his coworkers [37-46]. The properties of dense sprays, near the exit of pressure-atomizing injectors, and noncombusting and combusting dilute dispersed flows, in round-jet configurations, have been considered. Major conclusions of this work are as follows:

1. The properties of dense sprays exhibit structure and mixing properties similar to variable-density single-phase flows at high Reynolds numbers, within the atomization regime. The overall structure and mixing properties of the flow in this regime were estimated reasonably well using analysis based on the locally-homogeneous flow approximation. However, mixing properties deteriorate rapidly as the first wind-induced breakup regime is approached, which is not anticipated by predictions. Assessment of these effects has only been completed for a limited range of conditions and further study is needed to establish the effectiveness of the locally-homogeneous flow approximation for dense sprays.

2. The degree of development and turbulence levels at the injector exit have a surprisingly large effect on the structure and mixing properties of pressure atomized sprays – particularly when phase densities are large. This effect is probably largely responsible for past controversy concerning the properties of those flows and the effectiveness of the LHF approximation for analyzing them.

3. Contemporary stochastic analysis of dilute multiphase flows has provided encouraging predictions of turbulent dispersion for a wide variety of jet-like flows: particle-laden jets in gases and liquids; noncondensing and condensing bubbly jets; and nonevaporating, evaporating, and combusting sprays. However, present methods for executing stochastic simulations are crude and, although effects of turbulence modulation have been observed experimentally, existing methods of prediction have shown little capability to reliably estimate turbulence modulation phenomena. Examination of turbulence spectra in multiphase flows and consideration of particle response and multiscale effects are needed to gain a better understanding of this phenomenon.

4. Several aspects of interphase transport to drops need further study, as follows: high relative turbulence intensities are encountered in these flows, particularly when phase densities are comparable, existing methods for treating transport at these conditions are not highly developed [3]; effects of virtual mass and the Basset history force are important when phase densities are comparable, existing information concerning these phenomena is very limited [49]; and combusting sprays involve circumstances where differences in transport due to the presence or absence of envelope flames must be considered, reliable criteria for the existence of envelope flames are needed for conditions representative of combusting sprays [46].

ACKNOWLEDGEMENTS

The research described here was sponsored in part by the Air Force Office of Scientific Research, Grant No. AFOSR-85-0244, under the technical management of J. M. Tishkoff; the Army Research Office and the U. S. Army Armament Research, Development and Engineering Center, Contract No.

DAAL03-86-K-0154, under the technical management of D. M. Mann and P.-L. Lu; the Office of Naval Research, Contract No. N00014-85-C-0148, under the technical management of G. Roy; and the National Aeronautics and Space Administration, Grant No. NAG 3-190, under the technical management of R. Tacina.

REFERENCES

1. F. V. Bracco, in *Recent Advances in Gas Dynamics* (Plenum Publishing, New York, 1983.).

2. F. V. Bracco, SAE Paper No. 850394 (1985).

3. R. Clift, J. R. Grace and M. E. Weber, *Bubbles, Drops and Particles* (Academic Press, New York, 1978), p. 185.

4. C. T. Crowe, *J. Fluids Engr.* 104, 197 (1982).

5. D. A. Drew, *Ann. Rev. Fluid Mech.* 15, 261 (1983).

6. C. K. Law, *Prog. Energy Combust. Sci.* 8, 171 (1982).

7. W. A. Sirignano, *Prog. Energy Combust. Sci.* 9, 291 (1983).

8. G. M. Faeth, *Prog. Energy Combust. Sci.* 3, 191 (1977).

9. G. M. Faeth, *Prog. Energy Combust. Sci.* 9, 1 (1983).

10. G. M. Faeth, *Prog. Energy Combust. Sci.*, in press (1988).

11. A. M. Al Taweel and J. Landau, *Int. J. Multiphase Flow* 3, 341 (1977).

12. Ranz, W. E., *Can. J. Chem. Engr.* 36, 175 (1958).

13. R. D. Reitz and F. V. Bracco, *Phys. Fluids* 25, 1730 (1982).

14. R. E. Phinney, *J. Fluid Mech.* 60, 689 (1973).

15. B. Chehroudi, Y. Onuma, S.-H. Chen and F. V. Bracco, SAE Paper 850126 (1985).

16. H. Hiroyasu, M. Shimizu and M. Arai, *Proceedings of the 2nd International Conference on Liquid Atomization and Spray Systems* (Madison, Wisconsin, 1982).

17. S. L. Soo, *Fluid Dynamics of Multiphase Systems* (Blaisdell, Waltham, MA, 1967).

18. M. W. Thring and M. P. Newby, *Fourth Symposium (International) on Combustion* (Williams and Wilkins, Baltimore, MD, 1953), p. 789.

19. K. J. Wu, C.-C. Su, R. L. Steinberger, D. A. Santavicca and F. V. Bracco, *J. Fluids Engr.* 105, 406 (1983).

20. K. J. Wu, A. Coghe, D. A. Santavicca and F. V. Bracco, *AIAA J.* 22, 1263 (1984).

21. A. J. Shearer, H. Tamura and G. M. Faeth, *J. Energy* 3, 271 (1979).

22. C.-P. Mao, G. A. Szekely, Jr. and G. M. Faeth, *J. Energy* 4, 78 (1980).

23. C.-P. Mao, Y. Wakamatsu and G. M. Faeth, *Eighteenth Symposium (International) on Combustion* (The Combustion Institute, Pittsburgh, PA, 1981), p. 337.

24. G. A. Ruff, A. D. Sagar and G. M. Faeth, First Annual Conference, ILASS-Americas, Madison, WI (1987).

25. G. A. Ruff, A. D. Sagar and G. M. Faeth, AIAA Paper No. 88-0237 (1988).

26. T. W. Lee, J. P. Gore, G. M. Faeth and A. Birk, *Comb. Sci. Tech.* 57, 95 (1988).

27. S.-M. Jeng and G. M. Faeth, *J. Heat Trans.* 106, 721 (1984).

28. F. C. Lockwood and A. S. Naguib, *Comb. Flame* 24, 109 (1975).

29. A. Birk and P. Reeves, Ballistic Research Laboratory Report No. BRL-TR-2780, 1987.

30. W. F. McBratney, Ballistic Research Laboratory Report No. ARBRL-MR-03018, 1980.

31. W. F. McBratney, Ballistic Research Laboratory Report No. ARBRL-MR-03128, 1981.

32. K.N.C. Bray, *Seventeenth Symposium (International) on Combustion* (The Combustion Institute, Pittsburgh, PA, 1979), p. 223.

33. K.N.C. Bray, in *Turbulent Reacting Flows* (Springer, Berlin, 1980), p. 115.

34. S. Gordon and B. J. McBride, NASA Report No. SP-273, 1971.

35. R. W. Bilger, *Comb. Flame* 30, 277 (1977).

36. S. K. Liew, K.N.C. Bray and J. B. Moss, *Comb. Sci. Tech.* 27, 69 (1981).

37. J.-S. Shuen, L.-D. Chen and G. M. Faeth, *AIChE. J.* 29, 167 (1983).

38. J.-S. Shuen, L.-D. Chen and G. M. Faeth, *AIAA. J.* 21, 1483 (1983).

39. J.-S. Shuen, A.S.P. Solomon, Q.-F. Zhang and G. M. Faeth, *AIAA J.* 23, 396 (1985).

40. D. L. Bulzan, J.-S. Shuen and G. M. Faeth, AIAA Paper No. 87-0303 (1987).

41. R. N. Parthasarathy and G. M. Faeth, *Int. J. Multiphase Flow*, in press (1988).

42. T.-Y. Sun and G. M. Faeth, *Int. J. Multiphase Flow* 12, 99 (1986).

43. T.-Y. Sun, R. N. Parthasarathy and G. M. Faeth, *J. Heat Trans.* 108, 951 (1986).

44. A.S.P. Solomon, J.-S. Shuen, Q.-F. Zhang and G. M. Faeth, *J. Heat Trans.* 107, 679 (1985).

45. A.S.P. Solomon, J.-S. Shuen, Q.-F. Zhang and G. M. Faeth, *AIAA J.* 23, 1548, 1724 (1985).

46. J.-S. Shuen, A.S.P. Solomon and G. M. Faeth, *AIAA J.* 24, 101 (1986).

47. D. Modarress, H. Tan and S. Elghobashi, *AIAA J.* 22, 624 (1984).

48. A. D. Gosman and E. Ioannides, AIAA Paper No. 81-0323 (1981).

49. F. Odar and W. S. Hamilton, *J. Fluid Mech.* 18, 302 (1964).

50. J. C. Lopes and A. E. Dukler, *AIChE J.*, in press (1987).

51. R. D. Reitz and R. Diwakar, SAE Paper 870598 (1987).

52. S. E. Elghobashi and T.-W. Abou-Arab, *Phys. Fluids* 26, 931 (1983).

NOMENCLATURE

c	reaction progress variable
C_i	empirical constants in turbulence model
d	injector diameter
f	mixture fraction
g	mixture fraction fluctuations squared
k	turbulence kinetic energy
L	injector passage length
L_e	characteristic eddy size
r	radial distance
Re	injector Reynolds number
$S_{p\phi}$	dispersed-phase source term
t_e	characteristic eddy lifetime
u,v,w	streamwise, radial, and tangential velocities
x	streamwise distance
α_f	liquid volume fraction
ε	rate of dissipation of turbulence kinetic energy
ρ	density

Subscripts

c	centerline quantity

p   particle or dispersed-phase property

o   initial condition

Superscripts

(~),(·)   Favre- and time-averaged mean property

(~)",(·)'   Favre- or time-averaged root-mean-squared fluctuating property

LAGRANGIAN SIMULATION OF PARTICLE DISPERSION

A. BERLEMONT, G. GOUESBET, P. DESJONQUERES

Laboratoire d'Energétique des
Systèmes et Procédés
UA CNRS 230
INSA de Rouen - BP 08
76130 Mont-Saint-Aignan
FRANCE

I. Introduction

Eulerian and Lagrangian approaches have been studied at Rouen for the prediction of particle dispersion in turbulent flows. A first computer code DISCO (DISpersion COmputing) has been developed during the past years for predicting turbulence and the dispersion of discrete particles in the framework of an Eulerian approach. In that method, the particles are considered as a continuum and satisfy a transport equation which involves a dispersion tensor linked to particles and fluid characteristics. The Tchen theory (on discrete particle displacement) and the Batchelor theory (for diffusion of fluid particles) are used to determine the dispersion tensor, and a correction factor is defined to take into account for the crossing trajectories effects.

Although that approach can be used for several engineering purposes, the assumptions of the underlying theories are limiting new developments of the code to handle more complex situation.

In the Lagrangian approach which is now developed and here presented, the particles are considered individually : a number of particle trajectories are simulated and statistics are carried out to obtain mean quantities for comparisons with experimental data.

The particle trajectory simulation is different when either a fluid particle or a discrete particle is considered. For a fluid particle, its instantaneous velocity is determined using the mean values of the turbulence obtained with a (K-ϵ) model supplemented with algebraic relations deduced from a second order closure scheme for the Reynolds tensor, and taking random values with respect to spatial and time fluid velocity correlations for fluctuating velocities.

These correlations are expressed through a Frenkiel family of correlation functions and introduced in a correlation matrix linked to the particle trajectory and thus taking into account for most of the particle history.

For a discrete particle the forces acting on the particle are expressed following the RILEY equation and classical relations for the drag coefficient. To get the instantaneous fluid velocities at the discrete particle location, a fluid particle is simultaneously followed with the discrete particle and, as the two trajectories are different, spatial correlations are introduced to take into account for the lack of coincidence between the trajectories. The crossing trajectory effects are also simulated by this method, by considering a new fluid particle when the distance between the discrete and the fluid particle becomes too large.

The interaction between particles and the turbulent flow due to non negligible mass loading is expressed using extra source terms in the governing equations of a ($K-\epsilon$) model, namely by computing momentum and energy exchanges between both phases and using an iteration procedure.

As the whole Lagrangian approach is presented elsewhere more extensively (DESJONQUERES (1987), GOUESBET et al (1987)) the opportunity is taken to point out the specific procedure used to simulate a fluid particle trajectory.

II. Fluid particle trajectory

The Lagrangian simulation of a fluid particle trajectory is obtained by the quite simple relation :

$$x_i(t + \Delta t) = x_i(t) + \left(U_i(x_i(t)) + u_i(x_i(t))\right) \Delta t \qquad (1)$$

where x_i is the particle position, U_i its mean velocity and u_i its fluctuating velocity, and Δt the time step.

The mean velocity can be deduced from previous predictions of the turbulent field, by use of a (K-ϵ) model for example, but the fluid particle trajectory simulation requires the knowledge at each time step of the fluctuating velocity u_i. The problem is to obtain statistics which are describing the turbulence.

A first condition to satisfy is the fluid velocity r.m.s. which can be obtained, for instance, by supplementing the (K-ϵ) model with algebraic relations deduced from a second order scheme as done by RODI (1979). We thus need to assume that the fluid velocity p.d.f. is Gaussian, which is a quite classical assumption but does not depend on the method we are presenting.

The second condition we decide to take into account in the random process is that the fluctuating velocities must comply with the Lagrangian time correlation $R_L(T)$, which is an important quantity to characterize the turbulence under study. A method , which is used by GOSMAN and IOANNIDES (1981), DURST et al (1984), SHUEN et al (1984), ORMANCEY and MARTINON (1984), among others, is based on the definition of an "eddy Life time". That process mainly consists in keeping constant the fluctuating velocity during a characteristic time linked to the turbulence properties, such as the Lagrangian integral time scale T_L, and then to generate another random value, and so on they thus obtain a Poisson's scheme which leads to a Lagrangian correlation $R_L(T)$ expressed by :

$$R_L(T) = \exp -T/T_L \tag{1}$$

But following HINZE (1975), it can be shown that the Lagrangian correlation $R_L(T)$ must involve negative loops. We therefore first introduced such correlations in the code DISCO, in the framework of the Eulerian approach (PICART et al (1986), GOUESBET et al (1984), with the aid of a Frenkiel family of correlation functions, namely (FRENKIEL (1984)) :

$$R_L(T) = \exp \frac{-T}{(m^2+1)T_L} \cos \frac{mT}{(m^2+1)T_L} \tag{2}$$

where m is the loop parameter linked to the number and the importance of the negative parts in the correlation.

The results have been found much more satisfactory when we used a value of m not equal to 0, which leads to an exponential form for the correlation ; indeed, we choose the value m = 1, since the corresponding predictions with the code DISCO showed a good agreement with various experimental situations (grid turbulence or pipe flow, for example). Thus we decided to use such Lagrangian correlations in the Lagrangian approach. But a Poisson's scheme is no more valid for the random generation process and we introduced a correlation matrix which is now presented.

III. Fluctuating velocities

III.1. Random generation process

The procedure to estimate the fluctuating velocities is presented in an 1D-formulation.

Let be $u(n \Delta t)$ the value of the fluctuating velocity at the time $n \Delta t$. From a mathematical point of view, the problem is to determine a vector U of correlated random variables :

$$U = (u(0), u(\Delta t), ..., u(i \Delta t), ..., u(n \Delta t))$$

which complies with the given Lagrangian correlations and the Gaussian p.d.f.

We define the positive definite, symetric correlation matrix by (DESJONQUERES (1987), GOUESBET et al (1987)) :

$$A = (a_{ij}) = \begin{vmatrix} \overline{u(0)^2} & & & & \\ \overline{u(0)u(\Delta t)} & \overline{u(\Delta t)^2} & & & \\ \vdots & \vdots & & & \\ \overline{u(0)u(i\Delta t)} & \overline{u(\Delta t)u(i\Delta t)} & \vdots & \overline{u(i\Delta t)^2} & \\ \vdots & \vdots & \vdots & \vdots & \vdots \end{vmatrix} \quad (3)$$

which is used under a reduced from $R(r_{ij})$:

$$r_{ij} = \frac{\overline{u(i\Delta t)\,u(j\Delta t)}}{\sqrt{\overline{u^2(i\Delta t)}}\sqrt{\overline{u^2(j\Delta t)}}} = \overline{u^*(i\Delta t)\,u^*(j\Delta t)} \qquad (4)$$

that is, with the Frenkiel family of correlation functions :

$$r_{ij} = \exp\frac{-|i-j|\,\Delta t}{(m^2+1)\,T_L}\;\cos\frac{m|i-j|\,\Delta t}{(m^2+1)\,T_L} \qquad (5)$$

To determine the vector $U^*(u^*_i)$ and then the vector U, we start from a vector $Y(y_i)$ of uncorrelated, Gaussian random variables, namely

$$\overline{y_i} = 0 \qquad \overline{y_i y_j} = \delta_{ij} \qquad (6)$$

We then define a matrix B^*, such as :
$$U^* = B^*Y \qquad (7)$$
And using a Cholesky scheme, the matrix B^* is deduced from :

$$R = B^* B^{*t} \quad \text{where} \quad b^*_{ij}{}^t = b^*_{ji} \qquad (8)$$

The above method has been extended to 2D-problems, including cross correlations of the kind $\overline{u_x(t)u_y(t+i\Delta t)}$ and $\overline{u_y(t)u_x(t+i\Delta t)}$, and proves very satisfactory for all the numerical tests which have been carried out (DESJONQUERES (1987)).

III.2. Numerical simplifications

The method which has been presented could appear as time consuming and with a quite important computer storage, since the involved matrix are growing with the time and are linked to the time step. But the following simplifications (numerically checked) can be stated :

i) The matrix size can be physically limited when considering a diffusion time larger than, for example, ten times the Lagrangian integral scale, as the correlation is nearly equal to zero. The average number of elements in the matrix is thus, for 1.D problems, of the order of $10\,T_L/\Delta t$.

ii) Using the Cholesky scheme permit to define the matrix step by step, without the necessity of a whole storage of every elements (DESJONQUERES (1987)).

iii) An important result is that the time step Δt can be choosen greater with our method than with a Poisson's process. For instance, for the first case described in the next section, a similar precision is obtained with a time step equal to $0.2\, T_L$ with the correlation matrix, compared with a time step of the order of $0.01\, T_L$ for the Poisson's scheme.

More precisely, it can be exemplified in that case by the running time which is for the Poisson's process twice than with the method under discussion.

IV. Results

A number of numerical tests have been carried out to validate the method we are using. But any complete validation can not be stated without comparisons against physical situations. Two simulations are now presented.

IV.1. Point source in homogeneous and isotropic turbulence

In the fundamental case of the diffusion of a tracer from a point source in homogeneous and isotropic turbulence, an analytical expression for the tracer concentration is given by HINZE (1975) :

$$C(x,r) = \frac{S}{4\pi \epsilon x} \exp - \frac{U_1 r^2}{4 \epsilon x} \qquad (9)$$

for a position far away from the point source and near the axis. S is the source strength, U_1 the mean velocity and ϵ the diffusion coefficient of the tracer, x the distance from the source and r the radial coordinate.

Figure (1) presents the tracer concentration profile, for a distance $x = 500$ cm from the source, versus r. The 2D-turbulence field is given by $U_1 = 655$ cm/s, $U_2 = 0$, $\overline{u_1^2} = \overline{u_2^2} = 171$ cm^2/s^2 and $\epsilon = 15.6$ cm^2/s that is $T_L = 91$ ms ($\epsilon = \overline{u^2} T_L$ for long diffusion time). The theoretical results are compared with simulations using either a Poisson's process or our method with m = 1, which shows a very good agreement with HINZE's results, by contrast with the exponential form for the correlation which exhibit some discrepancies of the order of 25 % on the center.

IV.2. Comparisons with experimental results

We then apply the fluid particle trajectory simulation to the experiment described by TAYLOR and MIDDLEMAN (1974) for the mean square displacement $\overline{Y^2}$ of a tracer in a turbulent pipe flow of water. The Reynolds number is 35000, with a mean velocity on the axis equal to 85 cm/s and the pipe diameter is 5.08 cm. The Lagrangian integral time scale is estimated to 80 ms and the turbulence intensity is 3.2 % on the centerline velocity.

The simulation is carried out with the assumption that the turbulence is homogeneous in the flow region where dispersion is studied, since the diffusion times under study are small compared with the Lagrangian integral time scale (less than $10\, T_L$).

The results are presented on the Figure (2), and the agreement is very satisfactory (m = 1).

V. Conclusion

The Lagrangian approach we are developing requires fluid particle trajectory simulations. In order to take into account negative loops in the Lagrangian correlation, a matrix is defined linked to the fluid particle and evolving with the turbulence characteristics, using a Frenkiel family of functions to express the Lagrangian correlations. The method has been numerically tested, compared with experimental results and proved very satisfactory.

Discrete particle trajectories can be afterwards simulated, using the above fluid particle simulations to estimate the influence of the turbulent field on the particle behaviour, as described in DESJONQUERES (1987) and GOUESBET et al (1987).

REFERENCES

DESJONQUERES P.
    Modélisation Lagrangienne du comportement de particules discrètes en écoulement turbulent. Thèse de 3ème cycle. ROUEN. 28 Janvier 1987

DURST F., MILOJEVIC D., SCHONUNG B.
    Eulerian and Lagrangian predictions of particulate two-phase flows : a numerical study. Appl. Math. Modeling 1984

FRENKIEL F.N.
    Etude statistique de la turbulence. Fonctions spectrales et coefficients de corrélation. Rapport Technique, O.N.E.R.A, n 34, 1948

GOSMAN A.D. et IOANNIDES E.
    "Aspects of Computer Simulation of Liquid-Fuelled Combustors", AAIA Journal, 81, 0323, 1981

GOUESBET G., BERLEMONT A., PICART A.
    Dispersion of discrete particles by turbulent continuous motions. Extensive discussion of the Tchen's theory using a two parameters family of Lagrangian correlation functions. Phys. Fluid 27, 827-837, 1984

GOUESBET G., DESJONQUERES P., BERLEMONT A.
    Eulerian and Lagrangian approaches to turbulent dispersion of particles. Seminaire International "Transient Phenomena in Multiphase flow", I.C.H.M.T. Dubrounik. 1987

HINZE J.O.
    Turbulence. McGraw-Hill, New-York, 1959 (Réédition 1975)

ORMANCEY A. and MARTINON A.
    Prediction of particle dispersion in turbulent flows. Physico-Chemical Hydrodynamics, vol.5, n 314, 1984

PICART A., BERLEMONT A., GOUESBET G.
    Modelling and Predicting Turbulence Fields and the Dispersion of Discrete Particles Transported by Turbulent Flows. Int. J. Multiphase Flow 12, n 2, pp 237-261, 1986

RODI W.
    Turbulence models for environmental problems. V.K.I. Lect. series 2979-2, 1979

SHUEN J.S., SOLOMON A.S.P., ZHANG Q.F., and FAETH G.M.
    Structure of Particle-Laden Jets : Measurements and Predictions, AIAA-84-0038

TAYLOR A.R. and MIDDLEMAN S.
    Turbulent Dispersion in Drag. Reducing Fluids. AICHE Journal, Vol.20, n 3, 1974

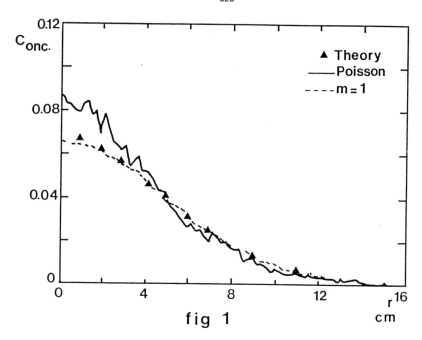

fig 1

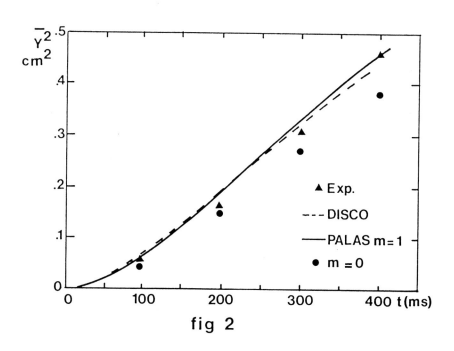

fig 2

NUMERICAL MODELLING OF DEVOLATILIZATION IN PULVERISED
COAL INJECTION INSIDE A HOT COFLOWING AIR FLOW

O. SIMONIN - P.L. VIOLLET
E.D.F. - Laboratoire National d'Hydraulique
6 quai Watier - 78400 Chatou (France)

Abstract - A two-dimensional separated flow model has been developed for predicting fluid-particles turbulent recirculating two-phase flows. Thus, separate conservation equations are formulated for both phases including interphase transfer terms (mass, momentum and enthalpy). Turbulence is modeled by means of a $q^2-\varepsilon$ eddy viscosity model in the continuous phase which does not, at the present, take account for the interaction between the two phases.

The model, including pyrolysis of coal and combustion of volatile matter, is used to compute an axisymetric injection of pulverised coal in a hot coflowing air flow, for two different particle sizes and two different inlet temperatures of the coflowing air flow. A first attempt is made to model the heterogeneous gaseous combustion process due to the local distribution of matter around particles.

1. INTRODUCTION

For the purpose of steam-generator thermal-hydraulic analysis, a two-dimensional separated flow model has been developed at the "Laboratoire National d'Hydraulique" for predicting recirculating turbulent bubbly flows and computations has been carried out for flows across an inline tube bundle, as described by Simonin and al. (1987). As a matter of fact, such a numerical modelling can be used for most of the turbulent two-phase flows with dispersed inclusions (bubbles, drops or particles) which occur in processing applications.

Basically two approaches have been employed to model these flows : Lagrangian and Eulerian.

In the Lagrangian formulation a large number of particle trajectories are calculated using a previously computed fluid flow field, several iterations can be necessary to get a solution which accounts for the mutual interaction of the two phases. In the Eulerian approach used there, the two-phases are treated as separate interpenetrating continua, and mean equations with interaction terms are solved for each phase.

In this paper, we present the application of the two-dimensional separated flow model -named MELODIF- to a monodisperse pulverised coal injection inside a hot coflowing air flow with an emphasis on the effect of the particle size and the temperature of the coflowing air flow on the global rate of devolatilization.

2. THE MODEL

Basic equations

In the separated flow model formulation used there, the field equations for each phase are derived from the local instant conservation equations in single-phase flow by density-weighted averaging with in addition local instant balances of mass, momentum and energy at the interface (Delhaye 1974 , Ishii 1975). Closure of the averaging equations requires constitutive relations to describe the exchange of quantities between the two phases and is finally achieved by modelling the turbulent correlations with the help of a $q^2 - \varepsilon$ eddy viscosity model in the continuous phase.

In the following presentation of the differential equations, the dependant variables will be denoted by subscript 1 for the gaseous phase and subscript 2 for the solid phase.

In summary, the main hypothesis upon which these equations are based are listed below :

(1) The flow is turbulent enough, so molecular diffusion and intergranular stress are negligible.

(2) The pressure gradient term in the momentum equation for each phase is written as the partial gradient of a single mean pressure.

(3) Gravity effects are negligible.

(4) Heat transfers due to radiation are negligible.

(5) The specific heat of particles is constant.

The field equations for the two-fluid model are:
(a) The mass-conservation equations:

$$\frac{\partial}{\partial t}(\alpha_k \rho_k) + \frac{\partial}{\partial x_i}(\alpha_k \rho_k U_{k,i}) = \Gamma_k \qquad (1)$$

where, $U_{k,i}$ is respectively the mean velocity for gaseous phase (k=1) and dispersed particles (k=2).
α_k is the volumetric fraction, ρ_k the mean density and Γ_1 (= $-\Gamma_2$) the rate of mass transfer per unit volume from the solid to the gaseous phase due to the devolatilization process of the coal.

(b) The conservation of momentum equations:

$$\alpha_k \rho_k \frac{\partial}{\partial t} U_{k,i} + \alpha_k \rho_k U_{k,j} \frac{\partial}{\partial x_j} U_{k,i} = -\alpha_k \frac{\partial P}{\partial x_i} + I_{k,i}$$
$$+ \frac{\partial}{\partial x_j} \alpha_k \rho_k \nu_k^t \frac{\partial}{\partial x_j} U_{k,i} \qquad (2)$$

where P is the mean pressure, ν_k^t the eddy viscosity coefficient and $I_{k,i}$ the interfacial momentum transfer between phases.

(c) The conservation of energy equations:
Gas-phase energy equation:

$$\alpha_1 \rho_1 \frac{\partial}{\partial t} H_1 + \alpha_1 \rho_1 U_{1,j} \frac{\partial}{\partial x_j} H_1 = Q_1 + \Gamma_1 (h_v - H_1)$$
$$+ \frac{\partial}{\partial x_j} \alpha_1 \rho_1 K_1^t \frac{\partial}{\partial x_j} H_1 \qquad (3)$$

Particle-phase energy equation:

$$\alpha_2 \rho_2 C_{p,2} \frac{\partial}{\partial t} T_2 + \alpha_2 \rho_2 U_{2,j} C_{p,2} \frac{\partial}{\partial x_j} T_2 = Q_2 - \Gamma_1 h_{dv}$$
$$+ \frac{\partial}{\partial x_j} \alpha_2 \rho_2 K_2^t C_{p,2} \frac{\partial}{\partial x_j} T_2 \qquad (4)$$

where H_1 is the total enthalpy per unit of mass of the gaseous mixture, defined from the mass fraction $X_{1,m}$ and the enthalpy h_m of each species m (with molar mass W_m) by :

$$H_1 = X_{1,m} h_m \qquad (5)$$

$$h_m = h_{mo} + \int_o^T C_{p,m} dT$$

where h_{mo} is the enthalpy of formation of the specie m ;
h_v is the enthalpy per unit of mass of volatile matter produced by pyrolysis of the coal and supposed to be at the temperature of the particle, and h_{dv} is the corresponding heat of sublimation ;
K_k^t is the eddy diffusivity coefficient and Q_k the rate of interfacial heat transfer between phases.

(d) The equations of state for gaseous species :

$$P = \rho_1 R T_1 \sum_m X_{1,m}/W_m \qquad (6)$$

The turbulence model

The nature of turbulence in two-phase flows is rather complex and not well understood at the present time. There have been, however, several attempts to predict the behaviour of particle-laden turbulent flows in the Eulerian formulation with models which take account of the influence of particles on the turbulence of the continuous phase (S. Elgobashi and al.
1984 , Gouesbet and al. 1987). In the model used there, no extra terms are added to the standard single-phase equations of the turbulent kinetic energy q^2 and its dissipation rate ε (Launder and Spalding 1974).

Comparisons of the predictions of the model with experimental results (Modarress and al. 1984) show that the addition of heavy particles in turbulent jets flows tends to reduce the mean eddy characteristic scale and thus the dispersion effect due to turbulence. In the industrial process studied, the mass loading ratio of the particles is high enough to modify strongly the turbulence properties and the computations probably over-estimate the mixing of the injection with the coflowing flow.

The kinematic eddy viscosity ν_1^t and the eddy diffusivity K_1^t of the continuous phase are obtained directly from the turbulent kinetic energy q^2 and its dissipation rate ε :

$$\nu_1^t = C_\mu \frac{q^4}{\varepsilon}$$
$$K_1^t = \nu_1^t / P_r^t \tag{7}$$

where $\alpha_k \rho_k q^2 = \frac{1}{2} < \rho u''_i u''_i >_k \qquad k=1$

The transport equations allowing to compute q^2 and ε are written in the conventional form :

$$\alpha_1 \rho_1 \frac{\partial q^2}{\partial t} + \alpha_1 \rho_1 U_{1,j} \frac{\partial q^2}{\partial x_j} = \frac{\partial}{\partial x_j} \left(\alpha_1 \rho_1 \frac{\nu_1^t}{\sigma_q} \frac{\partial q^2}{\partial x_j} \right) - < \rho u''_i u''_j >_1 \frac{\partial U_{1,i}}{\partial x_j} - \alpha_1 \rho_1 \varepsilon \tag{8}$$

$$\alpha_1 \rho_1 \frac{\partial \varepsilon}{\partial t} + \alpha_1 \rho_1 U_{1,j} \frac{\partial \varepsilon}{\partial x_j} = \frac{\partial}{\partial x_j} \left(\alpha_1 \rho_1 \frac{\nu_1^t}{\sigma_\varepsilon} \frac{\partial \varepsilon}{\partial x_j} \right) - \frac{\varepsilon}{q^2} \left[C_{\varepsilon 1} < \rho u''_i u''_j >_1 \frac{\partial U_{1,i}}{\partial x_j} + C_{\varepsilon 2} \alpha_1 \rho_1 \varepsilon \right] \tag{9}$$

An expression for ν_2^t, the turbulent particle momentum eddy diffusivity, should account for the interaction between phases so that, for instance, the larger particles are not as much affected by the turbulence of the continuous phase as the smaller ones. In the model, the diffusivity ν_2^t is derived from the kinematic eddy viscosity ν_1^t as follows :

$$\nu_2^t = \nu_1^t / (1 + t_{12}/t_q) \tag{10}$$

the interaction relaxation time t_{12} is related to the interfacial drag force coefficient F_D defined by (16) :

$$t_{12} = 1/\left[\alpha_2 \rho_1 F_D \left(\frac{1}{\alpha_2 \rho_2} + \frac{1}{\alpha_1 \rho_1}\right)\right] \tag{11}$$

and t_q is the time scale of the energetic turbulent eddies given by :

$$t_q = \frac{3}{2} C_\mu \frac{q^2}{\varepsilon} \tag{12}$$

The interfacial momentum transfer

The interfacial transfer terms in the basic equations (1) to (4) are obtained from the analysis of the exchange between an isolated particle and the surrounding fluid. The pratical expressions for Γ_k, $I_{k,i}$ and Q_k, respectively the rate of interfacial transfer of mass, momentum and heat per unit volume, derive by averaging from the late results with in addition attempts to take account for the interaction between particles, and must be related to the mean computed variables.

According to Wallis (1969), the drag force induced by the relative motion of a spherical particle in a steady fluid velocity field can be written :

$$\vec{f}_D = -\frac{\pi d^3}{6}\frac{3}{4} \rho_F \frac{C_D}{d} \left|\vec{U}_P - \vec{U}_F\right|(\vec{U}_P - \vec{U}_F) \tag{13}$$

where d is the diameter and $\vec{U}_P$ the velocity of the particle, ρ_F is the density and $\vec{U}_F$ the mean velocity of the surrounding fluid, C_D is the drag coefficient :

$$\begin{aligned}R_e \leq 1000 \quad & C_D = \frac{24}{R_e}(1 + 0.15\, R_e^{0.687})\\ R_e > 1000 \quad & C_D = 0.44 \quad \text{where } R_e = \frac{\left|\vec{U}_P - \vec{U}_F\right| d}{\nu_F}\end{aligned} \tag{14}$$

The corresponding term in the conservation of momentum equations can be obtained by averaging on the dispersed phase :

$$I_{2,i} = -I_{1,i} = -<\frac{3}{4}\rho_F \frac{C_D}{d}|\vec{U}_P - \vec{U}_F|(U_{P,i} - U_{F,i})>_2 \quad (15)$$

And we approach the previous evaluation in the following way :

$$I_{2,i} = -I_{1,i} = -\alpha_2 \rho_1 F_D V_{r,i} \quad (16)$$

where $F_D = \frac{3}{4}\frac{C_D}{d}|\vec{V}_r|$ and $\alpha_2 V_{r,i} = <U_{P,i} - U_{F,i}>_2$

But in general V_r, obtained by averaging of the local relative velocity, is not equal to the simple difference between the respective mean velocities of the two-phases. This gap is due to the correlation between the instantaneous distribution of particles and the turbulent fluid velocity field, and can be calculated with the help of the eddy diffusivity concept from the relation :

$$(U_{2,i} - U_{1,i}) - V_{r,i} = -\frac{\nu_2^t}{\sigma_p}\frac{1}{\alpha_2}\frac{\partial \alpha_2}{\partial x_i} \quad (17)$$

where σ_p is the particle turbulent Schmidt number.

Using the above equality in the mass-conservation equation of the dispersed phase, we obtain :

$$\frac{\partial}{\partial t}(\alpha_2 \rho_2) + \frac{\partial}{\partial x_i}\alpha_2 \rho_2 (U_{1,i} + V_{r,i}) = \Gamma_2 + \frac{\partial}{\partial x_i}(\rho_2 \frac{\nu_2^t}{\sigma_p}\frac{\partial \alpha_2}{\partial x_i}) \quad (18)$$

This expression, which is used in the numerical solution procedure, yields to the conventional form of the scalar transport equation as the diameter of the particles tends to zero.

The interfacial heat transfer

On the same way, the rate of interfacial heat transfer per unit volume can be obtained by averaging :

$$Q_2 = - Q_1 = - \left\langle \frac{6}{d^2} \lambda_F \cdot Nu \, (T_P - T_F) \right\rangle_2 \qquad (19)$$

where T_P is the temperature of the particle, λ_F is the thermal conductivity and T_F the mean temperature of the surrounding fluid, Nu is the Nusselt number given after Ubhayakar (1977) by :

$$Nu = \frac{q_v \, d}{\lambda_F} \Big/ \left(\exp\left[\frac{q_v \, d}{2 \cdot \lambda_F} \right] - 1 \right) + 0.6 \, R_e^{1/2} \, P_r^{1/3} \qquad (20)$$

where $P_r = \dfrac{\rho_F \, \nu_F}{\lambda_F}$

The first term in the previous expression takes account for the influence of the flux of volatile matter q_v due to the devolatilization process which tends to reduce the conductive heat transfer, the second term represents the advective heat transfer.

Finally, neglecting the correlation between the instantaneous distribution of particle and the turbulent fluid temperature field, we approach the interfacial heat transfer term in the following way :

$$Q_2 = - Q_1 = - \alpha_2 \, \rho_1 \, N_T \left[T_2 - T_1 \right] \qquad (21)$$

where $N_T = \dfrac{6}{d^2} K_1 \, Nu \qquad K_1$ is the molecular diffusivity of gas

The devolatilization model

The coal reacting particle is assumed to be composed of three components : coal, char and ash. Ash is defined as that part of coal particle that is inert in the devolatilization and combustion processes.

Char is the residue left in the particle when the volatiles are released which reacts heterogeneously, after diffusion of the reactant to the particle surface. Characteristic time scale of such a phenomena is larger than residence time of particles in the computed domain so no modelling of char combustion is carried out.

The raw coal is assumed to pyrolyse according to the parallel reactions : raw coal $\longrightarrow$ Y_n volatiles + $(1 - Y_n)$ coke $n = 1,2$.

The reaction rate constants are expressed by Arrhenius formulas with stoichiometric coefficients Y_n, frequency factors A_n and activation energies E_n given by Kobayashi (1976). The total mass of a particle decreases according to the equation :

$$\frac{d m_p}{dt} = - m_p X_R \sum_{n=1}^{2} Y_n A_n \exp\left[-\frac{En}{R T_p}\right] \qquad (22)$$

where X_R is the instantaneous concentration of raw coal in the particle which is given by :

$$\frac{d m_p X_R}{dt} = - m_p X_R \sum_{n=1}^{2} A_n \exp\left[-\frac{En}{R T_p}\right] \qquad (23)$$

$X_R = (1 - X_A)$ at the injection

where X_A is the initial mass fraction of ash in the particle.

The rate of interfacial mass transfer per unit volume can be obtained by averaging :

$$\Gamma_2 = - \Gamma_1 = - \left\langle \rho_p X_R \sum_{n=1}^{2} Y_n A_n \exp\left[-\frac{En}{R T_p}\right] \right\rangle_2 \qquad (24)$$

neglecting the non-linearity of the reaction rate constant depending of the local distribution of particle temperature :

$$\Gamma_2 = - \Gamma_1 = - \alpha_2 \rho_2 X_R \sum_{n=1}^{2} Y_n A_n \exp\left[-\frac{En}{R T_2}\right] \qquad (25)$$

The mean concentration of raw coal (X_R) is needed to calculate the interfacial mass transfer (Eq. 25) and can be obtained from the Eulerian mass-conservation equation :

$$\frac{\partial}{\partial t}(\alpha_2 \rho_2 X_R) + \frac{\partial}{\partial x_i}(\alpha_2 \rho_2 X_R U_{2,i}) = -\alpha_2 \rho_2 X_R \sum_{n=1}^{2} A_n \exp\left[-\frac{E_n}{RT_2}\right]$$
$$+ \frac{\partial}{\partial x_i} \alpha_2 \rho_2 K_2^t \frac{\partial X_R}{\partial x_i} \quad (26)$$

According to experimental results, the diameter of the particles is assumed to remain nearly constant during devolatilization process, so particles become porous and the mean density (ρ_2) decreases.

The corresponding Eulerian volume-conservation equation is used, in addition to the particle mass-conservation equation (Eq. 18), to obtain the mean density (ρ_2) and the volumetric fraction (α_2) :

$$\frac{\partial}{\partial t}\alpha_2 + \frac{\partial}{\partial x_i}(\alpha_2 U_{2,i}) = \frac{\partial}{\partial x_i}\alpha_2 \rho_2 K_2^t \frac{\partial}{\partial x_i}\left(\frac{1}{\rho_2}\right) \quad (27)$$

The combustion model

A significant portion of the reaction process in pulverised coal injection takes place in the gaseous phase. As the coal undergoes devolatilization, the volatiles react further with the oxidizer and the heat generated by the combustion process may increase considerably the temperature and consequently the devolatilization rate itself.

The combustion model must provide a method of evaluating the mean mass fractions $X_{1,m}$ of each species present in the gas phase, and in addition allows the calculation of mean fluid temperature and density with the help of equations (5) and (6). In the fast chemistry approximation invoked here, all reactions are assumed to go to completion as soon as the reactants are mixed, and proceed via a single-step irreversible reaction of the form :

fuel (volatiles) + oxidant (air) ⟶ products

The local-instantaneous composition consists either fuel (x_1) plus products (x_3) or oxidant (x_2) plus products (x_3) according to the value of the gas mixture fraction f :

$$\begin{cases} f \geq f_s \\ x_1 = (f - f_s)/(1 - f_s) \\ x_2 = 0 \\ x_3 = (1 - f)/(1 - f_s) \end{cases} \quad \begin{cases} f \leq f_s \\ x_1 = 0 \\ x_2 = (f_s - f)/f_s \\ x_3 = f/f_s \end{cases} \quad (28)$$

where f_s is the mixing rate value in the pure products obtained from the stoichiometry of the studied reaction, and the mixing rate f is defined as :

$$f = \frac{Y_m - Y_{m,2}}{Y_{m,1} - Y_{m,2}} \quad (29)$$

Y_m is the mass fraction in the gas mixture of any element m (C, H, O etc...), $Y_{m,1}$ the mass fraction of element m in the pure fuel (volatiles) and $Y_{m,2}$ the mass fraction of element m in the pure oxidant (air). With these conventions, $f = 1$ in the pure fuel and $f = 0$ in the pure oxidant. Finally, with reasonable accuracy, it can be assumed that all species (and heat) have equal diffusion coefficients (molecular or turbulent). The transport equation for F_1, the density-weighted average of f in the gaseous phase, is written :

$$\alpha_1 \rho_1 \frac{\partial}{\partial t} F_1 + \alpha_1 \rho_1 U_{1,j} \frac{\partial}{\partial x_j} F_1 = (1 - F_1) \Gamma_1 + \frac{\partial}{\partial x_j} \alpha_1 \rho_1 K_1^t \frac{\partial}{\partial x_j} F_1 \quad (30)$$

Because of the strong non-linearity of the relations (Eq. 28) which give the local-instantaneous composition of the gas mixture, it is necessary to take account of the fluctuations in mixture fraction f due in the same time to the turbulence of the flow and to the local distribution of matter around particles.

If the relative motion of particle is slow enough, and translating in an oxidant environment, an envelope diffusion flame is established.

The concentric distribution, governing the mixture fraction around a single particle which undergoes steadily pyrolysis in an infinite calm atmosphere, can be easily generalized to a bounded domain if spherical symmetry is respected an takes the form :

$$f(r) = 1 - C(1 - \bar{f}) \exp\left[-\frac{q_v}{2\lambda_F} \frac{d}{2} \frac{d}{r}\right] \quad (31)$$

where r is the distance to the center of the particle, and the mean value $\bar{f}$, in a spherical domain of diameter D, is defined below :

$$\bar{f} = \frac{24}{D^3 - d^3} \int_{d/2}^{D/2} f(r) r^2 \, dr \quad (32)$$

So, an attempt has been made to take account for the local heterogeneous distribution of matter in the calculation of the mean thermochemical state. Substituting F_1 and $L = d/\alpha_2^{1/3}$ for $\bar{f}$ and D in the late equations, we can obtain the local-instantaneous composition of the gas mixture according to equation (28) and by integration calculate the corresponding mean mass fractions $X_{1,m}$ (L is the particle mean free path).

In parallel, in order to take account for the turbulent fluctuations, the probability density function for mixture fraction $P_1(f)$ is introduced making it possible to write :

$$X_{1,m} = \int_0^1 x_m(f) P_1(f) \, df \quad (33)$$

The p.d.f. is modelled as described by beta functions, whose parameters a and b can be determined explicitly from the values of F_1 and $\overline{f''^2}$:

$$F_1 = \int_0^1 f P_1(f) \, df \qquad \overline{f''^2} = \int_0^1 (f - F_1)^2 P_1(f) \, df$$

$$P_1(f) = \frac{f^{a-1}(1-f)^{b-1}}{\int_0^1 f^{a-1}(1-f)^{b-1} df} \qquad (34)$$

The transport equation for $\overline{f''^2}$, neglecting the fluctuations of the interfacial mass transfer, is written :

$$\alpha_1 \rho_1 \frac{\partial \overline{f''^2}}{\partial t} + \alpha_1 \rho_1 U_{1,j} \frac{\partial \overline{f''^2}}{\partial x_j} = \frac{\partial}{\partial x_j} \alpha_1 \rho_1 K_1^t \frac{\partial \overline{f''^2}}{\partial x_j} \qquad (35)$$

$$- <\rho u''_j f''>_1 \frac{\partial F_1}{\partial x_j} - \frac{2}{C_f} \alpha_1 \rho_1 \frac{\varepsilon}{q^2} \overline{f''^2}$$

Solution procedure

The governing differential equations were solved using an orthogonal finite-difference numerical solution procedure based on the fractional step method and developped at the Laboratoire National d'Hydraulique for the purpose of reactor thermal-hydraulic analysis (Viollet 1987).

All transport equations, are solved in the same way, using an unsteady algorithm, written in the general form as :

$$\frac{G^{n+1} - G^n}{dt} = - U_j^n \frac{\partial}{\partial x_j} G^n + \frac{\partial}{\partial x_j} K^n \frac{\partial}{\partial x_j} G^{n+1} + S_x^n + S_i^n G^{n+1} \qquad (36)$$

where G^n and G^{n+1} are the values of the variable G at the nth and $(n+1)^{th}$ time steps.

The solution of (36) is divided in two steps :

a) Advection step, solving :

$$\frac{\hat{G} - G^n}{dt} = - U_j^n \frac{\partial G^n}{\partial x_j} \qquad (37)$$

The solution $\hat{G}$ is obtained using a two dimensional characteristics method, after Esposito (1981), with third order interpolation allowing to minimize numerical diffusion.

b) Diffusion step, solving :

$$\frac{G^{n+1} - \hat{G}}{dt} = \frac{\partial}{\partial x_j} K^n \frac{\partial G^{n+1}}{\partial x_j} + S_x^n + S_i^n G^{n+1} \qquad (38)$$

For the computation of velocity components, a third step is required in order to prescribe the mass conservation equation (1). Mean pressure P and mass concentration $\alpha_2 \rho_2$ are computed on a staggered subgrid by solving the mass-conservation equations (1-18) respectively for gas mixture and particles.

The wall boundary conditions are written using wall function techniques for the continuous phase, and normal derivatives equal to zero for the dispersed phase. Accurate treatment of the wall grid nodes allows to describe oblique boundaries with respect to the orthogonality of the grid.

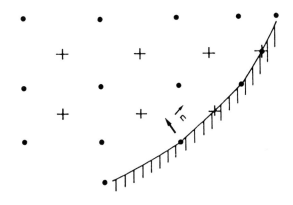

• grid for $U, H, T, q^2, \varepsilon$ .....
+ grid for pressure P, concentration $\alpha_2 \rho_2$

Figure 1 : Grid arrangement, and representation of a
boundary not parallel to grid axis

Table 1 - Physical Constants and Parameters

Value		Description
Physical constants		
P	$3.546 \cdot 10^5$ Pa	Mean pressure
ν_1	$0.937 \cdot 10^{-4}$ m²/s	Molecular viscosity of gas (1600°C)
K_1	$1.540 \cdot 10^{-4}$ m²/s	Molecular diffusivity of gas (1600°C)
ρ_2	1200 kg/m³	Density of raw coal
$C_{p,2}$	$1.80 \cdot 10^3$ J/kg.°C	Specific heat of coal
Turbulence model		
C_μ	0.09	Eddy viscosity coefficient
P_r^t	1	Turbulent Prandtl number
σ_{q^2}	1	Turbulent Schmidt number for q^2, the turbulence kinetic energy in the continuous phase
σ_ε	1.3	Turbulent Schmidt number for ε, the rate of dissipation of q^2
$C_{\varepsilon 1}$	1.44	Constant in the ε transport equation
$C_{\varepsilon 2}$	1.92	Constant in the ε transport equation
σ_p	1	Particle turbulent Schmidt number
C_f	1.6	Constant in the $\overline{f''^2}$ transport equation
Devolatilization model		

Subscripts 1 and 2 refer to parallel devolatilization reactions and parameters are taken from the work of H. Kobayashi and al. (1976).

Y_1	0.39	Stoichiometric coefficient
Y_2	0.80	Stoichiometric coefficient
E_1	$7.4 \cdot 10^4$ J/mole	Activation energy
E_2	$2.5 \cdot 10^5$ J/mole	Activation energy
A_1	$3.7 \cdot 10^5$ sec^{-1}	Arrhenius coefficient
A_2	$1.3 \cdot 10^{13}$ sec^{-1}	Arrhenius coefficient
h_{dv}	$1.724 \cdot 10^6$ J/kg	Heat of sublimation of the volatiles
X_A	0.15	Initial concentration of ash in the coal particles
Combustion model		
f_s	0.055	Stoichiometric mixing rate
W_1	$16 \cdot 10^{-3}$ kg/mole	Molar mass of the fuel (CH_4)
W_2	$28.9 \cdot 10^{-3}$ kg/mole	Molar mass of the oxidant (air)
W_3	$27.6 \cdot 10^{-3}$ kg/mole	Molar mass of the products
R	8.314 J/mole °C	Universal gas constant

3. PREDICTIONS

Application

Computations have been carried out for an axisymetric injection of pulverised coal in hot coflowing air flow. This work is only an aspect of a larger study of higher injection rates of coal in blast furnace tuyeres with the help of electric arc heaters. For a better inderstanding of the influence of overheating by plasma injection on the global devolatilization rate and char formation, numerical computations have been performed both without and with overheating, for inlet blast temperature T_1 = 1200°C and 1600°C respectively.

Table 2 - Input data used in numerical computations

Parameter	Value
Geometry	
Tuyere diameter	0.2 m
Tuyere length	0.5 m
Injection lance diameters (inner/outer)	16/20 mm
Coflowing air flow	
Mean velocity	120 m/s (1200°C)
	150 m/s (1600°C)
Pulverised coal injection and primary air	
Mean temperatures of gas and particles	40°C
Mean velocities of gas and particles	50 m/s
Coal mass flow rate	0.24 kg/s
Mass loading ratio of coal particles	6
Particle diameter	20 and 50 x 10^{-6} m

The calculated results were obtained with 36 x 59 grid nodes and are virtually grid-independant. The time step used in unsteady algorithm (dt $\simeq 10^{-5}$ sec.) corresponds to a reference Courant number ncou = 0.5 based on the inlet blast velocity and the minimum axial step size. Final steady state solution is believed to be reached after NT = 2000 time steps.

Results and discussion

From the aerodynamical point of view, the tuyere is too short to allow a complete mixing of the two-phase injection with the coflowing air flow. Moreover, the particle mean velocity and temperature are lower than the corresponding gas characteristics as the inertia of the particles is greater than the one of the gas. This effect is accentuated for larger particles, as shown by the predicted radial profiles of mean relative velocity for 20 μm and 50 μm coal particle injection (figure 2-3) and is physically due to the decrease in the surface-to-volume ratio, as the particle size increases, which reduces the interfacial transfer rate.

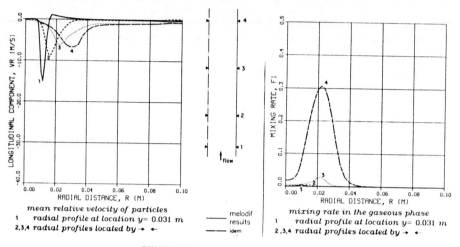

Figure 2 : PULVERISED COAL INJECTION IN HOT AIR FLOW
BLAST TEMPERATURE T1-1200C ; PARTICLE DIAMETER D-20 μm

The smaller particles heat up faster and thus devolatilize at an enhanced rate (figure 2-3) ; this effect is nearly explosive due to the overheating induced by combustion of emitted volatiles. So when the particle size decreases from 50 μm to 20 μm, the global rate of devolatilization at the tuyere exit increases from less than 0.5 % to 48 % of the whole coal injection. Moreover, the volatiles released by devolatilization process correspond respectively to 42 % and 65 % of the pyrolysed raw-coal, due to the competition of parallel devolatilization reactions.

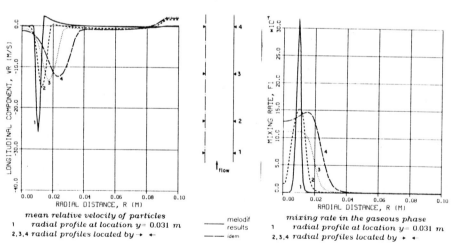

Figure 3 : PULVERISED COAL INJECTION IN HOT AIR FLOW
BLAST TEMPERATURE T1-1200C ; PARTICLE DIAMETER D-50 μm

Figures 4-5 show the predicted radial profiles of two-phase mean temperature for particle diameter d = 20 μm and 50 μm respectively. For smaller particles, it is obvious that combustion of the volatiles dominate the reaction process. A diffusion flame clearly appears in the middle of the tuyere with a peak temperature of 2500°C when the mixing rate reaches stoichiometric value. The gap between the mean temperatures of gas and particles increases in high devolatilization zone due to the energy absorbed by sublimation of the volatiles.

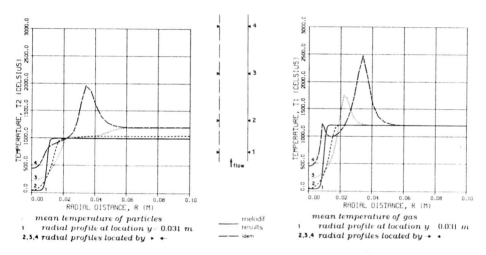

Figure 4 : PULVERISED COAL INJECTION IN HOT AIR FLOW
BLAST TEMPERATURE T1-1200C ; PARTICLE DIAMETER D-20µm

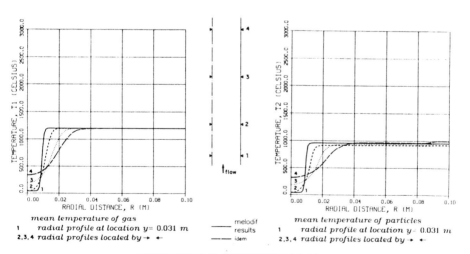

Figure 5 : PULVERISED COAL INJECTION IN HOT AIR FLOW
BLAST TEMPERATURE T1-1200C ; PARTICLE DIAMETER D-50µm

An increase of the inlet blast temperature involves an overheating of the coal particles and consequently leads to increase the rate of devolatilization (figure 6). It is an abrupt process due to both the non-linearity of reaction rate depending on particle temperature and the coupling with combustion of volatiles released in the gaseous mixture.

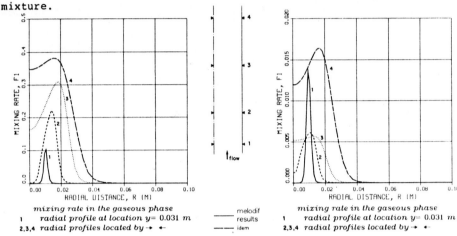

Figure 6 : PULVERISED COAL INJECTION IN HOT AIR FLOW
BLAST TEMPERATURE T1-1600C

PARTICLE DIAMETER D-20 µm   PARTICLE DIAMETER D-50 µm

When the inlet blast temperature increases from 1200°C to 1600°C, the global rate of devolatilization for 20 μm and 50 μm particles is multiplied by a factor about 1,5 and 5 respectively. The diffusion flame obtained for smaller particles also appears under these conditions but near by the injection lance and with a peak temperature of 2700°C (figures 7-8).

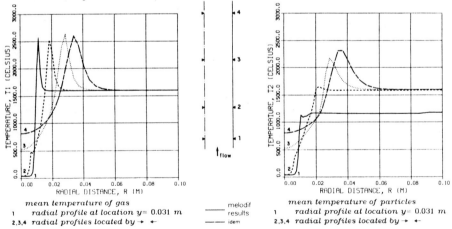

Figure 7 : PULVERISED COAL INJECTION IN HOT AIR FLOW
BLAST TEMPERATURE T1-1600C ; PARTICLE DIAMETER D-20 µm

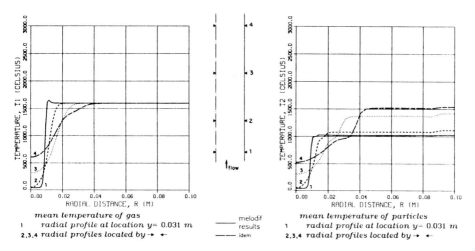

Figure 8: PULVERISED COAL INJECTION IN HOT AIR FLOW
BLAST TEMPERATURE T1-1600C ; PARTICLE DIAMETER D-50 μm

Previous results were obtained with the simplified laminar homogeneous model for the combustion process of the evolved volatile matter. To omit taking into account the local instantaneous distribution of mixing rate must involve an over-estimate of temperature in combustion areas.

Calculations have been carried out introducing the concentric distribution governing the gas mixture fraction around pyrolysing particles as defined by equations 31-32. Results seem not appreciably affected by this phenomena due to the fact that devolatilization and combustion generally occured in separated areas. Meanwhile, it was found a very light decrease in the global rate of devolatilization and also in the peak temperature near by the appearance region of flame where devolatilization occures in an oxidant environment (figure 9).

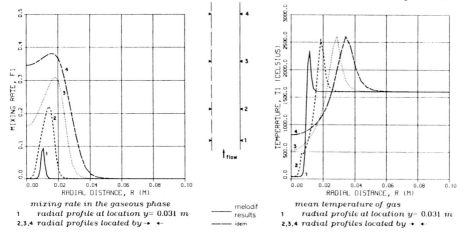

Figure 9: PULVERISED COAL INJECTION IN HOT AIR FLOW
INFLUENCE OF DISTRIBUTION OF MATTER AROUND PARTICLE
BLAST TEMPERATURE T1-1600C ; PARTICLE DIAMETER D-20 μm

On the other hand, turbulent combustion model, with presumed p.d.f. as defined by equations 33-34, is necessary to provide adequate description of the flow. Meanwhile, the global rate of devolatilization seems also no appreciably affected and shows a very light increase. The mean temperature profiles of gas are smoothed and the corresponding peaks for 20 μm particles, with an inlet blast temperature of 1200°C and 1600°C, decrease of about 200°C and 350°C respectively (figures 10 - 11).

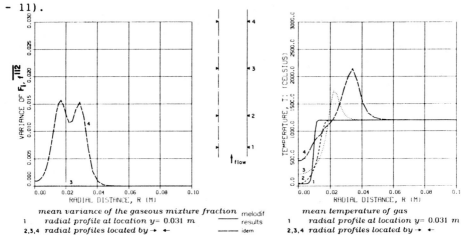

Figure 10: PULVERISED COAL INJECTION IN HOT AIR FLOW
INFLUENCE OF FLUCTUATIONS IN MIXTURE FRACTION (P.D.F)
BLAST TEMPERATURE T1-1200C ; PARTICLE DIAMETER D-20μm

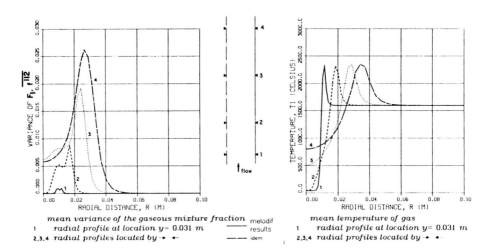

Figure 11: PULVERISED COAL INJECTION IN HOT AIR FLOW
INFLUENCE OF FLUCTUATIONS IN MIXTURE FRACTION (P.D.F)
BLAST TEMPERATURE T1-1600C ; PARTICLE DIAMETER D-20μm

4. CONCLUSION

A two-dimensional separated flow model -named MELODIF- including pyrolysis of coal particles and combustion of volatile matter has been used to predict an axisymetric injection of pulverised coal in a hot coflowing air flow. It was found that devolatilization process is nearly explosive, due to coupling with the combustion of volatiles released in the gaseous phase. This process is enhanced with a decrease in particle diameter and/or an increase in inlet coflowing air flow temperature.

The local gaseous combustion process, due to the heterogeneous distribution of matter surrounding individual pyrolysing particles, was considered in this model by introducing the concentric distribution governing the gas mixture fraction around a single particle which undergoes devolatilization in an infinite atmosphere. Results seem not appreciably affected by this phenomena due to the fact that devolatilization and combustion generally occured in separated areas. On the other hand, turbulent combustion model, with presumed p.d.f., seem necessary to provide adequate description of mean gas temperature field.

Further refinement of this model requires consideration of the interaction between particles and the turbulence of the continuous phase using extra source terms in the governing equation of the $q^2-\varepsilon$ eddy viscosity model. In addition, the non-linearity of the devolatilization rate must probably be taken into account for the computation of the mean rate of interfacial mass transfer. But firstly, validations against experimental data are needed to confirm the universality of the available modelling for predictive purposes.

REFERENCES

DELHAYE J.M., 1974, "Jump conditions and entropy sources in two phases systems. Local instant formulation". int. J. Multiphase Flow, Vol. 1 pp 395-409.

ELGHOBASHI S., T. ABOU-ARAB, M. RIZK & A. MOSTAFA, 1984, "Prediction of the particle-laden jet with a two-equation turbulence model", Int. J. Multiphase Flow, Vol. 10, 6, pp 697-710.

ESPOSITO P. 1981, "Résolution bidimensionnelle des équations de transport par la méthode des caractéristiques", EDF-LNH rep. HE 41/81.16.

GOUESBET G., DESJONQUERES P., & BERLEMONT A., 1987, "Eulerian and Lagrangian approaches to turbulent dispersion of particles", Seminaire International "Transient phenomena in multiphase flow", I.C.H.M.T., Dubrovnik.

ISHII M., 1975, "Thermo-Fluid Dynamics of two-phase flow", Eyrolles, Paris.

H. KOBAYASHI, J.B. HOWARD & A.F. SAROFIM, 1976, "Coal devolatilization at high temperatures", 16th Int. Symp. on Combustion. The Combustion Institute, Pittsburgh, pp 411-425.

LAUNDER B.E., SPALDING D.B., 1974, "The numerical computation of turbulent flows", Comp. Mech. in applied mech. and eng., 3.

MODARESS D., TAN H. & ELGHOBASHI S. 1983, "Two-component LDA measurement in a two-phase turbulent jet". AIAA 21st Aerospace Sci. Meeting. Reno. Nevada.

SIMONIN O., ROBINSON J. & BARCOUDA M. 1987, "Two-dimensional numerical simulation of turbulent bubbly flows across tube bundle", Proc. 22nd IAHR Congress (topics in industrial hydraulics) pp 98-103. Water Ressources Pub.

SIMONIN O., 1987, "Modélisation de l'injection de charbon pulvérisé dans la tuyère d'un haut fourneau", EDF-LNH rep. HE/44/87.05

SLEZAK S.E., 1985, "A model of flame propagation in rich mixtures of coal dust in air". Combustion and Flame 59 : 251-265.

UBHAYAKAR S.M., STICKLER D.B. & GANNON R.E., 1977, Fuel $\underline{56}$, 281.

VIOLLET P.L., 1985, "Modélisation numérique du mélange d'un jet de plasma en extinction dans un écoulement froid en conduite Bulletin DER N° 2 - 1985 (série A).

VIOLLET P.L. 1985, "The modeling of turbulent recirculating flows for the purpose of reactor thermal-hydraulic analysis", Nuclear Eng. & Design, 99, 365-377.

WALLIS G.B., 1969, "One-Dimensional Two-Phase Flow". MC Graw-Hill, New-York.

FLAME STABILIZATION IN A SUPERSONIC COMBUSTOR

M. Barrère
Office National d'Etudes et de Recherches Aérospatiales
BP 72 - 92322 Châtillon Cedex, France

1 - Introduction

The combustors using supersonic combustion are characterized by two fundamental problems:
- the stabilization of the flame in a supersonic air flow,
- the maximum combustion efficiency.

We will not discuss the second problem in connection with the mixing efficiency of the process. Figure 1 shows this point and underlines the strong dependence between combustion efficiency and mixing efficiency.

In this paper, we analyze different types of flame stabilization devices.

Our approach is a simple one and used ZELDOVITCH approach and the fact that, in a turbulent flow the similitude in transfer coefficients is, in general, verified [1,2].

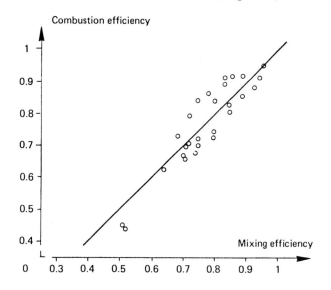

Fig. 1   *Strong dependence of the mixing efficiency on combustion efficiency (NASA-TMX 2895).*

2 - General flame stabilization

We discuss four processes:
1) the recirculating flow
2) the propagation of the flame ($u = v_{bn}$)
3) the conditions of spontaneous ignition
4) the production, in the flow, of gas pockets or vortices of fresh mixture, which burn.

2.1 - Recirculation flow

This is the most popular way of stabilizing a flame in a subsonic flow. It is shown in figure 2: a baffle is placed in the flow to produce a recirculating flow stabilizing the flame in the stream. This recirculation zone is analyzed like a "stirred reactor". The stability limits (lean and rich) are given by this analyzis, for instance, in the graph of figure 3, where we plot the mass flow rate in, as a function of the temperature T. The S shape curve shows three parts: an unstable area in the middle with two stable solutions to either side, limited by the temperatures : T_1 and T_2. The existence of the flame can be explained easily by this analyzis of the recirculating flow.

In figure 4, where we graph velocity against equivalence ratio, we see the influence for a given equivalence ratio a, on the rich side, of the velocity B at which the flame is blown off.

In this problem of stabilization, the pressure is an important parameter. For instance, the size of the volume of recirculating zone gives the altitude at which it is possible to reignite the combustor.

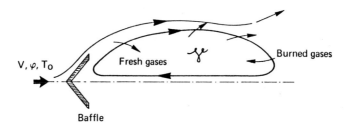

Fig. 2 Flame stabilization by a baffle to obtain a recirculating flow.

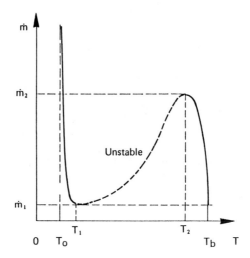

Fig. 3 Stabilization of the flow rate in the recirculating flow.

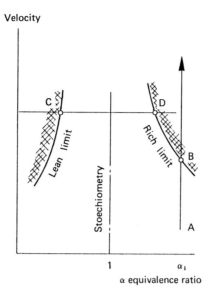

Fig. 4   Flame stability.

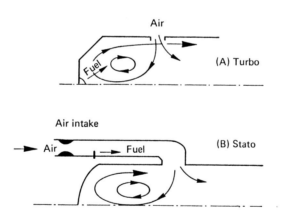

Fig. 5   Stabilization of the flame in practical combustors.

This device is used, at the same time, in the primary combustor of a turbojet and in the reheat part. The ramjet combustor also uses this device (figure 5).

2.2 - Stabilization by propagation condition

At the confluence of two flows (Air and Fuel for instance), the flame is stabilizezd when, in the mixing zone, the velocity of the flow u is equal to the normal flame velocity propagation v_{bn}. We assume a SCHMIDT number $S_c = v/D$ of (v, where is the kinematic viscosity (μ/ρ) and D the mass diffusion coefficient.

The velocity u in the mixing zone and the velocities of the two flows u_1 and u_2 are a function of the composition and are given by:

$$\frac{u - u_2}{u_1 - u_2} = f(Y_j)$$

Y_j is a mass fraction of species j.

In a diffusion flame, BAEV [4] gets the expression:

$$\frac{u - u_2}{u_1 - u_2} = \frac{1}{1 + ar_{st}}$$

where r_{st} is a stoechiometric coefficient (in the Air-Hydrogen mixture: $r_{st} = 34.5$), a is the excess Air. It is equal to unity at stoechiometry, a is less than one on the lean side and a greater than one on the rich side.

Representing the diffusion flame with an other way, we get [2, 3]:

$$\frac{u - u_2}{u_1 - u_2} = \frac{1}{1 + \Phi},$$

where Φ is the equivalent ratio.

So, either way we get the relation:

$$\frac{u - u_2}{u_1 - u_2} = \frac{1}{1 + m},$$

and m is given by the composition.

The normal velocity propagation $v_{bn} = e_b/\tau_c$, e_b is the thickness of the combustion zone and τ_c a characteristic combustion time. The stability condition, with ratio of the two velocities $\phi = u_2/u_1$, is given by:

$$\frac{u_1 \tau_c}{e_b} = \left(\frac{1 + m\phi}{1 + m}\right) = 1$$

The thickness δ_f of the boundary layer in the stabilization region depends on the abscissa with the axis of the figure 6.A and B:

$$\delta_f = C X_f = C L$$

C is a constant, for a first approximation; e_b is part of δ_f, $e_f \simeq \delta_f/n$, so we have a first approximation for the position of the flame L.

$$L = \frac{n u_1 \tau_c}{C} \times \left(\frac{1 - m\phi}{1 + m}\right);$$

if $C = C(\phi)$, a function of the ratio ϕ of the two velocities, then:

$$L_\phi = \frac{n u_1 \tau}{1 + m} \frac{1 + m\phi}{C(\phi)}.$$

From the experimental results, BAEV [4] gets the correlation (figure 7)

$$\frac{L_\phi}{L} = a \left|(1 + \phi m)\left(1 + \frac{1}{0.015}\right)^{\frac{1}{2}} - b\right|$$

In general a good correlation is obtained by comparing the momentum ρu^2 of the two streams $\rho_1 u^2_1/\rho_2 u^2_2$:

$$\frac{L_\phi}{L} = f\left(\frac{\rho_1}{\rho_2} \times \frac{u^2_1}{u^2_2}\right) = f(R\,\phi^2)$$

R is the ratio ρ_2/ρ_1 of densities.
This result is well verified for different Fuels and various conditions (figure 8).

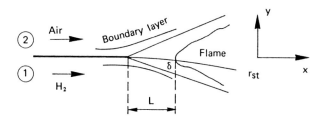

Fig. 6.A   Stabilization of the flame in the case of two parallel flows.

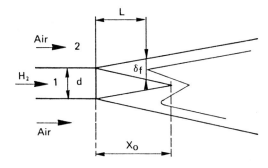

Fig. 6.B   Diffusion flame stabilized at the exit of a Fuel tube.

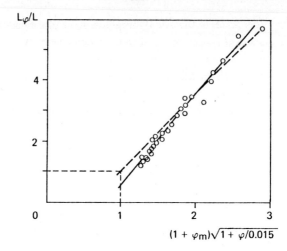

Fig. 7   *Mixing effect on the detachment length [4].*

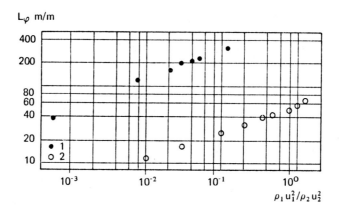

Fig. 8   *Detachment length of the flame:*
            1) Exp. FERRI - 2) Exp. BAEV [4].

2.3 - Spontaneous ignition

In this case, the ignition conditions are:

$$X_i = u\, \tau_i,$$

where X_i is the distance at which spontaneous condition occurs, u is the flow velocity and τ_i the induction time.

With this approach, the combustion should be modelled by two characteristic times:
- the *induction time*, $\tau_i \simeq B\, exp(T_A/T)$ which is a function of the temperature and is given by the first chemical reactions,
- the *heat release time* τ_r, which is the time during which the main reactions of combustion occur.

In the case of the combustion of the Air-Hydrogen system in a flow having velocity $u = 1000$ m/sec, a temperature of $T = 1000$ K and an induction time of 10^{-4} sec, the value of $X_i = 10$ cm and stabilization is possible at a realistic length.

The analyzis is the same with the assumption of the *flame propagation* $(L_o)_p = u\, \tau_c$ and with the *spontaneous ignition* $(L_o)_S = u\, \tau_i$.

The choice depends on the temperature level of the Air stream: at high temperature, for instance, $\tau_i < \tau_S$.

2.4 - Formation of pockets

An other way of stabilizing the flame is to form pockets of gases at the confluence of:
a) two streams of Air and Fuel or
b) two streams of a mixture oa Air and Fuel and hot gases (figure 9).

Each pocket has, in the relative motion, a life of its own.

For instance, with reference axis in the relative motion of the pocket, each pocket has an ignition process and a normal velocity propagation v_{bn}. The ignition condition is given by:

$$\left(L_o\right)_a = u\, \tau_a$$

u is the velocity of the pocket and τ_a is an ignition lag that is a function of the square of the diameter Δ of the pocket ($\tau \propto \Delta^2$), and also a function of the external conditions.

In conclusion, we have two types of devices:
1) recirculation zone
2) propagation : τ_p
    spontaneous ignition: τ_i
    pocket formation: τ_a

with an ignition lag which defines the process τ_p, τ_i or τ_a.

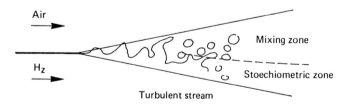

Fig. 9   Pocket formation of fresh gases.

3 - Flame stabilization in supersonic combustor

We give here four techniques used in supersonic combustors.

3.1 - Stabilization by parietal injection

This device is given in the figure 10.A. Fuel is injected at the surface of the wall, at a certain angle from the surface. In a supersonic flow, this injection induces an oblique shock and, upstream, a separation flow. In this part of the flow, we note two recirculating zones on either side of the jet. A recompression shock wave appears after this bubble and the combustion occurs after the induction time τ_i. This device has two principal disadvantages:
1) the stabilization depends on the volume of the recirculating flow and this volume is small.
2) the combustion in the boundary layers increases the heat transfer to the wall, and the integrity of this device is questionable.

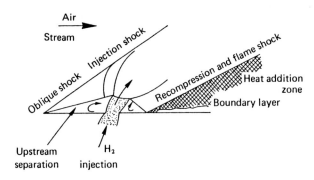

Fig. 10.A   Flame stabilization by parietal injection.

3.2 - Stabilization by a step

This process is shown in figure 10.B; a step or many steps are placed in the wall of the combustion chamber, with injection of Fuel in the step parallel to the surface and in the recirculating flow. The supersonic stream parallel to the wall is deflected by decompression upstream and compression downstream, thereby forming a shock wave. The Fuel is mixed with Air and a flame is stabilized in the boundary layer. The same problem occurs, as with the previous process: a large heat transfer at the wall.

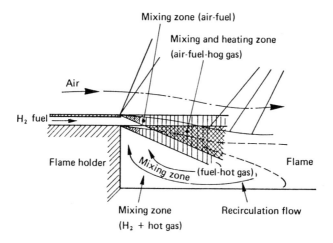

Fig. 10.B Stabilization by a step and parietal injection.

3.3 - Stabilization by a normal shock

This process is like a detonation. A normal shock is produced at the exit of a supersonic nozzle (figure 10.C). The premixed stream of Air and Fuel passes through the normal shock and after the induction time a spontaneous ignition occurs and the combustion is stabilized by the normal shock. This process is characterized by a large loss in the normal shock and the flux is limited.

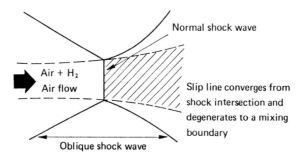

Fig. 10.C Combustion behind a normal shock.

3.4 - Stabilization by pocket formation

This process is described in figure 11. A nozzle of Hydrogen is placed in a supersonic flow of Air. This device is shown in figure 16. The flame is stabilized at a distance from the nozzle in the mixing flow as in figure 6.B. The experiment of figure 17 shows a setup with a nozzle of Hydrogen in the center of a nozzle of Air. Figure 12.B shows the stabilization of a flame in the Air stream at a velocity of Mach 2.5. The stagnation Air temperature is 2300K. The central nozzle of Hydrogen is at the same velocity $u_1 = u_2$. The flame is stabilized at

$L\phi/d \simeq 1.5$, d is the Hydrogen nozzle exit diameter. The values of $\rho_1 u^2_1/\rho_2 u^2_2 \simeq 0.25$ and $L\phi/d \simeq 30$ mm are of the same order of magnitude as in BAEV experiments (figure 8).

In general, in this type of stabilization, we need to have information on the length of flame detachment $L\phi$, the length of the lift off or the breakway of the flame and we have to know the stability of $L\phi$.

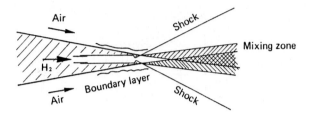

Fig. 11   Stabilization at the exit of a Hydrogen nozzle.

In the case of a turbulent flame the propagation velocity

$$v_{bu}^2 = \frac{u' l_t}{\iota_c}$$

where u' is the fluctuating velocity, l_t the scale of the turbulence.

The stability condition is:

$$\frac{v_{bu}^2}{u_1^2} = \left(\frac{1+\phi m}{1+m}\right)^2$$

We introduce the height of the potential cone X_o (figure 6.B) to qualify the mixture. The length L_S at the lift off of flame is the following. We introduce a dimensionless parameter by dividing the length by the diameter d of the Hydrogen nozzle

$$x_o = \frac{X_o}{d}, \quad l_S = \frac{l_S}{d},$$

so the value of l_S is:

$$l_S = \frac{1}{4x_o^2}\left(\frac{1+m}{1+\phi m}\right)^2$$

BAEV [4] gives the relation

$$l_S = \frac{(1+m)^2}{4 x_o^2} \times \frac{1}{1 + R\,\Phi^2|\sigma^2 - 1|}$$

where $\sigma = d_2/d_1$.

For an Hydrogen-Air mixture, he gets the expression:

$$l_S = \frac{378}{x_o^2\left|1 + R\,\phi^2(\sigma^2 - 1)\right|}$$

The fluctuation of L_ϕ in 1D and 2D configuration (figure 13) is a difficulty of this type of stabilization.

The frequency variations observed are of 10 sec^{-1} to 100 sec^{-1}, and are a function of the nature of the Fuel, the geometry and the turbulent structure of the flow.

For a 1D analysis, the maximum fluctuation velocity u'_{max} and the amplitude A of these oscillations for a sinusoidal type oscillation is given by:

$$A = u'_{max} \int_0^{\frac{1}{4n}} sin(2n\omega t)\, dt = \frac{u'_{max}}{2n\omega};$$

The coefficient of turbulent transfer is

$$D_t = 2n\omega A^2$$

The maximum gradient of the velocity fluctuations u'_{max}/d is:

$$\frac{u'_{mlax}}{d} = \frac{2n\omega A}{d} \simeq \frac{1}{t_c}$$

This result underlines the importance of the characteristic combustion time:

$$t_c = \frac{d}{2n\omega A} = \frac{1}{2n\omega A}$$

When the Fuel is liquid in a supersonic stream, the configuration of the flame changes. Figure 14 shows this configuration. The Kerosene is injected in liquid phase at three points:
a) along the axis
b) near this axis and
c) near the wall, the white trails are drops in the stream.

The deformation of drops by the flow is an important problem but it is sure that the liquid Fuel does in fact vaporize and burn in the supersonic stream. The mechanism of combustion of drops is the same as in pocket combustion when the Fuel is vaporized, and produces rich gas pockets.

Fig. 12.A Stabilization with two coaxial nozzles (H_2 in the center and Air outside).

Fig. 12.B   Hydrogen flame stabilization, Air at M = 2.5.

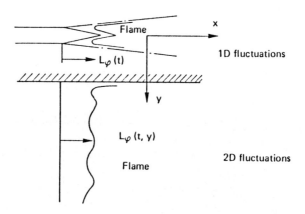

Fig. 13   Oscillations of the detachment length.

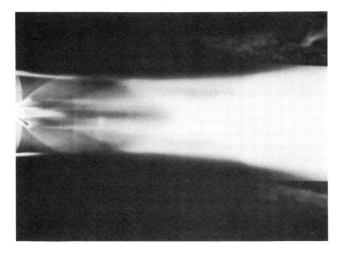

Fig. 14   Kerosene combustion in a supersonic Air flow.

4 - The supersonic combustor

Different types of combustors are planned or already built and have been published.

We have chosen five combustors:

a) a classical combustor proposed by SWITHENBANK is represented in figure 15.A. Steps stabilize the flame with parietal injection [5].

b) NASA scramjet combustor, with struts and flame stabilization by steps with parietal injection of Hydrogen, is given in figure 15.B [6].

c) a combustor with stabilization by a pilot flame is shown in figure 15.C; a part of the Air flow mixed with the Fuel at subsonic state burns classically in a subsonic combustion chamber of a ramjet and the sonic jet burned gases stabilize the supersonic part of the Air-Fuel flow mixture.

d) the ESOPE combustor is diagrammed in figure 15.D. This combustor has the posibility to operate in subsonic and supersonic combustion, with two sections of Hydrogen injection: one downstream for subsonic combustion, the other upstream (section 1*) for supersonic combustion [7,8,9,10].

e) In certain designs, parietal injection has the disadvantage of releasing high heat, near the wall, leading to a large heat transfer. We propose an other way, represented in figure 15.E, with Fuel injection in the wake of a profile. The main problem in this solution, is the possible fluctuations of the flame and the low mixing efficiency in a supersonic combustor.

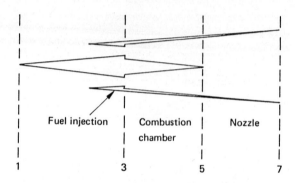

Fig. 15.A Classical supersonic combustor [5].

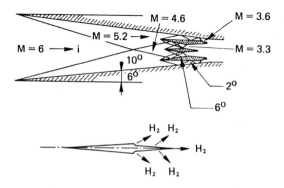

Fig. 15.B Supersonic combustor of the NASA Mach = 7.6 - Strut element [6].

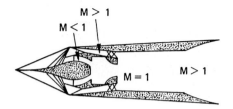

Fig. 15.C Combustor with a pilot flame for stabilization.

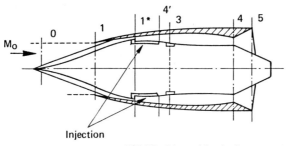

ESOPE with combined sub- and supersonic combustion

Fig. 15.D ESOPE combustor with subsonic and supersonic regimes /9/.

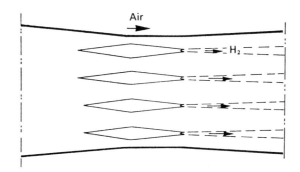

Fig. 15.E Proposed 2D supersonic combustor with Hydrogen injection in the tail.

5 - Conclusion

We would like to make three remarks in conclusions:
1) we have to choose between two types of stabilization device:

    a) organization of recirculating flow at high velocity,

    b) a stabilization of the flame at a certain distance from the nozzle exit that depends on a characteristic time:

    - τ_c for propagation,
    - τ_i for spontaneous ignition
    - τ_a for pocket ignition.

    We have two main problems:
- Heat transfer for (a) and flame fluctuation for (b).

2) We underline the advantages of a 2D combustor:
- stabilization by struts with minimum losses
- simpler modelling
- possible modular combustor to save time and money
- configuration well fitted for a 2D nozzle, with or without variable geometry.

3) at the time, ONERA ended its studies of supersonic combustion, we recognized the difficulties of supersonic combustion: to get for instance good flame stabilization and to have good mixing and good combustion efficiency.

A great deal of effort is now devoted to the computational approach. We support this orientation but :
- more fundamental Physics is necessary
- new methods are required for complex phenomena
- computer capacity must be increased.

Certain alternatives that were difficult to study in the past are now coming within reach, in particular with new high temperature composites; but supersonic combustion remains a complicated and unsolved problem of fluid mechanics.

Bibliography

[1] Y. ZELDOVITCH: Théorie Mathématique de la Combustion. Publication de l'Académie Soviétique, Moscou, 1980.

[2] F.A. WILLIAMS: Combustion Theory. Addison-Wesley, 1965.

[3] R.W. BILGER: The Structure of Diffusion Flames. Combustion Science and Technology, Vol. 13, n° 1-6, pp. 155-168, 1976.

[4] V.K. BAEV, W.A. YASAKOV: Stability of a Diffusion Flame in Single and Mixed Jets. Fizika Goreniya i Wzryva, Vol. 11, n° 2, pp. 163-178, March, April 1975.

[5] M.T. JACQUES, R. PAYNE, J.SWITHENBANK: Studies Leading to the Realization of Supersonic Combustion in Propulsion Applications. H/C. 174, 1972.

[6] J.R. HENRY, G.Y. ANDERSON: Design Considerations for the Airflame. Integrated SCRAMJET. NASA, 1er ISABE, Marseille, 1972.

[7] A. MESTRE, L. VIAUD: Combustion Supersonique dans un Canal Cylindrique. AGARD, p. 93-111, London, 1963.

[8] F. HIRSINGER: Optimisation des Performances d'un Statoréacteur Supersonique. ISABE, 1972. T.P. ONERA n° 1106.

[9] R. MARGUET: Propulsion des Véhicules Hypersoniques. N.T. ONERA n° 196, 1970.

[10] 0. LEUCHTER: Problèmes de Mélange et de Combustion Supersonique d'Hydrogène dans un Statoréacteur Hypersonique. D.G.L.R., Trauen, mai 1971.

MIXING PROBLEMS IN SUPERSONIC COMBUSTION

J. Swithenbank, F. Boysan, B. C. R. Ewan,
L. Shao and Z. Y. Yang

ABSTRACT

The attainment of satisfactory supersonic combustion requires that the fuel and air be mixed to a sufficient extent *at the molecular level*, at a high enough temperature, so that the combustion reaction is largely complete before the gases leave the combustor. The macro-mixing must be achieved through turbulence, and molecular diffusion effects are only significant within the dimensions of the viscous eddies. The effects of macro-mixing only extend for the dimensions of the macro-scale eddies, hence the fuel injectors must be separated by a distance not exceeding twice the macro-scale. Since the turbulence is produced at the expense of the kinetic energy of the flow, there is an engine thrust penalty as the level of turbulence is increased. On the other hand, combustion efficiency and therefore thrust is limited by the degree of mixing. Hence there is an optimum level of turbulence, mixing and combustion efficiency in a scramjet engine.

The design of practical supersonic combustors entails arranging the geometry so that this optimum can be achieved. Various systems have been investigated experimentally at Sheffield University and it should be noted that interaction between the pressure gradients due to heat release and the flow must be taken into account since there may be large differences between the flows with and without combustion. Thus calculation procedures are required to accurately predict the supersonic (or mixed supersonic-subsonic) turbulent reacting flows. One of the most successful practical geometries which we have tested involves the generation of vortices within the flow by injectors spanning the stream, with their trailing edges swept back behind the local Mach angle. These produce multiple vortices within the flow, and in common with other vortex flows will have turbulence which is highly anisotropic. We have already developed a very successful Reynolds stress code for subsonic swirling flow calculations and this technique may be extended to supersonic flows.

(NB Parts of this paper were covered in Ref. 10. However, since that document is of very limited availability, is was considered that its inclusion here would be helpful).

NOMENCLATURE

A,B   constants

D	dissipation
h	static enthalpy
I	specific impulse
K.E.	kinetic energy
KE_{def}	wake K.E. deficit
K_d	process efficiency
l_e	average size of energy containing eddies
M	Mach number
η_c	combustion efficiency
η_m	mixing efficiency
η_d	kinetic energy of diffuser
p	static pressure
P	total pressure
Q	heat release rate
q	$\rho U^2 / 2$
ρ	density
T	temperature
τ_s	reactor stay time
τ_d	diffusion time
u'	r.m.s. turbulence velocity
U	mean velocity
W	mass flow rate
X	reactor volume
λ	characteristic dimension
ψ	fraction of oxygen unreacted

Subscripts

1, 3, 5, 7	Engine stations (see Fig. 1)
f	fuel
def	deficit

INTRODUCTION

Hypersonic air breathing propulsion systems are proposed for both cruise and acceleration missions. From low supersonic speeds up to about Mach 5 the subsonic combustion ramjet provides an attractive propulsion system however as the flight Mach number increases, the diffusion losses in the intake rise to the point where the performance is inferior to the supersonic combustion cycle. The supersonic combustion ramjet, or SCRAMJET is the system principally considered within this paper, however the approach given is completely fundamental and generally applicable to any turbulent reacting flow system.

As the flight speed increases, the enthalpy of the air passing into the engine increases, until at Mach 5, the enthalpy of the air is approximately equal to the enthalpy of the fuel. Since the kinetic energy of the air is increasing with the square of the velocity, it follows that the energy of the air is dominant throughout the hypersonic range and the flow energy in the combustion system must be monitored carefully. This is particularly important when we produce turbulence to promote mixing in these high velocity flow systems.

At a flight Mach number of ten the Mach number in the scramjet combustor would be about three but the velocity is only a few percent below the flight velocity. The change in Mach number in the combustor is due to the change in the speed of sound in the flow caused by the increase in temperature through the intake diffusion process.

ENERGY CONSIDERATIONS

The above introduction highlights the role of energy flows in hypersonic propulsion systems and it is useful to expand on this fundamental aspect to provide insight into the objectives to be achieved in a scramjet combustor.

First of all, we should recall that temperature is the random mechanical motion of molecules ($T = mv^2/2k$) in which the r.m.s. velocity of the molecules v is of the same order as the speed of sound in the gas. The **velocity of the gas** is actually a directed motion superimposed on this random motion. Turbulence on the other hand is also a random motion but in this case groups of molecules are concerned. Exothermic chemical reactions within a gas are due to a rearrangement of the atoms, in which the excess energy appears as the rapid separation velocity of two or three molecules After a few collisions, this motion is randomised into temperature. Energy transfers between these forms of molecular motion occur by various mechanisms;

a) shock waves, which achieve a change from directed to random motion (accompanied by an increase in density) within two or three molecular collisions.

b) turbulence dissipation, in which the kinetic energy of turbulence decays, leading to an increase in gas temperature. At subsonic conditions this temperature rise is extremely small, but at hypersonic conditions the temperature rise is significant. In this case the energy flow is from the large eddies to the small eddies, until their Reynolds number (i.e. the ratio of their inertial to viscous forces) gradually falls towards unity, at which point the eddy only survives for approximately one revolution.

c) shear layers, in which turbulence is generated provided that the size of the eddy with a Reynolds number of unity is significantly smaller than the thickness of the shear layer. If the unity Reynolds number eddy is larger than the shear layer thickness, the turbulence which is tending to be formed by the velocity gradient is dissipated before it is formed and the flow remains laminar. These shear layers may be either on a surface, or free as in a wake. Ref. 1.

THE VEHICLE

Starting with the overall concept of a body moving at a constant high speed through the atmosphere, energy considerations show that the energy of the fuel is eventually transferred to increased molecular motions, or heating, of the atmosphere both directly and through the decay of the velocity profile in the wake. It will be noted that conservation of momentum in the wake is not violated by this energy transfer to thermal (random) motions of the molecules.

THE INTAKE

If we next turn our attention to the integrated engine-airframe vehicle itself, Fig 1, there are large energy flows occurring within a typical system. Starting with the lifting surface, there is direct transfer of some of the directed motion of the air molecules (velocity) to random motion (heat) in the shock wave structure. At the surface, a boundary layer is formed and some of the directed kinetic energy is converted to heat and also possibly to turbulence. Any heat which is transferred to the surface and radiated away is lost from the system, however heat which is transferred to an internal coolant can be recycled through the system and can contribute significantly to the system efficiency. The magnitude of this contribution can be estimated from Reynolds analogy which relates the heat transfer to the skin friction.

From the outline design shown in Fig. 1, the intake is located in the flow field of the lifting surface. This achieves several purposes, in particular the intake is less susceptible to changes in vehicle incidence, and the pressure rise across the shock wave contributes to the intake compression. The purpose of the intake is to diffuse the flow as nearly isentropically as possible. Any increase in entropy represents a performance loss. However, the diffusion process must be achieved over a wide range of Mach number, particularly in a vehicle designed for acceleration. This leads to well known design problems such as achieving starting of the intake, and maintaining high intake efficiency. The compression is achieved most efficiently by a large number of weak shocks emanating from the surface. If these shocks intersect within the capture flow, a shear layer will be formed, which can become turbulent if the Reynolds criterion discussed above is met. The lifetime of these eddies is important since they could contribute to turbulent mixing within the combustion chamber if they persist for the correct distance. The amount of energy which is transferred from the stream via turbulence to heat represents a increase in entropy and hence a loss to the intake, but since turbulent mixing is required in the combustor, the overall effect may be beneficial. The main point to note at this stage is that the optimisation of the intake and combustor must be carried out simultaneously.

The effects which take place in the wall boundary layer are very similar to those already considered in the discussion on the lifting surface. Nevertheless, the opportunity for recycling heat from the boundary layer into the system should be noted.

THE NOZZLE

Earlier studies, Ref. 2 have shown that expansion of the exit flow into the full base area of a hypersonic cruise or accelerating vehicle still leaves an excess of exit pressure above ambient. The purpose of the nozzle is therefore to expand the flow into the base area of the vehicle whilst maximizing the pressure on the surfaces so as to maximize the thrust. As shown in Ref. 2 there is also some benefit, on cruise vehicles, in deflecting the flow downwards by a small amount (10^0 to 14^0) in order to maximize the cruise range.

In recent work, Stalker has reported that interaction between the expansion waves in the nozzle and the Mach number profile across the flow can result in a beneficial pressure distribution on the wall. This interesting mechanism suggests that practical scramjet systems may not utilise uniform nozzle exit conditions, however the performance attained in practice can not exceed that governed by overall energy considerations.

THE COMBUSTOR

The principal topic for discussion in this paper is the mixing problem in supersonic combustion, and hence major emphasis is given to the scramjet combustor. At the present time we normally design subsonic combustion systems with the aid of a computational fluid dynamics code such as FLUENT.

Codes of this type evaluate the governing equations of the reacting flow field using a finite volume technique. The major equations used are; the conservation equations for mass, momentum, energy and species with a turbulence model such as the well known k,ε model. Since the combustion is usually mixing limited, a model such as the eddy break-up model is used to represent the mixing process. In supersonic flow, the equations are parabolic, however there will usually be some subsonic areas of the flow where re-circulation is present, necessitating a mixed flow analysis. The use of such codes ensure that factors such as energy balance are satisfied, however they can conceal some of the fundamental features of the reacting flow field. In this paper, a simple approximate method will first be used in order to draw attention to these fundamentals, then the need for a general computational fluid dynamic code for supersonic combustion design dynamic code for supersonic combustion design will be highlighted.

SCRAMJET CYCLE OPTIMISATION

Early analysis of the overall scramjet cycle allowed for intake losses, losses due to heat addition, and nozzle losses, but invariably assumed negligible mixing loss. Although it is well known that subsonic combustors require a significant pressure drop across the baffle (e.g. about 4% of the total pressure in gas turbine combustors), in the case of supersonic combustion it was assumed that hydrogen fuel from downstream or cross-stream injectors would mix readily in the jets. Early photographs of the appearance of flame at the fuel/hot air interface seemed to confirm the zero mixing pressure drop concept, and the eddy viscosity models used to compute the flow field did not highlight the pressure losses actually occurring. In fact, as shown in Ref. 1, the hydrogen jet carries negligible mixing energy and behaves more like a wake behind a bluff body. The mixing energy is thus extracted from the air stream and appears as a total pressure loss in which velocity energy is converted first to turbulence which then decays to heat. By comparison, the intake shock losses also convert total pressure to heat, although without the intermediate formation of turbulence, and both processes result in a loss of overall engine performance.

In this study, energy balance principles are applied to the mixing and combustion process to determine the effect of varying the total pressure loss due to mixing (i.e. the degree of mixedness) and hence combustion efficiency on the overall engine cycle performance. The trade off between high total pressure loss leading to high combustion efficiency but low cycle performance, and low pressure loss causing low combustion efficiency again leading to low cycle performance, is investigated and **optimised**. The practical realization of an optimum combustor requires not only the correct amount of mixing but also the attainment of this mixing in a short length since wall friction losses in ducted supersonic flow are extremely large. Turbulence theory can again be used to determine the length of the mixing region and hence the required scale of turbulence. Experimental evidence of the validity of this concept is obtained at simulated Mach 10 flight conditions, and also reported herein.

A schematic arrangement of a scramjet is shown in Fig. 1 where various engine stations are identified. An important parameter in the design of scramjets is the performance of the intake, where some compromise is usually necessary between the amount of intake diffusion and the efficiency with which the air is diffused. The amount of diffusion is normally specified by either the intake velocity ratio (V_3/V_1) or the diffusion factor (M_1/M_3). The latter has the advantage that its optimum value remains almost constant throughout the hypersonic speed range.

The intake efficiency may be specified by two kinds of parameters which are either independent of the amount of diffusion, or relate the efficiency to the amount of diffusion. An example of the former is the kinetic energy efficiency η_d which may be defined by

$$\eta_d = \frac{h_t - h_1'}{h_t - h_1}$$

where h_1' is the hypothetical enthalpy obtained by isentropic expansion from station 3 to ambient conditions.

This may be compared with the process efficiency K_d which relates efficiency to diffusion and is defined as

$$K_d = \frac{h_3 - h_1'}{h_3 - h_1}$$

$$= \frac{\eta_d - (V_3/V_1)^2}{1 - (V_3/V_1)^2}$$

Fig. 2 illustrates the difference between these definitions and attention is drawn to Refs. 2 and 3 for further discussions on these parameters.

Some typical performance curves are shown in Fig. 3 which are plotted for four values of η_d spanning the range of likely intake efficiencies. These performance curves were calculated by the exact analysis outlined in Ref. 2 and assume an optimum diffusion factor of $(M_1/M_3) = 3$. A simple two shock inlet was also assumed with stoichiometric hydrogen fuelled combustion at 100% efficiency and an isentropic nozzle expanding to $p_7/p_1 = 2$. These results were calculated for a height of 40,000m but are not considered to be very sensitive to altitude. Data such as that given in Fig. 3 provides a starting point from which design studies can be evolved, since the results can be re-calculated to incorporate component performance detail as it becomes available.

In order to incorporate combustion chamber performance into an overall engine cycle calculation it is necessary to define further efficiency parameters which describe the mixing and combustion processes. If the combustion chamber is considered to consist of a duct into which a mixing device (e.g. a centre-body fuel injector) has been placed, then a kinetic energy efficiency, η_m due to the turbulent mixing processes may be defined as Ref. 4.

$$\eta_m = \frac{h_t - h_{ms}}{h_t - h}$$

where h_t is the total enthalpy
h is the static enthalpy
h_{ms} is the static enthalpy in mixed region when expanded isentropically to the initial pressure.

This definition is then compatible with the intake kinetic energy efficiency η_d defined previously. If now a **constant pressure** mixing system is assumed then the mixing efficiency parameter can be shown to reduce to

$$\eta_m = (U_m/U)^2$$

where U_m denotes the mean velocity in the mixed region.

If the turbulence generating device is designed with an aerodynamically shaped upstream surfaces, then the momentum of the fluid which is lost in wall fraction is negligible, and the energy loss in the wake (due to base drag) appears first as turbulence. An energy balance across the turbulence producing region then gives

$$3/2 \, (\rho \, u'^2_{max}) = \Delta P$$

where u'_{max} is a reference r.m.s. turbulence velocity assuming that the turbulence generated is isotropic and all the energy becomes turbulence i.e. no dissipation. At hypersonic flow conditions this pressure loss can also be equated to the overall loss in mean kinetic energy

$$3/2 \, (\rho u'^2_{max}) = \Delta P = 1/2 \, \rho U^2 - 1/2 \, \rho U_m^2$$

hence

$$U_m/U = (1 - 3(u'_{max}/U)^2)^{1/2}$$

The mixing efficiency parameter then becomes

$$\eta_m = 1 - 3(u'/U)^2_{max}$$

which can then be incorporated into engine cycle calculations in the usual manner.

To show what this means in terms of loss of specific impulse we can write approximately

$$\Delta I_f = (W \Delta U)/(g W_f)$$

$$\simeq (3 U / 2 \alpha g)(u' / U_{max})^2$$

This function is plotted in Fig. 4 where a loss in specific impulse amounting to a few hundred lb_f sec/lb_m is apparent. It is interesting to note that this loss of impulse increases linearly with flight velocity in the same way that the loss of impulse due to skin friction increases linearly with flight velocity, as shown in Ref. 2. Thus turbulence generating devices must be designed to produce only low turbulence intensities to avoid large thrust penalties and they must also be designed to release the turbulence efficiently.

In scramjet systems currently envisaged for flight at M > 8, the intake processes will generally result in static temperatures at the combustion chamber entry in excess of 1000°K, such that spontaneous ignition of hot hydrogen (or other) fuel is possible. Thus the chemical kinetic requirements of combustion can usually be met providing the fuel and air can be mixed satisfactorily. In such a mixing limited system the maximum achievable combustion efficiency must therefore be related to the turbulence dissipation mechanism. A study of stirred reactor theory allows the **maximum** completeness of combustion to be expressed in terms of the dimensionless time ratio τ_{sd} (Ref. 5).

$$\eta_c = 1 - \Psi = 1 / (1 + 1/\tau_{sd})$$

where Ψ = fraction of oxygen untreated
τ_{sd} = τ_s/τ_d
τ_s = residence or stay time
τ_d = mixing or turbulent diffusion time

The stay time τ_s is determined by the reactant throughput and is defined as

$$\tau_s = \frac{\text{Reactor Volume}}{\text{Volumetric flowrate of reactant}}$$

$$\simeq X / U$$

where the reactor volume is confined to the region where turbulence is being dissipated. In the present simplified model, mixing rate is assumed to be directly proportional to the dissipation rate of the turbulent fluctuations by Reynolds analogy. It can be shown (Ref. 5) that the diffusion time τ_d is given by

$$\tau_d \simeq l_e / u'_{max}$$

where le is the average size of the energy containing eddies and is directly proportional to some characteristic dimension (λ) of the turbulence producing device. Thus

$$\tau_{sd} = (X/U)(u'_{max}/le)$$

In order to ascribe values to τ_{SD} it is reasonable to assume at this stage that for a simple shear layer turbulence generation system, $le \simeq 0.2\lambda$ and $X \simeq 10\lambda$ (Ref. 6), so that

$$\tau_{sd} = 50(u'/U)_{max}$$

and is independent of λ !. The mixing limited combustion efficiency is then

$$\eta_C = \frac{1}{1 + 1/50(u'/U)_{max}}$$

The two efficiency parameters η_m and η_C defined above may now be included in the engine cycle analysis. Both parameters relate the efficiency to the turbulence producing region through the turbulence intensity $(u'/U)_{max}$, which is a function of the energy loss across the turbulence generating device and for simple systems may be related directly to the size of the fuel injection step or baffle (Ref. 5). The effects of including η_m and η_C into the engine cycle calculation are shown in Figs. 5 to 8. The results are for a specific engine geometry at a specific flight condition and illustrate the effect on performance of varying the intensity of turbulence produced in the combustion chamber.

In Fig. 5 engine specific impulse is plotted as a function of turbulence intensity $(u'/U)_{max}$ for a flight Mach No. of 10, assuming an intake diffusion factor of 3. The results are plotted for four values of η_d covering the likely range of intake efficiencies. The curves show that an optimum value of $(u'/U)_{max}$ exists and that this value decreases as the intake efficiency is reduced. For turbulence intensities above the optimum value, so much mixing energy is taken from the air stream that the losses become prohibitive and the engine performance falls off. For values of $(u'/U)_{max}$ below the optimum there is insufficient mixing within the combustor and the combustion efficiency itself suffers, again reducing the performance. The optimum value of $(u'/U)_{max}$ thus represents a condition at which the mixing and combustion losses together have been minimised and engine performance maximised.

At this stage it should also be noted that not only is the intensity of turbulence important to the combustion process, but also that the scale of the turbulent fluctuations is important. Since the rate of mixing is here assumed to be directly proportional to the rate of turbulence dissipation (Ref. 5) then the position of the dissipation zone will depend upon the nature of the initial eddies. If the initial eddies are very large then they will travel a considerable distance before they have dissipated their turbulent kinetic energy, and the combustor will be long. The combustor must therefore be designed to generate the optimum turbulence intensity and to provide a scale of turbulence sufficiently low that the combustor can be kept as short as possible. On the other hand, the turbulence scale must initially be as large as the inhomogeneities which must be mixed. Thus reducing the scale leads to the requirement for many small fuel jets (or a narrow two-dimensional slot).

In Fig. 6 the engine performance is expressed in terms of the combustion efficiency η_c by relating η_c to $(u'/u)_{max}$. The curves suggest that the optimum efficiency is less than 80%, since the turbulence levels required to obtain efficiencies greater than this result in greater pressure losses and reduced overall performance. It would therefore appear that the highest attainable combustion efficiency is not the best in terms of overall engine performance.

The effect of diffusion on performance is illustrated in Fig. 7 for a scramjet operating at 4,900 m/s. The upper curve represents the performance of an intake having a process efficiency K_d of 0.9 and neglects combustion and nozzle losses. In the middle curve, optimum mixing and combustion losses have been incorporated and the resulting loss in performance is apparent. Including the combustor losses has no significant effect on the optimum value of (V_3/V_1) since the effect on the combustor of changing V_3 is included in the upper curve. In order to present a more complete picture, the effect of including friction and non-equilibrium losses in the nozzle, according to the scheme of Ref. 3, is illustrated in the lower curve. Since the nozzle losses increase with decreasing velocity ratio the optimum (V_3/V_1) has increased above the value previously indicated. The important feature of Fig. 7 however is the **large loss in performance incurred by incorporating the mixing and combustion losses**. At the flight condition of Fig. 7 these losses are so large that the final performance value is so low that it is matched by conventional chemical rockets and makes scramjet systems apparently uneconomic at this speed. This observation is based on a very crude analysis and more exact study is required to validate the conclusion and provide a firm estimate for the upper speed limitation of the scramjet.

The variation of engine performance with flight Mach number is shown in Fig. 8 which assumes a constant diffusion factor of $M_1/M_3 = 3$ throughout the flight range. The optimum overall combustor efficiency for each flight condition has been chosen and the performance expressed as $I_{sp(optimum)}$ is the maximum attainable at that condition. Nozzle losses are neglected in this figure. As can be seen from Fig. 8 the engine performance decreases approximately linearly with increasing flight Mach number >5, finally reducing to zero between Mach 19 and 22 depending upon the intake performance. If nozzle losses are taken into account and reasonable values of intake performance assumed then it would appear that scramjet performance approaches that of the chemical rocket at a Mach number of about 17. It would appear then that for scramjets to be an economic proposition their flight speed should be confined to less than about Mach 14.

The data presented above thus show the general trends in engine performance with varying combustor performance and indicate the large impulse losses for scramjets which inevitably operate with mixing limited combustion systems. The significant features of the data are that

1. an optimum level of turbulence exists at which mixing and combustion losses, taken together, are minimised;
2. that the best combustion efficiency in terms of engine performance is considerably less than 100%;
3. that when combustor losses are included in the theoretical scramjet cycle, engine performance is a linear function of the flight Mach number.
4. the crude theory used directly relates the combustor performance to the geometrical arrangement and pressure loss and may be extended to include droplet evaporation, chemical kinetics, and flow features such as re-circulation, so that it can be applied to many other combustor configurations.

EXPERIMENTAL APPROACH

An investigation into the effects of initial turbulence generation on the position and extent of the mixing and combustion zone was carried out using our shock tunnel test facility. For the tests, this was fitted with a 2-D Mach 3.5 test section, and supplied air at stagnation conditions of approximately 6000°K at 200 bars. This permits simulation of scramjet combustor inlet conditions for flight at Mach 10 at an altitude of 25-30 Km. These conditions are obtained using the combustion driven hypersonic shock tunnel operated at the tailored interface condition ($10.2 < M_s < 10.6$) to give a contamination free test time of 2 msec.

In general the time required for the heat release process to go to completion, and hence the length of the combustor, is dependent upon the time needed to mix the fuel and air, and the time needed for it to react. In practice these two processes are occurring simultaneously, however with the test section conditions used in this study (T_{static} = 2000°K, P_{static} = 1 atm., M = 3.5) the chemical kinetic time for the hydrogen/air reaction is of the order of 10^{-6} secs. Consequently once reaction is initiated, complete combustion can occur within a very short distance (approx. 3mm). However in order to obtain complete combustion all the fuel and air must be mixed in the correct proportions down to the molecular level. The size of the mixing region can be estimated as follows:-

As stated previously the mixing rate is directly proportional to the turbulence dissipation rate. Structural turbulence theory (Ref. 6) shows

$$dD/dt = \rho A u'^3 / le$$

Consequently the dissipation rate and hence the mixing rate at any axial position for a given system is dependent upon the turbulence level (u') and a constant (le or λ). Integration of the above equation gives (Ref. 5)

$$\text{Mixing rate} \sim (\Delta D/q)/(\Delta x/\lambda)$$

$$\sim B \left(3(u'/U)^2/4\right)^{3/2} \qquad *$$

The mixing rate and hence combustion rate can be related to the geometry by carrying out an aerodynamic energy balance. This reasoning shows that for a given amount of free stream energy available for turbulence production, and a given throughput of fuel, the shortest combustor will be obtained when the turbulence dissipation rate is maximised in as short a distance as possible.

The experimental study investigates the effects of initial turbulence generation on the position and extent of the turbulence dissipation rate for a non-reacting hydrogen/nitrogen system, and compares the results with the observed positions and extent of the combustion zones obtained using hydrogen/air. Two separate midstream injector configurations were tested, having equal geometric blockage ratios (and similar drag) in order that the total pressure energy, theoretically available for turbulence generation, was the same in each case. One injector consisted of a plain 12° wedge with fuel injected from the base. The other injector was specifically designed to be an efficient turbulence generator, and consisted of a plain 12° wedge with backward facing deltas attached to the trailing edges. These deltas are swept back behind the Mach angle and produce swiss-roll type vortices which interact in the near wake. The hydrogen was injected from the base into this highly turbulent region.

The nature and duration of the test flow precludes conventional velocity, turbulence and concentration traverses in the wakes. However midstream pitot pressures were obtained using a probe designed specifically to detect changes in test gas molecular weight. The aerodynamic energy balance across the injector is:-

$$p_1 + KE_1 = p_2 + KE_{wake} + KE_{turb} + D$$

Non-dimensionalising throughout by $q = \rho U^2/2$ gives

$$\frac{\Delta P}{q} + \frac{KE_{def}}{q} = 3 \left(\frac{u'}{U}\right)^2 + \frac{D}{q}$$

From this equation it is possible to obtain information about the turbulence generation and dissipation, and hence about the mixing zones in the wakes of the injectors if the total pressure drop ($\Delta P/q$) across the injector is known, together with the axial distribution of the wake K.E. deficit (KE_{def}/q).

The pressure drop across the injector appears first as base drag $\Delta P = \text{Drag} / A_{ref}$, where A_{ref} is a reference area, which ideally should be taken as the initial area of the stream tube within which all the subsequent turbulence generation and mixing occur.

The base drag is given by; $\text{Drag} = C_D q A_B$, where C_D is an experimentally determined drag coefficient and A_B is the base area

$$\frac{\Delta P}{q} = C_D \frac{A_B}{A_{ref}}$$

The photographic results of the combustion zones (Plate 1) indicate that the reacting wake does not spread appreciably, hence A_{ref} is assumed equal to A_B, and then

$$\frac{\Delta P}{q} = C_D$$

For a 12° wedge the experimental value of $C_D \sim 0.5$ (Ref 7). KE_{def}/q is the non-dimensionalised total K.E. of the wake velocity deficit at any given axial position. For a 2-D wake the axial decay of the velocity deficit is $\sim 1/x^{1/2}$ (Ref. 8). Hence $KE_{def}/q \sim 1/x$

The constant of proportionality in the preceding equation differs for different wakes and in general needs to be determined experimentally. Values of this constant were determined for the two injector configurations using measured wake pitot pressures.

Knowing both $\Delta P/q$ and the axial dependence of KE_{ref}/q it is possible to determine the sum of the turbulence kinetic energy ($3(u'/U)^2$) and the dissipation energy (D/q) from the energy balance equation. Absolute values of these two terms are then found from the relationship between these two terms from Equation * where an initial guess at the value of (u'/U) results in a corresponding value of (D/q). Hence for a given ($\Delta x/\lambda$) and assuming $l_e = 0.2\lambda$, an iterative routine is followed until the overall energy balance equation is satisfied numerically.

The resultant curves of (D/q), (KE_{ref}/q) and ($3(u'/U)^2$) are shown in Figs. 9 and 10 for the wakes of both the vortex generating injector and the plain wedge injector. In each case it can be seen that the turbulent K.E. is only a very small proportion of the total available energy, and the dissipation energy rises rapidly, asymptotically approaching the total energy in the far wake. The turbulence dissipation rates and hence the mixing rates for the two wakes are shown in Fig. 11 where it can be seen that the maximum mixing rate for the vortex generator is over twice that for the plain wedge and occurs at an $x/\lambda \sim 2.5$ compared to $x/\lambda \sim 5$ for the plain wedge. It is clear that in both cases the mixing rate decreases rapidly after reaching a maximum value. It is the shape of these mixing curves which shows that the near wake region closely approximates to a well stirred reactor containing the mixing rate maximum, followed by a plug flow region in the far wake where the mixing rate has fallen to negligible proportions.

An estimate of the amount of mixing occurring in the well stirred region can be obtained from the area under the dissipation rate curve. Assuming a minimum dissipation rate beyond which mixing is negligible (taken to be < 0.015) the stirred reactor zone for the wake of the vortex generator is seen to extend from an $x/\lambda \sim 0$ to 8 with a mean dissipation rate of about 0.035. Similarly for the wedge wake the mixing region extends from $x/\lambda \sim 3$ to 11 with a mean rate of about 0.025. This shows that the mixing in the well stirred region is 30% more efficient for the vortex generator. These estimated positions of the well stirred reactor zones agree well with the observed positions of the combustion zones shown in Plate 1. These photographs show that for the wedge the reaction zone starts at $x/\lambda \sim 8$ and is well established by $x/\lambda = 11$, whilst for the vortex generator it begins at $x/\lambda = 5$ and is well established by $x/\lambda \sim 8$.

The combustion efficiency of the system can be estimated from the equation

$$\eta_c = 1 - \psi = 1/(1 + 1/\tau_{sd})$$

hence η_c increases with increasing τ_{sd} which, for a mixing limited system is given by

$$\tau_{sd} = (x/l_e)(u'/U)$$

Taking values of X and u'/U from Figs. 9 and 10 and assuming $l_e = 0.2\lambda$, η_c is found to be about 86% for the vortex generator and about 84% for the plain wedge. This close agreement is not surprising when it is realised that the total pressure drop available for both configurations is equal, and as the total dissipation curve becomes asymptotic to $\Delta P/q$ then the total mixing will be the same. However the important difference between the two from the propulsive efficiency viewpoint is the extent of the mixing zone X and the turbulence intensity u'/U. It is clear that large X and small u'/U will give good combustion efficiency but large friction, heat transfer and weight penalties on the overall system efficiency. Small X and large u'/U will offset these penalties for the same combustion efficiency but if u'/U is too large, losses in propulsive efficiency can be expected as has been shown in the overall cycle analysis. The optimum combustor must combine small X and optimise u'/U to give the required combustion efficiency for maximum specific impulse.

EXPERIMENTAL RESULTS

Qualitative interpretation of wall static pressure measurements and direct glow photographs supports the assumption that mixing rate is approximately proportional to dissipation rate and shows the importance of the initial turbulence generation on the subsequent mixing and reaction zones. Fig. 12 shows;
(a) measured wall static pressure distributions for the vortex generator,
(b) the computed pressure increase due to wall friction alone, and
(c) the final equilibrium combustion pressure as calculated from one dimensional Mollier chart calculations. It can be seen that the measured wall static pressure closely approaches the final pressure calculated for complete combustion. It is worth noting that at these conditions the wall friction effects produce pressure rises comparable to the theoretical combustion pressure rise.

The steep compression zone following injection is due to the combined effects of heat and mass addition and the presence of the fuel injector. The theoretical equivalence between heat and mass addition (Ref. 9) shows that if dQ/dx is large, steep compression of the external flow field will result. In order to separate the effects of heat and mass addition, nitrogen replaced air as the test gas. A comparison of the pressures obtained with equivalent hydrogen flow rates using air and N_2 is shown in Fig. 13. These results show that there is a considerable pressure rise with mass addition alone, however the increased pressure obtained with combustion indicates that mixing and combustion occur within a very short distance of injection.

The general shape of the wall static pressure distribution can be explained qualitatively in terms of the stirred reactor and turbulence concepts already presented. The most important factor affecting the external pressure field is the rate of heat release per unit length (dQ/dx), the absolute magnitude of which is determined by the mean flow velocity and the "reaction rate". The reaction rate is mixing limited and hence depends upon the turbulence dissipation rate. In the near wake of either injector the mean axial velocity will be less than the free stream value. This relatively low velocity, highly turbulent region corresponds to the well stirred reactor zone. Consequently if combustion occurs in this region, dQ/dx will be large and the resultant pressure field will give rise to a high wall static pressure. Further downstream the turbulence has all been dissipated and the mean axial velocity increases. These conditions approximate to those of a Plug Flow region in which dQ/dx will be relatively small, resulting in lower wall static pressure. Comparison of wall static pressures obtained from both injectors for equivalent stoichiometry is shown in Fig. 14 where it can be seen that the pressure induced by the well stirred region is lower at station 2 for the plain wedge, indicating that dQ/dx is lower in this region. Pitot pressure measurements showed that the axial velocity in the near wake of the wedge was less than that for the vortex generator and hence the residence time was higher in this region. Consequently the inferred reduction in dQ/dx must be due to a decreased reaction rate brought about by lower mixing rates in this near wake region.

FINITE DIFFERENCE MODELLING

Space precludes a presentation of the application of computational fluid dynamic (C.F.D.) modelling to scramjet systems herein. However, the purpose of this document is to draw attention to the **governing principles** underlying scramjet combustor performance. Nevertheless, inspection of the graphics presentation from the C.F.D. calculations will show that the turbulence kinetic energy, turbulence dissipation and mixing rates are all concentrated in the near wake of the turbulence generating system. Thus the same conclusions could be drawn from the application of these modern codes and they should of course be used to quantify actual designs.

CONCLUSIONS

Fundamental principles of turbulence generation, dissipation and mixing limited combustion have been applied to the scramjet with particular emphasis on the combustor. It has been shown that severe penalties may be incurred in overall engine performance in order to achieve adequate mixing and hence good combustion efficiency in an acceptable length of combustor. Cycle optimization studies show that the highest performance, in terms of fuel specific impulse, is obtained at a combustion efficiency of about 80% (at M = 10). Optimum intake diffusion factors are slightly smaller than those estimated in previous studies in which mixing losses were neglected. The average performance loss due to mixing amounts to about 6 Ns/g (600 lb_f sec/lb_m) throughout the hypersonic speed range. As a result, the rocket may be preferred over the scramjet at Mach numbers above 16 (or perhaps less when system weights are taken into account).

In experimental studies it has been shown that it is possible to deduce the mean turbulence characteristics and mixing rates in the wake region of fuel injectors using only mean flow measurements. The predicted axial positions of the mixing zones agrees well with the observed combustion zones. These results support the basic assumption of the mixing model that the mixing rate is approximately proportional to the turbulence dissipation rate. They also show that propulsive efficiency will be optimized when the required amount of turbulence needed to produce the optimum combustion efficiency, is dissipated in as short a distance as possible. The basic turbulence theory indicates that this can only be achieved by producing initial turbulent eddies which are as small as possible (consistent with the number of injection points). Consequently the injection system must be designed with this in view, as well as the other problems of penetration, interaction and low temperature ignition.

References

1. Swithenbank, J. and Chigier, N.A. "Vortex Mixing for Supersonic Combustion", 12th Int. Symp. on Combn., 1969 pp. 1153-1162.

2. Swithenbank, J., "Hypersonic Air Breathing Propulsion", Prog. in Aeronautical Sci., Vol. 8, ed. D. Kuchemann, Pergamon Press, 1967.

3. Curran, E.T. and Swithenbank, J. "Really High Speed Propulsion by Scramjets", Aircraft Engineering, Jan., 1966.

4. Swithenbank, J., "Combustion Fundamentals", Dept. of Chem. Eng., Sheffield Univ., Report H.I.C. 150.

5. Swithenbank, J., "Flame Stabilization in High Velocity Flow", COMBUSTION TECHNOLOGY Ed. Palmer and Beer, pp 91 -127, Academic Press, 1974.

6. Hinze, J.O., "Turbulence", McGraw Hill, 1959.

7. Hoerner, S.F., "Fluid Dynamic Drag", Published by Author, 1958.

8. Schlichting, H., "Boundary Layer Theory", McGraw Hill, 1968.

9. Rues, D., "Concerning the Equivalence between Heat, Force and Mass Sources", R.A.E. Library Translation No. 1119, July 1965.

10. Swithenbank, J., Jaques, M. T., and Payne, R. "The Role of Turbulence in Scramjet Combustor Design" Proceedings of the 1st International Symposium on Air Breathing Engines, Marseilles, June 1972.

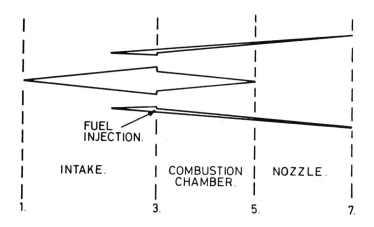

FIG. 1. SCRAMJET ENGINE STATIONS.

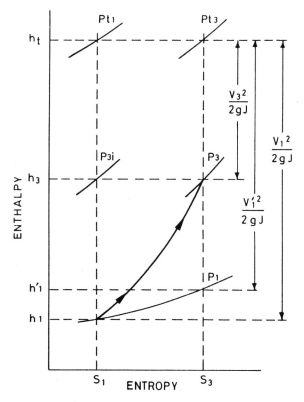

FIG. 2. SCRAMJET INLET PROCESS.

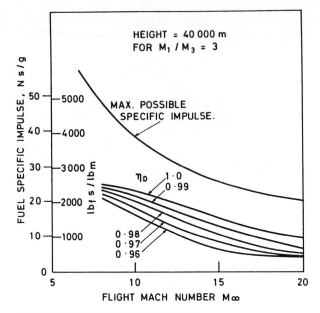

FIG. 3. SCRAMJET PERFORMANCE.

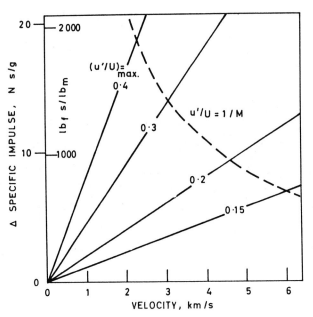

FIG. 4. SPECIFIC IMPULSE LOSS DUE TO MIXING.

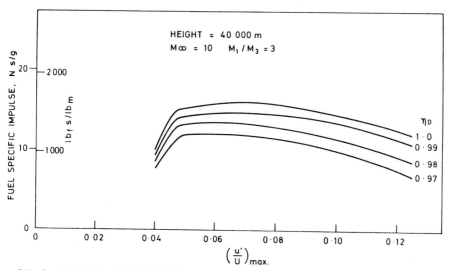

FIG. 5. VARIATION OF ENGINE PERFORMANCE WITH TURBULENCE INTENSITY.

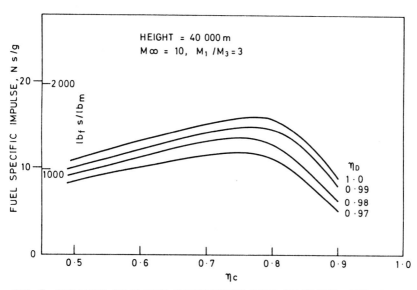

FIG. 6. VARIATION OF ENGINE PERFORMANCE WITH COMBUSTION EFFICIENCY.

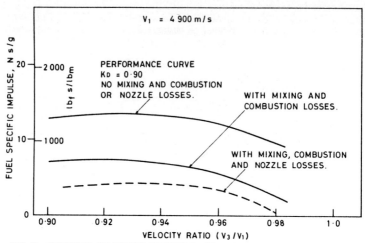

FIG. 7. VARIATION OF ENGINE PERFORMANCE WITH INTAKE DIFFUSION ILLUSTRATING THE EFFECT OF MIXING AND COMBUSTION AND NOZZLE LOSSES.

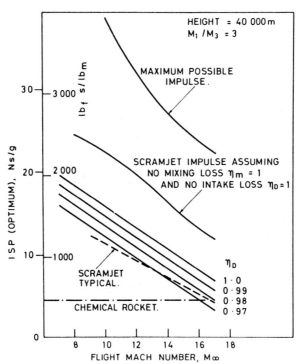

FIG. 8. VARIATION OF ENGINE PERFORMANCE (ISP) WITH FLIGHT MACH NUMBER, NEGLECTING NOZZLE LOSSES.

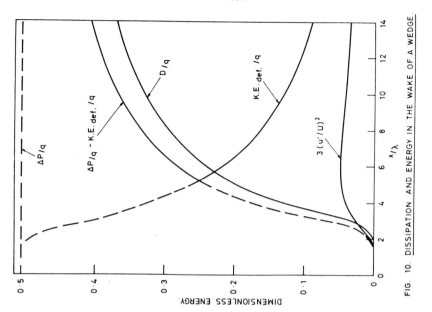

FIG. 10. DISSIPATION AND ENERGY IN THE WAKE OF A WEDGE.

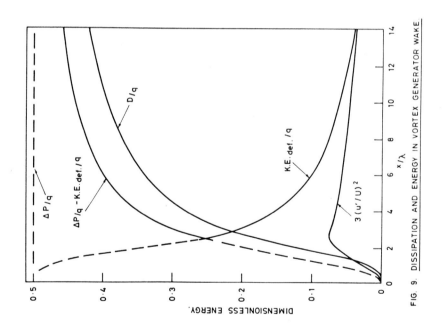

FIG. 9. DISSIPATION AND ENERGY IN VORTEX GENERATOR WAKE

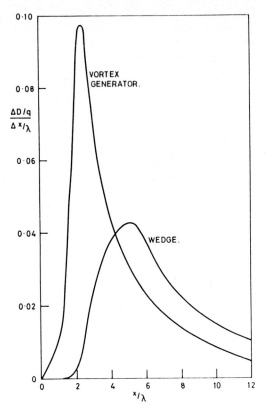

FIG. 11. COMPARISON OF MIXING RATES IN THE WAKES OF THE TWO INJECTORS.

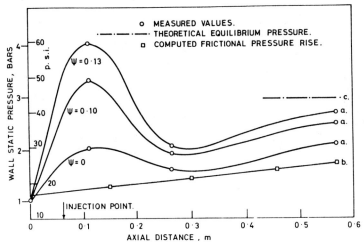

FIG. 12. COMBUSTOR WALL STATIC PRESSURES — VORTEX GENERATOR.

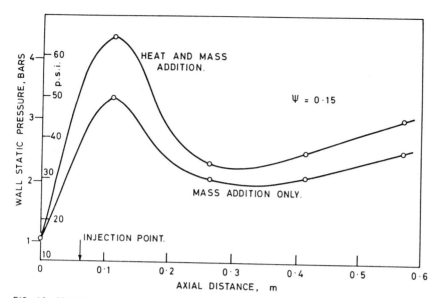

FIG. 13. STATIC PRESSURES WITH AND WITHOUT COMBUSTION – VORTEX GENERATOR.

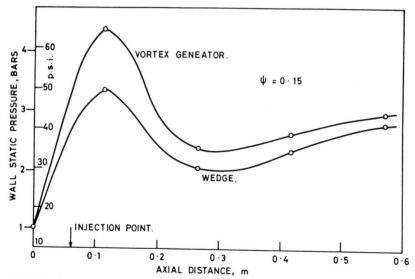

FIG. 14. COMPARISON OF STATIC PRESSURES FROM VORTEX GENERATOR AND PLAIN WEDGE.

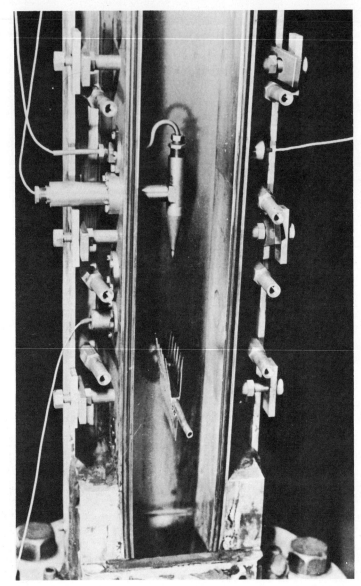

PLATE 1. VORTEX GENERATING INJECTOR AND PROBE MOUNTED IN TWO DIMENSIONAL MACH 3·5 TEST SECTION.

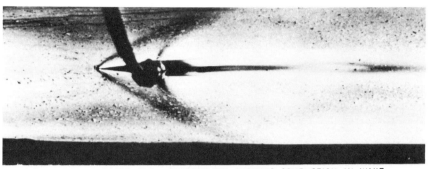

(a) <u>VORTEX INJECTOR GLOW PHOTOGRAPH SHOWING COMBUSTION IN WAKE.</u>

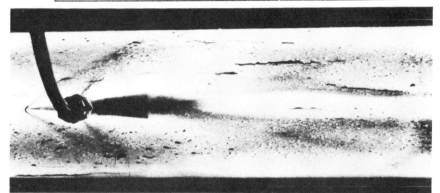

(b) <u>BLUFF BASE WEDGE INJECTOR GLOW PHOTOGRAPH WITH COMBUSTION.</u>

(c) <u>VORTEX INJECTOR GLOW PHOTOGRAPH USING NITROGEN TEST GAS.</u>

MORPHOLOGY OF FLAMES SUBMITTED TO PRESSURE WAVES

T. Poinsot, A. Trouve, D. Veynante, S.M. Candel and E. Esposito
E.M2.C. Laboratory, CNRS and Ecole Centrale des Arts et Manufactures
92295 Châtenay-Malabry, FRANCE

Abstract

The influence of pressure waves on the mean structure of a turbulent flame is described. Two cases are presented :

(1) When the pressure wave is **generated by the flame itself**, the system is submitted to combustion instability. In some extreme circumstances, the coupling between combustion and acoustics can lead to a flame structure characterized by the **shedding of large vortices** behind the flame holders ; for this regime, the mean flame structure definition is a result of a cyclic formation and destruction of the reactive pattern at the frequency of the pressure oscillation. The characteristics of this cycle are studied by a phase average imaging method of the local reaction rate. This low frequency combustion instability is described in the first part of this paper.

(2) Pressure waves can also be <u>externally</u> imposed to the flame. Arrays of driver units (loudspeakers) can be used to excite the flame at a chosen frequency and with a predetermined acoustic structure. With this technique it is possible to excite the flame in the transverse sloshing mode. In the case of a multiple flame combustor, this excitation induces strong modifications of the flame pattern : small vortices generated by the sloshing motion interact and set the jets of fresh mixture into a flapping motion. The flame structure appears as the result of the **nonlinear interaction** between small vortices generated by high frequency excitation and a low frequency induced flapping motion of the jets. Phase average measurements cannot be used in such circumstances because the fluctuations are the result of the superposition of many different modes. A spectral imaging method is developed to analyse these nonlinear coupling effects and its application is described in the second part of the paper.

1. INTRODUCTION

Many recent studies have indicated that pressure waves may strongly interact with turbulent flames (Keller et al. 1981, Darabiha et al. 1986, Bray et al. 1983). For example, it has been observed that combustion instabilities coupled with an acoustic resonant mode are often associated with large flame-front motions and coherent periodic patterns. An early indication of such pressure wave interactions with combustion may be found in a study of high-frequency oscillations in ramjets and afterburners due to Rogers and Marble (1956). Under "screeching" conditions, a transverse acoustic mode is excited in the duct. This acoustic wave produces large sloshing velocities in the vicinity of the flameholder lips. Vortices are generated by these fluctuations and transported downstream and after a certain characteristic time, these vortices burn and feed energy into the transverse acoustic mode. When certain phase conditions are satisfied the cycle becomes self sustained. More recently, Smith and Zukoski (1985) have shown that combustion instabilities in a dump combustor are caused by a coupling between longitudinal acoustic modes of the duct and vortex shedding from the rear edge of the flame holder.

Other types of large scale flame front motions in the presence of pressure waves have been found in other experiments carried on ducted flames. In some cases the amplitude of the motion becomes so large that the flame separates into lumps (Yamaguchi et al. 1985).

These experiments indicate that in many circumstances pressure waves modify the flame structure by triggering large scale fluctuations and by enhancing vortex pattern formation.

To improve our understanding of the coupling mechanism it is particularly interesting to study regimes dominated by vortex structures and obtain information on how these structures form, grow, interact, burn and couple to the pressure wave. This kind of analysis is carried in Section 3. It concerns a low-frequency vortex driven acoustically coupled instability of a multiple inlet combustor. This regime is characterized with instantaneous spark schlieren photographs and with a novel phase average imaging method (PAI) providing the local instantaneous rate of reaction. A mechanism for this self-sustained oscillation involving vortices and acoustic feedback is put forward.

Another situation of current interest is that in which an external pressure wave couples with the natural hydrodynamic modes of the turbulent flame. According to some recent studies, certain types of oscillation may be related to turbulent shear layer instability modes (Keller et al 1981, Pitz and Daily 1981). This view is already contained in the work of Blackshear (1956) and it is explored in the theoretical studies of Flandro (1985). This last author indicates that hydrodynamic instabilities may lead to combustion oscillations in solid rocket engines if the hydrodynamic frequency is close to an acoustic eigenfrequency of the combustor. Our own investigations indicate that the coupling between hydrodynamic and acoustic modes may take more complex forms. In some cases, it is nonlinear and involves small and large scale motions. In others, it features collective interaction phenomena similar to those observed in excited non-reactive shear layers (see a review of this problem by Ho and Huerre 1984). Mechanisms which give rise to this complex behaviour are examined in Section 4. The analysis is based on an original spectral imaging method and it is conducted on the same multiple inlet combustor. It indicates how a high-frequency pressure wave may trigger a low frequency flame response.

2. EXPERIMENTAL CONFIGURATION

The experimental configuration is displayed in Fig. 1. A mixture of air and propane is injected through a long duct into a dump combustor. The combustor has a square cross-section of 100 x 100 mm^2 and is 200 mm long. The injection plane comprises five narrow injection slots separated by four backward facing steps (Fig. 2). Two half-steps separate the first and last slots from the lateral walls. Each slot has a rectangular cross-section of 3 x100 mm^2 and the blockage is 85 %. The combustor walls are made of ceramic and they are surrounded by a layer of insulating material. A stainless steel structure holds these parts together. In designing this combustor our objective was to dispense with the use of water cooling and to approach adiabatic combustion conditions. The combustor side walls are equiped with two air cooled circular quartz windows. These windows have a diameter of 40 mm and their axis is located on the centerline at 21 mm from the inlet plane. They allow a direct examination of the flame structure and they are also used for schlieren imaging and C_2-radical emission detection.

Acoustic measurements are performed on the test duct located on the upstream side of the combustor. A Bruel & Kjaer 4135 microphone is employed to measure the pressure signal in the test duct. This microphone is placed at 0.38 m from the combustor inlet plane. It provides the reference signal for the phase average imaging. Two JBL 2482 driver units plugged on the duct allow external acoustic excitation of the system.

3. PHASE AVERAGE IMAGING OF LOW-FREQUENCY COMBUSTION INSTABILITIES

Phase average

In studies of combustion instabilities, it is important to obtain informations on the non-steady energy release process. Exact time resolved measurements of the local energy release are not available but many observations indicate that the light emitted by certain radicals like C_2, OH or CH is directly related to the reaction rate or equivalently to the heat release rate. Our estimation of the local heat release relies on the assumption that a monotonic relation exists between the heat release and the emission of C_2 radicals in the (0,0) band. The corresponding wavelength (λ = 5165 A) is isolated with a narrow-band interferential filter ($\Delta\lambda$ = 50 A). The optical arrangement is shown in Fig. 3. The light emitted by the flame is collected by an f = 54 mm convex lens and filtered around 5165 A. The light intensity is then detected by a "La Radiotechnique" XP 1002 photomultiplier through a ϕ = 1 mm pinhole to provide a **local** measurement : the emitting area effectively seen by the photomultiplier has a diameter of 0.8 mm. The detector output is amplified, low-pass filtered to prevent aliasing and transmitted to a PDP 11/23 computer through an 8 channel 12 bit A/D converter (Fig. 4). Acquisition and displacement of the optical system are controlled by the computer. It is convenient to extract the fluctuating components from the complete signal : the fluctuating heat release corresponds to the combustion of vortices while the continuous heat release corresponds to the steady combustion process. This extraction is achieved by substracting the "pedestal" of the photomultiplier output.

The photomultiplier is displaced on a predetermined grid of 41 x 41 = 1681 points. At each point of this grid, the data acquisition is synchronized with respect to the pressure signal detected by microphone M1. The zero crossings of this signal are recognized and used as triggering signals for the PDP 11/23 clock. If T_K designates the k-th triggering moment, then the fluctuating intensity data samples acquired at a constant rate from this initial moment are

$$f_k(\ell) = f(\ell \Delta t + T_k)$$ for k = 0,..., N-1

and ℓ = 0,..., L-1

In this expression, Δt is the sampling period, N is the number of cycles examined and L is the number of samples acquired in each cycle. A phase-average or conditionned average can then be obtained by summing samples corresponding to the same moment in each cycle :

$$\langle f(\ell) \rangle = \frac{1}{N} \sum_{k=0}^{N-1} f_k(\ell)$$

The measurements described in this paper are performed by acquiring L = 9 data points per cycle at a sampling period Δt = 0.23 ms and averaging these samples over N = 100 cycles. Spark schlieren photographs are also synchronized with respect to the same pressure signal. Photographs are obtained at nine different moments of the cycle separated by a constant time interval of Δt = 0.23 ms thus allowing a direct comparison with the phase-average images. The spark duration is less then 30 ns in order to "freeze" the flow structure (at the mean jet speed of 50 m/s, the displacement of a particle is less than 2 m during the spark).

Experimental results

Let us now consider an unstable regime at 530 Hz. The air flow rate is 74 g/s corresponding to an inlet jet velocity of 40 m/s and the equivalence ratio is = 0.92. Other modes can be obtained for different operating conditions and a simple one dimensional analysis of the longitudinal acoustic modes of the system shows that all unstable modes correspond to acoustic modes (Poinsot 1987). These instabilities are mainly due to interactions between acoustics and combustion. The 530 Hz mode is the strongest unstable mode. Pressure signals recorded during unstable combustion at 530 Hz are almost sinusoidal, allowing conditional measurements and phase-average imaging. A complementary description of unstable modes of the combustor can be found in Poinsot (1987).

Schlieren visualization

Figure 5a presents a schlieren visualization of a stable regime. In this case, no vortex is present. The mean flow structure is not periodic : the two lower jets interact and combine, the two inner jets form another single jet while the upper jet remains isolated. The flow structure is greatly modified when the 530 Hz instability appears : the jets are not deviated, and the mean flow structure is nearly periodic. Direct observation of the flow shows that the recirculation regions become luminous indicating the presence of chemical reactions. Schlieren visualization reveal the existence of large vortices formed behind the flameholders (Fig. 5b). Figure 6 shows seven schlieren photographs taken at different instants of the 530 Hz cycle with a time step of 0.23 ms. At the first instant t_1 (Fig. 6h) the jet is nearly straight. At moment t_2 (Fig. 6i) a corrugation develops on the central jet at about 1 cm from the inlet plane. At the next instant t_3 (Fig. 6d), the bulge has started to roll into a vortex and after another time step (Fig. 6k), the vortex has grown into a large "mushroom" with a rounded cap. Vortices growing on neighbouring jets become visible. A turbulent wake is formed behind the cap. At moment t_5, the central and side vortices have grown and they strongly interact (Fig. 6ℓ). Intense mixing with production of small scale turbulence takes place in the interaction regions. Although schlieren pictures cannot be used to distinguish between burning and nonburning gases it may be infered that intense reactions are taking place in the interaction regions. At the next instant t_6 (Fig. 6m) mixing and burning continue and extend over the whole vortex wake. At instant t_7 (Fig. 6n) turbulence decays and vortices and wakes are convected downstream. This evolution continues until the end of the cycle : at time t_9 (not presented here), the cycle is almost completed (t_9-t_1 = 1.84 ms while the period of the cycle is 1.88 ms). The flow configuration is similar to that observed at time t_1 and an other cycle begins.

Phase average images

Figure 6 also provides phase average images of C_2-radical emission at the same moments t_1, ... t_7 used for schlieren photographs. As indicated before, this data may be interpreted as representative of the local heat release source term. The intensity maps are plotted on a scale of grey levels. The upper and lower limits of this scale are fixed to allow direct comparisons between images taken at various moments in the cycle. The darkest symbol represents the maximum light intensity. At the initial moment t_1 (Fig. 6a), some light is radiated on the left of the window, indicating that some residual burning still takes place in the wake of the previous structure. At the second moment, light emission is weak (Fig. 6b). Combustion begins on the sides of the vortex corrugation (Fig. 6i). At the next moment t_3 (Fig. 6c), four spots of combustion are made evident. Central spots are located on the sides of the central vortex displayed in Figure 6j. The side spots correspond to vortices developing on the side jets. After another time step at t_5 (Fig. 6d), combustion takes place near the middle of the window. The size of the spots and their intensity have augmented. Vortices of the central jet begin to interact with vortices of the side jets.

The maximum combustion intensity is obtained between times t_5 (Fig. 6e) and t_6 (Fig. 6f). The coalescence between adjacent spots is almost achieved at time t_5 and intense combustion takes place behind the vortex side branches in the regions where the central vortex interacts with the side vortices. Comparison of schlieren photographs and phase average images at times t_5 and t_6 confirms that the combustion intensity is maximum when small scale turbulence appears.

Combustion continues after time t_6 but its intensity decreases in the spots. Burning proceeds in the wakes and in the central jet (Fig. 6g). During the last moments of the cycle, combustion spots slowly drift towards the left side of the window. The maximum light emission becomes weak indicating that most of the combustible material introduced during the cycle has burnt. At time t_9, the C_2 emission map is similar to that observed at time t_1.

A comparison between the schlieren photographs and the C_2 radiation maps reveal additional features of the combustion process. From the photographs taken at instants t_4 and t_5 and from the corresponding emission maps, it is clear that little combustion takes place on the vortex cap boundary. This may be explained by the high values of the strain rate reached in this region. An estimate of the local strain rate is obtained by calculating the average stretch factor (1/A) (dA/dt) imposed to the cap during the initial growth of the structure (times t_2 to t_5). A direct measurement of the interfacial length gives an estimation of the stretch factor of 2 10^3 to 10^4 s^{-1}. Such large values have the effect of nearly extinguishing any flame which would tend to develop at the vortex cap. This situation is very close to that examined in some recent studies of premixed strained flames (Darabiha et al. 1986) : there it is shown that when the premixed flame is highly strained it enters a regime of partial extinction.

Similar results are obtained by Cattolica and Vosen (1984) in a study of the evolution of a flame developing in a vortex. From observations of laser induced fluorescence, these authors show that the concentration of OH radicals is very low near the vortex cap and much larger in the side branches and wake where the flame develops.

Discussion of the instability mechanism

Let us now examine the relations between the pressure variations and the vortex shedding and burning processes. It is first important to locate the moments $t_1, ..., t_9$ with respect to the pressure cycle. This will enable us to describe the entire instability mechanism. Figure 7 displays the variations of pressure and velocity at the inlet plane of the combustor along with the global heat release fluctuations. This last quantity is obtained by integration over the scanning window of the instantaneous local heat release given by C_2 emission measurements. Moments t_1 to t_9 are also indicated in Figure 7. The global heat release reaches a maximum near time t_6. Now, it is worth looking at some implications of Rayleigh's criterion. Locally, instability is amplified if the instantaneous heat release and the pressure oscillation are in phase. It is damped if these two signals have opposite phase. These situations and intermediate cases may be observed if one considers the time histories of pressure and heat release at various points in the combustor. Because the acoustic wavelength λ in the combustor is large compared to the window dimension d ($\lambda/d \sim 20$), we will suppose that the pressure fluctuation is homogeneous over the scanning window and equal to its value at the inlet plane. Let us select as observation points those which correspond to a maximum heat release at times t_3, t_5, t_7 and t_8. Figure 8 displays the time variations of the local heat release at points A, B, C, D along with the pressure fluctuations at the inlet plane of the combustor. At point A, close to the inlet plane, little energy is fed into the acoustic field (the phase φ between pressure and heat release fluctuations is $\pi/2$). Point B corresponds to a damping zone : heat is released when the pressure is minimum ($\varphi = \pi$). At points C and D, located on the left side of the window, heat is released when the pressure is positive ($\varphi > 3\pi/2$) and these regions amplify the instability.

Now, if one considers the acoustic energy balance for the whole combustor cavity, local amplification and damping zones compensate each other. Globally, the total heat release and the pressure signals should be in quadrature if acoustic losses from the cavity are small. Figure 7 shows that the "global" Rayleigh criterion is indeed satisfied. Figure 7 presents the variations of velocity at the inlet plane, too. This quantity is also of interest because it certainly plays a role in the vortex shedding process. The vortex is formed at instant t_2 (Fig. 6b) : this moment corresponds to a nearly maximum value of the velocity fluctuation. Vortices formed during a 530 Hz cycle are generated by large positive excursions of the velocity at the inlet plane. The time lag between the pressure maximum and the vortex formation is about T/4. After its formation, the vortex grows and burns when it interacts with vortices developing on the side jets. The time lag between the formation of the vortex and the maximum heat release is about T/2. These results allow us to propose the following mechanism for the vortex driven combustion instabilities studied in this paper :

Step 1 : A large positive excursion of velocity at the inlet plane of the combustor causes the shedding of a vortex. In our experiment, this takes place at time t_2.

Step 2 : The vortex grows and is convected downstream (t_2 to t_5). Maximum heat release is produced when two neighbouring vortices interact (time t_6).

Step 3 : The heat pulse generated by the vortex coalescence and combustion feeds energy into the acoustic oscillation. This acoustic wave will induce a large positive excursion of velocity at the inlet plane at time $t_1 + T$ where T is the period of the cycle and the cycle is completed.

This oscillation will be self sustained if (1) the acoustic fluctuation corresponds to an acoustic mode of the system, (2) the amplitude of the velocity oscillation is large enough to induce vortex shedding, (3) the time lag between pressure and heat release satisfy Rayleigh's criterion : heat is added when the pressure becomes positive. In our case, this time lag is about 3T/4 and the global heat release maximum occurs at $3\pi/2$ after a positive pressure maximum.

4. SPECTRAL IMAGING OF THE COMBUSTOR RESPONSE TO HIGH FREQUENCY ACOUSTIC EXCITATION

Spectral imaging method

The spectral imaging method described in this section extends and enhances the commonly used spectral analysis of time varying signals. In standard practice, the signal $\phi(\underline{x},t)$ delivered by a local probe like a hot wire, laser Doppler velocimeter, microphone, thermocouple etc... is conditionned, sampled and stored and its spectral content is determined. The power spectral densities of the signal ϕ are usually estimated at a set of points in the flow and the spectral data are displayed in the form of one dimensional curves : $S_{\phi\phi} = S_{\phi\phi}(f)$. The analysis of large amounts of spectral data is tedious and when the physical mechanisms are complex, the characteristic features are difficult to extract without additional processing.

Now, consider a typical signal $\phi(\underline{x},t)$ obtained from a local probe and let $S_{\phi\phi}(x,f)$ designate the spectral density corresponding to the measurement point $\underline{x}$. By scanning the flow with the probe, the local signals may be collected and stored into a large time signal data base DB_1 (Fig. 9). A typical 2D exploration grid may have 200 points and 10000 samples may be acquired at each grid point. When the flow exploration is over, the data collected are used to estimate the local power spectral densities at each point of the grid. All the spectral estimates are stored in a second data base DB2 and this information is then processed to form the final displays : it is for example possible to gather the maximum spectral amplitudes belonging to a particular frequency band (f_1, f_2) and corresponding to all the grid points. A spectral image is constructed with these informations. Such an image reveals the spatial extent and intensity of a particular process if it is characterized by a specific frequency. The spectral imaging principle may be used without changes to process cross spectral or coherence informations. An application of this concept to combustion noise source location is presented in Poinsot et al. (1986). The source location is based on the coherence level between an outer microphone placed in the far field and the signal delivered by an optical (laser Schlieren) probe scanning the reacting flow.

The potential of the spectral imaging method will appear clearly in the next sections. The method does not provide phase informations like conditional methods do, but its range of application is much broader. It may be used in a variety of flows to determine the dominant frequencies and the spatial extent of the corresponding oscillations.

In most cases, spectral imaging must be complemented by other measurements. Spark Schlieren visualizations are used in this paper to distinguish the relevant flow structures. The spectral imaging method itself is applied to the signals obtained from a laser Schlieren set up and from a C_2 radical light emission system. The laser Schlieren probe measures the deviations of a laser beam crossing the flame. This deviation is caused by the local gradients of refraction index and it provides informations on the mixing of the fresh and burnt gases. The C_2 radical emission system uses the light emitted by radicals with a short life duration to characterize the chemical reaction process. As already indicated in Section 3 the light emitted by such radicals as C_2, CH and OH may be related to the local reaction rate. The laser Schlieren and C_2 light emission measurements are performed simultaneously by making use of the optical arrangement shown in Fig. 3.

Non-linear response of the multiple flame combustor under external excitation

Let us now consider the behavior of the combustor shown in Fig. 2 under external acoustic excitation. This device exhibits numerous modes of instability with or without external excitation (Darabiha et al. 1986, Poinsot et al.1987). In this section, we will only consider excited modes of the combustor : the upstream flow is set into oscillation by the driver units placed on the upstream duct. As explained by Zikikout et al. (1986), it is possible to select acoustic modes of oscillation and obtain a transverse sloshing motion in the combustor cavity. If the excitation frequency is close to the preferred hydrodynamic frequency of the jets, the combustor behaviour takes a complex form (the preferred frequency f_i is estimated by a simple Strouhal value established for non reacting flows, see Ho et Huerre 1984).

The high-frequency excitation (f_i = 3820 Hz) induces a sinous motion of the jets of fresh mixture. This motion causes vortex roll up at the same frequency (Fig. 10) but spectral analysis of sound pressure reveal the presence of a strong low frequency oscillation at a frequency f_2 = 470 Hz (Fig. 11). This oscillation vanishes when the external excitation is cut off.

Spectral imaging is now used to examine this nonlinear coupling. The two optical probe signals (laser Schlieren and C_2 light emission) are examined in the two frequency ranges centered on f_1 = 3820 Hz and f_2 = 470 Hz. The first frequency corresponds to the vortex roll-up and burning while the second is the induced low frequency oscillation.

Spectral images of the laser Schlieren signal are displayed in Fig. 12a for the high frequency range and Fig. 12b for the low frequency range. The two maps are similar and suggest that the vortices generated at the jet boundaries at a frequency f_1 also oscillate with the jets at a frequency f_2. The asymmetric structure of the spectral images of the laser Schlieren signal is not completely understood but similar features are also observed in the Schlieren visualizations (Fig. 10).

Spectral images of the C_2 radical emission signal convey additional information. In the high frequency range (Fig. 13a), little combustion takes place. The vortices are formed at the excitation frequency f_1 but most of them do not burn at this frequency. Now, the spectral images corresponding to the low frequency range show high combustion levels in the interaction zones between the jets (Fig. 13b) These regions are the source of the nonlinear coupling mechanisms. This process can be described as follows :

(1) Assume that the jets oscillate at the low frequency f_2. When the jet velocity is high, the transverse motion at this frequency has a small amplitude and the vortices formed at the frequency f_i are convected on the jet boundaries and burn at this frequency f_i near the jet axis (Fig. 13a). A typical Schlieren photograph of the flow at this moment is given in Fig. 10a.
(2) When the velocity decreases, the jet deviation from the horizontal axis increases. At one point, the jet deviation becomes sufficiently large to allow interactions between vortices issued from neighbouring jets (Fig. 10b). This interaction takes place at the frequency f_2, produces small scale turbulence and intense non-steady combustion (Fig. 13b).

(3) This non-steady combustion at frequency f_2 feeds energy into a longitudinal acoustic mode of the cavity (close to the quarter wave mode in this case).
(4) The low frequency acoustic mode sustains the jet oscillation at the frequency f_2 and the cycle is completed.

Direct Schlieren observations of the flow at various times reveal that the jet of fresh mixture flaps up and down, explaining the two combustion spots observed in Fig. 13b. The movement of the central jet cannot be reduced to a simple sinuous mode. All the jets are involved in the oscillation and this coupled movement explains the non linear response of the combustor.

The coupling mechanisms described in this section between high frequency vortices and low freuency acoustic modes may occur in many practical systems, where multiple flames are used. In this study, high frequency vortices were generated by an external acoustic excitation but natural coupling between high frequency vortices and low frequency modes may occur without external excitation. In fact, it has been observed that high frequency instabilities in afterburners (**"screech"**) usually trigger low frequency oscillations (Rogers and Marble 1956).

CONCLUSION

Pressure waves may strongly modify the structure of turbulent flames. This point is illustrated in this paper with two generic examples. In the first case a low frequency vortex driven acoustically coupled self-sustained oscillation is considered. The mechanism suggested for this regime is as follows : (1) vortex shedding occurs when the velocity fluctuation at the combustor inlet reaches a maximum positive vlue (2) vortices produced on neighbouring jets convect downstream and interact, producing small scale turbulence and intense combustion (3) the resulting heat pulse feeds energy into the acoustic oscillation and induces large velocity excursions which cause the vortex shedding at the inlet.

In the second case, an external pressure wave couples with a hydrodynamic instability mode of the flow. Hydrodynamic vortices formed on neighbouring jets induce a low frequency oscillation of the combustion intensity and a flapping motion of the fresh mixture jets.

Acknowledgements

We wish to acknowledge helpful discussions with Pr Frank Marble and Edward Zukoski of the California Institute of Technology.

This study was financially supported by a DRET contract monitored by Jacques Besnault.

References

Blackshear, P.L (1956). Growth of disturbances in a flame generated shear region NACA Rep. 1360.

Bray, K.N.C., Campbell, I.G., Lee, O.K.L. and Moss, J.B. (1983). An investigation of reheat buzz instabilities. Aeronaut. and Astronaut., AASU Memo 83/2. University of Southampton.

Cattolica, R.S. and Vosen, S.R. (1984). Two-dimensional fluorescence imaging of a flame-vortex interaction. Sandia Rep. Sand 84-8704.

Darabiha, N., Poinsot, T., Candel, S.M. and Esposito, E. (1986). A correlation between flame structures and acoustic instabilities. Presented at the tenth ICODERS, Berkeley 1985. Progr. in Astronaut. and Aeronaut., AIAA, 283-295.

Darabiha, N., Candel, S.M. and Marble F.E. (1986). The effect of strain rate on a premixed laminar flame. Comb. Flame 64, 203.

Flandro, G.A. (1986). Vortex driving mechanism in oscillatory rocket flows. J. of Propulsion 2, 3, 206-214.

Ho, C.M., and Huerre, P. (1984). Perturbed free shear layers. Ann. Rev. Fluid Mech. 16, 365-424.

Ho, C.M., and Huang, L.S. (1982). Subharmonics and vortex merging in mixing layers. J. Fluid Mech. 119, 443-473.

Keller, J.O., Vaneveld, L., Korsheld, D., Hubbard, G.L., Ghoniem, A.F., Daily, J.W., and Oppenheim, A.K. (1981). Mechanism of instabilities in turbulent combustion leading to flashback. AIAA J. 20, 254.

Keller, J.O., and Westbrook, C.K. (1986). Response of a pulse combustor to changes in fuel composition. 21st Symposium on Combustion. Munich Aug. 3-8.

Pitz, R.W., and Daily, J.W. (1981). Experimental study of combustion in a turbulent free shear layer formed at a rearward facing step. 19th Aerospace Sci Meeting, St Louis, Missouri. AIAA paper 81-0106.

Poinsot, T., Hosseini, K.M., Le Chatelier, C., and Candel, S. (1986). An experimental analysis of noise sources in a dump combustor. Prog. in Astronaut. and Aeronaut. 105, 333-345.

Poinsot, T. (1987). Analyse des instabilités de combustion de foyers prémélangés turbulents. Doctorat ès Sciences, University of Paris XI.

Poinsot, T., Trouvé, A., Veynante, D., Candel, S., and Esposito, E. (1987). Vortex driven acoustically coupled combustion instabilities. J. of Fluid Mech. 177, 265-292.

Rogers, D.E., and Marble, F.E. (1956). A mechanism for high-frequency oscillations in ramjet combustors and afterburners. Jet Propulsion 26, 456-462.

Smith, D.A. and Zukoski, E.E. (1985). Combustion instability sustained by unsteady vortex combustion. 21st Joint Propulsion Conference, Monterey, California, AIAA Paper n° 85-1248.

Yamaguchi, S., Ohiwa, N. and Hasegawa, T. (1985). Structure of blow-off mechanism of rod stabilized premixed flame. Comb. Flame 62, 31.

Zikikout, S., Candel, S.M., Poinsot, T., Trouvé, A. and Esposito, E. (1986). High combustion oscillations produced by mode selective acoustic excitation. 21st Symposium (International) on Combustion, The Combustion Institute, Pittsburgh.

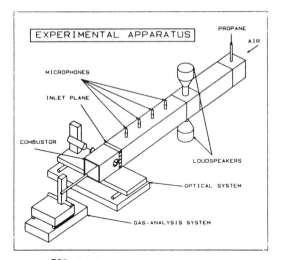

FIG. 1. Experimental arrangement.

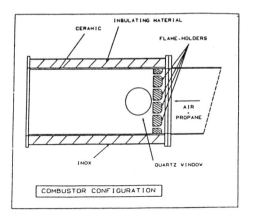

FIG. 2. Combustor configuration.

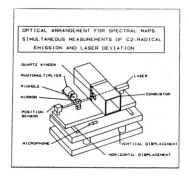

FIG. 3. Optical system used to simultaneously measure C_2 radical light emission and laser deviations.

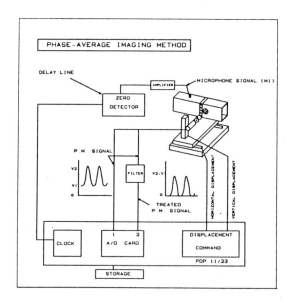

FIG. 4. Phase-average imaging method.

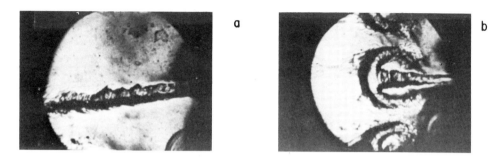

FIG. 5a. Schlieren visualization of a stable regime. $\dot{m}_{air}$ = 87 g/s , ϕ = 0.72.
FIG. 5b. Schlieren visualization of an unstable regime. $\dot{m}_{air}$ = 74 g/s , ϕ = .92.

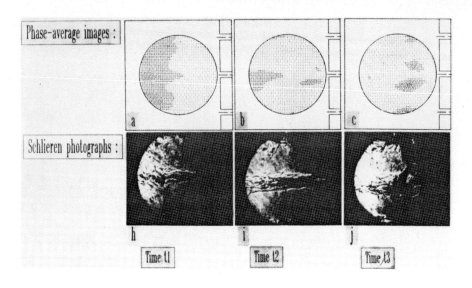
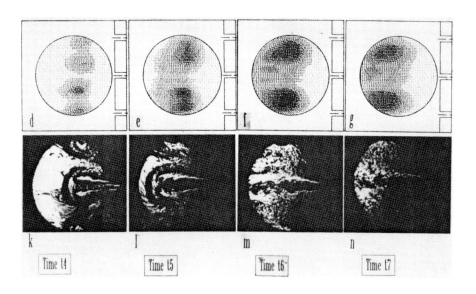

FIG. 6. Schlieren visualization (6h to 6n) and phase average image (6a to 6g) of an unstable combustion regime. Oscillation at 530 Hz. $\dot{m}_{air}$ = 74 g/s , ϕ = .92.

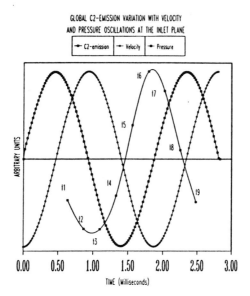

FIG. 7. Pressure, velocity and global heat release fluctuation during a 530 Hz cycle

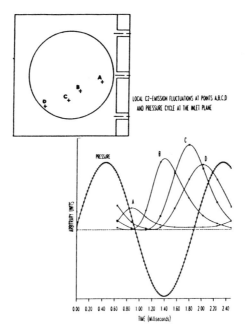

FIG. 8. Pressure and local heat release fluctuations at four points of observation A, B, C, D during a 530 Hz cycle.

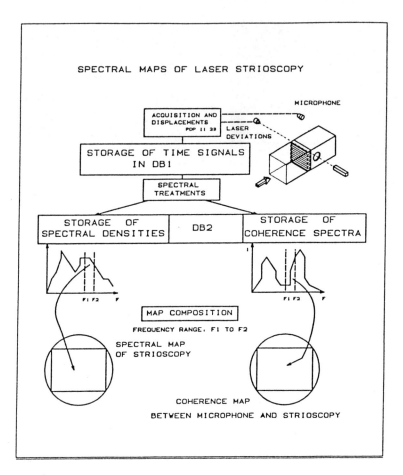

FIG. 9. Data processing performed during spectral imaging.

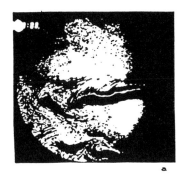 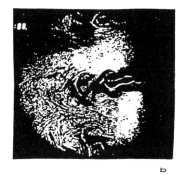

FIG. 10. Vortex formation and interaction in the combustor shown in Fig. 2.
a) Instant of the cycle with high jet velocity and small deviation.
b) Instant of the cycle with low jet velocity and large deviation.

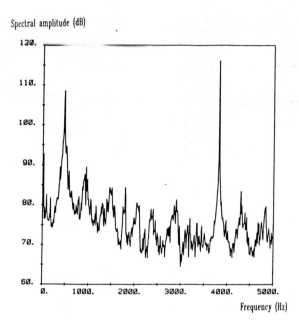

FIG. 11. Sound spectral analysis of an acoustically excited combustion oscillation (inner microphone). f_{exc} = 3820 Hz.

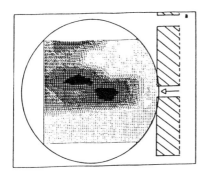

Spectral map of laser deviation signal

Frequency range : 3800-3840Hz

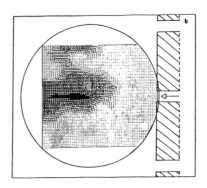

Spectral map of C_2-emission signal

Frequency range : 3800-3840Hz

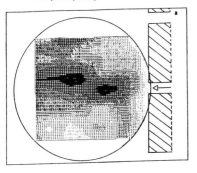

Spectral map of laser deviation signal

Frequency range : 460-510Hz

Figure 12

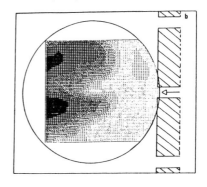

Spectral map of C_2-emission signal

Frequency range : 460-510Hz

Figure 13

FIG. 12. Spectral maps for laser Schlieren signal
a) in the high frequency range : f_1 = 3820 Hz
b) in the low frequency range : f_2 = 470 Hz.

FIG. 13. Spectral maps for C_2 emission signal
a) in the high frequency range : f_1 = 3820 Hz
b) in the low frequency range : f_2 = 470 Hz.

CONTROL OF TURBULENCE IN COMBUSTION

K. C. Schadow, E. Gutmark, T. P. Parr, D. M. Parr, and K. J. Wilson
Research Department
Naval Weapons Center
China Lake, CA 93555-6001

ABSTRACT

Passive shear-flow control methods are being investigated to enhance turbulence for subsonic combustion. In this paper, two control methods are discussed. These methods were developed in subsonic nonreacting flows and verified in subsonic combustion processes to avoid combustion instabilities in a dump combustor. Both methods, which are based on the detailed understanding of flow instability mechanisms in shear layers, were used to change the initial conditions of the air flow (jet) issued from the inlet duct into dump combustor, in an attempt to decouple combustion from the large-scale structures in the flow. Two different jet nozzles located at the dump plane are discussed: jet nozzles having sharp corners (these jets are characterized by large azimuthal variations of the flow field), and a multi-step nozzle with numerous sources for turbulence production yielding highly turbulent homogeneous incoherent flow fields. These nozzles augment turbulence or fine-scale (molecular) mixing and, at the same time, reduce large-scale mixing. Both methods were shown to be able to modify the flow field mixing pattern and inner eddy structure and to provide the potential for minimizing combustion instabilities.

NOMENCLATURE

D	Exit diameter
D_e	Equivalent diameter
f	Frequency
f_F	Forcing frequency
f_j	Preferred mode frequency
H_{ST}	Step height
L_{ST}	Step length
$\dot{m}_a$	Air mass flow
P	Mean chamber pressure
r	Radial coordinate
R_e	Equivalent radius

$R_{0.5}$	Jet width at the location of half centerline mean velocity
St	Strouhal number (= $f \cdot \theta/U$)
St_1	Preferred mode Strouhal number
U_{CL}	Centerline velocity
U_o	Exit velocity
u	Mean velocity
u'	Fluctuating velocity
$\overline{u'w'}$	Reynolds shear stress
x	Axial coordinate
θ	Momentum thickness
ΔP	Peak-to-peak pressure difference
$\lambda \approx U_o/2 \cdot f_F$	Wavelength
ϕ	Equivalence ratio

INTRODUCTION

In reaction systems, combustion rate and combustion stability are strongly related to the interaction of fluid dynamics with the combustion process. This paper discusses two methods of rationally controlling fluid dynamics to minimize combustion instabilities in a subsonic dump combustor.

Combustion instabilities have been observed in several dump combustors [1]. Different mechanisms have been postulated to be responsible for driving the pressure oscillations. The Naval Weapons Center (NWC) is exploring one possible driving mechanism that is related to the development of large-scale, coherent flow structures (vortices) in the shear flow downstream of the dump.

The dominant role of coherent structures in the evolution of shear layers has been recognized by many investigators, predominantly in laboratory-type flow at low Reynolds numbers [2-4]. Large-scale structures were also identified in the NWC coaxial dump combustor set-up in nonreacting tests at a maximum Reynolds number of $7 \cdot 10^5$. Specifically, the flow instability frequencies associated with initial vortex shedding at the dump and the subsequent vortex merging were determined [5]. Utilizing this knowledge of the shear-layer instability characteristics, the shear-flow development was actively controlled by excitation of the duct resonant acoustics. By matching the acoustic and shear layer instability frequencies, highly coherent large-scale structures were generated [5, 6]. The strongest effect on the flow was achieved when the acoustic frequency was near the jet preferred mode frequency, f_j [7]. This frequency is in the Strouhal number range $0.25 < St_j = f_j U_o/D < 0.5$ (normalized by the jet exit diameter, D, and jet exit velocity, U_o) depending on the experimental facility [8]. At this preferred-mode forcing, the shear layer rolled up into vortices with a wavelength of nearly the duct diameter. As a result, mixing was enhanced in the shear flow and pipe flow beyond the reattachment region in nonreacting and combustion experiments [6, 9].

While coherent structures are beneficial in enhancing large-scale mixing, they prevent fine-scale (molecular) mixing, particularly during the initial vortex development process. In the case of a dump combustor, fine-scale mixing between the hot reaction products from the recirculation zone and the "fresh" air/fuel mixture is reduced and therefore the reaction process is delayed. During the final stages of the vortex roll-up process, fine-scale mixing is amplified by vortex stretching and vortex break-up [10], which may lead to sudden heat release [11]. This interaction between fluid dynamics and combustion can lead to periodic heat release and to the driving of combustion instabilities when a certain phase relationship between the periodic heat release and the pressure oscillations exists (Rayleigh criterion) and/or when a feedback loop is established between the temperature waves associated with the vortex burn out and the pressure oscillations.

To prevent the driving of combustion oscillations during the described scenario, it is necessary to decouple the combustion process from the large-scale vortices generated when the flow instability frequencies match the acoustic pressure oscillation frequencies. In the NWC experiments the decoupling between combustion and coherent flow is achieved by combining passive shear flow control with controlled fuel injection. Two methods were investigated:

(1) Exit nozzles with sharp corners were used to alter the regular sequence of the flow instability processes, that lead to the generation of large-scale structures. From earlier studies [12, 13] it was conjectured that with triangular cross-sections fine-scale mixing can be enhanced in the flow issued from the corners, while large-scale mixing can be maintained at the flat sides. These tests suggest that fuel should be injected into the corner flow to avoid interaction between combustion and the organized flow of the flat side, thus avoiding periodic heat release.

(2) Fine-scale turbulence enhancement was also studied using an axisymmetric nozzle with a series of downstream facing steps inside the "multi-step nozzle." The idea is based on the theoretically proven relation between the number of unstable modes and the number of inflection points in the velocity profile. This relation was described by Howard [14] and was shown to be valid for instability problems governed by the Rayleigh stability equations.

The two methods to manipulate the conventional jet flow were researched with three experimental methods. Nonreacting free-jet tests were performed to study the shear-flow dynamics with hot-wire anemometry. Subsequently, the different shear-flow control methods were studied in a laboratory burner using Planar Laser Induced Fluorescence (PLIF) imaging for combustion dynamics visualization. Finally, laboratory ramjet dump combustor tests were performed to visualize the reactive shear-flow dynamics with the PLIF imaging technique and to establish the relationship between the shear-flow velocity characteristics and the pressure fluctuation amplitudes.

EXPERIMENTAL SET-UP

The free-jet facility using hot-wire anemometry is schematically shown in Figure 1. Orifices having square, equilateral-triangular, and isosceles-triangular cross sections were placed at the end of the circular pipe to study the shear-flow development for corner angles of 90, 60, and 30 degrees (Figure 2).

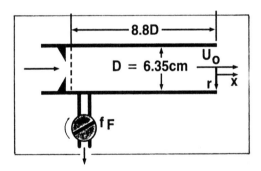

Fig. 1. Free-Jet Test Set-Up.

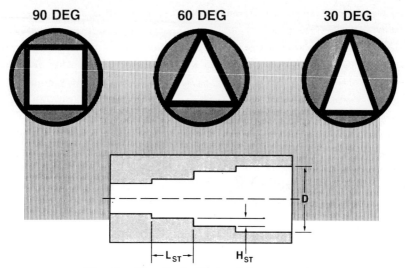

Fig. 2. Fine-Scale Mixing Enhancement in Jet Nozzles With Corners and Multi-Steps.

In the same set-up the multi-step nozzle was studied. This nozzle transitioned from a circular pipe to a circular exit with a diameter, D, via downstream-facing steps (Figure 2). The number of steps (N_{ST}) the step height (H_{ST}) and the step length (L_{ST}) were varied.

Acoustic excitation in the free-jet tests could be achieved by exciting the resonance frequencies of the duct upstream of the nozzles with a rotating valve (Figure 1).

The shear-flow characteristics of the propane/air burner were visualized by PLIF imaging of in situ OH radicals. In this technique, which is schematically shown in Figure 3 and described in detail in Reference 15, an excimer laser beam is expanded into a thin sheet that excites OH molecules in the annular diffusion flame of the burner. The fluorescence is imaged with an intensified diode array camera. The burner can be acoustically excited with varying phase shifts with respect to the laser. Changing the phase angle can be used as a means to stroboscopically map out the time dependence of the flame structure. Circular and triangular nozzles as well as the multi-step nozzle were used in the experiments.

The laboratory dump combustor has been described in detail in Reference 1. Circular and equilateral-triangular inlet ducts upstream of the dump were used to compare the pressure oscillation amplitude for fuel (ethylene) injection conditions: (1) from the circular inlet duct, (2) from the three flat sides of the triangular inlet duct, and (3) from the three vertex sides of the triangular duct. In addition, the shear-flow characteristics of the circular inlet duct were visualized using a quartz-wall combustor section to allow PLIF imaging.

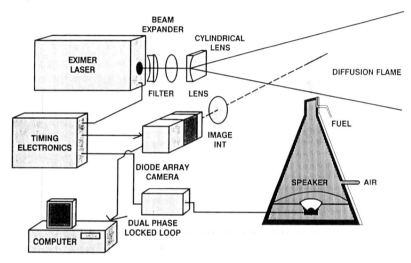

Fig. 3. Experimental Set-Up for PLIF OH-Imaging of a Burner.

RESULTS AND DISCUSSION

JET NOZZLES WITH CORNERS

The nozzles tested in this investigation had a range of corner angles from 30 to 90 degrees, thus the effect of these angles on the flow could be studied and compared to the flow issued at the flat sections of the nozzle. Mean velocity measurements [12] of these jets showed that the initial shear layer was thicker at the vertices relative to its thickness at the flat side. This results in different instability characteristics, which become a function of the azimuthal location in the jet flow. The highly coherent vortices, which are being shed from the flat sides of the nozzle, interact at the corners and produce intense three dimensional, incoherent shear layer flow [13]. Consequently, there is a substantial difference in the turbulence activity of the jet in the two sides. Figures 4 and 5 show the axial turbulence intensity profiles in various cross sections downstream of the nozzle, at the vertex and flat sides of the nozzle, respectively. The initial turbulence level at the vertex was about five times higher than on the flat side. Spectral analysis showed that the vertex turbulence contained a broader band of frequencies relative to the other side. The turbulence growth rate was much higher on the flat side. While at the vertex, the turbulence level remained almost constant in the downstream direction. The turbulence level at the flat side increased rapidly and reached the level of the other side at $x/D_e = 2$. This higher amplification is related to the larger ratio of the equivalent radius to the local shear layer momentum thickness (R_e/θ).

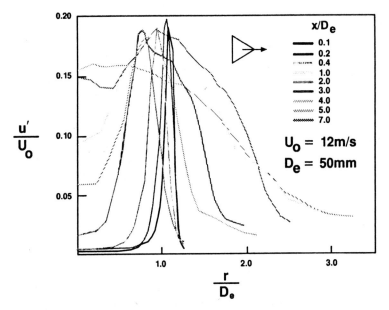

Figure 4. Turbulence Intensity Variation at the Triangular Jet Vertex Side.

With the higher R_e/θ there is an increased production of coherent Reynolds stresses. This process is usually accompanied by a modification of the shear layer, leading to a higher spreading rate of the flow. A comparison of the spreading rates at the flat and vertex sides of the triangular jet is given in Figure 6. It is indeed shown that the flat side spread much faster than the vertex side, which had a nearly parallel shear layer. At $x/D_e = 1$ both sides became equally wide and at $x/D_e = 6$ the jet was almost twice as wide at the original flat side than at the vertex side. The growth of the shear layer width at the flat side causes the local Strouhal-number ($St = f\theta/U$) to increase, and consequently the jet instability characteristics change. When the neutral Strouhal number is approached, the shear layer growth discontinues ($x/D_e > 6$).

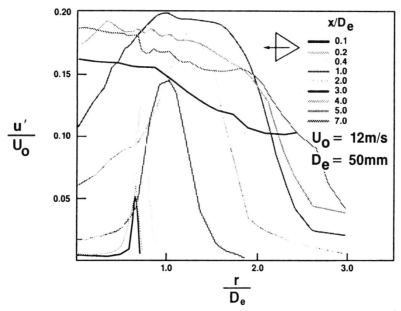

Figure 5. Turbulence Intensity Variation at the Triangular Jet Flat Side.

These flow dynamics are changing with the azimuthal locations. Consequently, the jet growth pattern in the downstream direction is very complex. Figure 7 shows the mean velocity contours at different downstream stations. The jet evolved from the initial triangular shape into a quasi circular shape at $x/D_e = 0.4$ and into a reversed triangle at $x/D_e = 2$. Even at a distance of seven equivalent diameters from the nozzle, the jet did not "recover" from the higher spreading rate of the flat side and had a reversed triangular shape relative to the original outlet.

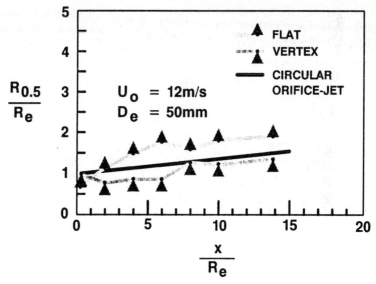

Figure 6. Spreading Rate of the Triangular Jet at the Flat and Vertex Sides.

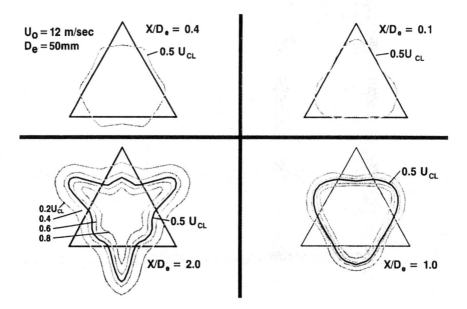

Figure 7. Iso-Velocity Contours of Triangular Jet.

Many observations that were made in the triangular jet were repeated in the square jet; except that in the case of the square jet the differences between the flat side and the vertex side were less distinct because of the larger vertex angle (90 degrees). Figure 8 shows the turbulence intensities of shear layers emanating from the 60-degree corner of the equilateral triangle compared to that of the square (90 degrees) and compared to that of the flat side of the triangle. Both initial shear layer intensities of the corner sides were much higher (3.5 to 4.5 times) than at the flat side. The intensity of the 60-degree corner shear layer was higher than that of the 90-degree corner shear layer by 30%. The intensity at a 30-degree corner triangle (not shown here) was even higher than at the 60-degree corner. These results show that, in the range of corner angles studied, the shear layer turbulence grew with decreasing corner angle. In all corner shear layers, no amplification of turbulence intensity was observed in the downstream direction. The growth rate at the flat side was much higher as a result of the large spreading rate, so that at $x/D_e > 4$ the turbulence intensity in all azimuthal locations was uniform.

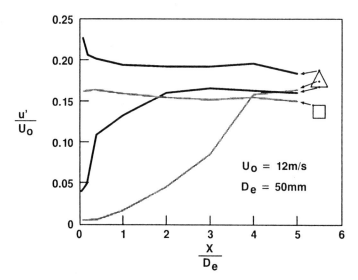

Figure 8. Turbulence Intensity Variation Along Shear Layers Emanating From Corners of Different Angles.

The basic differences between the circular and triangular jets, measured in non-reacting flows, were also studied in a reacting diffusion flame. The flames were acoustically excited at the preferred mode of the jets ($St_1 = f_1 D_e/U_o = 0.35$). An instantaneous picture (7 nsec) of the combustion process inside the vortices of a circular flame is given in Figure 9. It is shown here that following the initial roll-up of the vortices, the combustion was most intense in the vortex circumference. As the vortex was being convected downstream, the combustion proceeded into the vortex core, while the reaction at the circumference was completed. In a triangular burner flame (Figure 10) periodic flame structures were formed only at the flat side. The flame at the vertex side was highly diffusive and incoherent, as expected from the cold flow structure described earlier.

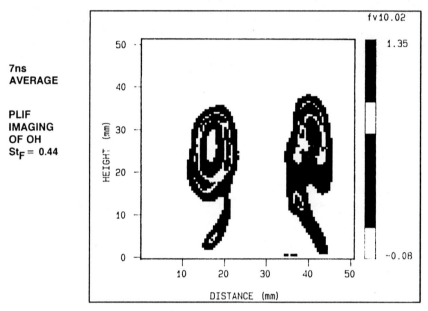

Figure 9. Instantaneous PLIF Picture of a Circular Diffusion Flame.

The breakdown of the flow coherence at the triangular jet vertex was proven beneficial in avoiding combustion instabilities that may develop under certain conditions in a dump combustor. Cold flow tests showed that the acoustic resonance of the combustion chamber can generate large-scale structures in the ducted jet being issued at the dump. This interaction is especially effective when the chamber acoustics match the flow instability frequencies. During combustion this interaction is further enhanced by the periodic heat release associated with the burning of the coherent large-scale structures. Figure 11 shows the periodic structures of the combustion inside a dump combustor with a circular inlet duct when high pressure fluctuations were present due to the previously described acoustic flow interaction.

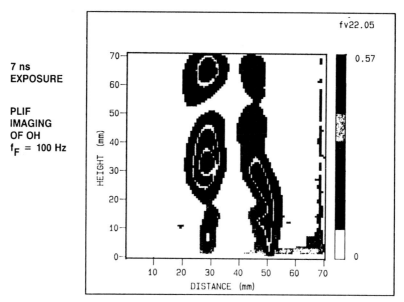

Figure 10. Instantaneous PLIF Picture of a Triangular Diffusion Flame.

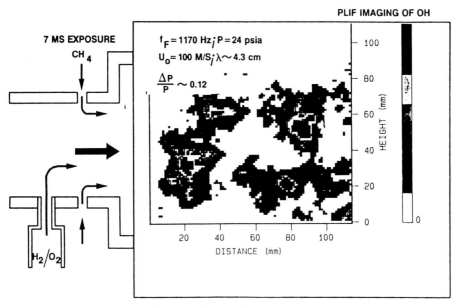

Figure 11. PLIF Imaging of Periodic Combustion Inside a Dump Combustor During Pressure Oscillations.

The pressure fluctuation amplitude varied with different fuel injection conditions. When fuel was injected into the flow at the locations where large-scale structures can be generated, i.e., around a circular jet outlet or at the flat sides of a triangle, the combustion was governed by the periodic heat release. This resulted in a high pressure oscillation inside the combustion chamber and led to combustion instability (Figure 12). When the fuel was injected into the incoherent and highly turbulent section at the triangle's vertex, the pressure oscillations were eliminated, as shown in the same figure.

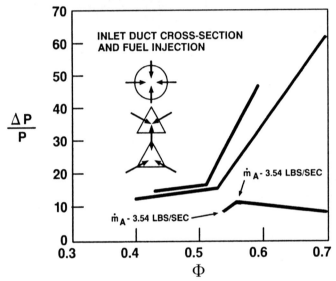

Figure 12. Pressure Oscillations in a Dump Combustor for Different Fuel Injection Locations.

MULTI-STEP NOZZLE

The concept of enhancing small-scale turbulence by altering the shear layer mean velocity profile to have multiple inflection points was tested in a circular jet. Figure 13 compares a regular jet having one inflection point in the shear layer and a modified jet that has two inflection points in the mean velocity profile. The same figure shows the corresponding turbulence intensity. The regular jet had a typical single peak turbulence intensity distribution across the shear layer. The other jet had two peaks corresponding to two sources of instability in the jet shear layer.

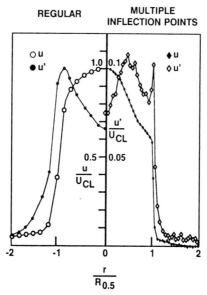

Figure 13. Comparison of Shear Layers With Regular Shape and Multiple Inflection Points.

This principle was extended by designing a nozzle with multiple backward facing steps with $N_{ST} = 3$ and $(L/H)_{ST} = 16$. The small sequential shear layers generated at each one of the steps modified the mean velocity profile to include numerous inflection points (Figure 14) and comprised many sources of fine-scale turbulence generation. The resulting turbulence intensity at the exit was up to six times higher than that of flow issuing from a regular circular nozzle (Figure 14).

In Figures 15 and 16 the contours of normal and shear Reynolds stresses, u' and $\overline{u'w'}$, of a regular circular jet are compared with a multi-step nozzle with $N_{ST} = 3$ and $(L/H)_{ST} = 7$. The regular circular jet had the highest level of turbulence activity in the jet shear layer. The intensity along the axis was low at the potential core and increased gradually towards the tip of this core. The jet issued from the multi-step nozzle had a uniform high turbulence activity across the jet from the jet exit plane. The intensity of the axial turbulence component, u', was twice that of the regular jet at the end of the potential core (Figure 15). The contours of $\overline{u'w'}$ for the multi-step nozzle were very irregular and homogeneously distributed in the entire flow field (Figure 16).

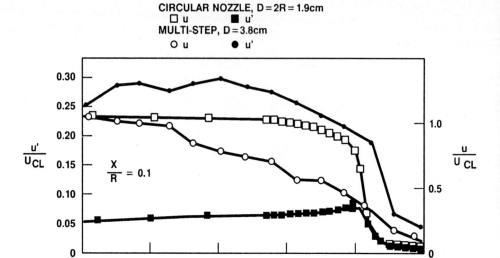

Figure 14. Turbulence Augmentation By Multiple Inflection Points in Mean Velocity Profile.

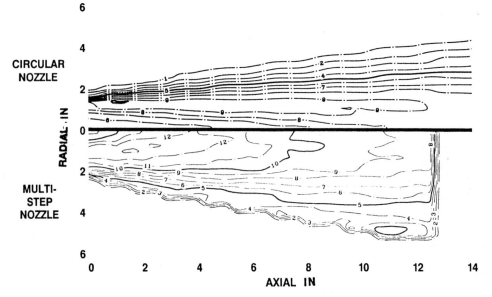

Figure 15. Comparison Between u' Contours of a Circular Jet and of a Multiple-Step Jet.

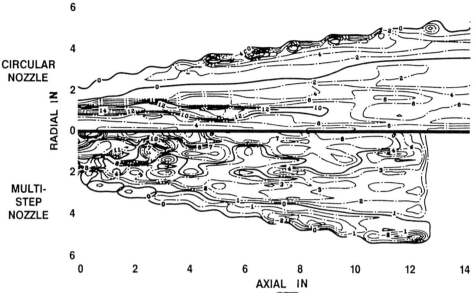

Figure 16. Comparison Between $\overline{u'w'}$ Contours of a Circular Jet and of a Multiple-Step Jet.

The fine-scale enhancement method with multiple steps was applied to combustion in free flames. The PLIF technique was used to compare a regular circular burner flame to a multi-step burner flame (N_{ST} = 2; $(L/H)_{ST}$ = 5) for both unforced and forced conditions. The concentration of the hydroxyl radicals in the flame was again used as an indicator of the combustion intensity. The combustion of the circular flame was confined to the thin shear layer surrounding the jet (left side of Figure 17). The multi-step flame was much more intense and was more uniformly spread throughout the jet (right side of Figure 17). The difference can be attributed to the homogeneous distribution of high intensity small-scale turbulence in the multi-step flame, which enhanced the combustion by promoting molecular mixing of fuel and air. In the circular flame these conditions existed only in the jet shear layer.

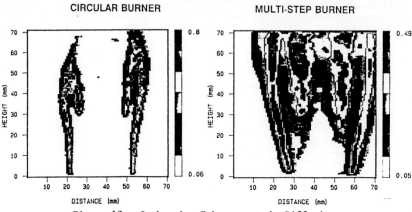

Figure 17. Combustion Enhancement in Diffusion Flames With Multi-Step Burner.

The forcing induced large-scale structures in the circular flame shear layer (left side of Figure 18). The flame was enclosed in the large-scale structures where the required small-scale mixing conditions are available. The flame was quenched in the small streaks connecting the vortices (the "braids") due to the high strain rates in the flow at this region. The flame of the multi-step nozzle was more homogeneously distributed without any trace of coherent structures (right side of Figure 18).

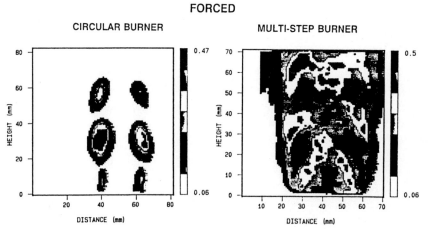

Figure 18. Effect of Forcing on Circular and Multi-Step Diffusion Flames.

The concept of enhancing fine-scale mixing and combustion using a series of downstream facing steps is presently being investigated in a dump combustor to avoid periodic heat release and therefore to reduce pressure oscillation amplitudes.

CONCLUSIONS

Two different methods were developed and investigated to enhance fine-scale mixing and to reduce the coherence of the large-scale structures in the flow. The suppression of these structures was shown to be important to avoid the onset of combustion instabilities:

(1) Sharp corners in the jet injector introduced high instability modes into the flow via the nonsymmetric mean velocity and pressure distribution around the nozzle. Both nonreactive and reactive flows showed the difference between the flow at the corner and at the flat side. Although highly coherent structures could be generated at the flat side, the corner flow was dominated by highly turbulent small-scale eddies. By injecting fuel into the corner flow, the periodic heat release, characteristic of combustion in the presence of coherent structures, was eliminated and combustion instability was avoided.

(2) Downstream-facing multiple steps inside a circular nozzle modified the initial mean velocity profile and added inflection points as sources of turbulence production. The resulting flow was highly turbulent with a homogeneous distribution in the entire flow field. Even high level excitation could not force the generation of large-scale structures in this flow. This concept is presently studied as a means of minimizing combustion instabilities in a dump combustor.

REFERENCES

1. J. E. Crump, K. C. Schadow, V. Yang, and F. E. C. Culick. "Longitudinal Combustion Instabilities in Ramjet Engines: Identification of Acoustic Modes," Journal of Propulsion and Power, Vol. 2, No. 2, March-April 1986, pp. 105-109.

2. G. L. Brown and A. Roshko. "On Density Effects and Large Structure in Turbulent Mixing Layers," Journal of Fluid Mechanics, Vol. 64 (1974), pp. 775-816.

3. C. D. Winant and F. K. Browand. "Vortex Pairing-the Mechanism of Turbulent Mixing Layer Growth at Moderate Reynolds Number," Journal of Fluid Mechanics, Vol. 63 (1974), pp. 237-255.

4. C. M. Ho and L. S. Huang. "Subharmonics and Vortex Merging in Mixing Layers," Journal of Fluid Mechanics, Vol. 119 (1982), pp. 443-473.

5. K. C. Schadow, K. J. Wilson, J. E. Crump, J. B. Foster, and E. Gutmark. "Interaction Between Acoustics and Subsonic Ducted Flow with Dump," AIAA 22nd Aerospace Sciences Meeting, AIAA Paper 84-0530, 9-12 January 1984, Reno, Nevada.

6. K. C. Schadow, K. J. Wilson, and E. Gutmark. "Characterization of Large-Scale Structures in a Forced Ducted Flow with Dump," AIAA 23rd Aerospace Sciences Meeting, AIAA Paper 83-0080, 14-17 January 1985, Reno, Nevada.

7. S. C. Crow and F. H. Champagne. "Orderly Structure in Jet Turbulence," Journal of Fluid Mechanics, Vol. 48 (1971), p. 547-591.

8. E. Gutmark and C. M. Ho. Physics of Fluids, Vol. 26 (1983), pp. 2932-2938.

9. K. C. Schadow, D. M. Parr, K. J. Wilson, E. Gutmark, and G. E. Ferrell. "Study of Flow Dynamics in Combustion Processes," 8th International Symposium on Airbreathing Engines, June 1987, Cincinnati, Ohio.

10. A.K.M.F. Hussein. "Coherent Structures and Turbulence," G. I. Taylor Symposium on Fluid Mechanics, Cambridge, England, March 1986.

11. F. E. Marble. "Growth of a Diffusion Flame in the Field of a Vortex," Recent Advances in Aerospace Sciences, Ed. C. Casci, 1985, pp. 395-413.

12. E. Gutmark, K. C. Schadow, D. M. Parr, C. K. Harris, and K. J. Wilson. "The Mean and Turbulent Structure of Noncircular Jets," AIAA Shear Flow Control Conference, AIAA Paper 85-0543, 12-14 March 1985, Boulder, Colorado.

13. K. C. Schadow, K. J. Wilson, D. M. Parr, C. J. Bicker, and E. Gutmark. "Reduction of Flow Coherence in Forced, Subsonic Jets," AIAA/SAE/ASME/ASEE 21st Joint Propulsion Conference, AIAA Paper 85-1109, 8-10 July 1985, Monterey, California.

14. L. N. Howard. J. De Mecanique, 3, 4, 1964.

15. E. Gutmark, T. P. Parr, D. M. Parr, and K. C. Schadow. "Planar Imaging of Vortex Dynamics in Flames," Presented at the National Heat Transfer Conference, Pittsburgh, Pennsylvania, 9-12 August 1987.

PROGRESS TOWARD SHOCK ENHANCEMENT OF
SUPERSONIC COMBUSTION PROCESSES

Frank E. Marble, Gavin J. Hendricks and Edward E. Zukoski
California Institute of Technology, Pasadena, CA 91125

ABSTRACT

In air breathing propulsion systems for flight at Mach numbers 7 to 20, it is generally accepted that the combustion processes will be carried out at supersonic velocities with respect to the engine. The resulting brief residence time places a premium on rapid mixing of the fuel and air. To address this issue we are investigating a mechanism for enhancing the rate of mixing between air and hydrogen fuel over rates that are expected in shear layers and jets.

The mechanism rests upon the strong vorticity induced at the interface between a light and heavy gas by an intense pressure gradient. The specific phenomenon under investigation is the rapid mixing induced by interaction of a weak oblique shock with a cylindrical jet of hydrogen embedded in air. The status of our investigations is described in three parts: a) shock tube investigation of the distortion and mixing induced by shock waves impinging on cylindric of hydrogen embedded in air, b) the molecular mixing and chemical reaction in large vortices, periodically formed in a channel, and c) two-dimensional non-steady and three-dimensional steady numerical studies of shock interaction with cylindrical volumes of hydrogen in air.

1. INTRODUCTION

Flight at hypersonic Mach numbers, between 6 and 15, appears as the next regime of major aeronautical research and development. The "trans-atmospheric aircraft," having a major component of air-breathing propulsion together with the capability of taking off from a runway and climbing to low earth orbit, is perhaps the most ambitious example. The operational flexibility of such an aircraft could offer a substantial advantage over current systems. The actual advantage that may be realized can be determined only by detailed investigations of the aerodynamic, cooling and propulsion problems which are unique to this Mach number range and by rather sophisticated studies of integration between the aerodynamic configuration, the propulsion system and the thermal control of the structure. The configuration and integration analyses which appear in Kirkham & Hunt (1977), Edwards, Small & Weidner (1975), Small, Weidner & Johnston (1974) and Nowak & Kelley (1976) provide substantial insight into both the importance and the formidable difficulty of the problem.

Among the conclusions of these and further studies is the clear confirmation that the problem of efficient, compact combustion stands as a central issue. In the first place, because of the severe penalties associated with excessive losses and heavy heat load, it is generally accepted that the combustion process must be carried out in a gas stream that is supersonic with respect to the engine. As a consequence, there is a very short time available for mixing and reaction and a high probability for internal shocks and losses. Because of the very high free-stream stagnation temperature associated with hypersonic velocities, the relative stagnation temperature rise resulting from combustion is rather small. Under this circumstance, internal aerodynamic losses or incomplete chemical reaction results in an unacceptable deterioration of performance.

In order to achieve a compact combustion process, considerable effort, e.g. Gross (1959), Gross & Oppenheim (1959) was devoted toward combustion in stabilized detonation waves. The results were not encouraging. Studies directed toward rapid mixing, Ferri (1968), Ferri (1973), Evans & Anderson (1974), McClinton (1978), have resulted in greater losses and longer combustors than desirable. Recent excellent reviews, Waltrup (1986) and Northam & Anderson (1986), report the current status of the combustion studies and confirm that the problem is still an open one, particularly at Mach numbers greater than 7. Therefore it is appropriate to investigate means for enhancing the rate of mixing between air and hydrogen fuel over that which may be achieved in shear layers and jets.

The mechanism to be described here depends essentially upon the Rayleigh-Taylor instability induced at the interface between a light and heavy gas by a strong pressure gradient. It is a property of combustion fields that the distribution of gas density is exceedingly non-uniform and this characteristic is accentuated when the fuel is hydrogen. One of the major effects of very strong pressure gradients on a field of non-uniform gas density is to accelerate the lighter gas at a rate several times that of the heavy gas. In addition to generating vorticity in the region of high density gradient, the large scale stability of the field is affected and flow reversal may be induced, a phenomenon which is of paramount importance in the supersonic mixing and combustion of hydrogen in air.

In an actual supersonic combustor there will exist oblique shock waves which constitute highly concentrated forms of the adverse pressure gradients discussed above. It is to be expected that the flow reversal, vortex formation and mixing observed in the steady phenomenon will appear, but in a more intense form. Therefore the distinct possibility exists of utilizing a set of "tailored" steady or unsteady shocks to induce very high intensity mixing and chemical reaction between the air and hydrogen fuel.

We are at present in the early stages of an extensive experimental program to investigate the detailed mechanism of the shock induced enhancement of mixing in non-uniform gas streams. The technological process described above has been broken down into two relatively independent phases for detailed experimental investigation, the first of which is being carried out in the Galcit 17-inch shock tube, the second in our unsteady combustion facility.

The work described here is supported by the United States Air Force Office of Scientific Research under Grant Number AFOSR-84-0286 and under URI Grant F49260-86-C-0113. Both programs are under the direction of Dr. Julian M. Tishkoff.

2. THE ENHANCEMENT MECHANISM AND BACKGROUND INFORMATION

When the pressure gradient in a gas is not aligned with the density gradient, vorticity may be generated in the gas. If the gas undergoes an isentropic change of state, the direct relationship between the pressure and the density guarantees the colinearity of these two gradients and no vorticity is generated. When, however, the density gradient is in part due to a variation of gas composition or the non-uniform conversion of chemical to sensible heat, the two gradients will not be aligned. Then, as shown in Fig.1, the displacement of the mass center from the geometric center of a gas element will allow the normal pressure stress on the surface of the element to generate a moment about the mass center and, hence, a rotation. A convenient analytical statement of this phenomenon is contained in Bjerknes theorem

$$\frac{d\Gamma}{dt} = \int \frac{1}{\rho^2} (\text{grad}\,\rho \times \text{grad}\,p) \cdot dA$$

where Γ is the circulation about a closed contour fixed to the fluid.

As applied to combustion problems, this vorticity generation mechanism was investigated explicitly by Fleming (1982) and that type of analysis was extended to relatively strong adverse pressure gradients by Marble & Hendricks (1986a). In this study a diffusion flame was situated, Fig. 2, along the horizontal axis and the gas flowed against a rising pressure field. The combustion products, which were of relatively low density, could not easily move against the pressure gradient and, as a consequence, reversed their flow direction and moved back upstream. The velocity and temperature profiles for a particular pressure gradient are shown in Fig. 3. In this circumstance the pressure gradient and the density gradient were nearly normal to each other, yielding the maximum rate of vorticity generation. This vorticity appears in the shear layers that bound the region of backflow, Fig. 3.

Consider now a cylindrical mass of low density gas (e.g. hydrogen) embedded in a higher molecular weight gas (e.g. air) situated in a pressure gradient as shown in Fig. 4a. The density jump between the hydrogen and the air constitutes a strong density gradient and this interacts with the imposed pressure gradient, according to Bjerknes theorem, to produce the vorticity distribution shown in Fig. 4b at the boundary of the hydrogen mass. The velocity field induced by the vorticity at the interface, distorts the boundary of the hydrogen mass, as shown schematically in Fig. 4c, and essentially creates a strong vortex pair.

Consider finally a shock wave passing over the cylindrical mass of hydrogen; this, as mentioned earlier, constitutes a very concentrated form of adverse pressure gradient. The shock passage, which can be considered very fast in comparison with other processes, deposits a distribution of vorticity over the boundary similar, but generally much stronger, than that associated with usual continuous gradients. Some of the earliest observations related to this phenomenon, reported by Rudinger (1958), Rudinger & Somers (1960), concerned the passage of a shock wave over bubbles and over flame surfaces. Subsequent computations by Evans et al (1962), Picone & Boris (1983) and Picone et al (1983) demonstrated the possibility of calculating the deformed surfaces observed by Rudinger (1958) with a relatively high degree of accuracy.

The earliest experiments which are directly related to the present investigations were performed by Sturtevant together with his graduate student, Haas (1983). These were shock tube studies, carried out at Caltech, concerning the interaction of relatively weak shock waves with regions of low density gas embedded in high density gas. The research was primarily aimed at the problem of shock diffraction in non-uniform atmospheres; fortunately the photographic data recording was carried to more than a millisecond after the shock impact and, as a consequence, preserved information valuable to the present investigation.

This interaction was found to generate a strong vortical motion, originating along the interface between the two gases. Severe and rapid distortion of the light gas region was observed followed by intense mixing. Figure 5 consists of three shadowgraphs taken during the passage of a Mach 1.22 shock wave over a 5 cm diameter cylinder helium embedded in air. The time elapsed after the shock impinges on the helium cylinder is given in milliseconds for each photograph. The results show, in a most striking manner, the generation of a vortex pair by the vorticity generated at the initial gaseous interface and the subsequent diffusion of the vorticity from that surface. This provides the time dependent development of the flow field described earlier in this section. The shock tube experiment does not, of course, include the density change associated with combustion, a factor which would further increase the intensity of the induced flow.

3. ANALYTICAL AND COMPUTATIONAL STUDIES

Referring specifically to the experiments shown in Fig. 5 the phenomenon may be divided, conceptually, into two phases. In the first phase, the vorticity is deposited at the interface between the helium and air and the ensuing motion generates an intense vortex pair. In the second phase these vortices entrain air into the helium and undergo a complex mixing process, leading to molecular mixing and chemical reaction. Although some of the results obtained by Marble (1985), Karagozian & Marble (1986), Karagozian & Manda (1987) are of interest with respect to the second phase, the discussion here will concentrate on the analysis of the initial distortion of the light gas region.

During the past few months we have carried out Euler code calculations, Marble & Hendricks (1986b), to investigate the distortion of a cylindrical mass of helium embedded in air, by the impingement of shock waves of various strengths. The code employed was one originally developed by Eric Baum of the Electronics and Defense Sector, TRW, for use in the analysis of shock tube experiments. We have modified the original code considerably and adapted it for the present calculations in non-uniform gases.

It is of interest first to examine the time period over which we are able to obtain a reasonable representation of the distorted helium cylinder observed in the experiments mentioned in the previous section, shown in Fig. 5; the photos were taken at times of 132, 378, and 724 microseconds respectively after the initial impingement of the shock on the helium. The wire hoop which appears in the photographs is one of those that supported the microfilm sheath which contained the helium before shock impingement. It is reasonable to assume that the wake of the ring support, particularly visible in Fig. 5b, is localized at the ends of the helium mass and has a small effect upon the larger portion of the gas into which the camera is looking. The boundary of the helium remains fairly well defined through this time span.

A reasonable test of our Euler code then is to make calculations of the field quantities at similar times and to compare the gas boundary contours. Three of these computed shapes, at comparable times after the shock impact are shown in Fig. 6a, 6b, and 6c. The boundaries of these computed contours tend to diffuse somewhat with time and these thicker boundaries have been reduced on the contours shown. Clearly the comparison between computed and experimental boundary shapes is adequate to assure us that other computed properties are correspondingly accurate. These will be presented subsequently.

Return for the moment to the technological issue of how such mixing augmentation could be employed if indeed the present program demonstrated that its use was warranted. Consider a single jet of hydrogen moving with a supersonic velocity parallel to the supersonic airstream toward or within the engine. Let this jet be intersected by an oblique shock wave, as shown schematically in Fig. 7. Then in each succeeding cross section of the jet, the shock appears moving across the jet in the manner suggested by the sketches on the figure. To the extent that this problem can be analyzed by conventional "thin body" theory, x/U can be exchanged with time in the unsteady calculations and the results applied accordingly. Thus the jet should develop a strong vortex pair parallel with the direction of flow as the oblique shock intersects it and will mix rapidly downstream of the shock.

The accuracy with which this thin body theory may be applied has been examined numerically by considering the steady problem of an oblique shock intersecting a cylindrical (non mixing) co-flowing jet of hydrogen. The results in successive cross sectional planes will be compared with the corresponding two-dimensional calculations.

Figure 8 shows the distortion of the jet when the shock intersection has passed to the point indicated. The two plots below the distorted jet contour give respectively the corresponding density and vorticity distributions. Notice particularly the distribution of vorticity building up strongly on the portions of the interface nearly normal to the shock wave, as suggested by the vorticity generation mechanism described earlier. Figure 9 provides similar information for a section further downstream along the jet, beyond the point where the shock passes out of the jet. Here the complete vorticity distribution is developed and it provides a very clear intuitive picture of how the jet distortion is induced.

Figures 10 and 11 present corresponding information computed from the two-dimensional unsteady model. The correspondence is extremely close, even in small details of the density distribution. It is possible, of course, that the comparisons will be less favorable for lower jet Mach numbers, but the differences should not be great so long as the hydrogen jet is supersonic.

This result is of great consequence with regard to the experimental program to be discussed in the next section, because it demonstrates that detailed measurements carried out in a two-dimensional unsteady flow have direct application to the three dimensional problem that is of technological interest. And it is clear that the experimental technique is greatly simplified in the two-dimensional unsteady analogy and that a wider range of conditions may be covered.

4. EXPERIMENTAL STUDIES IN PROGRESS

We are at present in the early stages of an extensive experimental program to investigate the detailed mechanism of shock induced enhancement of mixing and reaction in non-uniform gas streams. The process described earlier has been divided into two concurrent and relatively independent tasks, each of which concentrates on a critical portion of the physics. The first of these is being carried out in the Galcit 17-inch shock tube facility; the details of this facility are described by Liepmann et al (1962). The second investigation utilizes our unsteady combustion facility which is an outgrowth of several years of extensive research on combustion instability. The construction of this facility is described by Smith (1985) and by Smith & Zukoski (1985).

The shock tube program builds upon the previous experimental work of Sturtevant and Haas, Haas (1983). The initial portion of our program is concerned with the interaction between a shock wave propagating through air and a cylindrical mass of gas. The aims are to obtain a quantitative measure of the distribution of helium in a thin section normal to the axis of the cylinder and, further, to obtain local measurements of the degree of molecular mixing which has taken place. This latter measurement is of great importance because it will be the first reliable indication of the extent to which the shock interaction is able to enhance the combustion chemistry. In each case, it is the aim to carry the measurements through times longer than those available from the Sturtevant and Haas data.

The shock tube mixing studies will be based on the fluorescence of biacetal, a dye which, when illuminated with radiation in the visible range, produces radiation in the visible range. This technique has been employed effectively, particularly by

Epstein (1977). The dye will be added to the low density gas which, in turn, is contained within a microfilm membrane until the shock impingement fractures it. After an accurately determined time delay, a pulse of light from a 3 - Joule dye laser, in the form of a thin sheet, will be passed normal to the helium cylinder. The resulting radiation will be measured using an intensified video camera and the data examined subsequently to find the distribution of helium. This experiment will be performed with a sequence of values of time delay in order to provide an accurate record of the development of mixing. The minimum thickness of the laser sheet which will provide a sufficiently strong signal is about 0.1 cm. The video camera has an array of 244 by 388 pixels and this will limit the dimensions of the observed area also to about 0.1 cm. Because the original scale of the helium region will be in the range of 2 to 5 cm, the 0.1 cm resolution available will be quite adequate to study the gross mixing between the helium and air.

The question of molecular mixing is being addressed through a further development of the biacetyl technique. It is based upon the fact that, when appropriately excited benzene molecules collide with biacetyl molecules, a fraction of them transfer energy to the biacetyl molecules, leading them to phosphoresce. We shall employ this technique by placing the biacetyl dye in the helium, as before, and the benzene in the shock tube gas. After the shock wave has passed, the benzene will be excited by passing a sheet of pulsed laser light through the test region, the biacetyl dye will phosphoresce after collision with an excited benzene molecule, and the phosphorescence radiation will be recorded with the video camera.

A possible problem arises because the phospherescence radiation may be emitted over a period of milliseconds and an exposure time of microseconds is desirable to stop the motion of the gas. The question is then whether the integrated radiation over the exposure time gives a sufficiently strong signal. Although this technique has been used successfully, Cheng (1978), it is recognized as higher risk than the basic mixing experiment. However, the importance of such results is such as to make reasonable compromises acceptable.

The second portion of the experimental research program is concerned with the details of combustion within a vortex, carried out on a much larger physical scale and a longer time scale. This work is being performed utilizing our unsteady combustion facility shown in Fig. 12 in the configuration appropriate for the present experiments. A conventional blowdown supply system is used to furnish the combustible mixture from a 15.2 cm. diameter plenum chamber, through a converging nozzle with a 9:1 contraction ratio, into a two-dimensional, rectangular combustion chamber having a 15 cm by 7.6 cm cross section and a length of one meter. Gas will enter through an opening 2.5 cm by 7.6 cm in the top of the upstream wall. The velocity of the combustible gas mixture can be varied between 10 and 100 m/sec.

The combustion chamber permits access for pressure measurements, flow visualization, and optical measurements. The upper and lower walls are fabricated so that the inner steel liner of the combustor is cooled by water flowing inside an aluminum cooling jacket. The side walls are segmented and secured to the upper and lower walls by a window frame structure. Vycore glass walls are used for flow vizualization and optical experiments.

Velocity fluctuations will be forced at the inlet lip by producing large amplitude pressure fluctuations in the plenum chamber. These will be generated by a siren placed at the upstream end of the plenum chamber, adjusting the length of the plenum chamber to resonate with the desired acoustic frequency. Part of the combustable mixture will pass through the siren, the rest through a bypass system. The amplitude of the oscillation will be controlled by varying the flow fraction plenum chamber to resonate with the desired acoustic frequency. Part of the combustable mixture will pass through the siren, the rest through a bypass system. The amplitude of the oscillation will be controlled by varying the flow fraction through the siren. The system is designed to drive oscillations from 100 to 600 hz.

As a result of these velocity fluctuations, vortices are generated at the lip of the downstream facing step, Hendricks(1986), and develop in the shear layer which separates the unburned flow from the recirculation zone. The individual vortices are ignited by the hot combustion products in the recirculation zone. The time scale for production and growth of these vortices will be affected by the size of the apparatus, the gas velocity, as well as the strength of the disturbance imposed to form each vortex, Sterling & Zukoski (1987). In our system, we expect it to be in the range of 10 to 50 milliseconds. Although this is much longer than the time scale of the shock tube experiment, it does limit the instrumentation and techniques which can be used in the observation. The exposure time for the image intensified video camera, with which we plan to observe the chemiluminescence of the burning gas, can be as small as a few microseconds but the framing rate of 60 hz is too low to allow the development of a single vortex to be observed.

As a consequence, we have chosen to generate vortices periodically and to use phase averaging techniques which will permit determining the time resolved properties of the flow. The experiment involves producing a train of vortices with a repetition rate of 100 to 500 hz, making measurements at a frequency of about 60 hz which are timed to cover a complete cycle of the vortex. The complete experiment will take about one second and requires careful timing between exposure and the phase of vortex shedding.

5. CONCLUDING REMARKS

It is clear that any augmentation of the rate of mixing between hydrogen fuel and air in a supersonic combustion ramjet offers well-defined advantages so far as engine performance and engine-airframe integration are concerned. The theory and technique of shock induced mixing enhancement, which has been introduced and discussed here, offers a very attractive possibility for significant reduction in the time (length) required for the mixing process. It has been shown, moreover, that a major portion of the demonstration of this concept may be carried out through time-dependent, two-dimensional studies in which the gas conditions corresponding to the proposed flight corridor may be reproduced to a considerable degree. We anticipate that the shock tube investigation of mixing and reaction, the study of unsteady combustion in vortices and, finally, experiments under conditions of steady flow, will confirm the degree to which this technique merits incorporation into engine development programs.

6. REFERENCES

Cheng, W. K. (1978), "Turbulent Mixing in a Swirling Flow." Report No. 143, Gas Turbine and Plasma Dynamics Laboratory, Massachusetts Institute of Technology.

Edwards, C. L. W., Small, W. J., Weidner, J. P., and Johnston, P. J. (1975), "Studies of Scramjet/Airframe Integration Techniques for Hypersonic Aircraft." AIAA Paper 75-78.

Epstein, A. H. (1977), "Quantitative Density Visualization in a Transonic Rotor." ASME Journal of Power Engineering, pp. 460-475.

Evans, J. S., and Anderson, G. Y. (1975), "Supersonic Mixing and Combustion in

Parallel Injection Flow Fields." *Proceedings, AGARD Conference on Analytical and Numerical Methods for Investigation of Flow Fields with Chemical Reactions, Especially Related to Combustion* CP-164.

Evans, M. W., Harlow, F. H., and Meixner, B. D. (1962) "Interaction of Shock or Rarifaction with a Bubble." *Physics of Fluids* V. 5, No. 6, pp 651-656.

Ferri, A. (1968), "Review of SCRAMJET Propulsion Technology." *J.Aircraft* V. 5, No.1, 3-10.

Ferri, A. (1973), "Mixing-Controlled Combustion." *Annual Reviews of Fluid Mechanics* V. 5, pp. 301-338.

Fleming, G. C.(1982), "Structure and Stability of Buoyant Diffusion Flames." *Ph.D. Thesis, California Institute of Technology*.

Gross, R. (1959), "Research on Supersonic Combustion." *J. Amer. Rocket Society* V. 29, p 63.

Gross, R. and Oppenheim, A. K. (1959), "Recent Advances in Gaseous Detonation." *J. Amer. Rocket Society* V. 29, pp 173-179.

Haas, J-F. L. (1983), "Interaction of Weak Shock Waves and Discrete Gas Inhomogeneities." *Ph.D. Thesis* California Institute of Technology.

Hendricks, G. J. (1986), "Two Mechanisms of Vorticity Generation in Combusting Flow Fields." *PhD Thesis*, California Institute of Technology.

Karagozian, A. R. and Marble, F. E. (1986), "Study of a Diffusion Flame in a Stretched Vortex". *Combustion Science and Technology* V. 45, pp 65-84.

Karagozian, A. R. and Manda, B. V. S.,(1987), "Flame Structure and Fuel Consumption in the Field of a Vortex Pair." To appear in *Combustion Science and Technology*.

Kirkham, F. S., Hunt, J. L. (1977), "Hypersonic Transport Technology." *Acta Astronautica* Vol. 4, 181-199.

Liepmann, H. W., Roshko, A., Coles, D., and Sturtevant, B. (1962), "A 17-inch Diameter Shock Tube for Studies in Rarified Gasdynamics." *Rev. Sci. Instr.* V. 33. No. 6, pp. 625-631.

Marble, F. E. (1985), "Growth of a Diffusion Flame in the Field of a Vortex." *Recent Advances in Aerospace Sciences* Ed. C. Casci, pp 395-413.

Marble, F. E. and Hendricks, G. J. (1986a), "Behavior of a Diffusion Flame in a Flow Inducing a Pressure Gradient Along its Length." To appear in *Proceedings, 21st International Symposium on Combustion*.

Marble, F. E. and Hendricks, G. J. (1986b), "Distortion by Shock Impingement of Light Gas Regions Embedded in a Heavier Gas." Manuscript in preparation, unpublished.

McClinton, C. R. (1978), "Interaction Between Step Fuel Injectors on Opposite Walls in a Supersonic Combustor Model." *NASA Technical Paper 1174*.

Northam, G. B. and Anderson, G. Y. (1986), "Review of NASA/Langley Basic Research on

Supersonic Combustion." AIAA Preprint 86-0159.

Nowak, R. J. and Kelly, N. H. (1976), "Actively Cooled Airframe Structures for High Speed Flight." AIAA/ASME/SAE 17th Structures, Structural Dynamics and Materials Conference.

Picone, J. M. and Boris, J. P. (1983), "Vorticity Generation by Asymmetric Energy Deposition in a Gaseous Medium." Physics of Fluids V. 26, No. 2, pp 365-382.

Picone, J. M., Oran, E. S., Boris, J. P. and Young, T. R., Jr. (1983), "Theory of Vorticity Generation by Shock Wave and Flame Interaction." Progress in Astronautics and Aeronautics V. 94, pp. 429-448.

Rudinger, G. (1958), "Shock Wave and Flame Interactions." Combustion and Propulsion Third AGARD Colloquium, pp. 153-182.

Rudinger, G. and Somers, L. (1960), "Behavior of Small Regions of Different Gases Carried in Accelerated Gas Flows." J. Fluid Mech. 7, 161-176.

Small, W. J., Weidner, J. P., and Johnston, P. J. (1974), "Nozzle Design and Analysis as Applied to a Highly Integrated Research Airplane." NASA TM X-71972.

Smith, D. A. (1985), "An Experimental Study of Acoustically Excited, Vortex Driven, Combustion Instability Within a Rearward Facing Step Combustor." Ph.D. Thesis California Institute of Technology.

Smith, D. A. and Zukoski, E. E. (1985), "Combustion Instability Sustained by Vortex Combustion." 21st AIAA Joint Propulsion Conference.

Sterling, J. D. and Zukoski, E. E. (1987), "Longitudinal Mode Instabilities in a Dump Combustor." AIAA 25th Aerospace Sciences Meeting, Jan 12-15, Reno.

Waltrup, P. J. (1986), "Liquid Fueled Supersonic Combustion Ramjets: A Research Perspective of the Past, Present and Future." AIAA 86-0158.

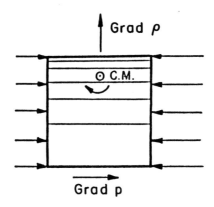

1. Vorticity Produced by Interaction of Gradients in Pressure and Density

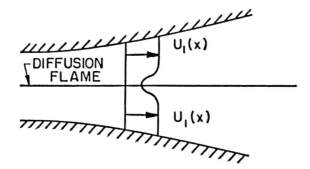

2. Diffusion Flame in Adverse Pressure Gradient

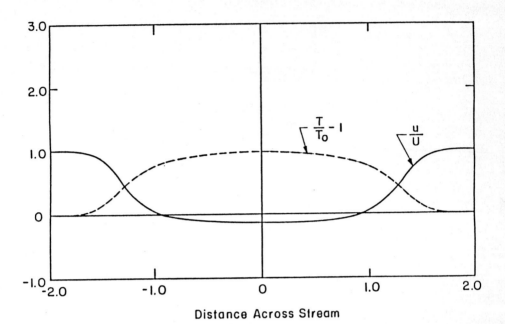

3. Backflow Induced in Diffusion Flame

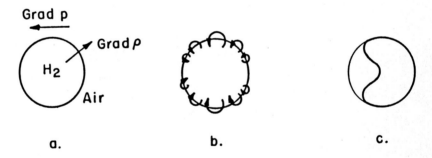

4. Vorticity and Distortion Induced by Shock Passage over Hydrogen Cylinder in Air

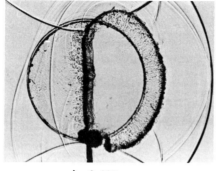

a) 0.132 ms

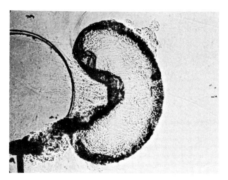

b) 0.378 ms

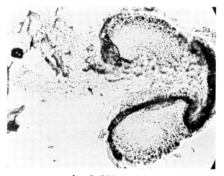

c) 0.724 ms

5. Shadowgraphs of Helium Cylinder Distortion by Mach 1.22 Shock

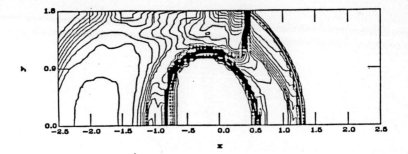

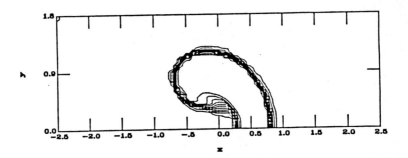

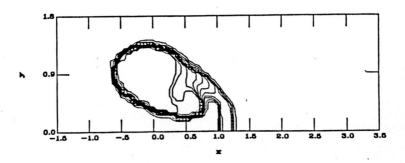

6. Computed Density Contours Corresponding to Shadowgraphs of Figure 5.

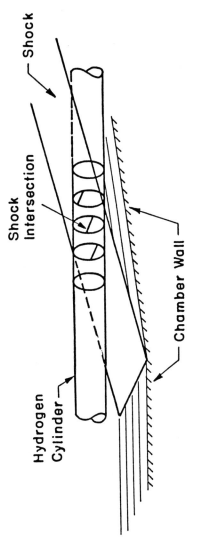

7. Schematic View of Oblique Shock Passage over Cylindrical Hydrogen Jet in Air

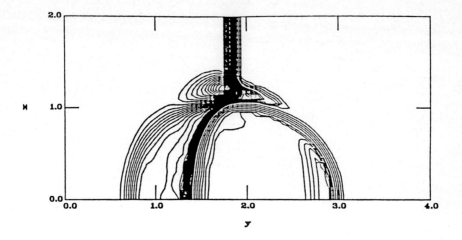

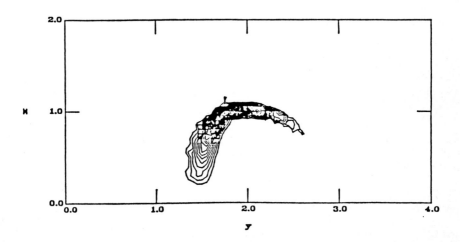

8. Density and Vorticity Distributions in Cylindrical Hydrogen Jet, Early Cross Section

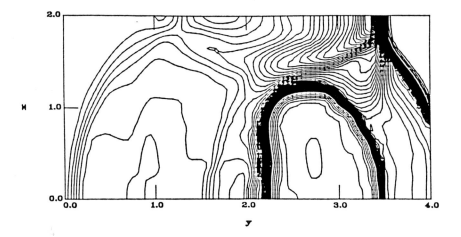
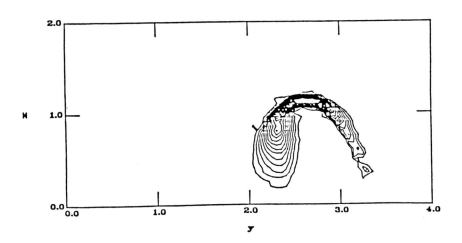

9. Density and Vorticity Distributions in Cylindrical Hydrogen Jet, Later Cross Section

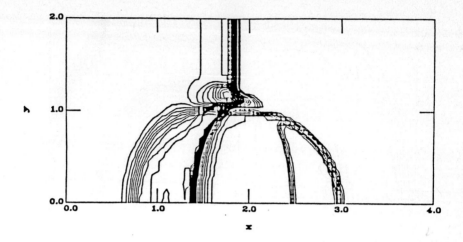

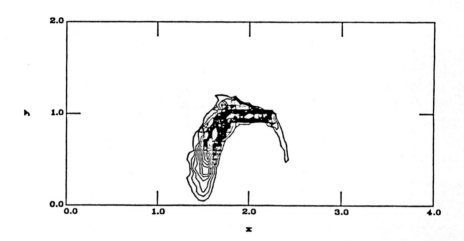

10. Two-Dimensional Computation of Density and Vorticity Corresponding to Three-Dimensional Calculations of Figure 8

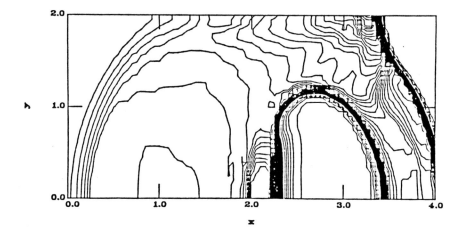

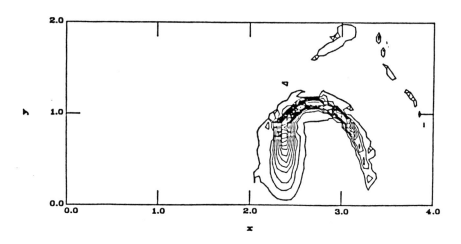

11. Two-Dimensional Computation of Density and Vorticity Corresponding to Three-Dimensional Calculations of Figure 9

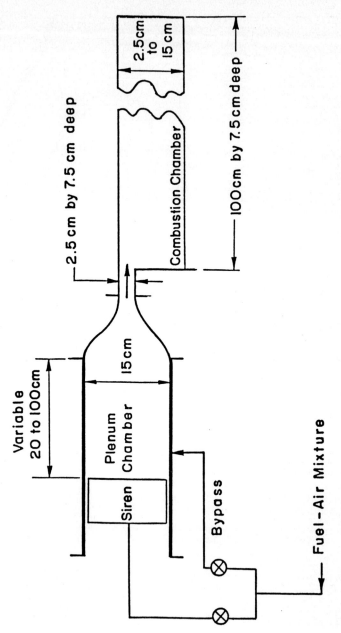

12. Schematic View of Unsteady Combustion Facility for Study of Vortex Combustion

Lecture Notes in Engineering

Edited by C. A. Brebbia and S. A. Orszag

Vol. 21: B. F. Spencer, Jr.
Reliability of Randomly
Excited Hysteretic Structures
XIII, 138 pages. 1986.

Vol. 22: A. Gupta, R. P. Singh
Fatigue Behaviour
of Offshore Structures
XXI, 299 pages. 1986.

Vol. 23: P. Hagedorn, K. Kelkel,
J. Wallaschek
Vibrations and Impedances
of Rectangular Plates
with Free Boundaries
V, 152 pages. 1986.

Vol. 24: Supercomputers
and Fluid Dynamics
Proceedings of the First
Nobeyama Workshop
September 3-6, 1985
VIII, 200 pages. 1986.

Vol. 25: B. Hederson-Sellers
Modeling of Plume Rise
and Dispersion –
The University of Salford
Model: U. S. P. R.
VIII, 113 pages. 1987.

Vol. 26: Shell and Spatial Structures:
Computational Aspects
Proceeding of the International Symposium
July 1986, Leuven, Belgium
Edited by G. De Roeck, A. Samartin Quiroga,
M. Van Laethem and E. Backx
VII, 486 pages. 1987.

Vol. 27: Th. V. Hromadka, Ch.-Ch. Yen
G. F. Pinder
The Best Approximation Method
An Introduction
XIII, 168 pages. 1987.

Vol. 28: Refined Dynamical Theories
of Beams, Plates and Shells and
Their Applications
Proceedings of the Euromech-Colloquim 219
Edited by I. Elishakoff and H. Irretier
IX, 436 pages. 1987.

Vol. 29: G. Menges, N. Hövelmanns,
E. Baur (Eds.)
Expert Systems in Production Engineering
Proceedings of the International Workshop
Spa, Belgium, August 18-22, 1986
IV, 245 pages. 1987.

Vol. 30: R. Doležal
Simulation of Large State Variations
in Steam Power Plants
Dynamics of Large Scale Systems
X, 110 pages. 1987.

Vol. 31: Y. K. Lin, G. I. Schueller (Eds.)
Stochastic Structural Mechanics
U.S.-Austria Joint Seminar, May 4-5, 1987
Boca Raton, Florida, USA
XI, 507 pages. 1987.

Vol. 32: Y. K. Lin, R. Minai (Eds.)
Stochastic Approaches
in Earthquake Engineering
U.S.-Japan Joint Seminar, May 6-7, 1987
Boca Raton, Florida, USA
XI, 457 pages. 1987.

Vol. 33: P. Thoft-Christensen (Editor)
Reliability and Optimization
of Structural Systems
Proceedings of the First IFIP WG 7.5
Working Conference
Aalborg, Denmark, May 6-8, 1987
VIII, 458 pages. 1987.

Vol. 34: M. B. Allen III, G. A. Behie,
J. A. Trangenstein
Multiphase Flow in Porous Media
Mechanics, Mathematics, and Numerics
IV, 312 pages. 1988.

Vol. 35: W. Tang
A New Transformation
Approach in BEM
A Generalized Approach for
Transforming Domain Integrals
VI, 210 pages. 1988.

Vol. 36: R. H. Mendez, S. A. Orszag
Japanese Supercomputing
Architecture, Algorithms, and Applications
IV, 160 pages. 1988.

Vol. 37: J. N. Reddy, C. S. Krishnamoorthy,
K. N. Seetharamu (Eds.)
Finite Element Analysis
for Engineering Design
XIV, 869 pages. 1988.

Vol. 38: S. J. Dunnett, D. B. Ingham
The Mathematics of Blunt Body Sampling
VIII, 213 pages. 1988

Vol. 39: S. L. Koh, C. G. Speziale (Eds.)
Recent Advances in Engineering Science
A Symposium dedicated to A. Cemal Eringen
June 20–22, 1988, Berkeley, California
XVIII, 268 pages. 1989